ANTARCTIC RESEARCH SERIES

American Geophysical Union

ANTARCTIC
RESEARCH
SERIES

American Geophysical Union

Volume 1 BIOLOGY OF THE ANTARCTIC SEAS
Milton O. Lee, *Editor*

Volume 2 ANTARCTIC SNOW AND ICE STUDIES
Malcom Mellor, *Editor*

Volume 3 POLYCHAETA ERRANTIA OF ANTARCTICA
Olga Hartman

Volume 4 GEOMAGNETISM AND AERONOMY
A. H. Waynick, *Editor*

Volume 5 BIOLOGY OF THE ANTARCTIC SEAS II
George A. Llano, *Editor*

Volume 6 GEOLOGY AND PALEONTOLOGY OF THE ANTARCTIC
Jarvis B. Hadley, *Editor*

Volume 7 POLYCHAETA MYZOSTOMIDAE AND SEDENTARIA OF ANTARCTICA
Olga Hartman

Volume 8 ANTARCTIC SOILS AND SOIL FORMING PROCESSES
J. C. F. Tedrow, *Editor*

Volume 9 STUDIES IN ANTARCTIC METEOROLOGY
Morton J. Rubin, *Editor*

Volume 10 ENTOMOLOGY OF ANTARCTICA
J. Linsley Gressitt, *Editor*

Volume 11 BIOLOGY OF THE ANTARCTIC SEAS III
Waldo L. Schmitt and George A. Llano, *Editors*

Volume 12 ANTARCTIC BIRD STUDIES
Oliver L. Austin, Jr., *Editor*

Volume 13 ANTARCTIC ASCIDIACEA
Patricia Kott

Volume 14 ANTARCTIC CIRRIPEDIA
William A. Newman and Arnold Ross

Volume 15 ANTARCTIC OCEANOLOGY
Joseph L. Reid, *Editor*

Volume 16 ANTARCTIC SNOW AND ICE STUDIES II
A. P. Crary, *Editor*

Volume 17 BIOLOGY OF THE ANTARCTIC SEAS IV
George A. Llano and I. Eugene Wallen, *Editors*

Volume 18 ANTARCTIC PINNIPEDIA
William Henry Burt, *Editor*

Volume 19 ANTARCTIC OCEANOLOGY II: THE AUSTRALIAN–NEW ZEALAND SECTOR
Dennis E. Hayes, *Editor*

Antarctic Oceanology II

The Australian–New Zealand Sector

FRONTISPIECE

USNS *Eltanin* in brash ice in antarctic waters (National Science Foundation photograph).

Volume 19 | ANTARCTIC RESEARCH SERIES

Antarctic Oceanology II
The Australian—New Zealand Sector

Dennis E. Hayes, *Editor*

Published with the aid of a grant from the National Science Foundation

PUBLISHER

AMERICAN GEOPHYSICAL UNION

OF THE

National Academy of Sciences—National Research Council

1972

Volume 19 | ANTARCTIC RESEARCH SERIES

ANTARCTIC OCEANOLOGY II: THE AUSTRALIAN–NEW ZEALAND SECTOR

DENNIS E. HAYES, *Editor*

1707 L Street, N.W.
Washington, D. C. 20036

Library of Congress Catalog Card No. 78-151300
International Standard Book No. 87590-119-0

List Price $32.00

Printed by
THE HORN-SHAFER COMPANY
DIVISION OF
Geo. W. King Printing Co.
Baltimore, Maryland

THE ANTARCTIC RESEARCH SERIES

The Antarctic Research Series is designed to provide a medium for presenting authoritative reports on the extensive and detailed scientific research work being carried out in Antarctica. The series has been successful in eliciting contributions from leading research scientists engaged in antarctic investigations; it seeks to maintain high scientific and publication standards. The scientific editor for each volume is chosen from among recognized authorities in the discipline or theme it represents, as are the reviewers on whom the editor relies for advice.

Beginning with the scientific investigations carried out during the International Geophysical Year, reports of research results appearing in this series represent original contributions too lengthy or otherwise inappropriate for publication in the standard journals. In some cases an entire volume is devoted to a monograph. The material published is directed not only to specialists actively engaged in the work but to graduate students, to scientists in closely related fields, and to interested laymen versed in the biological and the physical sciences. Many of the earlier volumes are cohesive collections of papers grouped around a central theme. Future volumes may concern themselves with regional as well as disciplinary aspects, or with a comparison of antarctic phenomena with those of other regions of the globe. But the central theme of Antarctica will dominate.

In a sense, the series continues the tradition dating from the earliest days of geographic exploration and scientific expeditions—the tradition of the expeditionary volumes which set forth in detail everything that was seen and studied. This tradition is not necessarily outmoded, but in much of the present scientific work one expedition blends into the next, and it is no longer scientifically meaningful to separate them arbitrarily. Antarctic research has a large degree of coherence; it deserves the modern counterpart of the expeditionary volumes of past decades and centuries which the Antarctic Research Series provides.

With the aid of a grant from the National Science Foundation in 1962, the American Geophysical Union initiated the Antarctic Research Series and appointed a Board of Associate Editors to implement it. A supplemental grant received in 1966, the income from the sale of volumes in the series, and income from reprints and other sources have enabled the AGU to continue this series. The response of the scientific community and the favorable comments of reviewers cause the board to look forward with optimism to the continued success of this endeavor.

To represent the broad scientific nature of the series, the members of the Board were chosen from all fields of antarctic research. At the present time they include: Avery A. Drake, Jr., representing geology and solid earth geophysics; A. P. Crary, seismology and glaciology; George A. Llano, botany and zoology; Martin A. Pomerantz, aeronomy and geomagnetism; Morton J. Rubin, meteorology and oceanography; David L. Pawson, biology; Waldo L. Schmitt, member emeritus; and Laurence M. Gould, honorary chairman. Fred G. Alberts, secretary to the U. S. Advisory Committee on Antarctic Names, gives valuable assistance in verifying place names, locations, and maps.

Morton J. Rubin
Chairman, Board of Associate Editors
Antarctic Research Series

PREFACE

The organization and production of *Antarctic Oceanology II: The Australian–New Zealand Sector* were motivated with several considerations in mind. It is obvious that the value of any scientific contribution treating a particular geographical area is greatly enhanced when the individual work is compiled and presented with other contributions that consider the scientific problems of the same area from different approaches. There is no doubt that the potential value of a collective volume far exceeds that of any individual contributions. Thus this geographical approach, which has been successfully used in many books, has also been used in this volume of the Antarctic Research Series. Moreover, the contributions in this volume have benefited from one important and basically different starting point, the research program of the USNS *Eltanin.* A large portion of marine geologic, marine geophysical, and oceanographic data for this sector of the ocean south of Australia and New Zealand has been collected simultaneously aboard the *Eltanin,* the antarctic research vessel of the National Science Foundation. Most of the *Eltanin* cruises spanning the period of the last 2–3 years have been devoted to reconnaissance surveying of this area. The surveys have been multidisciplinary by design, and the ship tracks have been systematically laid out as part of an over-all plan to survey the entire circumpolar ocean surrounding the antarctic continent.

The *Eltanin* has by no means provided the only data used in these studies, but there can be no question that without the *Eltanin* data this book would not have been realized. During the last several years, not only the *Eltanin* but a variety of other research vessels have operated in this ocean area south of Australia and New Zealand. The commendable attitude of cooperation existing among the scientists of those countries has led to many collaborative efforts. The participation of these scientists both in the field programs and in the data analysis has been an important contribution and provides additional and valuable credentials to insure the success of this book.

As previously mentioned, a collection of these studies is potentially far more valuable than any scattered individual contributions. The growing reluctance of many scientists to contribute their most contemporary research efforts for bound volumes is most understandable in view of the inevitably long delays associated with the production of a book of high quality. *Antarctic Oceanology II: The Australian–New Zealand Sector,* through the combined efforts of many individuals, has managed to circumvent this problem and has been distributed in less than 1 year from the time I received the first contributions to the volume. This prompt publication has been possible only through the conscientious efforts of the American Geophysical Union and its staff, the board of editors of the Antarctic Research Series, the section editors, the reviewers, and, most importantly, the contributors themselves. In spite of this rapid production and publication, the book has not suffered in quality. Quite the contrary; its value has been enhanced because the papers comprising the volume constitute the most current research efforts in the rapidly changing field of oceanographic science and do not suffer from the obsolescence that threatens many papers appearing in long-delayed publications.

In producing a volume of this type, the individual contributions are often closely related to one another. The conclusions of some inevitably should draw on the results of the others. Maximum communication between investigators is desirable, but this goal is generally mutually exclusive with a premise of rapid publication. A compromise approach has been tried in this book. The book is composed of three well-balanced sections, each of which has been edited by a scientist actively specializing in the field: (1) Physical Oceanography edited by Arnold Gordon, (2) Marine Geophysics edited by Dennis E. Hayes, and (3) Marine Sediments edited by John R. Conolly. These editors also reviewed each manuscript in their section and thus were well suited to write the brief chapters introducing each section. The introductions serve to synthesize the contributions of each section and to highlight both the important findings and the major unresolved problems within each subdiscipline. The introductory chapters also serve as a brief review but have the unique advantage of being able to review works published concomitantly. They are intended in part as a reasonable compromise to the extensive communication that is desirable between authors but that was not practical in this publication. Each introductory chapter may also be considered as an expanded preface to each section.

I am confident that the response to the book will further justify the enormous effort that has gone into its production, and hopefully the book may serve as a prototype for the oceanographic study of other sectors of the antarctic circumpolar ocean. In any event, the real joy for the authors must have already come in the process of solving many problems and formulating others. To quote William Faulkner, the scientist's voice, like that of the poet's, 'need not merely be the record of man, it can be one of the props, the pillars to help him endure and prevail.'

Dennis E. Hayes

CONTENTS

1

Physical Oceanography

INTRODUCTION: PHYSICAL OCEANOGRAPHY OF THE SOUTHEAST INDIAN OCEAN

ARNOLD L. GORDON

Lamont-Doherty Geological Observatory of Columbia University, Palisades, New York 10964

Abstract. The north-south constriction of the antarctic waters south of the Australian land mass connects the Indian and Pacific oceans. The interchange of water between these oceans contributes most of the deep and bottom water to the Pacific Ocean. This significant area is discussed in the papers in this section on physical oceanography.

The antarctic circumpolar water belt is clearly defined south of each of the Southern Hemisphere land masses. The major oceans of the world are connected by these zonal channels. Of the three passages, the one south of Australia and New Zealand has the greatest extent, the longitude totalling 60°. The north-south extent between adjacent shore lines varies from a maximum of 35° of latitude at 130°E to a minimum of 24° of latitude at 146°E and at 167°E. However, in considering the extent of the deep zone (deeper than 1000 meters), the minimum width passage is only 10° of latitude from Macquarie ridge to the topographic high of the Balleny Islands. This effective passage is very similar in width to the Drake passage and also lies at about the same latitude.

The basic eastward flowing antarctic circumpolar current (ACC) south of the Australian and New Zealand land masses supplies all the deep water of the entire Pacific Ocean. The deep reaching current structure produces a total transport of the ACC of nearly 235×10^6 m³/sec [*Callahan*, 1971]. The penetration of the ACC to the sea floor is clearly shown in the sediment distribution as discussed by *Watkins and Kennett* [this volume]. Most of the huge transport of the ACC continues across the Pacific sector of antarctic waters; only a relatively small amount turns northward into the Pacific as a western boundary undercurrent (WBUC). *Warren and Voorhis* [1970] estimate that the WBUC transport is 13×10^6 m³/sec. This water must eventually return to the Antarctic, so the Drake passage transport is expected to be similar to the transport south of Australia, as is apparently the case [*Reid and Nowlin*, 1971]. The salinity within the core layer of the circumpolar deep water drops by approximately 0.04‰ during its transit from the south Indian Ocean to the southwest Atlantic Ocean. Most of the drop in salinity occurs southeast of the Campbell plateau [*Gordon*, 1971*a*]. This feature may be related to the loss of the relatively high salinity circumpolar deep water to the WBUC.

The *Eltanin* physical oceanography program is now nearly completed in terms of a reconnaissance data network south of Australia and New Zealand. The work reported in the physical oceanography section of this volume stems mostly from the *Eltanin* program and Australian oceanographic cruises. The *Eltanin* data are reported by *Jacobs* [1965, 1966], *Jacobs and Amos* [1967], and *Jacobs et al.* [1970*b*], and the Australian data are contained in a series of reports by the *Commonwealth Scientific and Industrial Research Organisation of Australia* [1962, 1963*a*, *b*; 1966*a*, *b*; 1967*a*, *b*, *c*, *d*; 1968*a*, *b*; 1972]. Unfortunately, a full description of the general oceanography in the area south of Australia and New Zealand is not yet completed. However, data on the area are available in atlas form in *Tolstikov* [1966] and *Wyrtki* [1971], and the area is included in *Deacon* [1937] and *Ishino* [1963]. The reference lists of the papers in this section comprise a fairly complete bibliography on the oceanography of this area.

The basic lines of oceanographic singularities are shown in Figure 1. These lines or zones are also boundaries in the sound velocity patterns discussed by *Denham and Kibblewhite* [this volume] and earlier by *Kibblewhite and Denham* [1967]. The ACC axis closely follows the polar front zone as shown in the salinity profiles given by *Gordon* [this volume] in the Macquarie ridge study. The discordance in the Southeast Indian ridge discussed by *Hayes and Conolly* [this volume] near 120°–130°E does not seem to

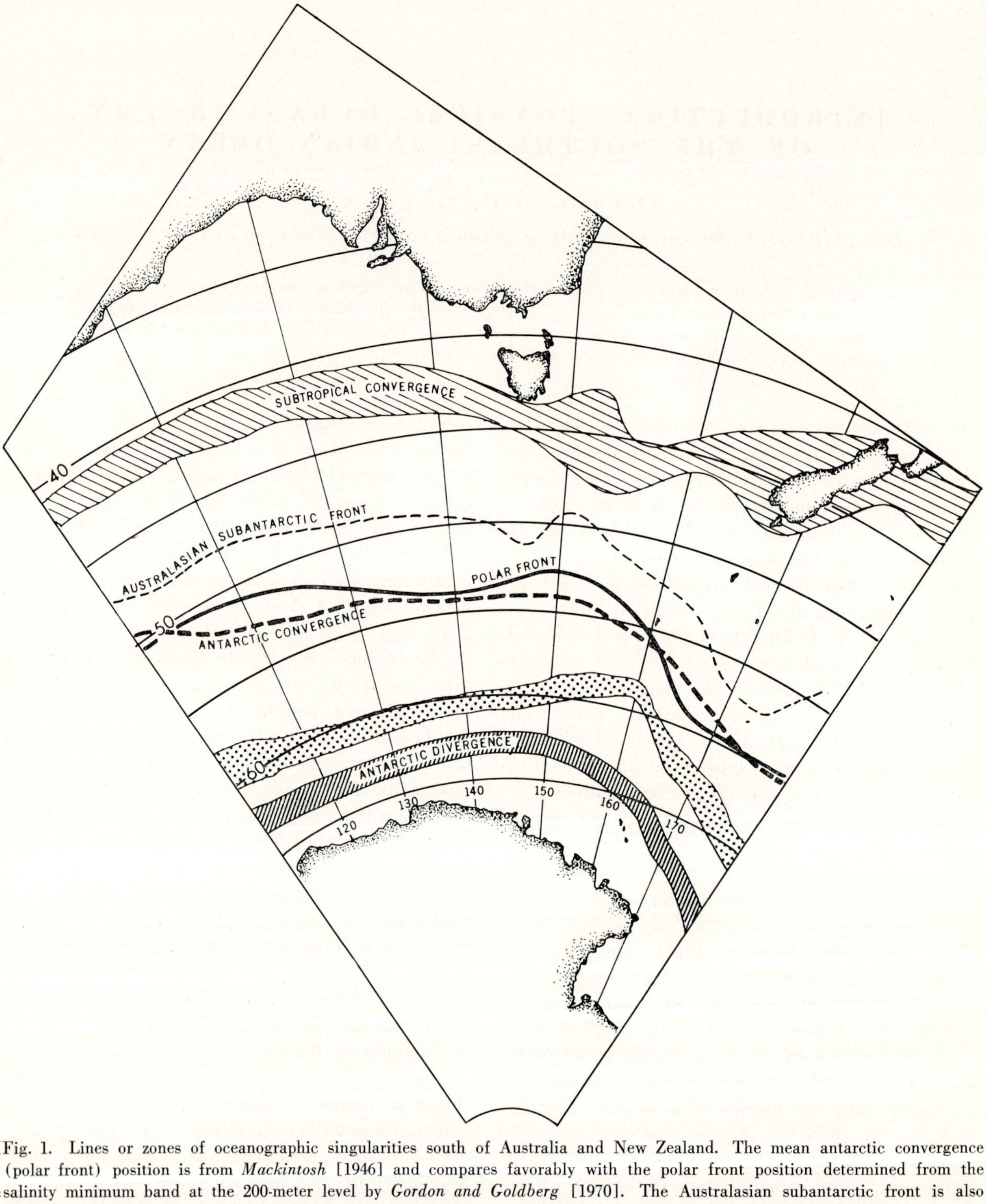

Fig. 1. Lines or zones of oceanographic singularities south of Australia and New Zealand. The mean antarctic convergence (polar front) position is from *Mackintosh* [1946] and compares favorably with the polar front position determined from the salinity minimum band at the 200-meter level by *Gordon and Goldberg* [1970]. The Australasian subantarctic front is also taken from the 200-meter salinity map of Gordon and Goldberg and follows the definition given by *Burling* [1961], i.e., the relatively large salinity gradient 'roughly between 150 and 400 meters.' The subtropical convergence zone is taken to lie in the region of strong salinity gradients at the 20-meter level (from *Gordon and Goldberg* [1970]); this position agrees with the positions given by *Deacon* [1937], *Burling* [1961], and *Garner* [1962]. The antarctic divergence zone is placed over the topographic crest of the salinity maximum core layer, also found from the 200-meter salinity distribution given by Gordon and Goldberg. The dotted pattern is a region of increased north-south gradients in both salinity and temperature.

have an influence on the general ocean circulation. Possibly the parallelism of the ACC path and the crest of the mid-ocean ridge is responsible. The basic water mass structure across the ridge produces a dynamic barrier to the northward flow of the colder southern waters. Perhaps if the discordance were broader and deeper, an influence would be evident.

The ACC undergoes an abrupt southward deflection at the Macquarie ridge. This anomalous path may be of significance to the energy balance of the current. The southward displacement of the ACC allows relatively warm water to flow over the ridge crest. As a result, possibly, the warm water eddies are shed from this feature and maintain their identity far into the Pacific Ocean. Such speculation is supported in the numerical solutions to the equations of motions presented by *Boyer and Guala* [this volume]. An interesting possibility is that transient flow patterns are generated through the interaction of a mean flow with bottom topography.

The Macquarie ridge area is likely to be active on a geologic time scale; hence alteration of the ACC in the geologic past is also likely. Perhaps the influence of the deep passage south of Macquarie ridge is similar to the interaction of the ACC and the Drake passage as discussed recently by *Gill and Bryan* [1971]; i.e., a slightly shallowed Drake passage actually increases the ACC transport. However, intuition would indicate that a drastic reduction in the dimensions of the deep passage south of Macquarie Island would eventually reduce the ACC transport and the deep water influx into the Pacific Ocean.

Speculations on the paleo-oceanography of antarctic waters and the interaction of these waters with the northern oceans are difficult to make. Paleo-oceanographic conditions may best be studied through an accumulation or a lack of accumulation of bottom sediments. However, alterations of bottom topography would strongly influence the course and transport of the ACC. This current would in turn influence the north-south slope of the isopleths and hence would have some control over surface characteristics at the southern regions of antarctic waters. Increased ACC velocities would increase the north-south slope of the isopleths, and this increase would in turn lead to higher near-surface salinity near Antarctica. This higher salinity produces an environment more susceptible to deep thermohaline convection, and hence more AABW may be produced.

The Australasian subantarctic front defined by *Burling* [1961] (see also *Garner* [1962] and *Houtman* [1964, 1967]) is a very pronounced feature extending from the Campbell plateau across the entire Indian Ocean. The salinity distribution at 200 meters suggests a westward intensification of this front [*Gordon and Goldberg*, 1970]. In the region south of Australia and New Zealand, the Australasian subantarctic front is approximately 3° of latitude north of the mean position of the polar front zone.

The surface expression of the Australasian subantarctic front is more pronounced than that of the polar front. The proximity of these fronts may lead to some confusion in their identification. I recommend [*Gordon*, 1971*b*] that the subsurface expressions of the polar front be used for identification, i.e., the termination of the temperature minimum layer and the salinity minimum at the 200-meter level.

The meridional flow of waters appears uninterrupted by the presence of the northern boundary of the Australasian land mass. The salinity profiles clearly show sinking of low salinity near surface water in the vicinity of the polar front. The sinking of water to the sea floor is evident in the Ross Sea region and Adélie Coast and is discussed by *Gordon and Tchernia* [this volume] and *Jacobs et al.* [1970*a*].

Approximately 500 km south of the low salinity band axis is another feature that often has a surface expression as an abrupt increase in the north-south temperature gradient. This feature has been found on many *Eltanin* cruises, is evident on the profiles along 115°E near 58°S (*Eltanin* cruise 45), and is given in Figure 2.

High and low salinity varieties of antarctic bottom water are produced in the Ross Sea [*Jacobs et al.*, 1970*a*; *Gordon*, 1971*c*]. The flow of high salinity bottom water derived from the Ross Sea is clearly observed at the base of the Adélie Coast continental margin. *Gordon and Tchernia* [this volume] present evidence that indicates the production of low salinity bottom water along the Adélie Coast. This bottom water intermixes with the high salinity Ross Sea bottom water. Neither the Ross Sea nor the Adélie Coast bottom water appears to be generated in large enough quantities to have a widespread influence. However, it is now fully established that the Weddell Sea is not the sole source of antarctic bottom water, although this sea does seem to be the most important source. In winter, open water (polynyas) in the coastal regions in the Ross Sea and along the Adélie Coast may be significant to bottom water formation. The mechanical removal by the wind of newly formed sea ice in these regions and the exposure of the water

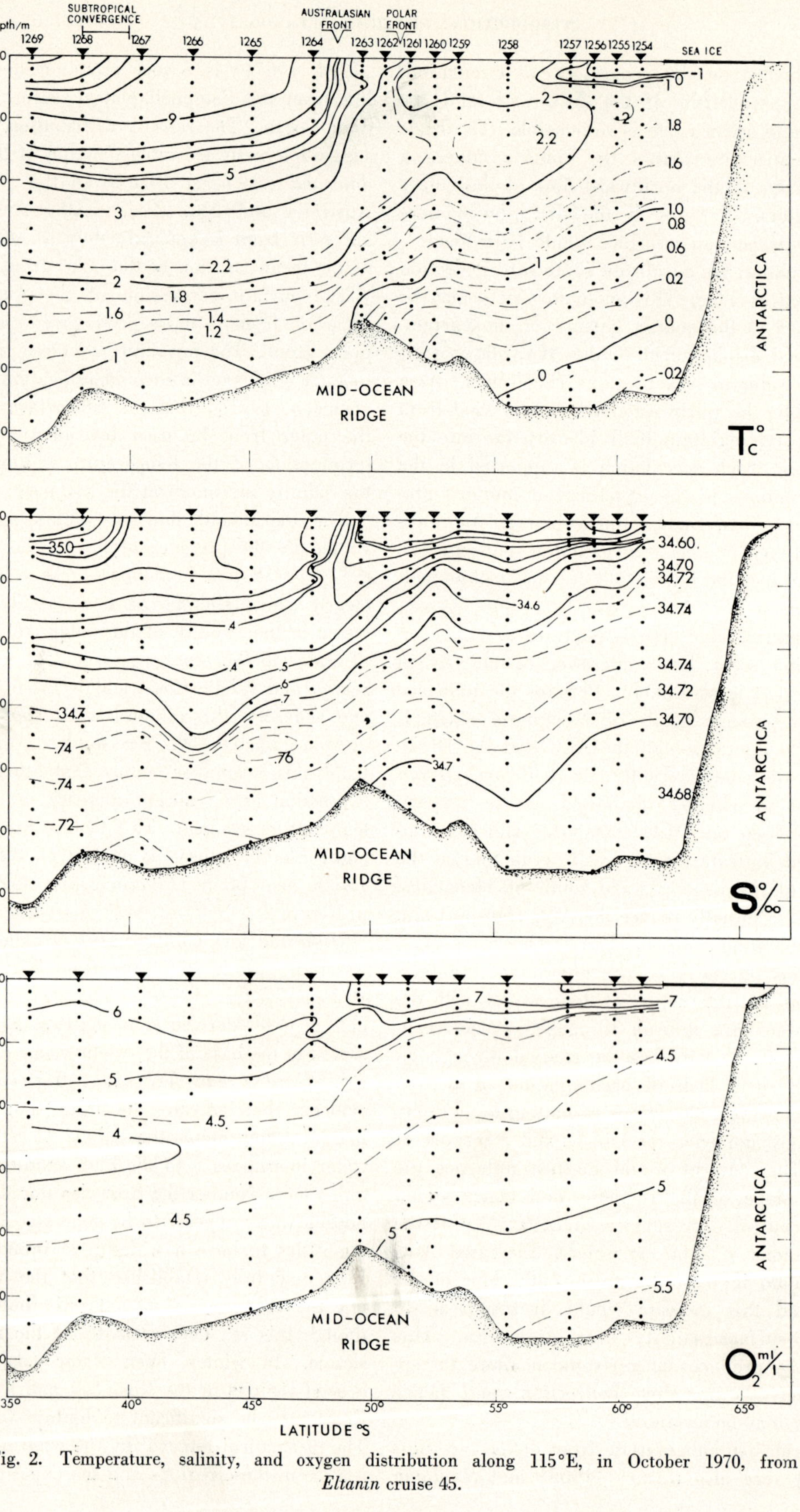

Fig. 2. Temperature, salinity, and oxygen distribution along 115°E, in October 1970, from *Eltanin* cruise 45.

surface to the atmosphere effectively produce great quantities of sea ice.

The bottom water mass derived from the Ross Sea is heavily laden with suspended material as discussed by *Eittreim et al.* [this volume], who believe that most suspended material settles from the water before completing one cycle of the general cyclonic bottom water flow within the South Indian basin. The bottom waters north of the mid-ocean ridge are derived in part from flow through a fracture zone at 140°–150°E in the mid-ocean ridge and perhaps in part are diluted bottom water derived from the Weddell Sea. Whereas the general pattern of flow is eastward, less intense westward currents are found along the antarctic coast throughout the water column and at times occur along the south shores of Australia. The antarctic coastal current is the east wind drift and the Australian coastal current is called the Flinders current and is discussed in *Bye* [this volume]. A westward flowing current may also occur immediately south of the Southeast Indian ridge as suggested by the salinity profiles given in the Macquarie ridge study and *Callahan* [1971]. *Thompson* [1971] notes that the ACC 'may be expected to be sharper when over a slope with deep water in the direction of the coriolis force on the current.' Hence a strong eastward flow would occur over the northern flank of the ridge, and a westward flow would occur over the southern flank.

The tidal characteristics of the water along 132°E during *Eltanin* cruise 41 and the results of that cruise are discussed in the study by *Irish and Snodgrass* [this volume]. They find inertial currents the magnitude of which is similar to that of the mean flow. The tidal currents are a mixture of diurnal and semidiurnal tides with a magnitude of less than 1 cm/sec, a magnitude far below the large value anticipated for open ocean circumpolar tides. The mean flow (frequency of less than 0.2 cycles/day) showed significant variability. Short, time series bottom current meter data collected south of Australia show an apparent random pattern [*Eittreim et al.*, 1971] that agrees with the findings of Irish and Snodgrass.

The structure of the temperature, salinity, and oxygen fields of the water along the western boundary of the area considered in this volume is shown in Figure 2. The northward slope of the isopleths is evident, the steepest slope occurring at all depths below the polar front zone situated (as it often is) over the mid-ocean ridge. The polar front is not simply a surface feature as is the subtropical convergence, which is confined to only the upper few hundred meters.

The water entering the region south of Australia eventually leaves the area south of New Zealand. Enroute the water is acted on at the sea–air boundaries by the antarctic and subantarctic atmosphere. The mean and seasonal sea-air interaction and atmospheric conditions south of Australia and New Zealand are given by *Zillman* [this volume]. Seasonal data arrays are rare in the remote antarctic waters; however, Zillman, using the data of *Eltanin* cruises 35–41, is able to inspect seasonal variability of the sea-air interchange in the region from 115° to 160°E. In addition to seasonal influences, the heat budget is strongly dependent on weather patterns, i.e., position of cyclones.

The summer temperature minimum layer, at the base of the antarctic surface water, behaves as a sound channel. The polar front zone represents a boundary between this shallow sound channel and the deeper Sofar channel to the north. The acoustical velocity pattern in the waters south of Australia and New Zealand is presented by *Denham and Kibblewhite* [this volume].

The structure of the salinity field immediately east of the Macquarie ridge is given by *Gordon* [this volume, Figure 9]. The slopes of the isopleths are somewhat increased over those shown along 115°E, this increase reflecting accelerated flow through the more restrictive passage south of the Macquarie ridge. The salinity in the lower part of the circumpolar current is only 0.004‰ below that of the water in the western boundary of the region. The difference may not be significant as this value is near the accuracy of the salinity determinations. There appears to be very little influence of turbulent mixing with the antarctic water column as the ACC transverses the region south of Australia and New Zealand.

Acknowledgments. The physical oceanographic studies at Lamont-Doherty Geological Observatory are supported by present grant GV-26230 from the Office of Polar Programs of the National Science Foundation.

Contribution 1771, Lamont-Doherty Geological Observatory.

REFERENCES

Boyer, D. L., and J. R. Guala, Model of the antarctic circumpolar current in the vicinity of the Macquarie ridge, in *Antarctic Oceanology II: The Australian–New Zealand Sector, Antarctic Res. Ser.*, vol. 19, edited by D. E. Hayes, pp. 79–93, AGU, Washington, D. C., 1972.

Burling, R. W., Hydrology of circumpolar waters south of New Zealand, *N. Z. Dep. Sci. Ind. Res. Bull., 149,* 66 pp., 1961.

Bye, J. A. T., Oceanic circulation south of Australia, in

Antarctic Oceanology II: The Australian–New Zealand Sector, Antarctic Res. Ser., vol. 19, edited by D. E. Hayes, pp. 95–100, AGU, Washington, D. C., 1972.

Callahan, J. E., Velocity structure and flux of the antarctic circumpolar current south of Australia, *J. Geophys. Res.*, *76*(24), 5859–5870, 1971.

Commonwealth Scientific and Industrial Research Organisation of Australia, Oceanographical observations in the Indian Ocean in 1960, HMAS *Diamantina* cruise Dm 1/60, *Cruise Rep. 2*, Melbourne, 1962.

Commonwealth Scientific and Industrial Research Organisation of Australia, Oceanographic observations in the Indian Ocean in 1960, HMAS *Diamantina* cruise Dm 1/61, *Cruise Rep. 7*, Melbourne, 1963*a*.

Commonwealth Scientific and Industrial Research Organisation of Australia, Oceanographic observations in the Pacific Ocean in 1961, HMAS *Gascoyne* cruise G 1/61, *Cruise Rep. 8*, Melbourne, 1963*b*.

Commonwealth Scientific and Industrial Research Organisation of Australia, Oceanographical observations in the Indian and Pacific Oceans in 1961, HMAS *Gascoyne* cruise G 2/61, *Cruise Rep. 10*, Melbourne, 1966*a*.

Commonwealth Scientific and Industrial Research Organisation of Australia, Oceanographical observations in the Indian Ocean in 1962, HMAS *Diamantina* cruise Dm 3/62, *Cruise Rep. 18*, Melbourne, 1966*b*.

Commonwealth Scientific and Industrial Research Organisation of Australia, Oceanographical observations in the Pacific and Indian Oceans in 1962, HMAS *Gascoyne* cruises G 2/62 and G 3/63, *Cruise Rep. 16*, Melbourne, 1967*a*.

Commonwealth Scientific and Industrial Research Organisation of Australia, Oceanographical observations in the Indian and Pacific Oceans in 1964, HMAS *Gascoyne* cruise G 3/64, *Cruise Rep. 35*, Melbourne, 1967*b*.

Commonwealth Scientific and Industrial Research Organisation of Australia, Oceanographical observations in the Indian Ocean in 1963, HMAS *Gascoyne* cruise G 2/63, *Cruise Rep. 22*, Melbourne, 1967*c*.

Commonwealth Scientific and Industrial Research Organisation of Australia, Oceanographical observations in the Indian Ocean in 1964, HMAS *Gascoyne* cruise G 2/64, *Cruise Rep. 34*, Melbourne, 1967*d*.

Commonwealth Scientific and Industrial Research Organisation of Australia, Oceanographical observations in the Indian Ocean in 1965, HMAS *Gascoyne* cruise G 2/65, *Cruise Rep. 43*, Melbourne, 1968*a*.

Commonwealth Scientific and Industrial Research Organisation of Australia, Drift bottle releases and recoveries in Bass Strait and adjacent waters 1958–1962, *Oceanogr. Sta. List 78*, Melbourne, 1968*b*.

Commonwealth Scientific and Industrial Research Organisation of Australia, Oceanographical observations in the Indian Ocean 1969, HMAS *Diamantina* cruise Dm 7/69, *Cruise Rep.*, Melbourne, in press, 1972.

Deacon, G. E. R., The hydrology of the Southern Ocean, *Discovery Rep.*, *15*, 1–124, 1937.

Denham, R. N., and A. C. Kibblewhite, Sound velocity structure of the ocean south of Australia and New Zealand, in *Antarctic Oceanology II: The Australian–New Zealand Sector, Antarctic Res. Ser.*, vol. 19, edited by D. E. Hayes, pp. 41–49, AGU, Washington, D. C., 1972.

Eittreim, S., A. Gordon, M. Ewing, E. Thorndike, and P. Bruchhausen, The nepheloid layer and observed bottom currents in the Indian-Pacific antarctic sea, in *Studies in Physical Oceanography—A Tribute to Georg Wüst on His 80th Birthday*, edited by A. L. Gordon, Gordon and Breach, New York, in press, 1971.

Eitteim, S., P. M. Bruchhausen, and M. Ewing, Vertical distribution of turbidity in the South Indian and South Australian basins, in *Antarctic Oceanology II: The Australian–New Zealand Sector, Antarctic Res. Ser.*, vol. 19, edited by D. E. Hayes, pp. 51–58, AGU, Washington, D. C., 1972.

Garner, D. M., Analysis of hydrological observations in the New Zealand region, 1874–1955, *N. Z. Dep. Sci. Ind. Res. Bull.*, *144*, 45 pp., 1962.

Gill, A. E., and K. Bryan, Effects of geometry on the circulation of a three-dimensional southern hemisphere ocean model, *Deep Sea Res.*, *18*, 685–721, 1971.

Gordon, A. L., Recent physical oceanographic studies of antarctic waters, in *Antarctic Research*, edited by L. Quam, pp. 609–629, American Association for the Advancement of Science, Washington, D. C., 1971*a*.

Gordon, A. L., Antarctic polar front zone, in *Antarctic Oceanology I, Antarctic Res. Ser.*, vol. 15, edited by J. L. Reid, pp. 205–221, AGU, Washington, D. C., 1971*b*.

Gordon, A. L., Spreading of antarctic bottom waters, 2, in *Studies in Physical Oceanography — A Tribute to Georg Wüst on His 80th Birthday*, edited by A. L. Gordon, Gordon and Breach, New York, in press, 1971*c*.

Gordon, A. L., On the interaction of the antarctic circumpolar current and the Macquarie ridge, in *Antarctic Oceanology II: The Australian–New Zealand Sector, Antarctic Res. Ser.*, vol. 19, edited by D. E. Hayes, pp. 71–78, AGU, Washington, D. C., 1972.

Gordon, A. L., and R. D. Goldberg, Circumpolar characteristics of antarctic waters, *Antarctic Map Folio Ser.*, folio 13, edited by V. Bushnell, pp. 1–15, Amer. Geogr. Soc., New York, 1970.

Gordon, A. L., and P. Tchernia, Waters of the continental margin off Adélie Coast, Antarctica, in *Antarctic Oceanology II: The Australian–New Zealand Sector, Antarctic Res. Ser.*, vol. 19, edited by D. E. Hayes, pp. 59–69, AGU, Washington, D. C., 1972.

Hayes, D. E., and J. R. Conolly, Morphology of the southeast Indian Ocean, in *Antarctic Oceanology II: The Australian–New Zealand Sector, Antarctic Res. Ser.*, vol. 19, edited by D. E. Hayes, pp. 125–145, AGU, Washington, D. C., 1972.

Houtman, T. J., Surface temperature gradient at the antarctic convergence, *N. Z. J. Geol. Geophys.*, *7*(2), 245–270, 1964.

Houtman, T. J., Water masses and fronts in the Southern Ocean south of New Zealand, *N. Z. Dep. Sci. Ind. Res. Bull.*, *174*, 40 pp., 1967.

Irish, J., and F. Snodgrass, Australian-antarctic tides, in *Antarctic Oceanology II: The Australian–New Zealand Sector, Antarctic Res. Ser.*, vol. 19, edited by D. E. Hayes, pp. 101–116, AGU, Washington, D. C., 1972.

Ishino, M., Studies in the oceanography of the antarctic circumpolar waters, *J. Tokyo Univ. Fish.*, *49*(2), 73–181, 1963.

Jacobs, S. S., Physical and chemical oceanographic observations in the southern oceans, USNS *Eltanin* cruises 7–15, 1963–1964, *Tech. Rep. 1-CU-1-66*, 321 pp., Lamont-Doherty Geol. Observ. of Columbia Univ., Palisades, N. Y., 1965.

Jacobs, S. S., Physical and chemical oceanographic observations in the southern oceans, USNS *Eltanin* cruises 16–21, 1965, *Tech. Rep. 1-CU-1-66*, 128 pp., Lamont-Doherty Geol. Observ. of Columbia Univ., Palisades, N. Y., 1966.

Jacobs, S. S., and A. F. Amos, Physical and chemical oceanographic observations in the southern oceans, USNS *Eltanin* cruises 22–27, 1966–1967, *Tech. Rep. 1-CU-1-67*, 287 pp., Lamont-Doherty Geol. Observ. of Columbia Univ., Palisades, N. Y., 1967.

Jacobs, S. S., A. F. Amos, and P. M. Bruchhausen, Ross Sea oceanography and antarctic bottom water formation, *Deep Sea Res.*, *17*(5), 935–962, 1970*a*.

Jacobs, S. S., P. M. Bruchhausen, and E. B. Bauer, Hydrographic stations, bottom photographs, and current measurements, USNS *Eltanin* cruises 32–36, 460 pp., Lamont-Doherty Geol. Observ. of Columbia Univ., Palisades, N. Y., 1970*b*.

Kibblewhite, A. C., and R. N. Denham, Long-range sound propagation in the south Tasman Sea, *J. Acoust. Soc. Amer.*, *41*, 401–411, 1967.

Mackintosh, N. A., The antarctic convergence and the distribution of surface temperature in antarctic waters, *Discovery Rep.*, *23*, 177–212, 1946.

Reid, J. L., and W. D. Nowlin, Jr., Transport of water through the Drake passage, *Deep Sea Res.*, *18*(1), 51–64, 1971.

Thompson, R., Structure of the antarctic circumpolar current, *J. Geophys. Res.*, *76*(36), 8694, 1971.

Tolstikov, E. I. (Ed.), *Atlas Antarktiki*, vol. 1, G.U.C.K., Moscow, 1966. (English translation in *Soviet Geography: Reviews and Translations*, *8*(5-6), 225 pl., American Geographical Society, New York, 1967.)

Warren, B. A., and A. D. Voorhis, Velocity measurements in the deep western boundary current of the South Pacific, *Nature*, *228*, 849–850, 1970.

Watkins, N., and J. Kennett, Regional sedimentary disconformities and upper Cenozoic changes in bottom water velocities between Australasia and Antarctica, in *Antarctic Oceanology II: The Australian New Zealand Sector*, *Antarctic Res. Ser.*, vol. 19, edited by D. E. Hayes, pp. 273–294, AGU, Washington, D. C., 1972.

Wyrtki, K., *Oceanographic Atlas of the International Indian Ocean Expedition*, 531 pp., Nat. Sci. Found., Washington, D. C., 1971.

Zillman, J., Solar radiation and sea–air interaction south of Australia, in *Antarctic Oceanology II: The Australian–New Zealand Sector*, *Antarctic Res. Ser.*, vol. 19, edited by D. E. Hayes, pp. 11–40, AGU, Washington, D. C., 1972.

SOLAR RADIATION AND SEA–AIR INTERACTION SOUTH OF AUSTRALIA

JOHN W. ZILLMAN
Commonwealth Bureau of Meteorology, Melbourne, Australia

Abstract. Radiation measurements and bulk aerodynamic computations of the sea–air exchange of heat, water vapor, and momentum from data obtained on the USNS *Eltanin* cruises 35–41 provide estimates of the seasonal and latitudinal variation of the surface heat budget components and wind stress for the ocean south of Australia. Zonally averaged from 115° to 160°E, annual mean sensible and evaporative heat losses by the ocean show maxima of approximately 30 w m^{-2} and 100 w m^{-2}, respectively, between 45° and 50°S, and significant seasonal variation is apparent. In the spring and summer, oceanic heat loss to the atmosphere by eddy transfer processes is small south of 55°S, but quite large values occur at high latitudes during autumn and winter. Global radiation measurements suggest that in the most heavily clouded zone of the Antarctic trough less than one third of the extraterrestrial radiation normally reaches the surface. The measured global radiation from the *Eltanin* cruises accords well with data from Macquarie Island at 54.5°S. The diurnal variation of the heat budget components examined at 37.5°S and 60°S is dominated by the daily insolation cycle. In the vicinity of the Antarctic trough in summer, the solar radiation, despite the heavy cloud cover, is the only ocean surface heat budget component of significance. Synoptic influences on the sea–air energy exchange are considerable, large heat losses from the ocean on the order of 250 w m^{-2} occurring with strong southwesterly winds. In northerly flow ahead of intense cyclones, the ocean may take up heat from the atmosphere, but slight evaporation usually continues. A distinct relation between the latitude of maximum atmospheric heating and the belt of most frequent occurrence of atmospheric fronts and cyclogenesis appears to exist. Seasonal computations of mean wind stress on the ocean suggest peak values of the eastward zonal stress component of about 0.2 newton meters^{-2} (2 dyne cm^{-2}) between the latitudes 45° and 50°S.

This paper examines the energy and momentum exchange between the atmosphere and the ocean south of Australia principally on the basis of meteorologic data gathered during the USNS *Eltanin* cruises 35–41. The rationale for this kind of study lies in the far-reaching implications of the large, spatially and temporally varying interfacial energy source arising from the absorption of solar radiation in the upper 10 meters or so of the ocean on the dynamics of both the atmosphere and ocean. In particular for this geographical region, meteorologists have believed for a long time [e.g., *Taylor,* 1932; *Meinardus,* 1938; *Schmitt,* 1957; *Langford,* 1957; *Gray,* 1960; *Gibbs,* 1960] that, despite the seeming homogeneity of such a large expanse of ocean, a profound influence is exerted on the locales of preferred cyclogenesis, system tracks, and year to year shift in atmospheric circulation patterns through the mechanism of sea–air energy exchange by the various zonal asymmetries and meridional discontinuities in the temperature of the ocean surface. The subtropical convergence, the antarctic convergence (oceanic polar front), the edge of the pack ice, and the anomalously warm surface water where the antarctic circumpolar current deflects south of the Campbell plateau (Figure 1) have been attributed roles of varying importance from time to time. No less importantly, these spatial and temporal inhomogeneities in the ocean surface and their associated patterns of horizontal and vertical water movement extending even to the deep ocean have themselves been seen as largely forced by the solar radiation income and the exchange of momentum, heat, and water substance with the overlying atmosphere [*Schott,* 1935; *Deacon,* 1945; *Wyrtki,* 1960]. Recent and more quantitative studies of the atmospheric and oceanic products of this interaction have been made possible through the concerted data-gathering effort initiated by the International Geophysical Year and still more recently by the advent of meteorologic satellites. These studies have in some ways modified but in most respects tended to support the consensus of earlier views [e.g., *van Loon,* 1965; *Taljaard,* 1967; *Martin,* 1968; *Fletcher,*

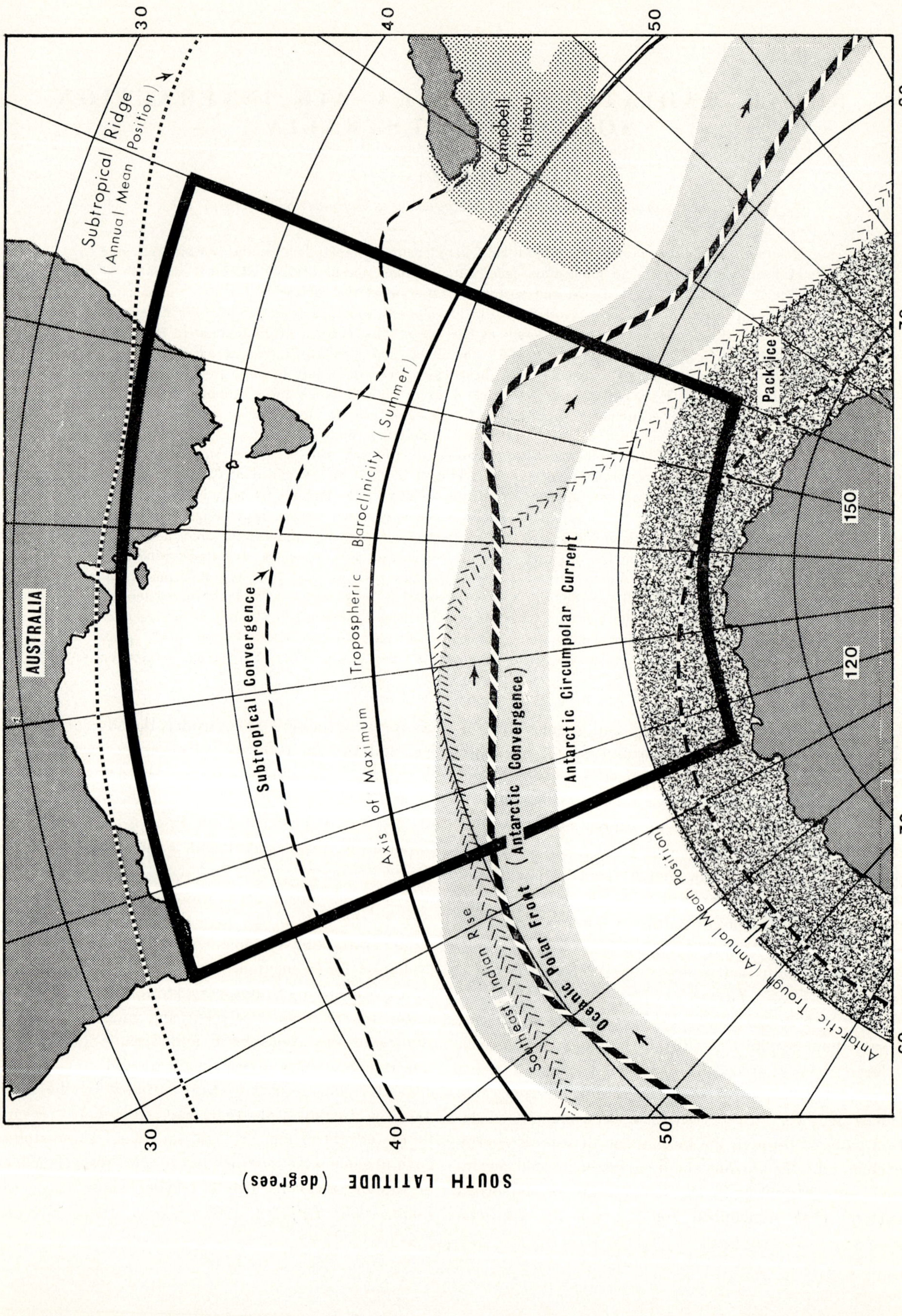
AUSTRALIA
Subtropical Ridge
(Annual Mean Position)
Subtropical Convergence
Axis of Maximum Tropospheric Baroclinicity (Summer)
(Antarctic Convergence)
Antarctic Circumpolar Current
Oceanic Polar Front
Southeast Indian Rise
Campbell Plateau
Pack ice
Antarctic Trough (Annual Mean Position)
30
40
50
60
70
120
150
70
60
SOUTH LATITUDE (degrees)

1969; *Streten*, 1969; *Gordon*, 1971*a*, *b*; *Bullock and Johnson*, 1971].

However, despite the accumulating evidence of the relevance of sea–air interaction to larger scale aspects of meteorology and oceanography (e.g., the dynamics of fronts and cyclones in the atmosphere, the formation of antarctic bottom water, and the maintenance of the circumpolar current and polar front in the ocean), the dearth of reliable conventional meteorologic data for this region has precluded any definitive study of the important sea–air interaction itself. The extreme data deficiency and the seasonal bias that results from whaling and antarctic relief operation schedules have not deterred several notable attempts to derive generalized patterns of extended period mean solar radiation, long-wave radiation exchange, sensible heat flux, evaporation, and vector wind stress. Recent examples treating the area of interest as part of a larger region include the monthly maps of *Budyko* [1963] and *Bernhardt and Philipps* [1958, 1966]; the seasonal maps of *Tolstikov* [1966], *Hellerman* [1967, 1968], and *Zillman* [1967]; and the annual maps of *Albrecht* [1960], *Privett* [1960], and *Landsberg et al.* [1963]. These maps have been based essentially on plausible climatologic fields of cloudiness, sea–air temperature difference, relative humidity, wind speed, and other variables. *Kangos* [1960] and *Viebrock* [1962] determined the meridional variation of evaporation and sensible heat exchange zonally averaged over longitude sectors embracing the region south of Australia but used seasonal mean values of the observables compiled from specific sets of data. *Radok* [1954] had earlier determined the instantaneous radiant, evaporative, and sensible heat fluxes during two meridional traverses between Australia and Antarctica. His results reflected the strong influence of the passing synoptic systems. *De Lisle* [1970] used data from surface synoptic ship reports over an 8-year period to compute sea–air heat transfer by season and 5° latitude and longitude squares for the Tasman Sea. *Zillman* [1968, 1970*b*] made use of radiation and conventional meteorologic measurements gathered during a 3-week, late summer cruise in 1961 of HMAS *Gascoyne* in the Great Australian Bight and Bass Strait to study diurnal factors in the oceanic heat budget. But, apart from these studies of rather restricted scope, some general conclusions based in part on the marine meteorologic measurements of the Soviet antarctic expeditions discussed by *Kort* [1962], and an attempt by *Loewe* [1957] to estimate the annual course of evaporation from meteorologic measurements at Macquarie Island (54.5°S, 159°E), a quantitative picture of the important sea–air exchange processes over the ocean between Australia and Antarctica has been lacking.

The presence of the *Eltanin* in these waters during 1968–1971 provided a unique opportunity to rectify this deficiency partly. The Australian Bureau of Meteorology assumed responsibility for the on-board meteorologic program during 1968 and the author's colleague, W. R. J. Dingle, undertook an augmented schedule of measurement of parameters pertinent to heat exchange computation. Solar and long-wave radiation measurements commenced in 1969. A brief note [*Zillman and Dingle*, 1969] sketched the aims of the sea–air interaction and heat budget study and presented a few preliminary results from cruises 35 and 36. The final account of the work will be published along with a full description of the processing techniques and basic data tabulation when the processing is completed. In the present contribution, some of the more significant results emerging from a first evaluation of the data from *Eltanin* cruises 35–41 in the sector 115°–160°E are presented together with some data from the 1961 *Gascoyne* cruise in the Great Australian Bight. Because of the preliminary nature of the *Eltanin* data treatment, several refinements to the bulk aerodynamic determination of the sensible and latent heat fluxes incorporated in the calculations described earlier [*Zillman and Dingle*, 1969] and to be

Fig. 1. Location map of the Australia-New Zealand sector of the Indian and Pacific oceans with the area under study outlined. The position of the Southeast Indian rise and the Campbell plateau (shown schematically) are from *Gordon and Goldberg* [1970]. The mean maximum extent of the pack ice in spring is based on the maps in *Tolstikov* [1966] and average brightness satellite photo composites (by courtesy of J. Kornfield at the University of Wisconsin). The subtropical convergence is from *Deacon* [1966] west of Tasmania and *Garner* [1959] east of Tasmania. The mean position of the oceanic polar front (antarctic convergence) is based on the location of the salinity minimum at 200 meters as given by *Gordon and Goldberg* [1970]. The antarctic circumpolar current is from *Tolstikov* [1966] and is based on the work of V. G Kort. The shaded channel carries one third of the integrated transport (surface to 3000 meters). The annually averaged latitude of the Antarctic trough is based on *Phillpot* [1968] and *Taljaard et al.* [1969], whereas the axis of maximum baroclinicity in the 1000–500-mb layer in summer is from *Taljaard et al.* [1968].

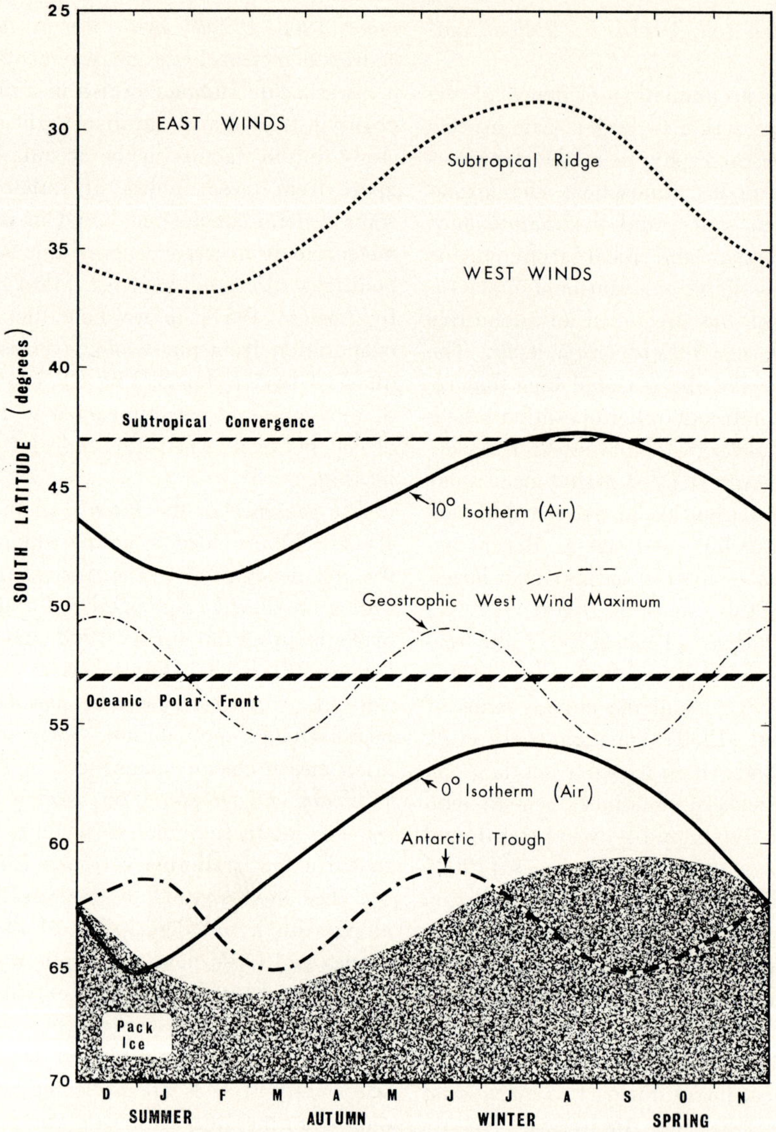

Fig. 2. The seasonal shift in the equatorward and poleward limits of surface westerly winds (taken as the mean position of the subtropical ridge and circumpolar trough, respectively) averaged in longitude over the octant 115°–160°E. Also shown is the average northern limit of the pack ice and the 0° and 10°C surface air temperature isotherms. The subtropical convergence and oceanic polar front also averaged between 115° and 160°E are shown as having no seasonal migration. This figure is compiled mainly from the same sources as Figure 1.

employed in revised form in the final treatment will be omitted here.

LARGE-SCALE SETTING

Meteorologic–Oceanographic Environment

Figure 1 shows the region studied. The most obvious characteristic of the region is a considerable departure from zonality extending from the ocean floor to the stratosphere. The general trend of submarine topography (the Southeast Indian rise), surface water movement and characteristics (the antarctic circumpolar current, the oceanic polar front, the edge of the pack ice, and so on), and atmospheric structure (the Antarctic trough, the belt of maximum tropospheric baroclinicity and cyclogenetic frequency, mean isotherms and isobars, and so on) is from west-northwest to east-southeast. A host of other geophysical and biogeographical parameters exhibit the same general orientation. Some of these parameters (e.g., the northern limits of various species of penguins [*Hedgpeth*, 1969]) are clearly directly related to one or another of the above, whereas some (e.g., the auroral zone [*Bond and Jacka*, 1962]) almost certainly are not.

Under the influence of periodic thermal forcing by solar radiation, the antarctic pack ice, the belt of surface westerly winds, and the atmospheric baroclinic zones undergo well-defined seasonal shifts in latitude and changes in strength, whereas, for example, the seasonally averaged position of the oceanic polar front appears to migrate only slightly if at all. These seasonal shifts of features pertinent to sea–air interaction are summarized in Figure 2. Of particular interest is the fact that, whereas the average northern limit of the westerlies as defined by the axis of the subtropical high pressure ridge reaches its most equatorward position in late winter, the southward extent of the westerlies (as defined by the position of the circumpolar pressure trough) is greatest at the time of their greatest strength around the equinoxes [*Schwerdtfeger*, 1970]. Further, there is no close association between the seasonal shift of the mean pressure trough and the ice edge [*van Loon*, 1967].

Heat Budget Climatology

The atlas type compilations of heat budget components [*Budyko*, 1963; *Privett*, 1960] mentioned earlier, based as they are of necessity on meager and heterogeneous data, provide only a very gross picture of the climatology of sea–air energy exchange in antarctic regions. The broad picture is, however, frequently useful in large-scale circulation studies and as a point of departure in evaluating the results from individual studies, such as those of *Kangos* [1960], *Viebrock* [1962], or the present investigation using *Eltanin* data. Rather than present a series of maps, we may summarize their essential features as isopleths of the heat fluxes (zonally averaged from 115° to 160°E) on a grid the co-ordinates of which denote latitude and time of year. These isopleths are given in Figure 3 and are based on an unpublished map series presented to the 40th Congress of the Australian and New Zealand Association for the Advancement of Science [*Zillman*, 1968]; they have been recomputed from climatologic means of the observables by using a lower bulk transfer coefficient (0.0015 in place of 0.0018) and have been extended to lower latitudes by using the data from the *Marine Climatic Atlas of the World* [*U.S. Navy*, 1957, 1959, 1965, 1969] and *Taljaard et al.* [1969].

Mean annual values of the components of the surface heat budget are shown schematically for 40°, 50°, and 60°S in Figure 3*e*. Compilations such as these qualify only as approximations, and the annual means of the heat balance $R - LE - H$ are very small differences between uncertain large numbers and cannot be regarded as significant. A number of published meteorologic heat budget climatologies for the high latitude oceans of the Southern Hemisphere were evaluated by the author [*Zillman*, 1970*b*], and these include one suggesting that the ocean transports significant amounts of heat out of antarctic regions. An alternative approach uses oceanic heat storage computations to estimate the heat given off during cooling and ice formation and taken up as the ice melts and the water warms in spring. This approach has limitations too [*Fletcher*, 1969].

The outstanding feature of Figure 3 is the very large sensible heat flux to the atmosphere that is implied between 55°S and the pack ice edge in winter. Unfortunately, the sea–air temperature difference patterns that largely specify this feature are poorly documented since the patterns are based on particularly fragmentary and heterogeneous data from scattered sources [e.g., *Vowinckel*, 1956]. This high latitude heat flux maximum is one feature on which data from the *Eltanin* cruises should eventually provide more definite information.

DATA AND COMPUTATIONAL PROCEDURES

Data

The basic data used in this study were obtained during the USNS *Eltanin* cruises 35–41 made between August 1968 and February 1970 and during the HMAS *Gascoyne* cruise G2/61 made between February 21 and March 15, 1961. Only the data between 115° and 160°E have been used, the restriction being imposed because of the much larger departures from zonality that characterize the New Zealand sector (Figure 1). The areal distribution of *Eltanin* data coverage between latitudes 35° and 65°S and longitudes 115° to 160°E is shown in Figure 4, whereas a summary of the distribution of *Eltanin* data by month and 5° latitude bands is shown in Figure 5. The *Gascoyne* cruise G2/61 took the form of an approximate figure eight between 35° and 41°S and 125° and 150°E.

Instrumental and observational details for the rele-

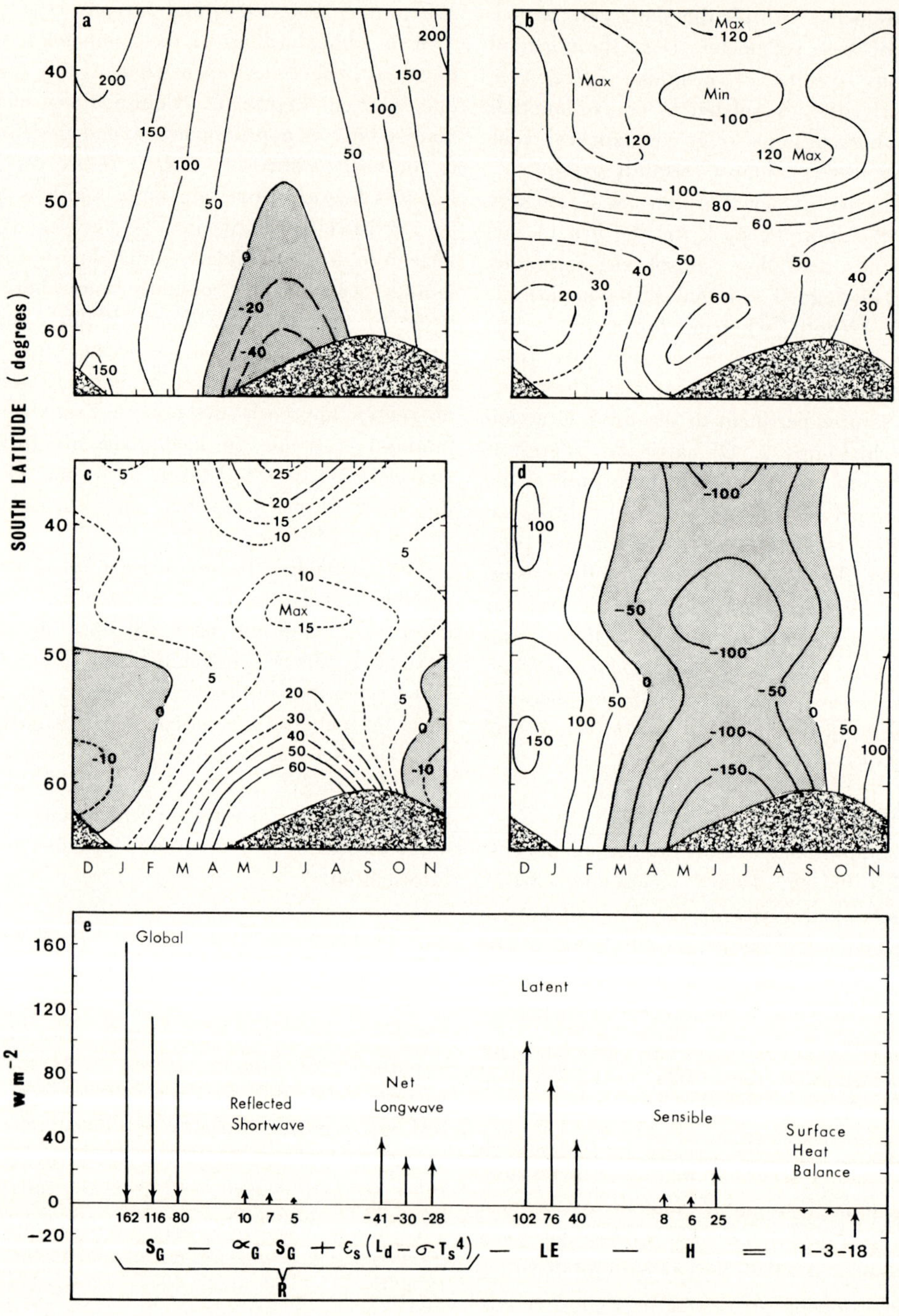

Fig. 3. Summarized surface heat budget climatology for the ocean south of Australia showing (a) net radiation R, (b) the latent heat flux LE, (c) sensible heat flux H, and (d) the heat available for storage and transport by the ocean, $R - LE - H$. The mean annual surface heat budget and its components at latitudes 40°, 50°, and 60°S (left to right for each component) are shown schematically in section *e*. Note that all heat fluxes are in watts per square meter and 1 w m^{-2} = 0.753 kcal/cm^2/year.

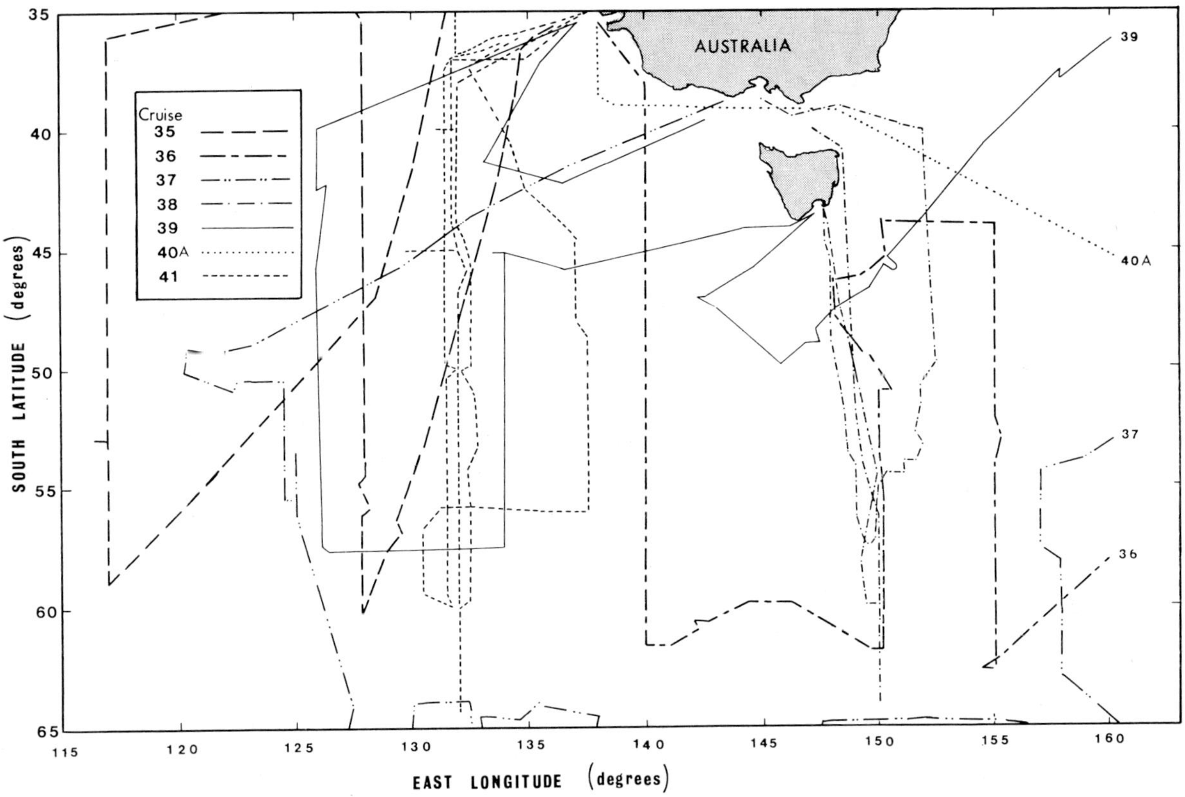

Fig. 4. *Eltanin* cruise tracks: cruise 35, August–October 1968; cruise 36, October–December 1968; cruise 37, January–March 1969; cruise 38, March–May 1969; cruise 39, June–July 1969; cruise 40A, December 1969; and cruise 41, December 1969–February 1970.

vant meteorologic parameters are summarized in the appendix, and the *Eltanin* radiation sensors are shown in Figure 6. Routine meteorologic observations were made every 6 hours (00, 06, 12, 18 UT) throughout all *Eltanin* cruises (1235 observations between 32° and 68°S), and these observations were supplemented by an additional 1298 hourly measurements made mostly between 50° and 60°S. Considerable resolution was thus possible in the vicinity of the oceanic polar front and in studying the variation of heat exchange with the passage of atmospheric synoptic systems. Measurements on the *Gascoyne* cruise were made approximately every 2 hours during daylight but less often through the night.

Computation of the Vertical Fluxes of Momentum, Heat, and Water Vapor

The direct measurement of the vertical eddy fluxes of momentum, heat, and water vapor within the constant flux layer over the sea requires sophisticated instrumentation and an extremely stable platform, such as an oil rig [*Hicks and Dyer*, 1970], a specially designed research tower [*De Leonibus*, 1971], or a gyrostabilized buoy [*Hasse*, 1970]. Lacking this capability we must resort to computational procedures that require only those parameters that can be readily obtained under cruise conditions. The relevant transfer formulas (see the notation list) are the following bulk aerodynamic equations:

$$\boldsymbol{\tau} = \rho C_D(|\mathbf{u}_b - \mathbf{u}_s|)(\mathbf{u}_b - \mathbf{u}_s) \quad (1)$$

$$H = \rho c_p C_H(\theta_s - \theta_a)(|\mathbf{u}_b - \mathbf{u}_s|) \quad (2)$$

$$E = \rho C_E(q_s - q_a)(|\mathbf{u}_b - \mathbf{u}_s|) \quad (3)$$

The derivation, capabilities, and shortcomings of these equations have been extensively discussed in the literature [*Manier and Möller*, 1961; *Deacon and Webb*, 1962; *Roll*, 1965; *Swinbank*, 1965; *Robinson*, 1966]. In the above and in slightly different forms, these equations have been widely used in both climatologic and synoptic studies of sea–air interaction [*Jacobs*, 1942; *Priestley*, 1951; *Budyko*, 1956; *Petterssen et al.*, 1962; *Laevastu*, 1965; *Garstang*, 1967]. Two problems seriously restrict the reliability of the approach: (1) the fact that the bulk transfer coefficients C_D, C_H,

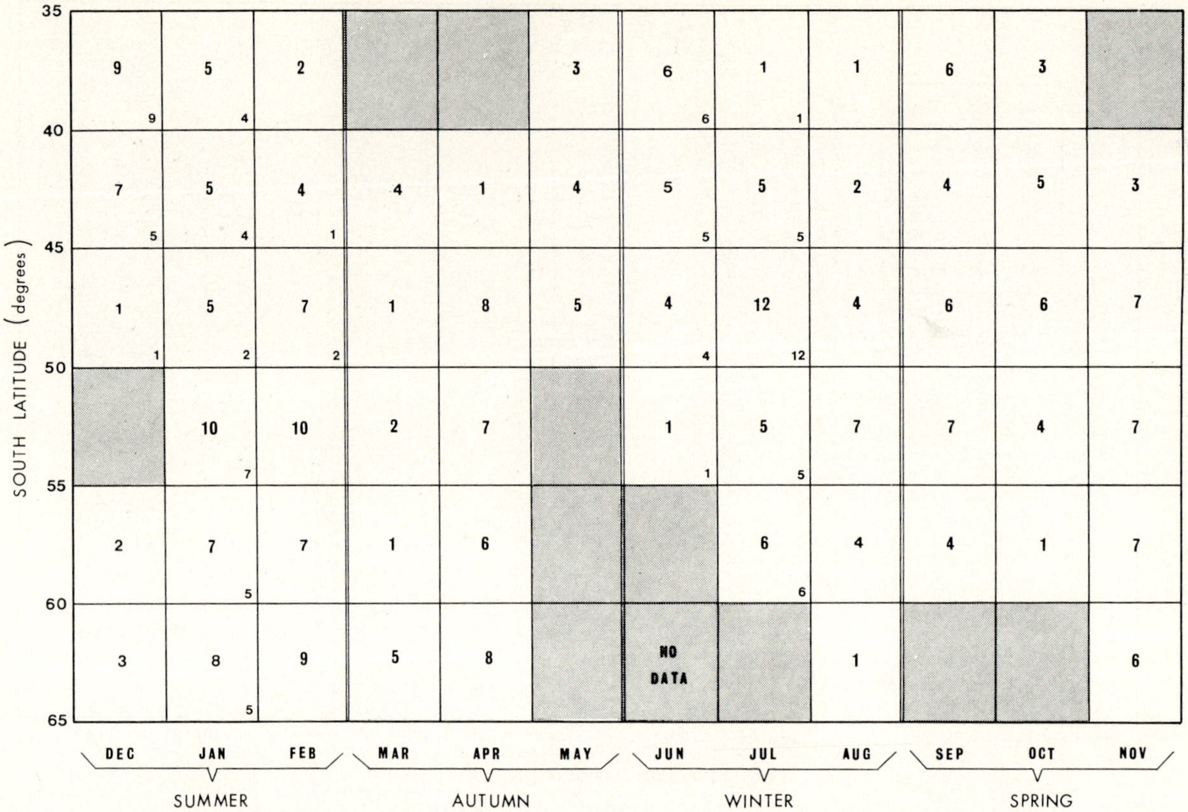

Fig. 5. Availability of *Eltanin* meteorologic data by month and 5° latitude bands. The number of days of data (with a minimum of four and a maximum of 24 observations per day) is given in the center of each box. In the small figures in the lower right of each box is the number of days with continuous radiation measurement.

and C_E are regarded as poorly known, are certainly dependent on atmospheric stability, and even under neutral stratification may vary with wind speed and other parameters (fetch, duration of the wind, and so on); and (2) the determination of true interface values for θ_s, q_s, and $\mathbf{u}_s$ as opposed, respectively, to conventional (e.g., dip bucket) sea water temperatures (corrected for atmospheric pressure), saturation specific humidity (corrected for salinity) at bucket temperatures, and zero.

The transfer coefficients C_D, C_H, and C_E appropriate to measurement levels at a (temperature and humidity) and b (wind) may be shown to be given for arbitrary stratification by

$$C_D = k^2[\phi_u(\frac{b}{L})]^2\,[f_u(\frac{b}{L}) - f_u(\frac{z_0}{L})]^{-2} \qquad (4)$$

$$C_H = k^2[\phi_u(\frac{b}{L})]^2\,[f_u(\frac{b}{L}) - f_u(\frac{z_0}{L})]^{-1} \cdot [f_\theta(\frac{a}{L}) - f_\theta(\frac{z_\theta}{L})]^{-1} \qquad (5)$$

$$C_E = k^2[\phi_u(\frac{b}{L})]^2\,[f_u(\frac{b}{L}) - f_u(\frac{z_0}{L})]^{-1} \cdot [f_q(\frac{a}{L}) - f_q(\frac{z_q}{L})]^{-1} \qquad (6)$$

where the functions f are given by $f = \int \phi d \ln (z/L)$ [*Sheppard*, 1970]. Explicit evaluation requires a knowledge of the form of the ϕ functions, the roughness lengths for the wind, temperature, and humidity profiles (z_0, z_θ, and z_q, respectively), and the Monin-Obukhov stability length L.

At least three recent treatments adopt empirical forms for ϕ for both stable and unstable stratification and account for the important stability effects explicitly. E. K. Webb (personal communication, 1969) and *Clarke* [1970] give, respectively, C/C_N (the ratio of the transfer coefficient to its value for neutral stability) and the nonneutral transfer coefficient itself in terms of the surface roughness length and stability as measured by a bulk Richardson number. *Deardorff* [1968] from a somewhat different approach generates

C/C_N in terms of the neutral drag coefficient and a different bulk Richardson number.

The problem of determining z_0, z_θ, and z_q is to some extent alleviated by recent evidence that all three values may be taken as identical with little error [*Sheppard,* 1970] and by the conclusion [*Brocks and Krügermeyer,* 1971] from a reanalysis of a large amount of North Sea profile data that for neutral stratification z_0 may be taken as 0.015 cm with negligible dependence on wind speed. These findings may be regarded as superseding the use of (1) theoretically inferred z_θ and z_q as recommended by *Swinbank* [1965], and (2) wind dependent z_0 from the widely adopted relation between drag coefficient and wind speed given by *Deacon and Webb* [1962]; both schemes were employed by *Zillman and Dingle* [1969].

For the present treatment of the *Eltanin* data, stability effects on the bulk coefficients have not been incorporated in the individual computations because the revised procedures had not been integrated into the data-processing routines at the time of preparation of this paper. However, on the basis of computations with the earlier scheme, the modification of the results by stability effects is discussed qualitatively. Further, the adoption here of a constant roughness length of 0.015 cm (or, equivalently, a 10-meter drag coefficient of 0.0013 for near-neutral stability) from the findings of *Brocks and Krügermeyer* [1971] permits us to revert to the simpler traditional approach of using a single constant transfer coefficient in each of equa-

Fig. 6. The *Eltanin* radiation sensors: (*right*) the gimbal-mounted Epply pyranometer and (*left*) modified Funk net radiometer. Photograph by W. R. J. Dingle.

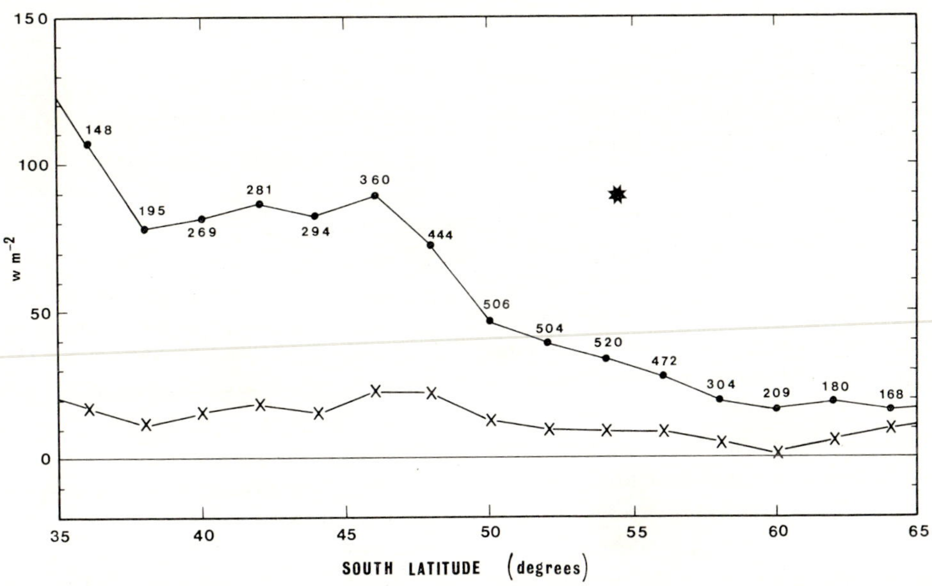

Fig. 7. Averages of the sensible and latent heat flux in overlapping 4° latitude bands plotted against latitude using all *Eltanin* data for all seasons. The numbers of individual computed flux values in each latitude band are shown adjacent to the latent heat flux data plot. The asterisk at 54.5°S indicates the annual global radiation average for Macquarie Island based on 1 year of data. Sensible heat flux H is marked by ×, and latent heat flux LE is marked by darkened circles.

tions 1–3. If 9 meters is adopted as a nominal measurement height for temperature and humidity (an average of the heights actually used) and 18 meters is adopted for wind, the transfer coefficients for near-neutral conditions ($z/L \approx 0$) are from equations 4–6:

$$C_D = k^2 \, (\ln 18 - \ln 0.00015)^{-2} = 0.00123 \qquad (7)$$

$$C_H, C_E = k^2 \, (\ln 18 - \ln 0.00015)^{-1} \cdot (\ln 9 - \ln 0.00015)^{-1} = 0.00130 \qquad (8)$$

In proceeding to use the bulk transfer equations with the coefficients from equations 7 and 8, the following simplifications and approximations are made:

1. The value $\mathbf{u}_s$ is neglected though it is believed [*Kraus*, 1967] to average around 4% of anemometer level wind speed. This approximation is not as bad as it appears since $\mathbf{u}_s$ is to some extent absorbed into the *Brocks and Krügermeyer* [1971] definition of C_D. The velocity of any general current (the antarctic circumpolar current in this case) is not, however, allowed for.

2. Potential temperature differences are replaced by temperature differences, the discrepancy being 0.12°C at most.

3. Bulk sea temperatures as measured are used for the sea surface temperatures. The discrepancy is not negligible [*Osborne*, 1964], the cool surface skin being believed to average about 0.5°C [*Roll*, 1965]. To the extent that this deviation between surface temperature and that of the water below is approximately constant, the resulting computed sensible heat fluxes may be simply modified; because of the nonlinear dependence of saturation vapor pressure on temperature, subsequent modification of evaporation estimates would be less simple. *Hasse* [1971], however, finds the deviation of the surface temperature from the bulk water temperature to vary with the heat flow through the sea–air interface and inversely with the wind speed. His results suggest the application of a small multiplicative correction factor.

4. The relation between specific humidity q and the more simply determined vapor pressure e, namely $q = 0.622e(p - 0.378e)^{-1}$, is approximated by $q = 0.625 \, (e/p)$ so that $q_s - q_a = 0.625 \, p^{-1} \, (e_s - e_a)$. Vapor pressure differences $(e_s - e_a)$ are derived from T_s, T_a, T_w, and p (1) by assuming e_s to be 0.98 of saturation vapor pressure over pure water at the bulk sea temperature T_s, (2) by adopting Tetens' formula for saturation vapor pressure as a function of temperature [*Dilley*, 1968], and (3) by applying the psychrometric formula for fully ventilated conditions. The working formula is

$$e_s - e_a = 0.98 \, \alpha \exp\left(\frac{\beta_s T_s}{\gamma_s + T_s} \right) - \alpha \exp\left(\frac{\beta_w T_w}{\gamma_w + T_w} \right) + Ap \, (1 + 0.00115 T_w) \, (T_a - T_w) \qquad (9)$$

where

$\alpha = 6.1078$;
$A = 0.000660$ for $T_w \geqslant 0°C$;
$A = 0.000594$ for $T_w < 0°C$;
$\beta_s = 17.269$;
$\beta_w = 17.269$ for $T_w \geqslant 0°C$;
$\beta_w = 21.875$ for $T_w < 0°C$;
$\gamma_s = 237.30$;
$\gamma_w = 237.30$ for $T_w \geqslant 0°C$;
$\gamma_w = 265.50$ for $T_w < 0°C$.

The values $(e_s - e_a)$ and p are in millibars, whereas T_s, T_a, and T_w are in degrees centigrade.

5. Values of air density, specific heat, and the latent heat of evaporation (by which the evaporation E must be multiplied to give the so-called latent heat flux) were computed from standard formulas [*List*, 1964] appropriate to the prevailing temperature, humidity, and atmospheric pressure.

The final forms of the bulk aerodynamic equations with the approximations and numerical values set out above are

$$\boldsymbol{\tau} = 0.00123 \rho (|\mathbf{u}|) \mathbf{u} \qquad (10)$$

$$H = 0.0013 \rho c_p (T_s - T_a) (|\mathbf{u}|) \qquad (11)$$

$$LE = 0.000813 (\rho L / p) (e_s - e_a) (|\mathbf{u}|) \qquad (12)$$

The international metric system of units (SI units) is employed (e.g., see *Grimson* [1970] and *Duncan et al.* [1970]) with the following partial exceptions:

1. Pressures and vapor pressures are quoted in millibars (1 mb = 10^2 newton meters^{-2}).

2. Wind stress is quoted in dyne cm^{-2} (1 dyne cm^{-2} = 0.1 newton meter^{-2}).

3. Roughness lengths are quoted in centimeters.

4. Cloud amounts are given in eighths (octas) of sky coverage.

5. Temperatures are in most instances quoted in degrees centigrade.

LATITUDINAL VARIATION OF HEAT FLUXES

The individual values for the sensible heat flux H and the latent heat flux LE were grouped in 4° latitude bands centered on each even degree of latitude. Thus except at the northern and southern limits, every in-

TABLE 1. Averages by Season in Overlapping 4° Latitude Bands of the Sensible and Latent Heat Fluxes Based on All Available Data (Upper Line) and 6-Hourly Data Only (Lower Line)

	Sensible Heat Flux, w m^{-2}					Latent Heat Flux, w m^{-2}					Number of Values				
Latitude Band, °	Summer	Autumn	Winter	Spring	All Seasons	Summer	Autumn	Winter	Spring	All Seasons	Summer	Autumn	Winter	Spring	All Seasons
32.0–35.9	23.0	...	15.2	25.4	23.5	164.3	...	57.0	115.1	138.6	46	...	5	33	84
	17.1	...	15.2	27.4	22.0	119.5	...	57.0	96.8	102.9	24	...	5	28	57
34.0–37.9	15.7	...	11.2	26.2	17.9	114.2	...	62.3	112.8	106.8	89	...	20	39	148
	8.9	...	11.2	27.8	15.6	83.3	...	62.3	115.5	89.9	49	...	20	34	103
36.0–39.9	10.7	5.7	12.4	16.8	11.3	81.7	63.8	65.1	84.3	77.7	121	19	30	25	195
	7.8	5.7	12.4	16.8	10.4	71.3	63.8	65.1	84.3	71.3	51	19	30	25	125
38.0–41.9	12.6	28.2	11.0	14.1	15.4	78.9	115.9	58.0	80.3	81.9	132	49	50	38	269
	18.3	13.8	14.4	12.3	15.4	100.5	82.0	69.1	79.4	83.9	46	38	38	29	151
40.0–43.9	10.2	52.6	12.7	13.9	18.1	72.2	172.1	62.1	80.8	86.9	123	44	66	48	281
	15.6	34.5	11.7	12.4	17.0	90.1	127.7	64.8	80.3	87.4	40	25	37	39	141
42.0–45.9	9.3	53.6	6.6	18.8	14.8	63.8	189.7	60.0	97.0	81.7	124	31	88	51	294
	9.3	60.2	9.0	18.8	17.5	58.5	199.5	58.5	97.0	85.8	34	14	45	51	144
44.0–47.9	17.8	24.7	19.9	34.7	22.3	80.7	112.3	77.3	115.4	89.3	119	44	135	62	360
	21.3	32.8	16.7	29.1	23.5	70.7	133.6	72.2	103.8	89.2	30	17	60	56	163
46.0–49.9	10.9	10.7	35.5	34.1	22.1	65.3	48.7	84.8	101.7	72.2	119	119	146	60	444
	15.5	12.7	33.8	27.4	23.4	52.3	50.4	85.8	87.5	71.3	32	56	62	54	204
48.0–51.9	4.5	9.4	28.7	30.4	13.7	35.9	31.7	66.8	81.4	45.9	208	138	111	49	506
	8.1	9.4	32.3	30.4	19.1	40.1	40.6	79.0	81.4	58.5	66	55	53	49	223
50.0–53.9	4.5	−0.2	24.5	26.8	9.8	32.7	21.7	61.2	60.8	39.0	251	102	92	59	504
	7.1	0.7	27.0	26.8	16.9	37.5	29.0	64.3	60.8	39.7	69	19	40	59	187
52.0–55.9	1.3	12.3	23.5	13.9	9.4	27.0	33.8	53.7	28.5	34.0	252	105	106	57	520
	2.3	18.4	23.3	13.9	13.2	25.3	46.5	53.1	28.5	35.7	52	22	41	57	172
54.0–57.9	−0.7	19.8	19.9	2.7	8.6	18.5	35.6	41.9	24.2	27.9	220	104	102	46	472
	1.9	20.7	23.3	2.7	11.6	20.6	38.9	42.3	24.2	31.1	47	37	46	46	176
56.0–59.9	−5.0	15.7	25.7	−2.8	4.8	7.4	29.4	42.8	18.0	19.5	145	67	49	43	304
	−1.6	14.5	26.3	−2.8	7.0	10.3	29.4	41.1	18.0	22.8	42	30	29	43	144
58.0–61.9	−4.9	14.7	23.1	1.1	1.3	12.4	27.2	38.2	13.3	16.5	117	41	8	43	209
	−2.1	12.1	23.1	1.1	3.1	14.8	20.9	38.2	13.3	16.9	54	23	8	43	128
60.0–63.9	0.4	14.8	31.2	6.7	5.9	15.7	23.9	47.9	17.6	18.9	103	50	4	23	180
	0.4	13.9	31.2	6.7	6.1	16.0	23.4	47.9	17.6	19.3	58	28	4	23	113
62.0–65.9	7.1	16.1	...	...	9.6	14.5	16.3	...	...	15.0	121	47	...	...	168
	8.1	15.2	...	...	10.0	15.9	16.9	...	...	16.2	93	33	...	...	126
64.0–67.9	9.8	19.3	...	...	12.2	17.2	16.8	...	...	17.1	77	26	...	...	103
	9.8	15.2	...	...	11.1	17.5	13.7	...	...	16.6	73	24	...	...	97

dividual value was included twice in a north-south profile. Averages were computed by season (summer includes December, January, February; autumn includes March, April, May; winter includes June, July, August; spring includes September, October, November) and for all data irrespective of season. With some reservations apparent from Figure 5, notably the lack of high latitude data in late autumn and early winter, the average of all data irrespective of season may be regarded as approximating an annual profile. All averages were computed by using all available data (6-hourly plus hourly observations) and were repeated by using 6-hourly data only. Both sets of averages are given in Table 1 along with the number of individual values represented in each 4° latitude band.

Figure 7 shows the overlapping 4° latitude averages of sensible and latent heat flux from all data irrespective of season. Profiles based only on 6-hourly data differ slightly. Both sensible and latent heat fluxes show maxima between 45° and 50°S. Whereas the latent heat flux (and hence also the evaporation) falls off steadily southward from the peak at 46°S, the sensible heat flux remains almost constant over 5° of latitude before falling to a minimum near zero at 60°S.

Figures 8, 9, 10, and 11 give the seasonal breakdown and include (for summer and winter only) the global radiation from *Eltanin* cruises 39–41 also averaged in overlapping 4° latitude bands. The small numbers accompanying the global radiation plots are

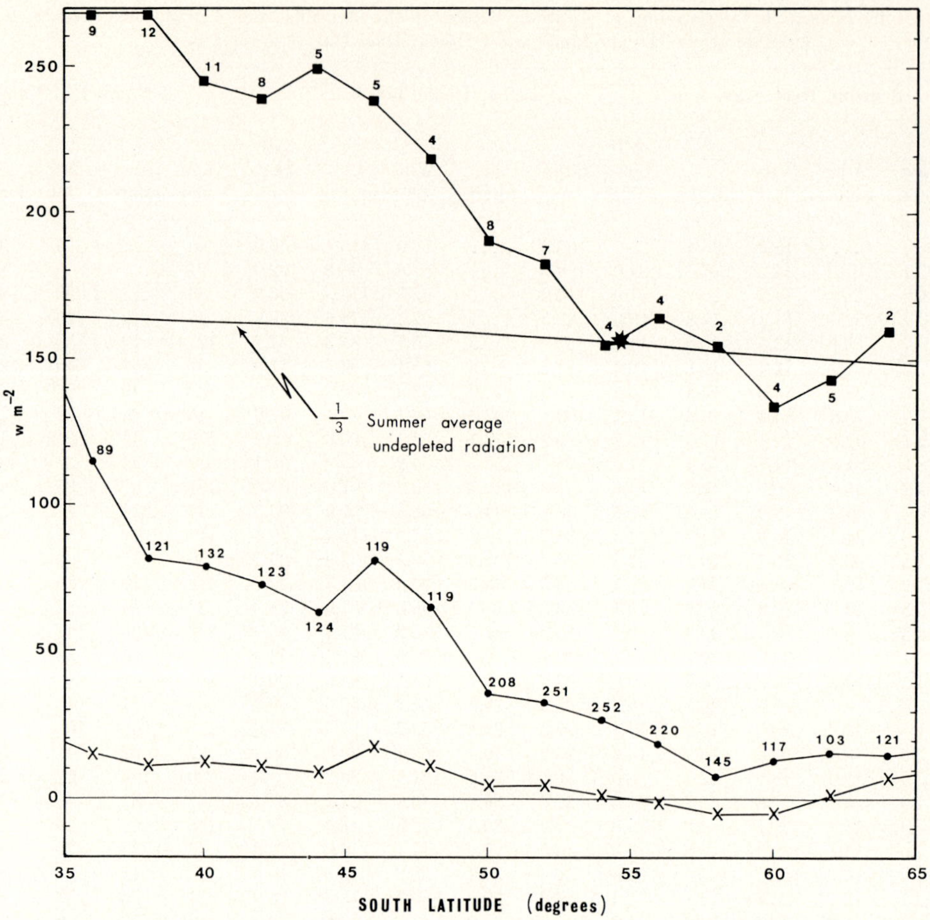

Fig. 8. Meridional profiles of global radiation and sensible and latent heat flux using all *Eltanin* data recorded during the summer months. The asterisk at 54.5°S gives the summer average global radiation for Macquarie Island. Global radiation is marked by the darkened squares, sensible heat flux H is marked by $\times$, and latent heat flux LE is marked by darkened circles.

the number of complete days of data for which the noon latitude of the *Eltanin* fell within the particular latitude band. The global radiation fluxes are not the actual recorded values, but have been adjusted for each latitude band by taking S_G (adjusted) as equal to the product of the seasonal mean extraterrestrial radiation times the ratio of the summation of the recorded daily global radiation totals to the summation of the daily total extraterrestrial radiation for the days of record.

In both summer and winter as the heavily clouded region of the Antarctic trough is approached from the north, the global radiation flux falls below one-third the amount that would be received if the atmosphere were absent. The asterisks (Figures 7–11) at latitude 54.5°S give the respective seasonal averages of the global radiation recorded at Macquarie Island (54.5°S, 159°E) during 1968–1969 as evaluated by the author in collaboration with G. E. Cook of the Commonwealth Bureau of Meteorology, Australia.

The broad features apparent in the annual profile are seen also in the individual seasons. The autumn peak in both sensible and latent heat loss by the ocean is much larger and at a lower latitude than in the other seasons, but note that all the autumn data were obtained during cruise 38 in which several north-south traverses were made approximately along 150°E, and this data may not be representative of the area westward of Tasmania. Other features in Figures 8, 9, 10, and 11 are the rather large autumn-winter sensible heat fluxes to the atmosphere in latitudes 55°–60°S as contrasted with the spring-summer values that are negative (air warming the sea) or near zero.

Although no standard deviations or other statistical measures of the variability of the H and LE values

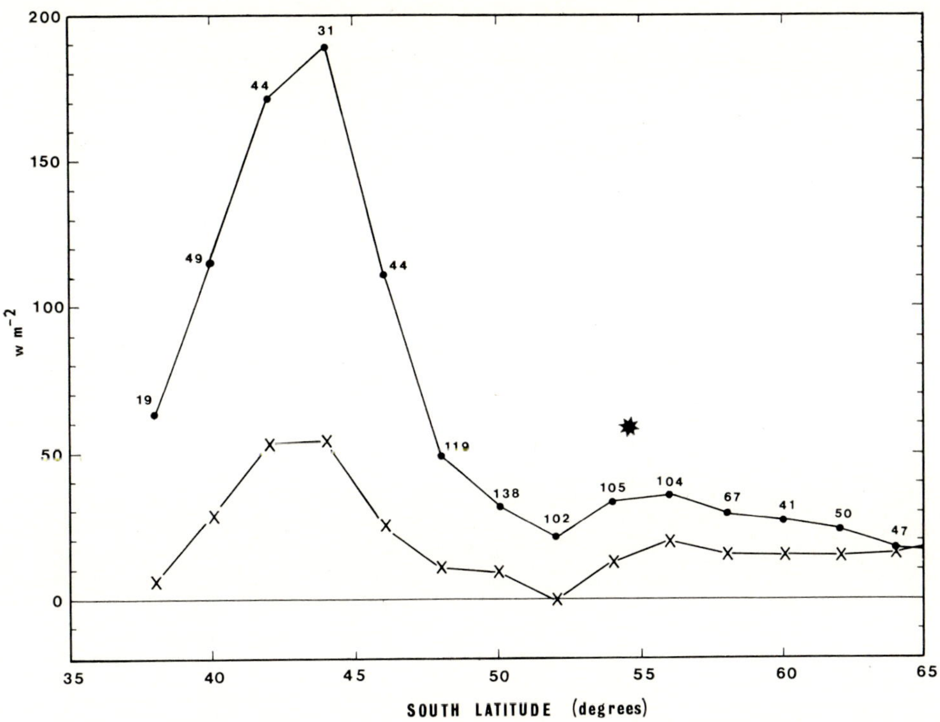

Fig. 9. Meridional profiles of global radiation and sensible and latent heat flux using all *Eltanin* data recorded during the autumn months. No global radiation was available. Latent heat flux *LE* is marked by darkened circles, and sensible heat flux *H* is marked by ×.

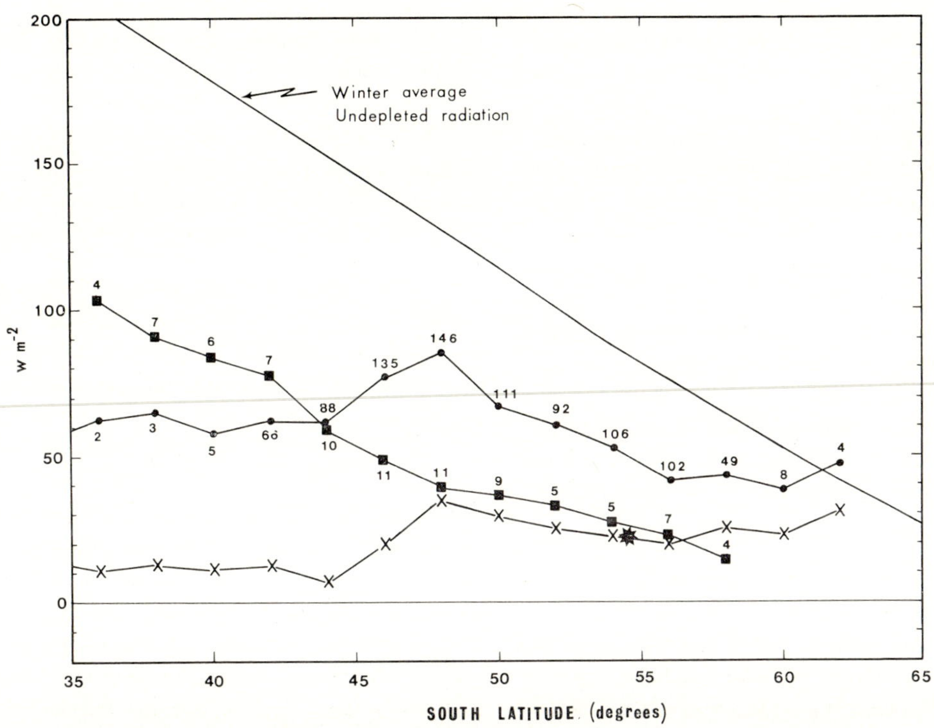

Fig. 10. Meridional profiles of global radiation and sensible and latent heat flux using all *Eltanin* data recorded during the winter months. Global radiation is marked by darkened circles, sensible heat flux *H* is marked by ×, and latent heat flux *LE* is marked by darkened squares.

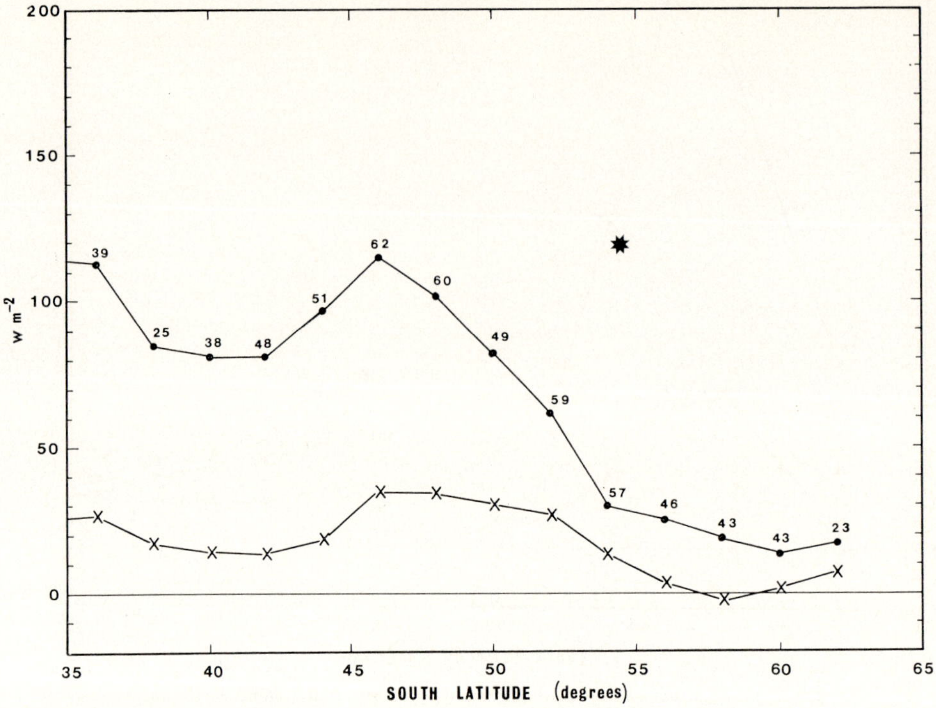

Fig. 11. Meridional profiles of global radiation and sensible and latent heat flux using all the *Eltanin* data recorded during the spring months. Sensible heat flux H is marked by $\times$, and latent heat flux LE is marked by darkened circles.

are included in Figures 7–11, the strong influence of specific synoptic situations (to be discussed in a later section) is evident if values of the Bowen ratio (H/LE) are computed from the seasonal averages of the heat fluxes in Table 1. The erratic behavior of the Bowen ratio indicates the limited confidence that can be placed in all but the most conspicuous features of the heat flux profiles derived from such limited data samples.

Table 2 presents average values of the sea–air temperature and vapor pressure differences and wind speed based on the same data (i.e., 6-hourly plus hourly) as used to compute the heat fluxes plotted in Figures 7–11. One of the important objects of this study was to derive, as functions of latitude and season, values for

$$\frac{\langle (T_s - T_a) u \rangle}{\langle T_s - T_a \rangle \langle u \rangle} \quad \text{and} \quad \frac{\langle (e_s - e_a) u \rangle}{\langle e_s - e_a \rangle \langle u \rangle}$$

If shown to be reproducible, these quantities would be the factors by which bulk transfer coefficients appropriate to instantaneous flux computations should be multiplied if an estimate of mean fluxes from the climatologic means of the observables is desired. No simple result emerges, and the point will not be discussed in detail here. Suffice it to say that in dealing with seasonal means of the 4° latitude group containing at least 40 individual observations, errors of up to 20% for the latent heat flux and 50% for the sensible heat flux may result if mean values for $(e_s - e_a)$, $(T_s - T_a)$, and u are used with the same transfer coefficients as those that apply for instantaneous flux computation.

DIURNAL VARIATION OF RADIATION AND HEAT BUDGET COMPONENTS

The radiation budget for the sea surface may be written

$$R = S_G - \alpha_G S_G + L_d - (1 - \epsilon_s) L_d - \epsilon_s \sigma T_s^4 \qquad (13)$$

The usually dominant term on the right-hand side is the global radiation S_G. It is zero through the night and, depending on cloud conditions, tends to reach its maximum around local noon.

Figures 12 and 13 show the measured diurnal variation of the four terms on the right-hand side of equation 13 in two quite dissimilar regimes. Figure 12 represents an average over 18 days of the *Gascoyne* cruise in the Great Australian Bight and Bass Strait in February–March 1961. The average noon latitude

TABLE 2. Averages by Season in Overlapping 4° Latitude Bands of the Sea–Air Temperature and Vapor Pressure Differences and Wind Speed Based on All 6-Hourly and Hourly *Eltanin* Cruise Data

Latitude Band, °	$T_s - T_a$, °C				$e_s - e_a$, mb				u, m s^{-1}			
	Summer	Autumn	Winter	Spring	Summer	Autumn	Winter	Spring	Summer	Autumn	Winter	Spring
32.0–35.9	1.44	...	1.78	1.34	6.36	...	4.49	4.28	10.65	...	4.64	11.69
34.0–37.9	1.22	...	1.26	1.25	5.98	...	4.39	3.82	7.93	...	5.97	11.71
36.0–39.9	1.05	0.85	1.21	0.95	5.57	4.49	3.92	3.40	6.09	5.86	7.14	9.43
38.0–41.9	1.15	2.21	0.80	0.67	5.10	6.19	2.51	2.67	6.16	6.94	10.04	12.94
40.0–43.9	0.80	3.33	0.86	0.68	4.05	7.43	2.55	2.59	7.30	8.99	10.39	13.32
42.0–45.9	0.69	2.65	0.51	0.88	3.47	6.16	2.26	3.00	7.99	12.91	13.12	13.21
44.0–47.9	1.20	1.11	0.94	1.33	3.94	3.45	2.35	3.02	9.10	13.23	13.72	14.68
46.0–49.9	0.65	0.62	1.82	1.58	2.45	1.75	2.77	3.10	12.37	11.59	11.31	13.00
48.0–51.9	0.33	0.29	1.71	1.72	1.31	1.15	2.50	2.94	11.52	10.92	10.15	11.44
50.0–53.9	0.41	−0.09	1.39	1.17	1.49	0.69	2.14	1.79	9.96	11.20	11.60	12.43
52.0–55.9	0.19	0.66	1.23	0.59	1.28	1.23	1.73	1.05	9.12	10.40	12.64	11.77
54.0–57.9	−0.02	1.50	1.23	0.26	0.79	1.69	1.47	0.94	9.12	7.87	11.43	10.52
56.0–59.9	−0.26	1.59	1.64	−0.02	0.34	1.83	1.77	0.76	9.29	6.68	9.10	10.13
58.0–61.9	−0.26	1.31	1.23	0.12	0.47	1.58	1.41	0.57	9.82	7.17	9.53	8.76
60.0–63.9	0.09	0.98	1.47	0.37	0.65	1.07	1.49	0.71	8.84	8.81	12.75	7.95
62.0–65.9	0.70	1.13	...	...	0.85	0.73	...	...	7.30	7.26	...	...
64.0–67.9	0.97	1.55	...	...	1.04	0.88	...	...	7.46	5.39	...	...

The number of observations in each category may be found in Table 1.

of the ship was 37.4°S, and average noon sea surface temperature and cloudiness were 20.0°C and 3.3 octas, respectively. Figure 13 gives the same information from *Eltanin* cruise 41 averaged over 6 days in January 1970 when the vessel was between 55° and 65°S. The average noon latitude of the *Eltanin* was 59.8°S, and the noon sea surface temperature and cloudiness were 2.4°C and 7.8 octas, respectively.

Comparing Figures 12 and 13 we see that, despite the longer hours of daylight and comparable values for the undepleted radiation in the high latitude summer, the heavy cloud cover reduces the global radiation to far below that recorded at lower latitudes. Also, with the air at deck level slightly warmer than the sea and with the overcast low cloud radiating downward as an effectively black body shell at temperatures comparable to the surface value, the downward long-wave radiation almost exactly balances the emission by the sea in the high latitudes. On the other hand, a net long-wave heat loss of about 60 w m^{-2} (W m^{-2} in S.I. unit notation) from the sea is found during the *Gascoyne* cruise in lower latitudes. The daylight decrease in the downward long-wave radiation in both sets of data is puzzling. Instrumental error cannot be ruled out completely (the L_d component is obtained as a difference between the all-wave downward (i.e., the global plus downward long-wave radiation) and global components on the *Eltanin* cruise and as a radiation balance residual with assumed sea surface albedo on the *Gascoyne* cruise) but is not considered likely. A similar pattern was found with different instrumentation over the tropical Indian Ocean [*Zillman*, 1970*b*], and *Paltridge* [1970] describes an essentially similar effect over land from the measurements of a net long-wave radiometer.

Figure 14 shows the net incoming radiation to the sea for the two sets of data. Its diurnal variation is seen to be almost exactly determined by the global radiation in the high latitude data and a close though not exact correspondence is evident at lower latitudes. Figure 14 also shows the other heat budget components and the surface heat balance $R - LE - H$. Both sensible and latent heat fluxes in the 6 days of *Eltanin* data (based on hourly measurements) are effectively zero, the evaporative heat flux averaging a few watts per square meter positive and the sensible heat flux averaging slightly negative with no perceptible diurnal variation. Thus whereas in the lower latitudes a large part of the absorbed solar radiation is used both to supply the latent heat for evaporation that continues steadily by night and day and to warm the lower atmosphere by infrared radiation, in the vicinity of the Antarctic trough almost all the incoming solar radiation is available for heating the surface water with no significant loss to the atmosphere. This observation is, of course, restricted to the particular short periods of data uses. Certainly there are comparable periods during the summer when quite large sensible and evaporative heat losses by the ocean occur in the environs of the Antarctic trough.

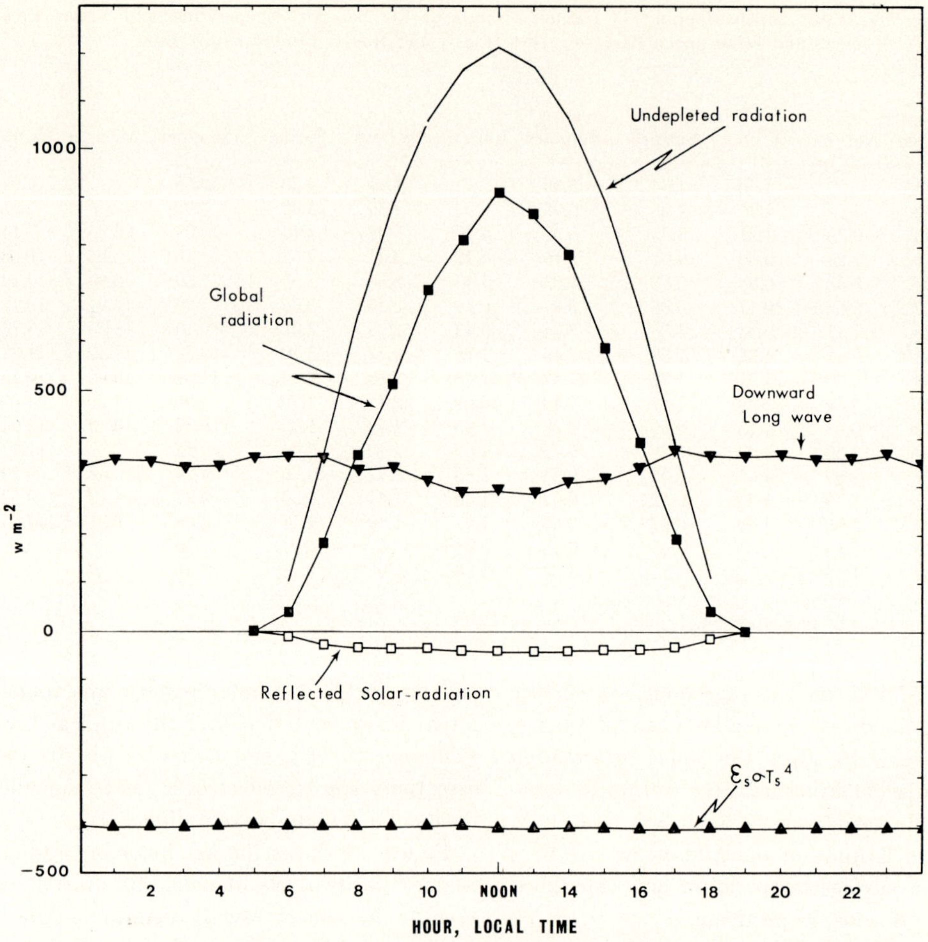

Fig. 12. Mean diurnal variation of the extraterrestrial solar radiation and sea surface radiation balance components during the *Gascoyne* cruise G2/61 in February–March 1961. The small reflected long-wave radiation term is not shown.

SYNOPTIC FACTORS IN SEA-AIR HEAT EXCHANGE

Dependence of Sensible and Latent Heat Fluxes on Vector Wind

Clearly from the form of the bulk aerodynamic equations, all other things being equal, the magnitudes of the heat fluxes will be greater with stronger winds. With essentially zonal sea surface isotherms, air moving northward will tend to be colder than the water, and a heat flux from sea to air will exist (i.e., H will be positive). With winds from the north, the air will tend to be warmer than the underlying ocean, and H will tend negative. Thus we are led to expect a definite dependence of the sensible (and also the latent) heat flux on the vector surface wind. A number of factors influence the exact form of the dependence (stability and past history of the air, orientation and spacing of sea water isotherms, and so on), and assessing their magnitude from physical considerations alone is difficult.

Enough data are available from *Eltanin* cruises 35–41 to examine this question empirically, at least for the more frequently occurring wind directions. Figure 15 shows the sensible and latent heat fluxes averaged in three wind speed categories and 30° wind azimuth intervals for all 6-hourly data irrespective of season between 40° and 50°S. Figure 16 gives the results of a corresponding analysis of data between 50° and 60°S. With strong to gale force south-southwest winds in both latitude bands, total heat losses to the atmosphere average about 250 w m^{-2}. On the other hand, with northerly winds of comparable strength, although the air cools by contact with the ocean, slight evaporation continues on the average and

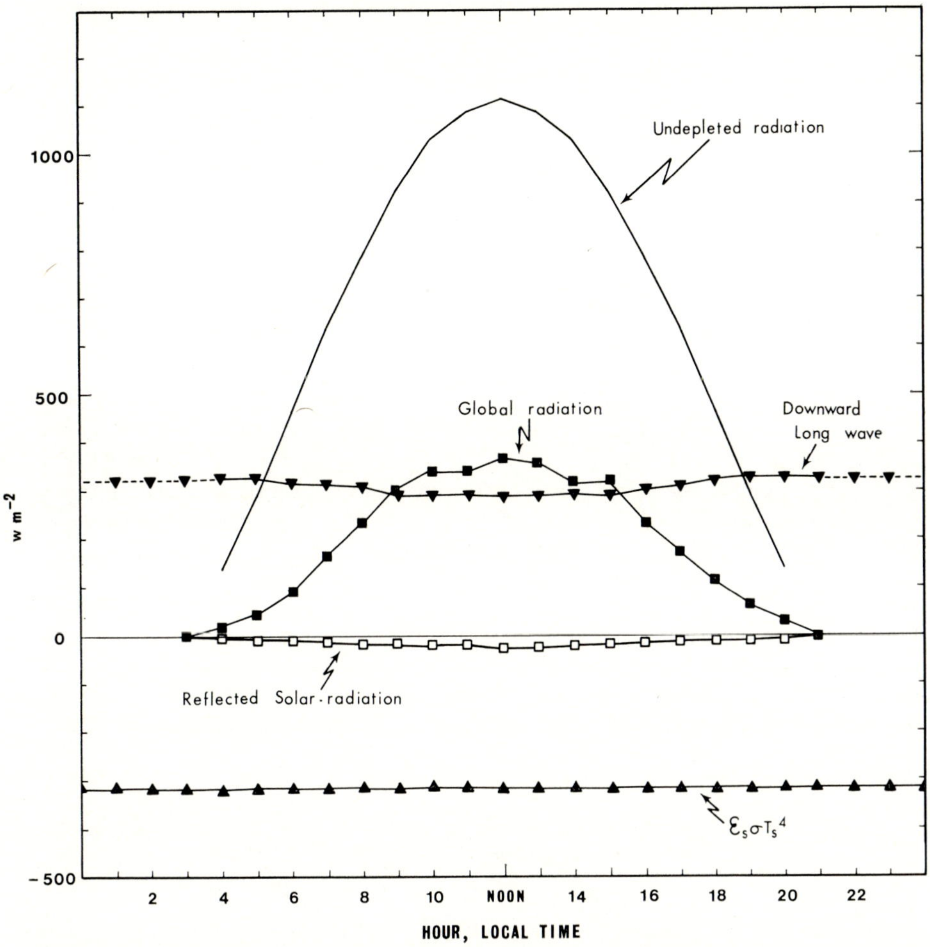

Fig. 13. Mean diurnal variation of the extraterrestrial solar radiation and surface radiation balance components during January 7 and January 10–14, 1970, from *Eltanin* cruise 41. The average noon latitude was 59.8°S. The small reflected long-wave radiation term is not shown.

the total heat loss by the ocean is near zero. We see immediately how the climatologic heat fluxes are positive (i.e., from sea to air) for most of the region despite the much greater frequency of winds with a component from the north. In fact, consideration of stability effects on the transfer coefficients suggests that an even greater tendency for positive average heat fluxes exists than is suggested by Figures 15 and 16. Application of *Deardorff*'s [1968] results implies a 5–10% increase in the already large positive heat fluxes with southwesterly winds, whereas the magnitude of the largest negative fluxes should be reduced by about 5%.

Winds from the south and southeast are relatively infrequent between latitudes 40° and 60°S. To the south of the Antarctic trough, however, the prevailing winds are from the southeast. Figure 17 compares the averaged dependence of the heat fluxes on wind azimuth (irrespective of speed) from *Eltanin* 6-hourly data between 60° and 70°S (summer months only) with a corresponding analysis for an area near 67°S, 41°E in the southwest Indian Ocean [*Zillman and Bell*, 1968]. Though the two sets of results are not completely comparable (a bulk transfer coefficient of 0.0018 was used with the Sōya data in the Indian Ocean, and fluxes were computed from grouped means of the observables; the first factor tended to overestimate the fluxes by comparison with the *Eltanin* computations, the second tended to reduce the peak values), we see that in this limited data sample the southerly winds off the antarctic coast south of Australia are undergoing much more intense heating than winds from the same direction off the west coast of Enderby Land. This result is plausible in view of the

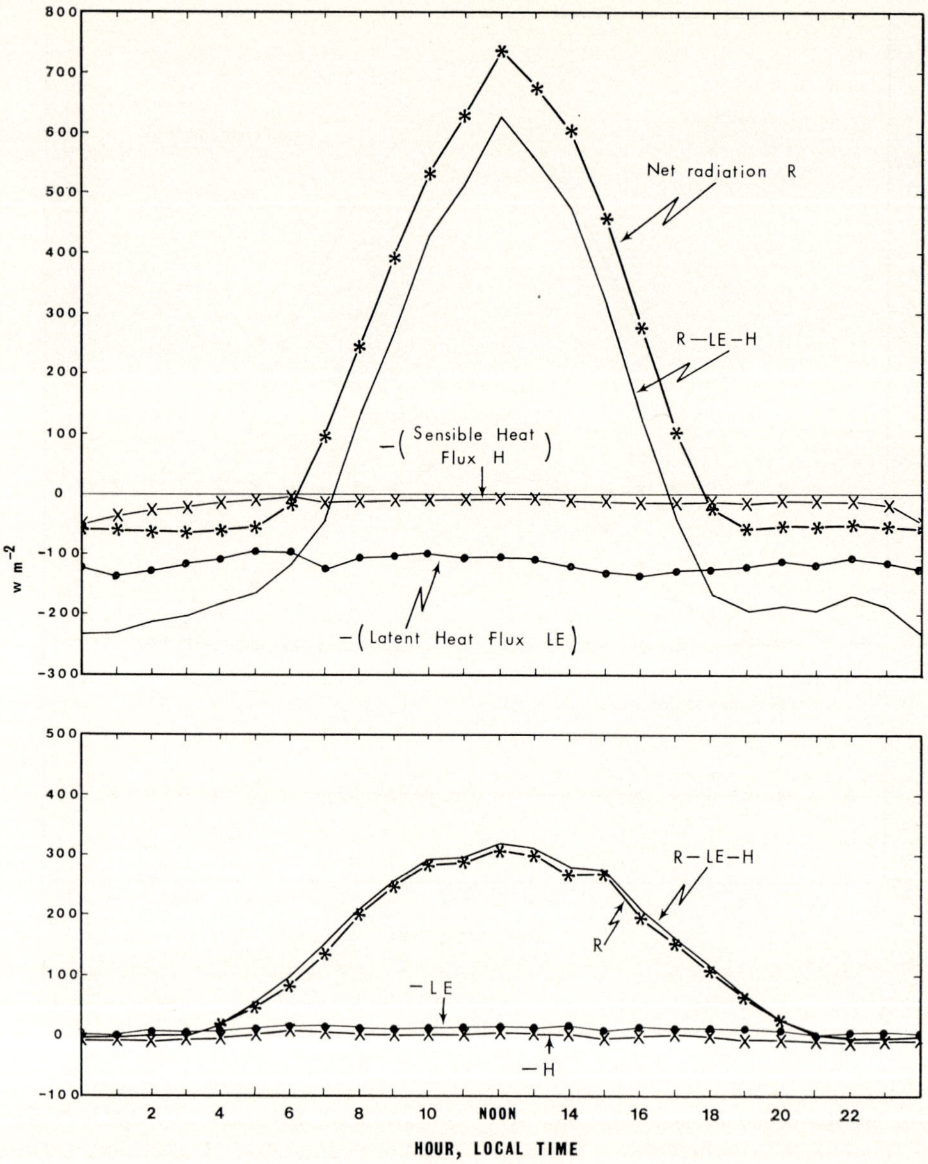

Fig. 14. The ocean surface heat balance and its components as a function of time of day during the *Gascoyne* G2/61 cruise (*top*) and averaged over January 7 and January 10–14, 1970, during *Eltanin* cruise 41 (*bottom*).

presence of quasi-isothermal surface water north of Enderby Land in summer in contrast with the still quite steep gradients in the Australian sector.

Effects of Fronts and Cyclones

Whereas the averaged values clearly show a dependence of sea–air fluxes on the surface wind direction (Figures 15 and 16), the individual values exhibit considerable scatter about the means and point to the significance of other characteristics of the surface air, in particular its past history and its relation to the fronts and cyclones that track eastward across the southern Indian Ocean. Satellite photographs of cyclones to the south of Australia provide spectacular evidence of the contrasting patterns of low level cooling in poleward flow ahead of the major 'frontal' cloud bands (evidenced by extensive areas of stratiform cloud and fog) and the heating of cold air flowing northeastward to the rear of the cyclone. In the cold air, cumulus and cumulonimbus clouds transport heat upward through a deep, relatively unstable layer, and large sea–air temperature differences are maintained at the surface. In regions of conventional meteorologic data, we can infer the pattern of sea–air temperature difference with considerable confidence from the alignment and structure of the satellite-viewed cloud patterns [*Zillman and Price,* 1972].

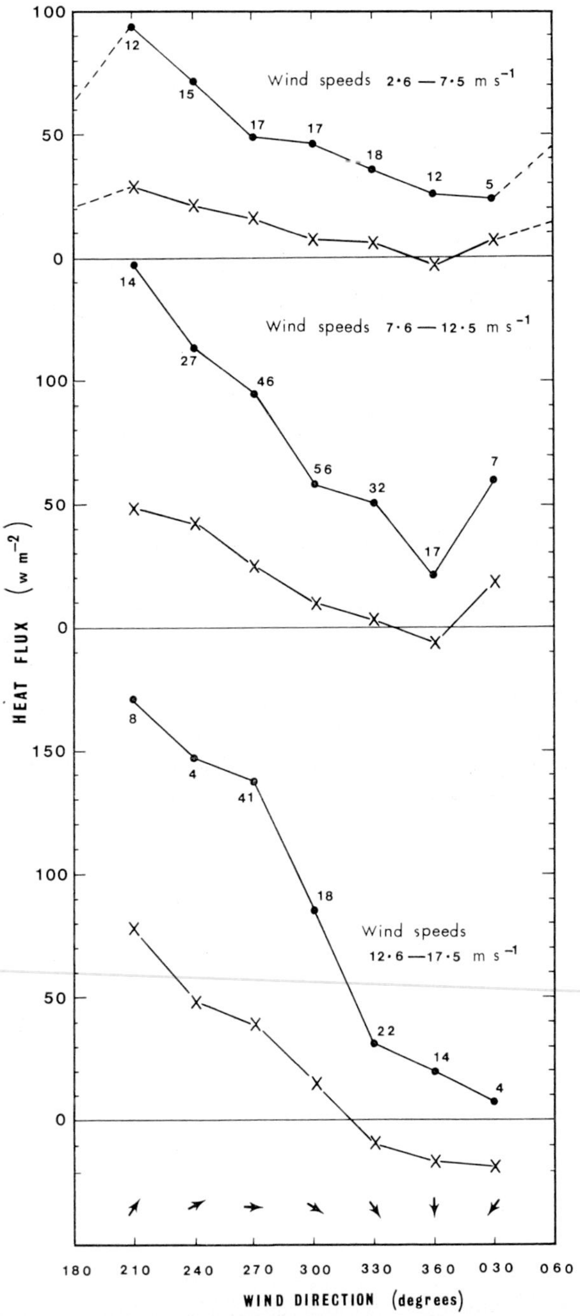

Fig. 15. Sensible and latent heat fluxes as a function of wind azimuth in three wind speed groupings averaged from the 6-hourly *Eltanin* data between 40° and 50°S irrespective of season. Sensible heat flux *H* is marked by X's, and latent heat flux *LE* is marked by darkened circles.

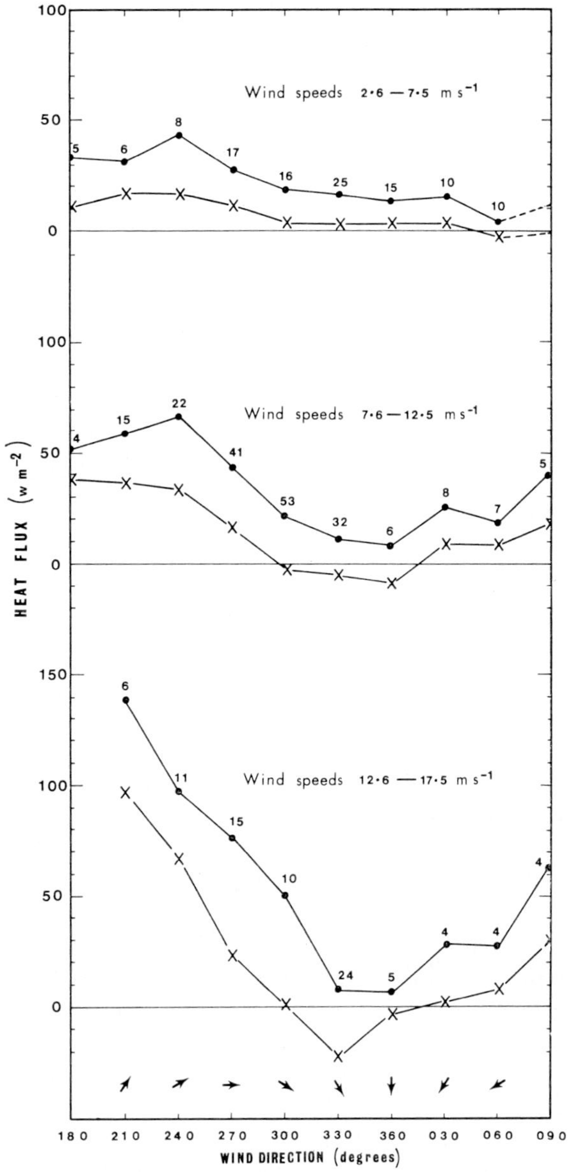

Fig. 16. Sensible and latent heat fluxes as a function of wind azimuth in three wind speed groupings averaged from the 6-hourly *Eltanin* data between 50° and 60°S irrespective of season. Sensible heat flux *H* is marked by X's, and latent heat flux *LE* is marked by darkened circles.

Figure 18 shows a particularly intense summer cyclone (February 17–18, 1969). The superimposed pattern of sea–air temperature difference is based on the analysis model of *Zillman and Price* [1972] but is adjusted slightly to accord with actual recorded values (*Eltanin* cruise 37) shown plotted along with cloud amount, wind, and barometric pressure in the time section format.

Figure 18 indicates clearly, if schematically, the type

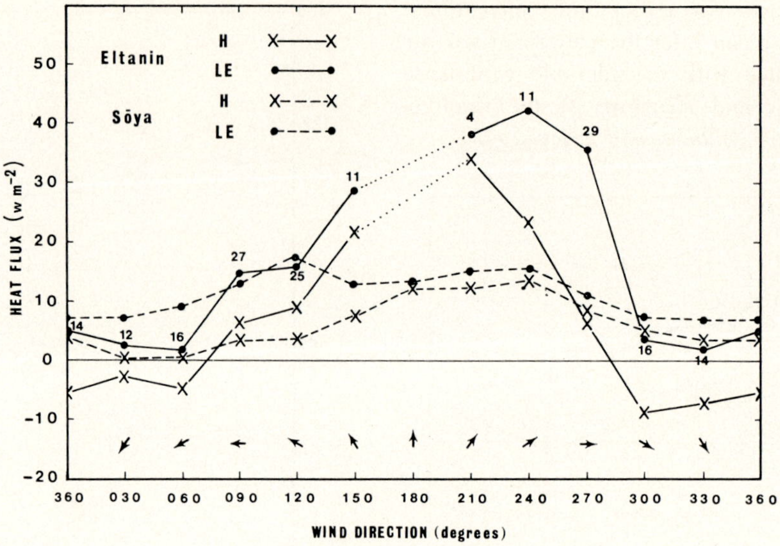

Fig. 17. A comparison of the sensible and latent heat fluxes as a function of wind azimuth from the *Eltanin* data 60° to 70°S during the summer months with a corresponding analysis from data of the Japanese ship *Sōya* in the vicinity of 67°S and 41°E.

of pattern of sea–air sensible heat flux likely to be associated with a mature cyclone over the high latitude oceans. In particular, the pressure gradient discontinuity and wind shift line usually found beneath the spiraling cloud band and analyzed on conventional weather maps as a cold front marks also the change from small or negative sea–air heat flux to positive values the magnitudes of which depend on the alignment of the front, the age of the cyclone and other factors. To derive an average cross-sectional picture of the changes in sea–air sensible and latent heat fluxes through fronts, five fronts were selected (three in winter, two in summer) that passed over the *Eltanin* between 40° and 50°S and for which hourly data were available from 9 hours before to 9 hours after frontal passage. Averages were taken of the heat fluxes, sea–air temperature differences, total cloud cover, atmospheric pressure, and vector wind; these are shown plotted in Figure 19. The vector mean winds, pressure, and cloud amount (partly interpolated) are shown in standard time section format in the lower part of Figure 19. The equivalent west-east cross section of heat fluxes is given in the upper part of the figure. In the northwesterly flow ahead of the average front, the sensible heat flux is negative, and the Bowen ratio (H/LE) is negative. As the wind shifts cyclonically and the air temperature falls with the frontal passage, the sea begins to lose heat at an increasing rate. As the frontal cloud band passes, the surface air becomes not only cooler but significantly drier, and the evaporative heat loss rises to values of 100–150 w m^{-2} 50 miles or so to the west of the front.

Figure 19 may be compared with the cross section through a single front given by *Zillman and Dingle* [1969]. Essentially the same features are evident, though smoother in the averaged case. Note that the absolute values of the fluxes given in Figure 19 have not been corrected for stability and the largest values are probably underestimates by about 10%.

Extreme Values of the Heat Fluxes

The largest positive sensible heat fluxes encountered during cruises 35–41 (between 115° and 160°E) occurred north of latitude 53°S. Under nine different synoptic situations, H values in excess of 100 w m^{-2} were computed, all but two of these being between 47° and 50°S. In all nine cases, the winds were in the southwest quadrant and the sea–air temperature differences ranged from 3.0 to 7.9°C. The largest individual value for the sensible heat flux, 280 w m^{-2}, occurred on July 22, 1969, near 48°S, 148°E with south-southwest winds in excess of 30 knots and sea–air temperature differences greater than 6°C persisting for 6 hours.

The extreme evaporative heat losses by the ocean were generally closely associated with the synoptic situations that produced sensible heat flux maxima. The largest individual value encountered was just

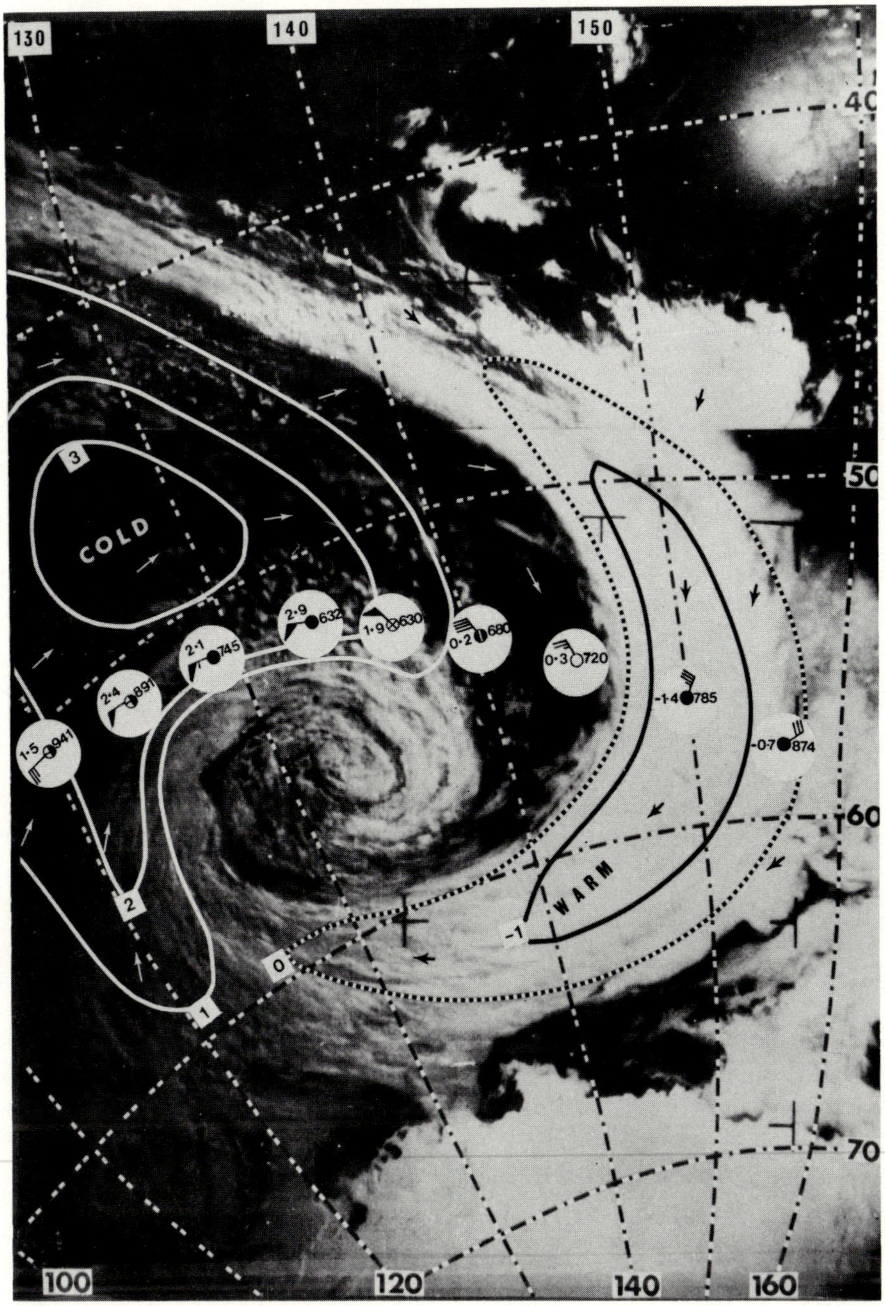

Fig. 18. A satellite photograph of an intense, mature cyclone south of Australia with superposed schematic depiction of low level wind flow and sea–air temperature difference. Note the tongue of warm air beneath the spiraling cloud band and the cold air to the northwest of the vortex center. The satellite photograph was taken at 2240 UT on February 17, 1969. The Dibble iceberg tongue with pack ice and fog to windward may be seen near 66°S, 135°E. The *Eltanin* observations as the cyclone moved southeastward to the south of the vessel are shown arrayed in time section format. Winds are plotted in standard form (each long barb equal to 10 knots), pressures are given in tenths of millibars with the hundreds figure omitted, and sea–air temperature differences are in degrees centigrade. Mean cloud cover in octas is in conventional form.

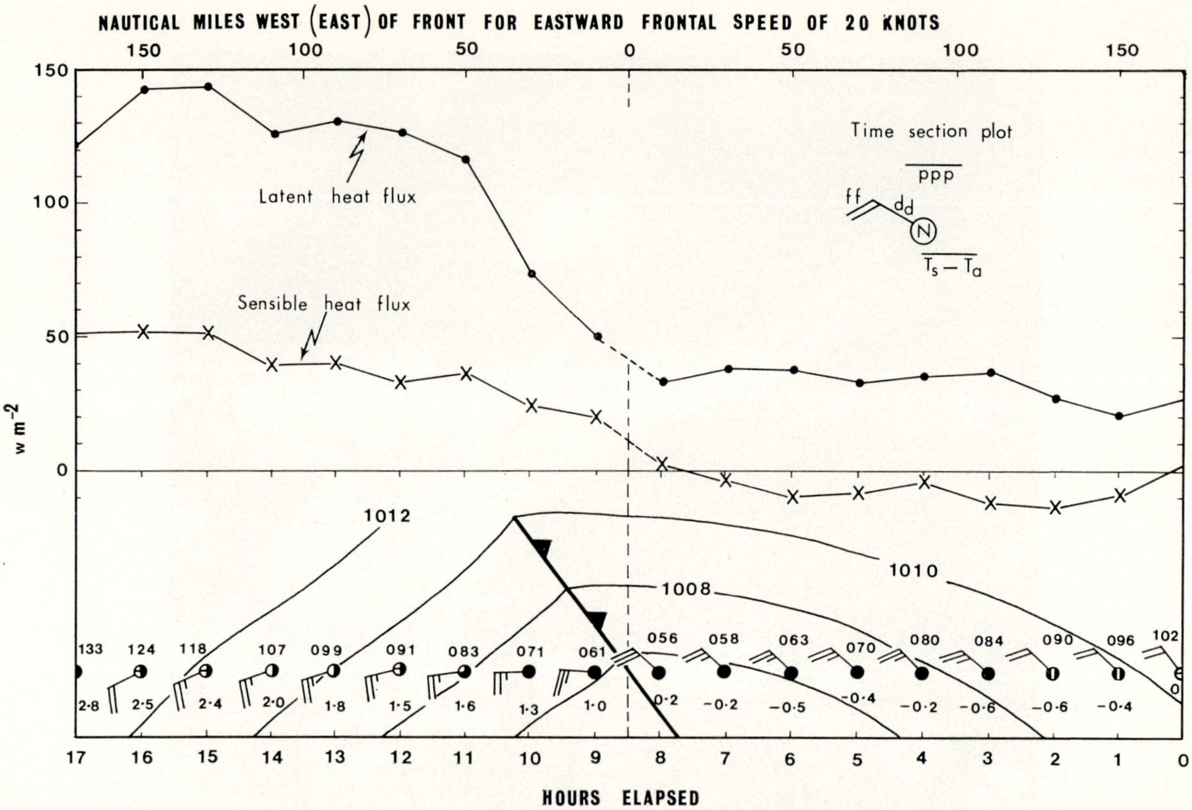

Fig. 19. Pressure, wind, cloud cover, sea–air temperature differences, and sensible and latent heat fluxes averaged at hourly intervals before and after the passage of five fronts (1) at 45.0°S on May 5, 1969; (2) at 41.7°S on June 24, 1969; (3) at 48.0°S on June 27, 1969; (4) at 45.0°S on February 1, 1970; and (5) at 49.1°S on February 10, 1970.

under 400 w m^{-2} with gale force westerly winds at 44.4°S, 148°E on April 12, 1969. The largest computed total (sensible plus latent) heat loss by the ocean (575 w m^{-2}) occurred with the south-southwesterly gales of July 22, 1969, south of Tasmania.

The largest negative heat fluxes (air warming the sea), excepting the occasions with offshore winds near the Australian coast line in summer, occurred mostly between 50° and 55°S with northerly or northwesterly winds. The extreme negative sensible heat flux computed was −70 w m^{-2}, and the extreme negative latent heat flux was −45 w m^{-2}. Although sea–air temperature differences between −1° and −2°C were common in northerly flow ahead of cyclones (see Figures 18 and 19), very few occasions occurred with values larger than −2.5°, the extreme being −3.0°C.

Some Apparent Climatologic Associations

Gray [1960], *Gibbs* [1960], *Karelsky* [1960], and others at the Melbourne Symposium on Antarctic Meteorology speculated on the existence of two belts of cyclones south of Australia, one consisting of migratory cyclones in latitudes 45°–50°S and the other consisting of quasi-stagnant systems near 60°S. *Gibbs* [1960], referring to the first group, pointed to the probable role of the antarctic convergence in producing conditions favorable for baroclinic cyclogenesis. *Vowinckel* [1957] produced evidence for a maximum in the poleward sea temperature gradient at about latitude 50° south of Australia, this latitude being rather north of the then accepted mean position for the antarctic convergence. The charts of *van Loon and Taljaard* [1958] and *Taljaard et al.* [1968] clearly show a tropospheric frontal zone at this latitude in the summer months.

Figure 20, a composite of *Eltanin* and other data, pursues the observational aspects of *Gibbs'* [1960] suggestion a little further. The top and bottom portions of the diagram are based on *Eltanin* data restricted to the longitude interval 115°–140°E and processed as overlapping 2° latitude averages. Care was taken to minimize spurious effects from individual synoptic systems and incomplete north-south traverses (that no doubt contaminated the seasonal average profiles to some degree) by using only those traverses that span the latitude interval 40°–60° and compiling the over-all average from individual averages for each north-south traverse. Figure 20*a* shows the poleward sea surface temperature gradient with a maximum at 50°S. The bottom section shows the averaged sensible and latent heat fluxes with maxima around 48°S and a tendency for slight minima just south of 50°S. These are the same features, though less pronounced, noted by *Zillman and Dingle* [1969] when the heat fluxes were averaged with respect to the observed position of the sharpest north-south temperature gradient on each individual north-south traverse. Supplementing the mean sea surface temperature gradient in Figure 20*a*, Figure 20*b* shows the distribution with latitude of observed positions of the middle of the zone of steepest poleward sea temperature gradient (at least 2.5°C in 90 miles) encountered in 26 crossings by the *Eltanin* and other ships between 115° and 140°E. The curve labeled atmosphere shows the frequency distribution with latitude of atmospheric fronts from daily synoptic analyses for the winter of 1957 extracted from the chart of *van Loon* [1965]. The summer pattern is similar. Figure 20*c* shows the distribution with latitude of the frequency of cyclogenesis loosely defined for present purposes as the first appearance (in satellite photographs) of a vortical cloud formation that, from analysis experience, is probably associated with a sea level cyclone. The frequency scale gives the number of new vortices per 100 days in overlapping 4° latitude bands and is based on once daily photo mosaics for the period from October to March 1967–1970. Numerous vortices are observed to move into the region poleward of 55° from the north and west, but the development of a completely new vortex there (either on a pre-existing band or within a field of cellular cloud) is quite rare. This coincidence of oceanic and atmospheric fronts and the latitude of most frequent cyclogenesis lends support to the ideas of *Gibbs* [1960] but provides only evidence of a plausible association and no real understanding of the dynamic process involved. A first attempt to isolate the role of the sensible heat flux in generating available potential energy on the cyclone scale was reported by *Bullock and Johnson* [1971]. Much work remains to be done.

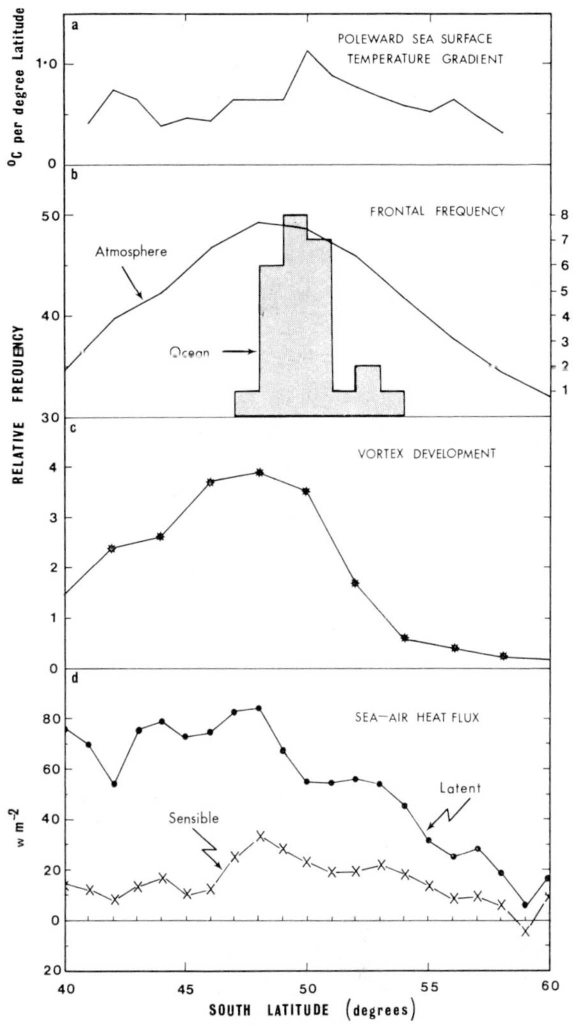

Fig. 20. (*a*) The mean poleward sea surface temperature gradient between 115° and 140°E from *Eltanin* cruise data. (*b*) The number of occurrences (right-hand scale) of oceanic 'fronts' within each 1° latitude interval in 26 north-south traverses. Most of the positions are from *Zillman* [1970*a*]. The left-hand scale refers to atmospheric fronts and gives an average over 120°, 130°, and 140°E of the number of times in 100 days when part of a front was situated within unit area of 400,000 km^2 during June, July, and August 1957 (adapted from the chart of *van Loon* [1965]). (*c*) Number of cases of development of a new vortex between 115° and 140°E per 100 days within the 4° latitude bands centered on the latitude of each data plot. This graph was compiled from daily satellite photos for the October to March period from October 1967 to March 1970. (*d*) Sensible and latent heat fluxes based on averages in overlapping 2° latitude bands from *Eltanin* cruise data (115–140°E) irrespective of season.

MERIDIONAL PROFILES OF MEAN WIND STRESS

The wind stress on the sea surface represents a momentum sink for the atmosphere and forces horizontal and vertical water movement in the ocean. The vertical velocity at the base of the divergent surface Ekman layer may be related directly to the curl of the wind

TABLE 3. West-East and South-North Components of the Seasonal Averages of Wind Stress Computed from 6-Hourly *Eltanin* Wind Data

Latitude Band, °	τ_{WE}, dyne cm^{-2}					τ_{NS}, dyne cm^{-2}				
	Summer	Autumn	Winter	Spring	All Seasons	Summer	Autumn	Winter	Spring	All Seasons
32.0–35.9	0.66	...	−0.09	0.96	0.74	0.54	...	0.01	−0.28	0.09
34.0–37.9	0.26	...	0.21	1.25	0.53	0.43	...	−0.14	−0.25	0.10
36.0–39.9	0.10	0.27	0.50	1.20	0.44	0.26	−0.41	0.03	−0.07	0.04
38.0–41.9	0.62	0.32	0.99	1.23	0.75	0.09	−0.14	−0.41	−0.47	−0.20
40.0–43.9	0.73	0.69	1.03	1.70	1.07	−0.22	0.28	−0.77	−0.64	−0.39
42.0–45.9	0.72	2.27	1.23	2.05	1.50	−0.20	0.34	−1.31	−0.77	−0.70
44.0–47.9	1.31	2.60	1.51	2.39	1.89	−0.01	−0.13	−1.23	−0.75	−0.72
46.0–49.9	2.16	1.32	1.10	2.04	1.58	−0.74	−0.71	−0.51	−0.28	−0.54
48.0–51.9	1.58	1.16	0.74	1.51	1.27	−0.72	−0.76	−0.35	0.02	−0.48
50.0–53.9	1.12	0.93	1.23	1.34	1.19	−0.59	−0.94	−0.26	−0.30	−0.47
52.0–55.9	0.93	0.44	1.50	1.05	1.04	−0.66	−0.29	−0.41	−0.55	−0.52
54.0–57.9	0.98	0.29	1.05	1.06	0.88	−0.64	−0.02	−0.18	−0.59	−0.38
56.0–59.9	0.55	0.26	0.54	1.27	0.67	−0.69	−0.32	0.25	−0.70	−0.43
58.0–61.9	0.49	−0.02	−1.52	0.81	0.29	−0.63	0.53	0.27	−0.52	−0.33
60.0–63.9	0.52	−0.63	−2.26	0.54	0.14	−0.38	0.59	0.80	−0.31	−0.09
62.0–65.9	−0.39	0.02	...	...	−0.29	−0.14	0.07	...	...	−0.08
64.0–67.9	−0.69	0.19	...	...	−0.48	−0.06	−0.02	...	...	−0.05

The distribution of data is given in Table 1.

stress with upwelling favored in regions of cyclonic vorticity. This relation has particular application in the vicinity of the Antarctic trough [*Gordon*, 1971*a*, *b*] where warm deep water from lower latitudes is brought to the surface by the action of the wind stress and gives up heat to the cold air masses moving off the antarctic continent. *Wyrtki* [1960] made a detailed theoretical analysis of the wind-driven circulation in antarctic waters and, in particular, offered several different interpretations for the polar front zone for different assumed profiles of zonal wind stress.

The charts of *Hellerman* [1967] suggest that nowhere in comparable latitudes is our present knowledge of the mean wind stress less securely based than over the ocean between Australia and Antarctica. *Eltanin*, 6-hourly wind data were thus examined by season with the zonal and meridional stress components computed from

$$\tau_{WE} = \rho C_D(u_{WE}^2 + u_{SN}^2)^{1/2}\, u_{WE} \tag{14}$$

$$\tau_{SN} = \rho C_D(u_{WE}^2 + u_{SN}^2)^{1/2}\, u_{SN} \tag{15}$$

The computed stress components were averaged in overlapping 4° latitude bands as for the heat fluxes, and the results are listed in Table 3 and are shown plotted in Figure 21. The maximum stress of around 0.2 newton meters^{-2} (2 dyne cm^{-2}) occurs between 45° and 50°S in all seasons though the winter profile shows a second peak near 55°S.

DISCUSSION AND CONCLUSIONS

In the introductory description of the region, we alluded to the significance of departures from zonal symmetry in several important characteristics of the ocean and atmospheric circulation. In discussion of the climatology of the Australasian region, the west-northwest to east-southeast trend of pressure and temperature fields is commonly attributed to the corresponding trend in ocean temperature, in particular the great relative warmth for its latitude of the Ross Sea sector. The oceanic polar front and in winter the edge of the pack ice are seen as tending to anchor the atmospheric baroclinic zones and axes of cyclogenesis. We have, however, no lack of evidence that, perhaps, we should reason the other way around: by frictional coupling the predominant west-northwesterly winds drive the antarctic circumpolar current carrying relatively warm subantarctic surface water far poleward in some areas (south of New Zealand) and transporting the antarctic surface water northward in others (central Indian Ocean). Divergence and convergence of the Ekman drift currents produce patterns of upwelling and sinking, and oceanic frontal zones align with features of the mean wind field.

It is difficult to avoid circular reasoning and to distinguish cause from effect with any certainty. There are, however, three factors that seem particularly relevant: (1) submarine topography, which *Callahan*

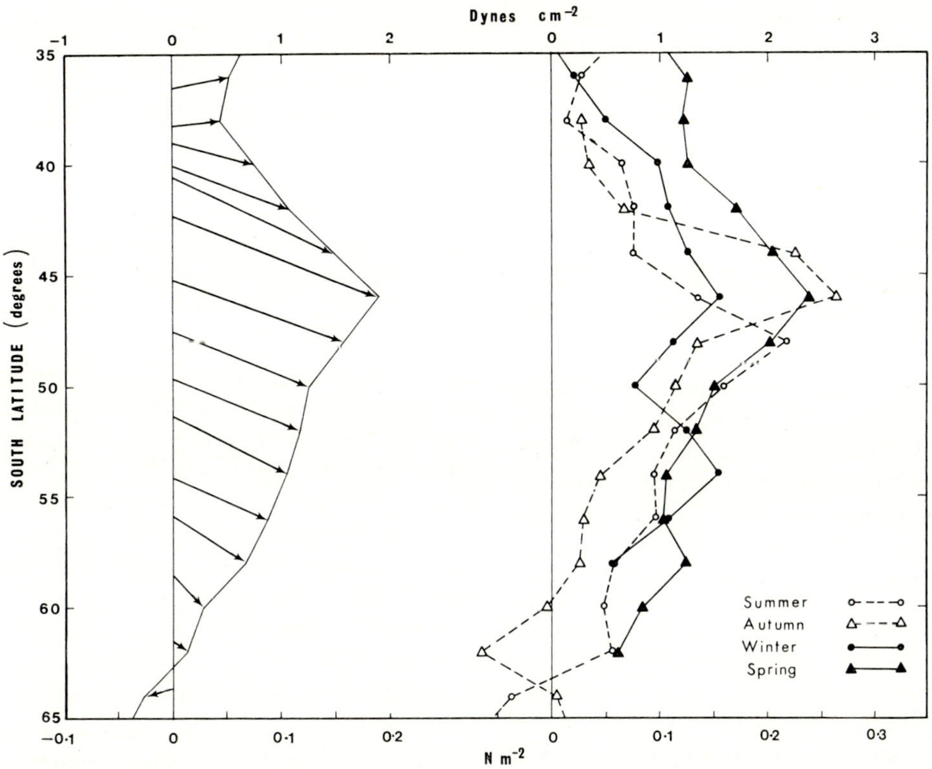

Fig. 21. The zonal component of the mean wind stress averaged in overlapping 4° latitude bands as a function of latitude by season (*right*) and based on all 6-hourly data (115°–160°E) irrespective of season (*left*). The arrows in the left portion of the diagram depict the vector mean stresses applicable at the latitude of the arrow head.

[1971], *Gordon* [this volume], and others suggest exerts a significant influence on the axis and structure of the antarctic circumpolar current; (2) the Australian continent and the topographic barrier of the New Zealand Alps with their important thermal and dynamic influences on the atmospheric circulation; and (3) the displacement of the center of the antarctic ice cap away from the geographical pole into East Antarctica, the displacement thus providing an asymmetrical elevated heat sink and so favoring the maintenance of an asymmetrical circumpolar vortex in the atmosphere. The first two factors are probably the most fundamental since we may argue that a preexisting asymmetrical circulation could have built up the antarctic ice cap to its present form. Little doubt exists, however, that under present conditions both surface and submarine topography and the asymmetrical ice cap exert significant control on the ocean and atmospheric circulation. Thus with both media partially responding to external forcing (solar radiation, orography) the role of sea–air interaction is the following:

1. In the vicinity of the Antarctic trough, the wind fields associated with the quasi-stagnant Antarctic trough cause upwelling of the warm deep water that flows in from lower latitudes. The upwelling warm water gives up heat (by long-wave radiation, sensible heat flux and evaporation) to cold air moving out from the antarctic continent with a mean northward component toward the circumpolar trough. Autumn and winter heat fluxes to the atmosphere by long-wave radiation, sensible heat flux, and evaporation appear to be of the order of 75–150 w m^{-2}. *Gordon and Goldberg* [1970] estimate an annual average heat loss of 14–19 kcal/cm^2/year (19–25 w m^{-2}) south of the polar front zone. Thus the winds produce upwelling by downward transfer of horizontal momentum to the surface, and the upwelling warm water provides an energy source for the winds by upward transfer of sensible heat and water vapor.

2. In the zone of the westerlies north of the Antarctic trough, the profile of mean wind stress and, in particular, the location of the eastward stress maximum (or maxima) in combination with the ocean

floor topography largely determine the flow pattern of the antarctic circumpolar current with wind-induced convergence and divergence patterns playing a role, and possibly a decisive role, in the formation of oceanic frontal zones. Cold air moving northward across the oceanic fronts is heated strongly by the warmer water, and this heating is believed [*Bullock and Johnson*, 1971] to constitute a partial energy source for the cyclones the wind fields of which in the mean maintain to some degree the thermal structure in the sea.

Thus an intimate coupling obviously exists between ocean and atmosphere. The interaction occurs over a wide range of time and space scales, and the time constants of the two media differ greatly. With this general picture in mind, we now summarize briefly some specific aspects of the interaction that have become apparent from the *Eltanin* data.

1. The position of the oceanic polar front is undoubtedly important in determining the location of maximum atmospheric heating, but the zone of steepest transition evident both in surface temperature records and the north-south profiles of time-averaged heat fluxes appears to be significantly (several hundred kilometers) north of the mean position of the oceanic polar front as based, for example, on the position of the salinity minimum at 200 meters. The apparent position of the maximum poleward sea surface temperature gradient based on 35 crossings made by the *Eltanin* and other ships was given by *Zillman* [1970*a*], but which of these positions represent surface expressions of the polar front in a more northerly position and which are simply sharp gradients within the subantarctic surface water are not known. G. E. R. Deacon (personal communication, 1971) has pointed out that he [*Deacon*, 1945] interpreted a similar observed increase in the north-south temperature gradient halfway across the subantarctic zone south of the Brazil and Agulhas currents as a boundary region between well-mixed water in the south of the zone and more stably stratified water to the north.

2. With the present choice of bulk transfer coefficients and with approximate correction for stability effects [*Deardorff*, 1968], the sensible heat flux averaged over the year and zonally across the Australian sector of the southern Indian Ocean shows a maximum between 45° and 50°S of approximately 30 w m^{-2}. This maximum is significantly larger than may be inferred from climatologic maps [*Privett*, 1960; *Budyko*, 1963] or from Figure 3 based on compatible bulk coefficients but atlas type compilations of the observables. The corresponding latent heat flux (inferred from *Eltanin* data) is broadly in agreement with the standard climatologic estimates, although the evaporation peak persistently evident in the *Eltanin* data does not appear on the maps.

3. The existence of a negative (i.e., air to sea) sensible heat flux in spring and summer near 55°–60°S and a corresponding minimum in the evaporation and annually averaged sensible heat flux appear to arise from two factors, the high frequency of northerly winds transporting warm air from lower latitudes toward the mean position of the Antarctic trough (poleward decrease of the heat fluxes in the westerly belt) and the high frequency of winds with a southerly component off the escarpment and pack ice moving out over the upwelling warm deep water (large positive fluxes near the ice edge). *Kangos* [1960] by using data zonally averaged from 20° to 180°E found an extremely pronounced minimum of the sea–air sensible heat flux near 50°–55°S, whereas *Viebrock* [1962] for the Ross Sea sector found the minimum near 60°S in the annual average with considerable seasonal shift. The interaction of the semiannual equatorward migration of the Antarctic trough (Figure 2) and the annual advance and retreat of the pack ice is an interesting question that has not been investigated.

4. The relation between the heat flux maximum and the oceanic polar front suggests that large values of sea–air energy exchange will be found in unusually high latitudes where the polar front dips poleward south of New Zealand. This relation is indeed apparent in the various heat budget climatologies (e.g., *Privett*'s [1960] map) and in *Viebrock*'s [1962] meridional profiles for the Ross Sea sector. Its oceanographic significance is discussed by *Gordon* [this volume].

5. Global radiation data available so far suggest that the lowest values of the relative insolation (the ratio of the global radiation to extraterrestrial radiation) of slightly less than 0.3 occur near latitude 60°S. Consideration of the net long-wave radiation in the form

$$R_L = \epsilon_s \sigma T_s^4 (\epsilon_a - 1) - 4\, \epsilon_s \sigma T_s^3 (T_s - T_a) \qquad (16)$$

from preliminary evaluation of the *Eltanin* long-wave radiation measurements [*Zillman*, 1970*b*] suggests that the first (black body defect) term is smaller than standard empirical formulas would suggest and that the main factor in the long-wave exchange is the sea–air temperature difference. In the heavily clouded region, average net losses as low as 10–20 w m^{-2} appear to be the norm in summer, but near the ice edge in autumn and winter radiative heat losses by the sea probably exceed 50 w m^{-2}.

6. The role of atmospheric synoptic systems in the oceanic heat budget is largely determined by the duration and frequency of occurrence of southerly flow in the western sectors of the migratory cyclones. Large sea–air heat fluxes in the mean can occur in regions of predominantly poleward flow provided a respectable number of cyclones with strong southerly flow to their rear traverse the region. Thus the tendency for a preferred cyclone track south of Australia [*Karelsky*, 1960; *Tolstikov*, 1966] to lie generally north of the zone of steepest sea temperature gradient is probably a contributing factor to the heat flux peak observed there. To what extent the heating maximum is a contributing factor to cyclogenesis cannot be determined from an observational study alone, but the observed sharp decrease in summer cyclogenesis (Figure 20) south of 50°S suggests that the heating maximum may be important.

7. The existence of two belts of cyclones, largely confirmed since satellite data became available, has important implications for the dynamics of the circumpolar current and the oceanic polar front. The geostrophic west wind maximum presented in Figure 2 is based on smoothed mean pressure profiles and thus may not provide a completely accurate picture. There is considerable evidence [e.g., *U.S. Navy*, 1957, 1959, 1965, 1969] for a double west wind maximum through the winter. The *Eltanin* mean wind and stress data tend to support this view but cannot be regarded as conclusive.

NOTATION

C_D, drag coefficient;
C_E, bulk transfer coefficient for water vapor;
C_H, bulk transfer coefficient for heat;
c_p, specific heat of air at constant pressure;
e, vapor pressure;
e_a, vapor pressure at level a meters;
e_s, vapor pressure at the sea surface;
H, sensible heat flux positive upward;
k, Karman constant, 0.41;
L, latent heat of evaporation, Monin-Obukhov stability length;
LE, latent heat flux positive upward;
L_d, downward long-wave radiation from the atmosphere;
N, total cloud cover, octas;
p, atmospheric pressure;
ppp, atmospheric pressure, tenths of millibars (hundreds and thousands digits omitted);
q_s, specific humidity of the air at the sea surface;
q_a, specific humidity of the air at level a meters above the sea surface;
R, net radiation positive downward;
S_G, global radiation;
T_s, sea water bulk temperature, degrees absolute (except where otherwise specified); also represents the true sea surface temperature;
T_a, air temperature at level a meters;
T_w, wet bulb temperature;
u_b, wind velocity at level b meters;
u_s, velocity of the interfacial current;
u_{WE}, west-east component of the wind velocity;
u_{SN}, south-north component of the wind velocity;
z_o, roughness length for wind profile;
z_θ, roughness length for potential temperature profile;
z_q, roughness length for humidity profile;
α_G, global radiation albedo of the sea;
ϵ_s, long-wave emissivity of the sea surface, 0.97;
ϵ_a, effective emissivity of the atmosphere;
θ_s, potential temperature at the sea surface;
θ_a, potential temperature at level a meters;
ρ, density of air;
σ, Stefan Boltzmann constant, 5.67×10^{-8} w m^{-2} $°K^{-4}$;
ϕ, function of z/L describing profile shape;
$\boldsymbol{\tau}$, vector stress;
τ_{WE}, eastward component of the wind stress;
τ_{SN}, northward component of the wind stress.

APPENDIX

Observational and instrumental details for the *Gascoyne* cruise G2/61 and the *Eltanin* cruises 35–41 are given below.

Sea surface temperature T_s.

Gascoyne: bucket thermometers and continuous recording towed thermistor.

Eltanin: bucket thermometer.

Air temperature. Dry bulb T_a and wet bulb T_w.

Gascoyne: Assman psychrometer on the windward side of the quarter-deck approximately 3 meters above sea surface.

Eltanin: (1) sling psychrometer (cruises 35, 36) operated on the quarter-deck (port or starboard) or the fantail approximately 6 meters above sea surface; (2) portable Stevenson screen (cruise 37 and subsequently) on the flying bridge approximately 12 meters above sea level.

Wind speed u.

Gascoyne: cup rotor anemometer at the stem of the vessel approximately 9 meters above the sea surface.

Eltanin: (1) cup rotor anemometer located on the forward ice house approximately 18 meters above the sea surface and used for winds coming from ahead of the beam; (2) Bendix impeller type anemometer located on the port aft King post approximately 18 meters above the sea surface and used for winds coming from aft of the beam.

Atmospheric pressure p.

Gascoyne: mercury barometer.

Eltanin: mercury barometer.

Cloud amount N.

Gascoyne: visual estimate and zenith viewing camera.

Eltanin: visual estimate.

Short-wave global radiation S_G.

Gascoyne: continuous recording Moll-Gorczynski pyranometer mounted high above foredeck.

Eltanin: continuous recording gimbal-mounted Eppley pyranometer on the starboard side of the helicopter deck approximately 10 meters above the sea surface.

All-wave downward radiation. That is, global plus downward long-wave radiation ($S_G + L_d$).

Eltanin: continuous recording CSIRO (Funk) net radiometer with the underside enclosed by a black body cavity gimbal-mounted on the starboard side of the helicopter deck approximately 10 meters above the sea surface.

Net radiation R.

Gascoyne: continuous recording CSIRO (Funk) net radiometer mounted 3 meters forward of the stem of the vessel approximately 7 meters above the sea surface.

Acknowledgments. This paper is published with the permission of the Director, Commonwealth Bureau of Meteorology, Australia. It was prepared while the author was attending the University of Wisconsin under an Australian Commonwealth Postgraduate Scholarship. The National Science Foundation provided the opportunity for Australian meteorologic observers to participate in the *Eltanin* cruises and the facilities that made this study possible. Thanks are due to E. E. Jesson, T. Mitchell, and D. Brooks who provided advice and assistance with the reduction of the radiation data; to my wife Dawn who contributed much of the computational effort in this paper; also to Dr. Arnold Gordon of the Lamont-Doherty Geological Observatory for helpful advice concerning the identification and location of the oceanic polar front; and to Dr. Gordon and Professor Werner Schwerdtfeger of the University of Wisconsin for reviewing the manuscript and making several suggestions for improvement. Finally I wish to acknowledge the dedicated effort of W. R. J. Dingle in recording and collecting the large amount of special meteorologic and radiation data for this study. The observational material is characterized by the consistent high quality that marks his work.

REFERENCES

Albrecht, F., Jahreskarten des Wärme-und Wasserhaushaltes der Ozeane, *Ber. Deut. Wetterdienstes, 9*(66), 1960.

Bernhardt, F., and H. Philipps, Die räumliche und zeitliche Verteilung der Einstrahlung, der Ausstrahlung und der Strahlungsbilanz im Meeresniveau, *Abhandlungen des Meteorologischen Dienstes der Deutschen Demokratischen Republik,* Teil 1, 227 pp., Akademie-Verlag, Berlin, 1958.

Bernhardt, F., and H. Philipps, Die räumlich und zeitliche Verteilung der Einstrahlung, der Ausstrahlung und der Strahlungsbilanz im Meeresniveau, *Abhandlungen des Meteorologischen Dienstes der Deutschen Demokratischen Republik,* Teil 2 and 3, 266 pp., Akademie-Verlag, Berlin, 1966.

Bond, F. R., and F. Jacka, Distribution of auroras in the Southern Hemisphere, *Aust. J. Phys., 15,* 261–272, 1962.

Brocks, K., and L. Krügermeyer, The hydrodynamic roughness of the sea surface, in *Studies in Physical Oceanography—A Tribute to Georg Wüst on His 80th Birthday,* edited by A. L. Gordon, Gordon and Breach, New York, in press, 1971.

Budyko, M. I., *The Heat Balance of the Earth's Surface,* translated from Russian by Nina A. Stepanova, 255 pp., U.S. Weather Bureau, Washington, D. C., 1956.

Budyko, M. I. (Ed.), *Atlas Teplovogo Balansa,* Gidrometeorologichesko Izdatel'skos, Leningrad, 1963.

Bullock, B. R., and D. R. Johnson, The generation of available potential energy by sensible heating in Southern Ocean cyclones, *Tech. Rep. 13,* 47 pp., Int. Antarctic Meteorol. Res. Centre, Commonwealth Bur. of Meteorol., Melbourne, 1971.

Callahan, J. E., Velocity structure and flux of the antarctic circumpolar current south of Australia, *J. Geophys. Res., 76*(24), 5859–5864, 1971.

Clarke, R. H., Recommended methods for the treatment of the boundary layer in numerical models, *Aust. Meteorol. Mag., 18*(2), 51–73, 1970.

Deacon, G. E. R., Water circulation and surface boundaries in the oceans, *Quart. J. Roy. Meteorol. Soc., 71*(307-308), 11–25, 1945.

Deacon, G. E. R., Subtropical convergence, in *The Encyclopedia of Oceanography,* edited by R. W. Fairbridge, Reinhold, New York, 1966.

Deacon, E. L., and E. K. Webb, Small scale interactions, in *The Sea,* edited by M. N. Hill, pp. 43–87, Interscience, New York, 1962.

Deardorff, J. W., Dependence of air-sea transfer coefficients on bulk stability, *J. Geophys. Res., 73*(8), 2549—2557, 1968.

De Leonibus, P. S., Momentum flux and wave spectra observations from an ocean tower, *J. Geophys. Res., 76*(27), 6506–6527, 1971.

De Lisle, J. F., Heat transfer from ocean to atmosphere in the Tasman Sea area, *N. Z. J. Sci., 13*(1), 166–176, 1970.

Dilly, A. C., On the computer calculation of vapour pressure and specific humidity gradients from psychrometric data, *J. Appl. Meteorol., 7*(4), 717–719, 1968.

Duncan, W. J., A. S. Thom, and A. D. Young, *Mechanics of Fluids, SI Units,* 725 pp., Edward Arnold, London, 1970.

Fletcher, J. O., Ice extent on the Southern Ocean and its relation to world climate, *RM-5793-NSF,* 108 pp., Rand Corporation, Santa Monica, Calif., 1969.

Garner, D. M., The sub-tropical convergence in New Zealand surface waters, *N. Z. J. Geol. Geophys., 2*(2), 315–337, 1959.

Garstang, M., Sensible and latent heat exchange in low latitude synoptic scale systems, *Tellus, 19*(3), 492–508, 1967.

Gibbs, W. J., Antarctic synoptic analysis, in *Antarctic Meteorology, Proceedings of the Symposium Held in Melbourne, February 1959,* pp. 84–95, Pergamon, New York, 1960.

Gordon, A. L., Oceanography of antarctic waters, in *Antarctic Oceanology I, Antarctic Res. Ser.,* vol. 15, edited by J. L. Reid, pp. 169–203, AGU, Washington, D. C., 1971*a*.

Gordon, A. L., Antarctic polar front zone, in *Antarctic Oceanology I, Antarctic Res. Ser.,* vol. 15, edited by J. L. Reid, pp. 205–221, AGU, Washington, D. C., 1971*b*.

Gordon, A. L., On the interaction of the antarctic circumpolar current and the Macquarie ridge, in *Antarctic Oceanology II: The Australian–New Zealand Sector, Antarctic Res. Ser.,* vol. 19, edited by D. E. Hayes, pp. 71–78, AGU, Washington, D. C., 1972.

Gordon, A. L., and R. D. Goldberg, Circumpolar characteristics of antarctic waters, *Antarctic Map Folio Ser.,* folio 18, edited by V. Bushnell, Amer. Geogr. Soc., New York, 1970.

Gray, T. I., Report of IGY Weather Central 1958, in *Antarctic Meteorology, Proceedings of the Symposium Held in*

Melbourne, February 1959, pp. 75–78, Pergamon, New York, 1960.

Grimson, J., *Mechanics and Thermodynamics of Fluids*, 343 pp., McGraw-Hill, London, 1970.

Hasse, L., On the determination of the vertical transports of momentum and heat in the boundary layer at sea, *Tech. Rep. 188*, pp. 70–72, Oreg. State Univ., Corvallis, 1970.

Hasse, L., The sea surface temperature deviation and the heat flow at the sea-air interface, *Boundary-Layer Meteorol.*, *1*(3), 368–379, 1971.

Hedgpeth, J. W., Introduction to antarctic zoogeography, *Antarctic Map Folio Ser.*, folio 11, Amer. Geogr. Soc., New York, 1969.

Hellerman, S., An updated estimate of the wind stress on the world ocean, *Mon. Weather Rev.*, *95*(9), 607–626, 1967.

Hellerman, S., Correction, *Mon. Weather Rev.*, *96*(1), 62–74, 1968.

Hicks, B. B., and A. J. Dyer, Measurement of eddy fluxes over the sea from an offshore oil rig, *Quart. J. Roy. Meteorol. Soc.*, *96*(409), 523–528, 1970.

Jacobs, W. C., On the energy exchange between sea and atmosphere, *J. Mar. Res.*, *5*, 37–66, 1942.

Kangos, J., A preliminary investigation of the heat flux from the ocean to the atmosphere in the antarctic regions, *J. Geophys. Res.*, *65*(12), 4007–4012, 1960.

Karelsky, S., The surface circulation over Southern Ocean, southern Indian Ocean, Australasia and southern Pacific Ocean regions during 1957 and 1958, in *Antarctic Meteorology, Proceedings of the Symposium Held in Melbourne, February 1959*, pp. 293–310, Pergamon, New York, 1960.

Kort, V. G., On heat exchange of the antarctic waters, in *Antarctic Research, Geophys. Monogr. Ser.*, vol. 7, pp. 163–167, AGU, Washington, D. C., 1962.

Kraus, E. B., Wind stress along the sea surface, in *Advances in Geophysics*, vol. 12, pp. 213–255, Academic, New York, 1967.

Laevastu, T., Synoptic scale heat exchange and its relation to weather, Proceedings of the Sea-Air Interaction Conference, Tallahassee, Florida, February 23–25, 1965, *Tech. Note 9-SAIL-1*, pp. 257–283, U.S. Dep. of Commerce, Washington, D. C., 1965.

Landsberg, H. E., H. Lippman, K. H. Paffen, and C. Troll, *World Maps of Climatology*, 28 pp., Springer-Verlag, Berlin, 1963.

Langford, J. C., Southern Ocean analysis with special reference to the period December 1954 to March 1955, *Aust. Meteorol. Mag.*, *16*, 1–22, 1957.

List, R. J., *Smithsonian Meteorological Tables*, 6th revised ed., Smithsonian Institute, Washington, D. C., 1964.

Loewe, F., Precipitation and evaporation in the antarctic, in *Meteorology of the Antarctic*, edited by M. P. van Rooy, pp. 71–89, Government Printer, Pretoria, 1957.

Manier, G., and F. Möller, Determination of heat balance at the boundary layer over the sea, final report, contract AF61(052)-315, 99 pp., Meteorologisch-Geophysikalisches Institut, Johannes Gutenberg-Universität Mainz, 1961.

Martin, D. W., Satellite studies of cyclonic developments over the Southern Ocean, *Tech. Rep. 9*, 64 pp., Int. Antarctic Meteorol. Res. Centre, Commonwealth Bur. of Meteorol., Melbourne, 1968.

Meinardus, W., Klimakunde der Antarktis, in *Handbuch der Klimatologie*, Band 4, Teil 5, pp. 1–133, Bornträger, Berlin, 1938.

Osborne, M. F. M., The interpretation of infrared radiation from the sea in terms of its boundary layer, *Deut. Hydrogr. Z.*, *17*(3), 115–136, 1964.

Paltridge, G. W., Day-time long-wave radiation from the sky, *Quart. J. Roy. Meteorol. Soc.*, *96*(410), 645–653, 1970.

Petterssen, S., D. L. Bradbury, and K. Pedersen, The Norwegian cyclone models in relation to heat and cold sources, *Geophys. Publ. Geophysica Norwegica*, *24*(9), 243–250, 1962.

Phillpot, H. R., A study of the synoptic climatology of the Antarctic, *Tech. Rep. 12*, 139 pp., Int. Antarctic Meteorol. Res. Centre, Commonwealth Bur. of Meteorol., Melbourne, 1968.

Priestley, C. H. B., A survey of the stress between the ocean and atmosphere, *Aust. J. Sci. Res., Ser. A*, *4*, 315–328, 1951.

Privett, D. W., The exchange of energy between the atmosphere and the oceans of the Southern Hemisphere, *Geophys. Mem. London*, *13*(104), 61 pp., 1960.

Radok, U., The voyage to Terre Adélie—Meteorological aspects, *Expedition en Terre Adélie 1951–1953, Extrait des Rapports Préliminaires*, vol. 24, Paris, 1954.

Robinson, G. D., Another look at some problems of the air-sea interface, *Quart. J. Roy. Meteorol. Soc.*, *92*(394), 451–465, 1966.

Roll, H. U., On the present state of knowledge in air–sea boundary layer problems, Proceedings of the Sea–Air Interaction Conference, Tallahassee, Florida, February 23–25, 1965, *Tech. Note 9-SAIL-1*, pp. 31–63, U.S. Dep. of Commerce, Washington, D. C., 1965.

Schmitt, W., Synoptic meteorology of the antarctic, in *Meteorology of the Antarctic*, edited by M. P. van Rooy, pp. 209–231, Government Printer, Pretoria, 1957.

Schott, G., *Geographie des Indischen und Stillen Ozeans*, 413 pp., Verlag von C. Boysen, Hamburg, 1935.

Schwerdtfeger, W., The climate of the antarctic, in *World Survey of Climatology*, vol. 14, pp. 253–355, Elsevier, Amsterdam, 1970.

Sheppard, P. A., The atmospheric boundary layer in relation to large-scale dynamics, in *The Global Circulation of the Atmosphere*, edited by G. A. Corby, pp. 91–112, Royal Meteorological Society, London, 1970.

Streten, N. A., A note on the frequency of closed circulations between 50°S and 70°S in summer, *Aust. Meteorol. Mag.*, *17*(4), 228–234, 1969.

Swinbank, W. C., Air-sea interaction, *Proceedings of the Symposium on Meteorological Results of the International Indian Ocean Expedition, Bombay, India, 22–26 July, 1965*, pp. 75–84, World Meteorol. Organ., Geneva, 1965.

Taljaard, J. J., Development, distribution and movement of cyclones and anticyclones in the Southern Hemisphere during the IGY, *J. Appl. Meteorol.*, *6*, 973–987, 1967.

Taljaard, J. J., R. L. Jenne, H. van Loon, and H. L. Crutcher, Seasonal range, anomalies and other aspects of sea level pressure, isobaric height, temperature and dewpoint at selected levels in the Southern Hemisphere, *Notos*, *17*, 63–140, 1968.

Taljaard, J. J., H. van Loon, H. L. Crutcher, and R. L. Jenne, Temperatures, dewpoints, and heights at selected levels, 1, *Southern Hemisphere,* vol. 1, *Climate of the Upper Air,* GPO, Washington, D. C., 1969.

Taylor, G., Climatology of Australia, in *Handbuch der Klimatologie,* Band 4, Teil 5, pp. 1–108, Bornträger, Berlin, 1932.

Tolstikov, E. I. (Ed.), *Atlas Antarktiki,* vol. 1, G.U.C.K., Moscow, 1966. (English translation in *Soviet Geography: Reviews and Translations, 8*(5-6), 225 pl., American Geographical Society, New York, 1967.)

U.S. Navy, *Marine Climatic Atlas of the World,* vol. 3, *Indian Ocean,* Washington, D. C., 1957.

U.S. Navy, *Marine Climatic Atlas of the World,* vol. 5, *South Pacific Ocean,* Washington, D. C., 1959.

U.S. Navy, *Marine Climatic Atlas of the World,* vol. 7, *Antarctic,* Washington, D. C., 1965.

U.S. Navy, *Marine Climatic Atlas of the World,* vol. 8, *The World,* Washington, D. C., 1969.

van Loon, H., A climatological study of the atmospheric circulation in the Southern Hemisphere during the IGY, 1, 1 July 1957–31 March 1958, *J. Appl. Meteorol., 4*(4), 479–491, 1965.

van Loon, H., The half-yearly oscillations in middle and high southern latitudes and the coreless winter, *J. Atmos. Sci., 24*(5), 472–486, 1967.

van Loon, H., and J. J. Taljaard, A study of the 1000–500 mb thickness distribution in the Southern Hemisphere, *Notos, 7*(3-4), 123–258, 1958.

Viebrock, H., The transfer of energy between the ocean and atmosphere in the antarctic region, *J. Geophys. Res., 67*(11), 4293–4302, 1962.

Vowinckel, E., Das Klima des Antarktischen Ozeans, *Arch. Meteorol. Geophys. Bioklimatol., Ser. B, 7,* 317–369, 1956.

Vowinckel, E., Climate of the antarctic ocean, in *Meteorology of the Antarctic,* edited by M. P. van Rooy, pp. 91–110, Government Printer, Pretoria, 1957.

Wyrtki, K., The antarctic circumpolar current and the antarctic polar front, *Deut. Hydrogr. Z., 13,* 153–174, 1960.

Zillman, J. W., The surface radiation balance in high southern latitudes, *Tech. Note 87,* pp. 142–171, World Meteorol. Organ., Geneva, 1967.

Zillman, J. W., Aspects of the energy exchange at the air–surface interface in the high southern latitudes, paper presented to 40th Congress, Aust. and N. Z. Ass. for the Advan. of Sci., Christchurch, New Zealand, 1968.

Zillman, J. W., Sea surface temperature gradients south of Australia, *Aust. Meteorol. Mag., 18*(1), 22–30, 1970*a.*

Zillman, J. W., The radiation and heat budget of the Southern Hemisphere oceans, M. Sc. thesis, 563 pp., University of Melbourne, 1970*b.*

Zillman, J. W., and J. A. Bell, Sea-to-air heat fluxes in the southwest Indian Ocean in summer, *J. Geophys. Res., 73*(22), 7057–7064, 1968.

Zillman, J. W., and W. R. J. Dingle, Southern Ocean sea–air energy exchange, *Aust. Meteorol. Mag., 17*(3), 166–172, 1969.

Zillman, J. W., and P. G. Price, On the thermal structure of mature Southern Ocean cyclones, *Aust. Meteorol. Mag., 20,* 34–48, 1972.

SOUND VELOCITY STRUCTURE OF THE OCEAN SOUTH OF AUSTRALIA AND NEW ZEALAND

R. N. DENHAM

Defence Scientific Establishment, HMNZ Dockyard, Auckland, New Zealand

A. C. KIBBLEWHITE

Physics Department, University of Auckland, Auckland, New Zealand

Abstract. An examination is made of the structure of sound velocity as derived from the hydrologic data obtained by the USNS *Eltanin* in the area south of Australia and New Zealand. Significant meridional changes in velocity structure are found to be associated with four different hydrologic boundaries that occur in the area: (1) the antarctic convergence or polar front zone, (2) the northernmost 34.2‰ isohaline enclosing the circumpolar band of minimum salinity in the 200-meter plane, (3) the Australasian subantarctic front, and (4) the southern boundary of the subtropical convergence region. South of the antarctic convergence, the velocity shows a generally monotonic increase with depth, whereas the velocity–depth profiles taken north of the boundary described in point 2 above always show a definite Sofar velocity minimum at depths greater than 500 meters. The geographical variation of Sofar axis velocity can be related to the characteristics of the general current circulation in the area.

The velocity of sound in the sea is a function of pressure, temperature, and salinity, and as a result the spatial and temporal variations of sound velocity may be related to the hydrologic characteristics of the area in which the velocity is measured. To calculate the sound velocity from temperature and salinity data an empirical formula must be used, and a formula given by *Wilson* [1960] has been applied extensively. This formula is rather complex. A simpler formula that is reported to yield values within 0.2 m/sec of the result given by Wilson's formula for depths less than 1000 meters and for the ranges of temperature and salinity encountered in the study area has been given recently by *Leroy* [1969]. This formula is

$$V = 1492.9 + 3(T - 10) - 6 \times 10^{-3}(T - 10)^2 - 4 \times 10^{-2}(T - 18)^2 + 1.2(S - 35) - 10^{-2}(T - 18)(S - 35) + Z/61 \qquad (1)$$

When the term involving the product of temperature and salinity is neglected, equation 1 is equivalent to

$$V = 1449.3 + 0.016Z + 4.56T - 0.046T^2 + 1.4(S - 35) \qquad (2)$$

In both equations 1 and 2, V is the sound velocity in meters per second, T is the temperature in degrees Celsius, S is the salinity in parts per thousand, and Z is the depth in meters. In this paper, however, all the velocities have been calculated by using the formula due to *Wilson* [1960].

A salient feature of the ocean in temperate and tropical regions is the presence of a strong permanent thermocline between depths of typically 100 and 500 meters; below the thermocline, the rate of decrease in temperature with depth is much smaller. As a result, a minimum in sound velocity occurs at a depth of around 1 km, the actual depth being that at which the temperature gradient is about −0.003°C/m (see equation 2). This minimum, which defines the Sofar axis, is very important in long-range sound propagation. Several studies have been made of the variation of Sofar velocity and axis depth in the Pacific, Atlantic, and Indian oceans [*Garner*, 1967*a*; *Johnson and Norris*, 1968; *Denham and Kibblewhite*, 1970; *Hunkins*, 1970], and one of the aims of the present study will be to extend the examination of this feature in the region south of Australia and New Zealand.

Two very important hydrologic features are present in the study area. The most significant of these is the antarctic convergence or polar front zone. This feature encircles the globe, lies between 50° and 60°S, and is farthest south at the longitude of New Zealand. The antarctic convergence has been discussed by various authors [*Deacon*, 1937; *Burling*, 1961; *Ostapoff*, 1962; *Houtman*, 1967; *Gordon*, 1967; *Gordon and*

Goldberg, 1970] and has been defined in various ways. The best definition is probably that given by Ostapoff: the polar front zone is associated with the circumpolar band of low salinity at 200 meters (see *Gordon and Goldberg* [1970, Figure 6, Plate 4]), the salinity at this depth being less than 34.2‰ within this band. The polar front zone is located at the absolute salinity minimum, which lies close to the more northerly of the 34.2‰ isohalines.

This salinity minimum is associated with the sinking at the polar front zone of relatively fresh antarctic surface water that contributes to the formation of antarctic intermediate water flowing northwest beneath warmer subantarctic surface water north of the zone. The difference in temperature between these two water masses results in the presence of a permanent thermocline north of the antarctic convergence that is sufficiently strong for the Sofar velocity minimum to be established. South of the polar front zone, the temperature variation with depth is such that the sound velocity shows an essentially monotonic increase with depth at all latitudes apart from a shallow sound channel, the antarctic duct [*Garner,* 1967*a*], which exists in the summer months at around 150–200 meters.

The other important hydrologic feature in the area under study is the subtropical convergence, which separates the subantarctic waters and subtropical surface waters of higher temperature and salinity; the maximum salinity is less than 34.9‰, and the density σ_t at 200 meters is greater than 26.8 south of this boundary [*Burling,* 1961]. East of New Zealand the subtropical convergence occupies a narrow band of latitude between 43° and 45°S, but west of New Zealand the subtropical convergence region is much more extensive and not so well defined. In the Tasman Sea, the northern limit of the region is probably given by the position chosen by *Wyrtki* [1962*a*], who defined a line running from about 45°S, 145°E to 40°S, 170°E; the southern limit of the region is probably given by the convergence as defined by *Garner* [1967*b*]. In this connection, note that several marked effects on long-range sound propagation in the Tasman Sea [*Kibblewhite and Denham,* 1967] occur at Wyrtki's boundary. Finally, south of Australia and west of 140°W, *Deacon* [1937] placed the subtropical convergence at around 40°S.

This study will deal mainly with data obtained during the cruises carried out by the *Eltanin* in the area of interest. These were, in chronological order, cruise 16 in February 1965 [*Jacobs,* 1966]; cruise 27 in February 1967 [*Jacobs and Amos,* 1967]; cruise 28 in March 1967 [*Scripps Institution of Oceanography et al.,* 1969]; cruises 32–36 in 1968 [*Jacobs et al.,* 1970]; cruises 37, 39, 42, 44, and 45 in 1969–1970 (A. L. Gordon, personal communication, 1971); and cruise 41 (J. A. T. Bye, personal communication, 1971).

TYPES OF VELOCITY-DEPTH PROFILE

Figures 1, 2, and 3 show examples of typical velocity–depth profiles encountered in the region south of Australia and New Zealand. In Figure 1, we show a selection of profiles derived from the data obtained by the *Eltanin* during cruise 16 near the 160°E meridian in February 1965. These profiles represent summer conditions along this meridian. For comparison, Figure 2 shows profiles that represent winter conditions along the same meridian and that were derived from data obtained during cruise 34 in June 1968. Finally, Figure 3 provides examples of the profiles obtained along the 117°E meridian during cruise 35 in September 1968; these profiles again represent winter conditions.

The profiles obtained at 59°S, 162°E; 60°S, 160°E; and 56°S, 117°E and shown in Figures 1, 2, and 3, respectively, are examples of antarctic profiles [*Denham and Kibblewhite,* 1970]. Apart from the antarctic duct at around 150 meters in the February profile, these profiles show the essentially monotonic increase with depth that is characteristic of the region south of the polar front zone.

In Figures 1, 2, and 3, the profiles obtained at 54°S, 160°E; 55°S, 160°E; and 49°S, 117°E, respectively, provide examples of the type of profile encountered in the vicinity of the antarctic convergence or polar front zone. The difference in character between these antarctic convergence profiles and the antarctic profile results from the sinking of antarctic surface water beneath the warmer subantarctic surface water at the antarctic convergence. As stated above, a permanent thermocline sufficiently strong to define a Sofar velocity minimum becomes established north of the antarctic convergence. In the examples shown, the Sofar axis is ill defined in comparison with the profiles existing farther north at 49°S, 160°E and at 46°S, 117°E. These latter velocity profiles are examples of subantarctic profiles. Apart from the well-defined Sofar axis the salient features of these profiles are a relative maximum in velocity at around 500 meters that may be defined as the subantarctic maximum and a velocity

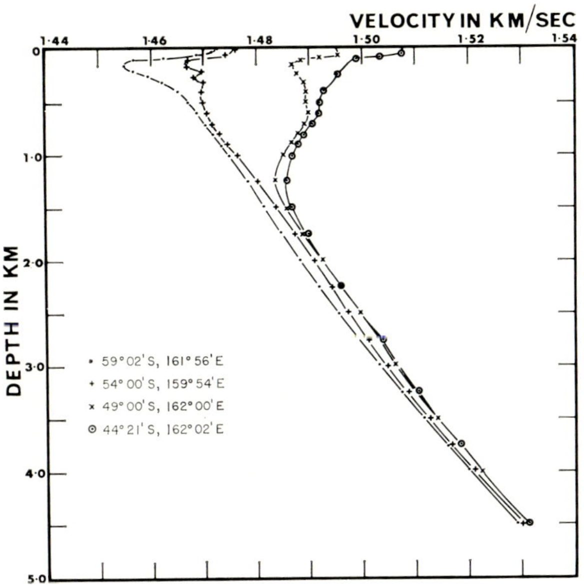

Fig. 1. Velocity–depth profiles obtained near the 160°E meridian in February 1965.

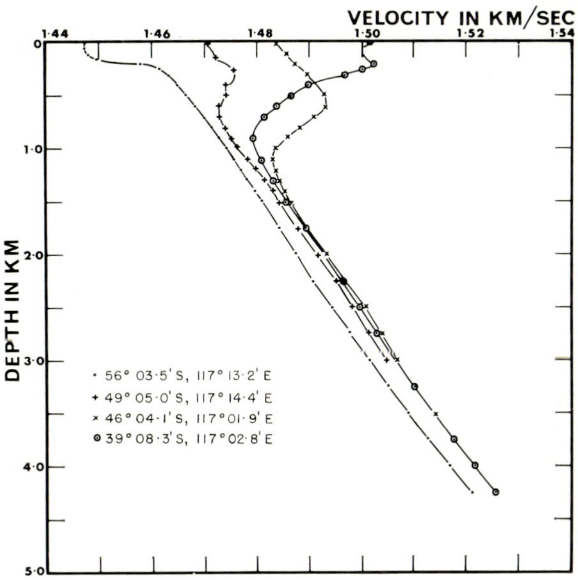

Fig. 3. Velocity–depth profiles obtained along the 117°E meridian in September 1968.

minimum above 500 meters that is present during the summer and defined as the subantarctic duct by *Garner* [1967*a*].

North of the subtropical convergence region, profiles usually exhibit a more or less smooth decrease in velocity between the bottom of the isothermal layer and the Sofar axis. Examples of this behavior are

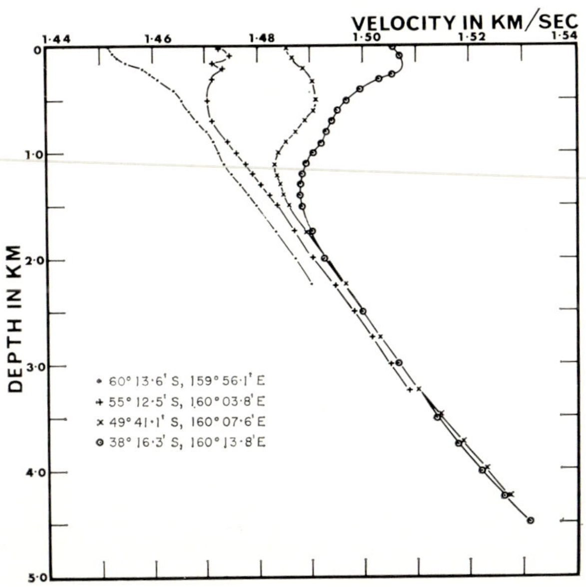

Fig. 2. Velocity–depth profiles obtained along the 160°E meridian in June 1968.

provided by the profiles obtained at 38°S, 160°E and 39°S, 117°E. The profile in Figure 1 obtained at 44°S, 162°E is a typical example of the profiles that occur in the subtropical convergence region, a characteristic feature of these subtropical convergence profiles being the knee at around 700 meters. A common feature of the three profiles obtained north of the boundary between the subantarctic and subtropical convergence regions is a more or less smooth decrease in velocity between 250 and 500 meters. These examples illustrate the profound change in character that takes place in the velocity–depth profile on proceeding north through the antarctic and subtropical convergences.

MERIDIONAL SECTIONS

A more detailed description of the variation of velocity structure with geographical position is provided by the meridional sections shown in Figures 4–6. These figures have been derived from data obtained during the numerous north-south traverses made by the *Eltanin* in the area. The corresponding tracks are shown in Figure 7. These traverses were made at various times: the traverse 398–404 was carried out in February 1965 [*Jacobs*, 1966]; the traverse 659–689 from 65°S, 161°E to Tasmania was made in February 1967 [*Jacobs and Amos*, 1967]; the traverses 1162–1167 and 1254–1269 were carried out in July 1969 and October 1970 (A. L. Gordon, personal communication, 1971); the traverse H2–H9 was carried out in the period from December 1969 to January 1970 (J. A. T. Bye, personal communication, 1971); and the remainder were made in the period from June to December 1968 [*Jacobs et al.*, 1970].

A salient feature of all these sections is the pronounced transition in velocity structure that takes place between the positions of the arrows labeled B and C on Figures 4, 5, and 6. The arrows labeled B represent the position of the more northerly of the 34.2‰ isohalines that enclose the circumpolar zone of minimum salinity in the 200-meter plane, whereas the arrows labeled C indicate the position of the polar front zone. This zone is defined here as the position of the absolute salinity minimum in the 200-meter plane. (Braces have been used on Figures 4*a*, 5*a*, 5*b*, and 6*b* where a salinity minimum could not be defined more closely.) North of boundary B a Sofar axis can be defined at depths greater than 500 meters in all cases. On the other hand, south of the polar front zone all sections show an essentially monotonic increase in velocity, except for the presence of an antarctic duct that is most noticeable in the sections representing the summer months and is situated at a depth of less than 250 meters. The transition in velocity structure that takes place between positions B and C can be associated with the sinking of antarctic surface water at the polar front zone. These sections show that north of the boundary B a permanent thermocline exists of sufficient strength to give rise to a Sofar velocity minimum below 500 meters.

Apart from the ubiquitous Sofar minimum, the most common feature north of boundary B is the pronounced maximum referred to earlier as the subantarctic maximum. This feature lies between depths of 200 and 700 meters. In Figures 4*a*–4*d*,, 5*b*–5*d*, 6*b*, and 6*c*, a decrease in depth of this feature is evident immediately to the north of the position B, and this movement can be associated with the presence of a boundary defined by *Burling* [1961] as the Australasian subantarctic front. This boundary, the position of which is shown by the arrows labeled A, defines the southern limit of Australasian subantarctic water, which is characterized above 500 meters by a salinity greater than 34.5‰ and a temperature greater than 8°C [*Houtman*, 1967; *Zillman*, 1970]. Burling referred to the subantarctic water south of the boundary where the maximum salinity is less than 34.5‰ as circumpolar subantarctic water. An additional consequence of the change in temperature structure associated with this front is that the Sofar axis lies at a

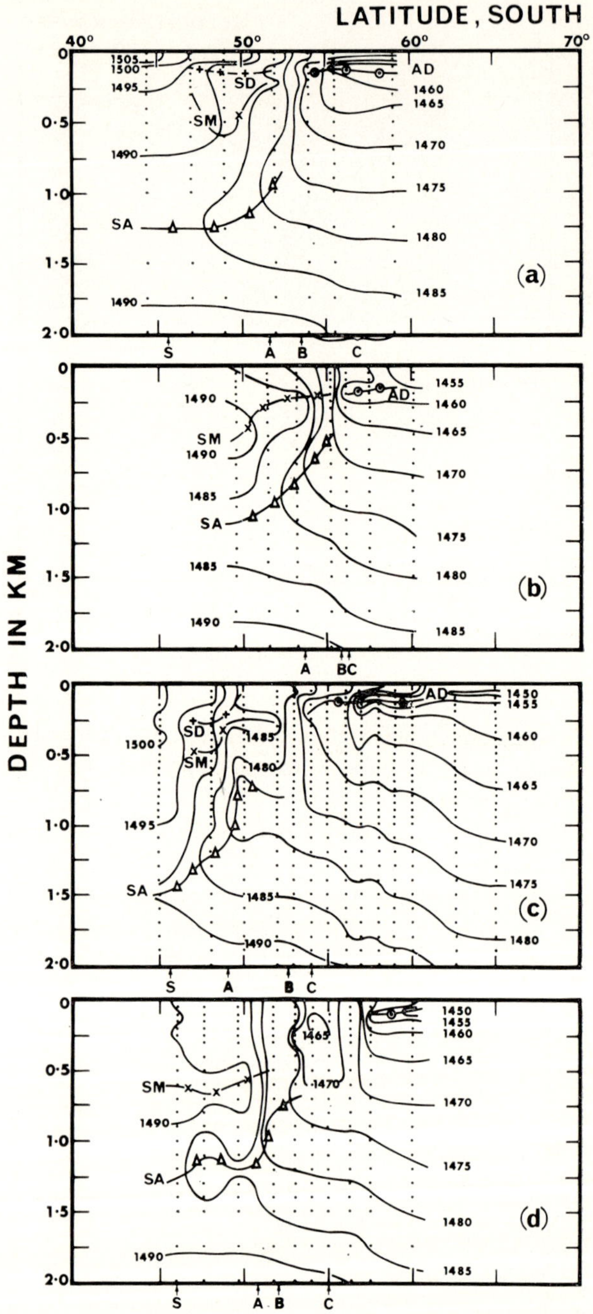

Fig. 4. Meridional sections of sound velocity obtained along (*a*) the 160°E meridian in February 1965 (stations 398–404); (*b*) the 160°E meridian in June 1968 (stations 836–842); (*c*) the track from 65°S, 161°E to Tasmania in February 1967 (stations 659–689); (*d*) the 155°E meridian in November 1968 (stations 935–951). The arrows S, A, B, and C show the positions of the hydrologic boundaries discussed in the text. The velocity is expressed in meters per second, and the contour interval is 5 m/sec. The triangles show the position of the Sofar axis (SA); the crosses, the subantarctic maximum (SM); the circles, the antarctic duct (AD); and the plus signs, the subantarctic duct (SD). The dots indicate the points at which the sound velocity was obtained.

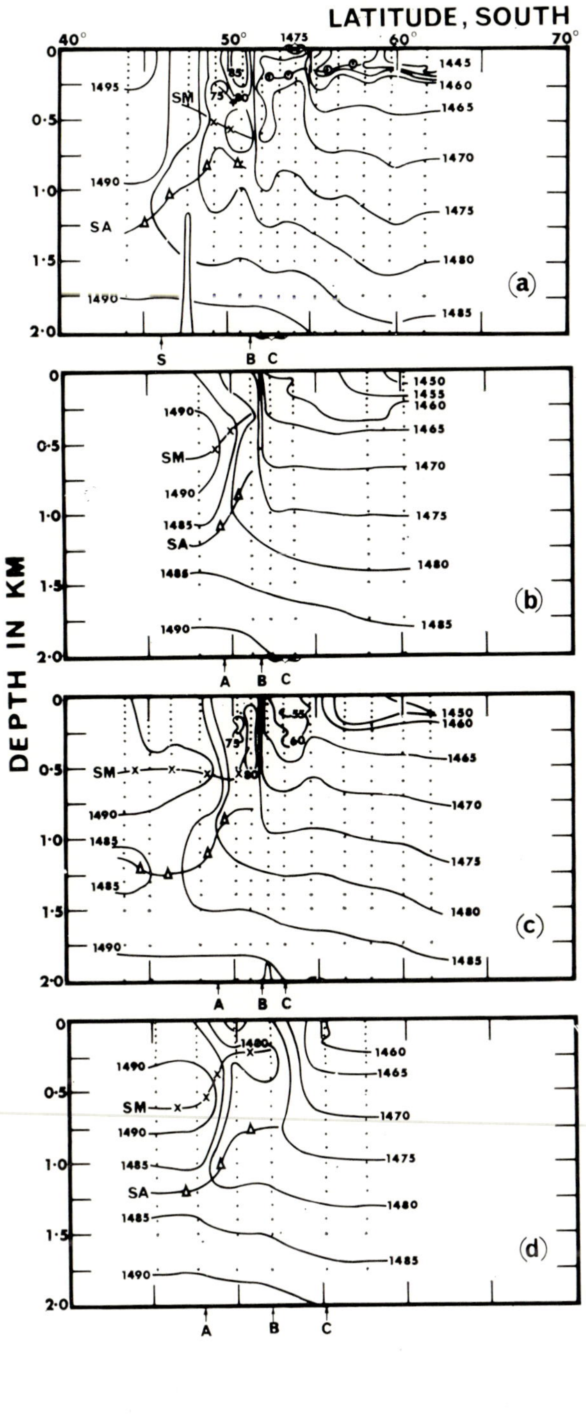

Fig. 5. Meridional sections of sound velocity obtained along (*a*) the 150°E meridian in November 1968 (stations 915–933); (*b*) the 145°E meridian in July 1968 (stations 844–849); (*c*) the 140°E meridian in October 1968 (stations 891–911); and (*d*) the 134°E meridian in July 1969 (stations 1162–1167). Other conditions are identical to those in Figure 4.

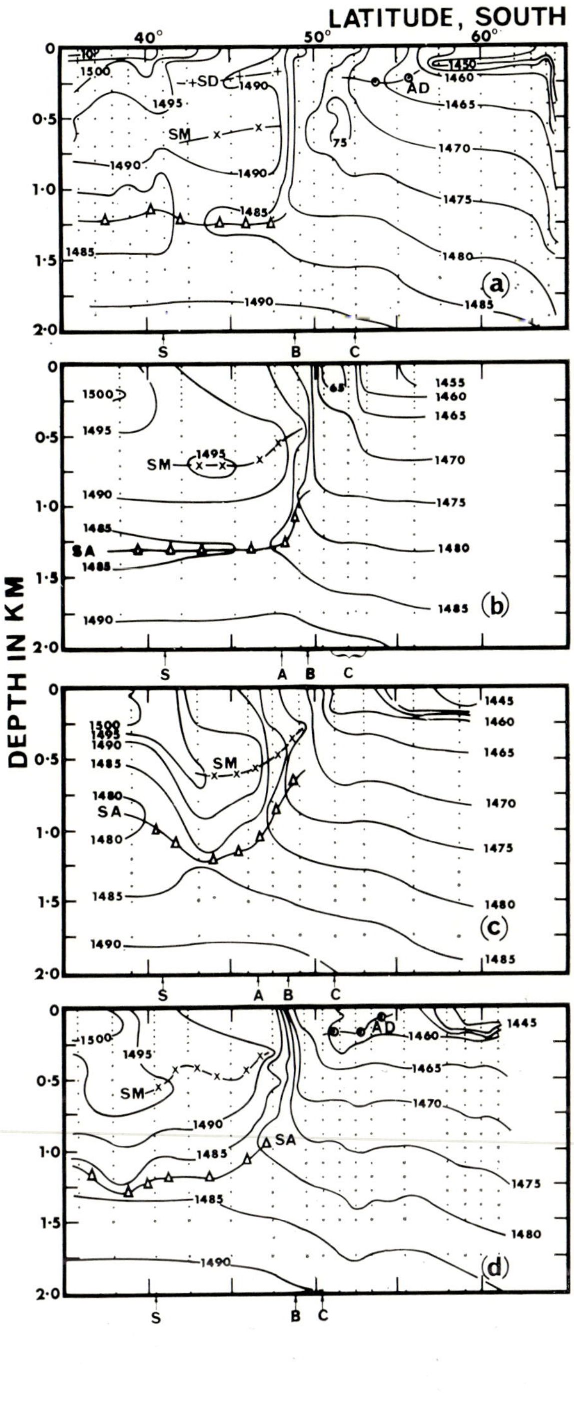

Fig. 6. Meridional sections of sound velocity obtained along (*a*) the 132°E meridian in the period December 1969–January 1970 (stations H2–H9); (*b*) the 128°E meridian in the period August–September 1968 (stations 869–877); (*c*) the 117°E meridian in September 1968 (stations 878–886); and (*d*) the 114°E meridian in October 1970 (stations 1254–1269). Other conditions are identical to those in Figure 4.

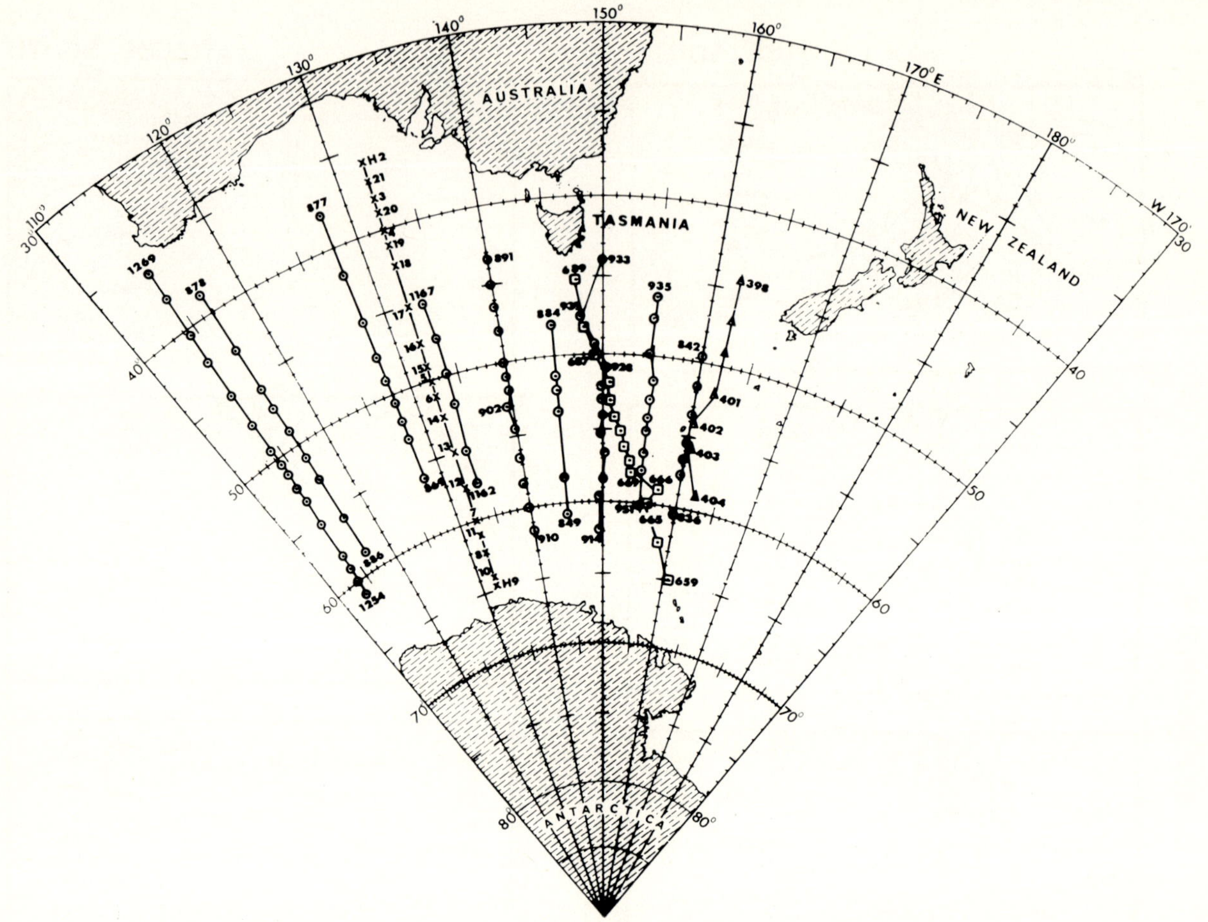

Fig. 7. Tracks corresponding to the meridional sections shown in Figures 4–6.

depth of less than 1 km, whereas the Sofar minimum velocity is less than 1480 m/sec at stations occupied south of this feature. The other notable feature in the subantarctic region is the presence in the summer months of a subantarctic duct (see Figures 4*a*, 4*c*, 6*a*).

Several of the meridional tracks crossed the southern boundary of the subtropical convergence region, this feature being indicated in Figures 4, 5, and 6 by the arrows labeled S. Except in Figure 4*a*, the position chosen corresponds to that given by *Deacon* [1937]. In Figure 4*a*, the position chosen for this boundary is the same as that given by *Burling* [1961]. In Figures 4*d* and 6*d*, the observed salinities indicate that the actual position of the boundary should be to the north of that assigned by Deacon. A noticeable feature of the sections that include the subtropical convergence region is the marked change in velocity of the upper layers at the southern boundary of this region. In particular, there is a marked change in the depth of the 1495-m/sec isopleth, and at 200 meters the velocity is less than 1495 m/sec at all stations occupied in the subantarctic region. In other words, the subtropical convergence region is close to the geographical position of the 1495-m/sec velocity contour in the 200-meter plane.

SOFAR VELOCITY CHART

The general pattern of the geographical variation in the Sofar axis velocity in the study area is shown in Figure 8. The line broken by ×'s shows the position of the northern boundary of the circumpolar zone where the salinity in the 200-meter plane is less than 34.2‰. This feature has been plotted since the previous section showed that the Sofar axis loses its identity south of this boundary. In the area south of the boundary, the broken contour lines shown were drawn by using the data taken during cruises 32–36 of the *Eltanin* in the period June–December 1968

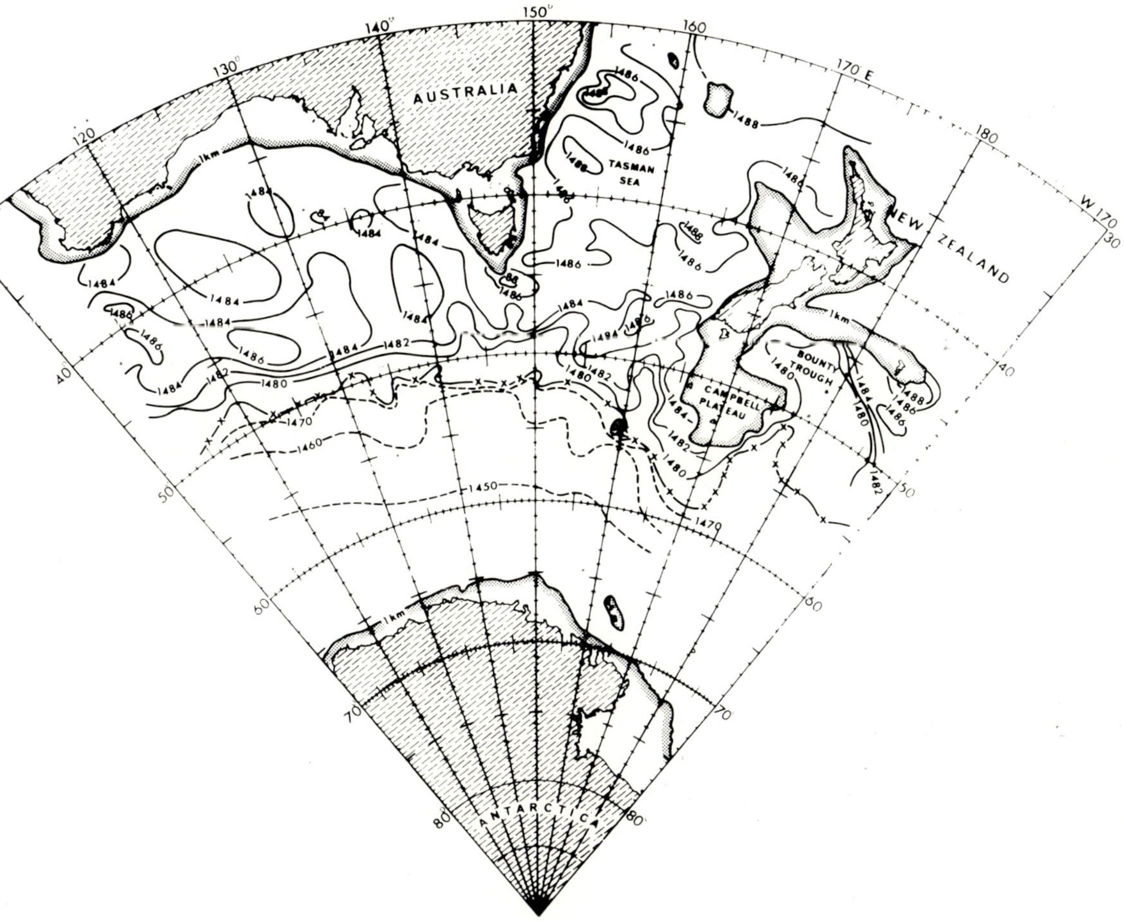

Fig. 8. Sofar velocity chart. The velocity is expressed in meters per second. The line broken by ×'s shows the northern 34.2‰ isohaline in the 200-meter plane. The broken lines south of this boundary represent minimum velocities obtained in the period June–December 1968. The stippled boundaries represent the 1-km depth contour.

[*Jacobs et al.*, 1970] and are thus approximately representative of southern winter conditions. The parameter presented in this case is the absolute minimum velocity at each station. North of the line broken by ×'s, the *Eltanin* data have been supplemented by data from cruises carried out by the *Discovery II* and the Commonwealth Scientific and Industrial Research Organisation of Australia.

The predominant current flow at all levels in the oceans south of Australia and New Zealand is toward the east. A major perturbation to the circulation at intermediate depths is produced by the Campbell plateau (see Figure 8). As a result, both the Sofar velocity contours and the polar front zone are deflected to the south and an area of low Sofar velocity prevails in the Bounty trough [*Garner*, 1967*a*].

In the Tasman Sea and in the seas south of Australia, there are numerous areas where the Sofar velocity is relatively high, being greater than 1486 m/sec in some places. In the Tasman Sea, the predominant current system is known to produce anticyclonic eddies [*Hamon*, 1965*a*; *Boland and Hamon*, 1970], and there is some evidence for the existence of similar eddies south of Australia [*Wyrtki*, 1962*b*; *Hamon*, 1965*b*]. The regions of high Sofar velocity may be associated with the presence of these anticyclonic eddies since isothermals always deepen toward the center of an anticyclonic gyral or eddy as a result of the effects of geostrophic balance [*Neumann and Pierson*, 1966].

SUMMARY

We have shown that as we proceed north from Antarctica into the oceans south of Australia and New

Zealand marked changes occur in sound velocity structure and these changes are associated with four different hydrologic boundaries.

1. The southernmost boundary is the antarctic convergence, or polar front zone, that has been defined here as the position of minimum salinity in the 200-meter plane. South of this boundary, the velocity shows an essentially monotonic increase with depth apart from the shallow minimum known as the antarctic duct, which is present in summer.

2. The circumpolar zone of minimum salinity in the 200-meter plane is enclosed by the 34.2‰ isohalines. In the vicinity of the position of the northern isohaline, a permanent thermocline exists of sufficient strength to give rise to a Sofar velocity minimum below 500 meters, and the Sofar axis can be defined everywhere north of this boundary.

3. In the area south of Australia and New Zealand there are two distinct types of subantarctic surface water present, defined by *Burling* [1961] as Australasian and circumpolar subantarctic waters, the maximum salinity of these water masses being greater than or less than 34.5‰, respectively. The southern boundary of Australasian subantarctic water has been defined as the Australasian subantarctic front. The marked change in temperature structure associated with this front results in a deepening of both the subantarctic velocity maximum and the Sofar axis on proceeding north. South of the Australasian subantarctic front the Sofar velocity is less than 1480 m/sec.

4. The final boundary of significance is that dividing the subantarctic and subtropical convergence regions. At this boundary, both the temperature and salinity of the upper layers have a sharp meridional gradient, the maximum salinity being less than 34.9‰ south of this boundary. A natural consequence of the relative increase in temperature of the upper layers is a marked change in velocity structure. North of the boundary the velocity tends to show a permanent and reasonably smooth decrease with depth between 250 and 500 meters; contrastingly, in winter between this boundary and the Australasian subantarctic front the velocity tends to show a monotonic increase with depth to between 400 and 600 meters. The data dealt with here indicate that the subtropical convergence region may be defined by the geographical position of the 1495-m/sec velocity contour in the 200-meter plane.

From the changes in velocity structure associated with these boundaries, they could possibly be investigated acoustically by means of suitably designed, long-range sound transmission experiments. Evidence of the profound effects that these boundaries can have on long-range acoustic propagation has already been obtained from several experiments carried out by the Defence Scientific Establishment in the South Pacific and Indian oceans [*Kibblewhite et al.*, 1965; *Kibblewhite and Denham*, 1967; *Kibblewhite and Denham*, 1971]. The relevance of these techniques to the study of major hydrologic discontinuities in the ocean needs more investigation, and further experiments are planned that involve transmission paths across the boundaries that are characteristic of the region south of Australia and New Zealand.

Acknowledgments. The authors are very grateful to Dr. A. L. Gordon of the Lamont-Doherty Geological Observatory and Dr. J. A. T. Bye of the Horace Lamb Centre, Flinders University, South Australia, for making available to us the unpublished data of cruises 37, 39, 41, 42, 44, and 45 of the *Eltanin*.

REFERENCES

Boland, F. M., and B. V. Hamon, The East Australian current, 1965–1968, *Deep Sea Res.*, *17*, 777–794, 1970.

Burling, R. W., Hydrology of circumpolar waters south of New Zealand, *N. Z. Dep. Sci. Ind. Res. Bull.*, *143*, 65 pp., 1961.

Deacon, G. E. R., The hydrology of the Southern Ocean, *Discovery Rep.*, *15*, 1–124, 1937.

Denham, R. N., and A. C. Kibblewhite, The sound velocity structure of the South Pacific Ocean, *N. Z. J. Geol. Geophys.*, *13*, 39–51, 1970.

Garner, D. M., Oceanic sound channels around New Zealand, *N. Z. J. Mar. Freshwater Res.*, *1*, 3–15, 1967*a*.

Garner, D. M., Hydrology of the south-east Tasman Sea, *N. Z. Dep. Sci. Ind. Res. Bull.*, *181*, 40 pp., 1967*b*.

Gordon, A. L., Structure of antarctic waters between 20°W and 170°W, *Antarctic Map Folio Ser.*, folio 6, edited by V. Bushnell, Amer. Geogr. Soc., New York, 1967.

Gordon, A. L., and R. D. Goldberg, Circumpolar characteristics of antarctic waters, *Antarctic Map Folio Ser.*, folio 13, edited by V. Bushnell, pp. 1–5, Amer. Geogr. Soc., New York, 1970.

Hamon, B. V., The East Australian current, 1960–1964, *Deep Sea Res.*, *12*, 899–921, 1965*a*.

Hamon, B. V., Geographic currents in the south-eastern Indian Ocean, *Aust. J. Mar. Freshwater Res.*, *16*, 255–271, 1965*b*.

Houtman, T. J., Water masses and fronts in the Southern Ocean south of New Zealand, *N. Z. Dep. Sci. Ind. Res. Bull.*, *174*, 40 pp., 1967.

Hunkins, K., Sound channels in antarctic waters, *Antarctic Map Folio Ser.*, folio 13, edited by V. Bushnell, p. 6, Amer. Geogr. Soc., New York, 1970.

Jacobs, S. S., Physical and chemical oceanographic observations in the southern oceans, USNS *Eltanin* cruises 16–21, 1965, *Tech. Rep. 1-CU-1-66*, 128 pp., Lamont-Doherty Geol. Observ. of Columbia Univ., Palisades, N. Y., 1966.

Jacobs, S. S., and A. F. Amos, Physical and chemical oceanographic observations in the southern oceans, USNS *Eltanin* cruises 22–27, 1966–1967, *Tech. Rep. 1-CU-1-67,* 287 pp., Lamont-Doherty Geol. Observ. of Columbia Univ., Palisades, N. Y., 1967.

Jacobs, S. S., P. M. Bruchhausen, and E. B. Bauer, Hydrographic stations, bottom photographs, and current measurements, USNS *Eltanin* cruises 32–36, 460 pp., Lamont-Doherty Geol. Observ. of Columbia Univ., Palisades, N. Y., 1970.

Johnson, R. H., and R. A. Norris, Geographic variation of Sofar speed and axis depth in the Pacific Ocean, *J. Geophys. Res., 73*(14), 4695–4700, 1968.

Kibblewhite, A. C., and R. N. Denham, Long-range sound propagation in the South Tasman Sea, *J. Acoust. Soc. Amer., 41,* 401–411, 1967.

Kibblewhite, A. C., and R. N. Denham, Low-frequency acoustic attenuation in the South Pacific Ocean, *J. Acoust. Soc. Amer., 49,* 810–815, 1971.

Kibblewhite, A. C., R. N. Denham, and P. H. Barker, Long-range sound propagation in the Southern Ocean—Project Neptune, *J. Acoust. Soc. Amer., 38,* 629–643, 1965.

Leroy, C. C., Development of simple equations for accurate and more realistic calculation of the speed of sound in sea water, *J. Acoust. Soc. Amer., 46,* 216–226, 1969.

Neumann, G., and W. J. Pierson, Jr., *Principles of Physical Oceanography,* pp. 161–164, Prentice-Hall, Englewood Cliffs, N. J., 1966.

Ostapoff, F., The salinity distribution at 200 meters and the antarctic frontal zones, *Deut. Hydrogr. Z., 15,* 133–142, 1962.

Scripps Institution of Oceanography, Woods Hole Oceanographic Institution, and Massachusetts Institute of Technology, Physical and chemical data from the Scorpio Expedition in the South Pacific Ocean, USNS *Eltanin* cruises 28 and 29, March 12–July 31, 1967, *SIO Ref. 69-15, WHOI Ref. 69-56,* 89 pp., La Jolla, Calif., 1969.

Wilson, W. D., Speed of sound in sea water as a function of temperature, pressure and salinity, *J. Acoust. Soc. Amer., 32,* 641–644, 1960.

Wyrtki, K., The subsurface water masses in the western South Pacific Ocean, *Aust. J. Mar. Freshwater Res., 13,* 18–47, 1962*a*.

Wyrtki, K., Geopotential topographies and associated circulation in the south-eastern Indian Ocean, *Aust. J. Mar. Freshwater Res., 13,* 1–17, 1962*b*.

Zillman, J. W., Sea surface temperature gradients south of Australia, *Aust. Meteorol. Mag., 18,* 22–30, 1970.

VERTICAL DISTRIBUTION OF TURBIDITY IN THE SOUTH INDIAN AND SOUTH AUSTRALIAN BASINS

Stephen Eittreim, Peter M. Bruchhausen, and Maurice Ewing

Lamont-Doherty Geological Observatory of Columbia University, Palisades, New York 10964

Abstract. A region of strong light scattering exists in the bottom few hundred meters of water over the antarctic continental slope and rise. The increase in light scattering occurs at about 2500 meters over the rise and is apparently associated with a current of high salinity antarctic bottom water derived from the Ross Sea that flows westward along the continental margin below this depth. In the region between Australia and Antarctica, the turbidity minimum observed at intermediate depths does not appear to be associated with any particular water mass, but marks the depth of overlap between a surface biogenic nepheloid zone and the bottom nepheloid layer. The bottom layer is apparently due to the interaction of moving water with the bottom. On the north side of the South Indian basin, the bottom nepheloid layer is considerably weaker than on the south side, and this variation indicates that most of the 'load' of particles has dropped out of suspension during the passage of this water around the western side of the basin.

Two *Eltanin* cruises, 35 and 37, obtained good quality nephelometer data on their traverses of the two basins separating Australia from Antarctica (Figure 1). The *Eltanin* 35 traverse provided a complete meridional section of antarctic waters, whereas the *Eltanin* 37 data were concentrated along the continental margin off the Adélie Coast, a region where significant antarctic bottom water spillover from the Ross Sea and local shelf region is likely [*Gordon and Tchernia*, this volume]. At certain *Eltanin* 37 stations, both nephelometer and *STD* profiles were obtained, and these profiles allowed detailed study of turbidity, temperature, and salinity through the water column. On the *Eltanin* 35 meridional section, standard Nansen casts were made, and they established the hydrographic structure along that section. The purpose of this paper is to describe the turbidity variations in the regions covered and to describe how these variations relate to the hydrographic structure.

The Lamont-Doherty nephelometer [*Thorndike and Ewing*, 1967] is a photographically recording instrument that measures relative light scattering of ocean water by using white light from an incandescent lamp at angles between about 7° and 30°. The nephelometer records continuously during a cast from surface to bottom to surface. Part of the film records light transmitted directly from the light source through an attenuator. The exposure along this part of the film is used as a monitor of light output. A bourdon tube sensor deflects a light beam that records the depth on the film. Variations in scattering intensity are recorded as exposure variations along the film. For analysis, optical densities are measured with a photodensitometer at points along the film. These densities are converted to log exposure by using the characteristic curve of film contrast, and the log exposures are plotted out as a function of depth. The characteristic curve is obtained by sensitometering the film on shipboard before lowering. The above operations result in a profile of relative turbidity from a depth of 250 meters to the bottom. Data are excluded from the near-surface waters, owing to daylight effects.

The profiles in this area, as well as those in most other regions of the world [*M. Ewing and Thorndike*, 1965; *Eittreim et al.*, 1969; *Hunkins et al.*, 1969; *M. Ewing and Connary*, 1970], show high turbidity values near the surface and near the bottom with a broad minimum in between. The near-bottom layer varies a great deal in its intensity of scattering but is almost always present to some degree. The broad expanse of clear water at intermediate depths has such low concentrations of suspended matter that measurement of these concentrations is difficult. In the North American basin, *Eittreim* [1970] found suspended matter concentrations of 0.01 ppm and less in the clearest intermediate water. If the assumption is made

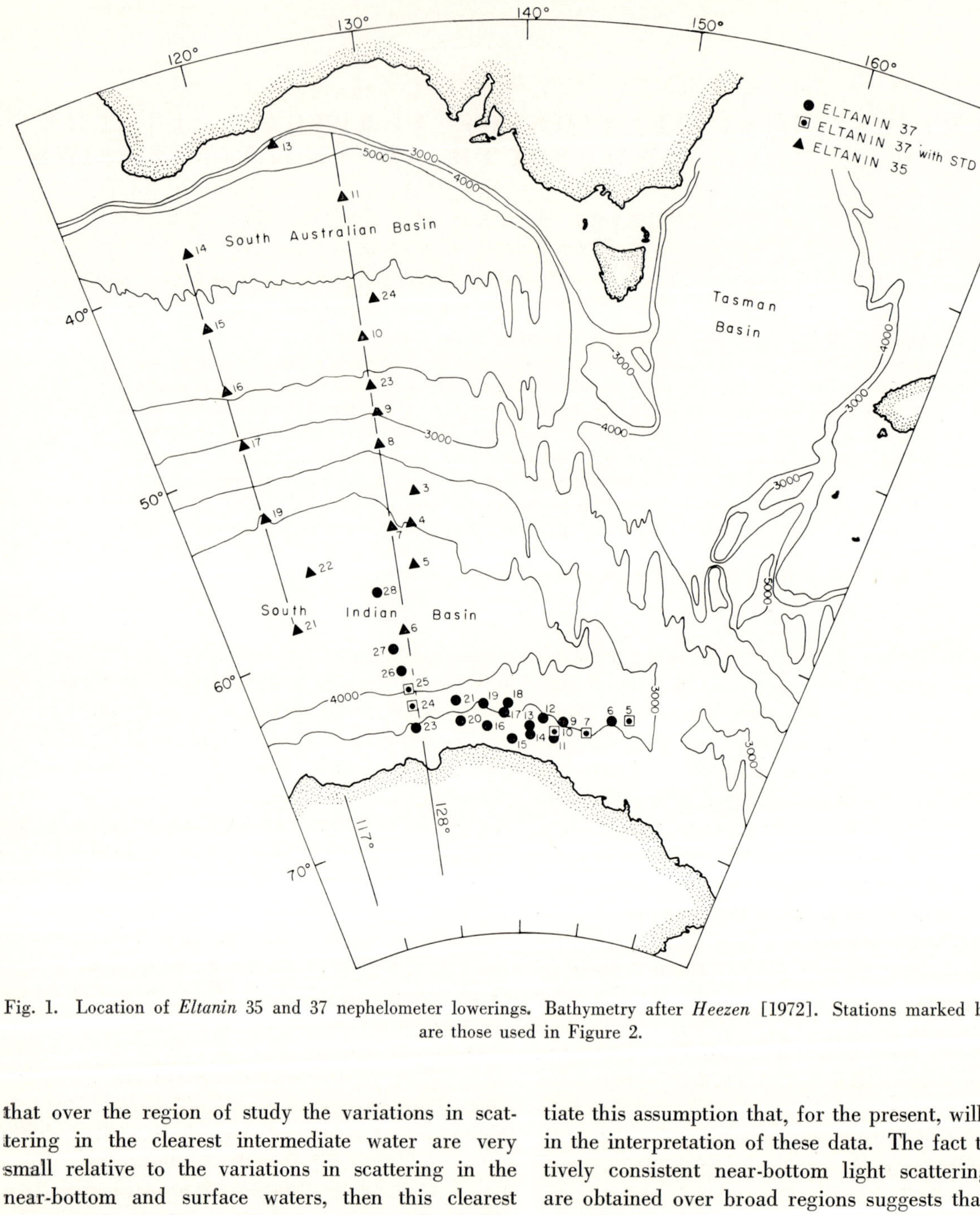

Fig. 1. Location of *Eltanin* 35 and 37 nephelometer lowerings. Bathymetry after *Heezen* [1972]. Stations marked by squares are those used in Figure 2.

that over the region of study the variations in scattering in the clearest intermediate water are very small relative to the variations in scattering in the near-bottom and surface waters, then this clearest water can be used as a reference scattering level. The film exposure units E used here are relative to the exposure in the clearest water E_0 for purposes of comparison of data from station to station. The units of scattering are thus log E/E_0. Quantitative sampling to establish absolute concentrations of particulate matter and their possible geographical variations in these clearest waters is obviously needed to substantiate this assumption that, for the present, will be used in the interpretation of these data. The fact that relatively consistent near-bottom light scattering values are obtained over broad regions suggests that the assumption is valid.

Theoretically, the variations in concentration of particulate matter should result in proportional variations in intensity of scattering or exposure of nephelometer film. The relationship can only be considered to hold in general because of the complicating factors that arise principally because of particle size, shape, and index of refraction variations. In the North

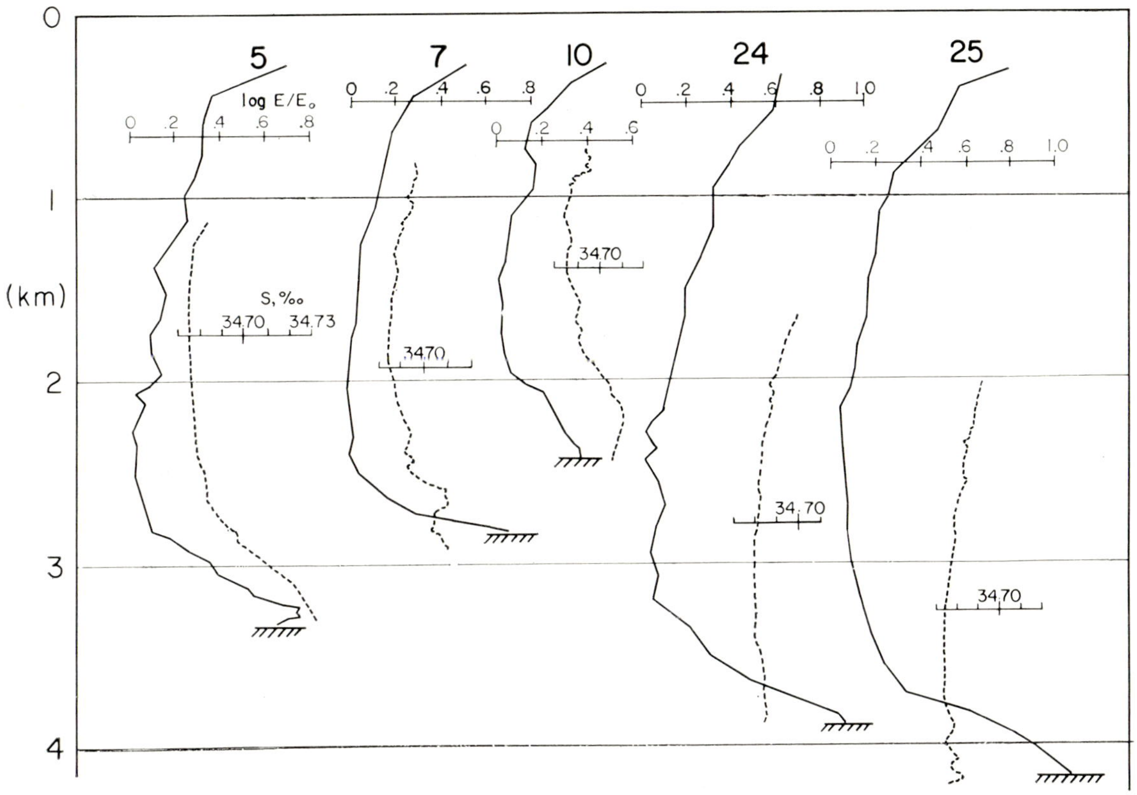

Fig. 2. Five nephelometer profiles and corresponding salinity profiles from stations at which both nephelometer and *STD* lowerings were made. Light scattering units are the log ratio of film exposure E to the exposure in clearest water E_0. Locations are indicated by those stations marked by squares in Figure 1.

American basin, where scattering intensities in the bottom nepheloid layer are a factor of 10 greater than those in the clearest water [*Eittreim et al.*, 1969], the concentrations of suspended matter collected in the most intense part of the layer were about 0.1 ppm. The concentrations represented by the regions of strong light scattering here (log $E/E_0 > 0.8$) are probably of the same order of magnitude.

CONTINENTAL MARGIN REGION

Seventeen nephelometer lowerings were made on the continental slope and rise off the Adélie Coast, five with accompanying *STD* lowerings. The turbidity profiles here are characterized by large increases ranging from factors of 3–10 in light scattering in the bottom few hundred meters. Turbidity increases of a factor of 10 and greater are commonly found in the Argentine and North American basins, particularly over the western parts [*M. Ewing et al.*, 1971; *Eittreim et al.*, 1969]. In the North Pacific, the bottom nepheloid layer is generally weaker in scattering and rarely exhibits factors of increase greater than 5 [*M. Ewing and Connary*, 1970]. Thus the intensity of scattering in the nepheloid layer on the slope and rise off the Adélie Coast appears to be intermediate in strength between the strong western Atlantic bottom layer and the weaker North Pacific layer. At present, however, this comparison must be taken with caution until it is substantiated that the absolute level of scattering in the clearest water in these three regions does not differ significantly.

In this region off the Adélie Coast, a type of antarctic bottom water (AABW) is found with higher salinities (>34.70‰) than the Weddell Sea variety. *Gordon* [1971] and *Gordon and Tchernia* [this volume] attribute this high salinity type to formation of AABW by a mixture of Ross Sea shelf bottom water with 'resident' slope water. A typical salinity profile through the deep and bottom waters in antarctic regions shows a salinity maximum associated with the circumpolar deep water at intermediate depths and then a continuous decrease in salinity to the bottom.

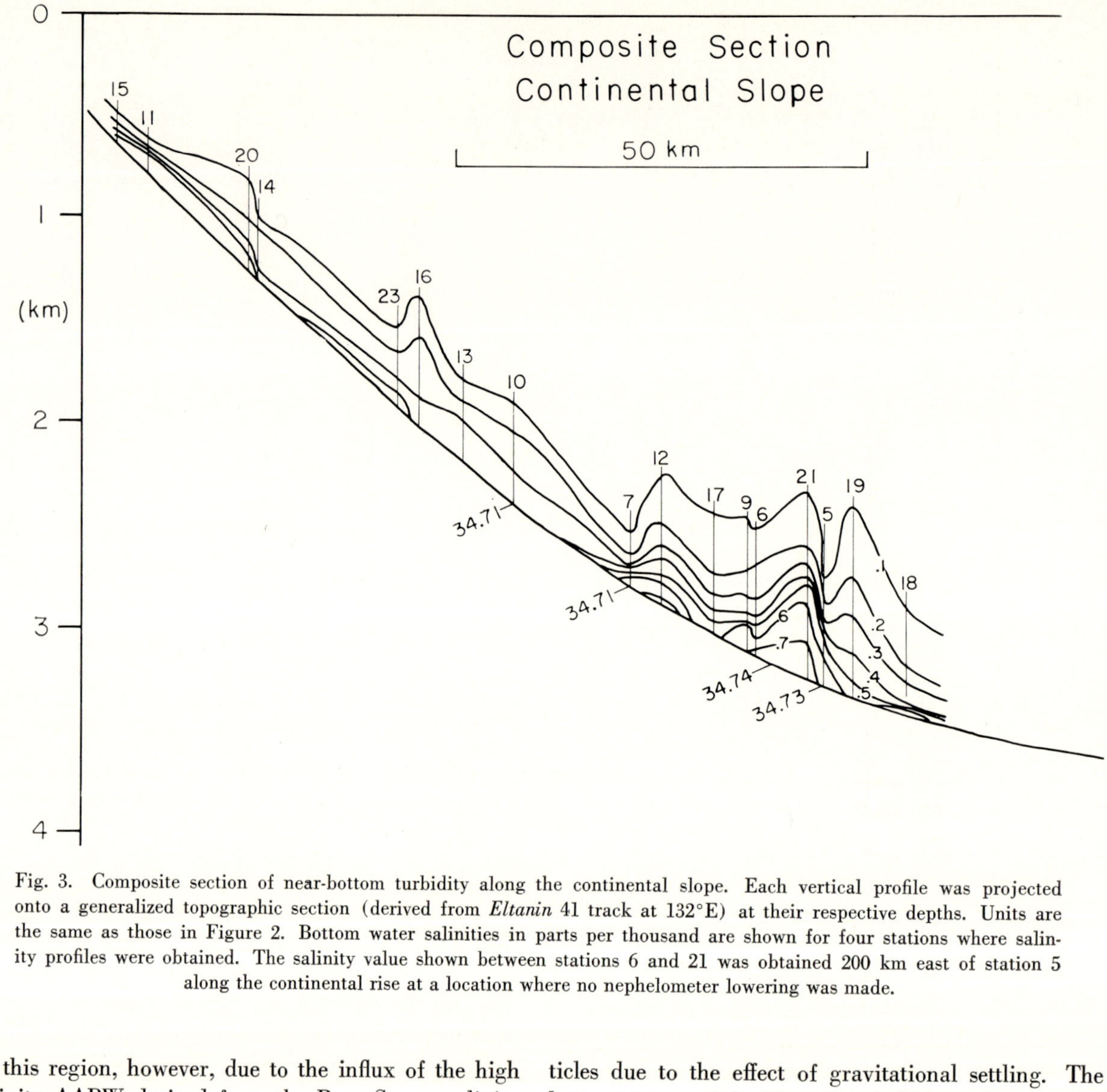

Fig. 3. Composite section of near-bottom turbidity along the continental slope. Each vertical profile was projected onto a generalized topographic section (derived from *Eltanin* 41 track at 132°E) at their respective depths. Units are the same as those in Figure 2. Bottom water salinities in parts per thousand are shown for four stations where salinity profiles were obtained. The salinity value shown between stations 6 and 21 was obtained 200 km east of station 5 along the continental rise at a location where no nephelometer lowering was made.

In this region, however, due to the influx of the high salinity AABW derived from the Ross Sea, a salinity minimum is found several hundred meters above bottom and salinity increases of up to 0.06‰ below this depth. The extent of this rise in salinity toward the bottom presumably marks the extent of injection of the Ross Sea type AABW.

Figure 2 shows that in this region the bottom nepheloid layer is confined to the zone of increased salinity identified with Ross Sea AABW. Figure 2 also shows that the salinity minimum tends to occur somewhat above the minimum in light scattering. If the particles causing the higher turbidity are indigenous to the AABW, this difference in the level of salinity and turbidity minimums would be expected from the preferential net downward diffusion of particles due to the effect of gravitational settling. The bottom water of the Ross Sea shelf was shown to be also characterized by an increase in turbidity [*Jacobs et al.,* 1970*a*], this increase suggesting that at least some percentage of the particles is indigenous to the water mass. On the other hand, this difference could also be explained by a more local source of particles, i.e., erosion from the sea floor along the path of AABW on the slope and rise.

All the continental slope nephelometer profiles have been combined in a composite section down the slope in Figure 3. A schematic topographic profile was obtained from *Eltanin* cruise 41, which crossed the slope at 132°E. The *Eltanin* cruise 37 nephelometer profiles were plotted on this section according to depth. This arrangement is not meant to necessarily

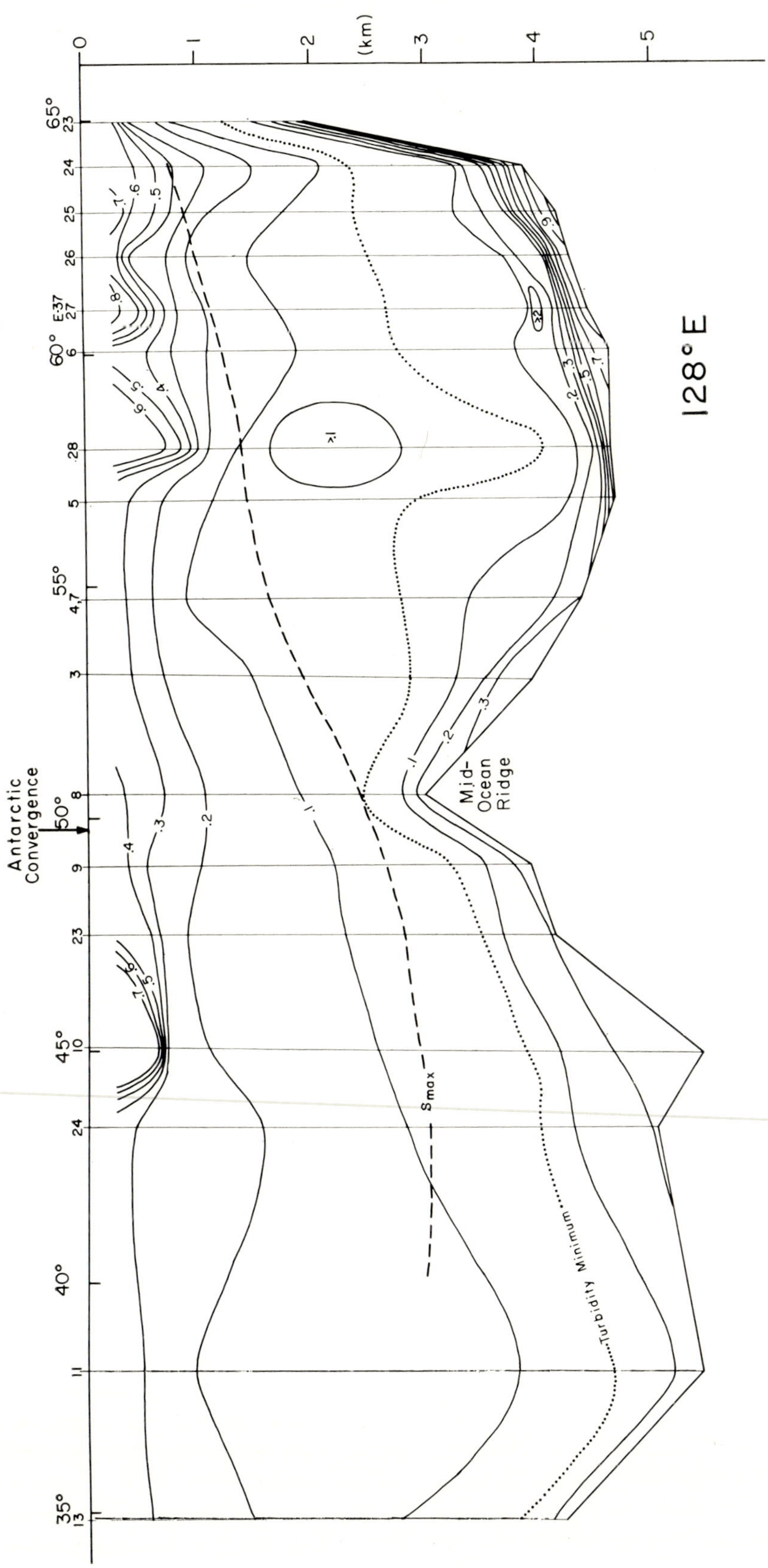

Fig. 4. Turbidity section along 128°E. Salinity maximum is from *Gordon* [this volume] and is derived from hydrographic measurements on *Eltanin* 35, the same cruise on which most of the nephelometer measurements were made. Units are the same as those in Figure 2.

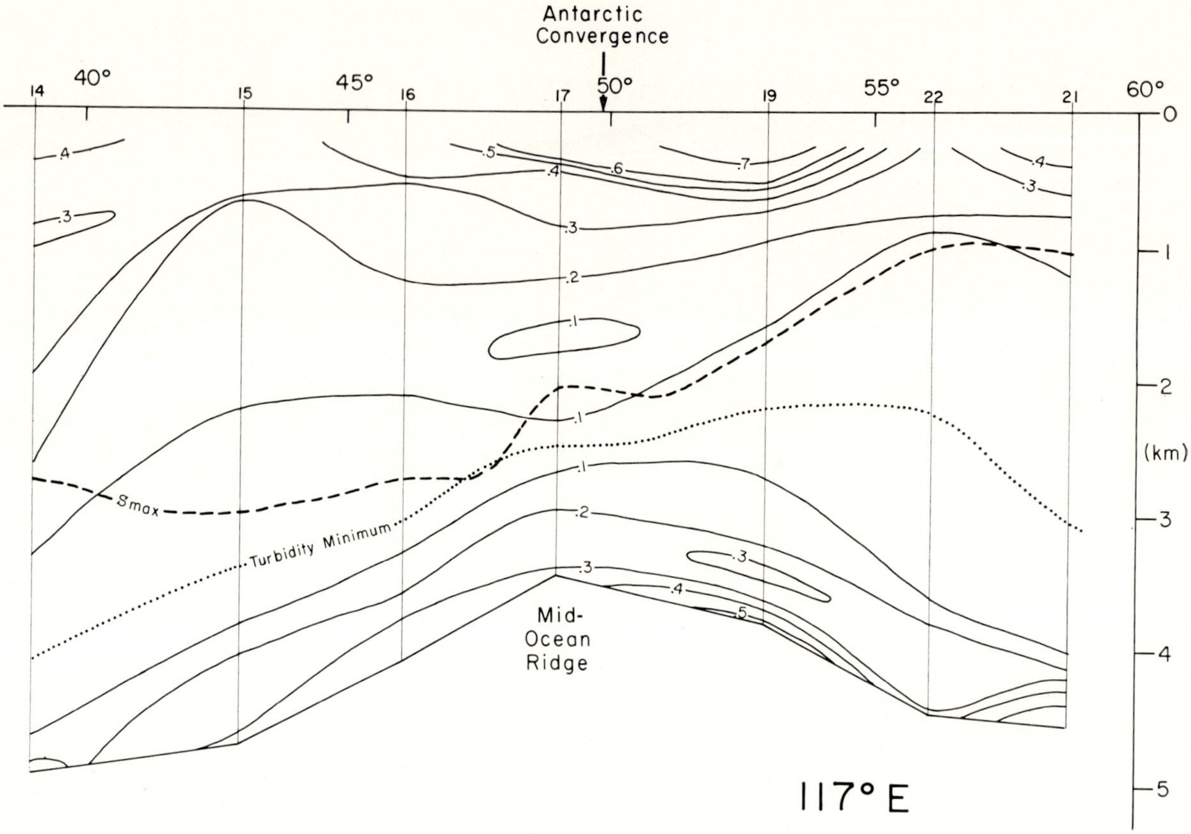

Fig. 5. Turbidity section along 117°E. Salinity maximum taken from the data of *Jacobs et al.* [1970*b*].

represent a real turbidity section across the slope but is only intended to illustrate any trends in thickness and intensity of the nepheloid layer that are dependent on depth. The light scattering values are relative to the clearest water in the water column. The increase seen at about 3000 meters is sufficiently strong and is represented by stations at sufficiently random longitudes to indicate that the increase is probably not simply an effect of changes in the reference water level, but is a real increase in turbidity at this level on the slope and is perhaps associated with Ross Sea AABW flowing westward along the continental margin. The higher salinity recorded at station 5 (Figure 3) indicates the main flow of this AABW below about 3000 meters.

SOUTH AUSTRALIAN AND SOUTH INDIAN BASINS

Turbidity values along the sections at 128°E and 117°E are contoured in Figures 4 and 5. The same general shape turbidity profile is found at all stations, i.e., a decrease downward from the surface and upward from the bottom with the minimum in turbidity generally closer to the bottom than to the surface. The surface layer is most likely due to the production of biogenic scatterers in the euphotic zone and their fallout below. Continental clay detritus is probably also present in the surface waters, and its gradual settling may also contribute to scattering in the upper layer, but its amount is probably small compared with the biogenic detritus in this area of high productivity. As the biogenic particles settle downward, they may aggregate and act as nuclei for the aggregation of clay particles [*Folger*, 1968]. This aggregation of particles plus the dissolution of particles with depth will reduce the total surface area of the particulate matter as it descends and thereby will reduce the light scattering.

The bottom nepheloid layer on the other hand is related to the circulation of bottom water and the interaction of this moving water with the bottom. Turbulence in the bottom water diffuses particles upward in the direction of negative concentration gradient and acts in a direction against their downward settling by gravity. Whether the particles are indigenous to the water mass or are suspended sediments from the bottom cannot be ascertained. The depth of the minimum in turbidity is then the depth at which the light scattering effects of the surface biogenic layer equal those of the bottom nepheloid layer, i.e., the depth of overlap of the upper and lower layers.

The 128°E section shows the association of the strong bottom nepheloid layer with the continental margin of Antarctica. The layer extends across the continental slope and rise but exhibits only low intensities on the north side of the deepest part of the South Indian basin. *Eittreim et al.* [1971] give evidence of clockwise circulation of AABW around the deep South Indian basin. On the basis of salinity and oxygen values in the near-bottom waters of the slope and rise (>2000 meters), *Gordon* [1971] and *Gordon and Tchernia* [this volume] postulate AABW flow westward along the continental margin here. Topographic constraints should cause such a bottom current to continue clockwise around the deep basin. Assuming that the bottom water does flow clockwise around the basin we clearly see from Figures 4 and 5 that in the course of one half turn around this basin the AABW has lost most of its suspended sediment load. This loss suggests that only minor amounts of particulate matter 'indigenous' to the AABW of the South Indian basin escape the basin. Apparently enough suspended load is maintained in the AABW to carry a significant amount to deposit eventually in the Tasman Sea after passage through the mid-ocean ridge via fracture zones south of the Macquarie ridge [*Eittreim et al.* 1971]. According to *Eittreim et al.* [1971, Figure 3], no comparable transport of turbid water appears to pass through the recently discovered fracture zones (called the Australian-Antarctic discordance and not included in the bathymetry of Figure 1) at 120°–128°E [*Hayes and Conolly*, this volume].

Figure 4 also indicates the salinity maximum, traced from *Gordon* [this volume], that is associated with the core of the circumpolar deep water (CDW). The turbidity minimum seems to have little relationship to the core of the CDW. This lack of relationship supports the idea that the turbidity minimum is simply a function of the location of overlap of the two turbid layers, the bottom nepheloid layer, and the surface biogenic nepheloid layer and is not a feature identified with a particular water mass.

North of the antarctic convergence, antarctic intermediate water (AAIW) descends and moves northward and eastward and is evidenced by a salinity minimum of less than 34.4‰ at a depth of 500–1000 meters [*Gordon and Goldberg*, 1970, Plate 18]. A turbidity maximum might be expected to be associated with AAIW, owing to the relatively recent residence of these waters in the euphotic zone south of the convergence. It is noteworthy that no such turbidity maximum was found in the nephelometer profiles north of the convergence (Figures 4 and 5). This finding suggests that the destruction, aggregation, or settling of organic particles occurs rapidly relative to the speed of descent of AAIW.

Acknowledgments. We thank A. Gordon and S. Jacobs for helpful discussions and assistance during this study. This research was supported by National Science Foundation grants GA-27281 and GV-26230.

Contribution 1772, Lamont-Doherty Geological Observatory.

REFERENCES

Eittreim, S., Suspended particulate matter in the deep waters of the northwest Atlantic Ocean, Ph.D. thesis, Columbia University, New York, 1970.

Eittreim, S., M. Ewing, and E. M. Thorndike, Suspended matter along the continental margin of the North American basin, *Deep Sea Res.*, *16*, 613–624, 1969.

Eittreim, S., A. L. Gordon, M. Ewing, E. M. Thorndike, and P. Bruchhausen, The nepheloid layer and observed bottom currents in the Indian-Pacific antarctic sea, in *Studies in Physical Oceanography—A Tribute to Georg Wüst on His 80th Birthday*, edited by A. L. Gordon, Gordon and Breach, New York, in press, 1971.

Ewing, M., and S. Connary, Nepheloid layer in the North Pacific, *Geol. Soc. Amer. Mem. 126*, edited by J. Hays, pp. 41–82, 1970.

Ewing, M., and E. M. Thorndike, Suspended matter in deep ocean water, *Science*, *147*, 1291–1294, 1965.

Ewing, M., S. Eittreim, J. Ewing, and X. Le Pichon, Sediment transport and distribution in the Argentine basin, 3, Nepheloid layer and processes of sedimentation, *Phys. Chem. Earth*, *8*, 49–77, 1971.

Folger, D. W., Trans-Atlantic sediment transport by wind, Ph.D. thesis, Columbia University, New York, 1968.

Gordon, A. L., Spreading of antarctic bottom waters, 2, in *Studies in Physical Oceanography—A Tribute to Georg Wüst on His 80th Birthday*, edited by A. L. Gordon, Gordon and Breach, New York, in press, 1971.

Gordon, A. L., On the interaction of the antarctic circumpolar current and the Macquarie ridge, in *Antarctic Oceanology II: The Australian–New Zealand Sector, Antarctic Res. Ser.*, vol. 19, edited by D. E. Hayes, pp. 71–78, AGU, Washington, D. C., 1972.

Gordon, A. L., and R. D. Goldberg, Circumpolar characteristics of antarctic waters, *Antartic Map Folio Ser.*, folio 13, edited by V. Bushnell, Amer. Geogr. Soc., New York, 1970.

Gordon, A. L., and P. Tchernia, Waters of the continental margin off Adélie Coast, Antarctica, in *Antarctic Oceanology*

II: The Australian–New Zealand Sector, Antarctic Res. Ser., vol. 19, edited by D. E. Hayes, pp. 59–69, AGU, Washington, D. C., 1972.

Hayes, D. E., and J. R. Conolly, Morphology of the southeast Indian Ocean, in *Antarctic Oceanology II: The Australian–New Zealand Sector, Antarctic Res. Ser.*, vol. 19, edited by D. E. Hayes, pp. 125–145, AGU, Washington, D. C., 1972.

Heezen, B. C., *Antarctic Map Folio Ser.*, folio 16, Amer. Geogr. Soc., New York, in press, 1972.

Hunkins, K., E. M. Thorndike, and G. Mathieu, Nepheloid layers in the Arctic Ocean, *J. Geophys. Res.*, *74*(28), 6995–7008, 1969.

Jacobs, S. S., A. F. Amos, and P. M. Bruchhausen, Ross Sea oceanography and antarctic bottom water formation, *Deep Sea Res.*, *17*, 935–962, 1970*a*.

Jacobs, S. S., P. M. Bruchhausen, and E. B. Bauer, Hydrographic stations, bottom photographs, and current measurements, USNS *Eltanin* cruises 32–36, 460 pp., Lamont-Doherty Geol. Observ. of Columbia Univ., Palisades, N. Y., 1970*b*.

Thorndike, E. M., and M. Ewing, Photographic nephelometers for the deep sea, in *Deep-Sea Photography*, edited by J. B. Hersey, pp. 113–116, Johns Hopkins Press, Baltimore, Md., 1967.

WATERS OF THE CONTINENTAL MARGIN OFF ADÉLIE COAST, ANTARCTICA

ARNOLD L. GORDON

Lamont-Doherty Geological Observatory of Columbia University, Palisades, New York 10964

PAUL TCHERNIA

Muséum National d'Histoire Naturelle Laboratoire d'Océanographie Physique, Paris V^e^, France

Abstract. The topographic depressions in the Adélie shelf most likely trap dense water formed during the winter. Therefore the winter water can be studied in the summer by sampling these trapped waters. The trapped shelf water is near the freezing point of sea water; however, each pocket of shelf water has slightly different salinities. Shelf waters sampled during *Eltanin* cruise 37 that were the highest in salinity and greatest in density were found in the 'Adélie depression' between 142° and 147°E. These waters measured —1.9°C, 34.6–34.7‰, and near 27.96 σ_t units. Shelf water results from a mixture of the surface water and brine released during sea ice formation. For each meter of sea ice, approximately 40–50 meters of shelf water is formed. The overflow of shelf water mixes with the upper parts of the slope water at approximately 0.5°C and 34.68‰ to form antarctic bottom water (AABW) with a potential temperature of —0.7°C and salinity of 34.67‰; this AABW is similar to the type reported by H. Mosby in the Weddell Sea. This type of AABW has been observed at one *STD* station along the floor of the continental slope. This dense water is probably sinking and deflecting toward the west under the influence of the Coriolis force and/or frictional interaction with the general east wind drift down the slope. This influence suggests a possible summer overflow of the trapped shelf water, although we expect that the shelf water would overflow mainly during the winter. On reaching the continental rise, the AABW sinking off the Adélie Coast mixes with the high salinity bottom water flowing from the Ross Sea. This mixing not only decreases the salinity of the Ross Sea water along its westward path but also increases the oxygen content of the Ross Sea bottom water by approximately 0.4 ml/l. The ratio of the change of salinity to oxygen is exactly the same as that expected when one considers the initial characteristics of the incoming Ross Sea water and the sinking Adélie bottom water.

The oceanographic processes operating within antarctic waters strongly influence the characteristics of abyssal waters of the entire world ocean. Much of this influence results from the northward spreading of the cold, highly oxygenated antarctic bottom water (AABW). This important water mass is believed to form near the antarctic continental margins by the intermixing of dense shelf water with components of circumpolar deep water (CDW). The process responsible for the production of the shelf water is not known, although four distinct possibilities exist:

1. The most widely accepted hypothesis calls for winter production during sea ice formation over continental shelf areas [*Brennecke*, 1921; *Mosby*, 1934].
2. A combination of sea ice formation and interaction of sea water with the thick floating glacial ice has also been proposed [*Jacobs et al.*, 1970; *Gordon*, 1971*a; Seabrooke et al.*, 1971].
3. The intense cooling and evaporation in near-coastal regions where katabatic winds are strong can produce the necessary cold salty shelf water.
4. *Gill and Turner* [1969] have suggested a mechanism dependent on the difference of the molecular diffusivity for salt and heat.

Measurement of the characteristics of the dense shelf water is difficult since we expect that most shelf water forms during the winter and is subsequently replaced by warmer water in the summer, the only season the shelf can be studied from research vessels. A unique opportunity exists to observe dense shelf water,

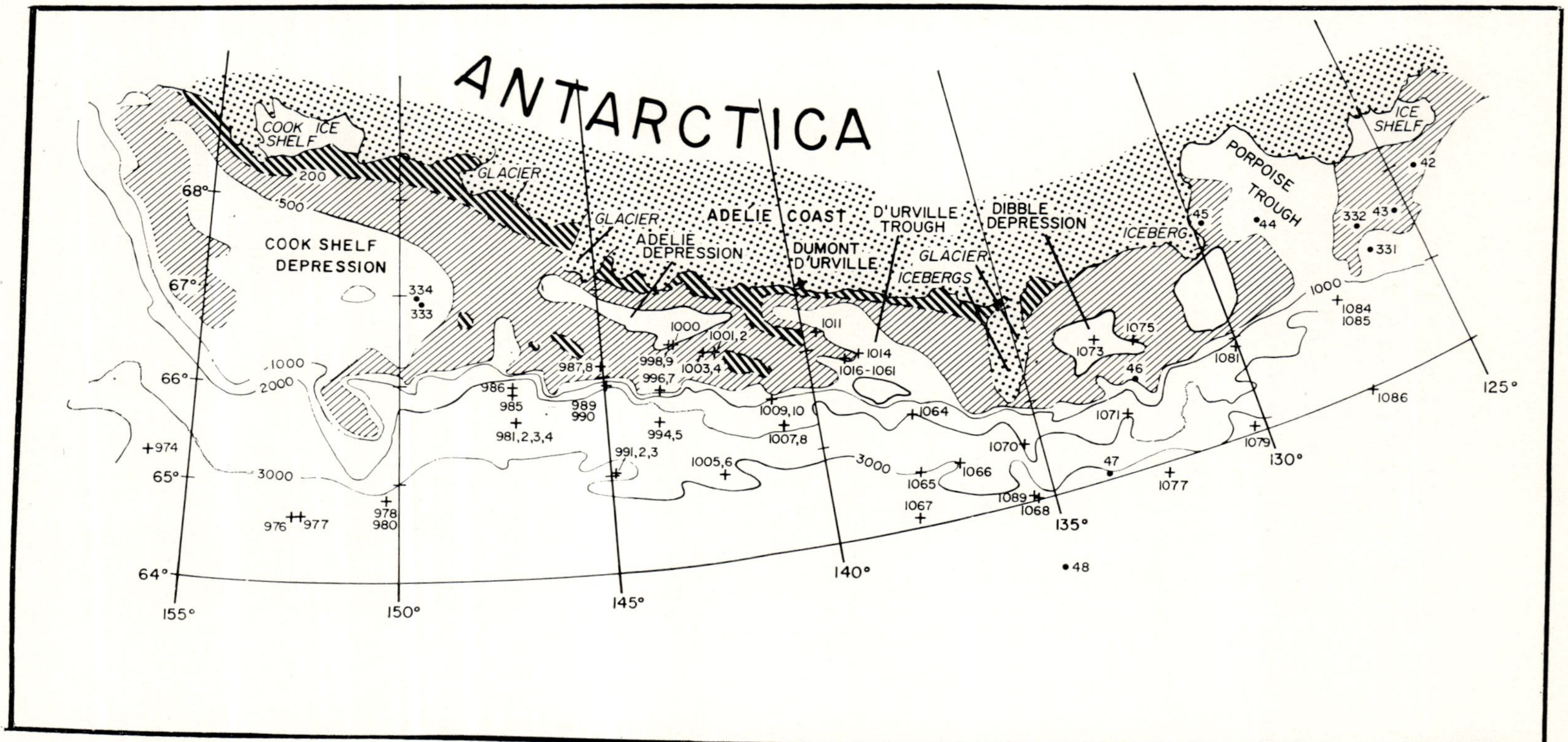

Fig. 1. The Adélie shelf region. Isobaths are taken from *American Geographical Society* [1970]. Unofficial names given to continental shelf features are explained in text. The crosses represent hydrographic stations from the *Eltanin* cruise 37 and the darkened circles represent *Ob* data for the International Geophysical Year.

even in the summer months, by sampling the waters within the numerous topographic deeps on the antarctic continental shelf.

The best known example of such a depression is within the Ross Sea. This depression extends to a maximum depth of 900 meters and is filled with water that has a salinity of 34.8–35.0‰ and a temperature of −1.95°C [*Countryman and Gsell*, 1966; *Jacobs et al.*, 1970]; these factors make this depression the most dense water mass yet observed within antarctic waters (as high as $\sigma_t = 28.17$). This dense Ross Sea shelf water produces a high salinity variety of AABW that flows into the South Indian and Southeast Pacific basins [*Gordon*, 1971*b*].

The shelf water measured during the summer in the Weddell Sea is also cold and salty (−2.1°C, 34.8‰), although the denser shelf water may have escaped by summer because this water rests on a flat broad continental shelf [*Seabrooke and Hufford*, 1970.] *Mosby* [1970] states that a deep depression exists near the edge of the Filchner ice shelf that isolates the water below 600–700 meters. The trapped shelf water is near freezing and has salinities between 34.66 and 34.77‰.

Deep depressions also occur on the continental shelf off the coast of East Antarctica in the vicinity of Adélie Coast. These depressions may also trap winter-produced shelf water, and this entrapment thus allows summer sampling [*Tchernia*, 1951*a*, 1951*b*.]

THE DATA AND AREA

From January 22 to February 10, 1969, the USNS *Eltanin* obtained oceanographic data along the Adélie Coast region of Antarctica. The hydrographic stations consisted of (1) standard serial data of nansen casts; (2) in situ recording of salinity, temperature, and depth (*STD*) with calibration data (Niskin sample bottles equipped with reversing thermometers attached to an *STD* frame; see *Gerard and Amos* [1968]); and (3) nepheloid measurements. The hydrographic station positions and basic submarine topography are shown in Figure 1 (see *Hayes and Conolly* [this volume]). Brief accounts of the cruise have been given (see *Antarctic Journal* [1969] and *Belliard* [1969]). The nepheloid results are discussed by *Eittreim et al.* [this volume].

The continental shelf topographic deeps are either large enclosed depressions isolated from the open ocean or elongate troughs open on one or both ends without large areas that trap significant quantities of dense water. We will refer to the closed variety as depressions and to the open variety as troughs. The names shown on Figure 1 for these depressions and troughs have been given by the authors for convenience; these names are thus unofficial. The continental shelf break occurs for the most part between the isobaths of 200–500 meters except at the mouths of troughs. The depressions and troughs extend to depths greater than 600 meters and often to depths greater than 1000 meters (e.g., 'Dibble depression').

Heavy pack ice conditions prevented the *Eltanin* from entering the 'Cook depression' and the western part of the 'Adélie depression.' The RV *Ob* gathered data in the 'Cook depression' during the International Geophysical Year [*Capurro*, 1964]. The *Ob* station positions are also shown on Figure 1.

THE SHELF WATER

The temperature–salinity relationship of the water column in each of the topographic deeps on the shelf is given in Figures 2 and 3. In the depressions, the near-bottom water, which will be referred to as shelf water, is near the 1-atm pressure freezing point for sea water given by *Knudsen* [1903]. The salinities of the section of the water column with temperatures near the freezing point vary from a low of 34.4‰ in the upper part of the 'Cook depression' to slightly over 34.7‰ in the bottom of the 'Adélie depression.' The shelf water filling the larger 'Ross Sea depression' has a higher salinity than any of the shelf water filling the depressions of the Adélie Coast, although the temperatures of all the shelf waters are near the freezing point for their respective salinities and some values are below the freezing point of sea water at 1-atm pressure. This cold water is not subfreezing at the observed depth, owing to the influence of the elevated pressure. The presence of the cold, relatively saline water can be explained either by the downwelling of ice crystals [*Countryman*, 1970] or as the residue following freezing to the underside of thick ice shelves [*Gordon*, 1971*a*].

The bottom waters in the troughs have salinities above that of surface water but are strikingly warmer than the bottom water in the depressions. The temperature minimum in the water column over the troughs occurs at depths between 100 and 200 meters and warms by 0.3–0.4°C to the bottom to reach a value of approximately 0.6°C above the freezing point.

We assume that only in the depressions can the dense water formed in a previous winter be trapped, the entrapment enabling summer sampling. The dense

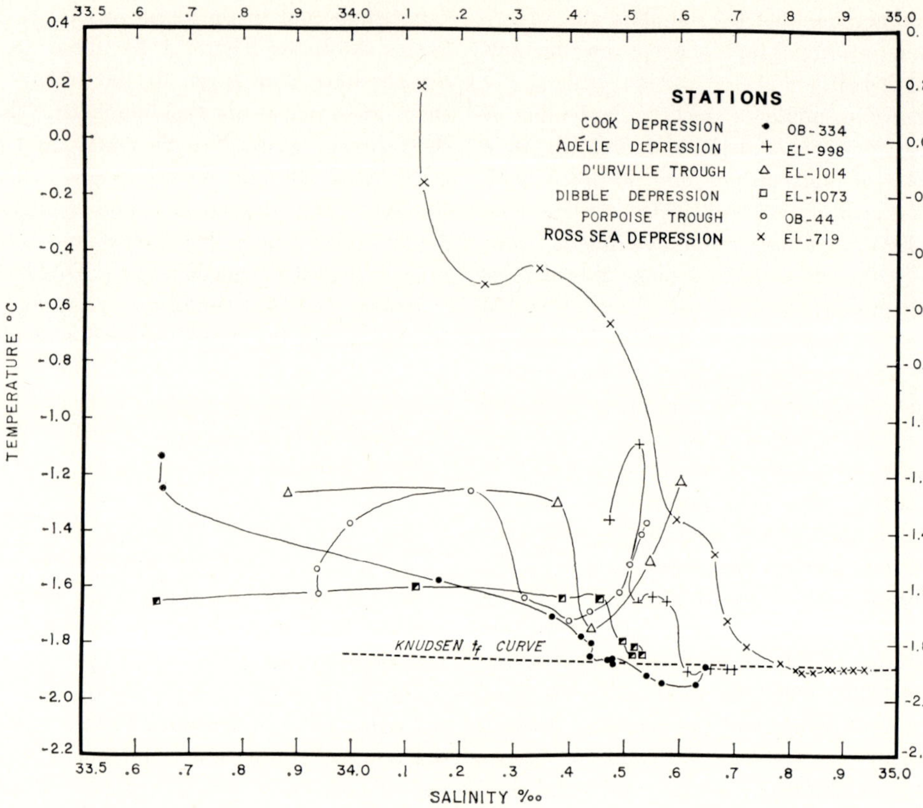

Fig. 2. Temperature–salinity relationship of water columns over the Adélie continental shelf and Ross Sea depression.

water reaching the bottom in the troughs and over the flat sections of the continental shelf is lost to the deep ocean or drains into the depressions. The summer bottom water within the troughs probably contains some modified circumpolar deep water.

The temperature–oxygen relationship of the continental shelf water columns is shown in Figure 4. The shelf water filling the depressions is high in oxygen and reaches 90% of saturation. The bottom oxygen of the water in the troughs is lower. The high oxygen of the trapped shelf water indicates that a significant portion of the shelf water has recently been in a near-surface position.

The variable salinity of dense shelf water filling the various depressions is puzzling, since one would expect each mass (unit) to be formed by a similar process, most probably sea ice formation. Perhaps geographical variations of the surface water salinity are responsible for this variation. At the time of *Eltanin* cruise 37, the surface salinity values varied from less than 33.600‰ (the result of local sea ice

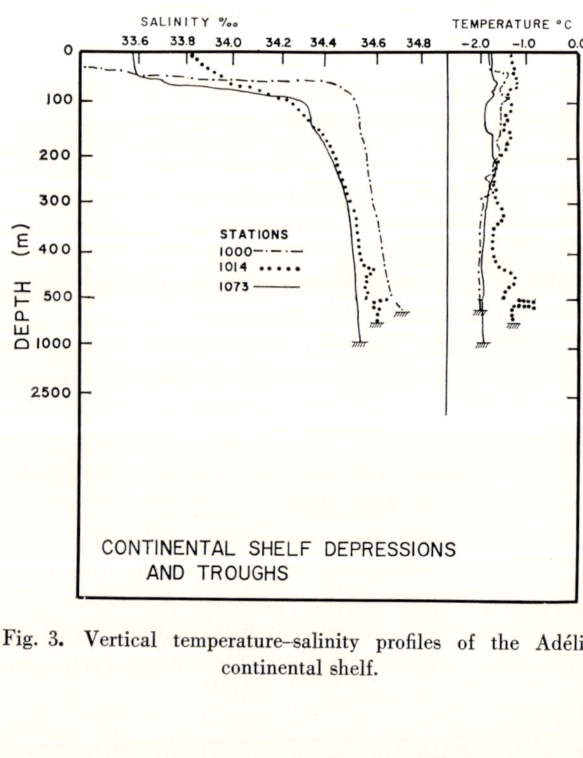

Fig. 3. Vertical temperature–salinity profiles of the Adélie continental shelf.

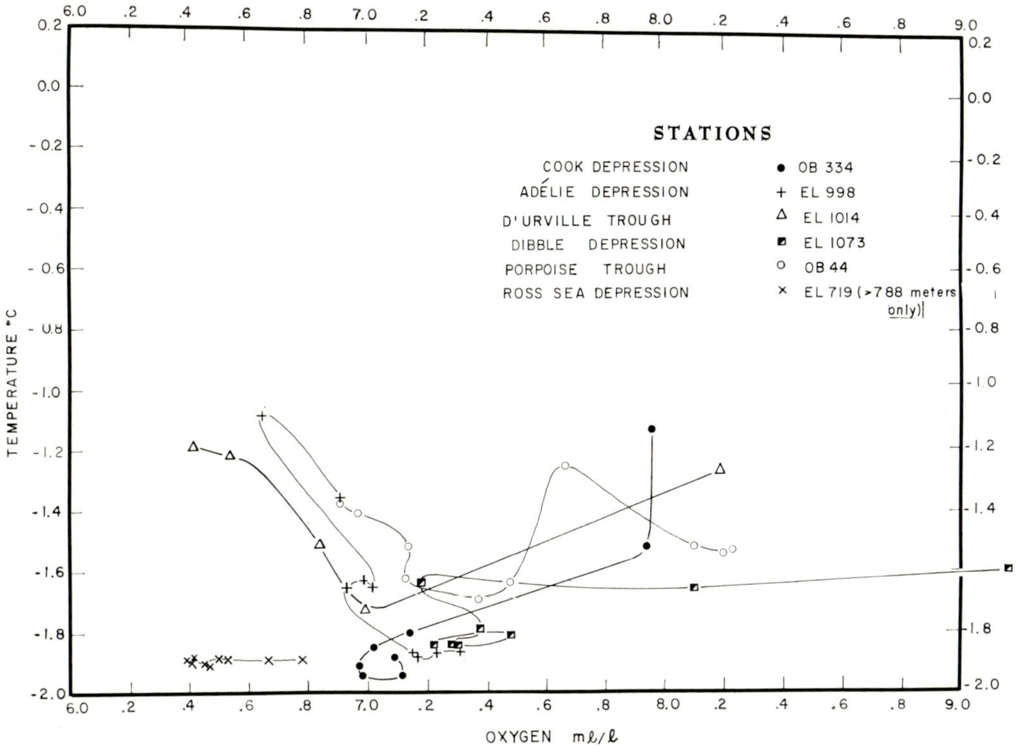

Fig. 4. Temperature–oxygen relationship of water columns over the Adélie continental shelf and Ross Sea depression.

melting) to a high of 34.475‰ over the 'Adélie depression.' The majority of values vary between 33.8 and 34.0‰. This range could result in variation to the shelf water salinity; however, the condition of the surface water at the beginning of the winter season (March–April) is not known and may be more uniform since ice melting would have ceased and local pockets of low salinity water would have mixed with the surrounding water.

A second cause for the salinity differences may be related to geographical variations in the amount of sea ice formation. Since approximately 3 grams of salt are released for each meter of sea ice formed, one would expect saltier water below areas with more extensive sea ice. The variation of the shelf water within depressions of such proximity as those along the Adélie Coast would not be expected at first. However, local meterologic effects may be of some importance.

A third possible cause of the variation in shelf water salinity is the depth of the shelf. We assume that as the brine leaving the immediate vicinity of the sea ice sinks this brine entrains the surrounding near-surface water that would most likely be near 34.0‰. The resulting mixture on reaching the shelf floor flows off the shelf and/or fills the depressions. The final salinity of the shelf water (composed of brine and entrained surface water) depends both on the salinity of the near-surface water and on the total distance of sinking, and this distance would be proportional to the amount of entrainment. The volume of shelf water depends on the quantity of ice formed and the ratio of the mixing of brine with surface water. The deeper the sill depth of the depression or the flat shelf floor, the more entrainment and hence the lower the salinity of the shelf water. In addition to the influence of the sill depth on the shelf water salinity is the degree of sloping of the isopycnals resulting from geostrophically balanced currents. The top of the Ross Sea shelf water (34.75‰ surface) slopes steeply up to the southwest corner of the Ross Sea to within 100 meters of the sea surface [*Jacobs et al.*, 1970]. This sloping effectively shortens the path of the sinking brine and reduces the amount of entrainment; the result is Ross Sea shelf water that is high in salinity.

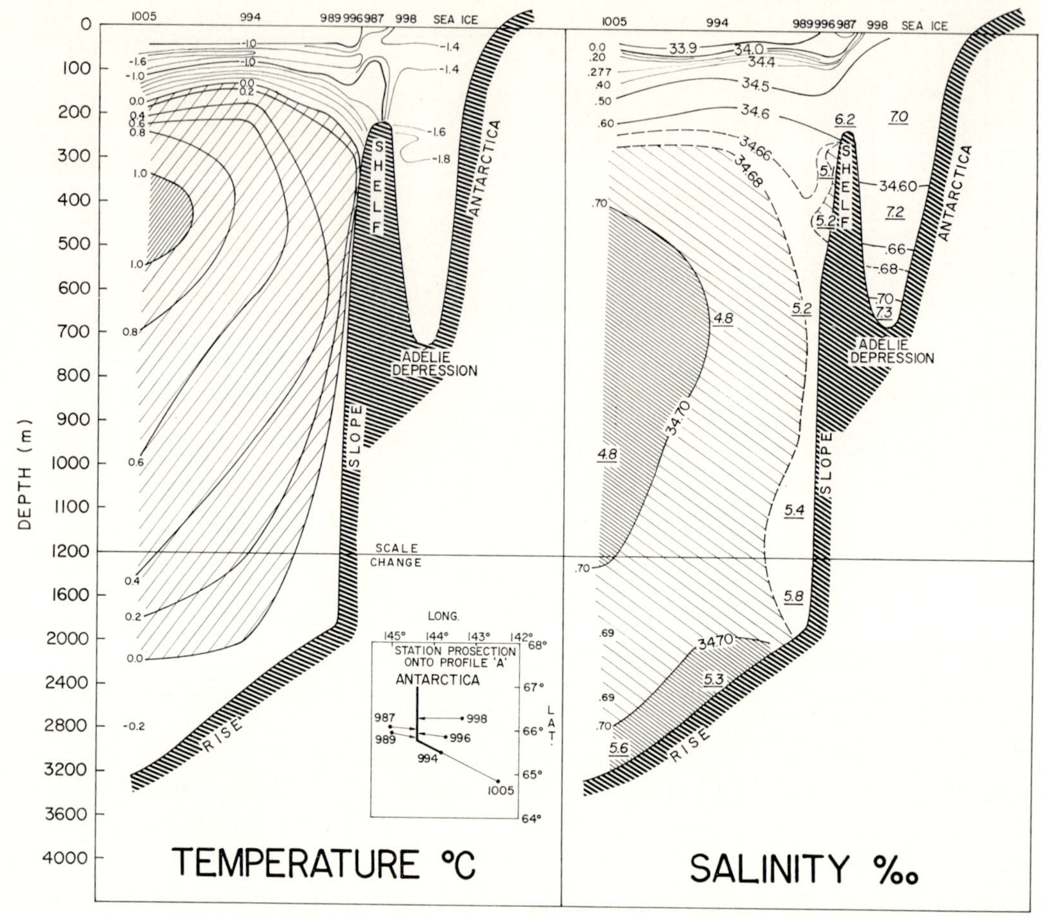

Fig. 5. Temperature and salinity section perpendicular to the continental shelf at 144°E. Underlined numbers in salinity section are oxygen values in milliliters per liter.

RELATION OF SHELF WATER TO DEEP WATER

The shelf water eventually escapes to the deep ocean. The rate of outflow may be directly proportional to the rate of production or perhaps a critical volume is attained and thus triggers a massive escape. Possibly, a relaxation of the westward flowing geostrophic coastal current would result in a decreased slope of the isopycnals and a subsequent release of shelf water to the deep ocean.

When the shelf water flows off the continental shelf, this water would sink because of its high density (σ_t = 27.96 in the 'Adélie depression'). It would first come in proximity with the upper circumpolar deep water found along the continental slope. This water is colder and slightly fresher than the main body of CDW found further out to sea and can be thought of as 'modified deep water' or 'slope water.' In this paper, the term slope water will be used. A general view of the water structure in the continental margin area may be gained from the north-south profiles of temperature and salinity near 144°E and 132°E; these sections are given in Figures 5 and 6, respectively. The circumpolar deep water is found near a depth of 500 meters north of the continental slope. The deep water is colder and fresher toward the south. In the 144°E section the deep water near the slope is 0.0–0.4°C and 34.68‰, whereas in the 132°E area the deep water is separated from the slope by a relatively thick, cold, low salinity zone. Typical continental slope water column characteristics are shown in Figure 7.

The *T/S* distribution of the slope water and upper and lower CDW, the *T/S* distribution of the trapped shelf water of the Adélie and Ross Sea depressions, and both varieties of antarctic bottom water are shown in Figure 8.

Note that mixtures of slope water and shelf water

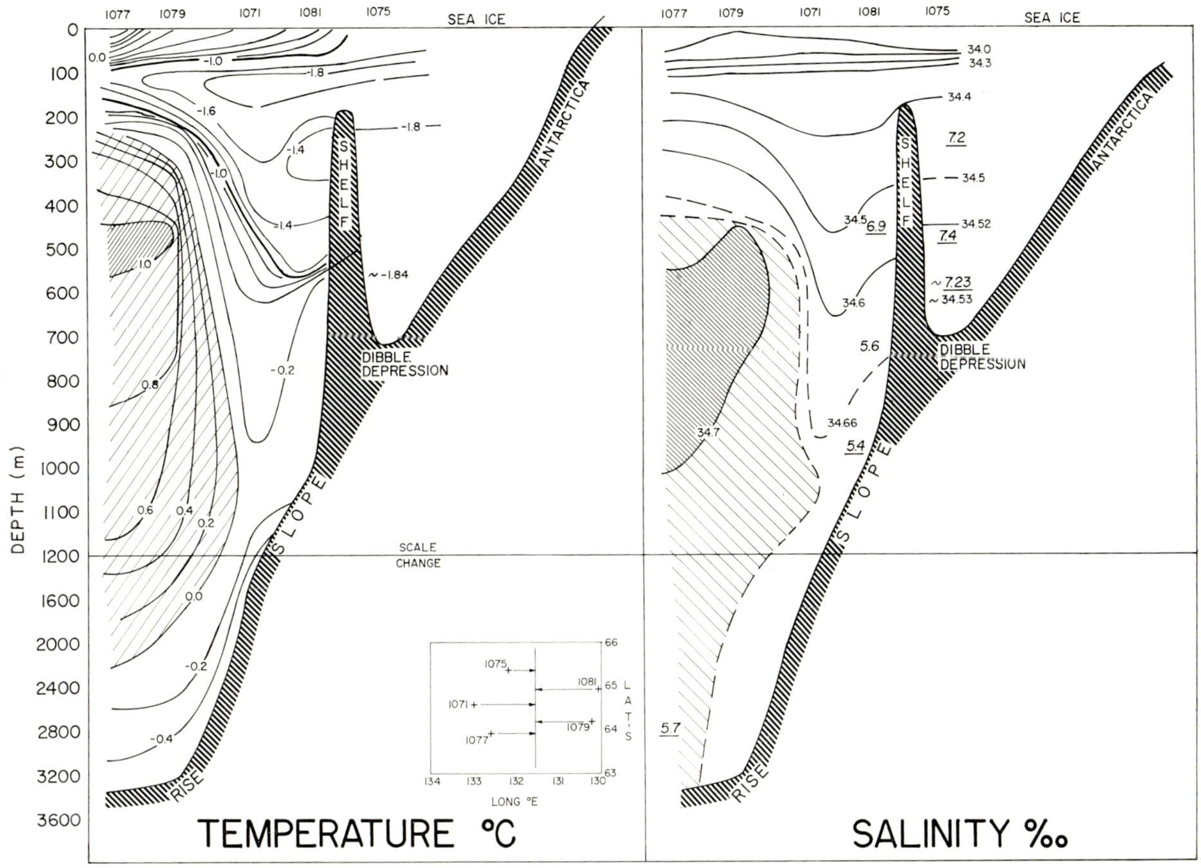

Fig. 6. Temperature and salinity section perpendicular to the continental shelf at 131°30′E. Underlined numbers in salinity section are oxygen values in milliliters per liter.

with salinity above 34.6‰ can produce a dense water capable of sinking to the sea floor with characteristics of the low salinity variety of antarctic bottom water as defined by *Mosby* [1934] and *Gordon* [1971*b*]. Note that Mosby used −0.4°C as the 'type' temperature; this temperature would reduce to −0.7°C potential temperature. The low salinity AABW variety falls in the area of approximate equal combinations of the shelf water within the 'Adélie depression' and slope water. The high salinity AABW variety falls in the region of mixtures of the Ross Sea shelf water and slope water.

Possibly some of this type of AABW was observed, as indicated by the lowest section of the *STD* traces of stations 985 and 1009. The *STD* traces for these stations are included in Figure 7. Both of these curves define thin, cold, relatively fresh bottom layers (indicated by the numbers 1 and 2 in Figure 7). In the case of station 985, the deepest *STD* value and water sample are about 60 meters (approximate since the pinger malfunctioned) from the sea floor. The near-bottom temperature, salinity, and oxygen are −1.08°C, 34.596‰, and 6.48 ml/l, respectively. The deepest water sample values at station 1009 are about 400 meters deeper than those of station 985 and about 33 meters from the sea floor (by pinger). The values are −0.62°C, 34.662‰, and 5.90 ml/l. A deeper *STD* measurement (approximately 30 meters from the bottom) gave values of −0.70°C and 34.650‰. The latter has a *T/S* point (shown as point 2 in Figure 8) that is very near the *T/S* point of the low salinity variety of antarctic bottom water. This measurement thus demonstrates that some AABW is produced from 'Adélie depression' shelf water and the slope water.

The colder, lower in salinity, and more highly oxygenated bottom water at station 985 may not yet have mixed with the deep water to a significant degree. This bottom water falls at point 1 on Figure 7. This bottom layer at station 985 appears to be derived

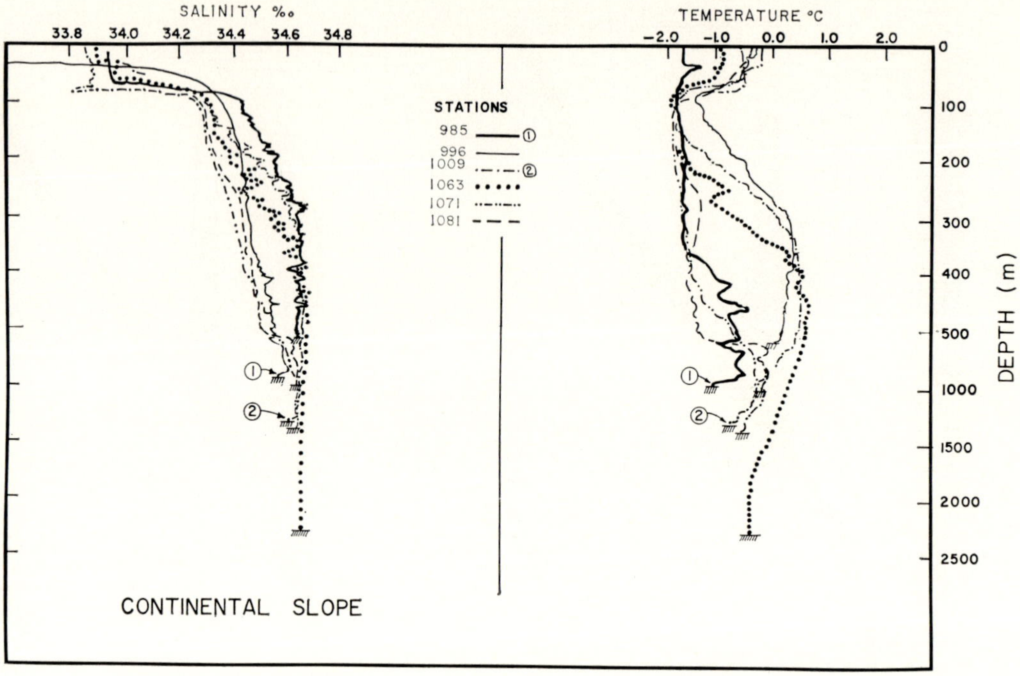

Fig. 7. Vertical temperature and salinity profiles over the Adélie continental slope.

from the low salinity shelf water of the 'Cook depression,' whereas at station 1009 the bottom layer is derived from the 'Adélie depression.' This situation is expected since the shelf water, on sinking down the slope, would turn westward under the influence of the Coriolis force.

The density of the bottom water found at station 985 is 27.85 σ_t, a value that is too low to allow sinking to the sea floor. The density of the bottom water at station 1009 is 27.88, a value that would permit sinking to the deep ocean floor. The low salinity bottom water has a density of 27.88.

The presence of the cold bottom layer at stations 985 and 1009 on the continental slope indicates the escape of some trapped shelf water during the summer months. Whereas the major overflow would be expected during periods of sea ice formation, we can reasonably expect that some summer overflow would result from internal waves or seiches within the trapped shelf water or perhaps from a relaxation of the geostrophically balanced east wind drift. The remarkable similarity of the AABW produced in the vicinity of the Adélie Coast to that produced in the southwestern Weddell Sea suggests an oceanographic process common to both areas. There are many logistic difficulties in working in the heavy ice cover of the Weddell Sea; therefore the vicinity of the Adélie Coast offers an attractive possible location for a practical study of the processes responsible for AABW because of the more manageable conditions of the Adélie Coast coupled with the proximity of the French antarctic station.

BOTTOM WATER OVER THE CONTINENTAL RISE

A cold, relatively salty bottom water derived from the Ross Sea [*Gordon*, 1971*b*] flows along the continental rise off the Adélie Coast. This water flows toward the west in a fairly narrow band and is recognized on the north-south profile of Figure 5 by a bottom salinity maximum below 2000 meters.

The isohaline and bottom oxygen values along the continental rise are shown in Figure 9. Salinity decreases along the path of spreading as a consequence of mixing with the surrounding fresher waters. Accompanying this freshening of the water is an increase in bottom oxygen concentration. Figure 10 shows the relationship of bottom salinity to oxygen on the continental rise. Thus the increased oxygen can be explained by an intermixing with the AABW sinking from the Adélie Coast. Mixing with the deep and bottom waters north of the continental rise would decrease salinity as observed, but would also decrease oxygen since the oxygen concentration decreases to

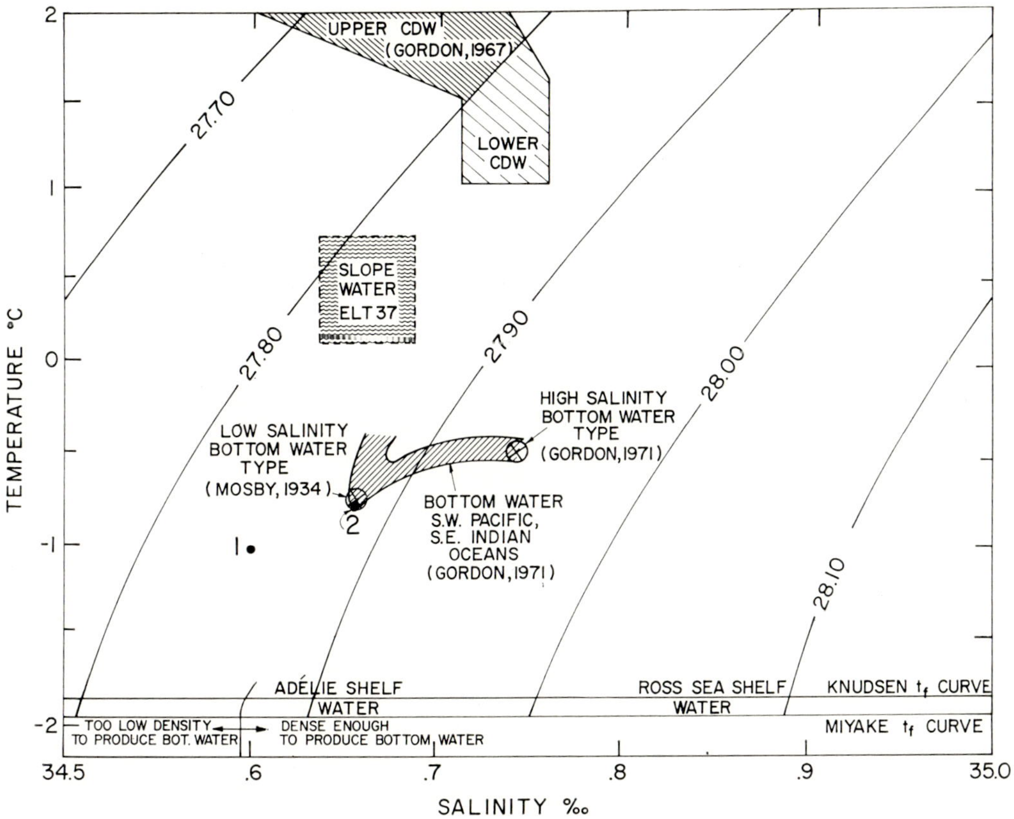

Fig. 8. Schematic potential temperature–salinity diagram showing core layer values of the antarctic water masses.

the north [*Gordon,* 1971*b*]. The bottom salinity–oxygen point at station 1009 (Figure 10) falls in line with the bottom salinity–oxygen points on the rise. Probably the locally produced AABW from the Adélie Coast flows down the continental slope and is then integrated into the westward flowing saltier bottom water derived from the Ross Sea.

The oceanographic data referred to in this paper were obtained during *Eltanin* cruise 37. Recently, however, during *Eltanin* cruise 50 (November–December 1971), many of the cruise 37 hydrographic stations on the continental rise were reoccupied. The low salinity variety AABW, believed to be derived from the Adélie Coast as discussed in this paper, was observed to be much more widespread during this latest cruise. This low salinity variety AABW formed a layer approximately 100 meters thick that separated the high salinity AABW from the sea floor. The most concentrated form of low salinity AABW on the continental rise was observed at 65°S, 143°30′E (potential temperature, −0.73°C; salinity, 34.676‰; oxygen, 6.06 ml/l). The water about 100 meters above this low salinity AABW was very similar to that shown in Figure 9 of this paper (potential temperature, 0.41°C; salinity, 34.720‰; oxygen, 5.52 ml/l). Possibly, the variation is seasonal, since the cruise 50 data represent late spring conditions and the cruise 37 data represent middle to late summer conditions.

SHELF WATER AND AABW PRODUCTION

The formation of the dense shelf water is probably related to sea ice formation with perhaps some influence from the high evaporation and cooling resulting from katabatic winds that are intense in the Adélie area [*Ball,* 1957; *Mather and Miller,* 1967]. When a sea ice salinity of 5‰ and a surface salinity of 34.0‰ at the end of summer are assumed, 1 meter of ice would produce by brine ejection and by mixing with surface water a layer of water approximately 41 meters thick that has a temperature of 1.9°C and a salinity of 34.7‰ on the shelf floor. Such water

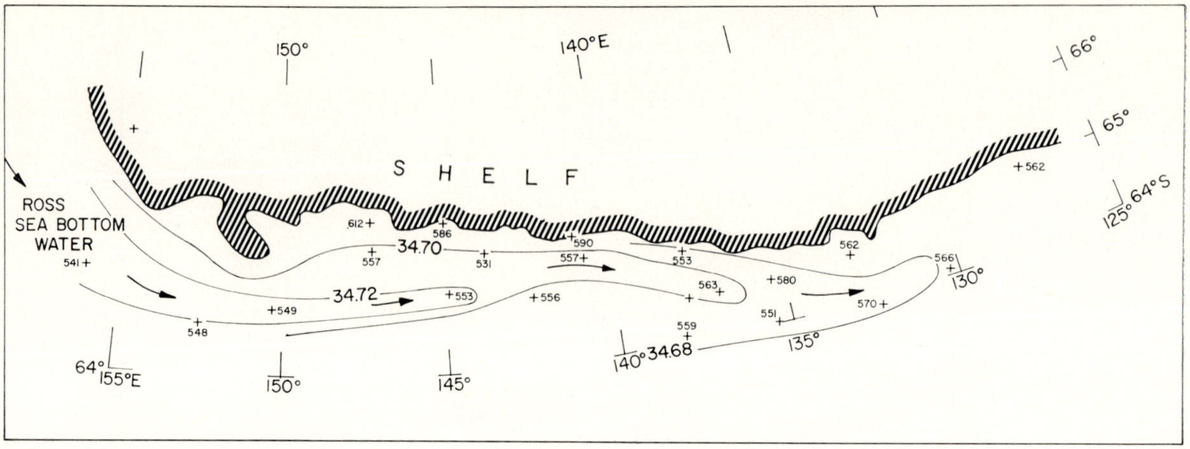

Fig. 9. Bottom salinity of the continental rise off the Adélie Coast. Numbers next to the crosses are bottom O_2 values in ml/l $\times$ 10^2.

would either flow off the shelf immediately to mix with the slope water to form AABW, or would drain into a depression and cause the less dense water already in the depression to overflow.

If the entire continental shelf of Antarctica (area 2.43×10^{12} m^2) were covered with 41 meters of water with a temperature of 1.9°C and a salinity of 34.7‰, the antarctic shelf, on mixing with equal amounts of deep water near the upper continental slope (0.5°C, 34.68‰), would produce 2.0×10^{14} m^3 of bottom water each year (an average value of 6.4×10^6 m^3/sec); this bottom water would be characterized by a temperature of 0.7°C and a salinity of 34.69‰. This water is slightly saltier and colder than the variety of the AABW described by *Mosby* [1934] but only slightly saltier than the type described by *Gordon* [1971*b*]. Naturally if a value of 34.65‰, which is closer to the average value of salinity for waters filling the depression of the Adélie Coast, were used for the shelf water, a thicker shelf water layer would result for each meter of sea ice (45 meters) and hence more AABW (7×10^6 m^3/sec) of −0.7°C and 34.665‰ would form. If 3 meters of sea ice are produced each winter, the value of AABW formed in this manner is approximately 21×10^6 m^3/sec.

In all likelihood, not only the sea ice of the continental shelf but also the brine released by the open ocean sea ice contributes to AABW production. The open ocean brine would most likely mix with surface water and upper CDW and would flow southward toward the continental margin where this open ocean brine may contribute to the formation of slope water. The dipping of the isohalines over the continental slope shown on Figure 6 may be related to slope water formation.

Other mechanisms for bottom water production related to the glacial ice shelf rather than to sea ice (called AABW 2 by *Gordon* [1971*c*]) may occur simultaneously and may increase the total AABW production. Possibly more than 3 meters of sea ice would form each year on the continental shelf.

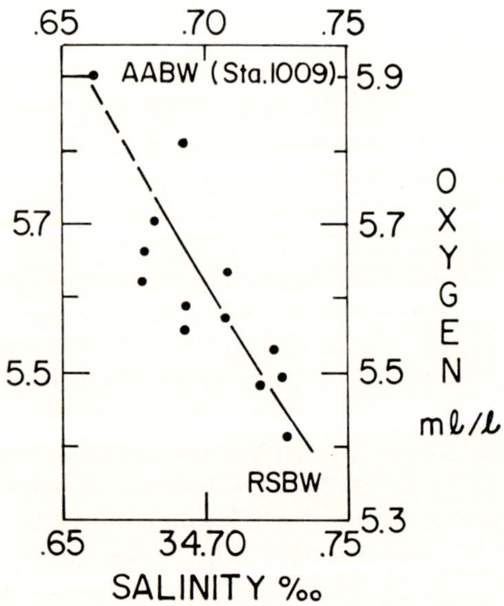

Fig. 10. Salinity-oxygen relationship of the bottom water on the continental rise off the Adélie Coast.

Antarctic bottom water is not produced uniformly about Antarctica. The Weddell Sea appears to be the major source region for AABW, and secondary sources appear to be the Ross Sea and the Adélie shelf. However, the westward flow of AABW near the continental margins may force the accumulation of the bottom water in the western parts of large embayments; this accumulation would give the appearance of a more localized source area than actually exists. Probably, however, most of the AABW is derived from a few major parts of the shelf, but small amounts may be produced in many areas.

Another factor in AABW production may be related to icebergs. During the *Eltanin* cruise 37, numerous icebergs were observed on the continental shelf, mostly to the east of each depression and trough. They were frequently grounded in shallow areas. Since the icebergs are approximately 300 meters thick, their lower portions contact the saltier, deeper waters observed in the troughs. This water may be cooled by contact (sensible heat exchange) with the glacial ice; if substantial bottom melting does not occur as may be the case during the winter, this resultant water may contribute to the dense shelf water observed in the depressions.

Acknowledgments. Grants from the Office of Polar Programs of the National Science Foundation supported most of the shipboard and laboratory work. The Atomic Energy Commission contract AEC-2663 contributed support for additional shipboard personnel.

Contribution 1773, Lamont-Doherty Geological Observatory.

REFERENCES

American Geographical Society, Antarctica, scale 1:5,000,000, New York, 1970.

Ball, E. K., The Katabatic winds of Adelie Land and King George V Land, *Tellus, 9*(2), 201–208, 1957.

Belliard, M., A bord du navire oceanographique USNS *Eltanin,* Supplement au numero 47 des TAAF, pp. 3–34, La Documentation Française, Paris, 1969.

Brennecke, W., Die ozeanographischen Arbeiten der Deutschen Antarktischen Expedition, 1911–1912, *Arch. Deut. Siewarte, 39*(1), 1–216, 1921.

Capurro, L., Hydrological observations in the southern oceans, *IGY Oceanogr. Rep. 2,* 386 pp., World Data Center, Washington, D. C., 1964.

Countryman, K. A., An explanation of supercooled waters in the Ross Sea, *Deep Sea Res., 17*(1), 85–90, 1970.

Countryman, K. A., and W. L. Gsell, Operation Deep Freeze '63 and '64 summer oceanographic features of the Ross Sea, *Tech. Rep. TR-190,* 193 pp., U. S. Navy Oceanogr. Office, Washington, D. C., 1966.

Eittreim, S., P. M. Bruchhausen, and M. Ewing, Vertical distribution of turbidity in the South Indian and South Australian basins, *Antarctic Oceanology II: The Australian–New Zealand Sector, Antarctic Res. Ser.,* vol. 19, edited by D. E. Hayes, pp. 51–58, AGU, Washington, D. C., 1972.

Gerard, R. D., and A. F. Amos, A surface-activated multiple sampler, *Mar. Sci. Instrum., 4,* 682–686, 1968.

Gill, A., and J. Turner, Some new ideas about the formation of antarctic bottom waters, *Nature, 224*(5226), 1287–1288, 1969.

Gordon, A. L., Oceanography of antarctic waters, *Antarctic Oceanology I, Antarctic Res. Ser.,* vol. 15, edited by J. L. Reid, pp. 169–203, AGU, Washington, D. C., 1971*a*.

Gordon, A. L., Spreading of antarctic bottom waters, 2, *Studies in Physical Oceanography—A Tribute to Georg Wüst on His 80th Birthday,* edited by A. L. Gordon, Gordon and Breach, New York, in press, 1971*b*.

Gordon, A. L., Recent physical oceanographic studies of antarctic waters, *Antarctic Research,* edited by L. Quam, American Association for the Advancement of Science, Washington, D. C., 1971*c*.

Hayes, D. E., and J. R. Conolly, Morphology of the southeast Indian Ocean, *Antarctic Oceanology II: The Australian–New Zealand Sector, Antarctic Res. Ser.,* vol. 19, edited by D. E. Hayes, pp. 125–145, AGU, Washington, D. C., 1972.

Jacobs, S. S., A. F. Amos, and P. M. Bruchhausen, Ross Sea oceanography and antarctic bottom water formation, *Deep Sea Res., 17*(5), 935–962, 1970.

Knudsen, M., Gejrierpunkttabellen fur Meerwasswe, *Publ. Circ. 5,* Copenhagen, 1903.

Mather, K. B., and G. S. Miller, The problem of the Katabatic winds on the coast of Terre Adelie, *Polar Rec., 13* (85), 425–432, 1967.

Mosby, H., The waters of the Atlantic antarctic ocean, *Sci. Res. Norwegian Antarctic Exped. 1927–1928 Rep. 11,* pp. 1–131, Norske Videnskaps-Akad, Oslo, 1934.

Mosby, H., South Atlantic bottom water, paper presented at the IAPSO/IABO/CMG/SCOR meetings, Tokyo, September 1970.

Seabrooke, J. M., and G. L. Hufford, Physical and chemical investigations of the Weddell Sea coastal current, *Antarctic J., 5*(4), 92–93, 1970.

Seabrooke, J., G. L. Hufford, and R. B. Elder, Formation of antarctic bottom water in the Weddell Sea, *J. Geophys. Res., 76*(9), 2164–2178, 1971.

Tchernia, P., Compte-rendu preliminaire des observations oceanographiques faites par le batiment polaire 'Commandant Charcot' pendant la campagne 1949–1950, *Bull. Inform. COEC, 3*(1), 13–22, 1951*a*.

Tchernia, P., Liste des stations hydrologiques exécutées par le batiment polaire 'Commandant Charcot' de la Marine Nationale dans l'Océan Indien et l'Océan Antarctique (campagnes 1948–1949 et 1949–1950), *Bull. Inform. COEC, 3* (10), 59–65, 1951*b*.

ON THE INTERACTION OF THE ANTARCTIC CIRCUMPOLAR CURRENT AND THE MACQUARIE RIDGE

ARNOLD L. GORDON

Lamont-Doherty Geological Observatory of Columbia University, Palisades, New York 10964

Abstract. South of Australia, the axis of the antarctic circumpolar current is near 50°S and is positioned near the mid-ocean ridge. On approaching the north-south trending Macquarie ridge, the current turns toward the south and returns abruptly to the north into the Emerald basin after passing east of the Macquarie ridge near 58°–60°S. The total southward deflection is near 1100 km. The flow around the southern segment of the ridge, which takes the form of a large permanent loop, recombines with the transport through two deep passages in the ridge (at 53.5°S and 56°S) and continues eastward as an intense flow along the southern slope of the Campbell plateau. The southward deflection of the streamline allows relatively warm water to move far south along the crest of the Macquarie ridge. This water, exposed to the cold antarctic atmosphere, releases heat at an accelerated rate. The temperature field on a north-south plane across the warm water feature exhibits a structure similar to the double polar front found east of the Macquarie ridge. In the Macquarie area, this structure is a permanent feature, but it is transient farther to the east. Speculations have been made that warm water eddies are shed from this feature, migrate far to the east, and produce the double polar front structure.

The antarctic circumpolar current (ACC) on its eastward course undergoes north-south shifts related to bottom topographic features. Mathematical considerations of such interactions have been summarized in *Neumann* [1960]. Though the various treatments differ in the details, they are essentially in agreement with most of the observed deflections of the current, i.e., northward migration over a shallowing sea floor and southward migration over a deepening bottom. Figure 1 shows the approximate axis of flow superimposed on the bottom relief. The apparent biaxial structure in the southwest portion of the south Indian Ocean sector is puzzling. Perhaps the structure is an artifact resulting from too few data in the region.

The major discrepancy of the path with topographic influences occurs south of New Zealand where the ACC deflects southward, although the topography is shallow. Qualitatively, such a deflection can be understood since a northward deflection would send the great volume of the ACC into the Tasman Sea where the shallow northern boundaries would prohibit easy exit. A generalized bathymetry is presented in Figure 2.

The mathematical consideration as well as the physical modeling of the Macquarie ridge problem is presented in the paper by *Boyer and Guala* [this volume]. The objective in this study is to better delineate the path of the ACC in the Macquarie ridge region by using the numerous hydrographic data gathered recently by the USNS *Eltanin*.

DATA

Most of the *Eltanin* data in the Macquarie ridge region are derived from cruises 36 and 44. The prime objective of the first half of cruise 44 (June–August 1970) was to determine the water characteristics on both sides of the Macquarie ridge and to clarify some problem areas in relation to bottom topography and geophysics [*Hayes and Talwani*, this volume]. A general account of the cruise is presented by *Gordon* [1971*a*]. In addition to the above mentioned cruises, hydrographic data in the Macquarie ridge area were also obtained on *Eltanin* cruises 16, 27, 35, and 37. Figure 3 shows the *Eltanin* data distribution. As shown by *Gordon and Goldberg* [1970, Plate 1], there are few data on the region except those collected from the *Eltanin*. The *Eltanin* data are available in the various data reports by *Jacobs* [1966], *Jacobs and Amos* [1967], and *Jacobs et al.* [1970].

The data consist of serial hydrographic stations and *STD* records calibrated against reversing thermometers and water sampling methods. In addition to temperature, salinity, oxygen, and nutrient data, nephelometer information is gathered as well as compass-

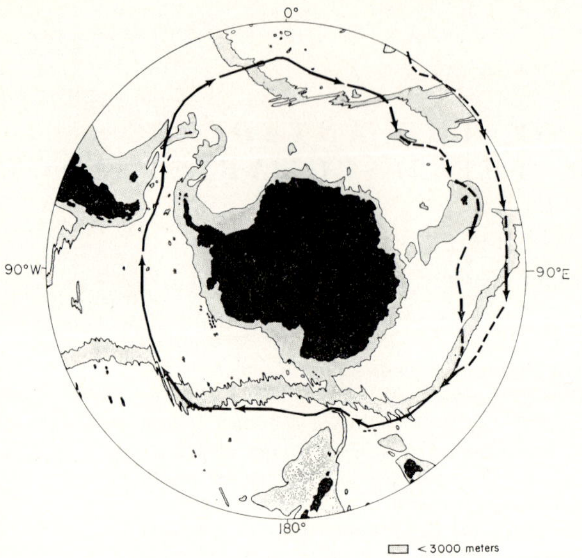

Fig. 1. Approximate axis of antarctic circumpolar current determined from sea surface dynamic topography relative to 3000-db level (from *Tolstikov* [1966, Plate 1031]) superimposed on bottom topography (from *Heezen et al.* [1972]).

oriented bottom photographs. The nephelometer information is included in a report by *Eittreim et al.* [1971].

TOPOGRAPHY

The generalized bathymetry shown in Figure 2 is presented in much more detail by *Hayes et al.* [1972] and *Hayes and Talwani* [this volume]. The Macquarie ridge extends to approximately 58°S. South of 56°S is the deep Hjort trench to the immediate west of the ridge. The ridge is cut by deep passes at 51°S, 53.5°S, and 56°S. South of the Macquarie ridge is a relatively deep ocean region and further south is the mid-ocean ridge and Balleny plateau (see *Hayes and Conolly* [this volume]). The system of ridges between New Zealand and Antarctica forms a relatively shallow barrier to the ACC.

The path of shallowest bottom topography from New Zealand to Antarctica is shown in Figure 4. From New Zealand to 55°S, the bottom averages nearly 1000 meters in depth (excluding the two deeper passes at 51°S and 53.5°S). South of 55°S, the shallowest path increases in depth and reaches approximately 3500 meters near 60°S with a deep passage (4500 meters) at 56°S. South of 60°S, the topography shallows to the Balleny Islands at 66°S. A 2350-meter passage exists between the Balleny Islands and Antarctica. This area usually has heavy sea ice conditions and has not been sampled by the *Eltanin*. During the International Geophysical Year, the USSR vessel *Ob* gathered data south of the Balleny Islands, but the salinity data of that cruise have numerous questionable values.

S_{max} CORE LAYER

The voluminous lower circumpolar deep water (CDW) is marked by a maximum in salinity [*Gordon*, 1967]. This S_{max} characteristic is derived from the influx of the relatively saline North Atlantic deep water. An interesting relation exists between the S_{max} core layer and the relative sea surface dynamic topography: the shallower the S_{max} core layer, the lower the sea surface dynamic topography [*Gordon*, 1967]. This relationship can be seen in Figure 5, which shows the relationship of the depth of the S_{max} core layer to the sea surface dynamic topography relative to the 4000-db (decibar) pressure surface. The relationship is linear when the S_{max} core layer is shallower than 2500 meters. When the core layer is deeper, the relation is not as clear, although a more gentle slope of the curve is suggested.

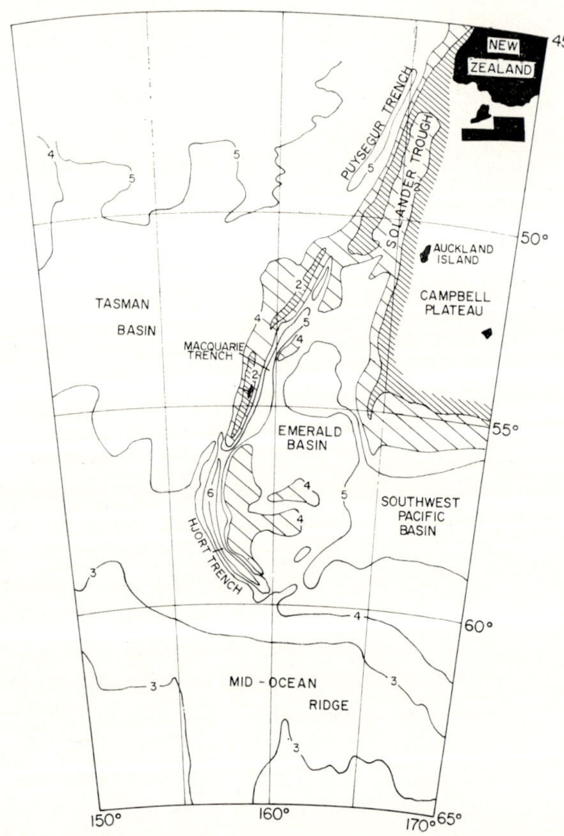

Fig. 2. Bathymetry of the region south of New Zealand (generalized from *Hayes and Talwani* [this volume]).

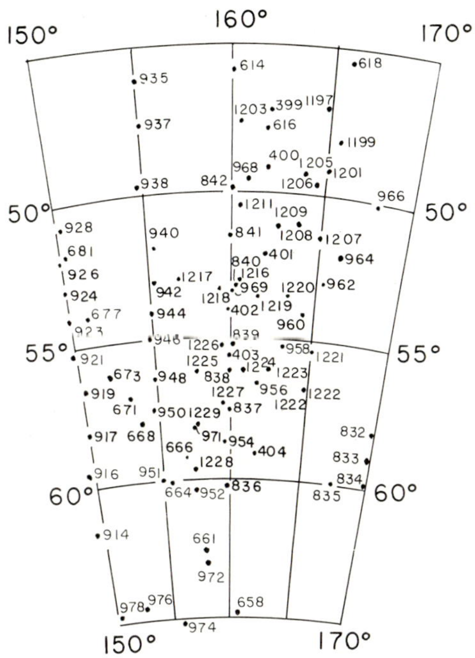

Fig. 3. *Eltanin* data distribution in the area shown in Figure 2.

The general north-south slope of the S_{max} core layer is such that the 2500-meter isobath of this layer occurs north of the mean position of the polar front zone. By assuming geostrophic balance within antarctic waters, the ACC occurs south of the 2500-meter isobath. Therefore the axis of the ACC is within the region of the linear relation of S_{max} core layer to dynamic topography of the sea surface.

This interesting relation allows determination of the ACC axis without the direct use of dynamic anomalies. The advantages of this method are: (1) the sea surface dynamic topography relative to 4000 db can be determined even in areas where the sea floor is less than 4000 meters or where hydrographic data do not extend to that depth, and (2) the sea surface dynamic topography can be determined from deep salinity data alone rather than by requiring the entire internal field of mass that would be subject to a greater number of possible errors.

Geostrophic velocities relative to a deep isobaric surface should not be considered to be approximately 'absolute.' This limitation results from the slow attenuation of the ACC with depth. For an area south of New Zealand, *Maksimov and Vorob'yev* [1962, Figure 2] show an attenuation between the sea surface to 3000 meters of only 40%. Therefore a 10-cm/sec current at the surface relative to 3000 meters would be 25 cm/sec in an absolute sense.

Another indicator of the axis of the ACC is the position of the axis of spreading of the CDW on the S_{max} core layer, identified as the maximum of salinity in this core layer. The coincidence of the axis of spreading and axis of flow is to be expected at positions far from the introduction of the saline water (which is essentially a tracer). The more rapid eastward advection in the ACC axis would transport more of the tracer and would so act as an axis of spreading.

AXIS OF THE ACC

Vertical profiles of the salinity are presented for 128°E, 140°E, 150°E, and for the profile immediately east of the shallowest path (Figure 4) in Figures 6–9, respectively. The axis of spreading of CDW is

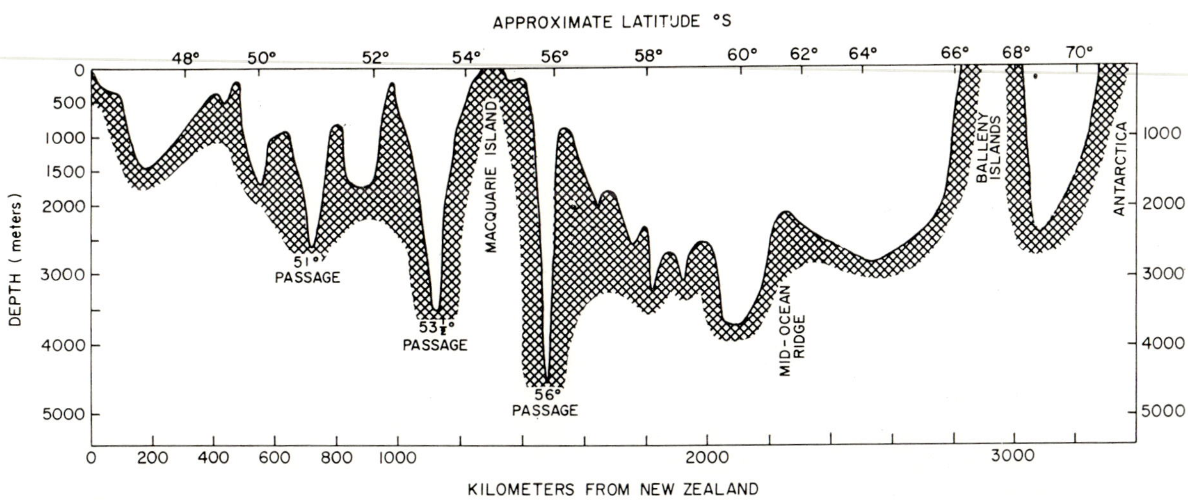

Fig. 4. Shallowest path from New Zealand to Antarctica along the Macquarie ridge and Balleny Island platform.

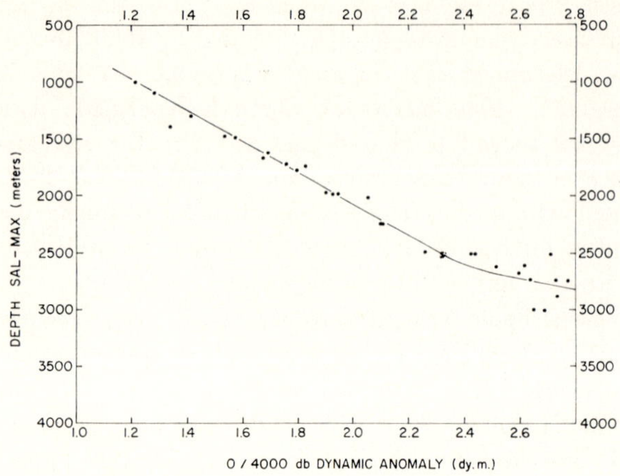

Fig. 5. Relation of the depth of the salinity maximum core layer to the sea surface dynamic topography relative to the 4000-db level in the area south of New Zealand and Australia.

marked, and this axis generally is also the axis of the ACC as determined by the relationship shown in Figure 5, i.e., the axis of surface geostrophic flow relative to the 4000-db level. In the first three sections, the axis of spreading is near or over the mid-ocean ridge. In the 140°E section, the lower CDW actually lies on the sea floor with a reversal in slope south of the mid-ocean ridge, the reversal indicating a westward counterflow.

The list below gives the latitude of the axis of spreading of CDW (axis of ACC) for the four sections.

49°S	128°E
52°S	140°E
55°S	150°E
59°S	Macquarie ridge

The total southward deflection from 128°E to a longitude east of the Macquarie ridge is approximately 1100 km, much of which (445 km) is accomplished to the east of 150°E.

THE ACC AT THE MACQUARIE RIDGE

The salinity structure immediately east of the Macquarie-Balleny ridge systems (Figure 9) shows that the highest salinity and steepest slope of the S_{max} core layer occur between 58° and 60°S. The slope of the S_{max} core layer indicates a counterflow (westward current) near 58°S and also exhibits an increased slope at the 53.5°S and 56°S passages. Therefore, whereas the main axis of spreading occurs south of the Macquarie ridge, the eastward flow appears to be intensified at the southern two passages through the ridge. The geostrophic flow relative to 4000 db at the 53.5°S and 56°S passages are 30 cm/sec and 40 cm/sec, respectively, as determined from the slope of the S_{max} core layer and the relationship shown in

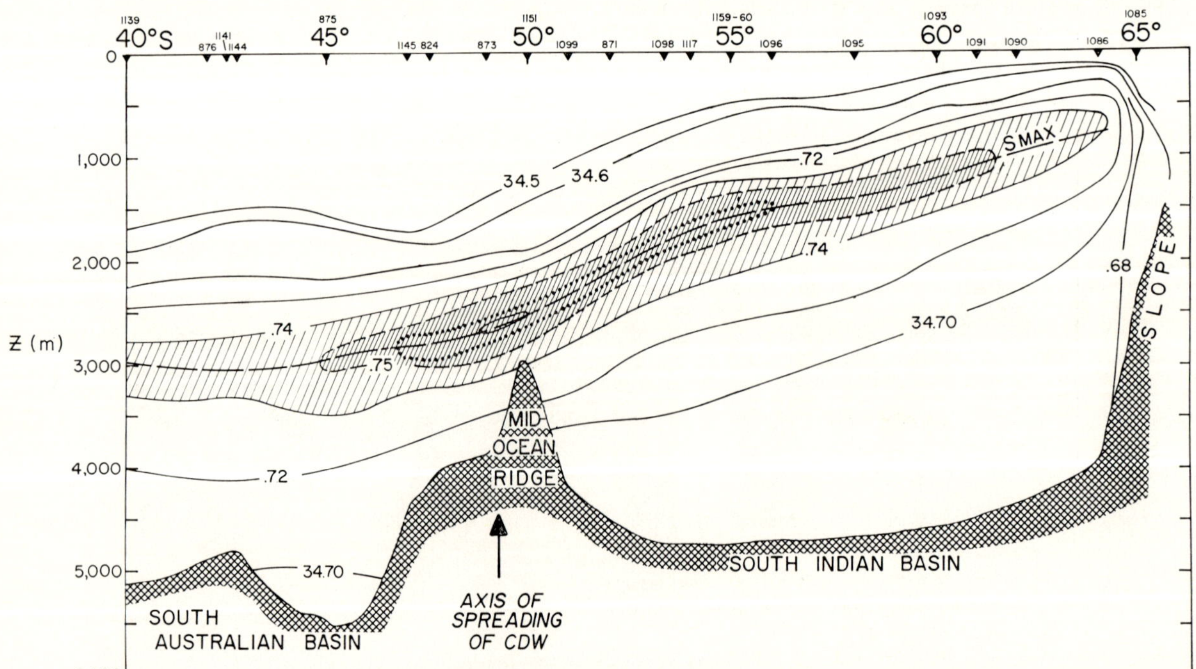

Fig. 6. Salinity structure along 128°E.

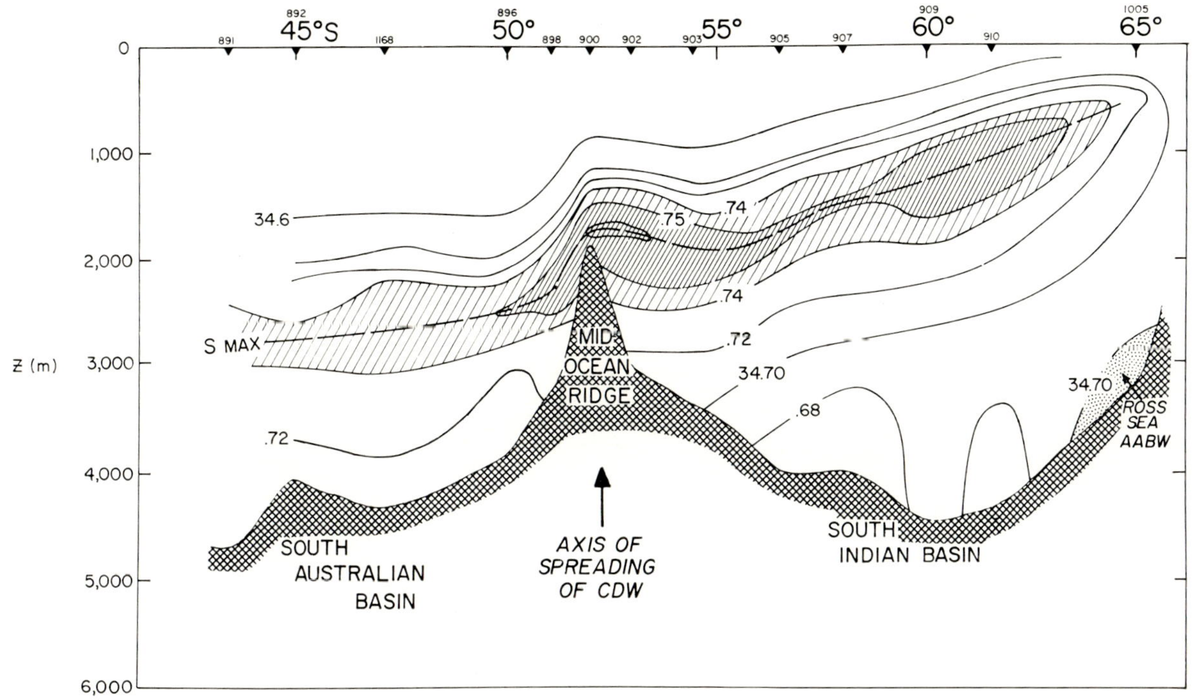

Fig. 7. Salinity structure along 140°E.

Figure 5. As noted earlier, the absolute flow may be substantially larger.

The areal distributions of the salinity and the depth of the S_{max} core layer are shown in Figures 10 and 11, respectively. The axis of spreading clearly outflanks the southern limbs of the Macquarie ridge. The 34.76‰ isohaline penetrates into the Emerald basin but does not continue east of the Campbell plateau. The salinity in the S_{max} core layer within the western boundary undercurrent [*Scripps Institute of Oceanography et al.*, 1969] is 34.739‰ and 34.73–34.74‰ for the current passing through the Drake passage [*Gordon*, 1967]. Thus only a decrease of 0.02–0.03‰ is observed in the transit of the ACC across the Pacific or in its offshoot in the western South Pacific.

The depth of the S_{max} core layer is proportional by the relationship shown in Figure 5 to the 0/4000-db dynamic topography; hence the sense of the slope

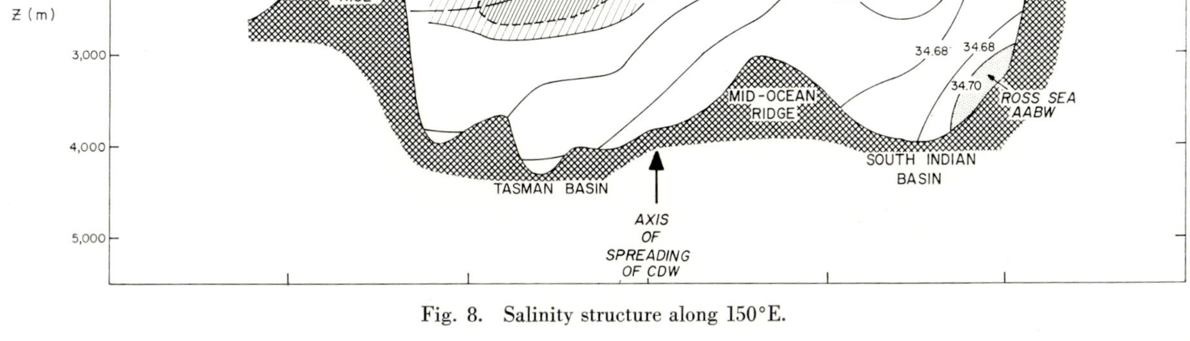

Fig. 8. Salinity structure along 150°E.

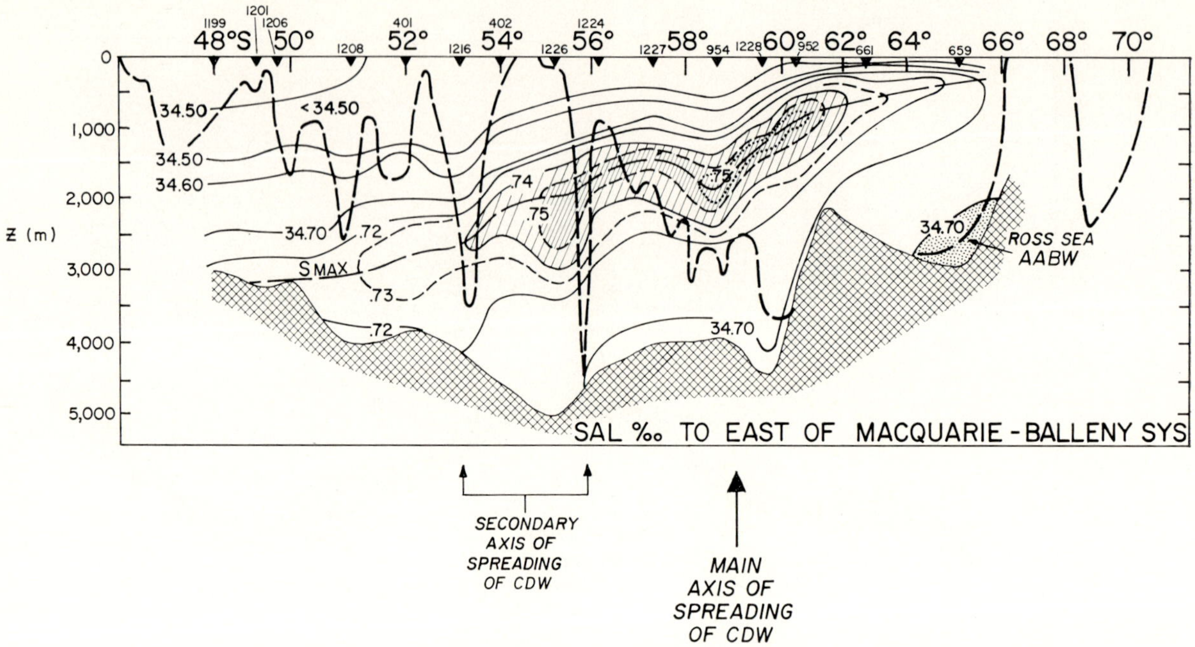

Fig. 9. Salinity structure immediately east of the shallowest path shown in Figure 4. The broken bathymetry trace is the shallowest path, whereas the solid trace is the actual bathymetry along the salinity section.

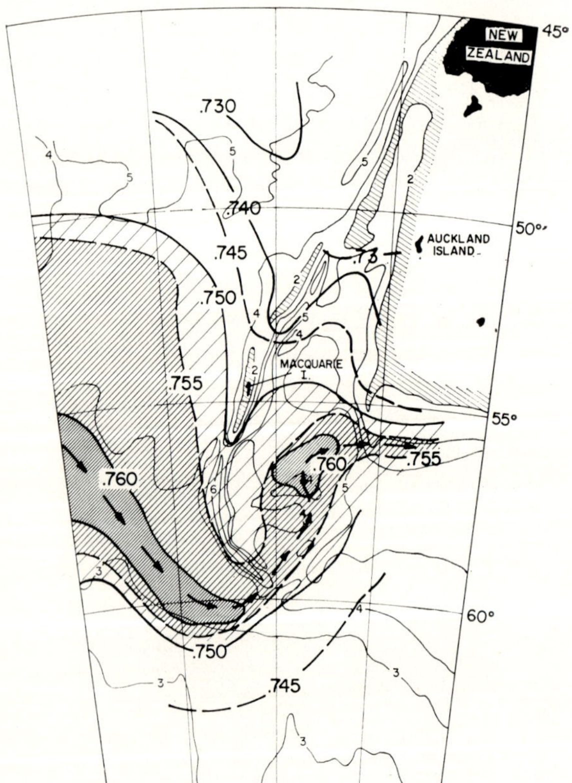

Fig. 10. Salinity on the salinity maximum core layer.

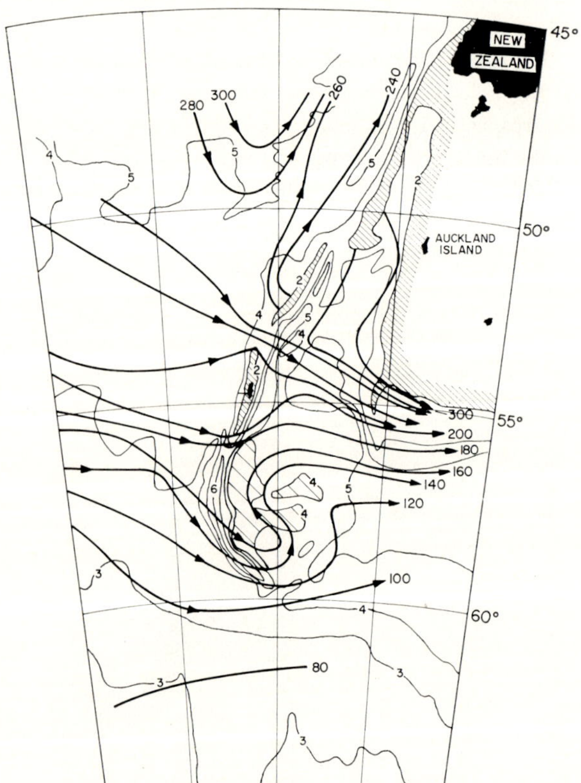

Fig. 11. Depth ×10 of the salinity maximum core layer.

gives the direction of flow, as shown by the arrows on the isobaths of Figure 11, and the spacing is related to the current magnitude; i.e., the more closely spaced the isobaths, the greater the flow.

THERMAL STRUCTURE OF UPPER KILOMETER OF THE OCEAN

The deflection of the axis of the ACC and the curling back of the current in the southern Emerald basin are also reflected in the temperature structure of the upper kilometer. Figure 12 shows the temperatures at 200 meters based on the hydrographic data shown in Figure 3. A line of relatively strong gradient (approximately 4°C/125 km) runs northwest-southeast from 52°S, 150°E to 59°S, 170°E. This line marks the northern boundary of the cold antarctic surface water called the antarctic convergence, or polar front zone. South of this feature is a warm water (greater than 3°C) intrusion over the crest of the southern section of the Macquarie ridge.

The warm water extends to the sea surface as shown in the bathythermograph section (Figure 13) along 159°30′E (cruise 44) where the sea surface temperatures reached 4°C at the end of July 1970, whereas the air temperature remained near 0°C. A heat loss by evaporation and sensible heat transfer in this situation (dew point near −1°C and wind speed near 40 knots) is 130 cal/cm^2/day. A value of approximately 70 cal/cm^2/day is expected for a wind of 20 knots, which is more the norm at those latitudes. The nonradiate heat loss with a more typical antarctic sea–air temperature difference of 1°C and a 20-knot wind would be about 17% lower than the value found for a temperature difference of 4°C with a 20-knot wind. The steady state nature of this feature (as indicated by its appearance on the hydrographic station construction of many cruises given in Figure 12) in light of the large heat flux to the atmosphere implies a constant renewal from the north.

The temperature structure for the upper few hundred meters along 159°30′E is strikingly similar to the temperature structure found in the double polar front zone [*Gordon*, 1971*b*]. However, in the case of the double polar front, the temperature structure seems to be a transient feature, and at the Macquarie ridge it appears to be more permanent. As suggested by Boyer and Guala, the double front is a permanent feature at the southern end of the Macquarie ridge where the flow follows the isobaths and a transient feature to the east associated with eddy shedding. Another interesting point is that the double polar

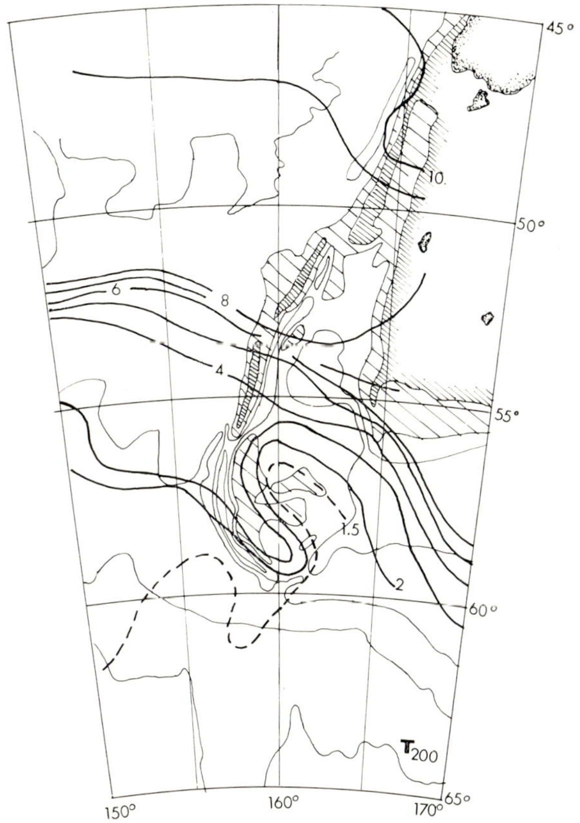

Fig. 13. Expendable bathythermograph profile along 159°30′E (cruise 44).

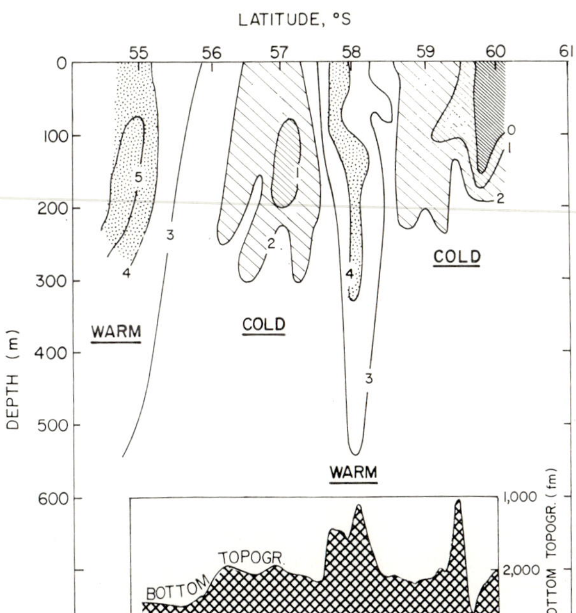

Fig. 12. Temperature at 200 meters.

front zone does not occur west of the Macquarie ridge [*Gordon,* 1971*b*]. Reasonable speculation is that the double polar front zone is created at the southern end of the Macquarie ridge by the shedding of warm water eddies. These eddies migrate across a major part of the Pacific Ocean (they are observed to 108°W [*Gordon,* 1971*b*]) before decaying by turbulence and sea–air exchange of heat. The problem of eddy shedding is discussed by *Boyer and Guala* [this volume].

CONCLUSIONS

The southward deflection of the ACC to pass south of the Macquarie ridge, rather than pass directly over it, has the effect of carrying water characteristics to latitudes nearly 10° farther south than they ordinarily are found. This effect results in accelerated sea to air transfer and pushes the limits of the pack ice to the south. Possibly, the deflection also acts as a drain on vorticity initially put into the water column by the wind. The wind transfers negative vorticity into the ocean south of the maximum westerlies. The counterclockwise flow around the southern Macquarie ridge would result in taking out negative vorticity by the frictional interaction with the ridge. Such extraction of negative vorticity would be most intense if the form of the current is jetlike. This form is suggested in the model studies of the accompanying paper by *Boyer and Guala* [this volume].

The shedding of eddies from the southern end of the Macquarie ridge presents the interesting possibility of the production of transient motion from interaction of mean flow and bottom topography. The speculation of the eddy shedding is supported by the model study, but for observational confirmation a multiship operation is needed, and/or (if the means of penetrating the cloud cover is developed that is capable of resolution to tens of kilometers) remote sensors from aircraft or satellite.

Acknowledgments. The data collection and processing were supported by National Science Foundation grant GV-26230 from the Office of Polar Programs. The analysis of the data was supported partially by the above grant and partially by Office of Naval Research contract TO-4.

Contribution 1774, Lamont-Doherty Geological Observatory.

REFERENCES

Boyer, D. L., and J. R. Guala, Model of the antarctic circumpolar current in the vicinity of the Macquarie ridge, in *Antarctic Oceanology II: The Australian–New Zealand Sector, Antarctic Res. Ser.,* vol. 19, edited by D. E. Hayes, pp. 79–93, AGU, Washington, D. C., 1972.

Eittreim, S., A. L. Gordon, M. Ewing, E. M. Thorndike, and P. Bruchhausen, The nepheloid layer and observed bottom currents in the Indian-Pacific antarctic sea, in *Studies in Physical Oceanography—A Tribute to George Wüst on His 80th Birthday,* edited by A. L. Gordon, Gordon and Beach, New York, in press, 1971.

Gordon, A. L., Structure of antarctic waters between 20°W and 170°W, *Antarctic Map Folio Ser.,* folio 6, edited by V. Bushnell, Amer. Geogr. Soc., New York, 1967.

Gordon, A. L., *Eltanin* cruise 44, *Antarctic J., 6*(1), 17–18, 1971*a*.

Gordon, A. L., Antarctic polar front zone, in *Antarctic Oceanology I, Antarctic Res. Ser.,* vol. 15, edited by J. L. Reid, pp. 205–221, AGU, Washington, D. C., 1971*b*.

Gordon, A. L., and R. D. Goldberg, Circumpolar characteristics of Antarctic waters, *Antarctic Map Folio Ser.,* folio 13, edited by V. Bushnell, Amer. Geogr. Soc., New York, 1970.

Hayes, D. E., and J. R. Conolly, Morphology of the southeast Indian Ocean, in *Antarctic Oceanology II: The Australian–New Zealand Sector, Antarctic Res. Ser.,* vol. 19, edited by D. E. Hayes, pp. 125–145, AGU, Washington, D. C., 1972.

Hayes, D. E., and M. Talwani, Geophysical investigation of the Macquarie ridge complex, in *Antarctic Oceanology II: The Australian–New Zealand Sector, Antarctic Res. Ser.,* vol. 19, edited by D. E. Hayes, pp. 211–234, AGU, Washington, D. C., 1972.

Hayes, D. E., M. Talwani, and D. A. Christoffel, Macquarie ridge complex, in *Antarctic Geology and Geophysics,* edited by R. J. Adie, Universitets forlaget, Oslo, in press, 1972.

Heezen, B. C., M. Tharp, and C. Bentley, *Antarctic Map Folio Ser.,* folio 16, Amer. Geogr. Soc., New York, in press, 1972.

Jacobs, S. S., Physical and chemical oceanographic observations in the southern oceans, USNS *Eltanin* cruises 16–21, 1965, *Tech. Rep. 1-CU-1-66,* 128 pp., Lamont-Doherty Geol. Observ. of Columbia Univ., Palisades, N. Y., 1966.

Jacobs, S. S., and A. F. Amos, Physical and chemical oceanographic observations in the southern oceans, USNS *Eltanin* cruises 22–27, 1966–1967, *Tech. Rep. 1-CU-1-67,* 287 pp., Lamont-Doherty Geol. Observ. of Columbia Univ., Palisades, N. Y., 1967.

Jacobs, S. S., P. M. Bruchhausen, and E. B. Bauer, Hydrographic stations, bottom photographs, and current measurements, USNS *Eltanin* cruises 32–36, 460 pp., Lamont-Doherty Geol. Observ. of Columbia Univ., Palisades, N. Y., 1970.

Maksimov, I. V., and V. N. Vorob'yev, Contribution to the study of deep currents in the antarctic ocean, *Sov. Antarctic Exped. Inform. Bull., 1*(4), 17–19, 1962.

Neumann, G., On the effect of bottom topography on ocean currents, *Deut. Hydrogr. Z., 13*(3), 132–141, 1960.

Scripps Institute of Oceanography, Woods Hole Oceanographic Institution, and Massachusetts Institute of Technology, Physical and chemical data from the Scorpio Expedition in the South Pacific Ocean, USNS *Eltanin* cruises 28 and 29, March 12–July 31, 1967, *SIO Ref. 69–15, WHOI Ref. 69–56,* 89 pp., La Jolla, Calif., 1969.

Tolstikov, E. I. (Ed.), *Atlas Antarktiki,* vol. 1, G.U.C.K., Moscow, 1966. (English translation in *Soviet Geography: Reviews and Translations, 8*(5-6), 225 pl., American Geographical Society, New York, 1967.)

MODEL OF THE ANTARCTIC CIRCUMPOLAR CURRENT IN THE VICINITY OF THE MACQUARIE RIDGE

DON L. BOYER

Department of Civil Engineering, University of Delaware, Newark, Delaware 19711

JOHN R. GUALA

Department of Mechanical and Aerospace Engineering, University of Delaware, Newark, Delaware 19711

Abstract. Both theoretical and laboratory models of the antarctic circumpolar current in the vicinity of the Macquarie ridge are presented. The models consider a homogeneous incompressible flow but neglect wind shear stress and β effects. The major phenomenon addressed is that of the deflection of the current due to the effects of bottom topography. The large-scale flow features indicated by the oceanographic data in this region are generally reproduced by the models. For example, for certain model parameters, large eddies are shed from the southern tip of the Macquarie ridge. The characteristics of this shedding lend support to A. L. Gordon's speculation that the double polar front zone found east of the Macquarie ridge system is formed by the shedding of eddies from the tip of the ridge.

In this paper we consider, both theoretically and by laboratory experiment, a model of the antarctic circumpolar current (ACC) in the vicinity of the Macquarie ridge. In particular, we investigate some of the effects of the ocean floor topography on the path of the current. The area to be considered is outlined by the broken line in Figure 1 [*Gordon*, this volume].

The present study is motivated by the results of some recent work on bottom topography effects on rotating flows done by *Boyer* [1971] and *Vaziri and Boyer* [1971]. These studies are theoretical as well as experimental. The experiments use a rotating water tunnel apparatus that is discussed in some detail in *Boyer* [1971]. The tunnel is a rotating channel of rectangular cross section through which water is recirculated. The basic capability of the tunnel is that it can provide a uniform rectilinear flow relative to a rotating observer. Topographies can then be placed in the flow, and the resulting motions can be observed by using dye tracer techniques.

In attempting to apply the theoretical and laboratory studies of *Boyer* [1971] and *Vaziri and Boyer* [1971] to real ocean current systems, one encounters numerous difficulties. In the first place, the fluid is assumed to be homogeneous and incompressible. This assumption is obviously not the case in the real ocean, but, relatively speaking, this approximation may be a reasonable one for the antarctic waters. Nonetheless, some stratification is present in the region being considered, and thus we can expect that the bottom topography effects will be exaggerated in the model.

The flows considered in *Boyer* [1971] and *Vaziri and Boyer* [1971] are laminar, whereas ocean current systems are clearly turbulent. In addition, wind shear and β effects are not considered in *Boyer* [1971] and *Vaziri and Boyer* [1971]. The ratio of the characteristic vertical to horizontal length scales in those studies is of order unity. For scales of motion in the real ocean for which the rotation of the earth is an important consideration, this ratio is very small.

The approach we take to justify, albeit in a limited way, the use of the physical system considered in *Boyer* [1971] and *Vaziri and Boyer* [1971] to model real ocean flows is to introduce a mathematical model in which the dissipative mechanism is defined by eddy viscosity coefficients (e.g., *Fofonoff* [1962]). In the theoretical development, we introduce the assumption that the vertical coefficient N_V is much smaller than the horizontal one N_H; for estimates of these quantities as obtained from oceanographic data, see *Sverdrup et al.* [1942]. By coupling this assumption with the requirement of a small vertical to horizontal length scale ratio, we show that the governing equations to lowest order for the mathematical model are identical to those considered in *Boyer* [1971] and *Vaziri and Boyer* [1971]. The rotating water tunnel, or labora-

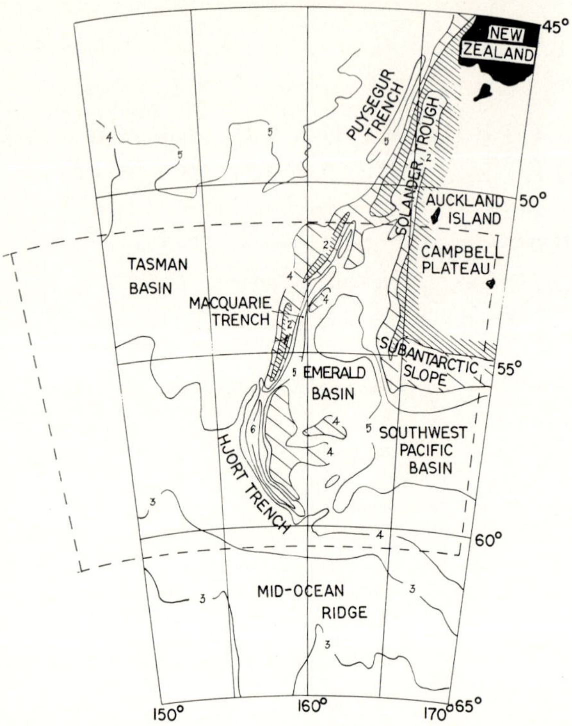

Fig. 1. Bathymetry in the vicinity of the Macquarie ridge. The numbers on the diagram represent the ocean depth in kilometers, and the broken line designates the region considered in the model studies. Figure taken from *Gordon* [this volume].

tory model, can thus be used as an analogue for the mathematical model even though the length scale ratios and dissipative mechanisms for the two systems differ.

With the exception of the eddy viscosity coefficients N_V and N_H, the various dimensional parameters of the mathematical model can be specified to a reasonable approximation. In the analysis, the horizontal coefficient N_H will not contribute to a determination of the lowest order motion. As such, the only parameter not easily prescribed is the vertical coefficient N_V, which will thus be treated as a 'free parameter'; i.e., the system behavior will be examined for a range of N_V.

In the next section on the laboratory model, we briefly review the general aspects of the physical system considered in *Boyer* [1971] and *Vaziri and Boyer* [1971]. Then we define the mathematical model and show its similarity to the Boyer-Vaziri system. The rotating water tunnel is briefly described in the fourth section. Various numerical runs and laboratory experiments are presented next, and the relation of these results to oceanographic measurements is given in the final section.

LABORATORY MODEL

The physical system considered in *Boyer* [1971] and *Vaziri and Boyer* [1971] is sketched in Figure 2. Identical topographies are mounted symmetrically on the upper and lower 'infinite' horizontal plane surfaces. The space between the bounding surfaces is filled with a homogeneous incompressible fluid. The entire system rotates at a constant angular velocity about a vertical axis, and upstream of the topographies and outside the Ekman boundary layers the fluid is in a uniform rectilinear motion relative to a rotating observer.

From the equations of motion and boundary conditions, we can easily show that the vertical velocity component and shear stress are identically zero on $z = 0$, the mid-plane. Either half (i.e., $z \geqslant 0$ or $z \leqslant 0$) of the system thus corresponds to the flow over a submerged topography above which is a horizontal surface on which no shearing stresses occur. In the experiments, photographs are taken from above by facing in the negative z direction and using the lower topography as a background; the upper bounding surface and topography are of course transparent. Thus anticipating the experimental configuration, we consider the $z \leqslant 0$ portion of the system in the following analytical development. Whereas the rotation is vertically upward in the experiments, the flow pattern for vertically downward, or Southern Hemisphere, rotation can be obtained by simply inverting the results, i.e., letting $x \rightarrow x$, $y \rightarrow -y$ (see paragraph on inversion and equation 16 on p. 84).

The dimensionless parameters that characterize the system are the Rossby number $Ro_l = U_l/2\Omega_l L_l$, the Ekman number $E_l = \nu/2\Omega_l L_l^2$, and the geometric ratios H_l/L_l and h_l/L_l where the subscript l denotes a

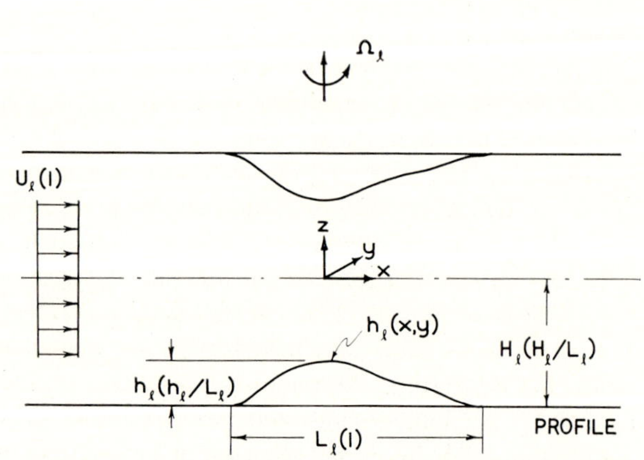

Fig. 2. The physical system considered in the laboratory experiments. The quantities in parentheses are dimensionless.

laboratory model quantity. Here U_l is the free stream speed, Ω_l is the rotation rate, and μ is the kinematic viscosity; the geometric quantities L_l, H_l, and h_l are defined in Figure 2. For general values of these parameters, the theoretical problem is intractable. By making the following a priori assumptions, however, the analysis is greatly simplified: (1) $E_l << 1$; (2) $Ro_l = k_l E_l^{1/2}$ where k_l is of order unity; (3) H_l/L_l is of order unity; and (4) $h_l(x, y)/L_l = h_{l0}(x, y)E_l^{1/2}$ where $h_{l0}(x, y)$ is of order unity.

Using these four assumptions, we can divide the flow field into an interior geostrophic region and Ekman layers along the bounding surfaces. The analysis proceeds by expanding the dependent variables in a power series of $E_l^{1/2}$ in each of the regions. One finds that the lowest order interior motion (i.e., of order unity) is horizontal, and one can thus define a stream function ψ_l, i.e.,

$$U_{l0} = -\psi_{ly} \qquad V_{l0} = \psi_{lx}$$

where U_{l0} and V_{l0} are the lowest order velocity components in the x and y directions, respectively. The governing equation for ψ_l is obtained by matching the interior and boundary layer solutions and is given by

$$k_l \frac{H_l}{L_l} [\zeta_{lt} + J(\psi_l, \zeta_l)] + \frac{1}{2^{1/2}} \zeta_l + J(\psi_l, h_{l0}) = 0 \qquad (1)$$

where ζ_l is the vertical component of relative vorticity and is related to ψ_l by

$$\nabla^2\psi_l = \zeta_l \qquad (2)$$

Equations 1 and 2 are then solved numerically subject to the appropriate boundary conditions (see *Vaziri* [1971]).

As noted previously, a rotating water tunnel is available that can be used to investigate experimentally the physical system sketched in Figure 2. In *Vaziri and Boyer* [1971], for example, the flow over conical obstacles is discussed in some detail. Experiments are presented in which the various dimensionless parameters strictly satisfy the four a priori theoretical assumptions given above. The streamline patterns in $z = 0$, or the mid-plane, are depicted by using dye tracer techniques, and the results so obtained are found to be in good agreement with the numerical runs. In addition, experiments are presented for cones in which the aspect ratios h_l/L_l are not small as required in the theoretical development. Nevertheless, the agreement between theory and experiment is still satisfactory.

MATHEMATICAL MODEL

The geometry considered in the mathematical model (Figure 3) is a greatly simplified version of the bathymetry in the vicinity of the Macquarie ridge (Figure 1). All the surfaces of the model topography in Figure 3 are plane. The ocean floor away from topographic features is taken to be horizontal and of depth H; specific values of the various model parameters used are the following:

$L = 7.5(10)^6$cm $\qquad U = 25$ cm/sec
$H = 5.0(10)^5$cm $\qquad \Omega = 7.29(10)^{-5}$ rad/sec
$h = 4.0(10)^5$cm $\qquad \Theta_0 = 56°$

In Figure 3, trenches are indicated by the dotted areas; the maximum depth of the trenches is $H + h/2$ below the free surface. Elevated features are indicated by the cross-hatched areas; the peaks of the Macquarie ridge, the Campbell plateau, and the seamount are $H - h$ below the free surface. The characteristic horizontal scale L of the topography is given by the width of the Macquarie ridge.

The following define other characteristics of the mathematical model:

1. The ACC is bounded on the northern and southern sides by vertical 'slip walls' ($u \neq 0$ at the boundary) representing the Tasman basin to the north and the mid-ocean ridge to the south.

2. The ACC is purely zonal upstream of the topography and is uniform and rectilinear across the stream; the free stream speed is given by U.

3. The fluid is homogeneous and incompressible.

4. The dissipative mechanism is defined by constant horizontal N_H and vertical N_V eddy viscosity coefficients.

5. The Coriolis parameter is constant and given by $f_0 = 2\Omega \sin \Theta_0$ over the entire domain, where Ω is the rotation rate of the earth and Θ_0 is the average latitude of the topography; i.e., we neglect the β effects.

6. Wind stress effects are neglected, and the free surface is taken to be horizontal.

Choose the rectangular Cartesian co-ordinate system (x, y, z) where x is to the east, y is to the north, and z is vertically upward (Figure 3). Let $\overline{v}$ (u, v, w) be the Eulerian velocity and t be the time. The governing equations for the model are then given by the three components of the momentum equation and by the equation for conservation of mass; i.e.,

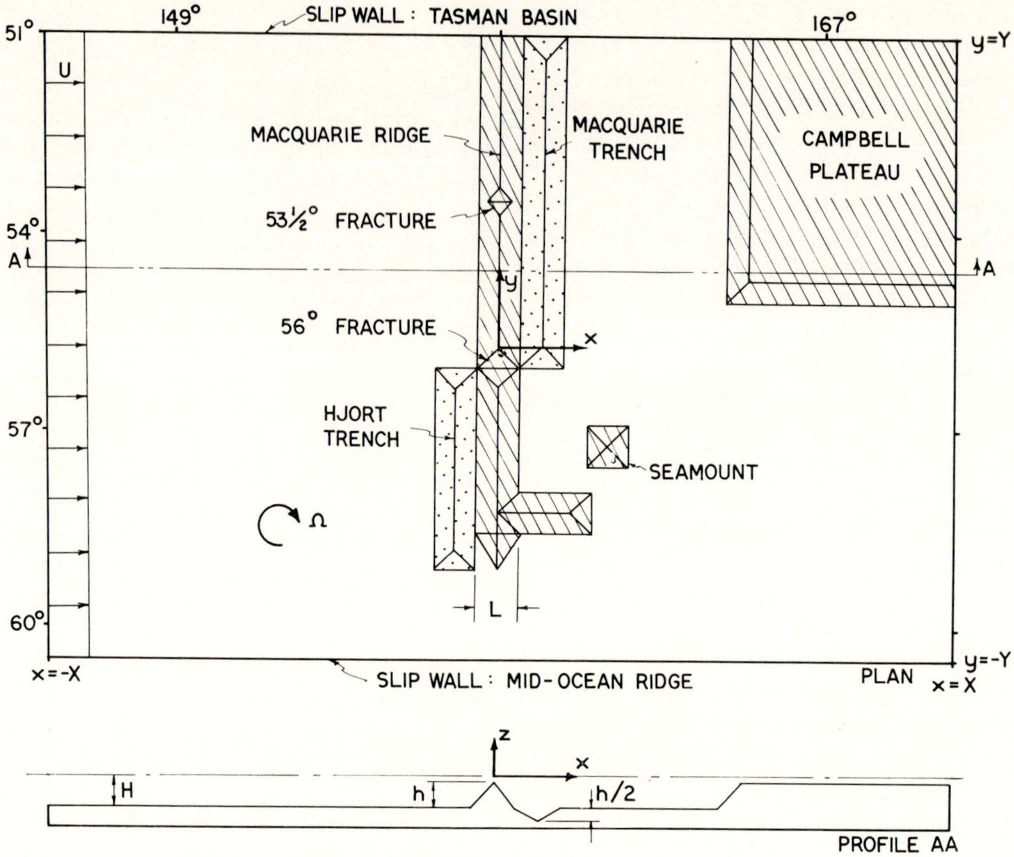

Fig. 3. The physical system considered in the mathematical model.

$$
\begin{aligned}
u_t + uu_x + vu_y + wu_z = &\\
&-\Phi_x - 2\Omega \sin \Theta_0 v - 2\Omega \cos \Theta_0 w \\
&+ N_H(u_{xx} + u_{yy}) + N_V u_{zz} \\
v_t + uv_x + vv_y + wv_z = &\\
&-\Phi_y + 2\Omega \sin \Theta_0 u + N_H(v_{xx} \\
&+ v_{yy}) + N_V v_{zz} \\
w_t + uw_x + vw_y + ww_z = &\\
&-\Phi_z + 2\Omega \cos \Theta_0 u + N_H(w_{xx} \\
&+ w_{yy}) + N_V w_{zz} \\
u_x + v_y + w_z = 0 &
\end{aligned}
\tag{3}
$$

respectively. Here $\Phi = p/\rho + \phi$ where p is the pressure, ρ is the density, and ϕ is the force potential; ϕ includes both gravitational and centrifugal effects.

We now nondimensionalize equation 3 as follows:

$$
\begin{aligned}
t &= (L/U)\, t^* \\
(x, y, z) &= (Lx^*, Ly^*, Hz^*) \\
(u, v, w) &= [Uu^*, Uv^*, (H/L)\, Uw^*] \\
\Phi &= 2\Omega \sin \Theta_0 UL\Phi^*
\end{aligned}
\tag{4}
$$

where the asterisks represent dimensionless quantities. Substituting equation 4 into 3, one obtains the following dimensionless equations:

$$
\begin{aligned}
Ro(u_t + uu_x + vu_y + wu_z) = &\\
&-\Phi_x - v - (H/L) \cot \Theta_0 w \\
&+ E_H(u_{xx} + u_{yy}) + E_V u_{zz} \\
Ro(v_t + uv_x + vv_y + wv_z) = &\\
&-\Phi_y + u + E_H(v_{xx} \\
&+ v_{yy}) + E_V v_{zz} \\
Ro(w_t + uw_x + vw_y + ww_z) = &\\
&-(L/H)^2\Phi_z + (L/H) \cot \Theta_0 u \\
&+ E_H(w_{xx} + w_{yy}) + E_V w_{zz} \\
u_x + v_y + w_z = 0 &
\end{aligned}
\tag{5}
$$

respectively, where the asterisks have now been omitted from dimensionless quantities. The dimensionless parameters are the Rossby number $Ro = U/f_0L$, the Ekman numbers $E_V = N_V/f_0H^2$ and $E_H = N_H/f_0L^2$, and the geometric ratios H/L and h/H.

The boundary conditions require that the flow far

upstream of the topography and outside the Ekman layers on the ocean floor is uniform and rectilinear. Far downstream we require the interior flow to be parallel to the x axis, i.e., purely zonal. The no slip (velocity is 0) condition is to be applied on the bottom boundary, and the vertical velocity component is to vanish on the free surface $z = 0$. The north-south velocity component v must be zero along the northern and southern boundaries. The general problem as defined by the governing equations (5) and these boundary conditions is intractable, and one is thus led to make simplifying assumptions.

By using the values of the various model parameters given on page 81, the dimensionless parameters of the system are given by

$$
\begin{aligned}
Ro &= 2.8(10)^{-2} \qquad & H/L &= 6.7(10)^{-2} \\
E_V &= 3.3(10)^{-8}N_V \qquad & h/H &= 0.8 \\
& & E_H = 1.5(10)^{-10}N_H &
\end{aligned}
\tag{6}
$$

We now make the following assumptions.

1. We assume that $E_V << 1$, $E_H << E_V^{1/2}$ so that the horizontal shear stresses can be neglected; i.e., $N_V << 10^8$ cm^2/sec and $N_H << 10^6 N_V^{1/2}$ cm^2/sec.

2. We assume that $Ro = kE_V^{1/2}$ where k is of order unity; then, for instance, $10^2 < N_V < 10^6$ cm^2/sec.

3. We assume that $H/L \sim 0(E_V^b)$ where $b \geqslant 1/2$.

4. We assume that $h(x, y)/H = h_0(x, y)E_V^{1/2}$ where $h_0(x, y)$ is of order unity. Note that $h(x, y)$ is the vertical co-ordinate of the topography relative to $z = -H$.

It must be emphasized that relation 4 above, which implies a topography of infinitesimal amplitude, is made for analytical convenience and is not strictly satisfied by the value of h/H given in equation 6. Restrictions 1 through 4 lead to an analysis that is very similar to that given in *Vaziri and Boyer* [1971] and also lead to an identical set of governing equations. As such, the derivation will be outlined rather than given in detail.

From equation 5 and relations 1–4, the flow field can be shown to consist of an interior geostrophic region and bottom Ekman boundary layers. In each of these regions the dependent variables can be expanded in power series in $E_V^{1/2}$; i.e.,

$$
\begin{aligned}
\bar{v} &= \bar{V}_0 E_V^0 + \bar{V}_1 E_V^{1/2} + \dots \\
\Phi &= \Phi_0 E_V^0 + \Phi_1 E_V^{1/2} + \dots
\end{aligned}
$$

where the order of the leading terms is dictated by the method of nondimensionalization and that of the first order terms is suggested by the structure of the Ekman layers.

Solutions of the zero order equations in the interior show that, to this order, the flow is horizontal and two-dimensional; i.e.,

$$
U_{0I} = U_{0I}(x, y) \qquad V_{0I} = V_{0I}(x, y) \qquad W_{0I} = 0 \tag{7}
$$

where the subscript I refers to the interior. Since the zero order motion is two-dimensional, one can introduce a stream function ψ defined by

$$
U_{0I} = -\psi_y \qquad V_{0I} = \psi_x \tag{8}
$$

To determine ψ, one must match the $0(E_V^{1/2})$ vertical velocity component as determined from the interior solution with that component obtained from the Ekman layer analysis. From the first order interior equations, one finds

$$
W_{1I} = -kz[\zeta_t + J(\psi, \zeta] \tag{9}
$$

where J is the Jacobian and ζ is the vertical component of relative vorticity defined by

$$
\nabla^2\psi = \zeta \tag{10}
$$

From an Ekman layer analysis, the vertical velocity component evaluated at the outer edge of the boundary layer is determined as

$$
W_{1E} = -[1/(2)^{1/2}]\,\nabla^2\psi + J(\psi, h_0) \tag{11}
$$

Since the topography is assumed to have an infinitesimal amplitude, $z = -1$ to lowest order at the bottom boundary for the interior flow. Thus matching equations 9 and 11 along $z = -1$, one obtains

$$
k[\zeta_t + J(\psi, \zeta)] + [1/(2)^{1/2}]\,\zeta - J(\psi, h_0) = 0 \tag{12}
$$

Relation (12) is a transport equation for the relative vorticity ζ. The combination of unsteady and advection terms represents the rate of change of relative vorticity following a fluid particle, i.e., a fluid column. This relative vorticity is dissipated by Ekman suction and is either enhanced or diminished by the topographic term.

The system (equations 10 and 12) can be treated as an initial value problem and solved iteratively by using a now standard numerical scheme (see *Vaziri* [1971] for details). In the present study, *Veronis'* [1966] Jacobian rather than *Arakawa*'s [1966] is used to represent the nonlinear terms in equation 12. Thus the vorticity on the lateral 'slip walls' has been

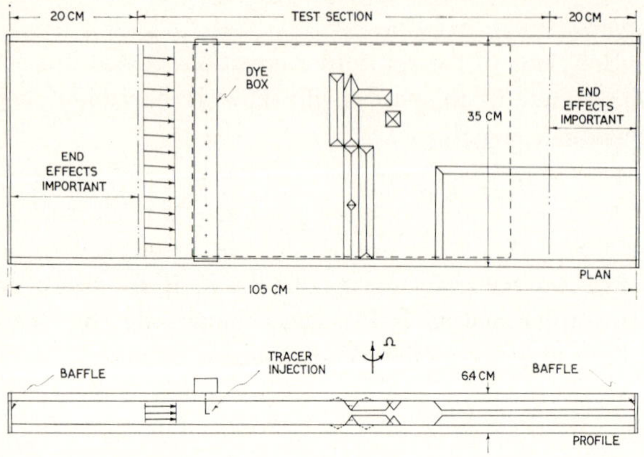

Fig. 4. Scale diagram of the rotating water tunnel. The broken line designates the region considered in the numerical model studies.

assumed unimportant in determining the interior flow field.

The boundary conditions require that

$$\begin{aligned} \psi(x \to -\infty, y, t) &= -y \\ \psi_x(x \to +\infty, y, t) &= 0 \\ \psi(x, -Y, t) &= Y \\ \psi(x, Y, t) &= -Y \end{aligned} \tag{13}$$

and the initial condition is taken as

$$\psi(x, y, 0) = -y \tag{14}$$

The first two equations in (13) are approximated by choosing a rectangular domain (Figure 3) of large streamwise extent and applying the following:

$$\begin{aligned} \psi(-X, y, t) &= -y \\ \psi_x(X, y, t) &= 0 \end{aligned} \tag{15}$$

Note that the x, y dimensions of Figure 3 are $2X$, $2Y$, respectively.

If we invert the flow field of the laboratory model by $x \to x$, $y \to -y$, and $\psi_l \to -\psi_l$, the lowest order equations for ψ_l and ζ_l (equations 1 and 2) are identical to those for ψ and ζ (equations 10 and 12) provided that

$$k = k_l \frac{H_l}{L_l} \to \frac{U}{(f_0 N_V)^{1/2}} \frac{H}{L} = \frac{U_l}{(2\Omega_l \nu)^{1/2}} \frac{H_l}{L_l} \tag{16}$$

$$h_0 = h_{l0} \to h \left(\frac{f_0}{N_V}\right)^{1/2} = h_l \left(\frac{2\Omega_l}{\nu}\right)^{1/2}$$

Thus the laboratory model can be used as an analogue for the mathematical model. The assumption of infinitesimal topographies is necessary to relate the laboratory studies to the mathematical model.

Assume that the characteristic dimensions of the laboratory model (L_l, H_l, and h_l) are specified and that L, H, h, U and f_0 for the mathematical model are given (say, by those values on page 81). Then after N_V is specified, the rotation rate Ω_l and free stream speed V_l for the laboratory flow are given by the equations in (16). Since Ω_l and U_l can be varied over only a very limited range of values with the present apparatus (i.e., typically $0.50 < \Omega_l < 2.00$ rad/sec and $0.15 < U_l < 0.40$ cm/sec), laboratory experiments corresponding to a wide range of N_V cannot be conducted. Large variations in N_V can be considered numerically, however.

ROTATING WATER TUNNEL

The details of the rotating water tunnel are discussed in some detail in *Boyer* [1971], but a brief description of the apparatus seems appropriate here. The working or tunnel section of the facility is sketched in Figure 4. This portion of the apparatus is immersed in a temperature bath to insure that the walls of the tunnel and the water remain at the same temperature during testing.

Fluid is pumped through the tunnel as indicated, and the entire channel rotates about a vertical axis. The purpose of the baffles at the entrance and exit of the channel is to approximate the cross-channel pressure gradient that occurs in the desired uniform test section flow. Without these baffles, a uniform flow cannot be obtained in a channel of the dimensions indicated. The baffle designs are approximate, and thus a certain portion of the flow field near the entrance and exits is not uniform. For the range of parameters considered, the test section length, which is determined under conditions for which the topographic features are absent from the tunnel, is approximately 65 cm. Because of the cross-channel transport in the Ekman layers on the horizontal surfaces, the free stream flow V_l is at a slight angle to the right and is facing downstream as indicated. This angle for the present experiments is at the most 2°–3°, and thus within experimental error this angle can be ignored for our purposes. For more details see *Boyer* [1971].

The various topographies being considered are placed in the channel in the positions shown in Figure 4. The outline of that portion of the tunnel corresponding to the numerical calculations is indicated by the broken lines. Observations of the flow in the midplane, or free surface, are made by releasing a neutrally buoyant dye tracer from the series of lateral positions shown, i.e., by observing a series of streak lines. Experimental runs are recorded photographically by both single shot and motion picture cameras. As previously noted, experimental runs are made by using a vertically upward rotation. Photographs are then reversed by letting $x \to x$ and $y \to -y$; this reversal thus gives runs corresponding to vertically downward, or Southern Hemisphere, rotation.

EXPERIMENTAL AND NUMERICAL RESULTS

The present model studies consider the flow behavior for the eddy viscosity range $10^3 \leqslant N_V \leqslant 10^6$ and for each of the following topographic configurations: (1) the Macquarie ridge, i.e., the L-shaped ridge in Figure 3 excluding the 53½° and 56° fracture zones; (2) configuration 1 plus the Macquarie and Hjort trenches and the 53½° and 56° fracture zones; and (3) configuration 2 plus the Campbell plateau and a seamount. A variety of topographies are considered so that the role of each of the various topographic features can be assessed.

Specify the laboratory model parameters $L_l = 2.54$ cm, $H_l = 1.90$ cm, and $h_l = 1.52$ cm, and take N_V *to* be given by $N_V = 2.5(10)^4$ cm^2/sec. For this choice of N_V, the flow pattern can be determined both experimentally and numerically. Using the values for L, H, h, U, and f_0 given on page 81, one determines the laboratory settings from the equations in (16); e.g., $U_l = 0.24$ cm/sec and $\Omega_l = 1.65$ rad/sec.

Figure 5*a* is a photograph of the free surface (mid-plane) streak lines obtained for flow past the basic Macquarie ridge (i.e., configuration 1). Figure 5*b* is a numerical computation of equations 10 and 12 for parameters corresponding to the experiment. The solution has converged and thus represents a steady state streamline pattern for the geostrophic flow. The kinks upstream of the ridge in Figure 5*b* have a wave length twice that of the grid spacing and are characteristic of numerical integrations of this type. They can be filtered out [*Shapiro*, 1970], but we have not done so except for the case $N_V = 10^3$ cm^2/sec discussed below.

The most striking feature of the streak line pattern in Figure 5*a* is the wavelike pattern that develops downstream of the ridge. Except for the two uppermost streak lines that enter the shear layers along the upper wall and, as described below, thus move out of the mid-plane, the remaining streak lines remain in the mid-plane and represent purely horizontal motions. The reason for the wavy pattern is the fact that the streak lines pass through a region of the flow field that is unsteady. The approximate boundary of this unsteady region is ascertained by viewing motion pictures and is indicated by the broken line α on Figure 5*a*.

To interpret Figure 5*a*, recall that streak lines are defined as the locus of points passing a fixed Eulerian position (i.e., the location of the tracer injection), whereas streamlines are defined as lines everywhere tangent to the velocity vector. Thus only for steady flows do streak lines and streamlines coincide.

Thus the streak lines that pass through α are streamlines only upstream of the unsteady region. Downstream and outside α, a smooth line drawn through the wavy patterns will approximate the corresponding upstream streamlines. Note that the increase in amplitude and decrease in wave length of the streak lines is due to the decreasing fluid velocities in the downstream direction (i.e., a stream tube phenomenon), whereas the rotation of the wave pattern is due to lateral shear, i.e., nonzero u_y.

The location of the unsteady region is associated with the large velocities near the tip of the ridge and the subsequent large deceleration required to return the flow to approximately the free stream level. The situation here is somewhat similar to the flow past a blunt object in a nonrotating system; i.e., unsteadiness first occurs in portions of the flow in which large decelerations of the fluid particles occur. In the rotating system as being considered presently Ekman suction is very effective in dissipating the relative vorticity generated near the tip of the ridge. Thus the unsteadiness dies out in a relatively short downstream distance as indicated by the broken line. In the nonrotating case, on the other hand, dissipation is by lateral diffusion and thus the unsteadiness will persist to a much greater distance.

The largest amplitude of the unsteady oscillations occurs near the tip of the ridge (arrow γ in Figure 5*a*). The frequency of the oscillations f_l within experimental error is the same as that of the turntable; i.e., $f_l = \Omega_l/2\pi$. Whereas the frequency of the oscillations is a very regular feature of the flow, the amplitude varies a bit with time as noted in the downstream pattern. These variations are thought to be due to small fluctuations in the rotation rate of the table.

The wavelike pattern near the tip of the ridge (arrow δ in Figure 5*a*) is a steady feature of the flow; i.e., the wave pattern is a standing one. By using a rather simple analysis, this pattern can be shown to be due to inertial oscillations. Letting U^* be a charac-

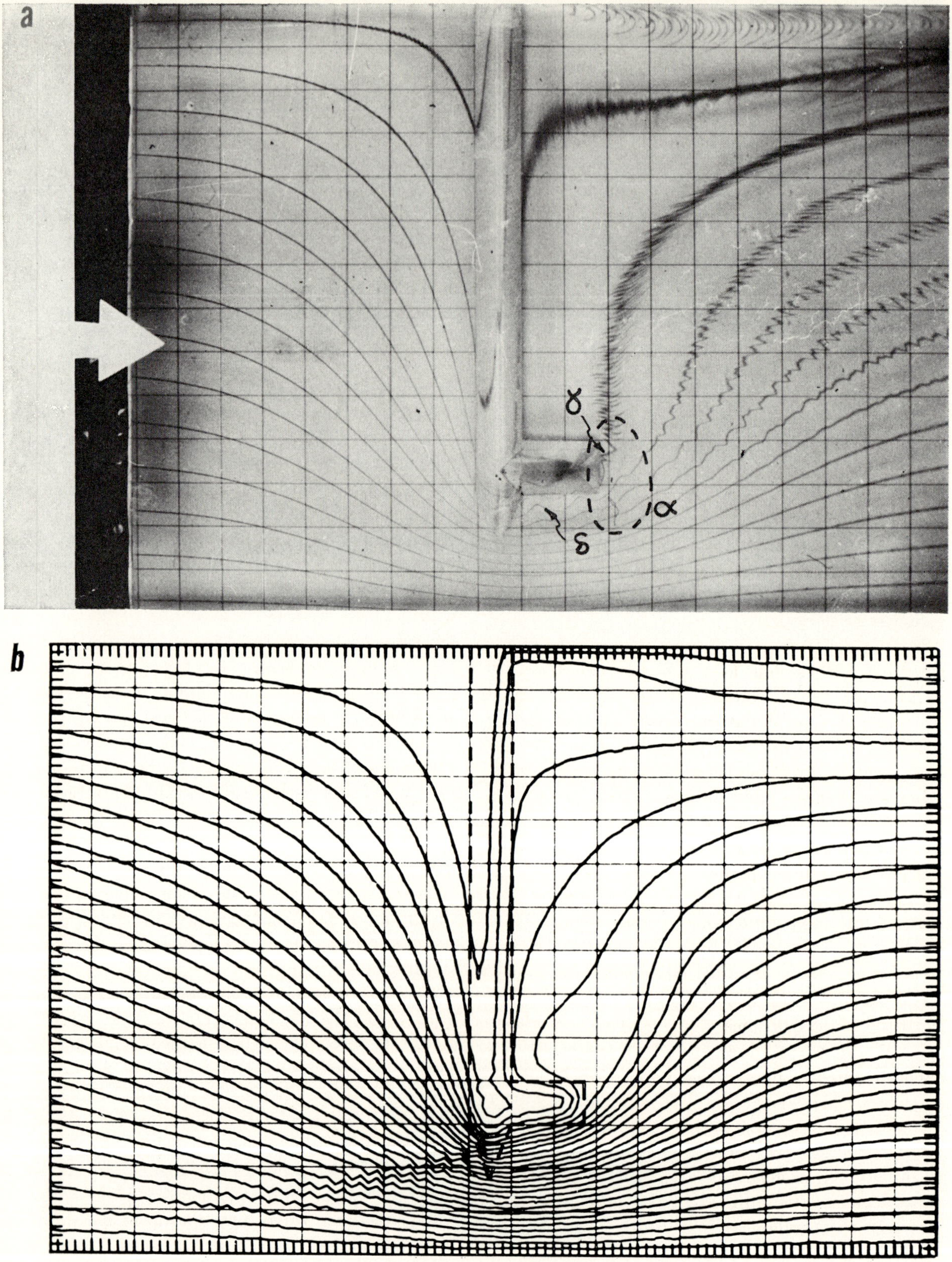

Fig. 5. Geostrophic flow for topographic configuration 1 where $N_V = 2.5(10)^4$ cm^2/sec, $E_V = 8.3(10)^{-4}$, and $Ro = 2.8(10)^{-4}$. The flow is from left to right, and the rotation is clockwise. (*a*) Mid-plane streak lines for an experimental run. (*b*) Numerical solution.

teristic velocity in the vicinity of the tip and then making a small perturbation on this field due to the topographic effect of the tip, we can show by considering an inertial ~ pressure ~ Coriolis balance (neglecting viscous effects) that the wave length λ_l of the streak lines should be given by

$$\lambda_l = U^*\pi/\Omega_l \tag{17}$$

Taking $U^* \cong 6U = 1.4$ cm/sec and recalling that $\Omega_l = 1.65$ rad/sec, one obtains $\lambda_l = 2.7$ cm. This value agrees reasonably well with the wave length noted at δ in Figure 5*a*.

The small ripples in the upstream portion of the uppermost streak line in Figure 5*a* are gravitational instabilities caused by the fact that the fluid velocities are small and that there is a slight difference in the tracer and working fluid densities.

The upper two streak lines in Figure 5*a* move into the vertical shear layers along this wall. Because of the slight density variation between the tracer and water, the tracer on reaching the vertical wall has dropped slightly below the mid-plane. We can show that the vertical velocities in the vertical shear layers are an order of magnitude larger than those in the remainder of the flow field. As such, the tracer is transported vertically downward and these two streak lines, downstream of the ridge, are significantly outside the mid-plane of the channel.

Considering the experimental streamlines to be an average line through the streak lines downstream of α, one notes good agreement between the numerical run and the laboratory experiment; note that the streak line spacing in the experiment is twice the streamline spacing in the numerical run. The numerical run does not exhibit the unsteady behavior noted in the laboratory. This result may be due to the coarse grid used in the computations or to the errors introduced by the approximations used in the theory, in particular, the assumption of infinitesimal topographies.

Figures 6*a* and 6*b* are experimental and numerical runs, respectively, for configuration 2. One notes that the streak line labeled *S*, being essentially smooth, can be considered a streamline. This streak line separates the flow field into two distinctly different unsteady regions α and β sketched on Figure 6*a*. The region α in the vicinity of the tip of the Macquarie ridge has the same general characteristics as those found for configuration 1 above; i.e., the frequency of the oscillations is the same as that of the turntable. Note that the amplitude of the oscillations is much less here and is associated with the much smaller velocities near the tip; these smaller velocities are of course due to the large transport now passing through the deep passages at 53½°S and 56°S.

The unsteady region β occurs in the vicinity of the Macquarie trench. The frequency of the oscillations in this region are, within experimental error, exactly one-half the frequency of the table. Whereas the frequency again is a very regular feature of the flow pattern, the amplitude was noted to vary somewhat in time.

The wave length of the inertial oscillations (arrow δ in Figure 6*a*) is smaller than that in Figure 5*a* because the local velocity near the tip ($U^* \cong 0.9$ cm/sec) is smaller in Figure 6*a*. From equation 17, $\lambda_l = 1.7$ cm. The result again is in reasonable agreement with experiment.

Note the relatively large velocities over the Hjort trench, through the 56° fracture zone, and past the tip of the Macquarie ridge. The three uppermost streak lines pass into the vertical shear layer along the upper wall and, for the reasons described above, are not in the mid-plane downstream of the topography.

Figures 7*a* and 7*b* give a laboratory experiment and numerical run, respectively, for configuration 3 at the same Rossby and Ekman numbers as those considered in Figures 5 and 6. The streamline labeled *S* again separates the field into two different unsteady portions as described above. The regions of unsteadiness are indicated by the broken lines α and β, and the frequencies are the same, respectively, as those for α and β in Figure 6. The characteristic velocity and corresponding wave length of the inertial oscillation near the ridge tip are approximately the same as those for configuration 2.

The streak line patterns here are much the same as those in Figure 6 except that one notes that the flow in the upper portion of Figure 7*a* now skirts along the Campbell plateau. In addition, the seamount tends to inhibit the flow from moving northward after passing the tip of the ridge. The uppermost streamline, downstream of the Macquarie trench, is in the Ekman layer on the floor of the tunnel.

Note that the flow fields in moving to more complex topographies as in Figures 5, 6, and 7 in that order are not affected to a great extent by the additional topographic features except in the immediate vicinity of the new features. For example, the addition of the Campbell plateau and a seamount in going from Figure 6 to Figure 7 had almost no effect on the flow except near these added features.

Before considering values of N_V that cannot be

a

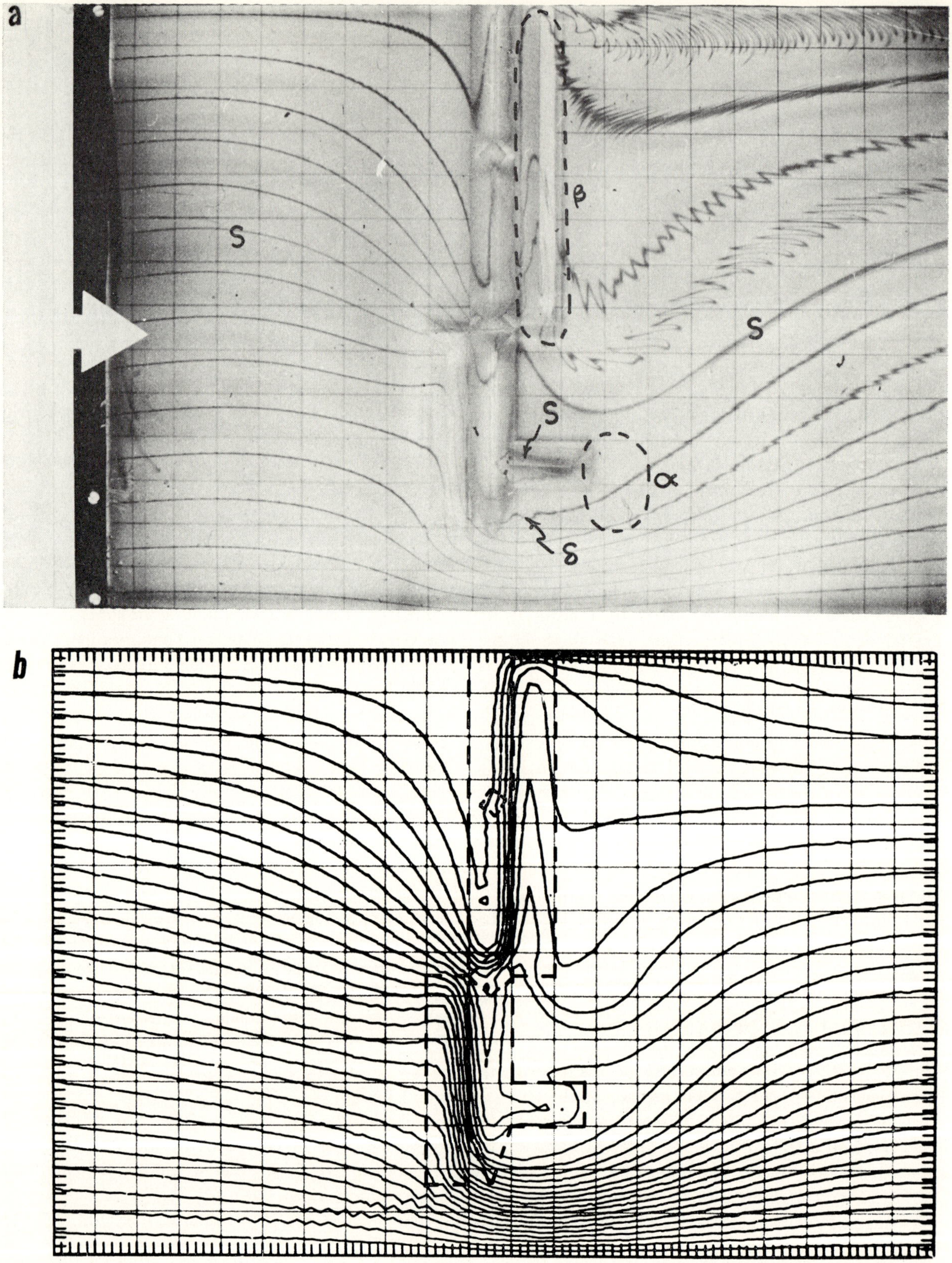

b

Fig. 6. Geostrophic flow for topographic configuration 2 where $N_V = 2.5(10)^4 cm^2/sec$, $E_V = 8.3(10)^{-4}$, and $Ro = 2.8(10)^{-4}$. The flow is from left to right, and the rotation is clockwise. (*a*) Mid-plane streak lines for an experimental run. (*b*) Numerical solution.

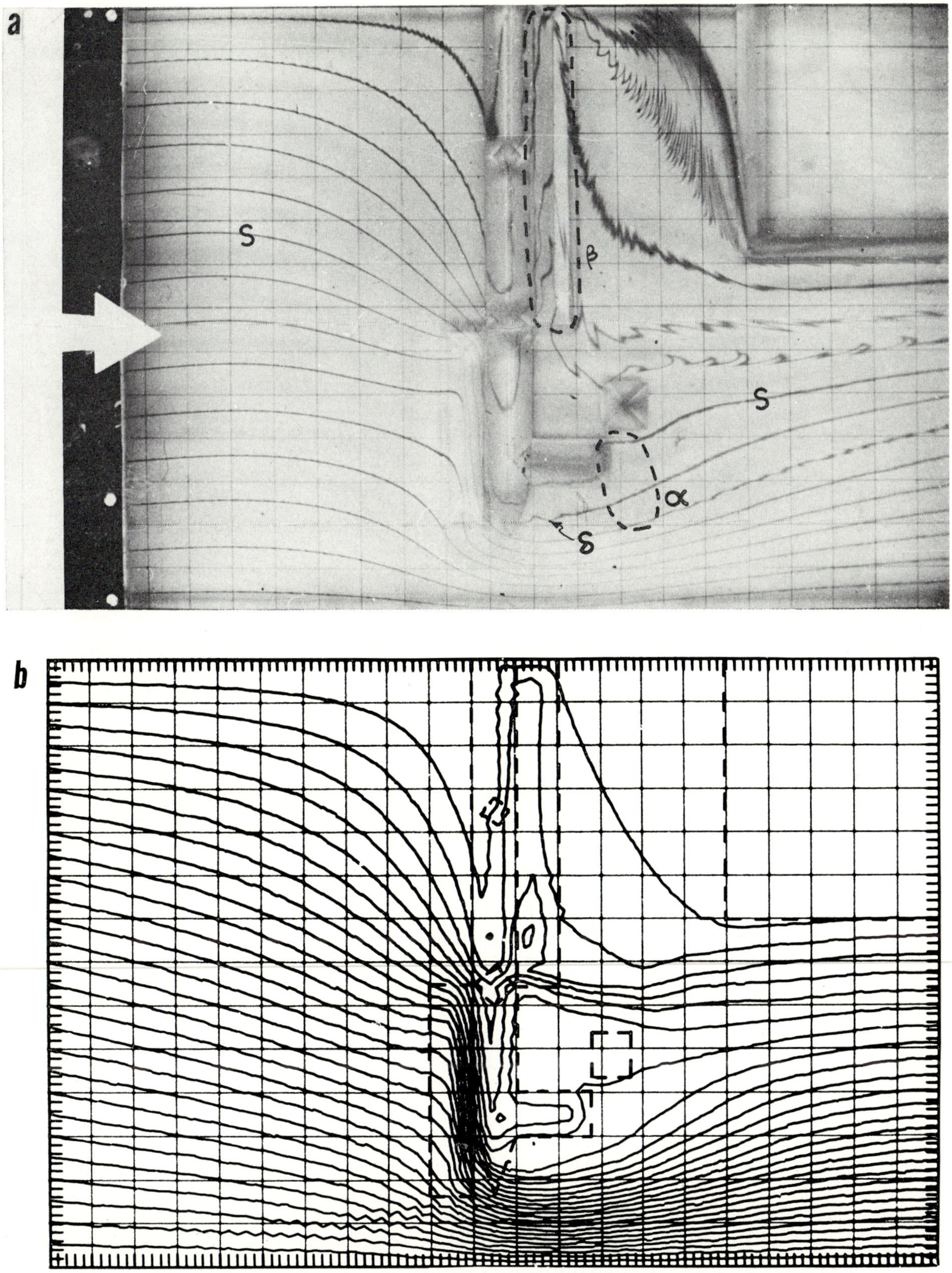

Fig. 7. Geostrophic flow for topographic configuration 3 where $N_V = 2.5(10)^4$ cm^2/sec, $E_V = 8.3(10)^{-4}$, and $Ro = 2.8(10)^{-4}$. The flow is from left to right, and the rotation is clockwise. (*a*) Mid-plane streak lines for an experimental run. (*b*) Numerical solution.

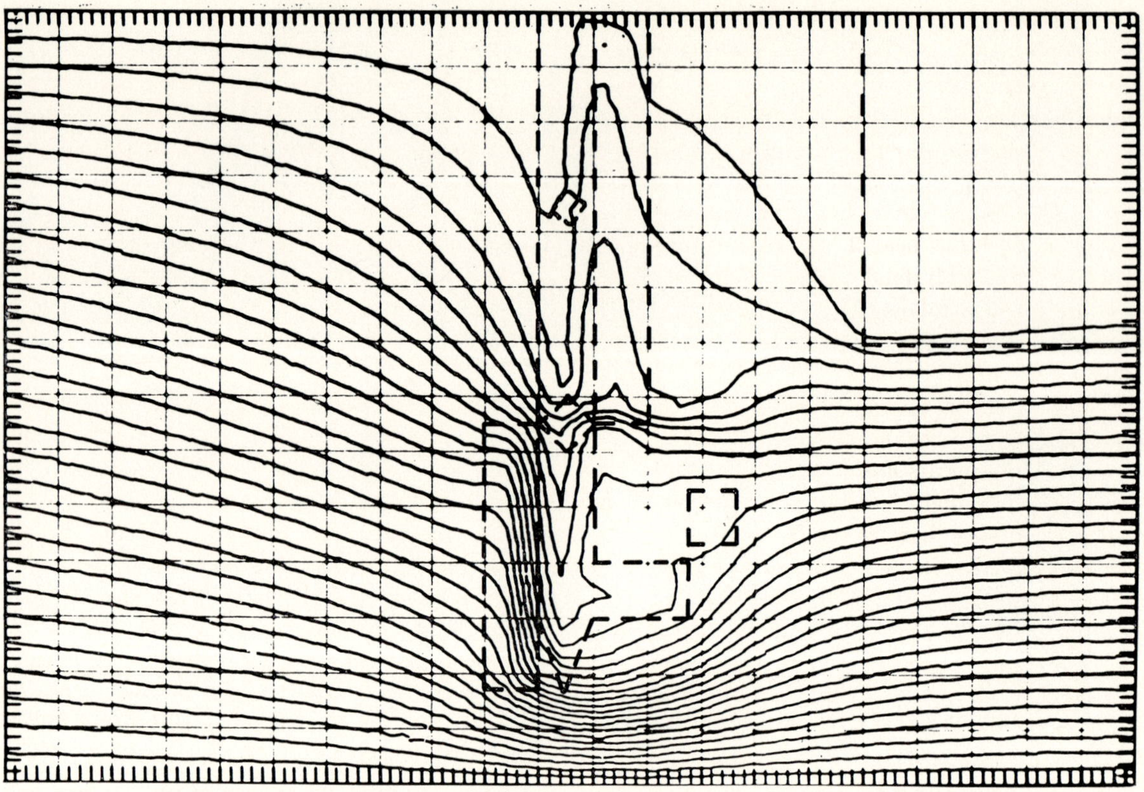

Fig. 8. Numerical solution for topographic configuration 3 where $N_V = (10)^5$ cm²/sec, $E_V = 3.3(10)^{-3}$, and $Ro = 2.8(10)^{-2}$.

investigated in the laboratory, we note that the behavior of configurations 1, 2, and 3 was investigated in the laboratory for the ranges of U_l and Ω_l given by $0.15 < U_l < 0.40$ cm/sec and $0.50 < \Omega_l < 2.00$ rad/sec; i.e., $4.0\ (10)^{-4} < E_l < 1.6(10)^{-3}$ and $1.5\ (10)^{-2} < Ro_l < 0.16$. One set of experiments was conducted at a fixed $E_l = 1.2(10)^{-3}$ and varying Ro_l. A second set was conducted at a fixed $Ro_l = 2.8(10)^{-2}$ and various E_l. Whereas the qualitative flow patterns for the three topographic configurations were the same as those described previously, the unsteady portions of the flow field became larger for increasing Ro_l at fixed E_l and for decreasing E_l at fixed Ro_l. This behavior should be expected since the Ekman suction or dissipative term $\zeta_l/(2)^{1/2}$ becomes smaller relative to the inertial effects in equation 1 for increasing Ro_l and decreasing E_l. Whereas the size of the unsteady regions changes with variations in Ro_l and E_l, the frequencies of the oscillations did not change for the range of parameters examined; i.e., the frequencies were that of the turntable for the α regions and half that of the turntable for the β regions.

For values of $N_V > 2.5(10)^4$ cm²/sec, the numerical flow fields are also steady. Figure 8, for example, is a solution for configuration 3 at $N_V = 10^5$ cm²/sec. We note that the flow pattern is very similar to that in Figure 7*b*, the major difference being that the characteristic speed in high velocity regions (e.g., near the tip of the Macquarie ridge and over the Hjort trench) is smaller in the larger viscosity case. A numerical run was also conducted for $N_V = 10^6$ cm²/sec. For this value, the streamlines experienced very little deflection due to the topographic features.

Another numerical run for configuration 3 at $N_V = 10^4$ cm²/sec was made. Because of the relatively small dissipation in the system, we have thus far been unable to obtain satisfactory results for $N_V < 10^4$ cm²/sec; work is continuing for these smaller viscosity cases. The integration of equations 10 and 12 subject to the initial condition (14) was carried out for 116 days of model time for $N_V = 10^4$. After approximately 40 days, the large-scale features of the flow developed into a periodic unsteady pattern. The major feature of the motion was the shedding of large eddies near the tip of the ridge. This shedding was observed over four cycles, and the period was approximately 18 days.

Figure 9 shows a time sequence of streamline patterns for a portion of the flow field near the tip of the ridge; the grid spacing and topographic locations are the same as those in other diagrams, e.g., Figure 8. An eddy clearly forms near the tip and is transported downstream. The eddy finally is dissipated by Ekman suction and is then followed by the formation of the next eddy near the tip and the process is then repeated.

Whereas, as noted above, conclusive results for $N_V = 10^3$ have not as yet been obtained, preliminary runs indicate that eddy shedding again takes place. The shedding period appears much smaller, and the eddies seem to maintain their identity over a much larger downstream distance than for the $N_V = 10^4$ runs. This difference of course should be expected because of the decreased dissipation.

RELATION TO OCEANOGRAPHIC DATA

We now compare the model studies with the oceanographic data presented by *Gordon* [this volume] in the preceding paper. In particular, we will refer to his Figures 11 and 13 on pages 76 and 77, respectively. Recall that Figure 11 in Gordon's paper gives isobaths of the depth of the salinity maximum S_{max} core layer. Gordon has shown that these depths are related in a rather simple way to the dynamic topography. Hence the isobaths are streamlines of the geostrophic flow with the separation of the contour lines being inversely proportional to the velocity. Gordon's Figure 13 shows an XBT section along 159°30′E and indicates a temperature structure similar to that of the double polar front zone discussed by *Gordon* [1971].

Except for the absence of eddy shedding near the tip of the Macquarie ridge, Figure 7*b* in this paper is typical, qualitatively, of the numerical flow patterns obtained for topographic configuration 3 in the range of eddy viscosities considered. As such, Figure 7*b* will be compared with the oceanographic data given in Gordon's Figure 13.

In making comparisons, the limitations of the model studies must be emphasized; e.g., none of the model restrictions delineated in the section on the laboratory model are strictly valid in the real ocean; the bottom topography has been greatly simplified, constant eddy viscosity coefficients have been introduced, the fluid has been assumed homogeneous and incompressible,

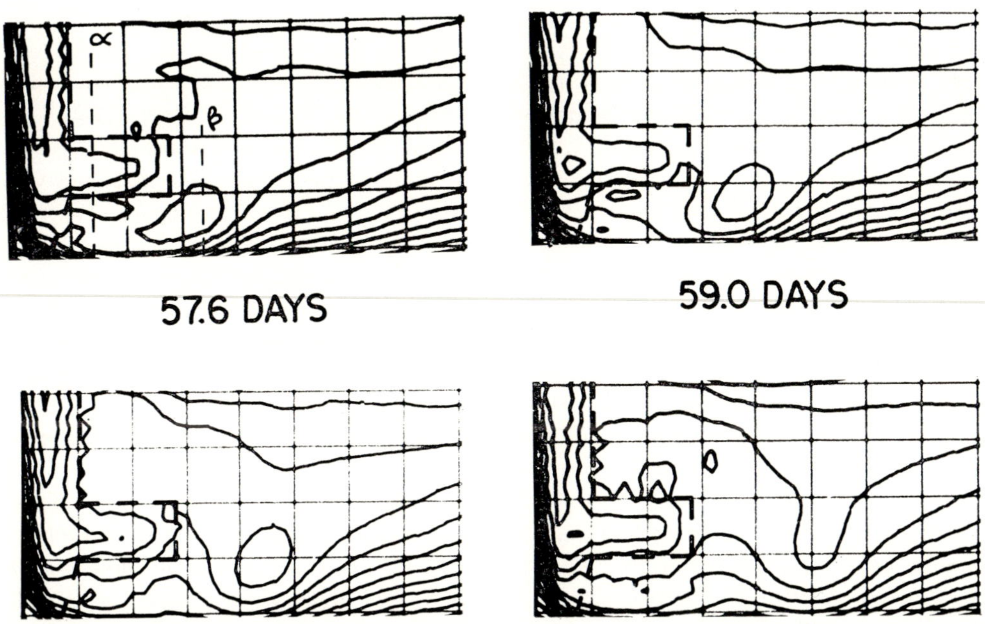

57.6 DAYS

59.0 DAYS

60.4 DAYS

61.8 DAYS

Fig. 9. Time sequence of an eddy shedding near the tip of the Macquarie ridge for topographic configuration 3 where $N_V = (10)^4$ cm^2/sec, $E_V = 3.3(10)^{-4}$, and $Ro = 2.8(10)^{-2}$.

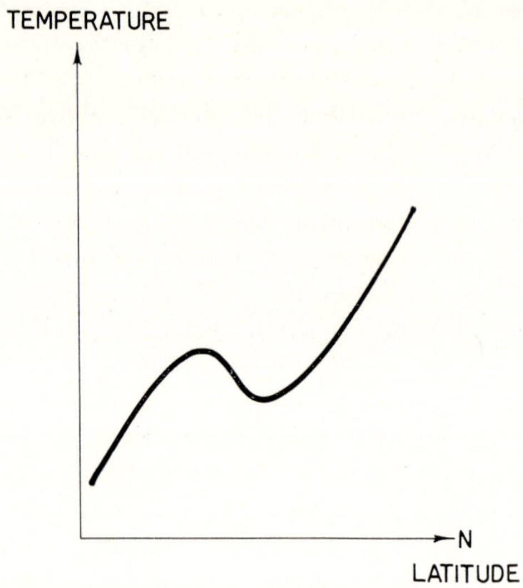

Fig. 10. Hypothetical temperature distribution for a south-north section across the snout of the Macquarie ridge or an eddy downstream of the ridge, i.e., either sections α or β from the 57.6-day graph of Figure 9.

and so on. As such, the comparisons to be made are of a purely qualitative nature.

From Figure 7*b* in this paper and Figure 11 of Gordon's paper, one notes that both the model and the oceanographic data indicate that a major portion of the flow is deflected southward on approaching the ridge system. Whereas the model indicates a strong southward flow over the Hjort trench, such an intense current is not evident in Gordon's Figure 11. The reason for this discrepancy is not evident, but more detailed oceanographic data may well indicate the presence of such a 'jetlike' current.

Both the model and prototype indicate an upstream (westward) current in the vicinity of the southern tip of the Macquarie ridge before the flow passes into the Emerald basin. This westward flow provides an explanation for the temperature structure given in Gordon's Figure 13. Assume that upstream of the ridge system in the model (or prototype) the temperature associated with fluid parcels increases monotonically with decreasing latitude. Further assume that the fluid parcels maintain their initial temperature; i.e., the material or Stokes derivative of the temperature is zero. Clearly then a south-north temperature profile across the snout of the ridge (broken line α in the 57.6-day graph of Figure 9) will look qualitatively like that given in Figure 10. Such a temperature distribution agrees with that of the surface profile given in Gordon's Figure 13. The permanency of this feature, as noted by Gordon, is also 'modeled' in the sense that a reverse flow is obtained for both the steady state cases (e.g., Figure 7*b*) and the eddy shedding regimes (Figure 9).

Both model and prototype indicate a large transport through the 56° fracture zone. The model (Figure 7*b*) predicts a relatively small transport through the 53½° fracture, whereas the oceanographic data (Gordon's Figure 11) shows a large transport in this region. We feel that the difficulty here is that the model fracture zone has been made too small. We also note that both our Figure 7*b* and Gordon's Figure 11 show that the flow skirts the Campbell plateau. One should observe, however, that the oceanographic data indicates a jet-like behavior along the southern boundary of the plateau. Jetting is not reproduced in this region in the model.

The model studies also seem to support Gordon's speculation that the double polar front zone observed east of the Macquarie ridge system is created by the shedding of eddies near the tip of the ridge. The reasoning here is similar to that noted above for the temperature profile expected for the section α (Figure 9) across the snout of the Macquarie ridge. In this case, however, we consider a temperature profile across an eddy as indicated by the broken line β in Figure 9. For such a section, again assuming a monotonically increasing temperature profile for decreasing latitude upstream of the ridge, one expects a qualitative temperature distribution as that sketched in Figure 10. Since the eddy advects downstream, such a temperature profile would not be a permanent feature for a given section. This result again is in agreement with Gordon's observations. Gordon also indicates that the double polar front zone may be seen several thousand kilometers downstream of the ridge system. This aspect also is modeled since the eddies shed in Figure 9, for example, maintain their identity far downstream of the topographic features.

The question arises as to which value of the vertical eddy viscosity coefficient N_V gives the best agreement with oceanographic data. The present status of the model work along with the limited oceanographic data available, however, does not allow for its determination. One might argue that N_V is probably less than $2.5(10)^4$ cm²/sec, since for this and larger values, no eddy shedding occurs. As noted earlier, both the eddy shedding period and the distance downstream for which the eddy retains its identity are functions of N_V. Thus if detailed, time dependent oceanographic data could be obtained, N_V might then be specified by matching the period and downstream characteristics of the model and prototype eddies. Whereas the model considered above is a rather simple one, it does demonstrate many of the large-scale characteristics of the ACC in the vicinity of the Macquarie ridge.

Acknowledgments. This work was supported by the Office of Naval Research under contract NR 083–259. The numerical computations were made on the CDC-6600 computer at the National Center for Atmospheric Research. The National Center is supported by the National Science Foundation.

REFERENCES

Arakawa, A., Computation design for long-term numerical integration of the equations of motion: Two dimensional flow, 1, *J. Comput. Phys., 1,* 119–143, 1966.

Boyer, D. L., Rotating flow over long shallow ridges, *J. Geophys. Fluid Dyn., 2,* 165, 1971.

Fofonoff, N. P., Dynamics of ocean currents, *The Sea,* vol. 1, pp. 323–395, Interscience, New York, 1962.

Gordon, A. L., Antarctic polar front zone, *Antarctic Oceanology I, Antarctic Res. Ser.,* vol. 15, edited by J. L. Reid, pp. 205–221, AGU, Washington, D. C., 1971.

Gordon, A. L., On the interaction of the antarctic circumpolar current and the Macquarie ridge, in *Antarctic Oceanology II: The Australian–New Zealand Sector, Antarctic Res. Ser.,* vol. 19, edited by D. E. Hayes, pp. 71–78, AGU, Washington, D. C., 1972.

Shapiro, R., Smoothing, filtering, and boundary effects, *Rev. Geophys. Space Phys., 8*(2), 359–387, 1970.

Sverdrup, H. U., M. W. Johnson, and R. H. Fleming, *The Oceans: their Physics, Chemistry, and General Biology,* 1087 pp., Prentice-Hall, Englewood Cliffs, N. J., 1942.

Vaziri, A., Rotating flow over shallow topographies, Ph.D. dissertation, University of Delaware, Newark, 1971.

Vaziri, A., and D. L. Boyer, Rotating flow over shallow topographies, *J. Fluid Mech., 50,* 79–95, 1971.

Veronis, G., Wind-driven ocean circulation, 2, Numerical solutions of the non-linear problem, *Deep Sea Res., 13,* 31–55, 1966.

OCEANIC CIRCULATION SOUTH OF AUSTRALIA

J. A. T. Bye

Horace Lamb Centre for Oceanographical Research, Flinders University of South Australia
Bedford Park, South Australia 5042

Abstract. New physical oceanographic observations south of Australia and north of 52°S are summarized and interpreted in terms of oceanic transports. The dynamic topography in the South Australian basin in 1967, 1968, 1969, and 1970 consisted of a ridge of high anomaly flanked on either side by troughs of low anomaly. The resulting anticyclonic circulation gave rise to a baroclinic transport toward the coast to the east of about 135°E and a transport away from the coast to the west of 135°E. This circulation that appears to transport about 15×10^6 m^3 sec^{-1} relative to 2000 db will be called the Flinders current. Farther to the south but north of the eastward circumpolar flow, a baroclinic transport of about 30×10^6 m^3 sec^{-1} to the west occurs, except to the south of Tasmania where large variations of dynamic topography are found.

Physical oceanographic observations between South Australia and just south of the antarctic front at 52°S have been discussed previously in the literature, notably by *Deacon* [1937], *Rochford* [1957, 1961], and *Wyrtki* [1962*a*, *b*]. Recently new data have been reported [*Jacobs and Amos*, 1967; *May and Tamuly*, 1968; *Bye*, 1967, 1971; *Bye and Tamuly*, 1970*a*, *b*; *Bye et al.*, 1968; *Bye et al.*, 1970]. These data have been obtained in the various programs of the *Eltanin* in the South Pacific and Indian oceans and in a repeat sampling program of the Horace Lamb Centre between 1967 and 1971. All together, 165 deep stations (sampling > 2000 meters) have been occupied in the region (Figures 1 and 2) and these include 76 stations of the Commonwealth Scientific and Industrial Research Organisation of Australia and 69 stations of the Horace Lamb Centre. This series of data enables the variability of the geostrophic flow to be assessed and new fields of the baroclinic transport to be drawn.

VARIABILITY OF THE BAROCLINIC TRANSPORT RELATIVE TO 2000 db

The baroclinic mass transport relative to a reference isobaric surface is represented conveniently by fields of anomaly of potential energy, i.e.,

$$\chi(p_0) = \frac{1}{g} \int_0^{p_0} p\delta dp$$

in which δ is the anomaly of specific volume, g is gravity, p is pressure, and p_0 is the reference pressure. Assuming geostrophic flow, we find the relative transport is related to $\chi(p_0)$ by the equations

$$U(p_0) = -(1/f)\ [\partial\chi(p_0)/\partial y]$$
$$V(p_0) = (1/f)\ [\partial\chi(p_0)/\partial x] \qquad (1)$$

where $0x$ and $0y$ are toward the east and north, $U(p_0)$ and $V(p_0)$ are the x and y components of baroclinic mass transport velocity in the water column relative to p_0, and f is the Coriolis parameter. Eight synoptic fields are presented (Figure 3) for a reference isobaric surface p_0 = 2000 db that occurs between the intermediate and deep circulations and corresponds closely to the level of no motion at 1750 db given by *Wyrtki* [1962*a*] for this region. The data in Figure 3 exhibit the following features:

1. The baroclinic circulation in the South Australian basin in 1961, 1967, 1968, and 1969 features a north-going baroclinic transport toward the coast of South Australia to the east of 135°E. For 1961, 1967, and 1970 another transport to the west of 135°E and away from the coast is also observed. This transport system will be referred to as the Flinders current. This current is a new feature of the general circulation of the oceans that is not resolved on previous charts. *Sverdrup et al.* [1942, p. 615] depict only a westward transport just south of Australia. The *Discovery II* data (Figure 3*a*), on which this chart is based, do not include stations in the South Australian basin. Alternatively, *Kort* [1962] from *Ob* data for 1956 [*Academy of Sciences of the USSR*, 1958] infers a cyclonic circulation to the north of the zonal west-

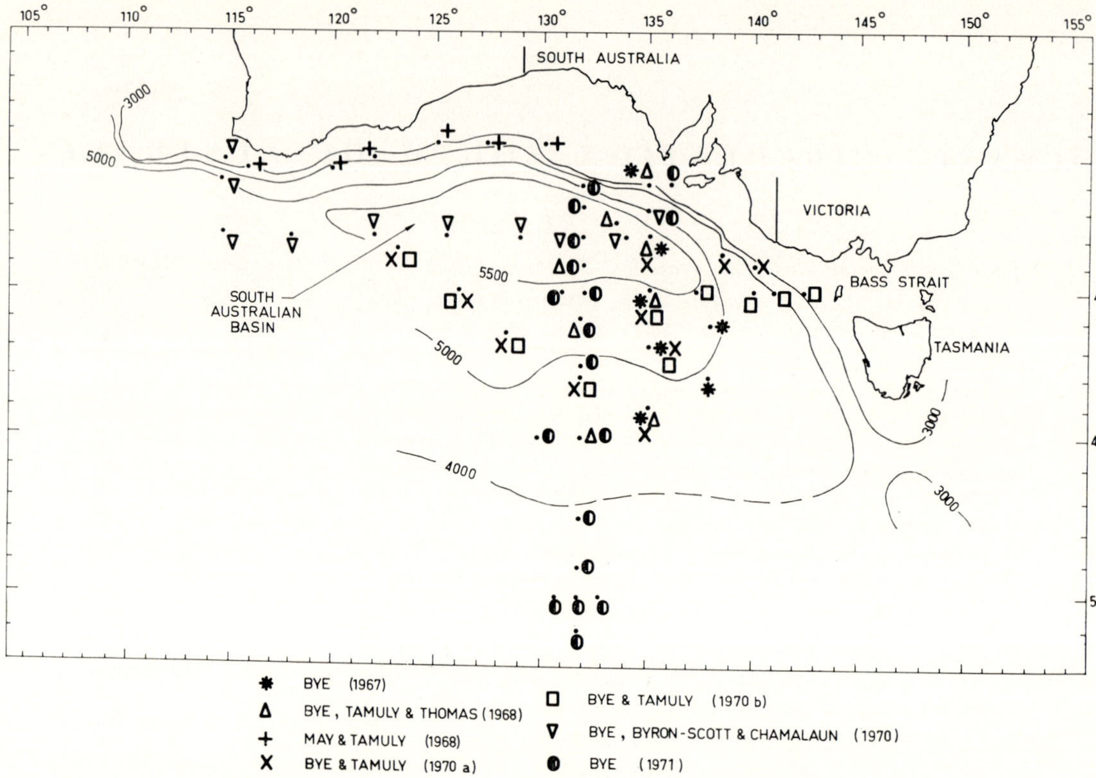

Fig. 1. Bathymetry, hydrographic deep stations (sampling > 2000 meters) occupied by Horace Lamb Centre.

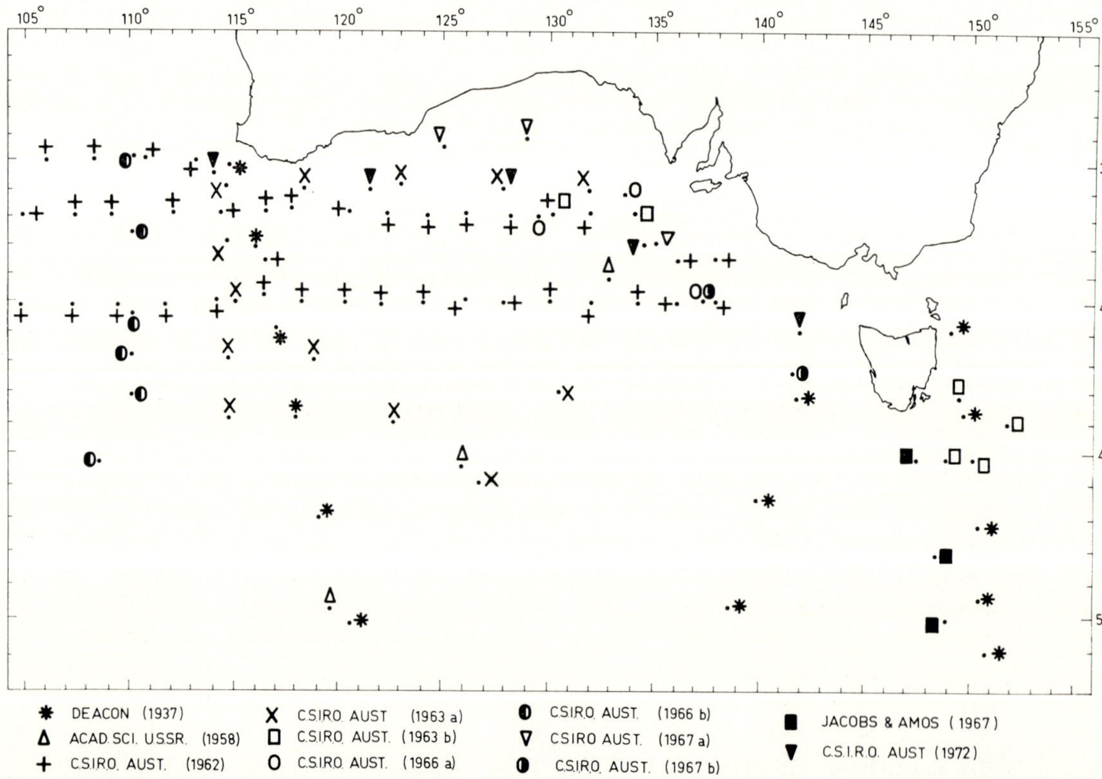

Fig. 2. Hydrographic deep stations (sampling > 2000 meters) other than those occupied by Horace Lamb Centre.

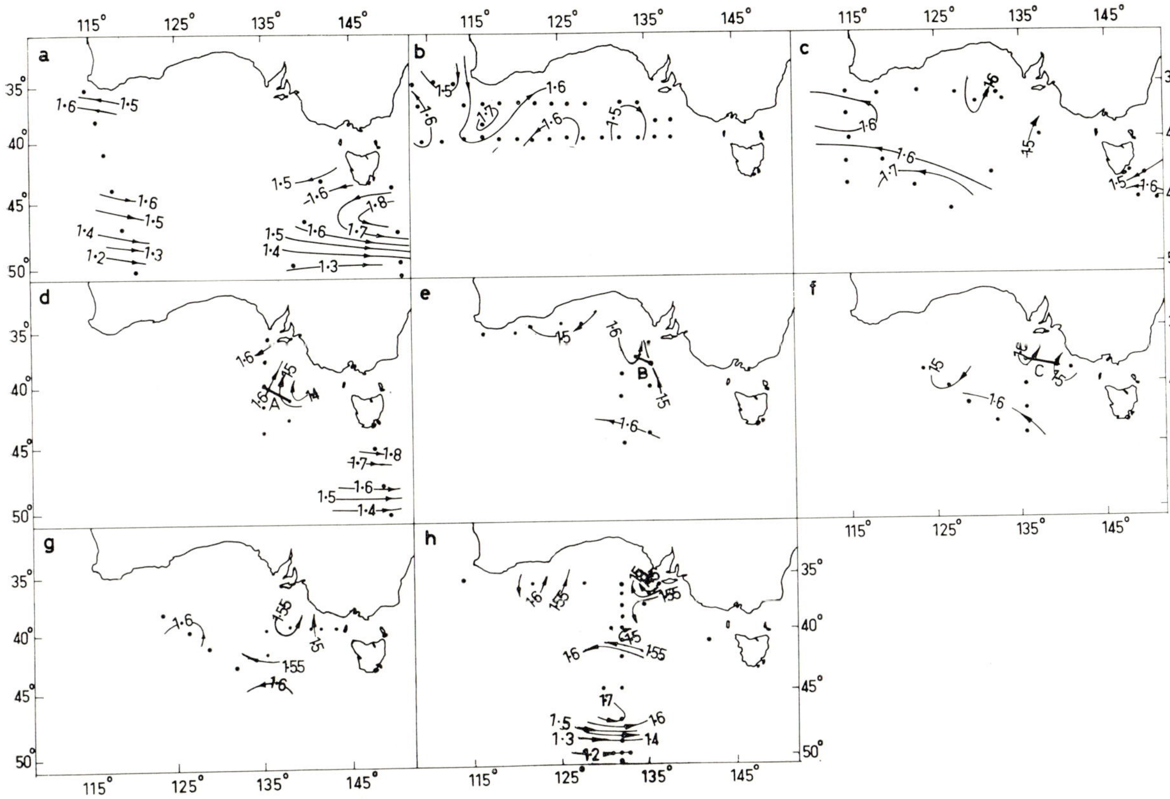

Fig. 3. Synoptic fields of $\chi(2000)$. The darkened circles indicate deep stations, and the contours are in dynamic meters squared. (*a*) Data recorded June–July 1932 [*Deacon*, 1937]. (*b*) Data recorded February–March 1960 [*Commonwealth Scientific and Industrial Research Organisation of Australia*, 1962]. (*c*) Data recorded February–March 1961 [*Commonwealth Scientific and Industrial Research Organisation of Australia*, 1963*a*, *b*]. (*d*) Data recorded February–March 1967 [*Bye*, 1967]; velocity section along line A is shown in Figure 5. (*e*) Data recorded in September–October 1967 [*Bye et al.*, 1968; *May and Tamuly*, 1968]; velocity section along line B is shown in Figure 5. (*f*) Data recorded in March 1968 [*Bye and Tamuly*, 1970*a*]; velocity section along line C is shown in Figure 5. (*g*) Data recorded in March 1969 [*Bye and Tamuly*, 1970*b*]. (*h*) Data recorded from November 1969 to January 1970 [*Commonwealth Scientific and Industrial Research Organisation of Australia*, 1972; *Bye*, 1971].

ward transport. Since only one deep station lies north of 45°S, $\chi(2000)$ is not contoured for 1956.

2. A flow of intermediate depth waters from the Pacific Ocean to the Indian Ocean south of Tasmania on the surface 80 cl ton^{-1} that occurs at approximately 1000 meters is indicated from water mass analysis [*Reid*, 1965]. This interpretation is in agreement with the more recent data [*Bye*, 1968]. However, the nature of the exchange is not clear. The data (Figure 3*a*, *c*, *d*) seem to imply an intermittent flow from the

TABLE 1. The Flinders Current

Date	Section From		Section To		Surface Salinity, %0	Dynamic Difference across Section, dyne cm	Geostrophic Surface Velocity, cm sec^{-1}	Transport Relative to 2000 db, 10^6 m^3 sec^{-1}
February–March 1967	40°S	135°E	41°S	138°E*	35.2	25	8	25
September–October 1967	37.5°S	133.5°E	38°S	135°E*	35.2	18	13	17
March 1968	38°S	135°E	38.5°S	138.5°E*	35.4	12	4	12
March 1969	40°S	137.5°E	40°S	141°E	35.2	12	4	10
January 1970	37°S	132°E	37°S	135°E	35.3	9	4	13

* Velocity sections along A, B, and C are shown in Figure 5.

east, possibly in the form of large anticyclonic eddies originating in the East Australia current [*Hamon*, 1965]. Clearly, regular and detailed sections south of Tasmania are required to monitor this interesting region adequately.

THE FLINDERS CURRENT

The basic feature of the hydrology of the region is an internal oscillation of the water column centered on the depth range of 500–1800 meters with a maximum amplitude at about 900 meters. The level of the occurrence of water of a given temperature and salinity may vary by as much as 250 meters (Figure 4). This oscillation gives rise to the characteristic topography of the Flinders current discussed in the previous section. Statistics of the geostrophic flow in sections across the current are summarized in Table 1, and typical velocity profiles are shown in Figure 5. Inshore of the 2000-meter contour, no marked changes in the density structure from that in the adjacent deep water have been observed [*Commonwealth Scientific and Industrial Research Organisation of Australia*, 1967*c*, *d*, 1968*a*].

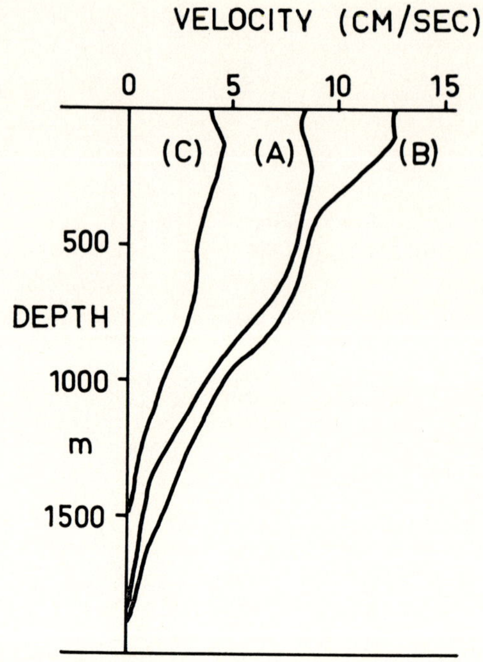

Fig. 5. Geostrophic velocity versus depth within the Flinders current as determined from sections A, B, C shown in Figure 3*d*, *e*, and *f*.

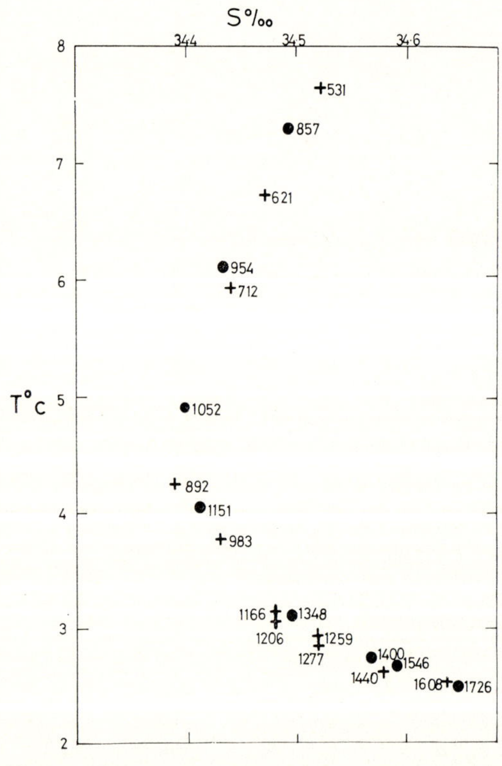

Fig. 4. Temperature–salinity diagram between 500 meters and 1800 meters for two stations in the northeast of the South Australian basin. Crosses indicate station taken from HLC 2 at 41°S, 138°E; darkened circles indicate station taken from Dm 2/68 (HLC 5) at 38°S, 135°E. Figures are sample depths in meters.

Measurements of drift velocity near the coast have also been made. From 1958 to 1962, a series of weighted drift bottles were released monthly from the following standard release positions [*Commonwealth Scientific and Industrial Research Organisation of Australia*, 1968*b*]: (1) 38°12′S, 140°40′E; (2) 39°15′S, 143°05′E; (3) 41°03′S, 144°23′E; (4) 41°56′S, 154°01′E; (5) 41°26′S, 144°31′E. These positions were to the west of Bass Strait and along the west coast of Tasmania. The results showed a sudden change along the Victorian and South Australian coasts from the south-going current in winter to a north-going current in summer. The change usually occurred about November, but the strength and duration of the north-going current varied widely from year to year. A lower limit on the average drift velocity was about 5 cm sec^{-1}, and individual velocities greater than 10 cm sec^{-1} were reported. This seasonal flow at the surface is not reproduced in the deeper layers (Figure 3) and appears to be an independent phenomenon probably confined to the Ekman layer and associated with the shifting of the wind direction from westerly in winter to southeasterly in summer.

BAROCLINIC TRANSPORT SECTIONS RELATIVE TO THE BOTTOM

The total (relative to the bottom) baroclinic transport except to the far south of the region in the circumpolar current is similar to that relative to 2000 db (Table 2), although there is evidence of a small, deep eastward flow in the South Australian basin.

The transports were calculated by invoking equation 1 at 2000 db or at the deepest common sampling level between pairs of stations. Between 132°E and 115°E the *Discovery II* and *Eltanin* cruise 41 sections agree on a total westward transport of about 30 × 10^6 m^3 sec^{-1}, but the variability south of Tasmania is demonstrated by the *Discovery II* and *Eltanin* cruise 27 sections.

TABLE 2. Baroclinic Transport Sections with Reference to the Australian Coast

Station	Latitude S	Longitude E	T, 10^6 m^3 sec^{-1}	T_{2000}
		Eltanin 41		
1	35°	132°	0	0
2	36°	132°	− 3	− 3
3	38°	132°	− 5	− 4
4	40°	132°	−10	− 2
5	50°	132°	47	30
6	51°	132°	47	29
23	36°	136°	0	0
22	37°	135°	3	3
21	38°	132°	−12	−10
20	39°	132°	−12	− 7
19	41°	132°	−14	− 6
18	42.5°	132°	−32	−21
17	45°	132°	−24	−19
16	47.5°	132°	−25	−26
15	49°	132°	35	13
14	52°	132°	58	29
		Discovery II		
877	35°	115°	0	0
878	38°	116°	−17	−23
879	41°	117°	−25	−17
880	44°	118°	−26	−19
881	47°	119°	0	4
882	50°	120°	58	36
883	53°	122°	101	53
898	44°	150°	0	0
899	47°	150°	−24	−24
900	49°	151°	38	28
901	51°	151°	58	45
903	54°	152°	...	52
		Eltanin 27		
689	45°	147°	0	0
686	48°	148°	25	23
683	50°	149°	52	50
682	52°	150°	45	47
679	53°	151°	128	58

Eltanin 41 data taken from *Bye* [1971], *Discovery II* data taken from *Deacon* [1937], and *Eltanin* 27 data taken from *Jacobs and Amos* [1967]. T is the cumulative baroclinic transport relative to the bottom and T_{2000} is the cumulative baroclinic transport relative to 2000 db. Transport is positive toward the east. Transports in each section are relative to the station nearest to the coast.

Acknowledgments. *Eltanin* cruise 41 data were collected jointly by a party from Scripps Institution of Oceanography led by Jeffrey E. Callahan and a party from the Horace Lamb Centre led by the author. A joint data report is in preparation. The author wishes to thank the chief scientists on *Eltanin* cruise 41, Professor Walter H. Munk and Dr. Frank E. Snodgrass, for making available time for the collection of the data, and the National Science Foundation for the use of facilities aboard the *Eltanin*. The assistance of many people, in particular A. Tamuly, B. Hughes, R. L. May, and my wife, in preparation of the work is gratefully acknowledged. The investigation has been made possible through the strenuous and continuing efforts of Professor J. R. M. Radok, Director of the Horace Lamb Centre at the Flinders University of South Australia.

REFERENCES

Academy of Sciences of the USSR, Hydrological, hydrochemical, geological, and biological investigations of research ship *Ob*, in *IGY Reports of the Complex Antarctic Expedition of the Academy of Sciences of the U.S.S.R.*, 214 pp., Hydro-meteorological Publishing House, Leningrad, 1958.

Bye, J. A. T., *Cruise Rep. HLC 2,* Horace Lamb Centre, Flinders Univ. of S. Aust., 1967.

Bye, J. A. T., Hydrology of the Southern Ocean, 1, Australian section–North, *Res. Pap. 26,* Horace Lamb Centre, Flinders Univ. of S. Aust., 1968.

Bye, J. A. T., *Eltanin* cruise 41, *Cruise Rep. HLC 10,* Horace Lamb Centre, Flinders Univ. of S. Aust., 1971.

Bye, J. A. T., and A. Tamuly, Cruise Dm 2/68, *Cruise Rep. HLC 5,* Horace Lamb Centre, Flinders Univ. of S. Aust., 1970*a*.

Bye, J. A. T., and A. Tamuly, Cruise Dm 1/69, *Cruise Rep. HLC 8,* Horace Lamb Centre, Flinders Univ. of S. Aust., 1970*b*.

Bye, J. A. T., A. Tamuly, and I. M. Thomas, *Cruise Rep. HLC 3,* Horace Lamb Centre, Flinders Univ. of S. Aust., 1968.

Bye, J. A. T., R. A. D. Byron-Scott, and F. H. Chamalaun, Cruise Dm 1/69, *Cruise Rep. HLC 9,* Horace Lamb Centre, Flinders Univ. of S. Aust., 1970.

Commonwealth Scientific and Industrial Research Organisation of Australia, Oceanographical observations in the Indian Ocean in 1960, HMAS *Diamantina* cruise Dm 1/60, *Cruise Rep. 2,* Melbourne, 1962.

Commonwealth Scientific and Industrial Research Organisation of Australia, Oceanographic observations in the Indian Ocean in 1960, HMAS *Diamantina* cruise Dm 1/61, *Cruise Rep. 7,* Melbourne, 1963*a*.

Commonwealth Scientific and Industrial Research Organisation of Australia, Oceanographic observations in the Pacific Ocean in 1961, HMAS *Gascoyne* cruise G 1/61, *Cruise Rep. 8,* Melbourne, 1963*b*.

Commonwealth Scientific and Industrial Research Organisation of Australia, Oceanographical observations in the Indian and Pacific Oceans in 1961, HMAS *Gascoyne* cruise G 2/61, *Cruise Rep. 10*, Melbourne, 1966*a*.

Commonwealth Scientific and Industrial Research Organisation of Australia, Oceanographical observations in the Indian Ocean in 1962, HMAS *Diamantina* cruise Dm 3/62, *Cruise Rep. 18*, Melbourne, 1966*b*.

Commonwealth Scientific and Industrial Research Organisation of Australia, Oceanographical observations in the Pacific and Indian Oceans in 1962, HMAS *Gascoyne* cruises G 2/62, and G 3/63, *Rep. 16*, Melbourne, 1967*a*.

Commonwealth Scientific and Industrial Research Organisation of Australia, Oceanographical observations in the Indian and Pacific Oceans in 1964, HMAS *Gascoyne* cruise G 3/64, *Rep. 35*, Melbourne, 1967*b*.

Commonwealth Scientific and Industrial Research Organisation of Australia, Oceanographical observations in the Indian Ocean in 1963, HMAS *Gascoyne* cruise G 2/63, *Rep. 22*, Melbourne, 1967*c*.

Commonwealth Scientific and Industrial Research Organisation of Australia, Oceanographical observations in the Indian Ocean in 1964, HMAS *Gascoyne* cruise G 2/64, *Rep. 34*, Melbourne, 1967*d*.

Commonwealth Scientific and Industrial Research Organisation of Australia, Oceanographical observations in the Indian Ocean in 1965, HMAS *Gascoyne* cruise G 2/65, *Rep. 43*, Melbourne, 1968*a*.

Commonwealth Scientific and Industrial Research Organisation of Australia, Drift bottle releases and recoveries in Bass Strait and adjacent waters 1958–1962, *Oceanogr. Sta. List 78*, Melbourne, 1968*b*.

Commonwealth Scientific and Industrial Research Organisation of Australia, Oceanographical observations in the Indian Ocean 1969, HMAS *Diamantina* cruise Dm 7/69, *Cruise Data*, in press, 1972.

Deacon, G. E. R., The hydrology of the Southern Ocean, *Discovery Rep. 15*, 1–124, 1937.

Hamon, B. V., The East Australia current 1960–1964, *Deep Sea Res., 12*, 899–921, 1965.

Jacobs, S. S., and A. F. Amos, Physical and chemical oceanographic observations in the Southern Ocean, USNS *Eltanin* cruises 22–27, *Tech. Rep. 1-CU-1-67*, Lamont-Doherty Geol. Observ. of Columbia Univ., Palisades, N. Y., 1967.

Kort, V. G., The antarctic ocean, *Sci. Amer., 207*(3), 113–128, 1962.

May, R. L., and A. Tamuly, *Cruise Rep. HLC 4*, Horace Lamb Centre, Flinders Univ. of S. Aust., 1968.

Reid, J. L., Jr., Intermediate waters of the Pacific Ocean, *Oceanogr. Stud., 2*, 85 pp., Johns Hopkins Univ., Baltimore, Md., 1965.

Rochford, D. J., The identification and nomenclature of the surface water masses in the Tasman Sea (data to end 1954), *Aust. J. Mar. Freshwater Res., 8*(4), 369–413, 1957.

Rochford, D. J., The intermediate depth waters of the Tasman and Coral seas, 2, The 26.80 σ_γ surface, *Aust. J. Mar. Freshwater Res., 12*(2), 129–149, 1961.

Sverdrup, H. U., M. W. Johnson, and R. H. Fleming, *The Oceans: Their Physics, Chemistry, and General Biology*, 1087 pp., Prentice-Hall, Englewood Cliffs, N. J., 1942.

Wyrtki, K., Geopotential topographies and associated circulation in the south-eastern Indian Ocean, *Aust. J. Mar. Freshwater Res., 13*(2), 1–17, 1962*a*.

Wyrtki, K., Geopotential topographies and associated circulation in the western South Pacific Ocean, *Aust. J. Mar. Freshwater Res., 13*(2), 89–105, 1962*b*.

AUSTRALIAN-ANTARCTIC TIDES

JAMES D. IRISH AND FRANK E. SNODGRASS

Institute of Geophysics and Planetary Physics, University of California at San Diego
La Jolla, California 92037

Abstract. To measure the Australian-antarctic tides, three deep-sea instrument capsules were dropped for 1 month between Australia and Antarctica along 132°E. Records of temperature, pressure, and current velocity were obtained. The velocity records were dominated by 3-cm/sec inertial currents, whereas tidal currents were less than 1 cm/sec. The mean flow showed great variability with changes as great as 6 cm/sec (reversing the direction of flow) over a period of 10 days. The capsule pressure records were compared with the antarctic and Australian coastal records. The observed diurnal tide can be interpreted as a Kelvin wave moving westward along the antarctic coast and a slightly smaller one traveling eastward along the south coast of Australia. Amplitude and phase vary smoothly from Australia to Antarctica. A Kelvin model does not fit the semidiurnal tide very well. The velocity is too great along the Australian coast and too small along the antarctic coast. The phase is smooth from Antarctica to the northernmost capsule and then changes abruptly at Australia, the change indicating a dissipative Australian coast with respect to the semidiurnal tide.

Until very recently, ocean tides were measured only at coastal and island stations. The offshore behavior of tides was inferred from these shore measurements given certain theoretical considerations. With the development of deep-sea instrument capsules, the measurement of offshore tides became possible. Our most ambitious program to date used a multistation array of deep-sea capsules off the coast of California. Observed tidal amplitudes and phases were then fit to a theoretical model that predicted a counterclockwise amphidrome in the M2 component of the tide midway between San Diego and Hawaii [*Munk et al.*, 1970]. Further measurements substantiated this prediction [*Irish et al.*, 1971].

For our second effort to study the deep-sea tides, the southeastern Indian Ocean was selected because it is experimentally and dynamically different from the area off California. Being a hostile environment, this ocean would thoroughly test our experimental method. Dynamically, the Indian ocean differs and is unique in that with the South Atlantic and South Pacific oceans it extends entirely around the earth, and at latitude 62°S the free shallow water wave velocity $(gD)^{1/2}$ equals the velocity of the sublunar point. Historically, the claim has been made that the tides would have enormous amplitudes at this latitude. However, this amplification seems unlikely since the 'resonance' occurs along a single latitude, whereas the tide wave averages over distances equal to wave length/2π (about 2000 km). Yet an examination of the tidal behavior between Australia and Antarctica with deep-sea capsules still seems appropriate. The antarctic research ship USNS *Eltanin* was made available to us from December 1969 through February 1970 (cruise 41) for the study of the mid-ocean tides [*Snodgrass*, 1971].

MEASUREMENTS OFFSHORE

Three deep-sea capsules were dropped to the sea floor between Australia and Antarctica along 132°E. Capsule Josie was located at 37°S off the Australian shelf on the abyssal plain, Kathy at 50°S just north of the crest of the South Indian ridge, and Flicki at 60°S off the coast of Antarctica on the abyssal plain (Figures 1 and 2). Josie was initially dropped at 40°S, but only 1 week of rather noisy data was obtained. She was then dropped at 37°S where 21 days of good data were obtained. Kathy at 50°S recorded pressure only; and at 60°S, Flicki's tape recorder failed after 21 days. Table 1 summarizes the records obtained from the expedition.

The deep-sea instrument capsule in Figure 3 [*Snodgrass*, 1968] is divided into two basic parts: the buoyant aluminum spheres containing the digital system, tape recorder, acoustical system, flashing light, and radio beacons; and the instrument frame that

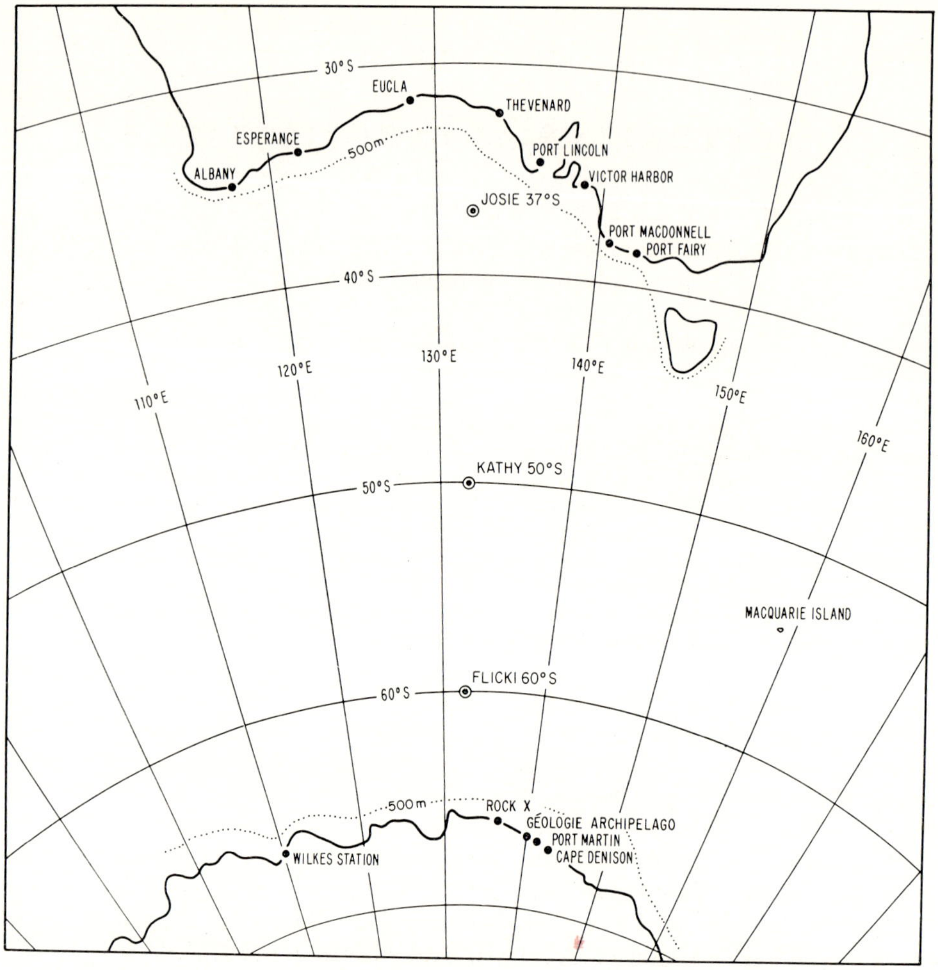

Fig. 1. The location of the three deep-sea capsule stations and of Australian and antarctic land stations. The reference station at Macquarie Island is also located.

sits on the battery pack and holds the sensors, tilt and direction indicator, and release mechanism. The lead acid battery pack not only supplies the power for the system but also acts as a ballast to hold the instrument frame firmly on the sea floor. Tethered on a 2-meter electric cable, the spheres float above the sensors to reduce the interference with the measurements.

The capsule is dropped freely to the sea floor where it remains unattended and records 2-min averages of temperature, pressure, current speed, and direction on computer compatible magnetic tape. The tape capacity and battery life allow recordings to be made for slightly longer than 1 lunar month. At the end of this time, the capsule is recalled acoustically. The instrument frame is separated from the battery frame by a solenoid or exploding bolt, and the capsule rises to the surface on the buoyancy of the spheres. The radio beacons and flashing light aid in recovery.

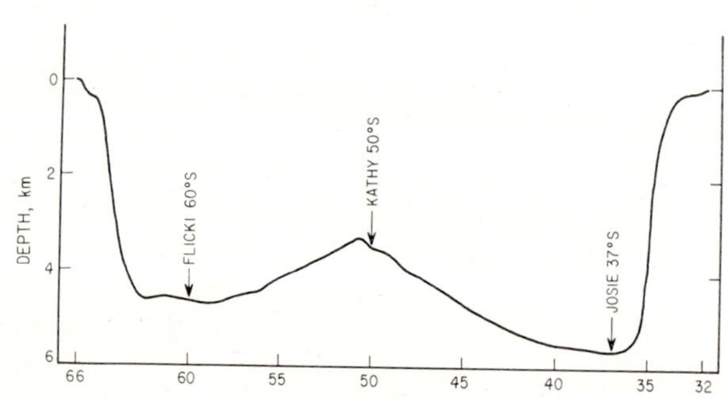

Fig. 2. The cross-channel section from Australia to Antarctica along 132°E is constructed from hydrographic charts and *Eltanin* cruise 41 data.

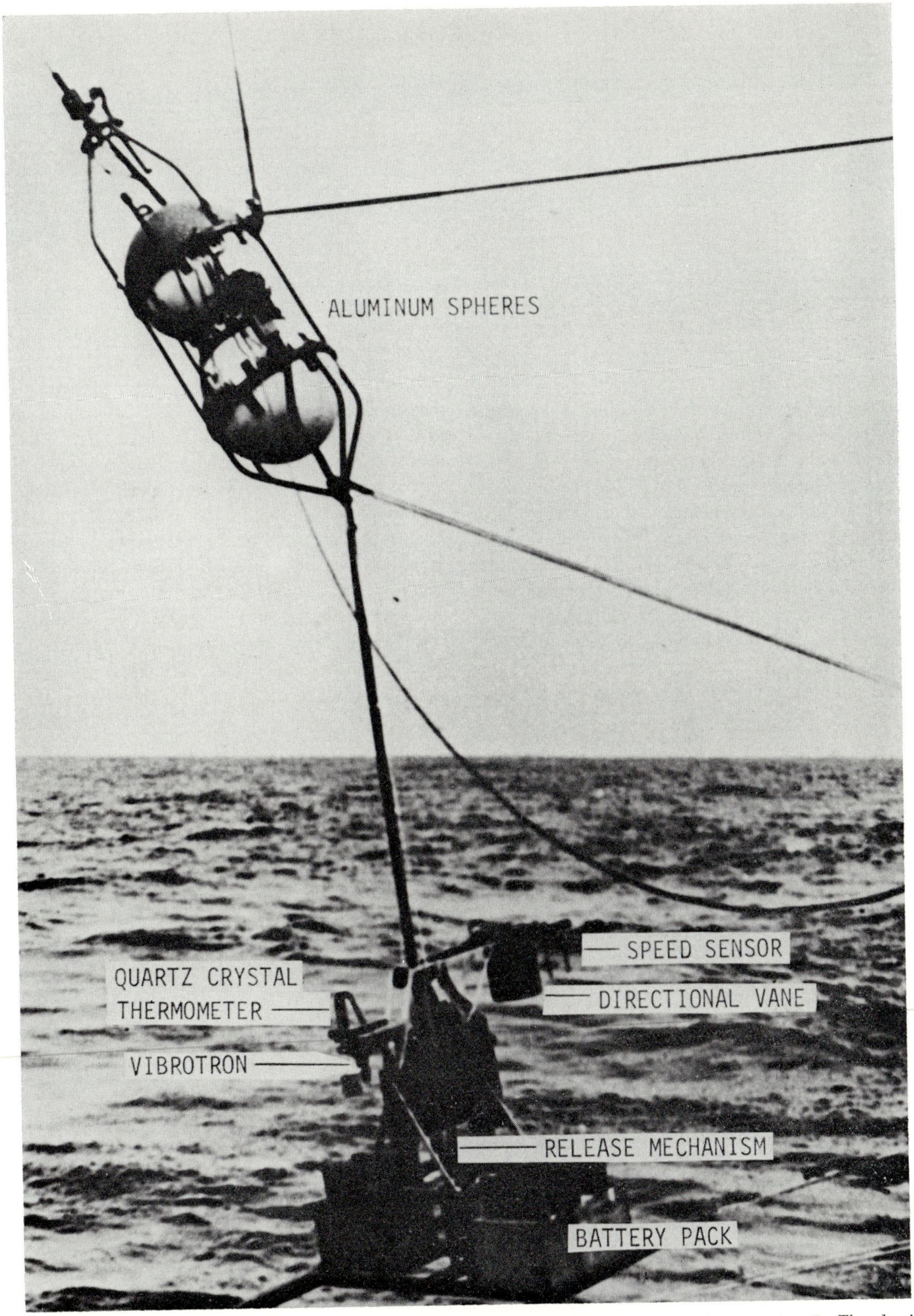

Fig. 3. A deep-sea instrument capsule is launched in the Antarctic from the *USNS Eltanin* during cruise 41. The aluminum spheres that contain the recording and release electronics are tethered to the instrument frame by a multiconductor cable. The sensors are attached to the instrument frame that sits on the battery pack.

TABLE 1. Summary of the Deep-Sea Tide Records

Station		Latitude, S	Longitude, E	Depth, meters	Record Length, hours		
					Temperature	Pressure	Velocity
Josie	37°S	37°01′	132°01′	5560	192	192	...
Josie	37°S	37°01′	132°01′	5550	501	501	498
Kathy	50°S	50°02′	132°09′	3440	765*	765	...
Flicki	60°S	60°01′	132°07′	4710	506	506	503

* This sensor showed a 0.1°C increase in temperature during the month that is probably nongeophysical. The sensor was damaged during recovery, so the calibration could not be checked.

Pressure Measurements

The tides were measured by recording the pressure at the sea floor with a Vibrotron pressure sensor [*Caldwell et al.*, 1969]. The ocean acts as a filter that attenuates surface waves by a factor of $\exp(-kz)$ where k is the wave number and z is the distance below the surface. At mid-ocean depths (typically 4 km), sea and swell are virtually eliminated since their wave lengths are short compared with the ocean depth. However, tides have wave lengths many thousands of kilometers long and so are hardly attenuated.

Relative temperature measurements of 10^{-3} °C were necessary to correct the Vibrotron records to the nearest millimeter of equivalent sea water pressure. These measurements were readily obtained with a quartz crystal thermometer [*Caldwell et al.*, 1969] that resolved relative changes of 10^{-5} °C. The time constant of the temperature sensor was made compatible with that of the Vibrotron by the addition of a convector. Besides being used to correct for the temperature sensitivity of the Vibrotron, the temperature data is also of general geophysical interest.

The records for the three deep-sea capsules plus coastal tide records at Australia and Antarctica (predicted from harmonic tidal constants) are plotted for visual comparison in Figure 4. The fortnightly alter-

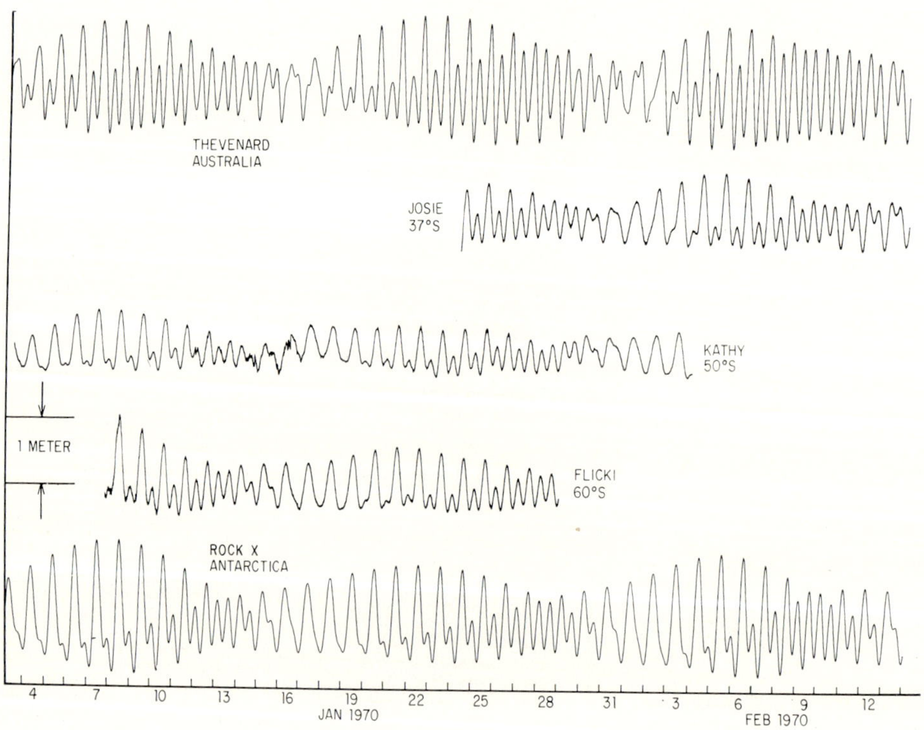

Fig. 4. The capsule pressure records are compared with two land records generated from harmonic tidal constants.

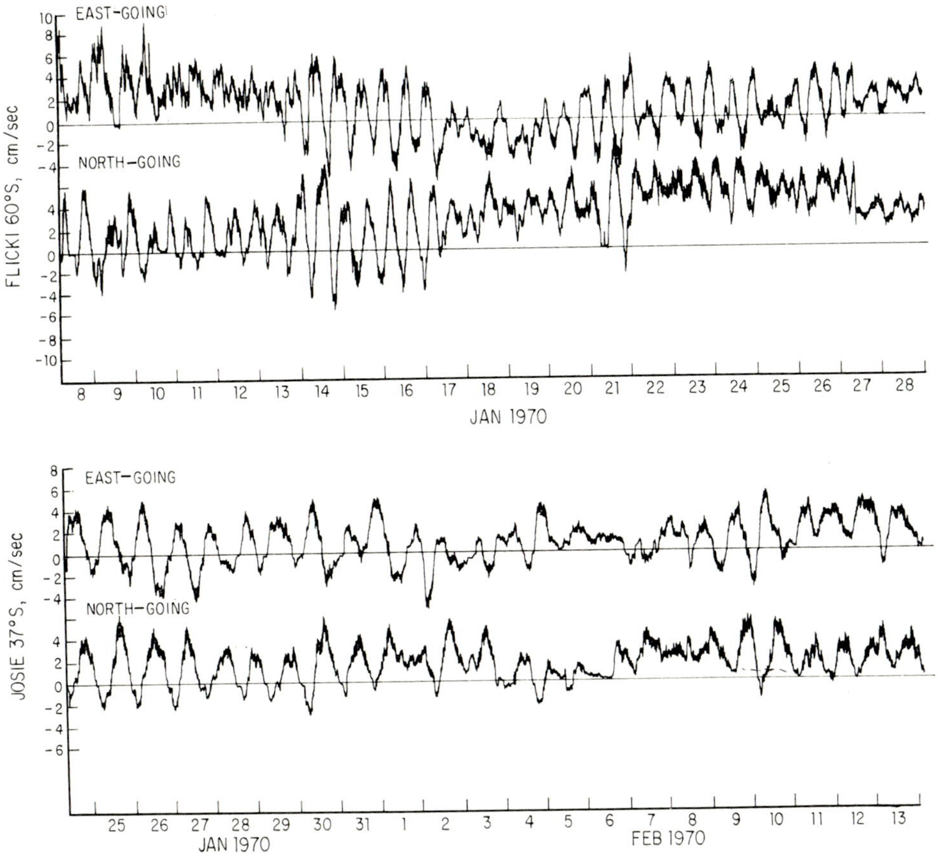

Fig. 5. The east-going and north-going components of velocity.

nation between predominately diurnal type tides (e.g., January 15–21) and semidiurnal type tides (January 27–30) can be followed from Australia to Antarctica.

In addition to the three deep-sea capsule records, several shallow water bottom pressure records were obtained on the Australian shelf by the Horace Lamb Centre for Oceanographic Research [*Nilsson*, 1971]. During and following *Eltanin* cruise 41, three shelf records were obtained 20 miles off the south coast of Australia. The first two records were less than 1 lunar month, whereas the last was of a 30-day duration.

Velocity Measurements

The magnitude of the abyssal current was measured with an indirectly heated thermister sensor [*Caldwell*, 1968]. This sensor, built for 1-month, low velocity measurements on the sea floor, operates on the same principle as a hot wire anemometer. The sensor performs best at velocities below 1 cm/sec where mechanical devices usually fail.

The direction of flow was measured with a standard oceanographic vane from Hydro Products that was adapted for use on the deep-sea capsule. In our previous work, multiple horizontal thermister sensors were used to determine the current direction; these had the advantage of no moving parts. However, this scheme led to occasional 180° ambiguities and was abandoned in favor of a traditional vane. The principal defect of the vane was that it stalled at velocities of less than 1 cm/sec, and these sections had to be patched out.

Figure 5 shows the east-going and north-going velocity components at Josie 37°S and Flicki 60°S that were calculated from the speed and direction records. The currents were measured in the bottom boundary layer, but the directions were not corrected for Ekman veering, and the velocities were not corrected to free stream velocitites. *Wimbush and Munk* [1970] measured the veering in the bottom boundary layer and found it almost nonexistent; the measured currents were about 80% of the free stream velocity.

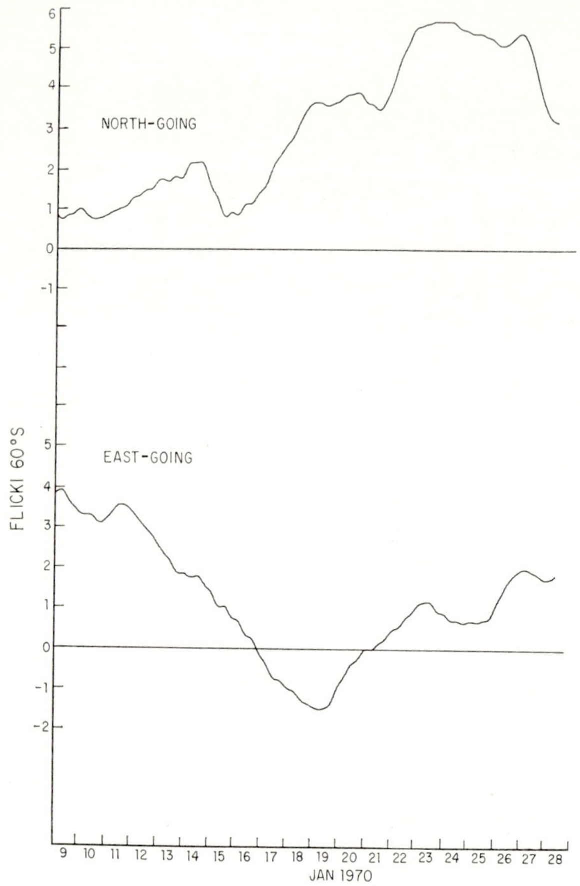

Fig. 6. The variation in mean flow for Josie 37°S.

MEAN CURRENTS

Figures 6 and 7 show the mean currents for Josie 37°S and Flicki 60°S, respectively; the means were obtained by low-passing the velocity records with a frequency cutoff at 0.2 cpd. The extreme variability of the flow is clearly evident. The circumpolar current would appear as a positive east-going velocity. In the Flicki 60°S record (Figure 7), the east-going velocity component changes from 4 cm/sec on January 9 to −2 cm/sec on January 19 and then back to 2 cm/sec by the end of the record on January 28, 1970. The north-going component increases from 1 cm/sec to over 5 cm/sec during the same period.

At 60°S, *Callahan* [1971] predicts a mean zonal east-going component of flow of 2.8 cm/sec and a northward component of 3.7 cm/sec by interpolating records from deep moored current meters as 55°S and 63°S. The mean of the entire Flicki 60°S velocity record indicates a 1.25 cm/sec east-going and 2.9 cm/sec north-going current. The deep moored records were of a 1- and 2-day duration and were taken near January 10 and 16, 1970, when the eastward velocity was at a high. If the records had been taken for 48 hours around January 19, the resulting mean would be 1 cm/sec west-going. Hence the danger of using short current records for geostrophic transport calculations is clearly evident.

The Flicki 60°S records show variations that would be expected from wave disturbances in zonal baroclinic current. Deep moored current records taken by J. L. Reid (personal communication, 1971) south of New Zealand showed similar variations in mean flow. *Bowen and Stommel* [1968] examined the geostrophic flow through the Drake passage and concluded that the circumpolar current is suprisingly steady but that this steadiness might be caused by the constriction of flow in Drake passage.

The currents at Josie 37°S (Figure 6) were not as strong or as variable as those at Flicki 60°S. The mean east-going component was 0.9 cm/sec, and the

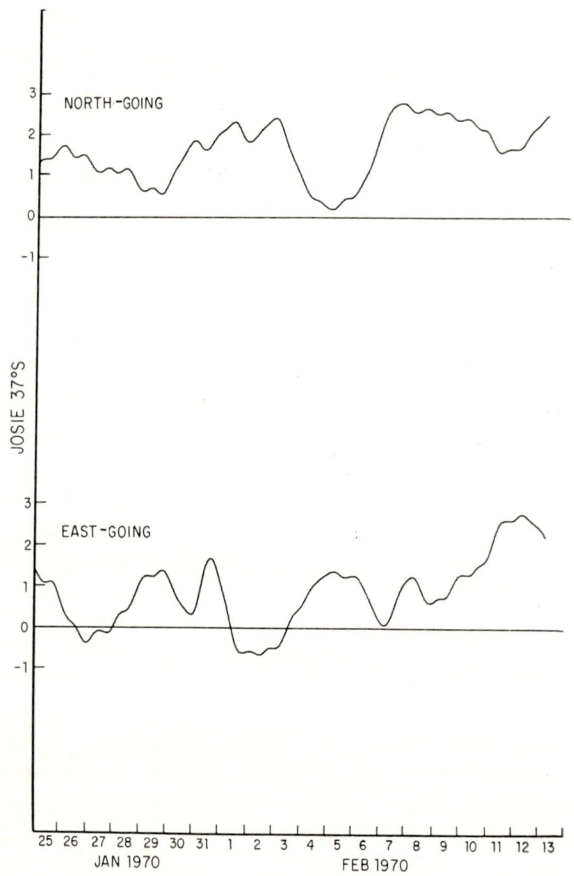

Fig. 7. The variation in mean flow for Flicki 60°S.

TABLE 2. Inertial Motion

Record Harmonic	Frequency, cpd	U Amp, cm/sec	U Phase, °	V Amp, cm/sec	V Phase, °	Relative Phase, °	Ellipticity, ϵ
			Josie 37°S (f_i = 1.203 cpd)				
24	1.1520	0.2505	49.19	0.3314	166.69	117.50	0.76
25	1.2000	1.0037	108.31	0.8118	−149.04	102.65	1.24
26	1.2480	1.5152	163.00	1.4468	− 97.81	99.19	1.05
27	1.2960	0.3576	41.24	0.2786	139.65	98.41	1.28
			Flicki 60°S (f_i = 1.732 cpd)				
35	1.6730	0.2302	−177.42	0.2167	− 92.29	85.13	1.07
36	1.7208	0.4522	− 72.26	0.7056	26.14	98.41	0.64
37	1.7686	0.7301	− 23.03	0.7042	82.10	105.13	1.04
38	1.8164	0.6353	114.61	0.7717	−144.29	101.10	0.82
39	1.8642	0.7060	− 75.93	0.9976	− 7.46	68.47	0.71

The relative phase equals the V phase minus the U phase (positive implies counterclockwise rotation); the ellipticity ϵ equals U amp/V amp; f_i is the inertial frequency.

north-going was 1.6 cm/sec. From standard hydrographic data, *Bye* [1968] discusses the existence of a large eddy in the South Australian basin with an eastern boundary current called the Flinders current flowing along Tasmania to South Australia. From this discussion, northwest-going currents would be expected, whereas the Josie 37°S currents were northeast. However, north of 42°S, *Bye et al.* [1968] have shown that the current is variable and sensitive to seasonal variations and at times can be absent.

INERTIAL MOTION

The velocity records (Figure 5) are dominated by inertial currents. The difference in inertial period between the two stations at different latitudes is clearly evident. In both cases, the observed inertial peak appears 'blue shifted' to higher frequencies. The shift appears greater for Josie 37°S, although the Flicki 60°S shift is somewhat confused by the semidiurnal tides. The Josie 37°S peak appears shifted about 0.027 cpd higher to a frequency of 1.230 cpd. *Munk*

TABLE 3. Bottom Pressure at a Depth of 5550 Meters at Station Josie 37°S (37°01'S, 132°01'E) for a Period of 21 Days (January 24–February 14, 1970) Contrasted with 'Predicted' Bottom Pressure from Macquarie Island

Admittances (Station/Reference)*					Principal Harmonic Constituents					
cpd	Real	Imaginary	R	ϕ, °	Symbol	cpd	Station H, cm	Station G, °	Reference H, cm	Reference G, °
0.8929346	1.2045	−0.5346	1.3178	−23.93	Q1	0.8932441	2.48	225.77	1.89	201.81
0.9295357	1.7696	−0.3647	1.8067	−11.65	O1	0.9295357	13.50	218.17	7.48	206.52
0.9661368	2.3494	−0.2993	2.3684	−7.26						
					P1	0.9972621	5.87	235.37	2.90	231.11
1.0027379	1.9700	−0.0881	1.9719	−2.56	K1	1.0027379	17.61	230.73	8.93	228.17
Recorded variance = 843.52 cm²										
Residual variance = 17.28 cm²										
1.8956725	−0.3783	−0.2337	0.4447	−148.30	N2	1.8959820	3.01	167.02	6.80	18.61
1.9322736	0.4862	−0.1567	0.5108	−17.86	M2	1.9322736	14.20	63.66	27.80	45.80
1.9688747	1.8409	−0.4861	1.9040	−14.79						
					S2	2.0000000	16.08	99.53	7.86	73.77
2.0054758	1.6342	−0.9450	1.8877	−30.04	K2	2.0054758	4.44	98.80	2.35	68.76
Recorded variance = 747.65 cm²										
Residual variance = 4.62 cm²										

* Interval is 1 cpm (0.0366011 cpd).
ϕ, station lead; G, Greenwich epoch.

TABLE 4. Bottom Pressure at a Depth of 3440 Meters at Station Kathy 50°S (50°02′S, 132°09′E) for a Period of 32 Days (January 3–February 4, 1970) Contrasted with 'Predicted' Bottom Pressure from Macquarie Island

Admittances (Station/Reference)*					Principal Harmonic Constituents					
cpd	Real	Imaginary	R	ϕ, °	Symbol	cpd	Station H, cm	Station G, °	Reference H, cm	Reference G, °
0.8929346	1.4004	−0.4450	1.4694	−17.63	Q1	0.8932441	2.78	219.52	1.89	201.81
0.9295357	1.5382	−0.3676	1.5815	−13.44	O1	0.9295357	11.81	219.96	7.47	206.52
0.9661368	1.4706	−0.0990	1.4739	−3.85						
					P1	0.9972621	4.13	234.88	2.90	231.11
1.0027379	1.4258	−0.1019	1.4294	−4.09	K1	1.0027379	12.76	232.26	8.93	228.17
Recorded variance = 656.13 cm²										
Residual variance = 5.43 cm²										
1.8956725	0.1140	−0.2081	0.2373	−61.30	N2	1.8959820	1.61	79.30	6.80	18.61
1.9322736	0.4173	−0.1485	0.4429	−19.58	M2	1.9322736	12.31	65.38	27.80	45.80
1.9688747	0.8228	−0.4675	0.9464	−29.60						
					S2	2.0000000	9.03	107.90	7.86	73.77
2.0054758	0.9256	−0.6203	1.1143	−33.83	K2	2.0054758	2.62	102.59	2.35	68.76
Recorded variance = 193.73 cm²										
Residual variance = 2.10 cm²										

* Interval is 1 cpm (0.0366011 cpd).
ϕ, station lead; G, Greenwich epoch.

and Phillips [1968] predict an increase in the frequency of the inertial peak depending on the zonal and vertical wave numbers. By selecting a vertical mode of 10 and using the observed frequency shift, the resulting east-west wave number of 220 cycles around the earth is not inconsistent with their reported results.

Table 2 lists the amplitude and phase for both components of velocity across the inertial band for the two capsules. The sense of rotation is counterclockwise as expected in the Southern Hemisphere, but the phase is about 100° in both cases instead of 90°. This difference is not understood. The amplitudes of

TABLE 5. Bottom Pressure at a Depth of 4710 Meters at Station Flicki 60°S (60°01′S, 132°07′E) for a Period of 21 Days (January 8–29, 1970) Contrasted with 'Predicted' Bottom Pressure from Macquarie Island

Admittances (Station/Reference)*					Principal Harmonic Constituents					
cpd	Real	Imaginary	R	ϕ, °	Symbol	cpd	Station H, cm	Station G, °	Reference H, cm	Reference G, °
0.8929346	1.7966	0.4705	1.8571	14.68	Q1	0.8932441	3.50	187.33	1.89	201.81
0.9295357	2.0306	−0.2853	2.0505	−8.00	O1	0.9295357	15.32	214.52	7.47	206.52
0.9661368	2.3573	−0.3602	2.3846	−8.69						
					P1	0.9972621	5.95	228.49	2.90	231.11
1.0027279	1.9484	0.1853	1.9572	5.43	K1	1.0027379	17.48	222.74	8.93	228.17
Recorded variance = 1051.47 cm²										
Residual variance = 8.44 cm²										
1.8956725	0.3955	−0.3866	0.5531	−44.34	N2	1.8959820	3.76	62.86	6.80	18.61
1.9322736	0.5559	−0.3602	0.6624	−32.95	M2	1.9322736	18.41	78.75	27.80	45.80
1.9688747	0.8370	−0.5811	1.0189	−34.77						
					S2	2.0000000	9.87	120.25	7.86	73.77
2.0054758	0.8198	−0.9636	1.2651	−49.61	K2	2.0054758	2.97	118.37	2.35	68.76
Recorded variance = 600.19 cm²										
Residual variance = 0.42 cm²										

* Interval is 1 cpm (0.0366011 cpd).
ϕ, station lead; G, Greenwich epoch.

TABLE 6. East-going Bottom Velocity at a Depth of 5550 Meters at Station Josie 37°S (37°01'S, 132°01'E) for a Period of 21 Days (January 24–February 14, 1970) Contracted with 'Predicted' Bottom Pressure at the Same Place (Table 3)

Admittances (Station/Reference)					Principal Harmonic Constituents					
cpd	Real, sec^{-1}	Imaginary, sec^{-1}	R, sec^{-1}	ϕ, °	Symbol	cpd	Station H, cm/sec	Station G, °	Reference H, cm	Reference G, °
0.8929346	0.00681	0.00914	0.01139	−126.69	Q1	0.8932441	0.028	352.46	2.48	225.77
0.9295357	0.00681	0.00914	0.01139	−126.69	O1	0.9295357	0.154	344.86	13.50	218.17
0.9661368	0.00681	0.00914	0.01139	−126.69						
					P1	0.9972621	0.067	2.06	5.87	235.37
1.0027379	0.00681	0.00914	0.01139	−126.69	K1	1.0027379	0.201	357.42	17.61	230.73
Recorded variance = 0.822 cm^2/sec^2										
Residual variance = 0.659 cm^2/sec^2										
1.8956725	0.00579	0.01200	0.01333	−115.77	N2	1.8959820	0.040	282.79	3.01	167.02
1.9322736	0.00579	0.01200	0.01333	−115.77	M2	1.9322736	0.189	179.43	14.20	63.66
1.9688747	0.00579	0.01200	0.01333	−115.77						
					S2	2.0000000	0.214	215.30	16.08	99.53
2.0054758	0.00579	0.01200	0.01333	−115.77	K2	2.0054758	0.059	150.34	4.44	98.80
Recorded variance = 0.432 cm^2/sec^2										
Residual variance = 0.296 cm^2/sec^2										

ϕ, station lead; G, Greenwich epoch.

the two components are not equal, as expected in circular motion, but a pressure gradient could cause this ellipticity. For Josie 37°S, the ellipticity ($\epsilon = U/V$) is about 1.1; for Flicki 60°S, the ellipticity is about 0.8.

TIDAL ANALYSIS

The three deep-sea pressure records were analyzed by the response method of *Cartwright et al.* [1969] by using a noise-free predictor series. This method is based on the use of a reference station with a record of a least 1 year. This station is chosen with tides as similar to the deep-sea records as possible.

Tables 3, 4, and 5 give the results for the three stations based on the Macquarie Island reference station. The admittances at cycle per month intervals are lis-

TABLE 7. North-going Bottom Velocity at a Depth of 5550 Meters at Station Josie 37°S (37°01'S, 132°01'E)) for a Period of 21 Days (January 24–February 14, 1970) Contrasted with 'Predicted' Bottom Pressure at the Same Place (Table 3)

Admittances (Station/Reference)					Principal Harmonic Constituents					
cpd	Real, sec^{-1}	Imaginary, sec^{-1}	R, sec^{-1}	ϕ, °	Symbol	cpd	Station H, cm/sec	Station G, °	Reference H, cm	Reference G, °
0.8929346	0.01631	−0.00223	0.01646	−7.79	Q1	0.8932441	0.041	233.56	2.48	225.77
0.9295357	0.01631	−0.00223	0.01646	−7.79	O1	0.9295357	0.222	225.96	13.50	218.17
0.9661368	0.01631	−0.00223	0.01646	−7.79						
					P1	0.9972621	0.097	243.16	5.87	235.37
1.0027379	0.01631	−0.00223	0.01646	−7.79	K1	1.0027379	0.290	238.52	17.61	230.73
Recorded variance = 0.762 cm^2/sec^2										
Residual variance = 0.570 cm^2/sec^2										
1.8956725	0.00113	−0.00348	0.00366	−72.09	N2	1.8959820	0.011	239.11	3.01	167.02
1.9322736	0.00113	−0.00348	0.00366	−72.09	M2	1.9322736	0.052	135.75	14.20	63.66
1.9688747	0.00113	−0.00348	0.00366	−72.09						
					S2	2.0000000	0.059	171.62	16.08	99.53
2.0054758	0.00113	−0.00348	0.00366	−72.09	K2	2.0054758	0.016	170.89	4.44	98.80
Recorded variance = 0.266 cm^2/sec^2										
Residual variance = 0.241 cm^2/sec^2										

ϕ, station lead; G, Greenwich epoch.

TABLE 8. East-going Bottom Velocity at a Depth of 4710 Meters at Station Flicki 60°S (60°01′S, 132°07′E) for a Period of 21 Days (January 8–29, 1970) Contrasted with 'Predicted' Bottom Pressure at the Same Place (Table 5)

Admittances (Station/Reference)					Principal Harmonic Constituents					
cpd	Real, sec^{-1}	Imaginary, sec^{-1}	R, sec^{-1}	ϕ, °	Symbol	cpd	Station H, cm/sec	Station G, °	Reference H, cm	Reference G, °
0.8929346	0.02662	0.01032	0.02855	−158.18	Q1	0.8932441	0.100	345.51	3.50	187.33
0.9295357	0.02662	0.01032	0.02855	−158.18	O1	0.9295357	0.437	12.70	15.32	214.52
0.9661368	0.02662	0.01032	0.02855	−158.18						
					P1	0.9972621	0.170	26.67	5.95	228.49
1.0027379	0.02662	0.01032	0.02855	−158.18	K1	1.0027379	0.499	20.92	17.48	222.74
Recorded variance = 0.833 cm^2/sec^2										
Residual variance = 0.058 cm^2/sec^2										
1.8956725	0.02974	0.00649	0.03044	−167.69	N2	1.8959820	0.114	230.55	3.76	62.86
1.9322736	0.02974	0.00649	0.03044	−167.69	M2	1.9322736	0.560	246.44	18.41	78.75
1.9688747	0.02974	0.00649	0.03044	−167.69						
					S2	2.0000000	0.300	287.94	9.87	120.25
2.0054758	0.02974	0.00649	0.03044	−167.69	K2	2.0054758	0.090	286.06	2.97	118.37
Recorded variance = 4.878 cm^2/sec^2*										
Residual variance = 4.176 cm^2/sec^2*										

* Includes some inertial energy.
ϕ, station lead; G, Greenwich epoch.

ted on the left, and the amplitude and phase lags for standard frequencies are listed on the right. Tables 6–9 give similar information for the two components of velocity at the two stations. These velocity constants are referred to the local predicted pressure that is calculated from the response function of Tables 3 and 5 and the Macquarie Island reference series. However, the restraint of constant admittance across each tidal band was imposed to reduce the effects of barotropic noise.

R. Radok (personal communication, 1971) has analyzed two shelf records for tides by conventional tech-

TABLE 9. North-going Bottom Velocity at a Depth of 4710 Meters at Station Flicki 60°S (60°01′S, 132°07′E) for a Period of 21 Days (January 8–29, 1970) Contrasted with 'Predicted' Bottom Pressure at the Same Place (Table 5)

Admittances (Station/Reference)					Principal Harmonic Constituents					
cpd	Real, sec^{-1}	Imaginary, sec^{-1}	R, sec^{-1}	ϕ, °	Symbol	cpd	Station H, cm/sec	Station G, °	Reference H, cm	Reference G, °
0.8929346	0.01566	−0.00955	0.01834	−31.39	Q1	0.8932441	0.064	218.72	3.50	187.33
0.9295357	0.01566	−0.00955	0.01834	−31.39	O1	0.9295357	0.281	245.91	15.32	214.52
0.0661368	0.01566	−0.00955	0.01834	−31.39						
					P1	0.9972621	0.109	259.88	5.95	228.49
1.0027379	0.01566	−0.00955	0.01834	−31.39	K1	1.0027379	0.321	254.13	17.48	222.74
Recorded variance = 0.445 cm^2/sec^2										
Residual variance = 0.128 cm^2/sec^2										
1.8956725	−0.00705	−0.00920	0.01159	−127.47	N2	1.8959820	0.044	190.33	3.76	62.86
1.9322736	−0.00705	−0.00920	0.01159	−127.47	M2	1.9322736	0.213	206.22	18.41	78.75
1.9688747	−0.00705	−0.00920	0.01159	−127.47						
					S2	2.0000000	0.114	247.72	9.87	120.25
2.0054758	−0.00705	−0.00920	0.01159	−127.47	K2	2.0054758	0.034	245.84	2.97	118.37
Recorded variance = 5.069 cm^2/sec^2*										
Residual variance = 5.029 cm^2/sec^2*										

* Includes some inertial energy.
ϕ, station lead; G, Greenwich epoch.

niques [*Noye and Easton,* 1967]. Both records are short, and the tide constants (Table 10) vary considerably. Perhaps better constants could be obtained by the response techniques used on the deep-sea capsule data.

MEASUREMENTS ONSHORE

For comparision with the deep-sea measurements, a summary of the results from Australian and antarctic coastal stations is useful. The amplitude and phase lag of the observed tides are shown along the antarctic coast south of Australia in Figure 8 and along the south coast of Australia in Figure 9. Refer to Figure 1 and Tables 11 and 12 for the locations of the stations and tidal constants for Antarctica and Australia, respectively.

The antarctic tidal constants are limited by record length and quality. There are no records of more than 1-month duration (the Cape Denison record is not continuous). The technical difficulties in obtaining good

TABLE 10. Horace Lamb Centre Stations

	34°39′S, 134°46′E January 31, 1970*	34°39′S, 134°46′E May 13, 1970⁺
	Component O1	
H, cm	13.38	11.61
G, °	217.3	232.7
κ, °	82.6	97.9
	Component K1	
H, cm	18.56	19.95
G, °	259.1	240.9
κ, °	124.3	106.1
	Component M2	
H, cm	16.12	19.48
G, °	79.3	77.7
κ, °	169.8	168.2
	Component S2	
H, cm	21.15	18.84
G, °	138.9	103.7
κ, °	229.4	194.2

* Record length is 19 days.
⁺ Record length is 11 days.

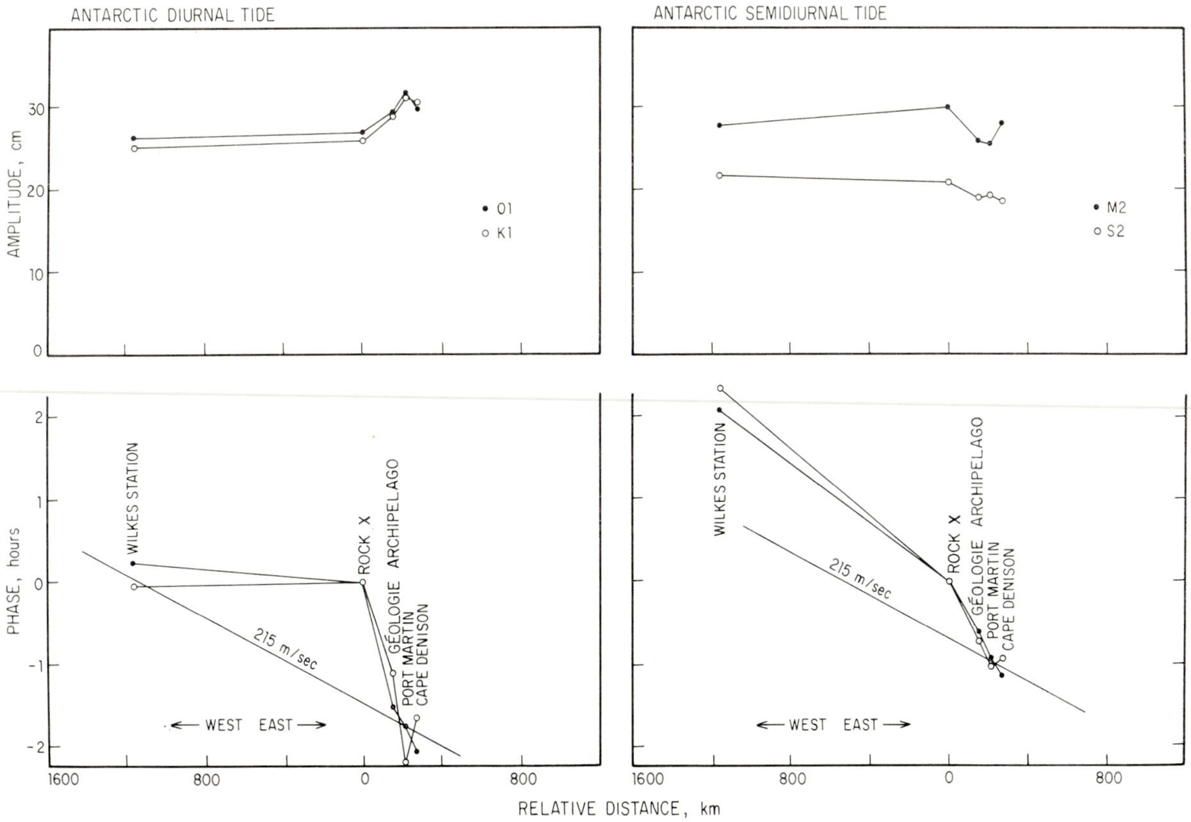

Fig. 8. The amplitude and phase of the tides along the coast of Antarctica. The phase lag is in hours later than Rock X. The antarctic Kelvin velocity (215 m/sec) is shown for reference.

continuous records are great. The gages are generally located on the ice shelf and use the ice itself as a float. The tide is then measured by recording the depth by a wire weighted to the sea floor. The effects of the ice are not well known, and the ice is moving and changing thickness with the season. The use of deep-sea instrumented capsules under the ice for periods of 1 year would solve these difficulties.

The Australian stations provide good continuous data of several years in duration. *Easton* [1968] has tabulated the tide gages of Australia, collected all the recent records, and analyzed them by Fourier techniques [*Easton*, 1970].

DESCRIPTION OF TIDES

The observed tide wave progresses westward along the antarctic coast at a speed equal to or less than the free shallow water velocity; $C = (gD)^{1/2} = 215$ m/sec, where $D = 4.7$ km is a representative depth off the antarctic shelf. This value is the direction (with the coast to the left) and velocity for a Kelvin wave trapped against the coast by the rotation of the earth. The observed phase change is not smooth, especially the reversal at Cape Denison and the 'hump' at Rock X. By ignoring such irregularities in phase, the propagation velocity along the coast becomes O1 = 192 m/sec = 0.9 antarctic Kelvin, K1 = 224 m/sec = 1.0 antarctic Kelvin, M2 = 135 m/sec = 0.6 antarctic Kelvin, and S2 = 118 m/sec = 0.6 antarctic Kelvin. The antarctic Kelvin number is in units of antarctic Kelvin velocity (215 m/sec).

Along the Australian coast, the diurnal tide progresses from west to east with a speed equal to the Australian Kelvin velocity ($C = 236$ m/sec where $D = 5.6$ km; i.e., O1 = 233 m/sec = 1.0 Australian Kelvin, and K1 = 219 m/sec = 0.9 Australian Kelvin).

The semidiurnal tide appears almost normally incident on the coast, and the phase is delayed and the height amplified by the shelf in the area of Port Lincoln in the Spencer Gulf. Eliminating the three stations centered at Port Lincoln gives a propagation velocity of M2 = −433 m/sec = 2.0 Australian Kel-

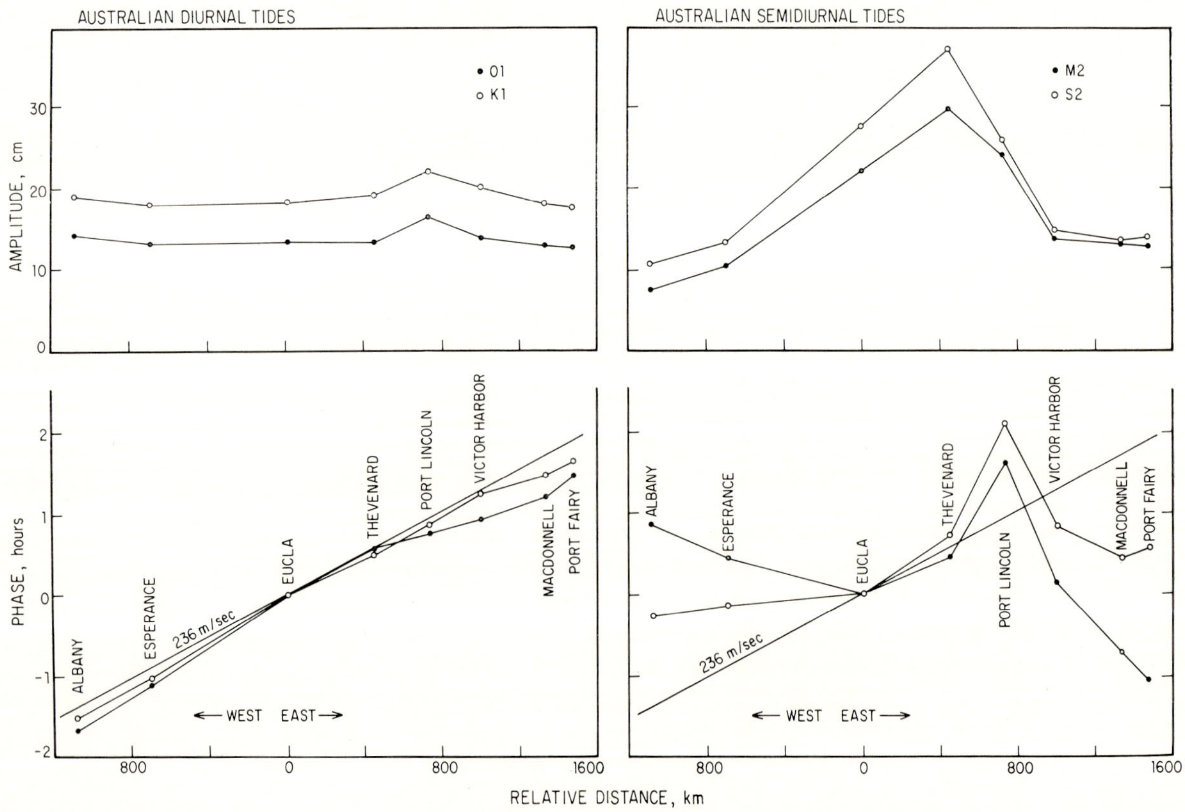

Fig. 9. The amplitude and phase of the tides along the south coast of Australia. The phase lag is in hours later than Eucla. The Australian Kelvin velocity (236 m/sec) is shown for reference.

TABLE 11. Antarctic Tide Stations South of Australia

	Wilkes Station 66°15′S, 110°31′E	Rock X 66°20′S, 136°42′E	Géologie Archipelago 66°40′S, 140°01′E	Port Martin 66°50′S, 141°25′E	Cape Dension 67°00′S, 142°40′E
			Component O1		
H, cm	26.33	27	29.4	31.7	29.9
G, °	232.4	229	208	204.6	200
κ, °	342.9	6	348	346.0	343
			Component K1		
H, cm	25.15	26	29	31.4	30.7
G, °	244.2	245	228	211.8	220
κ, °	354.7	22	8	353.2	3
			Component M2		
H, cm	27.71	30	25.9	25.5	28
G, °	140.3	80	62	52.3	47
κ, °	1.3	353	342	335.1	332
			Component S2		
H, cm	16.83	16	13.9	14.3	13.4
G, °	200.2	130	108	98.7	102
κ, °	61.2	43	28	21.5	27

Local epoch κ equals the lag in degrees behind local equilibrium tide; Greenwich epoch $G = \kappa + sl$; l is west longitude; and species s is 1 or 2 for diurnal and semidiurnal tides, respectively.

The record lengths, dates, and distances relative to Rock X for the above locations are as follows: Wilkes Station, 29 days, 1959, −1162 km; Géologie Archipelago, 2 × 29 days, 1952, 152 km; Port Martin, 3 × 29 days, 1950, 216 km; Cape Denison, 98 days, 1919, 273 km.

vin and S2 = 864 m/sec = 4.0 Australian Kelvin. The M2 component seems to travel in the opposite direction from the Kelvin wave, and both semidiurnal components move at several times the Kelvin velocity.

The amplitude and phase were then interpolated to 132°E along the antarctic and Australian coasts. The coastal and capsule constants then yield a cross-channel plot of amplitude and phase (Figure 10). Here the semidiurnal tide shows a sudden change in phase between the Josie 37°S station and the Australian coast. The amplitude and phase of the diurnal tide is fairly smooth across the channel.

THEORETICAL CONSIDERATIONS

For guidance in explaining the Australian-antarctic tides, a simple channel model with a constant Coriolis parameter and depth was chosen. Admittedly this model is poor, since the Indian Ocean is not continuously bounded on the north by Australia and the Coriolis parameter and depth vary considerably across the channel.

Consider a channel of constant depth with Antarctica and Australia as its southern and northern boundaries, respectively. Take the origin at the antarctic coast at 132°E, and let $\hat{x}$ be in the eastward and $\hat{y}$ in the northward direction. The variables U and V are the corresponding components of velocity; α is the eastward and β is the northward wave number. The equations of motion in a bottom of sea reference system [*Munk et al.*, 1970, equation 5] are

$$\begin{aligned} \partial_t U - fV + g\partial_x \eta &= g\partial_x \eta_E \\ \partial_t V + fU + g\partial_y \eta &= g\partial_y \eta_E \\ \partial_t \eta + \partial_y(Ud) + \partial_y(Vd) &= 0 \end{aligned} \tag{1}$$

where f is the Coriolis parameter and equals −2 sin (latitude) in cycles per day, η is the displacement of the sea surface relative to the sea bottom, and η_E is the equilibrium response of the sea surface to the tide potential. The solutions are found by adding free and forced solutions as if the sea floor were fixed and the tidal yield of the sea floor were absorbed in the forcing function $g \cdot \nabla \eta_E$.

Now consider the homogeneous equations and seek solutions of the form

$$\eta = \eta(y) \exp(i\alpha x - i\omega t) \tag{2}$$

where ω is the frequency, α and ω are real, and ω is positive. In terms of η, the velocity components become

$$\begin{pmatrix} iV \\ U \end{pmatrix} = \frac{g}{\omega^2 - f^2} \begin{pmatrix} \alpha f + \omega \partial_y \\ \alpha\omega + f\partial_y \end{pmatrix} \eta \tag{3}$$

TABLE 12. South Coast Australian Tide Stations

	Albany 35°02′S, 117°53′E	Esperance 33°52′S, 121°54′E	Eucla 31°43′S, 128°52′E	Thevenard 32°09′S, 133°39′S	Port Lincoln 34°43′S, 135°52′E	Victor Harbor 35°34′S, 138°38′E	Port MacDonnell 38°03′S, 140°42′E	Port Fairy 38°24′S, 142°15′E
				Component O1				
H, cm	14.30	13.32	13.38	13.47	16.64	14.05	13.14	12.89
G, °	195.5	203.0	218.7	225.9	229.3	231.6	235.6	239.4
κ, °	77.6	81.1	89.8	92.2	93.4	93.0	94.9	97.1
				Component K1				
H, cm	19.14	18.01	17.98	19.32	24.14	20.21	18.20	17.95
G, °	208.5	215.8	231.5	238.7	244.5	250.1	253.6	256.5
κ, °	90.6	93.9	102.6	105.0	108.6	111.5	112.9	114.2
				Component M2				
H, cm	7.53	10.45	22.37	29.90	24.05	13.75	13.01	12.92
G, °	97.5	84.7	71.8	85.1	118.8	76.0	51.1	41.2
κ, °	221.7	200.9	174.0	177.7	207.0	158.8	129.7	116.6
				Component S2				
H, cm	10.79	13.53	27.89	37.46	26.27	15.00	13.81	14.11
G, °	90.0	93.7	97.7	119.8	161.2	122.9	111.8	115.0
κ, °	214.2	209.9	199.9	212.4	249.4	205.7	190.4	190.4

Local epoch κ equals the lag in degrees behind local equilibrium tide; Greenwich epoch $G = \kappa + sl$; l is west longitude; and species s is 1 or 2 for diurnal and semidiurnal tides, respectively.

The distances relative to Eucla for the above locations are as follows: Albany, —1084 km; Esperance, —694 km; Thevenard, 454 km; Port Lincoln, 731 km; Victor Harbor, 1000 km; Port MacDonnell, 1331 km; and Port Fairy, 1472 km.

By assuming a constant depth $d = D$, equation 1 can be combined into

$$[gD(\partial_{yy} - \alpha^2) + \omega^2 - f^2]\ \eta = 0 \qquad (4)$$

By seeking cross-channel solutions of the form

$$\eta(y) = \tfrac{1}{2}(A - B)\ \exp(i\beta y) + \tfrac{1}{2}(A + B)\ \exp(-i\beta y) \qquad (5)$$

equation 4 becomes

$$gD(\alpha^2 + \beta^2) = \omega^2 - f^2 \qquad (6)$$

The boundary condition at $y = 0$ and $y = L$,

$$(\alpha f + \omega\partial y)\ \eta = 0 \qquad (7)$$

where L is the channel width, prevents flow into the channel sides. Substituting (5) into (7) yields

$$(\alpha^2 f^2 + \beta^2\omega^2)\,[\exp(i\beta L) - \exp(-i\beta L)] = 0 \qquad (8)$$

From equation 6, β is either pure real or pure imaginary, and this option determines whether $\eta(y)$ is trigonometric or exponential.

Consider β as pure real. Then the relevant solution is

$$\exp(i\beta L) - \exp(-i\beta L) = 0 \qquad (9)$$

Equations 6 and 9 yield

$$\beta = (\pi/L)n \qquad n = 0, 1, 2, \ldots$$
$$\alpha^2 = [(\omega^2 - f^2)/gD] - (\pi^2/L^2)n^2 \qquad (10)$$

This result is the Poincaré continuum of *Munk et al.* [1970] and is quantized by the channel. For the lowest mode ($n = 1$), a channel width of 3780 km, the frequency of the wave with $\alpha = 0$ must be $\geqslant$ 3.0 cpd. Conversely, if the wave frequency is 2.0 cpd, the channel must be $\geqslant$ 9247 km wide to support a Poincaré wave. Therefore, there are no Poincaré waves at tidal frequencies in the simple channel model.

Consider β as pure imaginary. Then the relevant solution is

$$\alpha^2 f^2 + \beta^2\omega^2 = 0 \qquad (11)$$

Equations 6 and 11 yield

$$\alpha^2 = \omega^2/gD$$
$$\beta^2 = -f^2/gD \qquad (12)$$

This result is the Kelvin solution for the simple channel. The wave progresses along the coast with a velocity of $\omega/\alpha = (gD)^{1/2}$, which is the free shallow water wave velocity. The wave decays exponentially toward the center of the channel at a rate of $f/(gD)^{1/2}$. For 60°S, this rate equals 1700 km or about 13° of latitude. Hence the wave decays to $1/e$ midway across the channel and to 0.11 at the opposite side, so little interaction occurs between the wave along Australia and the one along Antarctica. Figure 11 shows the $\alpha - \omega$ dispersion diagram for the channel model.

Since the simple channel theory predicts that only the Kelvin wave exists, the propagation velocity along the coast should equal the Kelvin velocity. For the

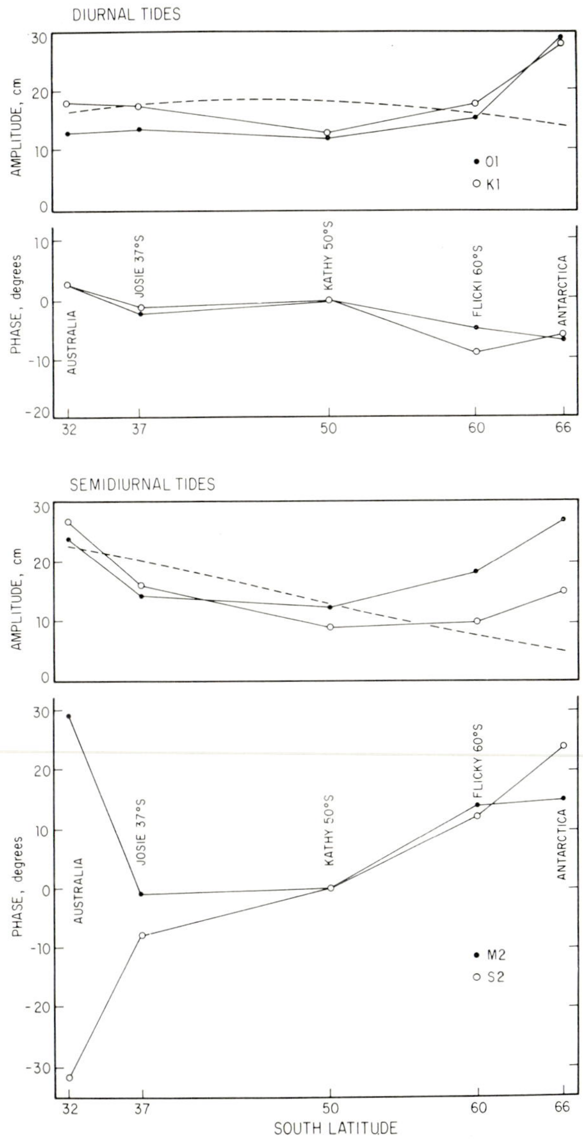

Fig. 10 The amplitude and phase of the tides on a section from Australia to Antarctica. The equilibrium tide for the K1 and M2 components is indicated by the broken line.

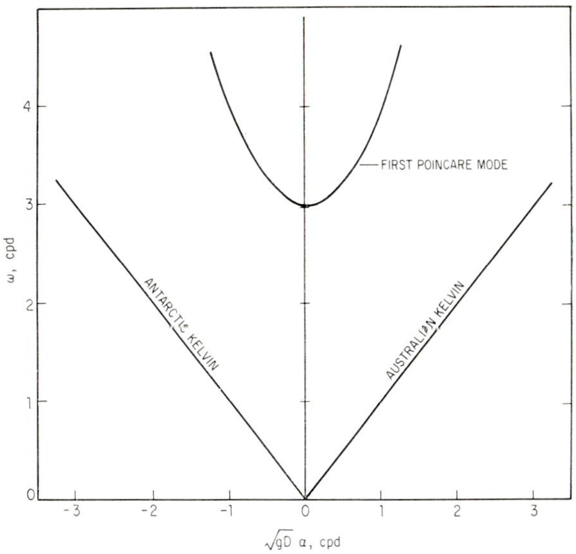

Fig. 11. The $\alpha - \omega$ dispersion diagram for the simple channel model.

diurnal tides along Australia and Antarctica (Figures 8 and 9), the propagation velocity almost equals the Kelvin velocity. If an antarctic shelf were added, Robinson-Rhines shelf waves would appear along the shelf edge. Since the observed velocity is nearly Kelvin, the shelf waves do not appear important at the antarctic coast.

For the semidiurnal tides (Figures 8 and 9), the observed velocities are too small along the antarctic coast and too great along the Australian coast. *Munk et al.* [1970] were able to combine a Poincaré and Kelvin wave to adjust the along shore propagation velocity. The channel quantizes the Poincaré wave, and no solution exists. The shelf waves have a frequency cutoff below the semidiurnal tides for both Australia and Antarctica. Therefore there are no additional solutions, and the simple model does not explain the semidiurnal tides.

The observed phase of the semidiurnal tide along the Australian coast (Figure 9) indicates that dissipation is occurring, this dissipation requiring a more complex theory than the diurnal case. This apparent difference in shelf dissipation cannot be ascribed to a small diurnal to semidiurnal ratio. Therefore, a model including dissipations seems necessary.

Acknowledgments. This work has been supported by the Office of Naval Research and by the Antarctic Research Program of the National Science Foundation.

REFERENCES

Bowen, J. L., and H. Stommel, On the steadiness of the antarctic circumpolar current, paper presented at the Symposium on Antarctic Research, Amer. Ass. for the Advan. of Sci., Dallas, Dec. 26–31, 1968.

Bye, J. A. T., Hydrology of the Southern Ocean, 1, Australian section–north, *Res. Pap. 26,* Horace Lamb Centre, Flinders Univ. of S. Aust., 1968.

Bye, J. A. T., A. Tamuly, and I. M. Thomas, *Cruise Rep. HLC 3,* Horace Lamb Centre, Flinders Univ. of S. Aust., 1968.

Caldwell, D. R., Indirectly heated thermister for measuring ocean currents, *Rev. Sci. Instrum., 39,* 1865–1869, 1968.

Caldwell, D. R., F. E. Snodgrass, and M. H. Wimbush, Sensors in the deep sea, *Phys. Today, 22,* 34–42, 1969.

Callahan, J. E., Velocity structure and flux of the antarctic circumpolar current south of Australia, *J. Geophys. Res., 76*(24), 5859–5870, 1971.

Cartwright, D. E., W. H. Munk, and B. D. Zetler, Pelagic tidal measurements, *Eos Trans. AGU, 50*(7), 472–477, 1969.

Easton, A. K., A handbook of selected Australian tide gauges, *Surv. Pap. 6,* Horace Lamb Centre, Flinders Univ. of S. Aust., 1968.

Easton, A. K., The tides of the continent of Australia, *Res. Pap. 37,* Horace Lamb Centre, Flinders Univ. of S. Aust., 1970.

Irish, J. D., W. H. Munk, and F. E. Snodgrass, M2 amphidrome in the Northeast Pacific, *Geophys. Fluid Dyn., 2,* 355–360, 1971.

Munk, W. H., and N. Phillips, Coherence and band structure of inertial motions in the sea, *Rev. Geophys. Space Phys., 6,* 447–472, 1968.

Munk, W. H., F. E. Snodgrass, and M. H. Wimbush, Tides off-shore, Transition from California coastal to deep sea waters, *Geophys. Fluid Dyn., 1,* 161–235, 1970.

Nilsson, C. S., Measurements of tides and swell on the continental shelf, *Res. Pap. 46,* Horace Lamb Centre, Flinders Univ. of S. Aust., 1971.

Noye, J., and A. K. Easton, The computation of tidal constants, *Comput. Programs 7,* Horace Lamb Centre, Flinders Univ. of S. Aust., 1967.

Snodgrass, F. E., Deep sea instrument capsule, *Science, 162,* 78–87, 1968.

Snodgrass, F. E., *Eltanin* cruise 41, *Antarctic J., 6,* 12–14, 1971.

Wimbush, M. H., Temperature gradient above the deep sea floor, *Nature, 227,* 1041–1043, 1970.

Wimbush, M. H., and W. H. Munk, The benthic boundary layer, in *The Sea,* vol. 4, part 1, edited by A. E. Maxwell, Interscience, New York, 1970.

2

Marine Geophysics

INTRODUCTION: MARINE GEOPHYSICS OF THE SOUTHEAST INDIAN OCEAN

DENNIS E. HAYES

Lamont-Doherty Geological Observatory of Columbia University, Palisades, New York 10964

Abstract. Three papers on morphology, magnetic anomalies, and sediment distribution constitute the geophysical reconnaissance studies of the southeast Indian Ocean. The explanation for asymmetric sea floor growth, major differential sedimentation rates, and the early rifting history of Australia, Antarctica, and New Zealand are the main unresolved problems of the area. The Pacific-Indian plate boundary south of New Zealand is examined in four papers, and the conclusions of these investigations are not in complete agreement. A study of the southwest Pacific defines many peculiar features in the tectonic and magnetic lineation pattern that may relate to similar features of the southeast Indian Ocean. The deep-sea drilling program planned for the near future should resolve many of the remaining problems in this area.

THE SOUTHEAST INDIAN OCEAN

The southeast Indian Ocean with its associated mid-oceanic ridge system has been regarded as one portion of the world ocean where the morphology and geophysical properties could be described and understood in terms of a simple history of sea floor spreading. The main section of the mid-oceanic ridge here, the Southeast Indian ridge, trends essentially east-west, and north-south spreading has resulted in the drifting apart of the Australian and antarctic continents. This area is one of the few where a major segment of the mid-oceanic ridge system trends east-west, and therefore this area is uniquely suited for examining subjects that have variables with strong latitudinal dependence, such as paleomagnetic inclinations, sedimentation patterns, oceanic and atmospheric circulation, biostratigraphy, and zones of primary productivity. The geologic history represented by the southeast Indian Ocean is significant on the geologic time scale (~60–100 m.y.) but is not so large as to be hopelessly indecipherable. The northern and southern margins of this area are essentially seismically inactive. Sparse seismic activity within the Australian continent does not define any major plate boundaries.

Three papers, one by *Hayes and Conolly* [this volume] on morphology, one by *Weissel and Hayes* [this volume] that treats the magnetic lineation pattern, and one by *Houtz and Markl* [this volume] that maps the sediment distribution, constitute the main reconnaissance geophysical studies of the southeast Indian Ocean (Figure 1). The eastern boundary of the study area is defined by an active plate boundary of unconfirmed nature. This boundary, defined by the Alpine fault, the Macquarie ridge complex, and the intersection of the mid-oceanic ridge systems of the Pacific and southeast Indian oceans, is the subject of four geophysical papers and two physical oceanography papers in this volume.

The history of sea floor spreading throughout the Cenozoic period is reasonably well established for the southwestern Pacific. Until this time, a similar history for the southeast Indian Ocean was defined only in a general way. Deciphering the detailed history of relative plate motions in the southeast Indian Ocean affords an opportunity to study the evolution of the Pacific-Indian plate boundary. A knowledge of the evolution of the southeast Indian Ocean may provide the best approach to the study of the complex history of the east and west Caroline basins, the Melanesian Archipelago, the south Fiji basin, and the Coral Sea.

The patterns of seismicity are most important in the definition and analysis of major plate boundaries, and the recent studies by *Sykes* [1970] and *Banghar and Sykes* [1969] relate directly to the geophysical investigations presented here. Prior to the acquisition of extensive marine geophysical data, earthquakes provided the only feasible means to examine the nature of global tectonics in this region.

The geophysical studies in this volume define at least two major features not yet observed or documented for any other portion of the world ocean. One, a major zone of disruption of the Southeast Indian

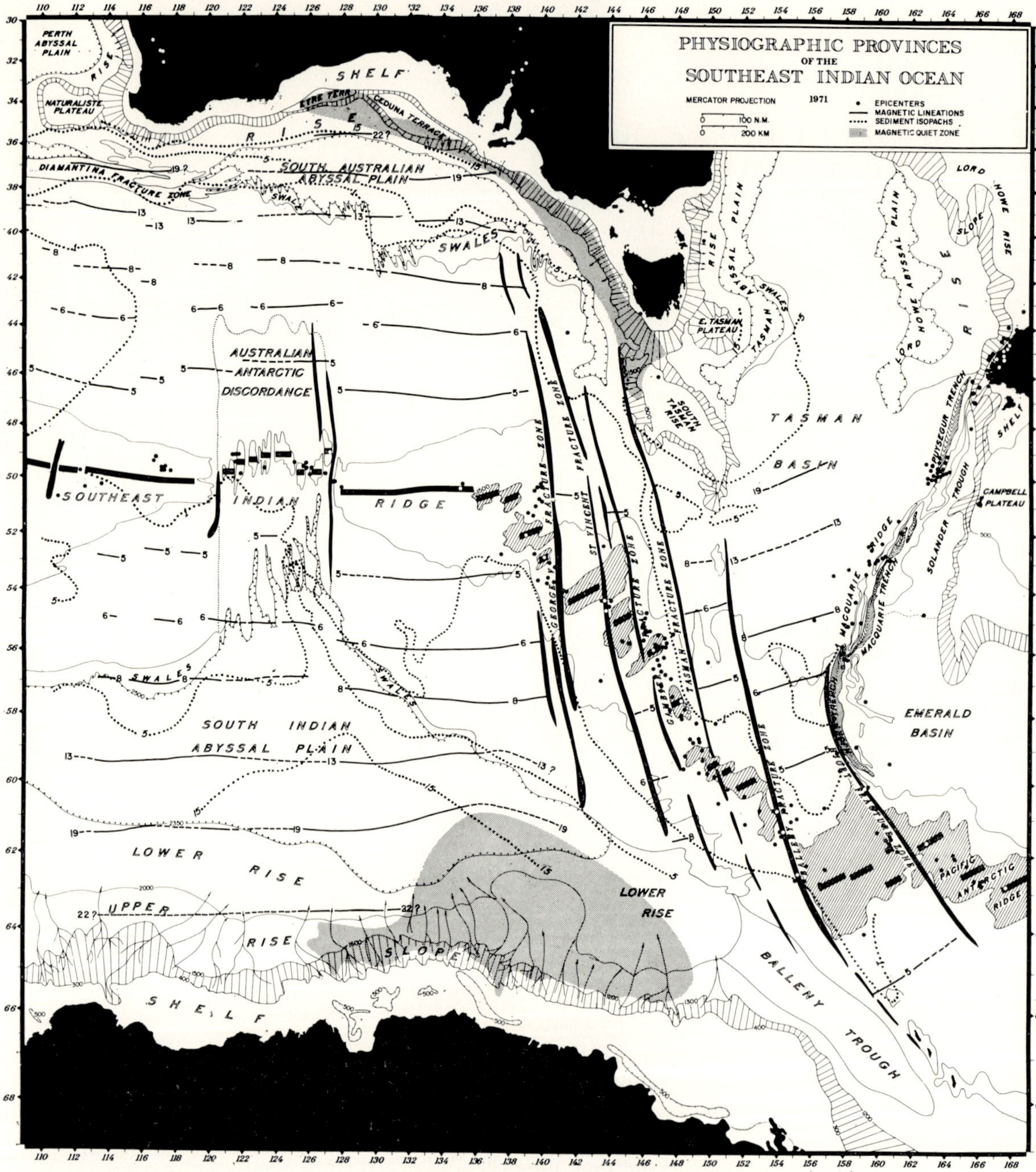

Fig. 1. Physiographic provinces of the southeast Indian Ocean modified from *Hayes and Conolly* [this volume]. Magnetic lineations are taken from *Weissel and Hayes* [this volume]. Selected sediment isopachs in tenths of a second of reflection time (0.1 sec ≈ 100 meters of sediment) are taken from *Houtz and Markl* [this volume]. Epicenters (darkened circles) are from the *Barazangi and Dorman* [1969] file. Mercator projection.

ridge and called the Australian-antarctic discordance, is discussed in the contribution by *Hayes and Conolly* [this volume]. This disrupted section of the Southeast Indian ridge is dominated by a topographic grain that trends at right angles to the axis of the ridge. The ridge morphology is so complicated that identification of a crestal province of the ridge system is difficult. The ridge crest in this region is assumed to be located near the broad linear belt of earthquake activity that extends through this area and coincides with the axis of the Southeast Indian ridge to the west. Probably similar zones of 'discordance' will eventually be found in other areas of the world ocean.

The disturbed morphology associated with the Australian-antarctic discordance is confined to a region within about 500 km on either side of the inferred ridge crest. *Weissel and Hayes* [this volume] have shown that the recognition of characteristic magnetic lineations is essentially impossible within this zone. Isolated anomalies are occasionally identified, but a coherent magnetic pattern cannot be mapped over any significant distance. Although the distribution of epicenters through the discordance is similar to that over the normal portion of the Southeast Indian ridge lying to the west, the eastern boundary of the discordance marks an abrupt seismic boundary. East of this boundary, the ridge morphology is simple, the pattern of magnetic lineations is clear, but the area is aseismic (Figure 1).

The morphology defined by *Hayes and Conolly* [this volume] and the distribution of magnetic anomalies presented by *Weissel and Hayes* [this volume] serve to define four longitudinal zones of the Southeast Indian ridge, each zone with unique characteristics. South of Tasmania, the ridge system undergoes a series of large right-lateral displacements. In the vicinity of 160°E, the ridge marks one branch of the Macquarie triple junction. Both the morphology and the offsets in the magnetic lineation pattern confirm the presence of the fracture zones previously inferred from the distribution of earthquake epicenters. Although the exact location and offset across individual fracture zones are not precisely known, the fracture zone azimuths are well determined by several independent geophysical parameters.

Within one section of the Southeast Indian ridge (~128°–140°E), a history of systematic asymmetric growth of the sea floor is convincingly recorded by the magnetic lineations as presented in the paper by *Weissel and Hayes* [this volume]. Although the symmetric distribution of magnetic lineations was viewed with great awe during the last 5 years, the systematic asymmetric growth of sea floor is even more perplexing. The zone characterized by asymmetric sea floor spreading is bounded on the west by a zone that during the same period of time was accreting crustal material symmetrically. This relationship implies that important modifications to the general concepts of global tectonics may be necessary [*Hayes*, 1971].

Weissel and Hayes [this volume] also show that there are pronounced differences in inferred natural remanent magnetism (NRM) properties of crustal rocks throughout the area. All available evidence suggests that Antarctica has remained relatively fixed in its present position with respect to the spin axis of the earth throughout the Cenozoic and that the Australian continent has migrated northward. The accreting zone has remained essentially median between the drifting continents and thus has also migrated northward at half the rate of separation of the antarctic and Australian continents. *Weissel and Hayes* [this volume] map magnetic quiet zones along portions of the Australian and antarctic continental margins. The quiet zones are further characterized by small areas of conspicuous 'piercement' structures as shown by *Houtz and Markl* [this volume]. The two areas of piercement structures were approximately contiguous according to the prerift reconstructions of *Weissel and Hayes* [this volume] and other workers. *Pitman and Talwani* [1971] have suggested that similar structures in the North Atlantic that also lie within magnetic quiet zones may represent salt diapirs. The Australian-antarctic quiet zone boundaries are subparallel to the continental margins. The magnetic lineations probably cut across the quiet zone boundaries and thus suggest that in this area the quiet zone boundaries do not represent isochrons.

The Diamantina fracture zone, a major zone of high relief lying subparallel to the Australian continental margin, defines another anomalous morphologic region. Its general configuration has been known from many earlier studies. The magnetic quiet zones are not recognized along the western portion of the Australian continental margin. The relationship of the Diamantina fracture zone to global tectonic processes is entirely speculative.

There is a pronounced nonuniformity in the distribution of sediments within the Diamantina fracture zone; this distribution is almost certainly controlled by bottom water circulation. The bottom water circulation, also discussed in the physical oceanography section, must be responsible for many other 'anomalies' in the distribution of the sediments mapped by *Houtz and Markl* [this volume]. The crestal portion of the Southeast Indian ridge is characterized by a thin sediment blanket. Large areas of nondeposition and current scour have been described in the marine geology section by *Watkins and Kennett* [this volume] and by *Conolly and Payne* [this volume]. The historic pattern of bottom water circulation must be regulated to a large degree by the major topographic barriers to the bottom water transport. These barriers almost certainly have been dynamic in nature (on a geologic time scale), and, in particular, the constraints imposed by the Macquarie ridge complex to the circumpolar flow may be of global importance.

A pronounced increase in the total sediment thickness is observed near the continental margins of Australia and Antarctica; the basement (layer 2) dips steeply toward the continent [*Houtz and Markl*, this volume] and geometrically resembles a buried trench. However, there is no evidence, particularly on the Australian continental margin, of tectonic deformation normally associated with crustal subduction. One may question whether the presence of a depressed oceanic crustal layer should necessarily be associated with 'trench structure' and lithospheric subduction. The observed geometry of the crustal layer is similar to that of many of the circum-Pacific trenches, whereas its genesis may be totally different.

The South Indian abyssal plain, lying adjacent to the antarctic continental margin, is underlain by sediments at least 1 km thicker than those of its counterpart to the north. The South Indian abyssal plain is extremely large (~2 million km^2), perhaps one of the largest in the world oceans. This area has only been available to receive sediment since the Eocene. Some controversy still remains as to the time of inception of the extensive continental ice sheet that now caps the antarctic continent. The ice sheet contains very little detritus, and little ice-rafted material is being contributed to the ocean floor at the present time. We are therefore left with a dilemma in explaining the pronounced contrast in total sediment accumulation north and south of the Southeast Indian ridge. If the continental ice sheet existed on Antarctica since the Late Cretaceous as suggested by some, or even since the Miocene as suggested by others, there is no obvious explanation for the 1.5 cm/1000 yr difference in the average Cenozoic accumulation rates north and south of the ridge. This difference constitutes one of the major unresolved problems of this area.

The magnetic lineations define rather concisely the spreading history of Australia and Antarctica back to 50 m.y. B.P. However, the oldest lineations still are about 200 km from their respective continental margins. We are left with a zone about 400 km wide for which the history of plate tectonics and of the earliest rifting is not defined by our marine geophysical observations. Geologic evidence exists on the Australian continent for the occurrence of pronounced rifting as early as Jurassic time, and marine transgressions have been recorded in the Cretaceous. The history of early rifting most probably spans a long period of time before the current phase of rapid sea floor spreading commenced in the early Eocene.

The cause of the observed asymmetric sea floor spreading is another major unresolved problem. The confirmation and understanding of asymmetric sea floor growth is extremely important. A large number of areas exist throughout the world ocean where the magnetic record has been studied from only one flank of the ridge, and it now seems apparent that a tacit assumption of a symmetric growth is no longer valid.

THE SOUTHWEST PACIFIC OCEAN

Christoffel and Falconer [this volume] in their study of the extreme southwest Pacific have been able to extrapolate the Cenozoic geomagnetic time scale to approximately 83 m.y. They have identified a major disturbed zone within the Pacific-Antarctic ridge that they have called the Pacific-antarctic fracture zone. Although the presence of the Australian-antarctic discordance was unknown to them, the Pacific-antarctic fracture zone may prove to be an analogue of the discordance. The morphology of the Pacific-antarctic fracture zone is extremely rough, and the magnetic pattern is not defined. Seismicity is uniform throughout the fracture zone area, but a sharp eastern boundary is defined that delineates the area where the magnetic pattern and the morphology are again simply developed. The southwest Pacific is a perfect example of an area where only one limb of the spreading system has been examined. I anticipate that many of the complications that *Christoffel and Falconer* [this volume] found on the north limb (e.g., magnetic lineations of different age offset by different amounts and the changing sense of offset across apparently continuous fracture zones) may be explained by the presence of discrete zones of asymmetric sea floor growth.

Christoffel and Falconer [this volume] have successfully mapped magnetic lineation patterns up to and adjacent to the continental slope south of the Campbell plateau. *Houtz and Markl* [this volume] present seismic data that show that the acoustic basement reflector is essentially lying flat as it abuts this continental slope. There is no evidence of a depressed crustal layer as along the margins of the southeast Indian Ocean. Similarly, there is no evidence of a magnetic quiet zone near the Campbell slope.

Christoffel and Falconer [this volume] suggest that a pronounced change in the pattern of motion occurred in the southwest Pacific at about anomaly 25 time (~63 m.y. B.P.). This change occurs slightly before the beginning of the major phase of sea floor spreading in the southeast Indian Ocean; however, the two phenomena are undoubtedly closely related. The Campbell plateau was separated from the antarctic continent by a large distance by early Cenozoic time. As Australia was not separated by a significant distance from Antarctica then, major differential motion between Australia and New Zealand must be postulated for late Mesozoic and early Cenozoic time. *Weissel and Hayes* [this volume] have suggested from the magnetic pattern and the sediment distribution that an ancient plate boundary defining a smaller Tasman subplate runs northeast between the South Tasman rise and Resolution ridge. A recent detailed *Eltanin* survey was conducted in the central Tasman basin. Preliminary analysis indicates that the central Tasman Sea evolved through a phase of spreading that was independent of that of the southeast Indian Ocean but that was intimately related to the earliest rifting of New Zealand and Antarctica (see also *Ringis* [1970] and *Hayes and Ringis* [1972]).

Christoffel and Falconer [this volume] have proposed a revised pole to describe the motion between the Pacific and antarctic plates. Their revised pole position affects the predicted motion and tectonic pattern along the Pacific-Indian plate boundary.

THE INDIAN-PACIFIC PLATE BOUNDARY

As previously mentioned, the nature of the southern Indian-Pacific plate boundary is the subject of considerable controversy. Several papers in this volume have examined parts of the boundary. *Hayes and Talwani* [this volume] concluded that the Macquarie ridge complex represents a zone of predominant transcurrent motion but with significant components of compression at the northern and southern extremities. A component of extension may have been active along the central part of the ridge that includes Macquarie Island. The pole describing the 'instantaneous' relative motion between the Pacific and Indian plates lies close to the complex, and inferences regarding the tectonics could change dramatically as a consequence of small errors in the determination of this pole and in the analysis of finite rotations.

Christoffel and van der Linden [this volume] and *Hayes and Talwani* [this volume] have offered explanations to reconcile suggestions of recent motion parallel to the northeast striking Alpine fault with evidence for a southeast dipping Benioff zone beneath Fiordland, New Zealand.

Williamson and Rubenach [this volume] and *Varne and Rubenach* [this volume] have examined, respectively, the geophysical and geologic properties of Macquarie Island. They arrive at slightly diverse conclusions. *Williamson and Rubenach* [this volume] suggest that the observed geophysical parameters of the island do not lead to the conclusion that the island was formed as a part of the mid-oceanic ridge system. They map northwest trending magnetic lineations that they view as unrelated to sea floor spreading processes. They believe that the lineation pattern reflects the degree of metamorphism of the mafic and ultramafic provinces that characterize the island. *Varne and Rubenach* [this volume] have shown that the tectonic pattern of the island is extremely complex, but that the petrologic suites there are characteristic of those of mid-ocean ridges. They suggest that the island was formed as part of the mid-oceanic ridge system. The only emergent part of the Macquarie ridge complex is Macquarie Island, a possible location for crustal extension as shown by *Hayes and Talwani* [this volume], and previously by *Christoffel* [1971] and by *Hayes et al.* [1972]. The inferred sense of crustal extension and possible sea floor growth is not consistent with the pattern of magnetic lineations deduced by *Williamson and Rubenach* [this volume]. Only the northernmost part of the island has been mapped in extensive detail. The northeast strike and attitude of the basalts that are transected by northwest striking ultramafic zones are consistent with the model of crustal extension proposed by *Hayes and Talwani* [this volume] if the ultramafics are associated with fracture zone elements. *Hayes and Talwani* [this volume] have suggested that northeast of Macquarie Island (in deep water) magnetic lineations lie subparallel to the complex. These anomaly correlations are tenuous, and the lineations have not yet been identified as part of the Cenozoic spreading pattern. A major problem remains in explaining the pattern of stresses consistent with sea floor spreading that also allows local uplift of the island by 3–4 km since late Miocene.

CONCLUSION

The resolution of the number of these outstanding problems is probably imminent. Plans are being formulated for the drilling vessel *Glomar Challenger* to drill several holes in the southeast Indian Ocean, the southwest Pacific, and the Macquarie ridge complex in the very near future. We anticipate that the results of the drilling will lead to an understanding of many aspects of the evolutionary history of this region not yet resolved by the papers in this volume.

Acknowledgments. This research has been supported by the Office of Polar Programs (grants GV-23334 and GV-27472) and the Oceanography Branch of the National Science Foundation (grant GA-27281) and by the Office of Naval Research (contract N00014-67-A-0108-0004).

Contribution 1775, Lamont-Doherty Geological Observatory.

REFERENCES

Banghar, A. R., and L. R. Sykes, Focal mechanisms of earthquakes in the Indian Ocean and adjacent regions, *J. Geophys. Res., 74*(2), 632–649, 1969.

Barazangi, M. W., and J. Dorman, World seismicity maps compiled from ESSA coast and geodetic survey epicenter data, 1961–1967, *Bull. Seismol. Soc. Amer., 59*(1), 369–380, 1969.

Christoffel, D. A., Motion of the New Zealand Alpine fault deduced from the pattern of sea floor spreading, in Recent Crustal Movements, edited by B. W. Collins and R. Fraser, *Roy. Soc. N. Z. Bull., 9,* 25–30, 1971.

Christoffel, D. A., and R. K. H. Falconer, Marine measurements in the southwest Pacific Ocean and the identification of new tectonic features, in *Antarctic Oceanology II: The Australian–New Zealand Sector, Antarctic Res. Ser.*, vol. 19, edited by D. E. Hayes, pp. 197–209, AGU, Washington, D. C., 1972.

Christoffel, D. A., and W. J. M. van der Linden, The Macquarie ridge–New Zealand Alpine fault transition, in *Antarctic Oceanology II: The Australian–New Zealand Sector, Antartic Res. Ser.*, vol. 19, edited by D. E. Hayes, pp. 235–242, AGU, Washington, D. C., 1972.

Conolly, J. R., and R. R. Payne, Sedimentary patterns within a continent–mid-oceanic ridge–continent profile: Indian Ocean south of Australia, in *Antarctic Oceanology II: The Australian–New Zealand Sector, Antarctic Res. Ser.*, vol. 19, edited by D. E. Hayes, pp. 295–315, AGU, Washington, D. C., 1972.

Hayes, D. E., Asymmetric sea-floor spreading; Some important consequences (abstract), *Eos Trans. AGU, 52*(11), 847, 1971.

Hayes, D. E., and J. R. Conolly, Morphology of the southeast Indian Ocean, in *Antarctic Oceanology II: The Australian–New Zealand Sector, Antarctic Res. Ser.*, vol. 19, edited by D. E. Hayes, pp. 125–145, AGU, Washington, D. C., 1972.

Hayes, D. E., and J. Ringis, Early opening of the central Tasman Sea, paper presented at the International Symposium on the Oceanography of the South Pacific, UNESCO, Wellington, New Zealand, February 1972.

Hayes, D. E., and M. Talwani, Geophysical investigation of the Macquarie ridge complex, in *Antarctic Oceanology II: The Australian–New Zealand Sector, Antarctic Res. Ser.*, vol. 19, edited by D. E. Hayes, pp. 211–234, AGU, Washington, D. C., 1972.

Hayes, D. E., M. Talwani, and D. Christoffel, The Macquarie ridge, in *Antarctic Geology and Geophysics*, edited by R. J. Adie, Universitets forlaget, Oslo, in press, 1972.

Houtz, R., and R. Markl, Seismic profiler data between Antarctica and Australia, in *Antarctic Oceanology II: The Australian–New Zealand Sector, Antarctic Res. Ser.*, vol. 19, edited by D. E. Hayes, pp. 147–164, AGU, Washington, D. C., 1972.

Pitman, W. C., III, and M. Talwani, Sea floor spreading in the North Atlantic, *Geol. Soc. Amer. Bull., 83*(3), 1971.

Ringis, J., Magnetic lineations in the Tasman, paper presented at International Conference on Geophysics of the Earth and the Ocean, Aust. Inst. of Phys., Sydney, Australia, January 1970.

Sykes, L. R., Seismicity of the Indian Ocean and a possible nascent island arc between Ceylon and Australia, *J. Geophys. Res., 75*(26), 5041–5055, 1970.

Varne, R., and M. J. Rubenach, Geology of Macquarie Island and its relationship to oceanic crust, in *Antarctic Oceanology II: The Australian–New Zealand Sector, Antarctic Res. Ser.*, vol. 19, edited by D. E. Hayes, pp. 251–266, AGU, Washington, D. C., 1972.

Watkins, N. D., and J. P. Kennett, Regional sedimentary disconformities and upper Cenozoic changes in bottom water velocities between Australasia and Antarctica, in *Antarctic Oceanology II: The Australian–New Zealand Sector, Antarctic Res. Ser.*, vol. 19, edited by D. E. Hayes, pp. 273–293, AGU, Washington, D. C., 1972.

Weissel, J. K., and D. E. Hayes, Magnetic anomalies in the southeast Indian Ocean, in *Antarctic Oceanology II: The Australian–New Zealand Sector, Antarctic Res. Ser.*, vol. 19, edited by D. E. Hayes, pp. 165–196, AGU, Washington, D. C., 1972.

Williamson, P., and M. J. Rubenach, Preliminary report on geophysical studies on Macquarie Island, in *Antarctic Oceanology II: The Australian–New Zealand Sector, Antarctic Res. Ser.*, vol. 19, edited by D. E. Hayes, pp. 234–249, AGU, Washington, D. C., 1972.

MORPHOLOGY OF THE SOUTHEAST INDIAN OCEAN

DENNIS E. HAYES

Lamont-Doherty Geological Observatory of Columbia University, Palisades, New York 10964

JOHN R. CONOLLY

Department of Geology, University of South Carolina, Columbia, South Carolina 29208

Abstract. A new bathymetric chart of the southeast Indian Ocean is presented. This portion of the Indian Ocean sea floor comprises the eastern half of the spreading Indian and antarctic plates. The bottom morphology is dominated by an east-west trending mid-ocean ridge that is symmetrically disposed between southern Australia and East Antarctica. The ridge is interrupted by a zone 250 miles wide of very rugged ridge topography herein called the Australian-antarctic discordance. The discordance has no recognizable counterpart elsewhere in the world ocean. It is characterized by north-south oriented topography lineations with relief of 200–800 fm; the lineations define a zone about 200 fm deeper than the adjacent ridge segments. Several large north-northwest trending fracture zones (the Hjort, Balleny, Gambier, George V, and St. Vincent fracture zones) offset discrete blocks of the mid-ocean ridge in a right-lateral sense, and the system is displaced southeastward toward the Balleny Islands region where it meets the Macquarie ridge complex at a triple plate junction. The Macquarie ridge complex is a zone 900 miles long of discontinuous ridges and trenches that defines a segment of the present Indian-Pacific plate boundary. Trenches lie both to the east and west of segments of the Macquarie ridge. The Tasman basin is characterized in the area north of 48°S by northeast-southwest oriented topography and a thick sediment cover and in the area south of 48°S by northwest-southeast oriented topography that parallels the present offsets of the Southeast Indian ridge. On the eastern edge of the Tasman basin, the Tasman rise system roughly marks the boundary between these two areas. The Tasman rise system consists of the East Tasman plateau and the South Tasman rise that were probably formed by downfaulted southeastward extensions of the Australian continent. Steep slopes separate Tasmania from the Tasman rise system. The continental margin of Australia forms a broad sedimented platform off southern Australia but is relatively narrow and steep off eastern Australia and off the Cambrian shield coast of Western Australia. Two prominent sediment wedges form the Eyre and Ceduna terraces off the Great Australian Bight. The Diamantina fracture zone is an east-west trending series of deep (3000 fm) en echelon trenches and ridges that separate the western part of the Southeast Indian ridge from the steep southern escarpment of the Naturaliste plateau and the continental margin of Western Australia. The antarctic continental shelf is wide (100–200 miles) and deep (300–900 fm) and has been sculptured by glacial processes that produced glacial troughs up to 900 fm deep within the depressed shelf platform. Fiord-like depressions up to 600 fm deep extend seaward across the shelf from the major glaciers of the coast. The continental slope and rise off Antarctica are also wide, but they are formed by normal margin sedimentation that has been most extensive off the Antarctic, this sedimentation building the level of the huge South Indian abyssal plain up to a level ~500 fm (1 km) higher than its northern counterpart, the South Australian abyssal plain.

The production of a bathymetric map (Figure 1; see foldouts at back of volume) of the ocean lying south of Australia and New Zealand was an essential prerequisite to serious marine studies in this region that consists of 25 million km^2 of essentially unmapped ocean floor. In some instances, the features defined in this area may prove unique for the world ocean.

The maps that were available to us at the undertaking of this project included the U.S. Navy Oceanographic Office charts, maps of the southwestern and southern continental margins of Australia [*Conolly*, 1969; *Conolly and von der Borch*, 1967], the New Zealand Oceanographic Office Oceanic Series maps, the Russian bathymetric map of the Pacific compiled by *Udintsev et al.* [1967], the bathymetric map of the antarctic circumpolar waters by Heezen that appeared in *Gordon and Goldberg* [1970], the French Monaco sheets, and the physiographic diagrams of the Pacific Ocean by *Heezen and Tharp* [1969] that appeared in *National Geographic.* With the exception of the continental margin maps of *Conolly* [1969] and *Conolly and von der Borch* [1967] and the New Zealand Oceanographic Office maps, all other charts represented highly generalized interpretations.

USNS *Eltanin,* the research vessel of the Office of Polar Programs of the National Science Foundation, has been operating in the area of interest almost continuously since early 1969 and had steamed approximately 100,000 miles in this area by early 1971. Precision 12 kHz echo-sounding data were collected almost continuously by the *Eltanin* during this period of time. These data have been digitally processed at the Lamont-Doherty Geological Observatory and consist of approximately 200,000 soundings.

Since 1967, navigation aboard the *Eltanin* has used the U.S. Navy's satellite navigation system, and the adjusted track has been determined by using the technique of *Talwani et al.* [1966] that employs the data from the electromagnetic log and gyrocompass of the ship to interpolate between fixed positions. The *Eltanin* surveys in this area have for the most part been conducted along north-south lines in a systematic reconnaissance fashion with track spacing at approximately each 5° of longitude or closer. Additional tracks were run along different azimuths in many areas (Figure 2).

The *Eltanin* data constitute the primary data source for this study, although many thousands of additional soundings were compiled from a variety of other sources (Figure 2). If unresolved sounding discrepancies were found between the *Eltanin* data and other data, the *Eltanin* data were assumed to be correct. The *Eltanin* data were combined with bathymetric data from the *Robert D. Conrad* and *Vema,* research vessels of the Lamont-Doherty Geological Observatory, and with the data of the bathymetric sounding libraries of the Australian Hydrographic Office, the New Zealand Oceanographic Institute, and the U.S. Navy Oceanographic Office. The *Vema* and *Conrad* data and the U.S. Navy Oceanographic Office data were also available to us in digital format and could thus be computer manipulated for presentation in a variety of sounding units.

The greatest amount of auxiliary data was provided by the Australian Hydrographic Office as master sounding sheets compiled into the standard Gebco (general bathymetric chart of the ocean) areas. These data were available to us only in this format in units of fathoms that were corrected for the variation of sound velocity in sea water according to the tables of *Matthews* [1939]. The Australian data constituted an important proportion of the total sounding data available to us, and, as these data could not be converted practically to alternative sounding units, all the additional data were compiled and plotted to conform to the format of the Australian Gebco sheets.

We recognize the primary disadvantage in using units corrected for the variation of sound velocity in sea water, since most shipboard echo-sounding recorders routinely record uncorrected or nominal fathoms or meters based on the assumed velocity of sound in sea water of 800 fm/sec, or 1500 m/sec. We wish to emphasize that the map (Figure 1; see foldouts at back of volume) was compiled in the units of corrected fathoms strictly from the practical considerations of normalizing and using all the significant data available to us. Nominal sounding units can be easily determined from our map by using the Matthews tables. Corrections for nominal water depths of 2500 fm are approximately +170 fm.

With the aid of digital computers and on-line plotters, all the data available in digital format were plotted at a scale of 1:1,000,000 to overlay the Australian master sounding sheets. In this manner, the data for a specific area could be simultaneously examined.

Initial contouring was done on the individual Gebco sheets at a scale of 1:1,000,000. The standard contour interval was 100 fm, although nonstandard contours were shown to describe abyssal plains and areas of steep bottom gradients where the inclusion of all contours was not practical. In some areas, notably in the extreme southeast corner of the map, the contour interval is greater than 100 fm primarily because of the sparsity of data.

The contoured Gebco work sheets (30 sheets) were then reduced to one common scale of 1:1,000,000 at 52° of latitude for drafting. These maps were then photographically reduced and spliced together by using computer-drawn geographic co-ordinates to minimize any accumulative distortion in the interior portions of the map that might otherwise result in the process of piecing together the individual segments.

Portions of the composite map draw heavily on existing detailed studies. These portions include the continental margin of south and southeastern Australia [*Conolly,* 1969; *Conolly and von der Borch,* 1967], the area immediately west of New Zealand [*van der Linden and Hayes,* 1972], the Macquarie ridge complex [*Hayes and Talwani,* this volume], and

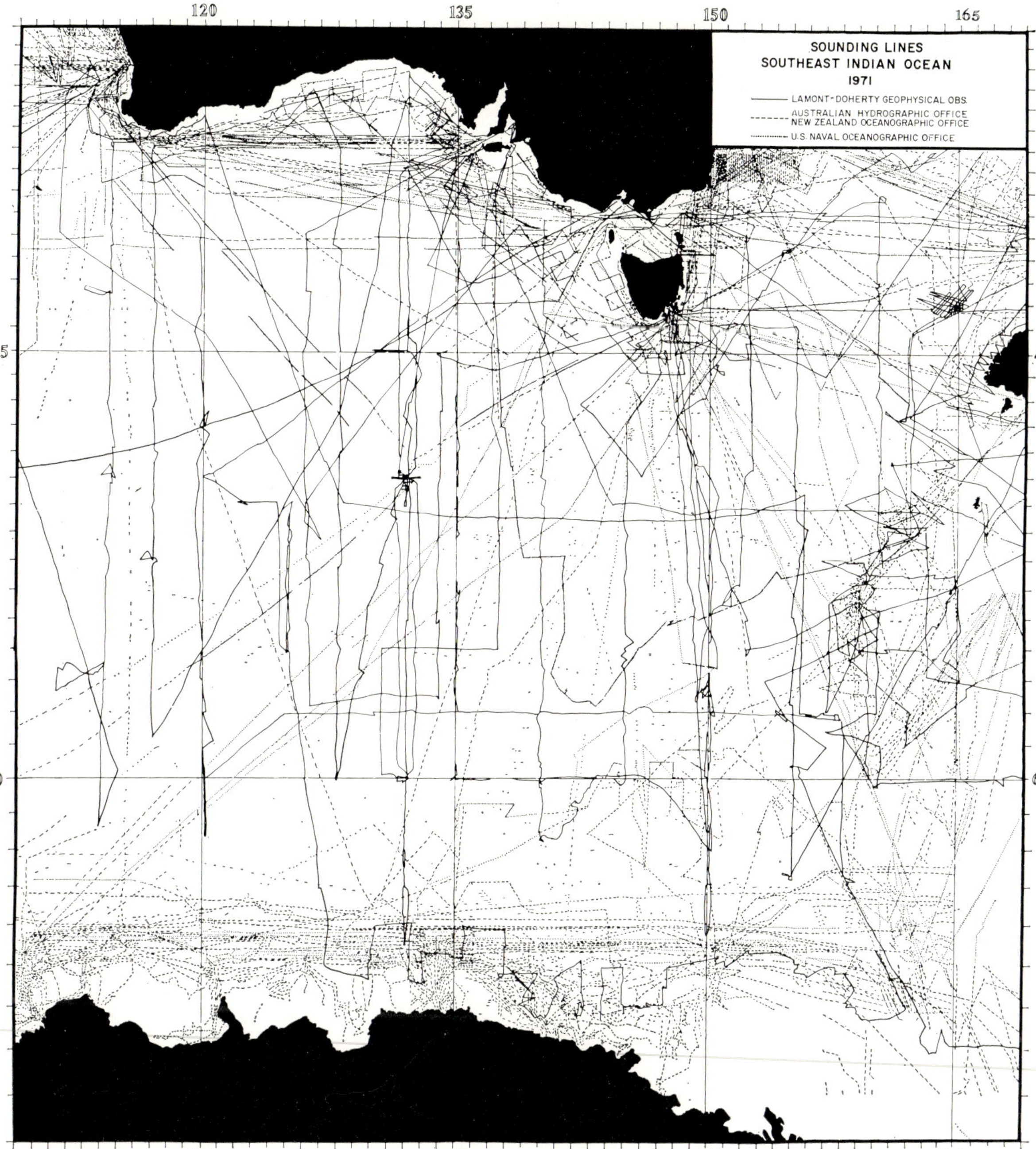

Fig. 2. Sounding lines by source used in the compilation of Figure 1. Isolated soundings are not shown. Australian Hydrographic Office and U.S. Naval Oceanographic Office sources overlap. Where this overlap occurs, control is designated as U.S. Naval Oceanographic Office.

the continental margin of the Wilkes Land coast [*Grinnell,* 1972]. The major portion of the map is entirely new, and this publication represents its first presentation.

In areas where the data density is poor, several alternative contour interpretations are possible. Auxiliary data, such as the distribution of earthquake epicenters and magnetic anomaly lineations and the disposition of marine sediments, were used in our interpretations. The reliability of these interpretations will be discussed in some detail in the discussion of the individual provinces. The distribution of track

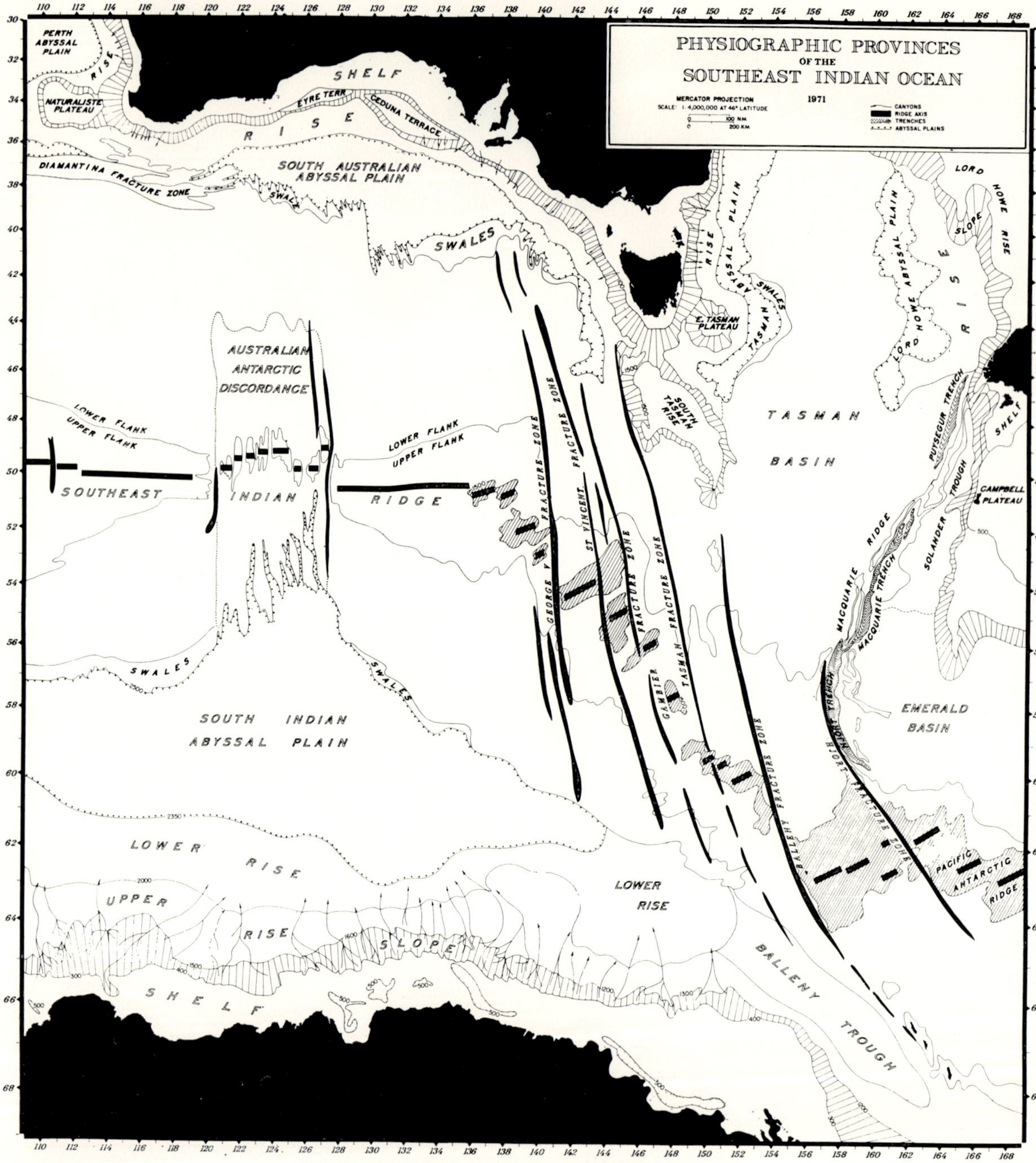

Fig. 3. Physiographic provinces of the southeast Indian Ocean as deduced largely from Figure 1. Selected generalized isobaths define most provinces. Fracture zones are highly schematic.

lines along which echo-sounding data were available for this study is shown in Figure 2. Although showing the control lines on the bathymetric map itself is generally desirable, the bottom topography was so complicated in some areas and the bathymetric data were so dense that the control data are given on the reverse side of Figure 1 to preserve the clarity of the bottom contours. Several of the more obvious characteristics of the ridge morphology south of Australia have been described by *Weissel and Hayes* [1971].

Figure 3 shows the major physiographic provinces as generalized from the detailed bathymetry shown on Figure 1. Selected isobaths are shown to help de-

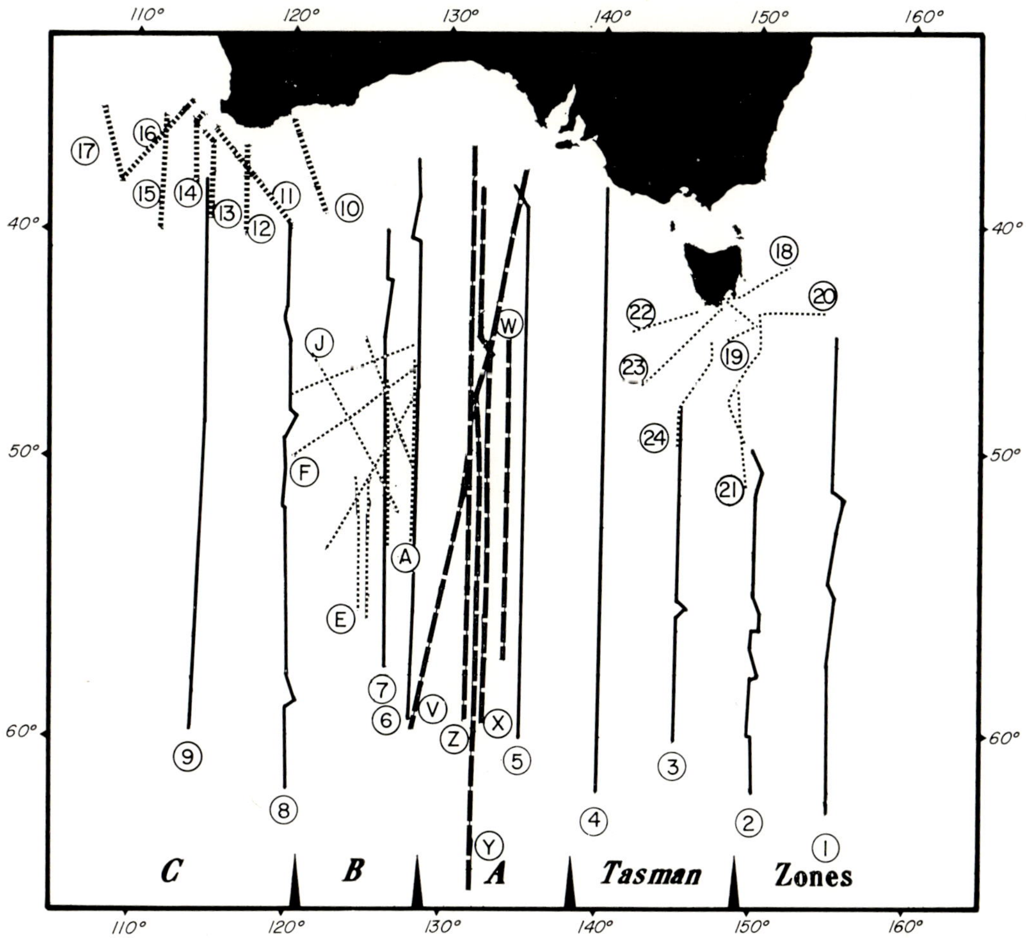

Fig. 4. Index map for profiles shown in Figures 5–9. Designations of zones A, B, and C are those of *Weissel and Hayes* [1971].

fine these provinces. The major morphologic features are (1) the ridge crest and ridge flank provinces, (2) the Australian and antarctic continental margins and their associated marginal plateaus and rises, (3) the South Australian abyssal plain and the South Indian abyssal plain, (4) the Tasman basin, (5) the Macquarie ridge complex, and (6) the Diamantina fracture zone.

RIDGE CREST AND RIDGE FLANK PROVINCES

The major portion of the sea floor consists of the ridge crest and ridge flank provinces. *Weissel and Hayes* [1971] have subdivided the Southeast Indian ridge west of 140°E into three longitudinal segments (Figure 4) called zones A, B, and C. These and a fourth segment, herein called the Tasman zone, will be described.

Australian-Antarctic Discordance (Zone B, 120°E–128°E)

The most striking feature of the sea floor is the highly fractured, northerly trending zone that interrupts the Southeast Indian ridge between approximately 120°E and 128°E. This zone is characterized not only by a highly fractured topography but also by the absence of magnetic anomalies that can be correlated. This zone exhibits an east-west band of seismic activity that ends abruptly at its eastern boundary.

Weissel and Hayes [1971] noted that this portion of the ridge province was characterized by regional depths several hundred meters greater than those of the ridge crest in adjacent zones. They noted the extremely high amplitude relief and suggested that the ridge crest could not be defined with confidence strictly on the basis of morphology. However, the locus of the minimum regional depth soundings

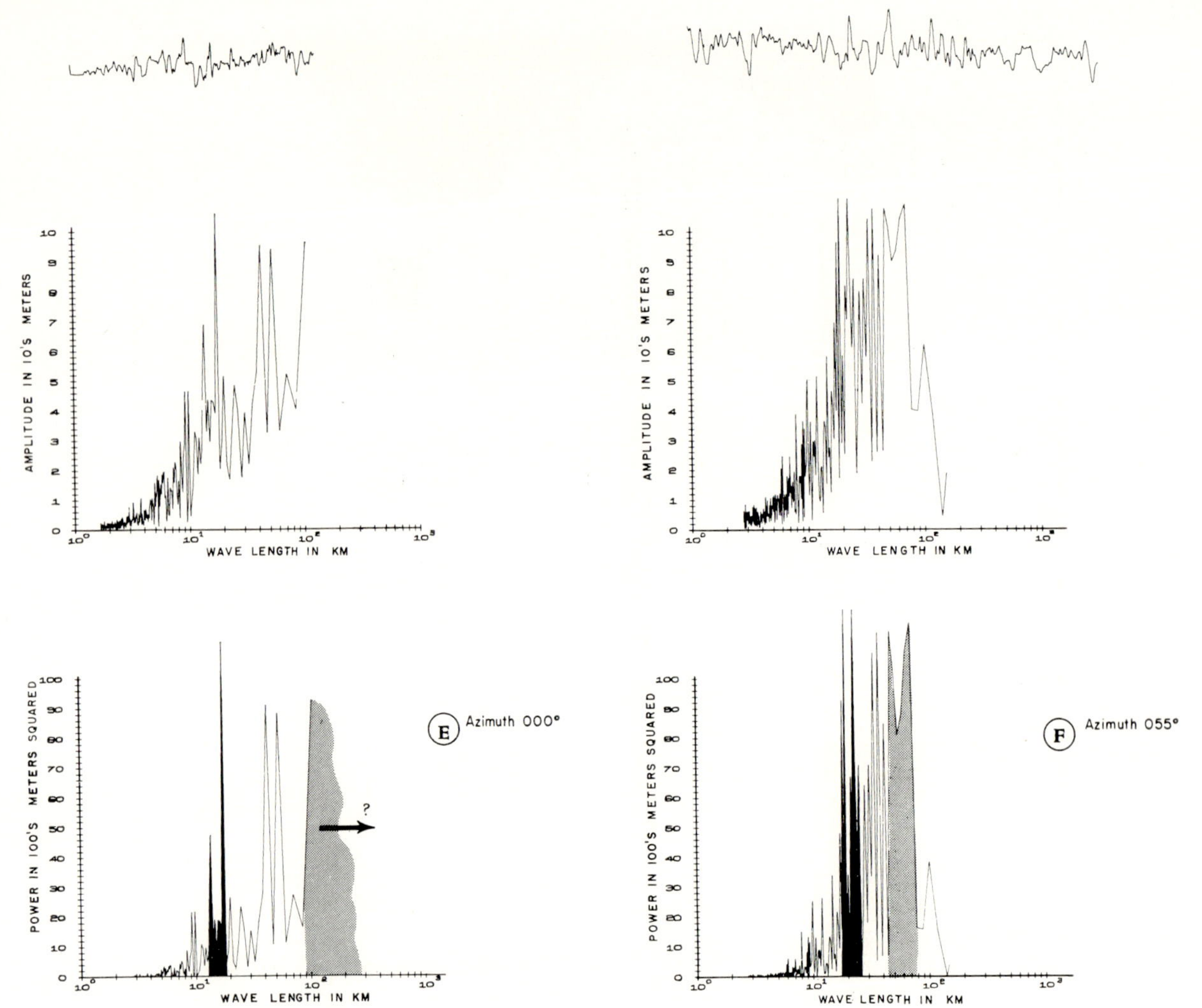

Fig. 5. Amplitude and power spectrum of bottom topography for selected profiles from the Australian-antarctic discordance. Profile of topography is given at the top. On the power spectrum plots, black areas define the short wave length components, stippled areas define the long wave length components. See Figure 4 for profile locations.

coincided closely with the rather broad and diffuse pattern of earthquake epicenters and also is positioned approximately midway between recognizable magnetic lineations. For example, anomaly 5, approximately 10 m.y. old, is located about 350–400 km both north and south of the position of the inferred ridge crest [*Weissel and Hayes,* this volume].

The major trends of the topographic fabric associated with the Southeast Indian ridge are east-west throughout most of the area between 105°E and 140°E. The exception, of course, is the Australian-antarctic discordance. In this area, the trend of the predominant morphologic fabric is essentially north-south. The orthogonal distribution of these major trends serves to highlight the extreme contrast between the discordance and the typical, almost classic morphologic expression of the ridge in the zones both to the east and to the west.

Throughout most of the region we have studied, the primary data control lines were oriented mostly north-south. The secondary control data throughout the entire region gave good azimuthal control, and thus we feel we have circumvented any major bias that might have been introduced into the contouring as a function of track distribution. However, within the discordance, we have a much better azimuthal distribution of tracks for the primary control data, and hence the trend of the fabric here is fairly well defined. Many of the details of the contours throughout the zone are highly interpretative. However.

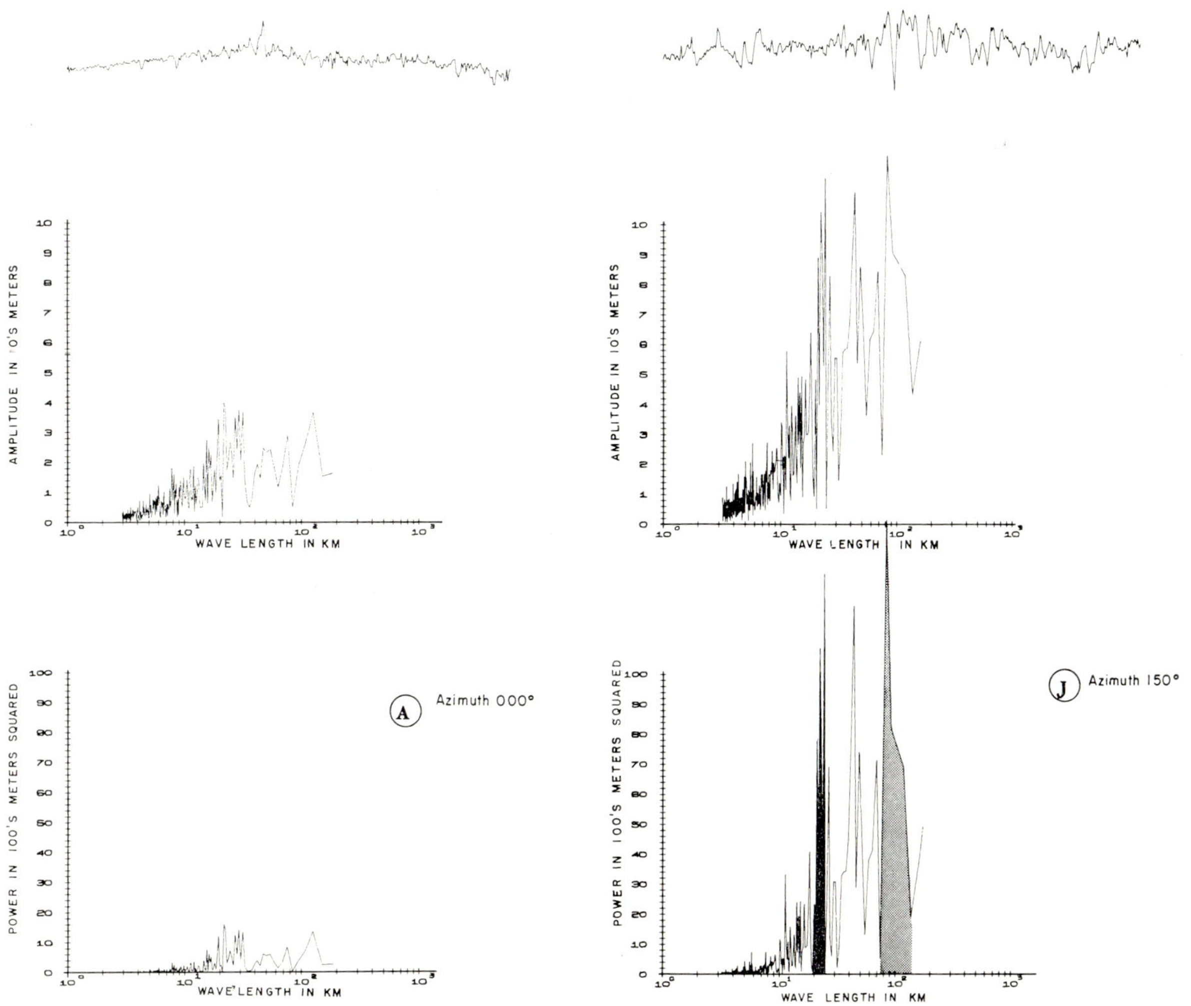

Fig. 5. (continued)

fingers of the South Indian abyssal plain and of the South Australian abyssal plain can be traced hundreds of miles into the highly fractured morphology of the discordance. Furthermore, prominent east-west trends cannot be correlated across closely spaced north-south topographic profiles.

Because of the prominence of this zone, the contrast in topographic trend, and the apparent uniqueness of such a major disruption of an otherwise normally developed ridge system, we attempted to demonstrate quantitatively the uniqueness of our general interpretation of the bottom morphology in this zone. Ten profiles intersecting near the center of this zone and distributed along a variety of azimuths were projected and examined in detail (Figure 4). Each of these projected topographic profiles was 225–400 n. mi. (~400–700 km) long. As the topographic relief along these profiles was very large, the profiles had been digitized originally at extremely close sampling intervals of approximately 1–3 minutes, or about 2 soundings per nautical mile. The topography was interpolated at equally spaced points at an interval of 0.5 to 1 km depending on the length of the profile. These interpolated data were then filtered to remove the components with wave lengths greater than 200 n. mi. (350 km). The spectral distribution of apparent topographic wave lengths was then computed for each profile. Published data on the ridge crest provinces of the world ocean show that there are two distinct topographic trends normally oriented approximately at right angles to each other. *Bhattacharyya and Ross* [1971] quantitatively demonstrated this general relationship for a portion of the Mid-Atlantic ridge near 45°N.

TABLE 1. Normalized Spectral Components of Bottom Topography within the Australian-Antarctic Discordance

Profile Identification	Track Azimuth	Apparent Wave Lengths			Normalized to	
		Short	Intermediate	Long	East-West (Short)	North-South (Long)
B	030°	21	40–60	85	18	42
C	000°	18–21	28–60	. . .	18–21	not defined
D	000°	18–22	25–45	>85	18–22	not defined
E	000°	15–18	40–55	>100	15–18	not defined
F	055°	20–27	30–40*	47–70	16–22	37–57
G	000°	14–18	30–70*	>100	14–18	not defined
H	070°	18–24	35–45	50–70	17–22	47–66
I	160°	16–22	32–50	80–160	15–21	28–55
J	150°	20–24	35–45	82	17–21	41

* Poorly determined.

Spectral Analysis of the Australian-Antarctic Discordance

Three dominant groups of wave lengths can be recognized on most of the spectral plots (Figure 5). These three groups account for almost all the 'energy' within the various projected topographic profiles. In an attempt to normalize the three groups, the spectral peaks were shifted to simulate north-south and east-west profiles and then were compared. Although a certain ambiguity exists in uniquely identifying each group, our technique did effectively normalize the data as presented in Table 1. Figure 5 shows several representative profiles of topography used in our analysis. The amplitude and energy of the spectral components of the topography are plotted against wave length. We interpret these data as follows:

1. The short wave length energy is associated with east-west trending relief with a true wave length of about 18 km (10 n. mi.) and typical amplitudes of 50–100 meters (27–55 fm).

2. The long wave length energy is associated with north-south trending relief with a true wave length of about 40–50 km (22–28 n. mi.) and typical amplitudes of 150–200 meters (84–110 fm).

Note that very little energy exists in these profiles for apparent wave lengths of less than 10 km (5.4 n. mi). By comparing the short wave length components of profiles E and F in Figure 5, one can easily see the apparent change in the wave length of these components for tracks oriented at azimuths 55° from one another. On profile F, the short wave length peaks lie between 20 and 27 km, whereas, on profile E, they lie between 15 and 18 km, the true wave length.

The trend of the intermediate wave length component is more difficult to assess. This component probably trends in a direction between north-south and east-west or, alternatively, may be randomly distributed.

Within the discordance, we have contoured the bottom features with a wave length of about 30–50 km or longer. Topographic relief associated with fracture zones is generally longer in wave length and higher in amplitude than that associated with the direction of the ridge crest. Approximately 400 or 500 km away from the inferred ridge crest, the longer wave length 'fracture zone' morphology is subdued, and here the magnetic anomalies can again be easily identified [*Weissel and Hayes*, 1971, this volume.].

We are confident that the generalized interpretation of the contours within zone B are reliable for the following reasons:

1. The topographic trends cannot be correlated across closely spaced, north-south profiles.

2. Fingers of the South Indian abyssal plain and of the South Australian abyssal plain can be traced several hundred miles along the north-south trends toward the position of the inferred ridge crest.

3. The spectral distribution of topographic components suggests that three predominant groups of wave lengths are present. The shorter wave length component is 'normalized' when associated with an east-west trend with a true wave length of approximately 18 km. A second and more predominant (higher amplitude) component is associated with a north-south trend with a true wave length of 40–50 km. The true trend of the third component that has an intermediate wave length cannot be resolved. The long wave length component has been contoured on Figure 1.

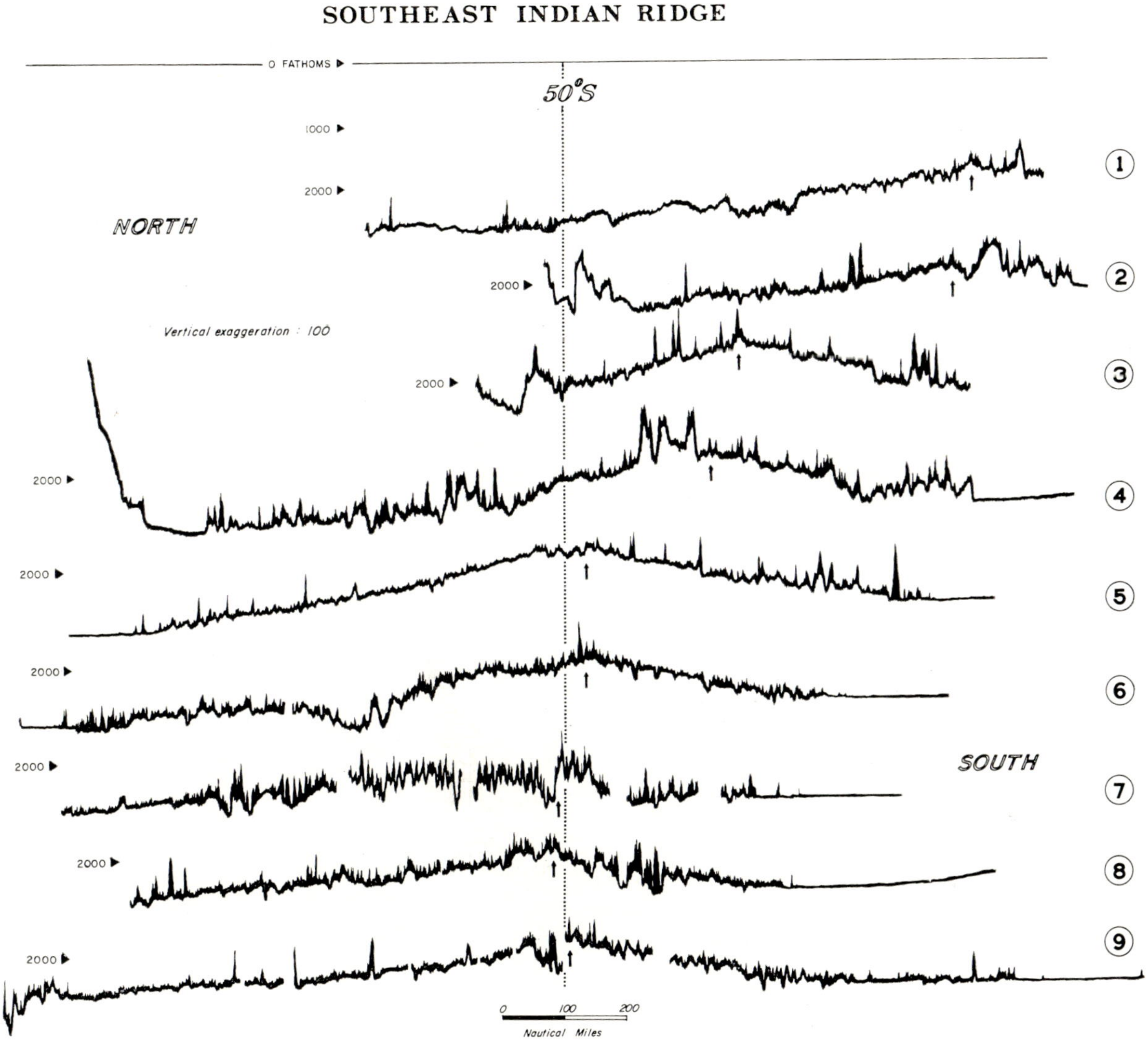

Fig. 6. Typical north-south transoceanic profiles of topography across the southeast Indian Ocean. Profiles are aligned on 50°S. The position of the ridge crest is indicated by the small arrows beneath the profiles, and the progressive north-south offset of the crest is apparent. Figure 4 shows the location of each profile and their grouping within zones A, B, C, and Tasman.

Western Zone (Zone C, ~105°E–120°E)

The ridge crest is well developed in this area and appears to be uncomplicated by major offsets as also suggested by the broad, linear, east-west zone of epicenters. The generalized isobaths that define the upper flanks of the ridge system trend about 20°–30° away from the east-west direction. Although no well-developed rift valley is associated with the crest of the ridge, there may be several significant east-west trending rifts near the crestal portion of the ridge that can be traced over distances of 60–100 n. mi. (Figure 1; Figure 6, profiles 8 and 9). Similar rifts are not recognized near the ridge crests within the other zones. The boundary between zone C and zone B is abrupt and is fairly well defined by a number of northerly trending lines. Note that the eastern limit of the Diamantina fracture zone coincides closely with the same boundary. The bathymetry of zone C is somewhat less well controlled than that of the other zones.

Eastern Zone (Zone A, 128°E–~139°E)

The crest and flanks of the mid-ocean ridge within zone A are easy to recognize. Profiles across the ridge show a simple morphology when compared with zone

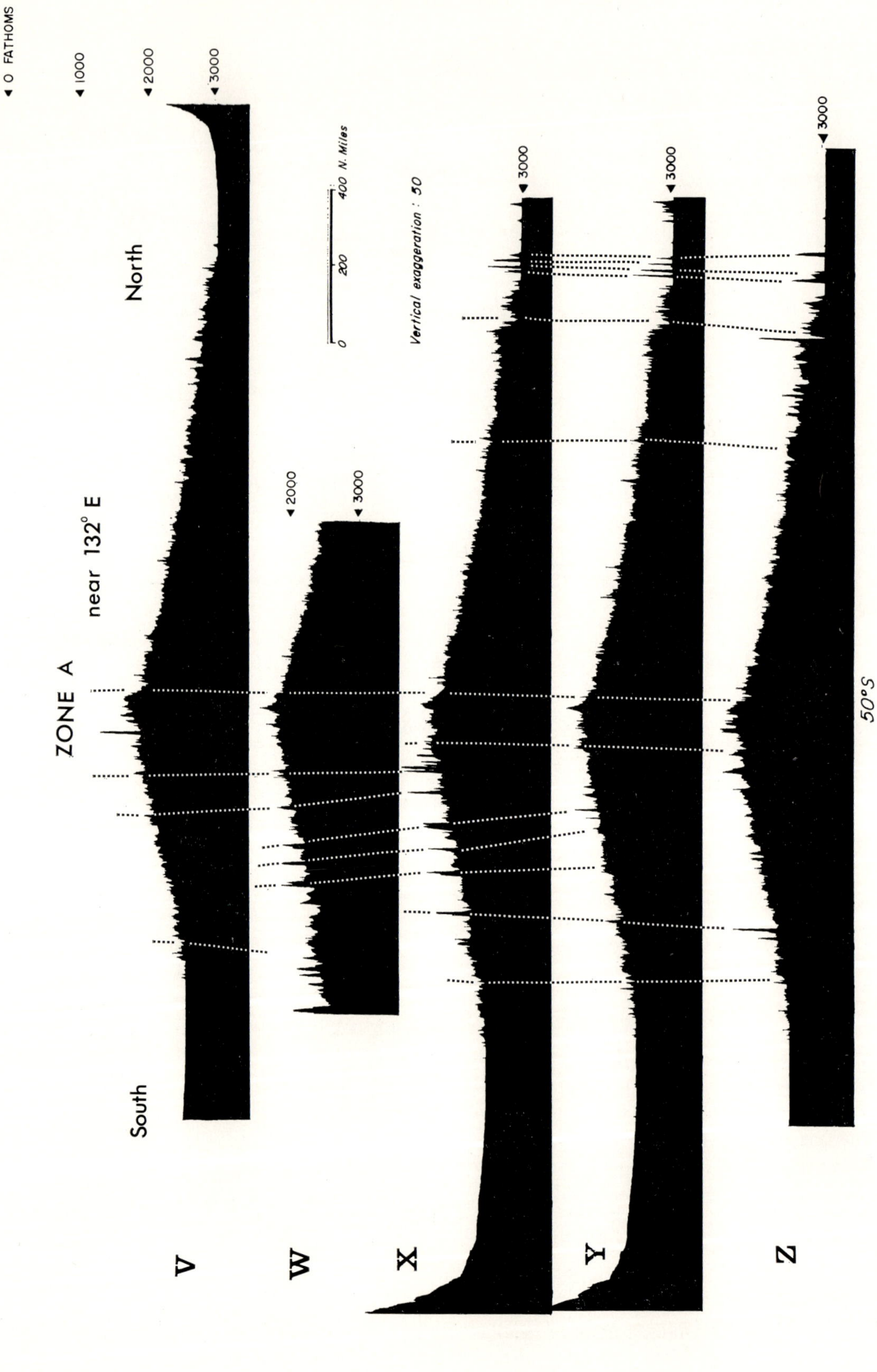

Fig. 7. Projected topographic profiles of the Southeast Indian ridge from closely spaced profiles from zone A. Correlations of bottom features across adjacent profiles are indicated. See Figure 4 for profile locations.

C and, particularly, with zone B (Figures 6, 7). A comparison of the spectral plots of Figure 5, profile A with those of profiles E, F, and J in the same figure quantitatively shows the pronounced contrast in relief between zones A and B. The ridge crest is almost always characterized by a small but distinct apex and by the absence of a rift valley. Its position agrees well with the distribution of magnetic anomaly 1 [*Weissel and Hayes*, this volume]. The local relief is quite small (<30 fm) in amplitude. This section of the ridge crest is essentially aseismic [*Weissel and Hayes*, 1971]. There is no evidence from the morphology or from the well-developed magnetic lineation pattern that any significant offsets of the ridge crest occur within this zone.

There appears to be slightly more relief on the southern flank of the ridge (in the vicinity of 50–59°S near 132–136°E) as apparent from Figure 1 where easterly and northeasterly trending topographic highs can be traced across several tracks. The continuity of these features can also be seen by examining Figure 7 that shows projected topographic profiles collected within a 3°-wide longitude band centered near 132°E.

Close examination of the projected profiles shown in Figure 7 allows us to define several fairly subtle features associated with the ridge in this zone. A small topographic step of about 200 fm in relief is consistently present on the north side of the ridge crest. The regional depth at approximately 400 n. mi. on either side of the ridge crest is nearly the same both north and south of the ridge. As a consequence of the 'crestal step,' the average gradient over 400 n. mi. is slightly steeper on the south side than on the north side. This difference gives rise to a subtle, but very real asymmetry in the cross-sectional profile of the ridge. The abyssal plain sediments lap almost 1 km (~500 fm) higher onto the south flank of the ridge crest than their counterparts lap onto the north flank. Seismic reflection profiler data [*Houtz and Markl*, this volume] show that the general slope of the south flank extends essentially uninterrupted under the abyssal plain sediments of the South Australian plain. Note that this zone is the zone that has been characterized by a period of asymmetric sea floor spreading as defined by magnetic anomaly lineations. However, this asymmetrical spreading development ceased approximately 10–15 m.y. ago [*Weissel and Hayes*, 1971] so that the asymmetry observed in the morphology at the present ridge crest may not be

TABLE 2. Major Fracture Zones of the Southeast Indian Ocean

Name	Midpoint Location	Amount of Offset of Ridges		Length of Topo Expression	
		n. mi.	km	n. mi.	km
Hjort	62°S, 163°E	30	55	450	825
Balleny	62°S, 156°E	150	275	1000	1830
Tasman	58.5°S, 151°E	120	220	700	1200
Gambier	57°S, 147.5°E	200*	370	800	1460
St. Vincent	55°S, 144°E*	100	183	1100†	2000
George V	53.5°S, 141°E	150	275	1100	2000
Zone A–B boundary	50°S, 127.5°E	110	200	not well defined	

All offsets are right-lateral in sense.
* These values may occur in two 'steps.'
† These values include the southern half and are uncertain.

related directly to the asymmetric pattern of sea floor generation.

A major, right-lateral offset of the ridge crest (~110 n. mi.) is indicated at 128°E along the boundary between zones A and B. The length of the transform zone represented by this offset has changed as a function of time as discussed by *Weissel and Hayes* [1971]. Therefore the present configuration of the accreting plate boundary (i.e., the Southeast Indian ridge axis) does not reflect the precise configuration of the initially rifted intracontinental boundary of Antarctica and Australia.

A suggested correlation of topographic peaks defining easterly trending ridges is shown by the broken lines on the south flank of Figure 7. These ridges do not extend as continuous features throughout the entire extent of zone A. They can be traced only for relatively short distances of the order of 50–100 n. mi. There are no obvious counterparts of these narrow ridges on the north flank. At least two additional topographic steps are defined on the lower flank on the north side of the ridge in zone A. These steps are correlated by the broken lines on Figure 7 and define fairly abrupt changes in depth of approximately 200–300 fm. Although the Diamantina fracture zone cannot be traced structurally eastward through the discordance (zone B), the trend and position of these topographic steps strongly suggest that they may eventually prove to be related to an eastern extension of the Diamantina fracture zone.

Tasman Zone

We will refer to that portion of the Southeast Indian ridge lying between about 139° and 158°E as the

Tasman zone. This zone is associated with the region of highest seismic activity [*Sykes,* 1970]. It is characterized by large offsets of the ridge crest of the order of 200 or 300 km. All of the major offsets are right-lateral in sense. The precise locations of these offsets are difficult to define by using any single type of data (seismic, topographic, magnetic). However, the data coverage within the Tasman zone is extremely good, and relatively little interpretation is required in constructing the bottom contours. Approximately five major offsets of the ridge crest are defined by the bottom contours. Additional small offsets are likely; their presence or absence correspondingly affects the exact positions and the amounts of offset associated with the major fracture zones (Table 2). Our interpretation of these major fracture zones is based primarily on the bottom morphology. Although the precise location and the amount of offset of individual fracture zones are subject to interpretation, the trends of these zones are firmly established. The trends of fracture zones as defined exclusively by epicenter belts have been subject to question. In many areas, the linear but broad distribution of epicenters could represent a series of small offset and en echelon fracture zones trending in a direction quite different from the over-all trend of the epicenter belt. Our bathymetric data suggest that for this area the trend defined by the earthquake epicenters very nearly represents the true trend of the fracture zones. The trend of the small blocks (~60–160 n. mi. in width) associated with the ridge crest changes gradually from east-west near 140°E to about N 70°E near 156°–158°E.

The fact that the major topographic lineaments associated with the extensions of the fractures can be traced over many tens and, in some cases, hundreds of miles beyond the actual transform zone (region between crests) adds a great deal of credibility to the specified fracture zone trends. The inferred location of the offset ridge crest elements within the Tasman zone is slightly different from that deduced from the magnetic lineation pattern [*Weissel and Hayes,* this volume]. Similarly, the amount of offset of these blocks as defined by the magnetic lineations is somewhat different from that defined by the morphology alone. Both interpretations are consistent with the seismic earthquake epicenter distribution. A comparison of the two interpretations provides an example of the reliablity of determining fracture zones from the two independent sets of data.

One of the major ridge crest offsets, the Tasman fracture zone (Figure 3), appears to extend and define a steep scarp along the western boundary of the South Tasman rise (Figure 8). The Balleny Islands lying between 66° and 68°S near 163°E have sometimes been related to a hypothetical southern extension of the Macquarie ridge south of its intersection with the mid-oceanic ridge system. Our map suggests that the Balleny Islands are most probably associated with the southeasterly extension of the major offset of the ridge crest near 155°E, herein called the Balleny fracture zone. Four more major offsets have been defined and are named the George V, Gambier, St. Vincent, and Hjort fracture zones (Figure 3, Table 2).

The predominant bottom fabric of the ridge province throughout the Tasman zone is that of the fracture zones and trends north-northwest. Recognition of the ridge crests within the Tasman zone is relatively easy as shown by profiles 2 and 3 of Figure 6. The true relief along the flanks of the ridge within this zone is difficult to ascertain in view of the large number of crestal offsets. This zone appears to be associated with considerably higher local topographic relief than that of zone A, but there is a high probability that much of this relief is only apparent and is associated with the fracture zones.

CONTINENTAL MARGINS AND ASSOCIATED MARGINAL RISES AND PLATEAUS

The Tasman rise system divides the continental margin of southern Australia into two distinct provinces (Figures 1 and 3: southeastern Australia, which flanks the Tasman Sea on the east, and southern Australia, which forms the northern continental border of the southeast Indian Ocean.) The physiography and bathymetry of both regions have been described in a series of recent papers [*Conolly and von der Borch,* 1967; *von der Borch,* 1967; *Conolly,* 1968; *von der Borch,* 1968; *Conolly,* 1969; *Conolly,* 1970; *Conolly et al.,* 1970; *von der Borch et al.,* 1970]. The additional track data provided from the *Eltanin* cruises do not radically alter the physiography as originally presented by *Conolly* [1968, 1969] in the western Tasman Sea or by *Conolly and von der Borch* [1967] and *Conolly et al.* [1970] along the southern margin of Australia, but the *Eltanin* data more clearly define the major provinces that are discussed below.

Southeastern Australian Margin

The southeastern continental margin of Australia is characterized by a narrow (10–40 miles) continental shelf and a steep (6°–8°) continental slope. Both the

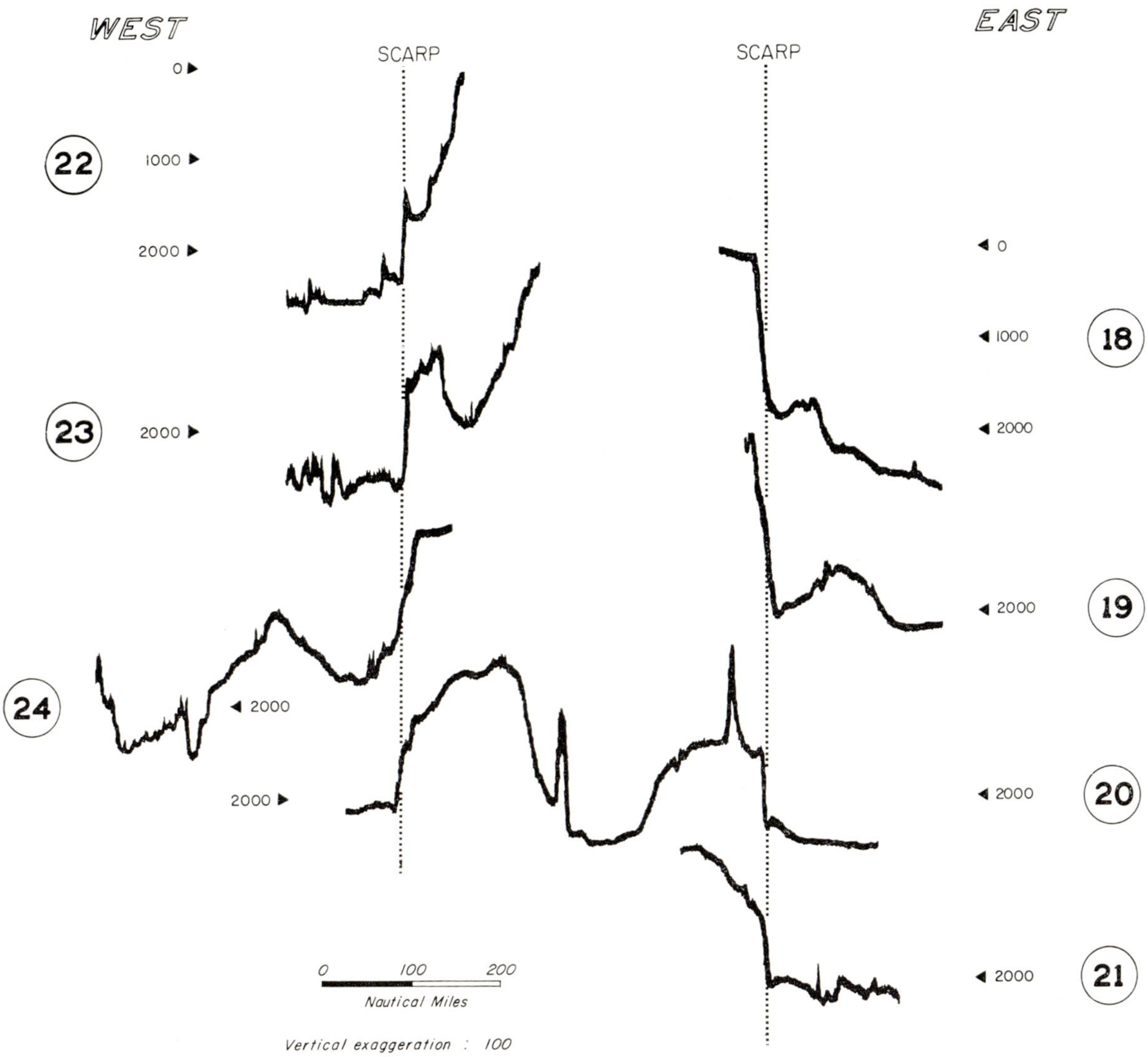

Fig. 8. Topographic profiles across the Tasman rise system. Depths are in fathoms. Those profiles on the right are associated with the East Tasman plateau; those on the left with the South Tasman rise. Figure 4 identifies the fanlike distribution of the profiles.

steepness and ruggedness of the continental margin of this part of Australia are reflected onshore by rocky headlands that merge inland with the 2000- to 3000-foot-high Great Dividing Range, which was uplifted during the Tertiary. The continental shelf is a narrow plain defined by an upper sediment surface that formed by a progradation from the local rivers and deltas from the Late Cretaceous to the Tertiary [*Conolly*, 1969]. Canyons have fed considerable amounts of sediment through the steep continental slope out onto the Tasman abyssal plain that forms the deepest parts of the western Tasman Sea. The most extensive canyon system occurs off the Bass Strait where several large canyons disperse from an amphitheatre-shaped part of the continental shelf. They extend to depths of 2000–2500 fm where they have built sediment fans and thus formed a continental rise [*Conolly*, 1968].

Tasman Rise System

The Tasman rise system is the southernmost extension of the Australian continental margin and is composed of the East Tasman plateau and South Tasman rise (Figure 3). Both these features are separated from the mainland of Tasmania by steep slopes and are characteristically bounded on the seaward side by steep scarps (Figure 8).

The South Tasman rise forms a plateaulike surface at depths from 800 to 1500 fm. It extends southeastward over a distance of 600 miles (1000 km) and forms the western margin of the Tasman basin. The South Tasman rise is bounded by an extremely steep scarp on its western margin between 46° and 48°S. This scarp has a vertical relief of up to 1200 fm with average slopes of up to 20° and is as steep as any slope described from the ocean floor. This scarp is analogous to the Blake escarpment of the Blake plateau off Florida [*Pratt and Heezen,* 1964]. The north-northwest–south-southwest orientation of the scarp is parallel to the general fracture zone trend of the mid-ocean ridge and presumably marks an offset in the initial pattern of rifting of the antarctic and Australian continents.

The East Tasman plateau also forms a plateaulike surface, but at depths of 1300–1600 fm. It is surmounted by at least one major volcanic spire (Figure 8). The plateau is separated from the Tasmanian mainland by a steep slope with a north-northeast–south-southwest orientation that is in line with the general direction of the eastern continental margin of southeastern Australia.

The shape and orientation of the Tasman rise system illustrate that it consists of two blocks (the East Tasman plateau and South Tasman rise) bounded by fault scarps and downfaulted 100–1500 fm from the Australian continent. The East Tasman plateau is separated from the South Tasman rise by a deep trough parallel to the faulted margins of the Bass basin of the Bass Strait [*Griffiths,* 1971] and the Jurassic-Tertiary volcanic trend of Tasmania [*Spry,* 1962]. The three features (the Bass basin, the Jurassic-Tertiary volcanic province of Tasmania, and the trough separating the East Tasman plateau and South Tasman rise) form a tectonic zone 100–150 miles wide that parallels the fracture zone trends of the Southeast Indian ridge. The trends on the continent (Bass basin and Tasmanian igneous province) have been active in the Jurassic and Cretaceous with sedimentation and downfaulting in the Bass basin and dolerite intrusions in Tasmania. Renewed basaltic lava extrusion and more downfaulting occurred along the Tasmanian province in the Tertiary. The South Tasman rise and East Tasman plateau are then part of an igneous horst and graben–fault complex, the development of which began with early rifting in the Jurassic and continued through renewed basaltic volcanic activity during the Tertiary.

South Australian Continental Margin and the Ceduna and Eyre Terraces

The continental margin of southern Australia can be subdivided into three major subprovinces that reflect the geologic framework of the adjacent continent. In the east, the Australian continent is younger geologically and rivers have fed detritus onto a tectonically active continental shelf since the Cretaceous [*Conolly,* 1970; *von der Borch et al,* 1970]. Here the continental shelf is generally narrow (30–50 miles), the continental slope is steep (5°–8°), and rugged canyon systems traverse the continental slope [*von der Borch,* 1968].

In the central portion of the southern Australian continental margin, however, the hinterland is flat and arid, and for almost 700 miles along the Great Australian Bight the present-day drainage is inland. Studies of the continental margin in this area that are based on numerous deep seismic traverses and some onshore well control [*Conolly et al.,* 1970] show that sedimentary progradation began in the south during the Late Cretaceous. Progradation continued into the Tertiary and built two distinct wedges of sediment 200–10,000 feet thick on the edge of the continental margin. Today, the upper surfaces of both these wedges of sediment form marginal terraces. The eastern wedge forms the Ceduna terrace (Figure 3) [*Conolly and von der Borch,* 1967; *Conolly et al.,* 1970]. The western wedge forms the Eyre terrace (Figure 3), which is much smaller than the Ceduna terrace and was derived from the erosion of the older and more subdued Pre-Cambrian mountains to the west of the Nullarbor plain.

Apart from the two marginal terraces lying at depths of 300–1500 fm, the continental slope off the Great Australian Bight is only moderately steep (1°–3°) and merges gradually with an extensive fan valley system that has built a large continental rise at depths of 2000–2900 fm. From a depth of about 2900 fm, the continental rise merges with the extremely flat surface of the South Australian abyssal plain.

The western part of the South Australian continental margin is characterized by a fairly narrow continental shelf (30–60 miles wide) and a steep and rugged continental slope incised by numerous deep canyons [*von der Borch,* 1967; *Conolly and von der Borch,* 1967]. The hinterlands are part of the crystalline Pre-Cambrian shield, and outcrops of these rocks occur at several localities across the continental shelf. The seaward extension of the margin abuts the east-west trending Diamantina fracture zone. The continental margin of southern Australia extends westward and forms the steep southern boundary of the Naturaliste plateau and eventually swings around the plateau where the canyons dominate the morphology of the continental slope off Perth.

Naturaliste Plateau

The Naturaliste plateau is a rectangular shaped block that forms a marginal plateau extending due west from southwestern Australia. Its upper surface lies at depths of 1200–1400 fm and is blanketed by at least 300–500 meters of sediment that has been dated as old as Upper Cretaceous [*Burckle et al.*, 1967]. The plateau is bounded on its three ocean-facing sides by steep (5°–12°) continental slopes that merge at depths of 2500 fm into a narrow, sediment-built continental rise. The troughs and ridges of the Diamantina trench parallel the southern margin of the Naturaliste plateau. The western margin is oriented north-south in the direction of the major tectonic lineaments of the western edge of southwestern Australia such as the Darling fault [*McWhae et al.*, 1957]. The northern margin of the plateau appears to be more irregular. The suggestion has been made that the Naturaliste plateau is a block of Pre-Cambrian crystalline rocks that has been downfaulted and separated from the Australian continent since the Early Jurassic or Early Cretaceous [*Conolly*, 1970].

Antarctic Continental Margin

The continental margin of Antarctica from the Balleny Islands west to 109°E includes a major portion of the East Antarctica coast and lies directly south of the Australian continent (Figure 1). In this region, the antarctic continent has a wide (70–200 miles) continental shelf. The shelf generally varies in depth from 100 to 300 fm, but many troughs and basins exist within the shelf (see *Zhivago* [1962]) with relief greater than 500 fm and with depths up to 900 fm. The contours describing these depressions have been extrapolated landward beneath near-coastal ice shelves.

The shelf lacks a general seaward gradient. The average depth of the break between the continental shelf and the upper continental slope lies between 200 and 300 fm and has a range from 100 to 400 fm (180–730 meters). In this part of Antarctica, the average width of the continental shelf is about twice the world average of 78 km [*Emery*, 1969], and the depth of the shelf–slope break here is three to four times the world average of 133 meters [*Emery*, 1969], or 73 fm.

In many places within about 75 km from the coast, the unusually great depths can probably be attributed to bedrock erosion during the periods of greater seaward extension of both present and previous coastal outlet glaciers. This hypothesis is substantiated by the fact that, where data exist, unusually great depths occur seaward of every glacial tongue along the coast of this part of Antarctica. Furthermore, in the coastal areas having outlet glaciers (where a significant number of reliable soundings exist), the deep areas seaward of the glacial tongues appear to be troughlike in shape. Their longitudinal axes coincide with those of the abutting outlet glaciers [*McCleod and Gregory*, 1967]. Fiords are also characteristically trough-shaped with deep inner basins and shallower sills at their seaward ends, and these features are probably submerged fiords or glacial troughs.

A similar relationship has been carefully documented and studied off the Vanderford Glacier on the continental shelf of East Antarctica in the Wilkes Land region [*Cameron*, 1965]. As in the case of the Adélie Coast drowned fiords, the Vanderford valley was shown to have its maximum depth, 1250 fm (2287 meters), immediately seaward of the glacial terminus.

The continental shelf ends abruptly seaward, gradients on the upper continental slope being as steep as 9°–10°. Submarine canyons begin in most cases at or somewhat below the shelf–slope break and continue down to the fan valleys that form the lower continental slope and rise. There is no pronounced morphologic difference between the slope and rise off the antarctic continent and those south of the Australian continent; except in the Wilkes Land region, there are no obvious marginal plateaus. The continental shelf of Antarctica is considerably deeper than average continental shelf depths elsewhere, presumably because it has been depressed by the load of the antarctic continental ice sheet. However, seaward of the continental shelf, normal submarine processes have been responsible for the movement of detritus from the edge of the continent resulting in a huge pile (up to 6 km) [*Houtz and Markl*, this volume] of sediment that comprises the continental rise. Average rates of sedimentation off the eastern antarctic continental margin have presumably been greater than those off the southern Australian margin, as the level of the South Indian abyssal plain is about 500 fm higher than its Australian counterpart.

ABYSSAL PLAINS

Abyssal plains are areas of the deep ocean floor (>2000 fm) in which the bottom is smooth and flat and the slope is less than 1:1000 [*Heezen et al.*, 1959]. The plains are essentially flat surfaces built by sediments infilling the basement relief of the deep ocean floor. They have a large areal extent in places where great amounts of detritus are derived from adjacent continents and where downslope transportation is effective.

The plains slope gently away from the continent. Their seaward boundary is generally marked by isolated sediment ponds that have not entirely filled the rugged mid-ocean ridge topography. Long flat-floored valleys extending from the flat major surface of the abyssal plain up to hundreds of miles into the mid-ocean ridge province are commonly found. Such sediment fingers (Figures 1 and 4) are extremely useful in delineating the topographic 'grain' of the ridge.

South Australian Abyssal Plain

The South Australian abyssal plain forms the flat sea floor measuring 1000 miles long south of the central part of southern Australia. It merges with the slightly more incised relief of the continental rise at depths of 2900 fm and has its deepest portions at depths of 3060 fm in the central part of the abyssal plain. One long arm of the plain extends southward between 130° and 131°E, a distance of up to 100 miles into the flank of the mid-oceanic ridge. On the west, fingers of the abyssal plain extend westward into the northern deeps of the Diamantina fracture zone and thus form long continuous flat-floored valleys that extend in an east-west direction.

South Indian Abyssal Plain

The South Indian abyssal plain is probably one of the largest abyssal plains in the oceans of the world. Its eastern margin is the mid-oceanic ridge near 150°E, and the plain extends westward to 86°E at the foot of the eastern margin of the Kerguelen plateau [*Heezen and Tharp*, 1965]. That part of the abyssal plain described herein (Figures 1 and 4) extends from the continental rise at depths of 2000–2400 fm to maximum depths of 2565 fm near 58°S, 129°E. Several major sediment ponds extend northward from this deep region 100–300 miles into the Australian-antarctic discordance. The South Indian abyssal plain is substantially shallower than the South Australian abyssal plain. We consider that this difference is caused by differential rates of sedimentation with the upper sediment surface off Antarctica being built up about 500 fm more than its Australian counterpart. Details of the nature of the upper sediment surface and sedimentation across both abyssal plains are given in companion papers [*Houtz and Markl*, this volume; *Conolly and Payne*, this volume; *Payne and Conolly*, this volume]. Several smaller abyssal plains (Figure 3) flank the continental margin of southernmost Australia (Perth abyssal plain, Tasman abyssal plain) and the Lord Howe rise–New Zealand region (Lord Howe abyssal plain).

Swale Topography

Sometimes the distal margin of the very flat surface of an abyssal plain is bounded by an area of fairly smooth but gently rolling swale topography. Areas of swale topography flank the Tasman abyssal plain off southeastern Australia [*Conolly*, 1969, Figure 3] and the Perth abyssal plain off southwestern Australia [*Conolly*, 1970, Figure 10]. One of the largest areas of such swalelike topography has been described by *Heezen and Tharp* [1965] on the distal margin of the Ganges abyssal plain in the Indian Ocean. Large areas of swale topography flank the South Indian abyssal plain in the west and the South Australian abyssal plain in the east. Examination of piston cores from these areas suggests that the swale topography is built from fine, suspended sediment that flowed over the distal margin of the abyssal plain, and then slowed, settled, and thus smoothed the ridge topography [*Payne and Conolly*, this volume].

TASMAN BASIN

The Tasman basin constitutes the major portion of the Tasman Sea and lies between the Australian and New Zealand continents, the Southeast Indian ridge and the Macquarie ridge complex (Figure 1). The western Tasman sea floor has been described in detail by *Conolly* [1969] and the Lord Howe rise–New Zealand region has been described by *van der Linden* [1969] and *van der Linden and Hayes* [1972]. Bathymetry of the region south of the latitude of Tasmania has not been previously discussed in the literature.

The northwesterly trend of the topographic fabric associated with the Tasman zone of the mid-oceanic ridge cannot be traced farther northward than about 48°S near the latitude of the South Tasman rise. This latitude also marks the approximate change in sediment character from relatively thin reverberant sediments to relatively thick acoustically transparent sediments. An obvious change also occurs here in the apparent wave lengths of magnetic anomalies along north-south tracks and serves to define the location of the hypothetical boundary of a pre-existing Tasman subplate (see *Weissel and Hayes* [this volume]).

The Tasman basin north of about 48°S is characterized by gently rolling, long wave length hills with relief of the order of 100–200 fm, and subtle topographic lineaments trend northeasterly. A very thick blanket (>1 km) of acoustically transparent sediments is associated with this portion of the basin as illustrated in *Houtz and Markl* [this volume].

The Tasman abyssal plain forms the flat sea floor flanking the Australian continent, and the Lord Howe abyssal plain forms a similar flat sediment apron on the western edge of the Lord Howe rise. The region between these two abyssal plains lies generally between 2500 and 2700 fm and forms the deep sediment-covered flank of old ridge topography. More recent (Tertiary) volcanic activity has probably been responsible for building the high Tasmantid seamounts that run north-south through the Tasman Sea in the north [*Vogt and Conolly*, 1971].

MACQUARIE RIDGE COMPLEX

The Macquarie ridge morphology is discussed in detail by *Hayes and Talwani* [this volume]. Only a brief discussion of the major morphologic features is repeated here. The Macquarie ridge complex, as the features are collectively named, is much more complicated than that depicted on previously existing bathymetric maps. The detailed morphology was defined primarily by the large collection of precise *Eltanin* echo-sounding data collected during several cruises that concentrated investigation on this feature.

The Macquarie ridge can be traced as a discrete unit from approximately 46°S to approximately 57°S where it branches into two major limbs. The ridge, however, is not continuous in the strict sense as it is offset at two or more localities. The two major offsets occur at approximately 51°S and 56°S. The offsets are also reflected by major gaps in the ridge that are of utmost importance in governing the path of meridional transport of water and the pattern of circumpolar flow across the Macquarie ridge (see *Gordon* [this volume]; *Boyer and Guala* [this volume]). Disruption and offset of the Macquarie ridge at these localities also coincide with radical changes in the distribution of the bounding trenches and also correspond to minor concentrations of earthquake activity.

The well-defined Puysegur trench lying west of Macquarie ridge terminates slightly north of the northern gap, although it may continue farther south as a structural feature. The Macquarie trench lies on the east side of the ridge; the trench begins approximately at the 51°S gap and continues to the 56°S gap as a continuous topographic deep at the 2500-fm isobath level. At the 56° gap, the Macquarie trench terminates and another pronounced arcuate deep, the Hjort trench, bounds the Macquarie ridge again on its western flank. The tectonic implications of this en echelon style of trench disposition with respect to Macquarie ridge are discussed in *Hayes and Talwani* [this volume]. Bounding the Macquarie ridge on the east is the Solander trough, the flanks of which are defined by narrow zones of normal faulting [*Houtz et al.*, 1971]. The grabenlike trough is characterized by thick accumulations of relatively flat lying sediments.

The Solander trough eventually merges into the Emerald basin near 53°S. The Emerald basin is interrupted by many pronounced irregularities that were previously undefined. The Solander trough and portions of the Emerald basin are bounded on the west by the steep western and southern flanks of the Campbell plateau. Between approximately 57°S and 59°S, the Emerald basin contains very deep closures at the 2700- to 3000-fm isobath levels. These topographic deeps trend in a northeasterly direction between approximately 47°S and 50°S. A northerly trending series of similar en echelon deeps lies west of the Macquarie ridge and the Puysegur trench.

The precise morphologic nature of the triple point junction defined by the intersection of the Macquarie ridge and Hjort trench, the Southeast Indian ridge, and the Pacific–Antarctic ridge is of great interest. Unfortunately, the existing data do not allow us to define the detailed nature at this intersection. However, the pronounced arcuate Hjort trench does not appear to continue its southeasterly trend into the axial portion of the spreading ridge system. This termination of the Hjort trench is considered to be well controlled. As previously mentioned, the Balleny Islands do not appear to be a logical extension of the Macquarie ridge complex south of the Pacific-Antarctic ridge system.

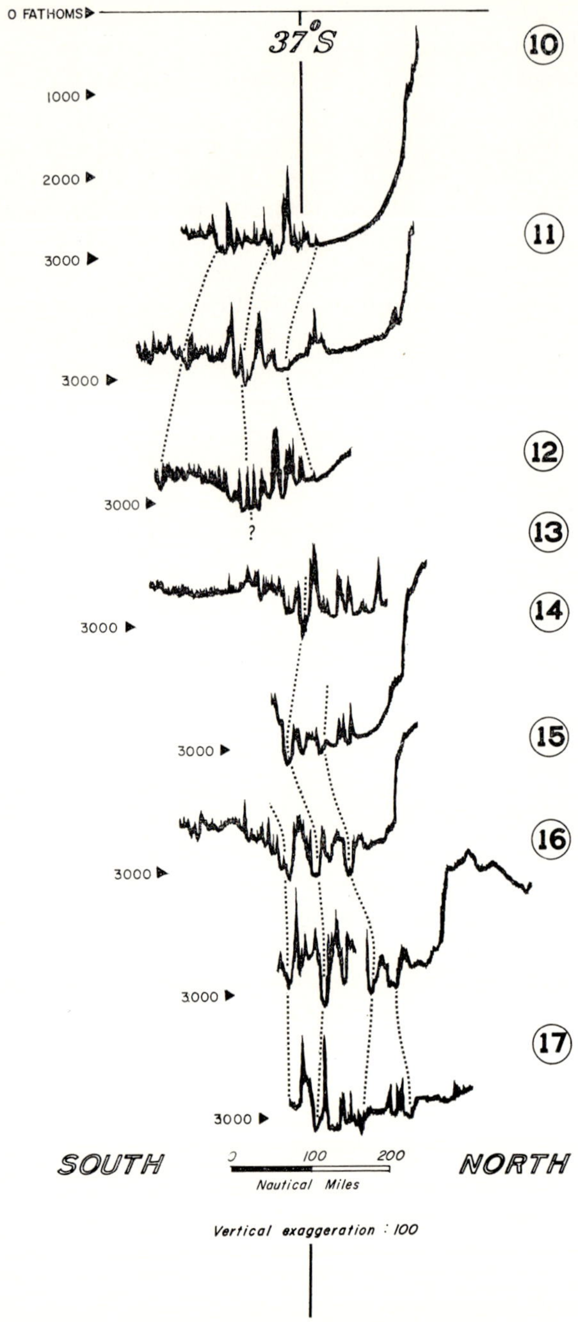

Fig. 9. Topographic profiles of the Diamantina fracture zone. See Figure 4 for locations. Correlations of major deeps are indicated. Profile 16 corresponds to the seismic profiler traverse shown in Figure 5, profile A of *Houtz and Markl* [this volume].

DIAMANTINA FRACTURE ZONE

The Diamantina fracture zone is an elongate region of abnormally high and rugged relief. The zone trends roughly east-west. It is about 100 n. mi. (180 km) wide and 1400 n. mi. (2500 km) long as shown schematically by *Heezen and Tharp* [1965]. The fracture zone defines the steep southern flank of Broken ridge between 90°–100°E and is terminated on the west where it abuts the Ninetyeast ridge [*Heezen and Tharp,* 1965]. Figure 1 shows the detailed bathymetry of only the eastern 400 n. mi. of this feature where it parallels the southern continental margin of Australia. Although the zone is fairly wide, it is composed of two or three major deeps (trenches), each separated by elongate ridges with typical relief of 500–1000 fm. The major deeps and ridges can be traced over several hundred kilometers (Figure 9). Near 116°E, an apparent right-lateral offset of the main deep of about 50 km occurs. Several smaller offsets may be present between 109° and 120°E. The sinuous pattern of the zone as it appears in plan view may be the result of a subtle en echelon distribution of the major trenches and ridges. Except for one or two profiles near the eastern limit of the zone, other profiles indicate that the trenches are not flat-floored. *Houtz and Markl* [this volume, Figure 5] show two seismic profiler traverses across the Diamantina fracture zone. There is a remarkable contrast in the disposition of sediments on adjacent trenches and ridges within the zone. In some cases the trenches are entirely devoid of sediment, whereas in others several hundred meters of acoustically transparent sediment reside within the deeps and along their bordering flanks. The sediments, where present, are not conformable with the basement relief. This unusual distribution of sediment must be controlled by the pattern of bottom water circulation, which would be dependent on the continuity of the individual trenches.

Aside from the South Australian and antarctic continental margins and the Southeast Indian ridge system, the Diamantina fracture zone defines the only prominent east-west trending bottom feature. The present connotation of the term 'fracture zone' is generic and implies that such a feature describes the pattern of relative motion between crustal plates. The name Diamantina fracture zone may be a misnomer in the sense that there is absolutely no evidence to suggest any east-west motion (present or past) that could be related to this 'fracture zone.' In addition, this zone is completely aseismic.

Although the Australian-antarctic discordance (zone B) does not extend as far north as the extension of the Diamantina fracture zone, the latter appears to be terminated near 120°E, the approximate western boundary of the discordance. The anomalous trend and character of the Diamantina fracture zone could be reconciled with concepts of plate tectonics if the zone represents the offshore locus of intensive fracturing and graben formation associated with the early rifting history and separation of Antarctica and Australia. Some ridges have prevented any large-scale transport of detritus into the major deeps and thus have preserved most of the initial morphologic expression of this zone.

A structural counterpart of the zone may be represented by the topographic step observed at the base of the north flank in the eastern portion of the ridge (zone A) near the 3000-fm isobath (Figure 7). If so, we assume the locus of offshore rifting was not as intense here as further west.

CONCLUSIONS AND SUMMARY

Evolution of the Australian-Antarctic Discordance

Weissel and Hayes [1971] presented two alternative explanations to account for the origin and evolution of this zone. One explanation called for major differences in the mechanics of crustal accretion along the east-west line of the Southeast Indian ridge through this area. The other explanation called for a second order intraplate deformation subsequent to the accretion and substantial lateral movement of material away from the ridge crest.

The discordance does not extend all the way to the Australian and antarctic continents. The clear development of the magnetic pattern [*Weissel and Hayes*, this volume] about 250 n. mi. (500 km) from the inferred crest substantiates this fact. We can thus conclude that whatever mechanism was responsible for creating the discordance, it did not exist or become effective until 40 m.y. or more after the initial rifting of the continents.

The western boundary of the discordance corresponds approximately with the eastern limit of the stable Pre-Cambrian shield of Western Australia. The zone thus correlates roughly with the intracratonic basins [*Brown et al.*, 1968] of central Australia. We point out this correlation but do not suggest that any causal relationship necessarily exists between these tectonic provinces. Australia is presently aseismic [*Doyle et al.*, 1968] within that portion that corresponds to the northward extension of the discordance boundaries. There is no evidence of any geologic or tectonic boundaries on Antarctica [*Craddock et al.*, 1970] that could be correlated with the east and west boundaries of the discordance.

Whatever the mode of emplacement of igneous material within the discordance, the observed changes in the relative height of the ridge system between the antarctic and Australian continents must provide an important clue to its origin. The two limbs of the Southeast Indian ridge become gradually higher away from the discordance and thus suggest that the build-up of the ridge, through basaltic igneous activity has become gradually more intensive away from the discordance toward the more open Indian and Pacific oceans. In other words, the discordance is situated in a region that received about 500 km^2/unit length less igneous rock over the past 10 m.y. than any other part of the ridge in that region. This relative deficiency of igneous rock could arise because of the presence of down-welling convective currents in the upper mantle that biased or inhibited the normal slow convective transfer of mantle material that tends to fill the rifted 'void.' In contrast, a massive ridge built up on the margins of the discordance where the mantle was effective in supplying considerably more rock through the last 10 m.y. The presence of the extremely sharp eastern and western boundaries of discordance is most difficult to understand.

There is good evidence that the antarctic plate has remained essentially fixed in its present position during the Cenozoic. As the Indian plate moved northward, the axis of the Southeast Indian ridge also must have migrated northward to maintain its median position. Since this locus of accretion moved with respect to the upper mantle, we do not have to invoke some unique mantle process that began 10 m.y. ago to explain the discordance. If the accreting zone passed over the position of an existing downward directed mantle plume, the coincidence of these features could give rise to the development of the discordance. Alternatively, the passing of the accreting zone over the type of local mantle 'hotspot' described by *Morgan* [1971] may, because of the elevated temperature there, have altered the mechanics of accretion within a fairly restricted zone. As less material was accreting within the discordance, this situation may have generated a preferential zone of weakness that could not sustain the development of large, discrete blocks of ridge crest. The discordance would then absorb most of the effects (strain) of small spreading motion changes.

The ridge crest appears to be highly fractured and dislocated in many places within the discordance, although identifying individual fractures or proving that they are normal transform faults is difficult. Whether the nature of these dislocations is similar to or different from that ascribed to ridge–ridge transform faults, the close spacing of the dislocations and the broad dimensions of the composite zone make the discordance a unique feature of the sea floor.

The discordance is certainly a wider (300 n. mi.) and more extensive (representing 10 m.y. of growth) zone of jumbled sea floor than any zone of the ridge system so far described from the ocean floor. One of the keys to understanding its origin must lie in the way it is symmetrically disposed between two continents that have drifted apart and also between two areas of normal sea floor ridge growth.

General

Zones A and C of the Southeast Indian ridge display rather simple ridge profiles and appear to define large blocks that are free of major transform faults. The fracture zones within the Tasman zone, as inferred from bathymetry, are consistent with those deduced from epicenter distributions and magnetic anomaly lineation offsets. The morphologic expression of these major fracture zones can probably be traced from Tasmania to the Victoria Land coast of Antarctica.

A change in the bottom morphology characteristics of the Tasman Sea occurs near 48°S and suggests (as does the magnetic pattern) that the sea floor in the areas north and south of this latitude evolved in quite different manners and probably at different times.

The Macquarie ridge complex is continuous as a composite feature over a distance of 900 n. mi. However, the individual ridges and trenches are discontinuous and disposed in a highly complicated fashion. The morphology of this province supports the contention of several authors that the features reflect a variety of tectonic regimes including transcurrent motion, extension and compression. The Balleny Islands platform appears to be the southeastern extension of a major offset of the Southeast Indian ridge. The major features of the depressed Wilkes Land continental shelf are the likely consequence of glacial erosion.

Acknowledgments. This research has been supported by the Office of Polar Programs (grants GV-23334 and GV-27472) and the Oceanography Branch of the National Science Foundation (grant GA-27281) and by the Office of Naval Research (contract N00014-A-0108-0004). Special acknowledgment is due the longstanding support and encouragement of the Office of Polar Programs (formerly the Antarctic Research Program) of the National Science Foundation. Support was received from National Science Foundation research grant GV-28803 for J. R. Conolly. We also thank Ed Haff who was responsible for a major portion of the computer-processing data and Robert Payne who spent considerable time working on preliminary bathymetric sheets and the data-processing in the early stages of the project. We thank Dan Grinnell for the use of material from his master's thesis and L. Newman for computing the spectral distributions.

The paper was critically reviewed by M. Ewing, W. Ryan, and R. Fisher, and many of their suggestions were incorporated. The map (Figure 1; see foldouts at back of volume) constitutes the major contribution of this project. Its production would not have been possible without the exceptional efforts of the draftsman, David Johnson.

Contribution 1776, Lamont-Doherty Geological Observatory.

REFERENCES

Bhattacharyya, P. J., and D. I. Ross, Detailed analysis of magnetics and bathymetry at mid-Atlantic ridge near 45°N (abstract), *Eos Trans. AGU, 52*(4), 237, 1971.

Boyer, D. L., and J. R. Guala, Model of the antarctic circumpolar current in the vicinity of the Macquarie ridge, in *Antarctic Oceanology II: The Australian–New Zealand Sector, Antarctic Res. Ser.*, vol. 19, edited by D. E. Hayes, pp. 79–93, AGU, Washington, D. C., 1972.

Brown, D. A., K. S. W. Campbell, and K. A. W. Crook, *Geologic Evolution of Australia and New Zealand,* 409 pp., Pergamon, New York, 1968.

Burckle, L. H., T. Saito, and M. Ewing, A Cretaceous (Turonian) core from the Naturaliste Plateau southeast Indian Ocean, *Deep Sea Res., 14,* 421–426, 1967.

Cameron, R. L., The Vanderford submarine valley, Vincennes Bay, Antarctica, *Geology and Paleontology of the Antarctic, Antarctic Res. Ser.*, vol. 6, edited by J. B. Hadley, pp. 211–216, AGU, Washington, D. C., 1965.

Conolly, J. R., Geology and submarine canyons, east Bass Strait, *Mar. Geol., 6*(6), 449–461, 1968.

Conolly, J. R., Western Tasman sea floor, *N. Z. J. Geol. Geophys., 12*(1), 310–343, 1969.

Conolly, J. R., Sedimentary history of the continental margin of Australia, *Trans. N. Y. Acad. Sci., Ser. 2, 32*(3), 364–380, 1970.

Conolly, J. R., and R. R. Payne, Sedimentary patterns within a continent–mid-oceanic ridge–continent profile: Indian Ocean south of Australia, in *Antarctic Oceanology II: The Australian–New Zealand Sector, Antarctic Res. Ser.*, vol. 19, edited by D. E. Hayes, pp. 295–315, AGU, Washington, D. C., 1972.

Conolly, J. R., and C. C. von der Borch, Sedimentation and physiography of the sea floor south of Australia, *Sediment. Geol., 1,* 181–220, 1967.

Conolly, J. R., A. Flavelle, and R. S. Dietz, Continental margin of the Great Australian Bight, *Mar. Geol., 8,* 31–58, 1970.

Craddock, C., et al., Geologic maps of Antarctica, *Antarctic Map Folio Ser.*, folio 12, edited by V. Bushnell, Amer. Geogr. Soc., New York, 1970.

Doyle, H. A., I. B. Everingham, and D. J. Sutton, Seismicity of the Australian continent, *J. Geol. Soc. Aust., 15*(2), 295–312, 1968.

Emery, K. O., The continental shelves, *Sci. Amer., 221*(3), 107–122, 1969.

Gordon, A. L., On the interaction of the antarctic circumpolar current and the Macquarie ridge, in *Antarctic Oceanology II: The Australian–New Zealand Sector, Antarctic Res. Ser.*, vol. 19, edited by D. E. Hayes, pp. 71–78, AGU, Washington, D. C., 1972.

Gordon, A. L., and R. D. Goldberg, Circumpolar characteristics of antarctic waters, *Antarctic Map Folio Ser.*, folio 13, edited by V. Bushnell, Amer. Geogr. Soc., New York, 1970.

Griffiths, J. R., Continental margin tectonics and the evolution of southeast Australia, *Aust. Petrol. Explor. Ass. J., 11*, 75–79, 1971.

Grinnell, D., Continental margin of the Wilkes Land coast, Master's thesis, University of South Carolina, Columbia, 1972.

Hayes, D. E., and M. Talwani, Geophysical investigation of the Macquarie ridge complex, in *Antarctic Oceanology II: The Australian–New Zealand Sector, Antarctic Res. Ser.*, vol. 19, edited by D. E. Hayes, pp. 211–234, AGU, Washington, D.C., 1972.

Heezen, B. C., and M. Tharp, Physiographic diagram of the Indian Ocean, Geol. Soc. of Amer., New York, 1965.

Heezen, B. C., and M. Tharp, Map of the Pacific Ocean floor (foldout), *Nat. Geogr., 136*(4), 1969.

Heezen, B. C., M. Tharp, and M. Ewing, The floors of the oceans, 1, The North Atlantic, *Spec. Pap. 65*, 122 pp., Geol. Soc. of Amer., New York, 1959.

Houtz, R. E., and R. G. Markl, Seismic profiler data between Antarctica and Australia, in *Antarctic Oceanology II: The Australian–New Zealand Sector, Antarctic Res. Ser.*, vol. 19, edited by D. E. Hayes, AGU, pp. 147–164, Washington, D. C., 1972.

Houtz, R., J. Ewing, and R. Embley, Profiler data from the Macquarie ridge area, in *Antarctic Oceanology I, Antarctic Res. Ser.*, vol. 15, edited by J. L. Reid, pp. 239–245, AGU, Washington, D. C., 1971.

Matthews, D. J., *Tables of the Velocity of Sound in Pure Water and Sea Water for Use in Echo-Sounding and Sound-Randing*, 2nd ed., 52 pp., H. M. Stationery Office, London, 1939.

McWhae, J. R. H., P. E. Playford, A. W. Lindner, B. F. Glenister, and B. E. Balme, The stratigraphy of Western Australia, *J. Geol. Soc. Aust., 4*, 1–161, 1957.

Morgan, W. J., Convection plumes in the lower mantle, *Nature, 230*, 42–43, 1971.

Payne, R. R., and J. R. Conolly, Turbidite sedimentation off the antarctic continent, in *Antarctic Oceanology II: The Australian–New Zealand Sector, Antarctic Res. Ser.*, vol. 19, edited by D. E. Hayes, pp. 349–364, AGU, Washington, D. C., 1972.

Pratt, R. M., and B. C. Heezen, Topography of the Blake plateau, *Deep Sea Res., 11*, 721–728, 1964.

Spry, A., Igneous activity, in The Geology of Tasmania, edited by A. Spry and M. Banks, *J. Geol. Soc. Aust., 9*, 255–284, 1962.

Sykes, L., Seismicity of the Indian Ocean and a possible nascent island arc between Ceylon and Australia, *J. Geophys. Res., 75*(26), 5041–5055, 1970.

Talwani, M., J. Dorman, J. L. Worzel, and G. M. Bryan, Navigation at sea by satellite, *J. Geophys. Res., 71*(24), 5891–5902, 1966.

Udintsev, G. B., et al., New bathymetric map of the Pacific Ocean, *Oceanol. Res. 9*, pp. 60–101, Inst. of Oceanol., Acad. of Sci. of the USSR, Leningrad, 1963.

van der Linden, W. J. M., Extinct mid-ocean ridges in the Tasman Sea and in the western Pacific, *Earth Planet. Sci. Lett., 6*, 483–490, 1969.

van der Linden, W. J. M., and D. E. Hayes, Resolution bathymetry, scale 1:1,000,000, Oceanic Series, N. Z. Oceanogr. Inst., Wellington, in press, 1972.

Vogt, P. R., and J. R. Conolly, Tasmantid guyots, the age of the Tasman basin, and motion between the Australia plate and the mantle, *Geol. Soc. Amer. Bull., 82*(9), 2577–2584, 1971.

von der Borch, C. C., South Australian submarine canyons, *Aust. J. Sci., 30*, 66–67, 1967.

von der Borch, C. C., Submarine canyons of southern Australia: Their distribution and ages, *Mar. Geol., 6*(4), 267–279, 1968.

von der Borch, C. C., J. R. Conolly, and R. S. Dietz, Sedimentation and structure of the continental margin in the vicinity of the Otway basin, southern Australia, *Mar. Geol., 8*, 59–83, 1970.

Weissel, J. K., and D. E. Hayes, Asymmetric spreading south of Australia, *Nature, 231*(5304), 518–521, 1971.

Weissel, J. K., and D. E. Hayes, Magnetic anomalies in the southeast Indian Ocean, in *Antarctic Oceanology II: The Australian–New Zealand Sector, Antarctic Res. Ser.*, vol. 19, edited by D. E. Hayes, pp. 165–196, AGU, Washington, D. C., 1972.

Zhivago, A. V., Outlines of southern ocean geomorphology, *Antarctic Research, Geophys. Monogr. Ser.*, vol. 7, edited by H. Wexler et al., pp. 74–80, AGU, Washington, D. C., 1962.

SEISMIC PROFILER DATA BETWEEN ANTARCTICA AND AUSTRALIA

ROBERT E. HOUTZ AND RUDI G. MARKL

Lamont-Doherty Geological Observatory of Columbia University, Palisades, New York 10964

Asbtract. Seismic profiler data from the area between Australia and Antarctica are used to compile a regional sediment isopach map and to discuss sediment distribution. Basement structure is deduced from sonobuoy data where the total sediment cover is too thick to be profiled. The Australian continental rise and South Australian abyssal plain sediments are acoustically similar to those of the antarctic continental rise and the adjacent South Indian plain. The underlying basement morphology is also similar in the two regions. These similarities suggest that similar depositional conditions have prevailed even though Antarctica has been capped by ice, probably since the Upper Miocene. Ice rafting has not contributed a significant volume of sediment to the post-Upper Miocene section. The South Indian basin contains a thicker sedimentary section, and this thickness makes it about 1 km shoaler than the South Australian basin; this characteristic implies more rapid erosion and sediment transport from Antarctica than from Australia before Upper Miocene time. Sonobuoy data from a north-south traverse at about 127°E and near the antarctic continental margin reveal that the basement dips steeply toward the continent and is covered by more than 4 km of sediment. This thick sediment sequence may represent a southern counterpart to the Otway basin, or rift, which extends along the southeastern coast of Australia. In marked contrast, the margins of the Tasman and Campbell plateaus and the Pacific sector of the antarctic margin are characterized by a flat basement surface up to the foot of the continental slope. A zone of basement highs having the appearance of diapirs is observed in the South Australian abyssal plain. A similar zone approximately equidistant from the mid-ocean ridge crest occurs in the South Indian abyssal plain.

The seismic profiler coverage is shown in Figure 1; the tracks shown are USNS *Eltanin* cruises 27, 34–37, 39–42, 45, and 47, as well as *Vema* cruise 18 and *Robert D. Conrad* cruises 8 and 9. With the exception of *Vema* 18, *Conrad* 8, and part of *Eltanin* 39, all the profiler data were obtained by using an air gun energy source. The deep-sea seismic profiler data from this region have not previously been published. The eastern edge of the area abuts the region described by *Houtz et al.* [1971], and the western edge adjoins the area studied by *M. Ewing et al.* [1969*a*]. Some profiler data from the Naturaliste plateau have been published by *Burckle et al.* [1967]. Numerous publications exist that deal with the southern Australian coastal zone and continental shelf. Profiler data from the continental rise and slope have been published by *Conolly et al.* [1970] and *von der Borch et al.* [1970]. Structure sections south of Cape Leeuwin and south of Adelaide from two-ship refraction data have been published by *Hawkins et al.* [1965].

The present work is the first attempt to analyze the sediment distribution of the entire region between Australia and Antarctica by using profiler data. The interpretation of the sediment distribution is complicated by the fact that modern depositional conditions near Antarctica are very different from past conditions when there was no continental ice sheet. Another factor that complicates the interpretation of the sediment distribution is the large number of offsets of the mid-ocean ridge in this region.

COMMENTS ON ANTARCTIC SEDIMENTATION

Antarctica has been in a polar position and has been sporadically contributing glacial marine sediments to the adjoining basins [*Margolis and Kennett*, 1970] since latest Mesozoic or earliest Cenozoic times. A general cooling of antarctic waters occurred during late Miocene time [*Hays and Opdyke*, 1967; *Bandy et al.*, 1971] and is thought to mark the beginning of the present glacial epoch that eventually resulted in the production of the present continental ice sheet. There is widespread evidence that temperate type glaciation occurred before the Upper Miocene, but firm evidence

Fig. 1. Seismic reflection profiler track chart between Australia and Antarctica. Tracks shown as solid lines are those on which an air gun energy source was used; tracks shown as broken lines indicate use of an explosive source; alternating dashes and dots indicate use of a sparker. The profiler sections shown in this paper are labeled as to figure and profile number and are marked by heavy lines. The region studied is broken into areas 1–7 along lines corresponding to Figure 3; the base of the Australian continental slope forms the northern boundary of areas 5–7. The two cross-hatched areas on 132°E mark zones of diapirlike features; the dotted line in the South Indian abyssal plain marks a change in reflector character.

for a pre-Upper Miocene ice sheet of continental dimensions during the Cenozoic is lacking (there is some evidence for an ice cap during the Permian [*Lindsay,* 1970]). According to *Le Pichon and Heirtzler* [1968], *Le Pichon* [1968], and *Dietz and Holden* [1970], Australia detached from Antarctica near the end of Cretaceous time and migrated northward to its present position. *Weissel and Hayes* [1971;

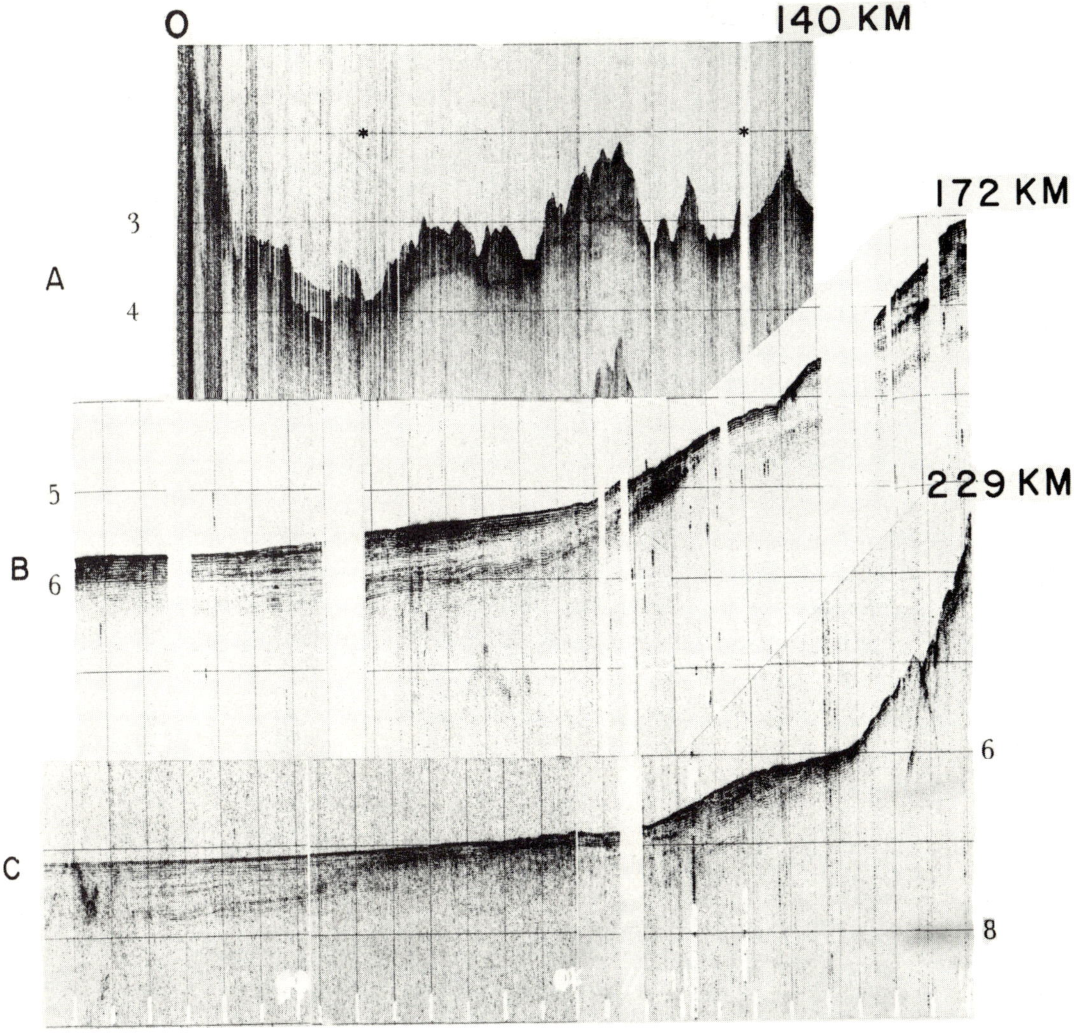

Fig. 2. Profile A: profiler section of the antarctic continental slope; the portion between the asterisks runs parallel to the coastline. Profile B: profiler section of the antarctic continental margin (north is to the left). Profile C: profiler section of the Australian continental margin (north is to the right). Profiles B and C were made on the same cruise.

this volume] trace this migration in detail and show that the separation rate increased 10 m.y. ago from about 5 cm/yr to about 8 cm/yr.

We often take for granted that glacial marine sediments are transported solely by the ice-rafting process. If this process contributed sediment at the rate of 2 cm/yr, the antarctic continental rise should have been covered by 100 meters of ice-rafted sediment since the Upper Miocene. However, sections of the antarctic continental rise and slope seem to be free of young sediments as shown in Figure 2, profile A, in which the submarine canyons appear to have been recently active. In other antarctic sections, the rise sediments appear to be unconsolidated (Figure 2, profile B) but are on-lapped by younger basin turbidites. This distribution suggests (as well as can be judged from the profiler data) that the main process of sediment transport to the South Indian basin is by turbidity flow and possibly also by bottom water circulation rather than by ice rafting. Figure 2, profile C shows a profiler section of the Australian slope and rise that is quite similar to the antarctic sections, even though no glacial deposition has occurred near Australia.

TREATMENT OF THE DATA

The following procedure was adopted in constructing the sediment isopach map (Figure 3; see foldouts at back of volume). The sediment thickness was measured (in units of 0.1 sec of reflection time) at three representative points on each profiler record, i.e., at approximately hourly, or 10-n. mi., intervals. The intervals were selected to indicate the average thickness of the over-all sediment cover rather than merely the average where sediment is actually present. As an example, small seamounts devoid of sediment were overlooked, whereas large ones were incorporated. In some areas, particularly those with an uneven draping of pelagic sediment, the data were smoothed by averaging the three measurements. In areas having sufficient profiler coverage, each measurement was used individually. We believe that if a bias exists in the measuring procedure used, the bias tends to maximize the sediment thickness rather than minimize it.

The data were contoured at intervals of 0.1 sec. As an approximation, 0.1 sec of two-way travel time equals 100 meters of sediment when a sediment velocity of 2.0 km/sec is assumed. Generally, sediment velocities range from about 1.7 km/sec at the sea floor to about 2.5 km/sec at depths of 1 km or more [*Houtz et al.*, 1968]. Hence this approximation is best for a moderate amount of sediment, the approximation being generally too high for very thin sediments and too low for very thick sediments. The percentage variation of true thickness from that obtained by using this approximation is equal to the amount by which the average velocity of the sediment differs from 2.0 km/sec; this amount is rarely more than 15% at reflection times of less than 1 sec.

In general, the contours used are 0.1 through 0.5, 1.0, 1.5, and so on. In some areas, selected contours were merged or deleted because of insufficient information or in the interest of clarity of presentation. For example, the 0.4 contour is not used at all in the northern half of the map. The small unlabeled closed contours within the 0.1-sec contour are 0.1–0.2 sec thick. Owing to the highly variable thickness of the pelagic deposits, contours in these areas represent trends rather than actual gradients between contours.

DISCUSSION OF THE ISOPACH MAP

The principal physiographic features shown in Figure 3 are the Southeast Indian mid-ocean ridge province and the bounding abyssal plains that have buried the northern and southern extremities of the ridge province, the South Australian abyssal plain near Australia and the South Indian abyssal plain near Antarctica, respectively. The abyssal plains represent thick accumulations of turbidites, whereas the area between the plains is characterized by pelagic sedimentation. The contoured area is delimited in the north by the inability to observe the basement beneath the Australian rise or by the abrupt termination of the sediments against the continental terrace. For ease of discussion, the region of study is divided into seven areas, each assigned a number 1–7 (see Figure 1). With the exception of Figure 2, all profiles shown are oriented such that right is north (or east in east-west profiles). All the profiles have a vertical exaggeration of about 25:1 and are uncorrected for sediment velocity.

Southeast Indian Ridge

The position of the crest of the Southeast Indian ridge was determined from topographic data where possible and by magnetic anomaly data elsewhere. The crestal region between 120°E and 128°E is apparently cut by numerous north-south fractures (see *Hayes and Conolly* [this volume, Figure 1]), though major lateral offsets do not appear to have occurred here. This zone is referred to as the Australian-antarctic discordance by Hayes and Conolly. The crest axis between 128°E and 139°E, determined to be aseismic by *Barazangi and Dorman* [1969], is discussed by *Weissel and Hayes* [1971]. The major offsets of the ridge crest between 139°E and 153°E have been discussed by *Le Pichon and Heirtzler* [1968] and *Weissel and Hayes* [1971; this volume]. Subsequent references to igneous basement or basement refer to buried mid-ocean ridge topography, a surface that underlies most of the area of study. Mid-ocean ridge province refers to that portion of this surface not covered by continental rise or abyssal plain.

The basement topography, although varying in roughness geographically [*Weissel and Hayes*, 1971], is visibly smoother on east-west profiles than on north-south ones. A dominant east-west trend was also obtained from plots of the average wave length of topography measured at intersections of three or more tracks. This direction, parallel to the crest, is expected within a given block of the ridge province, as the dominant linearity should be parallel to the spreading axis. The crestal zone between 120°E and 127°E and 44°S and 54°S (the Australian-antarctic discordance) has the greatest relief found in the ridge province in the area of Figure 3 (see foldouts at back of volume). Here the short wave length topography resulting from spreading of the sea floor is overshadowed by longer average wave lengths (see also *Hayes and Conolly* [this volume]). Determinations from track intersec-

Fig. 4. Seismic profiles of the boundary between the South Indian abyssal plain and the southern flank of the Southeast Indian ridge. The profiles are located in Figure 1; the vertical exaggeration is about 25:1. North is to the right.

tions in this region generally show north-south lineations, apparently a series of north-south fractures.

Area 1

With few exceptions, the mid-ocean ridge crest is devoid of sediment cover [*J. Ewing and M. Ewing*, 1967]. A prominent feature shown in Figure 3 is the vast area (area 1) of the ridge province that contains less than 100 meters of the sediment. The small sediment patches of 100–500 meters in thickness within the area are pelagic sediments generally draped over the rough basement topography. Some of the patches are elongate and show east-west trends. Several other patches (specifically those labeled 3, 4, 5 along 126°E) are small intermontane ponds, the sediment apparently being derived from the adjacent slopes.

The pelagic sediments in the area of study occur in one of two configurations. Most commonly they form 'caps' (Figure 4, profile A) having internal reflectors that in some cases are conformable to the underlying topography and that in other cases show bedding that is virtually horizontal. These deposits, which are commonly discontinuous or dissected, seem to occur more commonly on highs than in lows and rarely form ponds. The other configuration, shown in Figure 5, profile A and referred to as 'drift,' may not be truly pelagic. These sediments are acoustically transparent and show varying degrees of conformity to basement in the same locality. They tend to fill valleys asymmetrically (i.e., piling up against one side but not the other), and their deposition is clearly influenced by deep currents. In the area of study, drift occurs almost exclusively in area 5.

Area 2

This area is covered with about 100 meters of pelagic sediment with several poorly defined thicker patches. The northern boundary may be interpreted to extend even farther north than shown. A portion of the eastern boundary appears to be coincident with a major fracture zone (see *Hayes and Conolly* [this volume]).

Area 3

This area contains a thick pelagic cover with a maximum thickness at the western edge of the area. The contours represent considerable averaging, even though the control is reasonably good. The northeastern part of the area is within the Australian-antarctic discordance. Although most of the ridge province exhibits little or no sediment, the sediments in area 3 encroach to within 20 km of the apparent topographic crest near 121°E and may continue across it at 107°E.

Profiler data from the *Eltanin* cruise 47 along 50°S to Îles Kerguelen allow us to extrapolate westward from our study area. Pelagic sediments greater than 300 meters in thickness extend west to 100°E with a maximum of 700 meters occurring near 103°E. Between 100°E and 90°E, the thickness averages 200–300 meters, and beyond this area the thickness goes to less than 100 meters.

South Indian Abyssal Plain

As revealed acoustically by the seismic profiler, the abyssal plain turbidites are typical of abyssal plains elsewhere in the world. As is the case on the continental rise to the south, no particularly prominent or continuous internal reflectors are observed over most of the plain (e.g., Figure 4, profile E located in Figure 1). However, in the northern portion of the plain (north of the dotted line in Figure 1), discrete and continuous reflectors are observed (see Figure 4, profiles A–D). This change in reflector character is not an artifact of the profiler instrumentation; it appears to be related to distance from the antarctic margin. In both regimes, the reflectors are essentially horizontal and characteristically show no relation to the basement topography, though near the ridge conformity is seen locally, perhaps because of pelagic sedimentation beneath. The northern limit of the abyssal plain is quite easily determined from the seismic records, largely because of the manner in which turbidites are emplaced, i.e., near-bottom transport tending to create essentially horizontal layers. The turbidites fill in around isolated seamounts and lap against the rough topography of the ridge province.

A topographic profile along 150°E shows that the abyssal plain does not extend this far east and that here the antarctic continental rise abuts the ridge flank. Between the South Indian abyssal plain and area 2, the abyssal plain boundary essentially coincides with the 0.2 contour. The thin pelagic cover of area 2 is easily distinguished from the abyssal plain turbidites, which at 140°E (Figure 4, profile E) abut the ridge and thicken sharply toward the south. As shown in Figure 4, profile D, the boundary between the abyssal plain and the adjacent portion of area 1 (east of 127°E) is not so clear-cut. The relief of the ridge is substantially lower here and thus allows the turbidites to lap on the ridge over a broad area. The region west of 127°E is cut by numerous north-south fractures (the Australian-antarctic discordance) and consequently has great

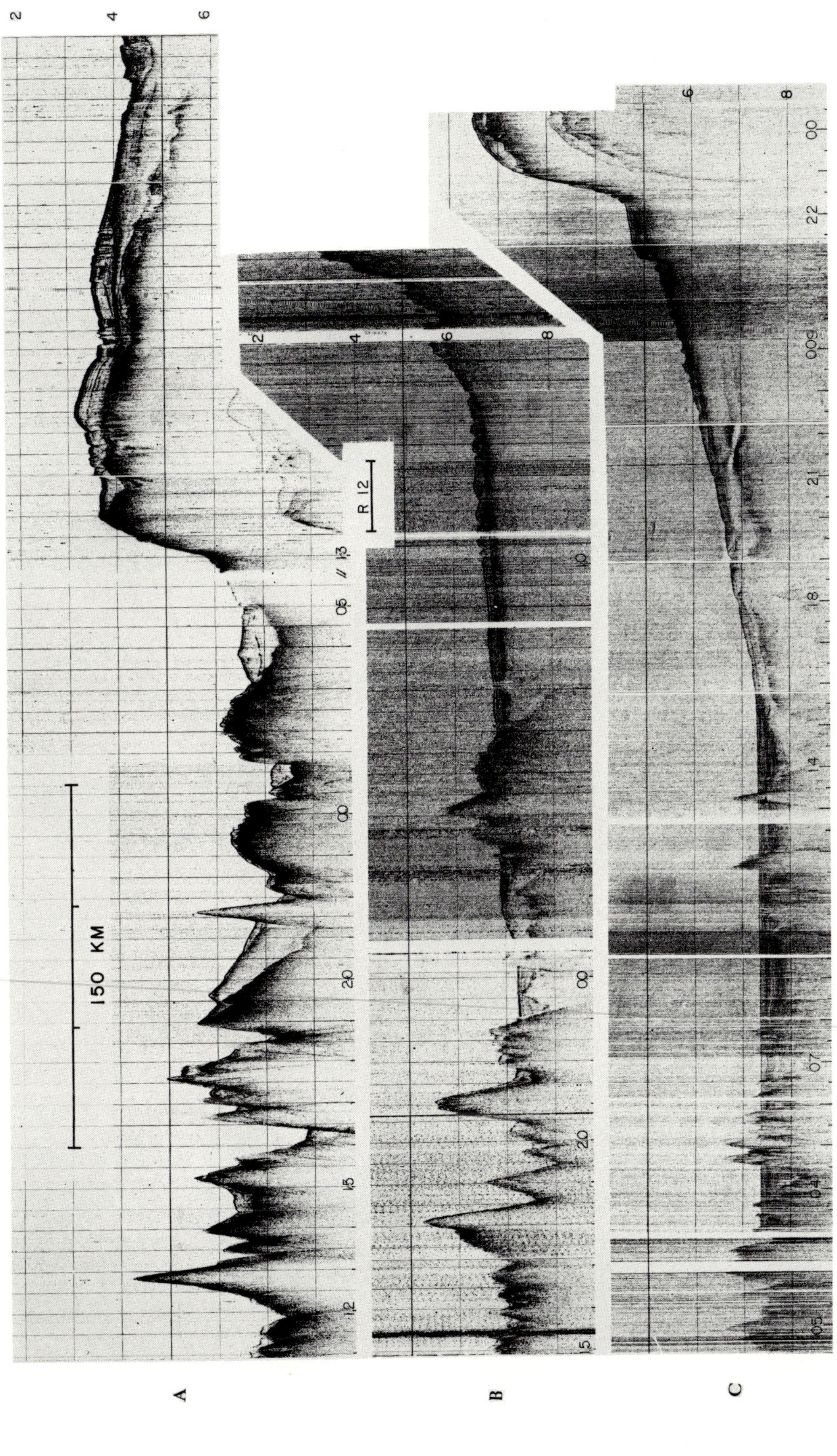

A B C

Fig. 5. Seismic profiles of the western Australian continental margin. The profiles are located in Figure 1; the vertical exaggeration is about 25:1. North is to the right.

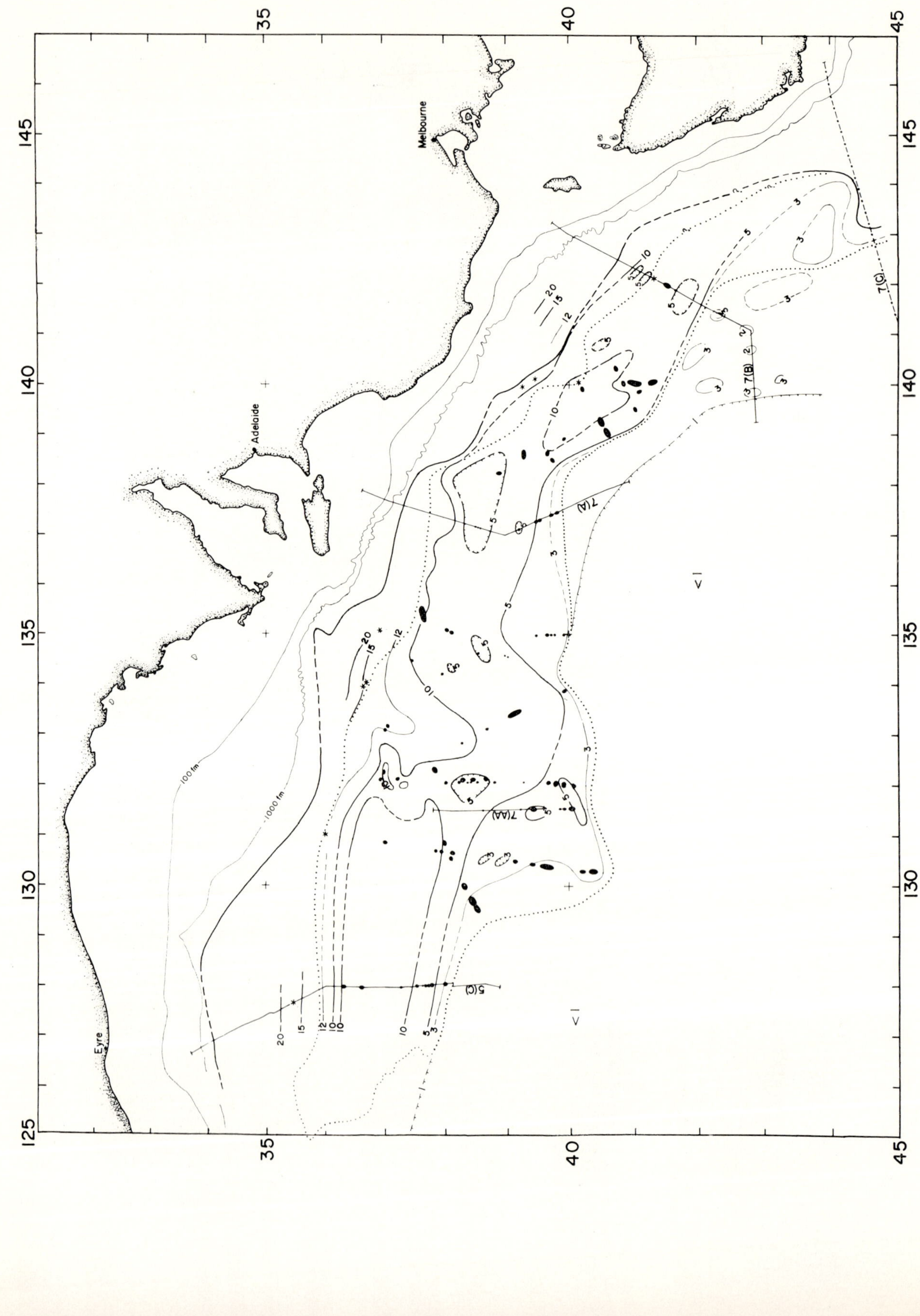

Melbourne
Adelaide
Eyre
100 fm
1000 fm
7(A)
7(B)
7(C)
7(AA)
5(C)
125
130
135
140
145
35
40
45

relief (800–1600 meters) as shown in Figure 4, profile C. The plain here is considerably deeper than in the other profiles shown. The sediments have filtered around basement highs and have left some seamounts protruding from the plain; the sediments have penetrated northward as far as 53°S, and this northward penetration supports the assumption that the Australian-antarctic discordance is composed of north-south fractures. Between the South Indian abyssal plain and area 3, the boundary is less sharp than elsewhere because the turbidites lie on older pelagics and are also intermingled with more recent pelagics on the ridge (Figure 4, profiles A and B).

The seismic profiler control in the South Indian abyssal plain is poor south of 60°S, owing to the difficulties of working in the polar sea ice. On the few tracks that penetrate south of 60°S, the igneous basement cannot be traced deeper than about 1.8 sec because it dips steeply under the antarctic continental margin (Figure 4 profile D). Therefore, the 2.0- and 2.5-sec contours are based primarily on sonobuoy measurements that are listed in Table 1. The choice of a southern boundary of the South Indian abyssal plain depends on the definition used. There is a perceptible change in bottom gradient at a depth of 2350 fm (at about 62°S along 128°E); the southern boundary shown in Figure 3 is the 2350-fm contour. The abyssal plain grades quite smoothly into the broad continental rise, which in turn rises gradually and meets the base of the continental slope at about the 2000-fm contour.

Area 4

The pelagic cover here is markedly thinner than that in area 3. The relatively thick circular patch on the margin of Figure 3 is well determined; however, the eastern boundary of the area marked by the broken line in Figure 3 reflects the poor control in area 4.

Area 5

The Naturaliste plateau lies west of the southwest corner of Australia. The plateau and environs are shown in Figure 5, profile A and have been discussed by *Burckle et al.* [1967]. Between the plateau and the 0.3-sec contour, the sediment (predominantly pelagic drift) overlies igneous basement that is comparatively smooth (relative to typical ridge topography). A few large basement highs are present here (crossing at an oblique angle in Figure 5, profile A); the sediments in the valleys in between the highs tend to be deposited asymmetrically against one side, though no consistent pattern is evident. The Diamantina fracture zone (relief of 1200–2000 meters) passes between the 0.3- and 0.2-sec contours. The sediment here also is drift that is quite thick locally (Figure 5, profile A), as indicated by isolated 1.0-sec contours. The ridge province begins between the 0.2-sec contour and area 1 (just south of the end of Figure 5, profile A); regionally, the ridge topography is smooth, but it has 'short wave length' roughness that is covered by pelagic caps. The profile coverage in area 5 is insufficient to define the sediments associated with this complex structure.

Area 6

This area, which includes the South Australian abyssal plain, is bounded on the north by the base of the continental slope, or escarpment (where no rise exists). The proximity of this boundary to the edge of the abyssal plain illustrates that the western and eastern portions of the area have a broad, well-developed continental rise, whereas the middle portion does not.

In the western part of area 6, as in area 5, the Diamantina fracture zone measuring 150 km wide separates the continental rise from the ridge province by major relief of 800–1800 meters (Figure 5, profile B). The northernmost peaks have tended to dam the continental rise. Sonobuoy 12E44 on the upper continental rise in Figure 5, profile B indicates a depth to the basement of 2.4 sec (3.3 km). North-south tracks in this region all reveal two or three major valleys in the fracture zone, the southernmost one being the deepest and containing little or no sediment. The other valleys contain ponds of stratified sediment that are filled to essentially the same level as the abyssal plain. If connected, the ponds form a sinuous pattern (stippled area in Figure 3) that may be a westward extension of the abyssal plain, or at least its equivalent in this area. The actual relation of this region to the abyssal plain is not clear, since we have no control

Fig. 6. Detailed isopach map of the South Australian abyssal plain; contours are in tenths of seconds of reflection time. The abyssal plain is outlined by a dotted line; the bold line north of the plain marks the base of the continental slope; the region south of the plain contains less than 100 meters of sediment. The bold track lines locate profiles shown in Figures 5 and 7. Seamounts within the abyssal plain are shown in black.

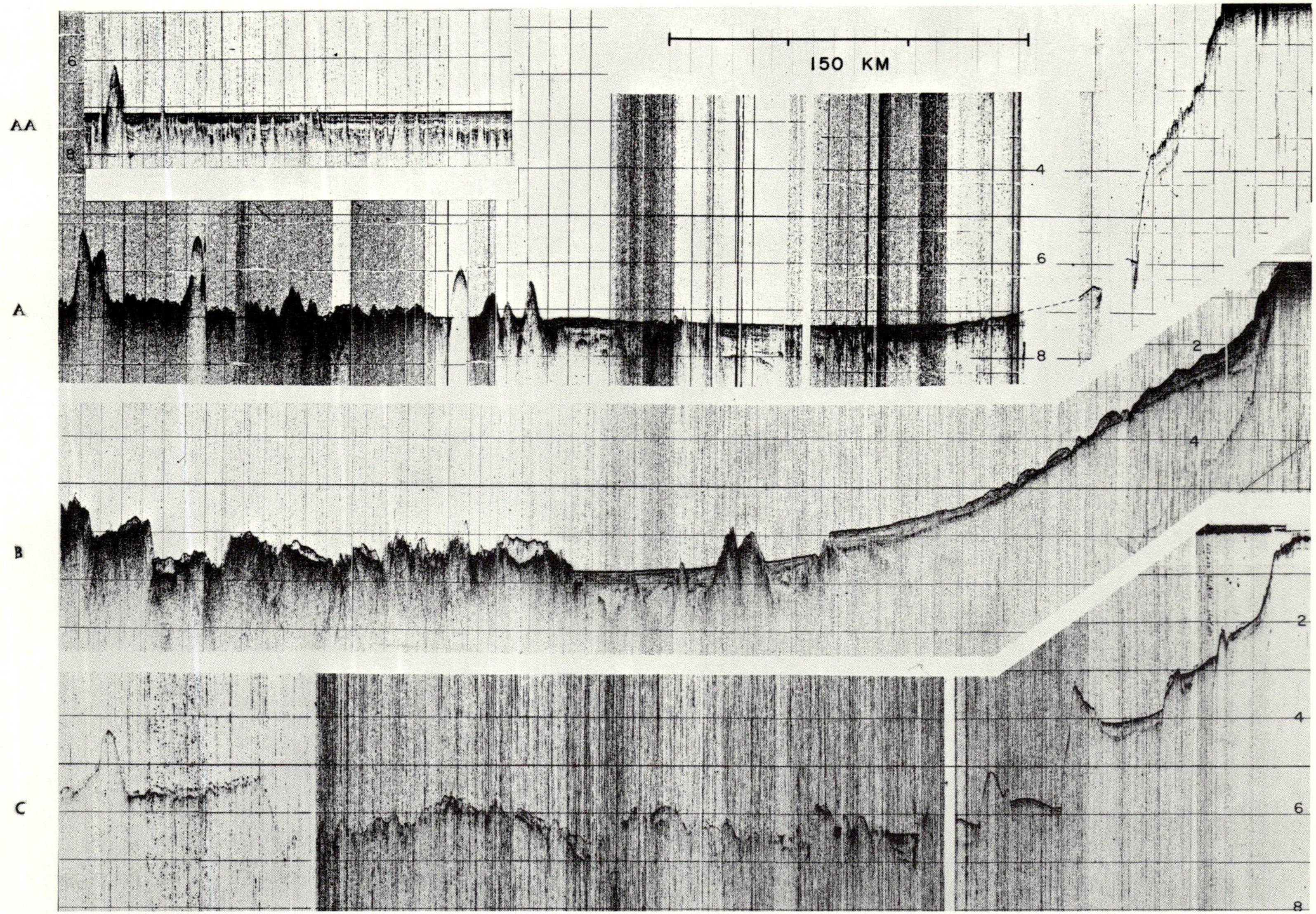

Fig. 7. Seismic profiles of the eastern Australian continental margin. The profiles are located in Figures 1 and 6; the vertical exaggeration is about 25:1. North (and east) is to the right.

between 121°E (shown by *Burckle et al.* [1967, Figure *3B*]) and 128°E (shown in Figure 5, profile C of this study).

In the central part of area 6, the South Australian abyssal plain reaches its maximum width and thickness. A detailed isopach map of the abyssal plain is shown in Figure 6. The western limit of the plain is taken from *Conolly and von der Borch* [1967]. Figure 5, profile C and Figure 7, profiles A, B, and C (shown in Figure 6) show the over-all relationships between the continental margin, abyssal plain, and mid-ocean ridge flank. The variation in development of the continental rise at different localities along the same continental margin is particularly striking. The upper rise is commonly cut by canyons and in some cases has been modified by slumping. The basement surface is not easily traced beneath the rise.

A basement high is frequently observed at the base of the Australian rise. In some crossings, such as that in Figure 7, profile B, these highs crop out and, having dammed the rise sediments, presumably are ridges. At 140°E, two crossings of a similar ridge show backfilling resulting in a hummocky plateau that is 40 km wide and is elevated 600 meters above the abyssal plain. Isolated peaks, which may be buried (as on Figure 5, profile C above the 1800-hours mark), occur on other crossings and may be equivalent to the ridges. In Figure 6, these basement highs are located with an asterisk; those on 131°E and 135°E are not well determined.

Of particular interest are diapirlike features that are similar acoustically to nearby buried basement peaks except for their narrower breadth, their greater number per unit area, and the fact that in some cases the surrounding strata are bowed up around them (Figure 7, profile AA). Similar features occur in the South Indian abyssal plain (see left end of Figure 4, profile D). The two areas in which these features occur are shown in Figure 1. Considering their setting (i.e., surrounded by bona fide basement peaks), one can most reasonably attribute these features to short wave length basement peaks that perhaps are accentuated through the defocusing of the incident energy by the overlying strata, as discussed qualitatively by *Collette and Rutten* [1970]. The two zones are similar in area, but the northern zone lies 150 km farther from the mid-ocean ridge than the southern zone; i.e., the northern zone lies between anomalies 13 and 18, whereas the southern zone lies between anomaly 10 and a point midway between anomalies 13 and 19 [*Weissel and Hayes*, this volume]. The two zones lie at the same longitude, whereas if they had a common origin at the ridge crest they should now be offset by nearly 20° of longitude. This offset would correspond to the longitudinal offset that accumulated between Australia and Antarctica since their breakup [*Dietz and Holden*, 1970].

The turbidites of the South Australian abyssal plain show many discrete, horizontal, and fairly continuous reflectors (Figure 5, profile C). Illustrating the area near the southern margin of the plain, Figure 7, profile A shows a broad high bounded by basement outcrops, and nearby profiles show other undulations. The plain is quite narrow farther east where section 7(B) in Figure 6 crosses. Here the rise is again well developed and is cut by canyons. A considerable amount of pelagic material occurs on the ridge in this region (see Figure 6). In Figure 7, profile C (a 5000-joule sparker profile), a sediment pocket at the base of the Tasmanian margin marks the southern extent of the abyssal plain. The block faulting on the Tasmanian margin aligns with the Tasman fracture zone [*Hayes and Conolly*, this volume], but may also represent the Otway rift valley (to be discussed). The escarpments at the left end of this profile and profile B in Figure 7 align with others along 140°E (south to at least 58°S) and define a major fracture called the George V fracture zone by *Hayes and Conolly* [this volume].

Area 7

The control in this area is inadequate to define the highly variable sediment distribution; consequently most of the contours shown are marked as broken lines. The structure of the southern margin of the South Tasman rise is similar to that of the eastern margin of Tasmania shown in Figure 10, profile A, though this structure is even more like that of the Campbell plateau (Figure 10, profile B). The thick (1.0 sec) unit of predominantly transparent sediment that abuts the southern edge of the South Tasman rise is interrupted by two separate west-northwest trending ridges of 1000–1800 meters in relief; however, the unit extends southward to the east-west profile shown in Figure 8. The sediment body, 0.5 sec thick in Figure 8, pinches out abruptly about 50 km south of that track. From this point southward to area 1, the basement has a rough surface with low relief and is overlain by thin pelagic caps. Antarctic bottom water (AABW) is observed to flow northward across the mid-ocean ridge in this region and probably merges with the southeasterly flowing antarctic circumpolar

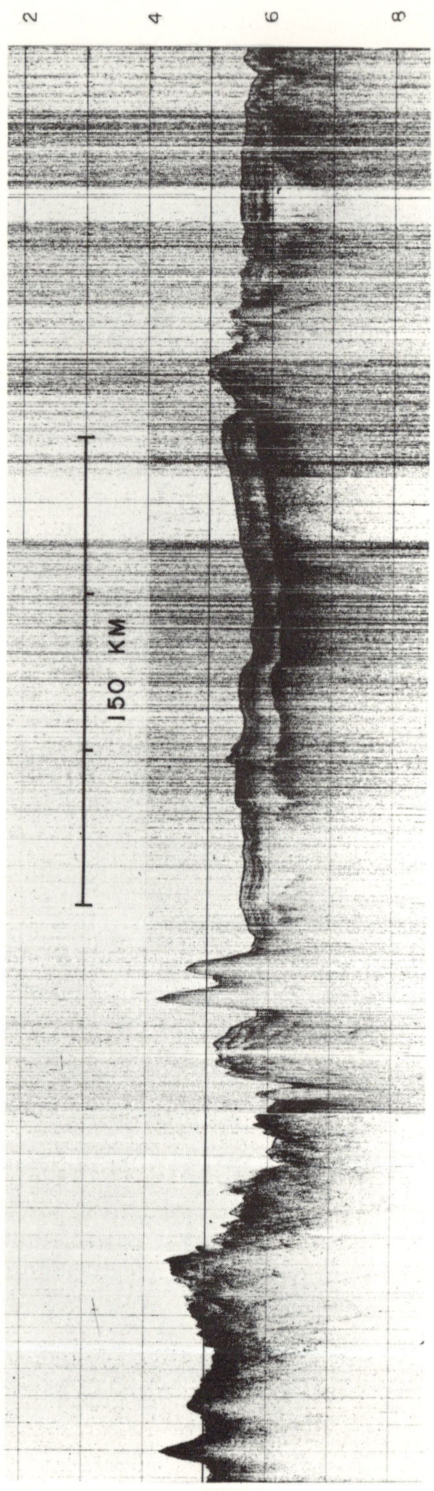

Fig. 8. An east-west profile south of the South Tasman rise. The profile is located in Figure 1. East is to the right.

current [*Gordon*, this volume]. We find no zone of increased sedimentation that might mark the path of the AABW. Since this water is thought to have previously circled the South Indian basin [*Eittreim et al.*, 1971] and in addition may increase in velocity when crossing the mid-ocean ridge, the amount of sediment being carried and deposited in this region may be insignificant.

Figure 8 crosses a major fracture zone (125 km from the left) that coincides with the boundary between areas 1 and 7 as far south as 58°S; this fracture (the Tasman fracture zone) has been delineated from magnetic anomaly data by *Weissel and Hayes* [this volume]. Note the difference in regional basement depth and basement roughness on either side of the fracture. Another north-south fracture (the Balleny fracture zone) is visible about 50 km from the right end of Figure 8; this zone also corresponds with a major offset determined from magnetic data. East of the Balleny fracture zone, the basement becomes rough again and stays that way to the Macquarie ridge. The sediments also extend to the Macquarie ridge and maintain conformity to the basement and a thickness of about 400 meters but show less stratification. *Houtz et al.* [1971] have described the sediment distribution in the Macquarie ridge area.

SEDIMENT PROVENANCE

Turbidites

Our data show that the sea floor in the South Indian basin is about a kilometer shallower than that in the South Australian basin and is largely the result of thicker sediments in the former basin. This difference is attributed to the low sedimentation rates that resulted from the subdued topography that has prevailed in Australia throughout much of the Tertiary. The two sediment bodies appear very similar in the profiler records, and this similarity may indicate that ice rafting has not contributed substantially to the sedimentation near Antarctica.

Crary et al. [1962] show that sea ice accretes to the glaciers that enter the Ross Sea, but rapid melting toward the seaward edge of the Ross ice shelf results in a net loss of bottom ice. Bottom melting therefore frees the particulate matter concentrated in the basal 10–20 meters of the glaciers. The larger particles would remain on the shelf, whereas the finer fraction would be transported to the open sea by the strong bottom currents on the shelf described by *Jacobs et al.* [1970]. The bottom water turns westward from the

Ross Sea shelf toward the South Indian basin, as shown by nephelometer data [*Eittreim et al.*, 1971] and hydrographic data [*Gordon*, 1971]. Crary and others have pointed out that most antarctic glaciers accumulate to form floating ice shelves (usually in embayments) where the ice may remain for hundreds of years before reaching the open sea. If other ice shelves likewise suffer a net loss of bottom ice, ice rafting cannot be a volumetrically significant process at great distances from the coast at the present time. Piston core data from the South Indian basin, as discussed by *Conolly and Payne* [this volume], show quite recent turbidity activity, which they attribute to the reworking of slope sediments deposited before the onset of continental glaciation. This interpretation is not in conflict with the evidence for typical turbidite deposition that appears in the profiler data. Probably, the South Indian basin has remained essentially unchanged since the Upper Miocene, the probable time when Antarctica entered ice age conditions.

Pelagic Sediments

Pelagic sediment (as interpreted from profiler data) refers to acoustically transparent material that has formed draped deposits that are essentially conformable to basement. These deposits are believed to have been formed by particles moving essentially vertically downward through the water column. The nature of the particles, whether detrital or biogenic, cannot be judged from the profiler data. In the absence of strong bottom currents, biogenic sediments should be accumulating most rapidly beneath the polar front. In our study area, the polar front overlies the mid-ocean ridge where the sediments are thinnest; several explanations are possible.

1. The polar front has only recently moved into its present position.
2. Accumulation of biogenic sediment is slow compared to its lateral transportation by sea floor spreading.
3. Deep currents redistribute biogenic fallout.
4. Pelagic components are mostly nonbiogenic.

Quite possibly all these factors come into play. *M. Ewing et al.* [1969*b*] noted similar discrepancies in the Pacific between the location of the present polar front and the location of the thickest pelagic sediments in the Pacific Ocean.

Although area 3 and area 2 of the ridge province contain a greater than average (for this region) pelagic sediment cover, there is a conspicuous zone between them that is relatively free of sediment. We know that the polar front does not loop southward here, nor are sea floor spreading rates higher here than elsewhere. Hence control of sedimentation by deep currents seems to be required.

Sedimentation Controlled by Bottom Currents

On the basis of nephelometer data, *Eittreim et al.* [1971] suggest that antarctic bottom water enters the South Indian basin from the Ross Sea and forms a clockwise gyre within the basin bounded on the north by the mid-ocean ridge. The eastward thinning sediments in area 3 could be interpreted as deposition from the gyre-controlled nepheloid layer; however, the bare spot between areas 3 and 2 remains a problem. Clear evidence of bottom current controlled sedimentation is not identified south of the mid-ocean ridge in this area. However, such evidence exists in the form of a sedimentary ridge near Îles Kerguelen to the west. North of the mid-ocean ridge, sedimentation controlled by the bottom current is present as the drift deposits near the Naturaliste plateau as discussed previously. Acoustically transparent sediments similar to drift, though not so locally variable or obviously current controlled, reach thicknesses of 1.0–1.5 sec on the eastern Tasmanian-Australian margin (Figure 10, profile A).

BASEMENT MORPHOLOGY

Compressional wave velocity and layer thicknesses obtained from sonobuoy records of variable angle reflections and refractions are listed in Table 1. The velocities and thicknesses in the table are not correlated; e.g., all the entries in the columns headed V_n are not necessarily from the same layer. Interval velocities based on variable angle reflection data appear in the table with their standard deviations (refraction velocities appear without standard deviations). At some stations (e.g., 6E37), a deep reflector was recorded that was too poorly defined to yield an interval velocity. If this reflector occurs below a refracting interface, the refraction velocity has been used as an interval velocity to estimate layer thickness. Some velocities were assumed, as indicated in the table.

The generalized sections showing depth to basement in kilometers (Figure 9) are plotted from sonobuoy data. These results show that the basement surface distinctly steepens beneath the continental rises of Australia and Antarctica. The sediment thickening that occurs beneath the Australian margin is well documented, notably by *Weeks and Hopkins* [1967]. It is interpreted by *Sproll and Dietz* [1969], *von der Borch*

TABLE 1. *Eltanin* Sonobuoy Solutions

Sonobuoy Stations	Water, km	V_1, km/sec	SD_1, km/sec	H_1, km	V_2, km/sec	SD_2, km/sec	H_2, km	V_3, km/sec	SD_3, km/sec	H_3, km	V_4, km/sec	SD_4, km/sec	H_4, km	Latitude S	Longitude E
1E37	0.67	1.70	...	0.45	2.30	...	0.56	2.94	...	0.35	6.10	...	...	49°04′	168°20′
2E37	2.87	2.50*	...	1.72	3.75	...	0.89	4.36	...	...	...	...	...	65°05′	143°47′
5E37	4.23	1.82	0.09	0.70	1.95	0.38	0.41	3.37	0.10	1.69	4.20	0.57	1.27	62°43′	126°54′
6E37	4.43	1.82	0.30	0.40	2.14	0.05	0.83	2.46	0.16	0.41	4.18	...	1.15	61°39′	126°43′
7E37	4.54	1.64	0.09	0.36	2.05	0.11	0.39	3.41	0.12	1.08	...	...	...	60°′38	126°19′
8E37	4.62	1.87	0.12	0.68	1.90	0.29	0.36	2.50*	...	0.37	...	...	...	59°10′	125°53′
9E37	4.66	1.75	0.17	0.40	...	...	...	...	...	...	...	...	...	57°06′	125°12′
10E37	4.74	2.05	0.15	0.31	...	...	...	...	...	...	...	...	...	54°13′	124°58′
13E37	4.41	2.37	0.10	0.48	2.47	0.04	1.26	...	...	...	...	...	...	39°53′	140°41′
14E37	3.76	2.26	0.10	1.03	2.70	0.03	0.99	3.50	...	2.45	...	...	...	39°41′	141°12′
15E37	2.93	2.50*	...	1.72	3.65	...	4.05	...	...	...	...	...	...	39°30′	141°45′
16E37	2.04	1.79	0.05	0.41	1.77	0.12	0.37	3.20	...	5.40	5.25	...	...	39°23′	141°59′
1E39	5.57	2.05	0.10	0.43	2.68	0.12	0.28	3.19	0.09	0.59	...	...	...	37°33′	132°04′
3E39	4.72	2.25	0.08	0.47	3.40	...	4.00	4.01	...	...	...	...	...	57°42′	127°28′
5E39	2.94	1.76	0.15	0.39	...	...	...	...	...	...	...	...	...	36°23′	163°06′
1E40	5.60	1.60	0.16	0.23	2.75	0.19	0.43	...	...	...	...	...	...	35°12′	132°04′
2E41	4.45	2.19	0.11	0.64	...	...	...	...	...	...	...	...	...	62°42′	132°01′
5E41	5.54	2.43	0.28	1.12	...	...	...	...	...	...	...	...	...	37°18′	132°11′
6E41	5.55	2.50	0.26	0.69	...	...	...	...	...	...	...	...	...	36°41′	133°04′
7E44	4.71	2.20*	...	...	4.43	...	2.84	6.50	...	...	...	...	...	58°01′	128°′42
8E44	4.69	1.64	0.09	0.65	2.50*	...	0.37	5.51	...	2.00	6.85	...	...	58°01′	126°21′
9E44	4.65	2.37	0.10	1.17	5.40	...	...	...	...	...	...	...	...	59°02′	122°23′
12E44	4.78	1.82	0.18	0.41	2.88	0.14	2.21	3.80*	...	0.72	...	...	...	36°09′	116°13′
13E44	0.14	2.00*	...	0.36	5.65	...	...	...	...	...	...	...	...	34°42′	114°55′
14E44	0.05	2.00*	...	0.28	4.41	...	0.82	6.48	...	...	...	...	...	33°48′	114°43′

* Assumed velocity.
SD, standard deviation of interval velocity.

et al. [1970] and *Griffiths* [1971] as a filled rift valley that formed between Australia and Antarctica just prior to their separation. Sonobuoy 12E44, well to the west of the section in Figure 9, reveals a fairly thick section that also indicates an abrupt steepening of the basement surface at the southwest corner of Australia. This solution agrees well with that of *Hawkins et al.* [1965] south of Cape Leeuwin. The basement outcrop at 56°S in the antarctic section (Figure 9) is peculiar to this part of the basin and is not found on other north-south tracks. The steepening dip of the basement surface is shown in Figure 4, profile D. Note that this increase in the dip of the basement surface is even steeper than it appears in the profile, owing to the greater sound speed associated with thickening sediments.

Unpublished *Eltanin* free air gravity minima of 50–60 mgals occur near the continental slope of Australia and Antarctica. Preliminary attempts to relate the dipping basement of the antarctic section (Figure 9) to observed shipboard gravity values show a reasonable correlation (L. Neuman, personal communication, 1971). The magnetically quiet zone reported by *Weissel and Hayes* [this volume] coincides approximately with the region where the basement surface steepens beneath the continental rises. Their work shows that the magnetically quiet zone is at least as old as Eocene.

The geophysical data from the Australian margin support the conclusions of numerous workers who describe the Otway basin or Otway rift valley (located in the inset of Figure 9) as an old submarine rift valley formed between Australia and Antarctica during the initial stages of their separation. *Sprigg* [1966], *Griffiths* [1971], and other workers have predicted that the southern half of the old submarine rift valley would be found near the antarctic continental margin. Our sonobuoy data near Antarctica are quite reliable (stations 5, 7, and 9 from *Eltanin* cruise 37 yield interval velocities down to the basement); but our data do not reveal a typical rift valley, although they do reveal a pronounced thickening of the sedi-

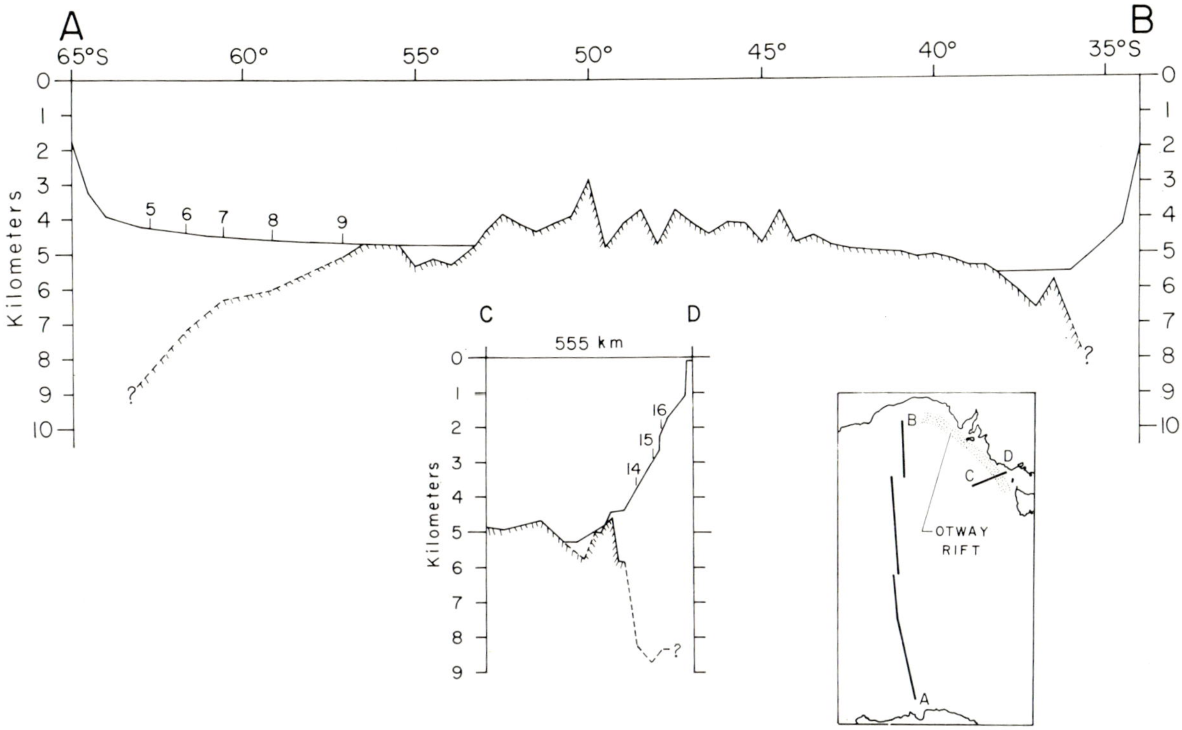

Fig. 9. Structure sections based on *Eltanin* sonobuoy solutions. Numbers represent stations listed in Table 1.

ments. Perhaps the original rift valley was asymmetrical or perhaps it broke apart unevenly. *Hawkins et al.* [1965] made five determinations of velocity in the deepest sedimentary layer resting on the basement along the Australian margin. These velocities measured 3.5, 3.8, 3.9, 3.9, and 4.1 km/sec. These values are similar to our interval velocities based on sonobuoy data from the same region and are also similar to our interval velocities obtained from the antarctic continental rise (see Table 1). The table also shows that the velocities in the sediments increase rather regularly as the overburden increases. Hence there is little correlative significance in the similar velocities that are obtained from the Australian and antarctic margins, the sediment sections of which are about the same thickness.

Unpublished profiler data from the Ross Sea show an irregular continental slope west of 175°W. This slope is apparently modified by normal faults (caused by rifting) and is therefore very similar to the western edge of the Tasmanian platform, as shown in Figure 7, profile C. East of 175°W, the continental slope in the Ross Sea is relatively undisturbed. Consequently, the Gondwanaland reconstruction of *Sproll and Dietz* [1969] and other workers that fits Tasmania into the Ross Sea west of 175°W is supported by the similar basement morphology of the two regions. Presumably, the formation of a rift valley often precedes the separation of cratons, as shown by modern rift valleys such as the Red Sea.

In contrast, other continental slopes such as the eastern margin of the Tasmanian platform (Figure 10, profile A) and the entire southeastern margin of the Campbell plateau (Figure 10, profile B) apparently lack marginal rifts. If Australia and New Zealand were formerly joined, the claim could be made that the thick sediments in the Emerald basin [*Houtz et al.*, 1971] have filled an old marginal rift formerly between Australia and New Zealand. This interpretation cannot be made with finality because more recent tectonic activity has greatly modified the western margin of southern New Zealand. If New Zealand were formerly joined to Antarctica (presumably in the Ross Sea sector, east of 175°W), the mode and/or the time of separation would have been quite different from the Australian-antarctic mode of separation. We therefore suggest that the Campbell plateau and the Ross Sea shelf were sheared apart as a result of a large

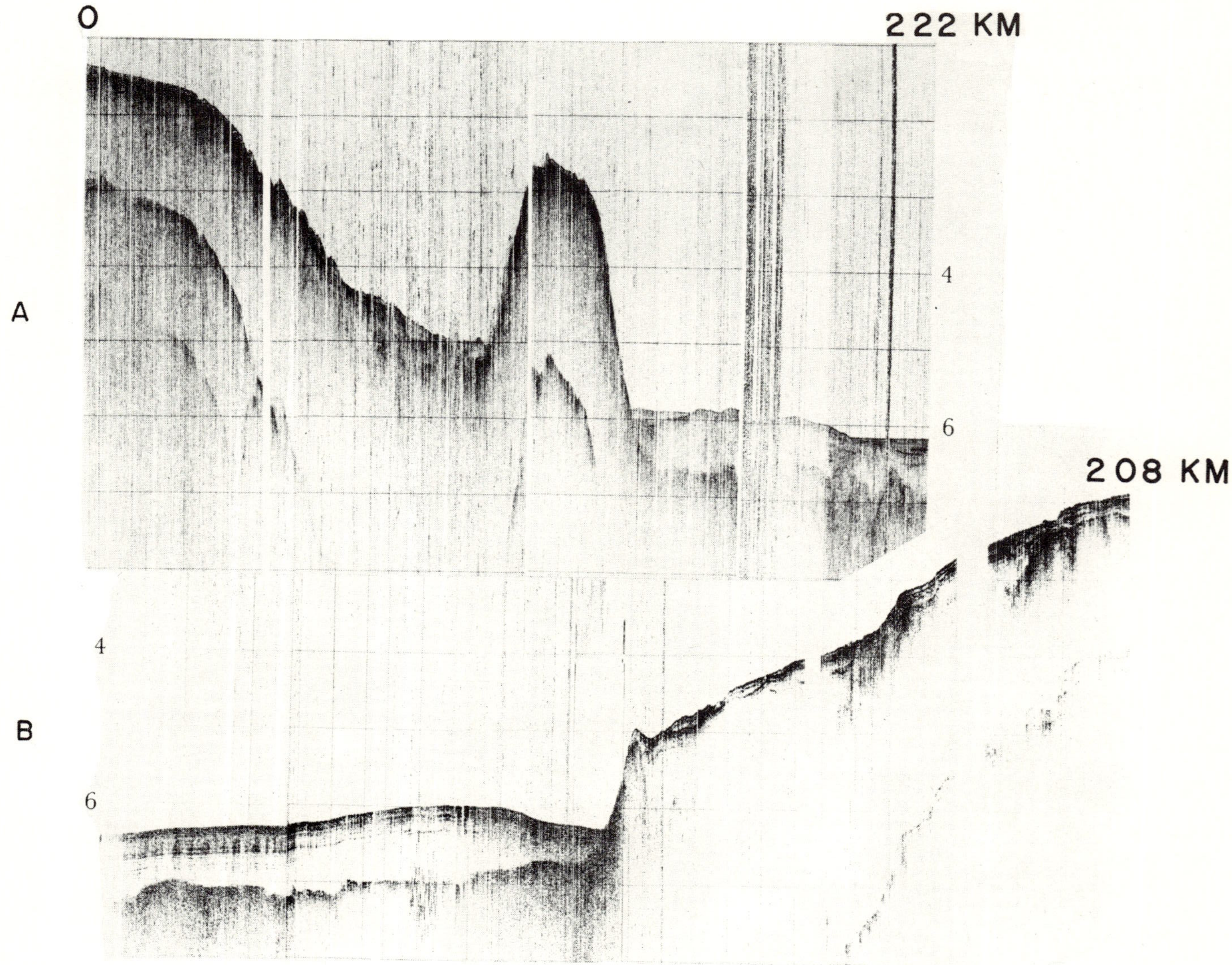

Fig. 10. Profile A: profiler section of the eastern margin of Tasmania. Profile B: profiler section typical of aseismic southeastern portions of the Campbell plateau; this section extends from 51°12′S, 177°34′E in a northwesterly direction to 50°06′S, 175°41′E on the right-hand side of the figure.

component of easterly translation of New Zealand relative to Antarctica in the initial stages of separation. This explanation may answer the question of why there is no steepening of the basement dip toward the continent, as there is between Australia and Antarctica where the separation was normal to their present coast lines.

SUMMARY

The profiler–sonobuoy data between Australia and Antarctica reveal the following:

1. The basement surface dips gently away from the mid-ocean ridge and then steepens abruptly near the two continents; this configuration suggests a crustal symmetry.
2. The acoustical character and inferred stratigraphy of the sediments from the basins and slopes of the two continents are quite similar; this similarity indicates that the dissimilar climates and modes of erosion that have existed in Australia and Antarctica since the Upper Miocene have had little effect on the appearance of the profiler records.
3. The distribution of pelagic sediments has been significantly controlled and modified by the action of bottom currents.
4. The marked lack of basement steepening in the Pacific sector of the antarctic margin and in the eastern margins of Tasmania and the Campbell plateau suggests a different type of separation (possibly shear) from the tension that occurred between Australia and Antarctica. An understanding of this difference would be a fundamental contribution.

Acknowledgments. The research work and the geophysics program aboard the USNS *Eltanin* were supported by the U.S. Antarctic Research Program and the Office of Polar Programs under National Science Foundation grants GA-375, GA-894, GA-1121, GA-1523, GA-12825, and GV-27472. RV *Robert D. Conrad* and RV *Vema* were supported by the National Science Foundation grant GA-550, the U. S. Navy Office of Naval Research grant NONR 266 (48), and the Bureau of Ships grant NOBSR 64547.

We acknowledge the efforts of the many participants in the geophysics program of the USNS *Eltanin* over the past 7 years. We are grateful to J. I. Ewing and D. Hayes for a critical reading of the manuscript.

Contribution 1777, Lamont-Doherty Geological Observatory.

REFERENCES

Bandy, O., R. Casey, and R. Wright, Late Neogene planktonic zonation, magnetic reversals, and radiometric dates, antarctic to the tropics, in *Antarctic Oceanology I, Antarctic Res. Ser.*, vol. 15, edited by J. L. Reid, pp. 1–26, AGU, Washington, D.C., 1971.

Barazangi, M., and J. Dorman, World seismicity maps compiled from ESSA coast and geodetic survey epicenter data, 1961–1967, *Bull. Seismol. Soc. Amer.*, *59*(1), 369–380, 1969.

Burckle, L., T. Saito, and M. Ewing, A cretaceous (Turonian) core from the Naturaliste plateau southeast Indian Ocean, *Deep Sea Res.*, *14*, 421–426, 1967.

Collette, B. J., and K. W. Rutten, Differential compaction vs. diapirism in abyssal plains, *Mar. Geophys. Res.*, 1, 104–107, 1970.

Conolly, J. R., and R. R. Payne, Sedimentary patterns within a continent–mid-oceanic ridge–continent profile: Indian Ocean south of Australia, *Antarctic Oceanology II: The Australian–New Zealand Sector, Antarctic Res. Ser.*, vol. 19, edited by D. E. Hayes, pp. 295–315, AGU, Washington, D. C., 1972.

Conolly, J., and C. von der Borch, Sedimentation and physiography of the sea floor south of Australia, *Sediment. Geol.*, *1*, 181–220, 1967.

Conolly, J., A. Flavelle, and R. Dietz, Continental margin of the Great Australian Bight, *Mar. Geol.*, *8*, 31–58, 1970.

Crary, A., E. Robinson, H. Bennett, and W. Boyd, Glaciological regime of the Ross ice shelf, *J. Geophys. Res.*, *67*(7), 2791–2807, 1962.

Dietz, R., and J. Holden, Reconstruction of Pangea: Breakup and dispersion of continents, Permian to present, *J. Geophys. Res.*, *75*(26), 4939–4956, 1970.

Eittreim, S., A. Gordon, M. Ewing, E. Thorndike, and P. Bruchhausen, The Nepheloid layer and observed bottom currents in the Indian-Pacific antarctic sea, in *Studies in Physical Oceanography—A Tribute to Georg Wüst on His 80th Birthday*, edited by A. L. Gordon, Gordon and Breach, New York, in press, 1971.

Ewing, J., and M. Ewing, Sediment distribution on the mid-ocean ridges with respect to spreading of the sea floor, *Science*, *156*, 1590, 1967.

Ewing, M., S. Eittreim, M. Truchan, and J. Ewing, Sediment distribution in the Indian Ocean, *Deep Sea Res.*, *16*, 231–248, 1969*a*.

Ewing, M., R. Houtz, and J. Ewing, South Pacific sediment distribution, *J. Geophys. Res.*, *74*(10), 2477–2493, 1969*b*.

Gordon, A. L., Spreading of antarctic bottom water, 2, in *Studies in Physical Oceanography—A Tribute to Georg Wüst on His 80th Birthday*, edited by A. L. Gordon, Gordon and Breach, New York, in press, 1971.

Gordon, A. L., On the interaction of the antarctic circumpolar current and the Macquarie ridge, in *Antarctic Oceanology II: The Australian–New Zealand Sector, Antarctic Res. Ser.*, vol. 19, edited by D. E. Hayes, pp. 71–78, AGU, Washington, D. C., 1972.

Griffiths, J., Continental margin tectonics and the evolution of southeast Australia, *Aust. Petrol. Explor. Ass. J.*, *11*, 75–79, 1971.

Hawkins, L., J. Hennion, J. Nafe, and H. Doyle, Marine seismic refraction studies on the continental margin to the south of Australia, *Deep Sea Res.*, *12*, 479–495, 1965.

Hayes, D. E., and J. R. Conolly, Morphology of the southeast Indian Ocean, in *Antarctic Oceanology II: The Australian–New Zealand Sector, Antarctic Res. Ser.*, vol. 19, edited by D. E. Hayes, pp. 125–145, AGU, Washington, D. C., 1972.

Hays, J., and N. Opdyke, Antarctic radiolaria, magnetic reversals, and climatic change, *Science, 158*(3804), 1001–1011, 1967.

Houtz, R., J. Ewing, and X. Le Pichon, Velocity of deep-sea sediments from sonobuoy data, *J. Geophys. Res., 73*(8), 2615–2641, 1968.

Houtz, R., J. Ewing, and R. Embley, Profiler data from the Macquarie ridge area, *Antarctic Oceanology I, Antarctic Res. Ser.*, vol. 15, edited by J. L. Reid, pp. 239–245, AGU, Washington, D. C., 1971.

Jacobs, S., A. Amos, and P. Bruchhausen, Ross sea oceanography and antarctic bottom water formation, *Deep Sea Res., 17*, 935–962, 1970.

Le Pichon, X., Sea-floor spreading and continental drift, *J. Geophys. Res., 73*(12), 3661–3697, 1968.

Le Pichon, X., and J. Heirtzler, Magnetic anomalies in the Indian Ocean and sea floor spreading, *J. Geophys, Res., 73*(6), 2101–2117, 1968.

Lindsay, J., Depositional environment of Paleozoic glacial rocks in the central Transantarctic mountains, *Geol. Soc. Amer. Bull., 81*(4), 1149–1172, 1970.

Margolis, S., and J. Kennett, Antarctic glaciation during the Tertiary recorded in sub-antarctic deep-sea cores, *Science, 170*(3962), 1085–1087, 1970.

Sprigg, R., Paleogeography of the Australian Permian in relation to oil search, *Australian Petrol. Explor. Ass. J., 6*, 1966.

Sproll, W. P., and R. S. Dietz, Morphological continental drift fit of Australia and Antarctica, *Nature 222*(5191), 345–348, 1969.

von der Borch, C. C., J. R. Conolly, and R. S. Dietz, Sedimentation and structure of the continental margin in the vicinity of the Otway basin, southern Australia, *Mar. Geol., 8*, 59–83, 1970.

Weeks, L. G., and B. M. Hopkins, Geology and exploration of three Bass Strait basins, Australia, *Amer. Ass. Petrol. Geol. Bull., 51*(5), 742–760, 1967.

Weissel, J. K., and D. E. Hayes, Asymmetric seafloor spreading south of Australia, *Nature, 231*(5304), 517–521, 1971.

Weissel, J. K., and D. E. Hayes, Magnetic anomalies in the southeast Indian Ocean, in *Antarctic Oceanology II: The Australian–New Zealand Sector, Antarctic Res. Ser.*, vol. 19, edited by D. E. Hayes, pp. 165–196, AGU, Washington, D. C., 1972.

MAGNETIC ANOMALIES IN THE SOUTHEAST INDIAN OCEAN

JEFFREY K. WEISSEL

Department of Applied Geology, University of New South Wales, New South Wales, Australia

DENNIS E. HAYES

Lamont-Doherty Geological Observatory of Columbia University, Palisades, New York 10964

Abstract. A detailed pattern of magnetic anomaly lineations has been mapped for the area south of Australia and for the south Tasman Sea. This area can be subdivided into four longitudinal zones on the basis of contrasts in magnetic, morphologic, and seismic properties. One zone is characterized by a long period of systematic asymmetric spreading (~40–15 m.y. B.P.) in which material was accreting about 30–40% faster on the north flank of the Southeast Indian ridge than on the south flank. Significant differences in the natural remanent magnetization of the oceanic crust are inferred for the Indian and antarctic plates in this area. The most striking difference is that the amplitudes of certain lineations are greater over the Indian plate than over the antarctic plate. Major differences in the amplitudes of the axial anomalies also occur along the Southeast Indian ridge. Paleoinclinations deduced from corresponding blocks on the Indian and antarctic plates older than 10 m.y. are not internally consistent. This inconsistency suggests that similar calculations from other areas must be viewed with extreme caution. Although large uncertainties exist for the inferred paleoinclinations, those inclinations associated with anomalies on the Indian plate appear to be about 20° lower than their counterparts on the antarctic plate. Magnetic quiet zones are identified along portions of the antarctic and South Australian continental margins and probably relate to special processes or conditions that occurred during the initial rifting history of the two continents. The quiet zone boundaries appear to cut across the magnetic lineations trends and are thus probably not isochrons. The history of relative motion of Australia and Antarctica is well documented and is described in terms of a series of sequential rotational poles that are calculated from magnetic lineations and complementary morphologic information. The poles of opening describing the Indian-antarctic motion are in good agreement with the independent studies of D. P. McKenzie and J. G. Sclater. Normal oceanic crust began to form between the Australian and antarctic continents about 55 m.y. ago in response to a major episode of sea floor spreading. Initial rifting likely took place long before that time. The relict transform directions associated with the initial phase of separation may have influenced the preferential location of sedimentary basins, deep marginal plateaus, and bathymetric offsets along the evolving continental margins. The results of the inferred Indian-antarctic plate motions are used to study the recent interaction between the Indian and Pacific plates along the section of their common boundary running from New Zealand to the Macquarie triple junction. The triple junction has probably migrated to the southwest about 100–200 km relative to the Southeast Indian ridge during the past 10 m.y.

We [*Weissel and Hayes*, 1971] carried out a preliminary appraisal of the marine magnetic data available from the area south of Australia. Although our previous report was very brief, it identified many of the problems and observations examined in detail in this paper. In the following text, frequent reference to the results of the *Weissel and Hayes* [1971] report is necessary, and we will do so without repeated references to our earlier work. The three longitudinal zones A, B, and C were designated (see Figure 1) within the Indian and antarctic plates by using magnetic, morphologic, and seismologic [*Barazangi and Dorman*, 1969; *Sykes*, 1970] parameters. Systematic asymmetric sea floor spreading as recorded by the pattern of magnetic lineations persisted exclusively in zone A from about 38 m.y. B.P. until 10 m.y. B.P. The recognition of this period of asymmetric sea floor spreading played an important part in predicting the evolution of the accreting plate margin between Australia and Antarctica and the subsequent separation of these continents.

This study of the area south of Australia and the

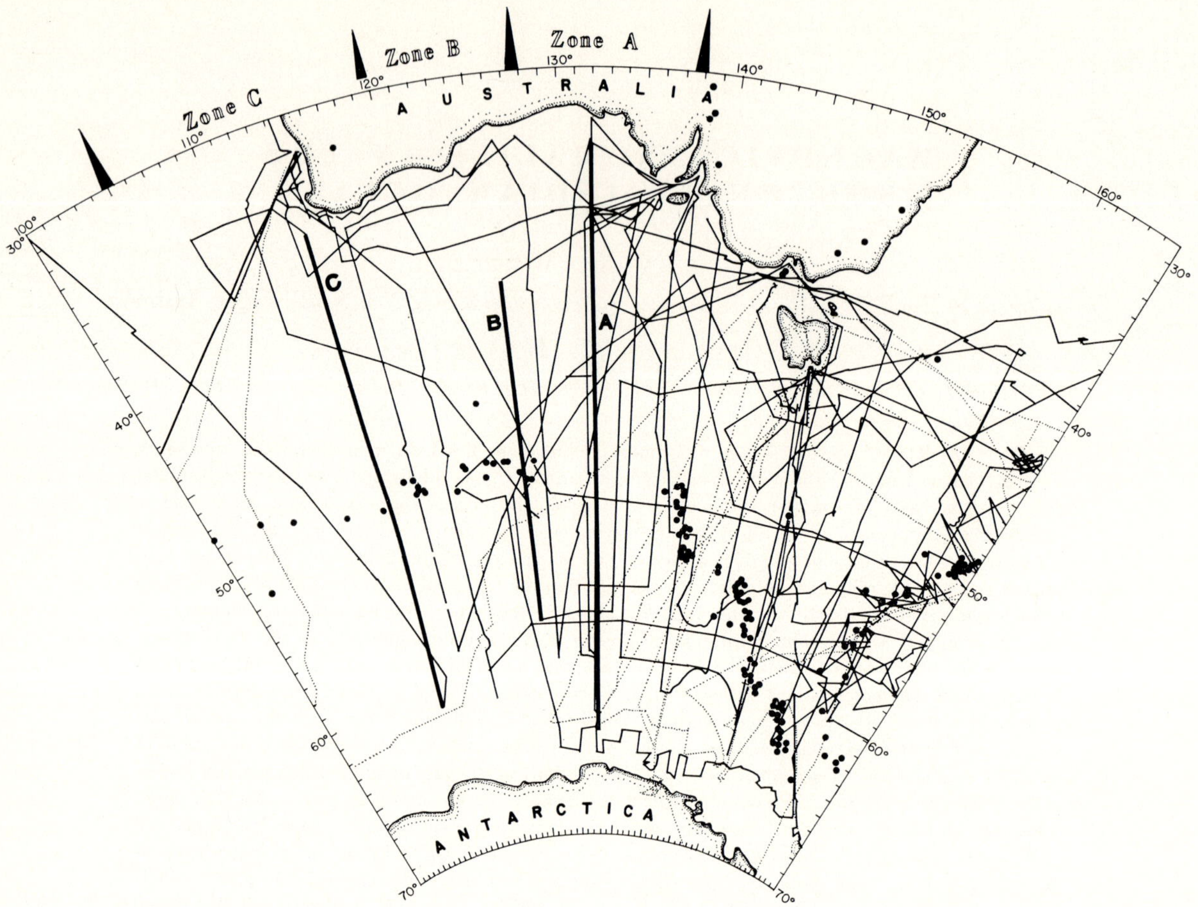

Fig. 1. Tracks along which magnetics data were available for this study as modified from *Weissel and Hayes* [1971]. Tracks of the *Eltanin, Vema,* and *Robert D. Conrad* are shown as solid lines; those of all other sources are shown as dotted lines. Epicenters are indicated by black dots and are taken from *Sykes* [1970]. Figure 4 shows A, B, C key profiles. Stereographic projection.

Tasman Sea (between 105°E and 165°E) incorporates all the magnetic data acquired by the *Eltanin,* the antarctic research vessel of the National Science Foundation, since it started concentrated operations in this part of the world in 1964. Additional data were acquired from vessels of Lamont-Doherty Geological Observatory, Australia, France [*Schlich and Patriat,* 1967], and Japan [*Tomoda et al.,* 1968] and from numerous aeromagnetic flight lines of the U.S. Naval Oceanographic Office. Figure 1 shows the available tracks and flight lines used and the relocated earthquake epicenters of *Sykes* [1970].

The correlation and areal distribution of magnetic anomaly lineations are presented in detail and discussed. The distribution of magnetic anomaly amplitudes and their relationship to geographical zones and to inferred age are described. The observed amplitude variations suggest that some areas have undergone a selective alteration of the initial natural remanent magnetization (NRM) acquired when the oceanic crust was formed at the accreting plate boundary.

The 'fidelity' with which a unique pattern of magnetic anomaly lineations can be generated and preserved is examined by comparing several closely spaced profiles (five north-south profiles collected within a 3° band of longitude).

Simple two-dimensional magnetic model calculations are presented to illustrate better the variations in spreading rates within the area. The models are compared with the observed shape and amplitudes of the sea floor spreading lineations in an attempt to determine the paleolatitudes of lithospheric plates in the manner of *Vine* [1968] and *Schouten* [1971, also personal communication, 1971].

By using the modern concepts of global plate tectonics [*Morgan*, 1968; *Le Pichon*, 1968; *Isacks et al.*, 1968; *McKenzie and Morgan*, 1969], the Indian plate (containing Australia) was rotated toward the antarctic plate (assumed fixed) by fitting anomaly boundaries together through the technique of *Pitman and Talwani* [1972]. The relative configuration of the two continents is thus reconstructed at various times during the Tertiary. This relative configuration can be determined confidently from the present back to about 50 m.y. B.P. The earliest history of rifting cannot be deciphered from the observed magnetic lineations; however, the presence of 'magnetic quiet zones' along presumed contiguous portions of the continents provide, together with auxiliary geologic and geophysical data, important constraints on the time and nature of the earliest rifting.

MAGNETIC ANOMALY AND FRACTURE ZONE PATTERN

The pattern of magnetic lineations in the southeast Indian Ocean and west of the Macquarie ridge complex is relatively easy to discern (see *Le Pichon and Heirtzler* [1968]). Magnetic anomalies from the standard Cenozoic spreading sequence of *Pitman et al.* [1968] have been identified and dated by using the *Heirtzler et al.* [1968] geomagnetic reversal time scale. Figure 2 (see foldouts at back of volume) shows the pattern of magnetic anomaly lineations and postulated fracture zones as indicated primarily by offsets in the magnetic pattern. Magnetic anomalies that can be confidently correlated are joined by the solid lines; somewhat tentative correlations are shown by broken lines. The confidence in the fracture zone identification and orientation is similarly indicated. The quality of correlation is good for most of the area east of 128°E (the zone A–zone B boundary) and is not as good for the area west of 128°E.

The oldest anomaly sequence identified in the entire region is 19–21. There is, perhaps, an indication of anomaly 22 in the Great Australian Bight between 36°S, 128°E and 36°S, 131°E and near the antarctic continent between 63.5°S, 127.5°E and 63.5°S, 132°E. Anomaly 21 is tentatively identified southeast of Tasmania at 48°S along 155°E. On the northern side of zone C, anomaly 19 abuts an east-west trending area of pronounced morphologic roughness known as the Diamantina fracture zone (see *Hayes and Conolly* [this volume]) that may cause obliteration of the oldest anomalies.

Important variations occur in the characteristics of the observed pattern throughout the region. *Weissel and Hayes* [1971] previously noted distinct amplitude differences among (1) magnetic anomalies of different ages, (2) magnetic lineations of the same age but located on opposite sides of the ridge axis, and (3) magnetic anomalies of the same age and on the same side of the ridge axis but from the different zones (A, B, and C). These variations are now discussed quantitatively and in more detail than before, and we have also incorporated data from the area east of 140°E.

The magnetic anomalies in the age range 0–10 m.y. are much larger to the east of 128°E than to the west (see Figure 4). Except for the seismicity associated with the large offset transform faults east of about 138°E, these large amplitude anomalies are associated with a ridge crest that is presently aseismic. Fracture zones associated with the transform faulting can be identified on the basis of magnetic pattern offsets and morphologic evidence as well as from the seismicity. There are at least six significant offsets of the magnetic pattern east of 138°E with the following magnitudes: 240 km at 140°E, 240 km at 141°E, 60 km at 144°E, 600 km at 146.5°–150°E (this may represent more than one offset), 100 km at 151°E, and 360 km at 153.5°–156°E.

These offsets occur south of the prominent southward bulge of the Australian mainland and Tasmania. At the eastern end of the area under discussion, the trend of the magnetic lineations gradually changes from an east-west to a northeast direction that is consistent with the inferred change in the azimuths of the fracture zones in this region. The boundary between old Tasman Sea crust and crust generated at the Southeast Indian ridge is tentatively suggested to run from the southern margin of the South Tasman rise in a northeasterly direction through the possible but uncertain observation of anomaly 21 at 48°S, 155°E and thence toward New Zealand. Although magnetic lineations with azimuths of 330°–340° have been identified by *Ringis* [1970] and *Hayes and Ringis* [1972] in the central Tasman Sea off the New South Wales and Victoria coasts, these lineations have not been traced farther south than about 44°S. Between this latitude and the inferred ancient plate boundary, the direction of magnetic lineation cannot be determined from the available data. The inferred position of this ancient plate boundary coincides with an increase in wave length of magnetic anomalies observed on a north-south profile along 155°E. Note that the observed amplitudes of magnetic anomalies in the central Tasman Sea are about half that of the older flank anomalies of zone A.

Seismic profiler results from *Eltanin* cruises 34 and 36 may substantiate the inferred ancient plate boundary. Seismically, this boundary is an area of rough basement. At least one track (along 155°E) suggests the basement roughness could be related to the extension of a fracture zone that intersects the Macquarie complex at 53°S, 159°E. The sediment seems thicker on the Tasman side of the questionable boundary. The character of the sediment also appears different on opposite sides of the inferred boundary. The flank of the Southeast Indian ridge is more scoured by erosion; the sediment is thin and sometimes drapes rough basement topography. The alternative explanation that observed rough basement morphology is associated with fracture zones cannot be discounted.

For anomalies abutting the Macquarie ridge complex, the along-strike length of lineations older than about anomaly 5 increases with age, and two explanations for the increase are suggested. One is that subduction of parts of the Indian plate may have taken place at the Macquarie ridge complex. Earthquake mechanism solutions [*Banghar and Sykes*, 1969; *Sykes*, 1970] support this hypothesis. This explanation is also supported by the observation that the along-strike length of the present ridge crest is slightly longer than the along-strike length of anomaly 5 on the Indian plate. The other explanation is that the evolution of the Macquarie triple junction has involved a shortening of the accreting boundary since about 10 m.y. ago. A similar change in the along-strike length of lineations may also be present south of the ridge axis according to data obtained by D. A. Christoffel (personal communication, 1970). From the apparent offset of anomalies 19–21 on the north flank, another fracture zone likely intersects the Macquarie complex at about 53°S, 159°E.

The location of fracture zones west of 128°E is highly speculative. The pattern of seismicity generally gives no guide to the strike of transform faults. For this reason, any offsets present in zone B and zone C are inferred to be small, although a discussion of discrete offsets within the central portion of zone B is not possible as explained by *Hayes and Conolly* [this volume]. This inference is supported by the occcurrence of only small offsets in the magnetic lineation pattern. *Weissel and Hayes* [1971] showed that the length of the transform fault boundary between zones A and B at about 128°E increased steadily with time during the interval from 38 to 10 m.y. B.P. An offset of about 190 km is tentatively identified at the zone A–zone B boundary from closely spaced crossings of the central anomaly, and this value is consistent with that inferred from spreading rates. If this mode of evolution occurs elsewhere, it occurs to a much smaller extent. Within zone A, the asymmetric growth of the ridge, which was considered to give rise to the change in the ridge crest offset distance, was not observed within zone C or zone B for this time interval. The data available for spreading rate calculations for zone B and zone C are more limited than those of zone A. Had asymmetric spreading occurred simultaneously in these zones and in the same sense, constant ridge offsets would have been maintained.

A gradual change in the trend of lineations from east-west to southeast is observed at about 110°E, and anomalies there have been discussed extensively by *McKenzie and Sclater* [1972].

For zone B, although a diffuse pattern of seismicity exists that suggests transform faulting, the very limited extent of magnetic lineations younger than about anomaly 6 (21.2 m.y.) precludes definite identification of individual offsets. Lineations older than anomaly 6 appear continuous on both flanks all across zone B, their appearance suggesting that any offsets and/or disturbances present are restricted to the areas of young anomalies. Offsets within zone B cannot be delineated from the magnetic data. The bathymetric map of *Hayes and Conolly* [this volume] shows that this zone (the Australian-antarctic discordance) is defined by a pronounced north-south topographic grain.

For zone C, there are a number of small offsets of anomaly lineations of all ages, but the data are insufficient to enable accurate location of these fracture zones. Those fracture zones postulated in Figure 2 were drawn by using the synthetic fracture zones that are shown in Figure 11 as guidelines. The bathymetric map of *Hayes and Conolly* [this volume] indicated one or more offsets between 110°E and 113°E with a total right-lateral displacement of about 100 km. The sediment distribution data of *Houtz and Markl* [this volume] also support this interpretation.

Figure 3 shows five magnetic anomaly profiles projected along a north-south direction. The traverses are located within 3° of longitude of one another and are centered near 132°E. These long (approximately 4500 km), closely spaced traverses are shown to highlight several important observations. Each of these profiles was recorded in zone A. This zone, described by *Weissel and Hayes* [1971], has the least ambiguous

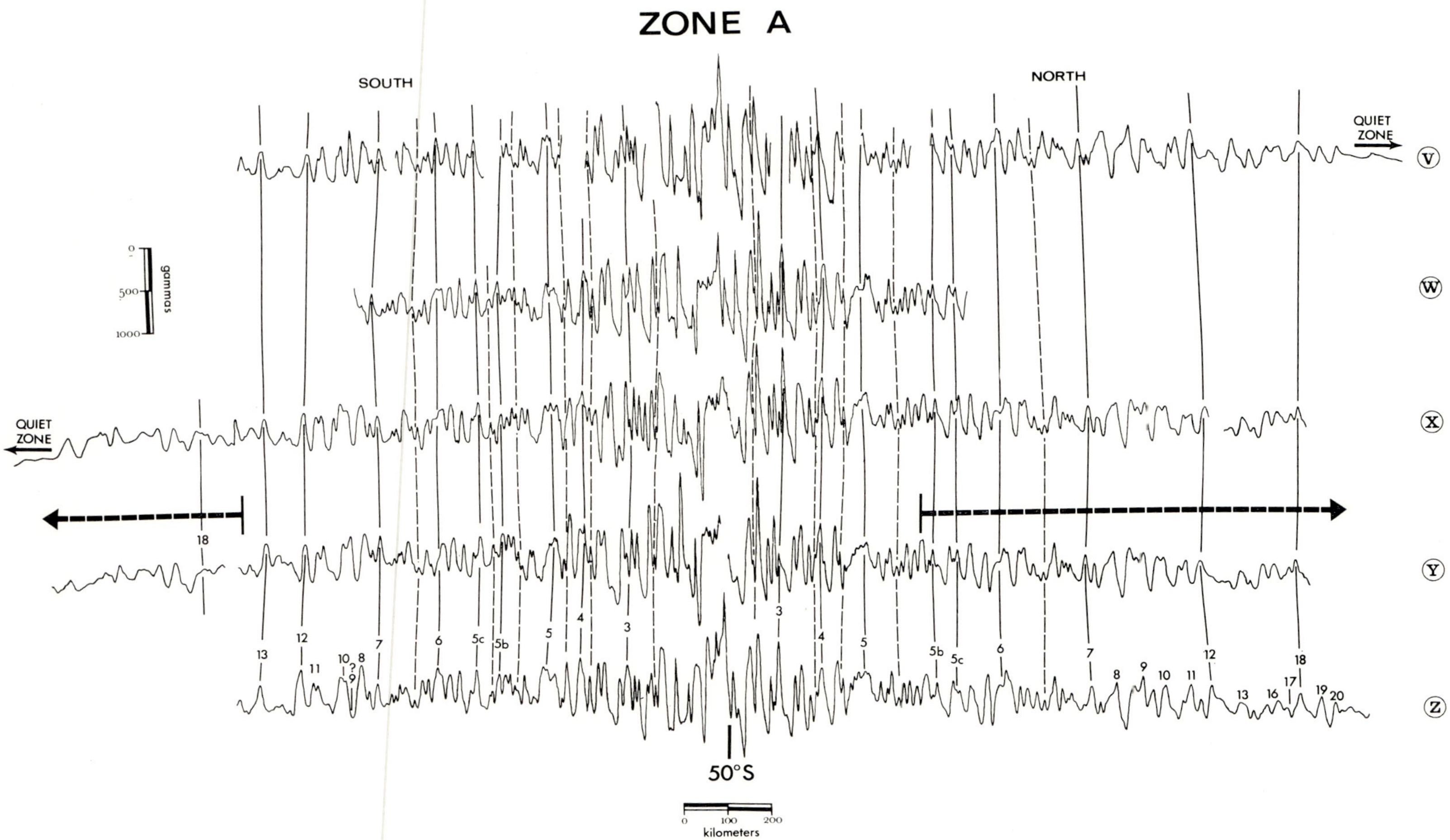

Fig. 3. North-south projected total intensity magnetic anomaly profiles recorded within zone A. Correlations of well-established lineations of the *Pitman and Heirtzler* [1966] Cenozoic sequence are shown (solid lines), as well as minor peaks or troughs (broken lines) not yet correlated on a world-wide basis. The heavy broken lines on the left-hand and right-hand sides of the figure indicate portions of the profiles X and Y that are nearly coincident. Profiles are all located near line A shown in Figure 1. Topographic profiles corresponding to these magnetic profiles are shown in Figure 7 of *Hayes and Conolly* [this volume].

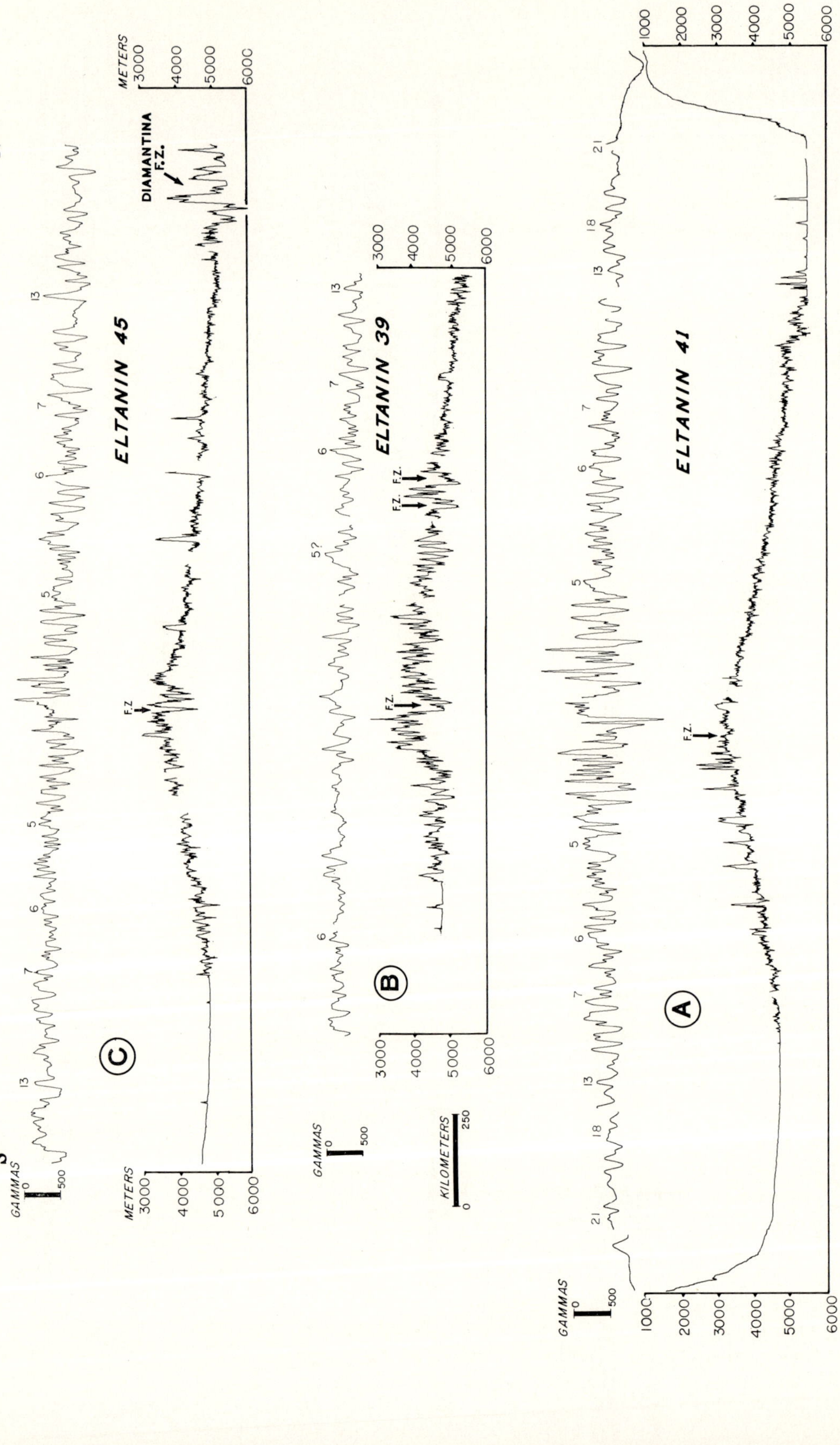

Fig. 4. Representative topographic and magnetic anomaly profiles of zones A, B, C taken from *Weissel and Hayes* [1971]. Lines A, B, and C of Figure 1 mark profile locations.

pattern of magnetic anomaly lineations and a well-developed mid-oceanic ridge with an easily recognizable, well-developed topographic crest uncomplicated by any major fracture zones.

Figure 3 demonstrates unequivocally that the unique character of the magnetic pattern that has recognizable counterparts on both sides of the ridge is part of the same pattern recognized in all the major world oceans. These data allow an opportunity to assay the degree of variability in the magnetic pattern as recorded in the crustal rocks over a relatively short distance. These closely spaced profiles allow us to correlate relatively minor features of the pattern, i.e., those that are small when compared with the major anomaly lineations of *Pitman and Heirtzler* [1966]. Some of the details of this pattern have previously been identified by *Chase et al.* [1970]. As noted by *Pitman et al.* [1968], the sequence of anomalies between anomaly 5 and anomaly 6 and also the sequence between anomaly 6 and anomaly 7 are characterized by short wave lengths that can be correlated between the North and South Pacific areas. Although the general character of this sequence of anomalies has been reproduced in the model proposed by *Heirtzler et al.* [1968], they made no attempt to model the details of these anomaly sequences.

Several other aspects of the magnetic lineation pattern within zone A are clearly illustrated by Figure 3. One is the asymmetric distribution in space of magnetic anomaly lineations of the same age on opposite sides of the ridge. The asymmetry can be readily evaluated by comparing the distance of corresponding anomalies on opposite sides of the ridge for those anomalies older than anomaly 5 (10 m.y.). T. Atwater (personal communication, 1971) has pointed out to us that anomaly 9 may either be missing entirely or severely compressed on the south side of profiles X, Y, and Z of Figure 3. This configuration is not typical of the areas immediately to the west or east but may mark the site of a small, local jump in the paleoposition of the ridge crest. The presence of an unusually wide anomaly 9 on the north side of profiles X, Y, and Z supports this contention. Note that profiles X and Y are virtually coincident along the 7–13 anomaly sequence and profile Z is located only about 50 km away.

Although the composite pattern of lineations is reproduced with remarkable fidelity across each of the profiles, that portion of the magnetic anomaly pattern younger than anomaly 3 (about 5 m.y.) shows the greatest variability across adjacent profiles. This variability cannot be attributed to differences in the bottom topography, as such differences are very slight (see *Hayes and Conolly* [this volume, Figure 7]). Also of interest is that the apparent skewness of anomaly 5 (see Figure 3, profiles W and Z) is opposite in sense to the skewness representative of most other portions of the anomaly sequence.

A number of interesting problems arise on close examination of the over-all distribution of magnetic anomalies. These groups of profiles also illustrate the contrast in the average magnetic anomaly amplitudes (larger on the north) on the opposite sides of the ridge crest. The variation of anomaly amplitude both as a function of location relative to the ridge axis and also as a function of age has been examined for zone A, zone B, and zone C. Figure 4, taken from *Weissel and Hayes* [1971], shows representative profiles from each of these zones. Average peak to peak magnetic anomaly amplitudes for selected anomalies are given in Figure 5.

Ridge Crest

The average amplitude of the central anomaly in zone A is almost twice that of the anomaly in zone C. There are no positive identifications of the central anomaly for zone B. The depths to the ridge crest (and hence the minimum depth of the magnetic layer) are not significantly different between zones A and C. Likewise, the inclination, declination, and magnitude of the present magnetic field and of the axial dipole field of the earth are about the same for zones A and C. Hence the difference in average amplitude may be due to a thinner magnetic layer under the ridge crest of zone C and/or a contrast in the magnetic properties of the crust accreted along the plate boundary for the different zones. For the second case, the amount of magnetization of the magnetic material at the axis of zone C would be about half that of the material at the axis of zone A. A third possibility is that the thermoremanent magnetization of the material in zone C was destroyed or partially inhibited by some process localized at the ridge crest. The diffuse seismicity occurring near the crest in zones B and C might be related to the decrease in magnetization.

Ridge Flanks

An examination of the different anomaly characteristics as they relate to constant spreading rate episodes and to intervals of asymmetric spreading may provide insight into the cause of the variable magnetic properties. Figure 6, taken from *Weissel and Hayes* [1971],

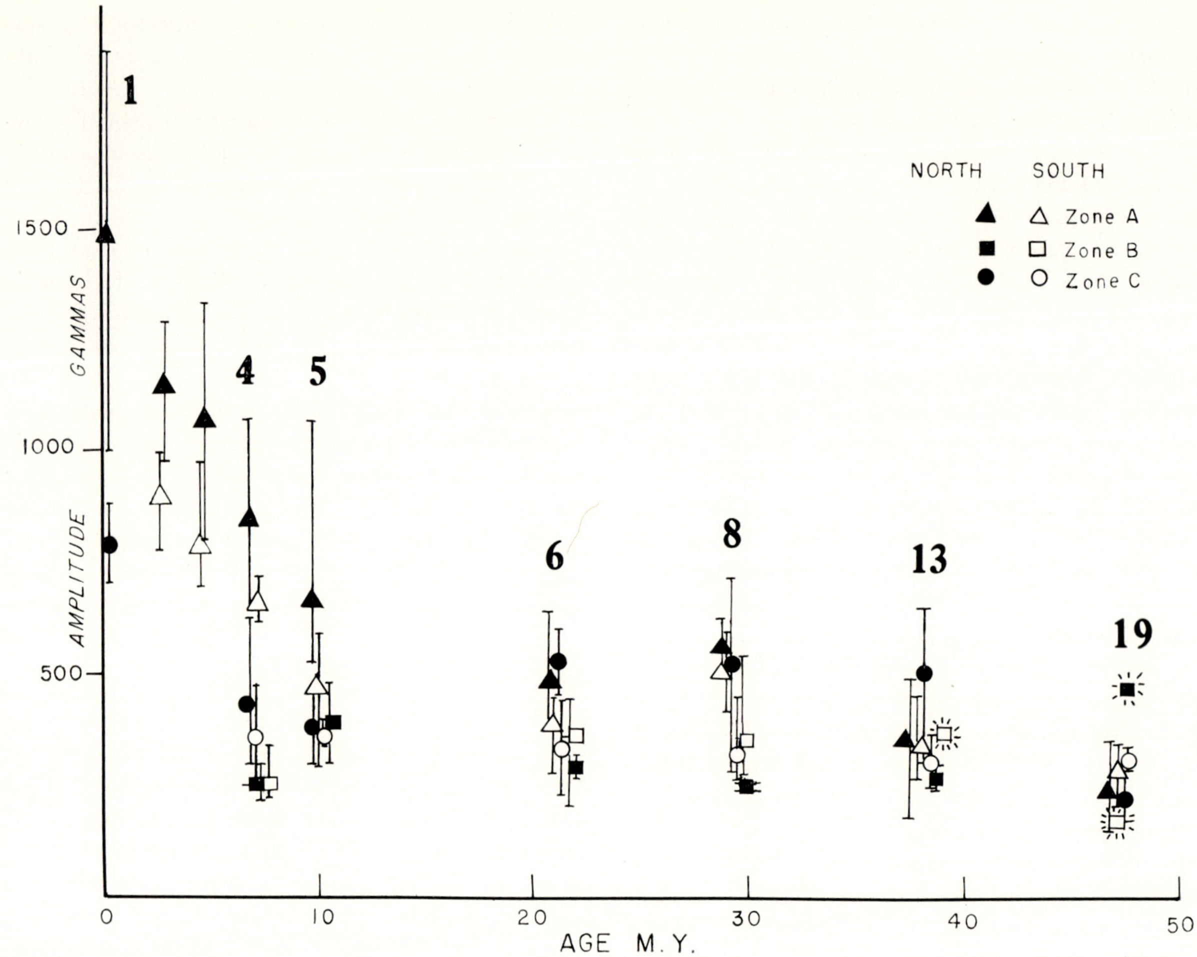

Fig. 5. Average peak to peak amplitudes of key magnetic anomalies in zones A, B, and C. The scatter of the individual observations is indicated by the range bars. Single observations are highlighted.

illustrates the inferred 'half' opening rates for each side of the ridge and for each zone.

The unusual relationship between the amplitudes of anomalies in zone A and the other two zones persists markedly to at least anomaly 5 for both flanks of the ridge. The average amplitude of anomaly 4 in zone B is significantly lower than that in zone A by at least 400 γ and is somewhat lower than the average amplitude in zone C. Both flanks in zone B show about the same amplitude for anomaly 4. The minimum depth to the magnetic layer in zone B is slightly greater ($\sim$400 meters) than that in zone A, but this increase in depth alone could not account for the observed difference. Figure 5 shows that anomaly amplitudes for zone B are systematically lower than those in zone A from anomaly 13 time (38.2 m.y. B.P.). This time corresponds to the onset of the period of asymmetric sea floor spreading observed in zone A. The systematic difference in amplitudes is most marked for anomalies that are younger than anomaly 6 (21.2 m.y. B.P.), and we suggest that the difference in amplitudes decreases with age. This amplitude change is paralleled by an increase in the roughness of the morphology of zone B as the crestal zone is approached.

As shown in Figure 5, the average amplitude of the first peak of the Gauss anomaly in zone A shows different magnitudes on the north and south flanks. Those on the north flank are 250 γ higher than those on the south. This relationship is also shown for the last peak of anomaly 3. For anomalies 4 and 5, the difference is about 200 γ (also see Figure 3). This disparity in anomaly magnitudes within zone A persists out to about anomaly 6. Anomalies older than about anomaly 8 show almost equal amplitudes on

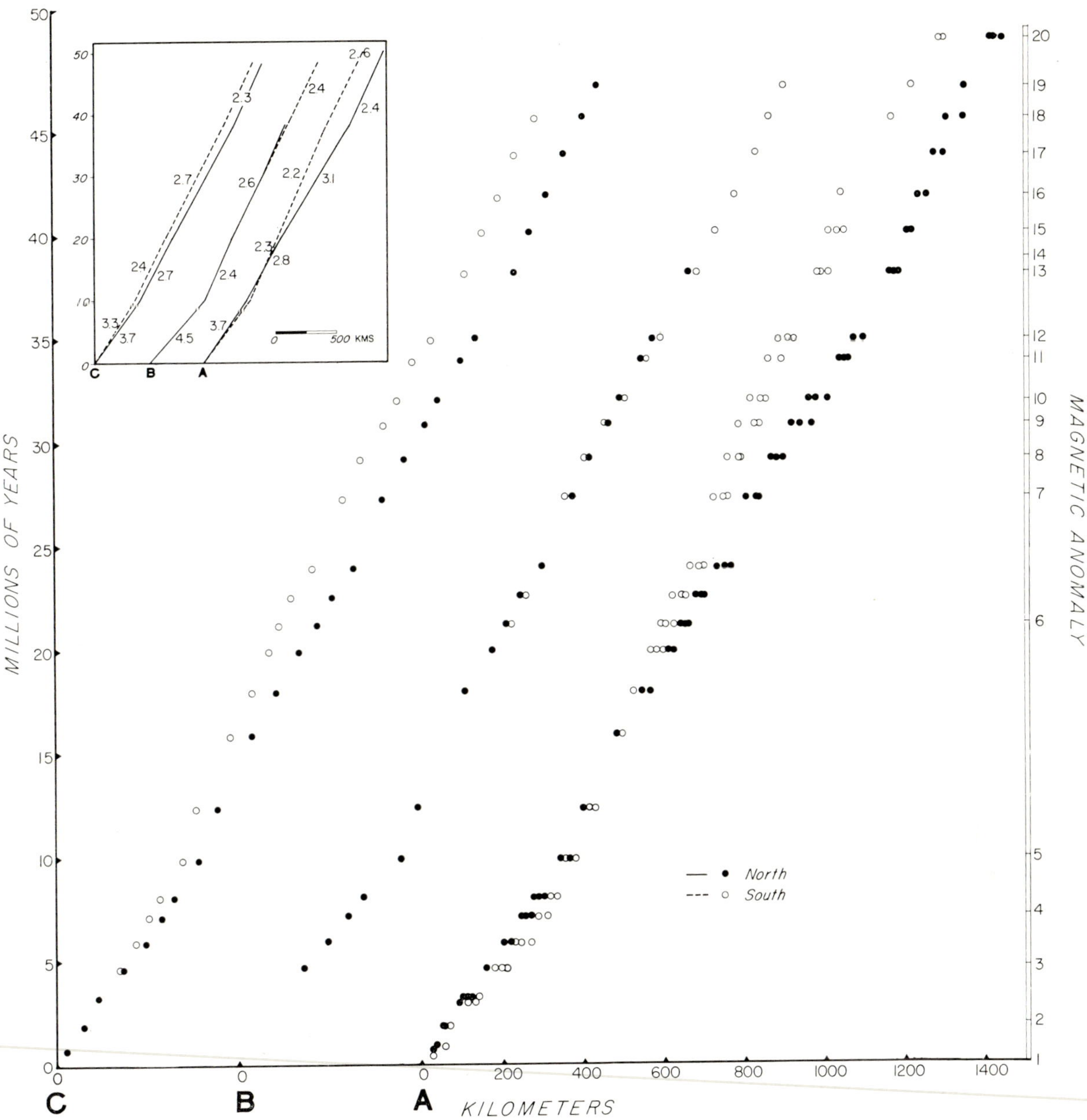

Fig. 6. Constant spreading rate episodes of zones A, B, and C taken from *Weissel and Hayes* [1971]. Note large difference in rates on opposite limbs indicated for zone A for the period of approximately 15–40 m.y. B.P. The spreading rate values determined by linear least squares for each limb are shown in the inset. Note that total opening rates for all zones during any time period are essentially equivalent.

both ridge flanks, although there is a great variation in the amplitude of anomaly 13 north of the axis as shown by the range bar in Figure 5.

From Figure 4, profile C, the anomalies on the north flank of zone C appear to be larger than those on the south. This difference is quite marked for anomalies older than about anomaly 5 (see Figure 5) where amplitudes on the north flank are about 200 γ higher than those on the south. This relationship persists out to at least anomaly 13 and is not readily explainable by different depths to the magnetic source layer on each flank. Hence *Weissel and Hayes* [1971] suggested that second order, intraplate deformation subsequent to the formation of the crust containing this sequence of anomalies may have to be invoked to explain the observed difference in average amplitudes.

The possible relationship between low amplitude axial anomalies and belts of seismic activity cannot be extrapolated to explain the observed amplitude differences in zone C, because the corresponding parts of an anomaly of a given age were together when the material acquired its thermoremanent magnetization. Thus equal magnetic properties should be found on both flanks regardless of the conditions operating when that crust was being formed. The decrease in thermoremanent magnetization of the material on the south flank of zone C most likely occurred after the anomalies had been formed and migrated away from the ridge crest.

The amplitude relations from the segments of the ridge flanks east of 140°E are similar to those found in zone A, although information from the south flank is very limited. The inference can be made that these parts of the plates are similar in geophysical and morphologic properties to Zone A, except that the spreading center is offset progressively southward by large, seismically active transform faults.

Continental Margins and Adjacent Areas

Adjacent to the continental margins of Australia and Antarctica are regions where the amplitudes of local magnetic anomalies are extremely small. These two areas are labeled 'quiet zone' in Figure 2 (see also Figure 16). The quiet zone area adjacent to Antarctica does not appear to be offset by the large fracture zones near 140°E. We suggest that magnetic lineations 12–19 bend around the northern edge of the antarctic quiet zone and that lineations 20 and 21 also show bending but appear, although strongly attenuated, to be present within the quiet zone. The antarctic plate quiet zone occurs mostly in the deep basin, presumably over an area of typical oceanic crustal thickness. Adjacent to the Australian margin, the quiet zone occurs both over oceanic crust and transitional crust beneath the lower continental slope. The quiet zone along the Australian margin was not observed consistently west of about 125°E. Its absence may be due to magnetic anomalies associated with the Diamantina fracture zone, which transects the lower continental margin and adjacent basin in the western region.

Perhaps the oceanic crust loses its thermoremanent magnetization by some type of thermal demagnetization as it becomes depressed along the continental margin of Australia. The anomalously high heat flow observed in southeast Australia may be indicative of some dynamic process at depth that could explain the loss in thermoremanent magnetization of crust immediately adjacent to this part of the Australian continent. This suggestion is entirely speculative, and a number of alternative explanations are also plausible.

The presence of a magnetic quiet zone is not unique to this area. *Heirtzler and Hayes* [1967] described similar zones in the North Atlantic. The origin of the quiet zones has been a subject of considerable speculation [e.g., *Heirtzler and Hayes*, 1967; *Taylor et al.*, 1968; *Vogt et al.*, 1970; *Drake and Nafe*, 1972; *Pitman and Talwani*, 1972]. Obviously, magnetic quiet zone boundaries observed in a number of oceans do not represent one single isochron as they have been mapped adjacent to dated magnetic lineations of divergent ages. At present, the most feasible explanation of the quiet zones that bound continental margins is that they represent some phenomena or geometry associated with the earliest stages of continental rifting. These conditions alone are not sufficient, since no quiet zone is observed in the northern sector of the South Atlantic and since probably none exists in the western sector of the southeast Indian Ocean. Possibly, the quiet zone represents a pre-existing boundary between the rifted continents similar to that suggested for the North Atlantic by *Drake and Nafe* [1972]. Alternatively, the quiet zone may be underlain by foundered continental crust. The relationship of the quiet zones to the early rifting history is considered in more detail in a later section. For the region containing the quiet zones (from about 128° to 145°E), the accreting plate boundary is not centrally placed in the Indian Ocean as is apparent from the reconstruction of the continents 38.2 m.y. ago, the time of formation of anomaly 13 (see Figure 14). The boundary is closer to the Australian margin. Furthermore, the trend of the axis at anomaly 13 time is not perpendicular to the direction of motion for the initial rifting stage (55–38 m.y. B.P.) or parallel to the Australian continental margin. Anomalies 18–22 appear truncated along the margin northwest of Tasmania. These anomalies south of Australia do not exhibit the marked bending of their south flank counterparts. Rotation of anomaly 13 north of the axis back onto anomaly 13 south of the axis results in a poorer coincidence at the eastern end of zone A than elsewhere along this fossil plate boundary because of this slight bending. The present geometry of anomaly 18 or older lineations on the north and south flanks precludes their fitting together on a single curve. One must suspect that intraplate deformation of the south flank anomalies took place subsequent to the formation of anomaly 13 or even anomaly 12. Neither the mechanism nor the duration of the deformation is known at this time.

The precise history of the ancient accreting plate boundary for the period 55–38 m.y. B.P. at the eastern end of zone A is difficult to confidently recreate. The margin that truncates the anomaly lineations 18–22 on the north flank defines an area of overlap in the continental reconstruction obtained by *Sproll and Dietz* [1969]. They speculate that postrift relaxation in the form of block faulting of the margin and sedimentation could have extended the margin southward, this extension thus explaining at least part of the overlap. If the trend of the anomaly 22 lineation is extrapolated eastward to just south of Kangaroo Island (~36°S, 138°E), this anomaly lineation would occur about 100 km on the continental side of the 2000-meter isobath.

MODEL STUDIES

Synthetic magnetic anomaly profiles were calculated to study further the variations in magnetic properties of the Indian and antarctic plates. An additional objective was to obtain an estimate of the past movement of these plates relative to the geocentric axial dipole by determining parameters of the remanent magnetic intensity vector needed to give the best correspondence between the model profiles and the observed profiles.

We assume that the magnetic properties of the oceanic crust can be modeled by two-dimensional blocks aligned parallel to the ridge axis with alternately normal and reversed magnetization. The geomagnetic reversal time scale of *Heirtzler et al.* [1968] is used in conjunction with the spreading rate data of *Weissel and Hayes* [1971, Table 1] to determine the distribution of the model blocks. A schematic representation of the regional sea floor morphology was taken from Figure 4.

Talwani et al. [1971] found that a thickness of the magnetic layer of a few hundred meters was sufficient to reproduce observed anomaly amplitudes over the Reykjanes ridge in the North Atlantic. *Herron* [1972] used a magnetic layer thickness of 0.5 km combined with a magnetization of 0.01 emu to reproduce observed magnetic profiles in the east central Pacific. We have therefore assumed the thickness of the magnetic layer to be 0.5 km. The assumption of a relatively thin magnetic layer tends to minimize the effects of varying depth to the upper surface. Any variations inferred for the magnetic intensities of different blocks would be greater if a thicker layer were consistently assumed for the model.

Only the natural remanent magnetization (NRM) of the basalts in the oceanic crust was considered in these model studies. The ridge flank morphology for zones A and C is relatively smooth, and induced magnetization arising from topographic irregularities should be small. Also, although the morphology encountered in zone B is much rougher than that in the other zones, the generally high Koenigsberger (Q) ratio common for oceanic basalts [e.g., *Opdyke*, 1970] would still require any induced components present to be relatively insignificant. *Hayes and Conolly* [this volume] have demonstrated that the topographic grain in zone B is oriented north-south rather than parallel to the ridge crest. The assumption of a two-dimensional model oriented east-west is thus invalid in the central part of this zone. The synthetic magnetic anomalies obtained there are subject to additional errors and must be treated with caution.

Note that east-west trending magnetic lineations cannot be used to determine uniquely the declination of the NRM vector but only the inclination and the magnitude of the cross-strike component of magnetization. However, if we assume that in the area of study the spreading center maintained an east-west trend during the Cenozoic and that the magnetic field of the earth was described by a geocentric axial dipole, then the declination of the ambient field should have been near zero. Consequently we have examined only the effects of variable inclination of the NRM vector in the model studies and have assumed that the NRM declinations are near zero. Anomalies for inclinations of 85°, 74°, and 59° were computed for each sequence of anomalies. By assuming an axial dipole field, these inclinations correspond to latitudes of formation of 80°, 60°, and 40°, respectively.

The reader is reminded that only a few selected anomaly profiles are presented in this paper. However, a large number of profiles have been examined in selecting representative profiles for comparison with the synthetic magnetic anomalies shown in Figures 7, 8, and 9. Almost every observed magnetic anomaly sequence longer than about 300 km contains one or more minor irregularities that do not correspond well with the idealized synthetic anomalies. Such a lack of correspondence can be attributed to several possible causes: (1) the possibility of the profiles crossing small offset fracture zones, (2) the

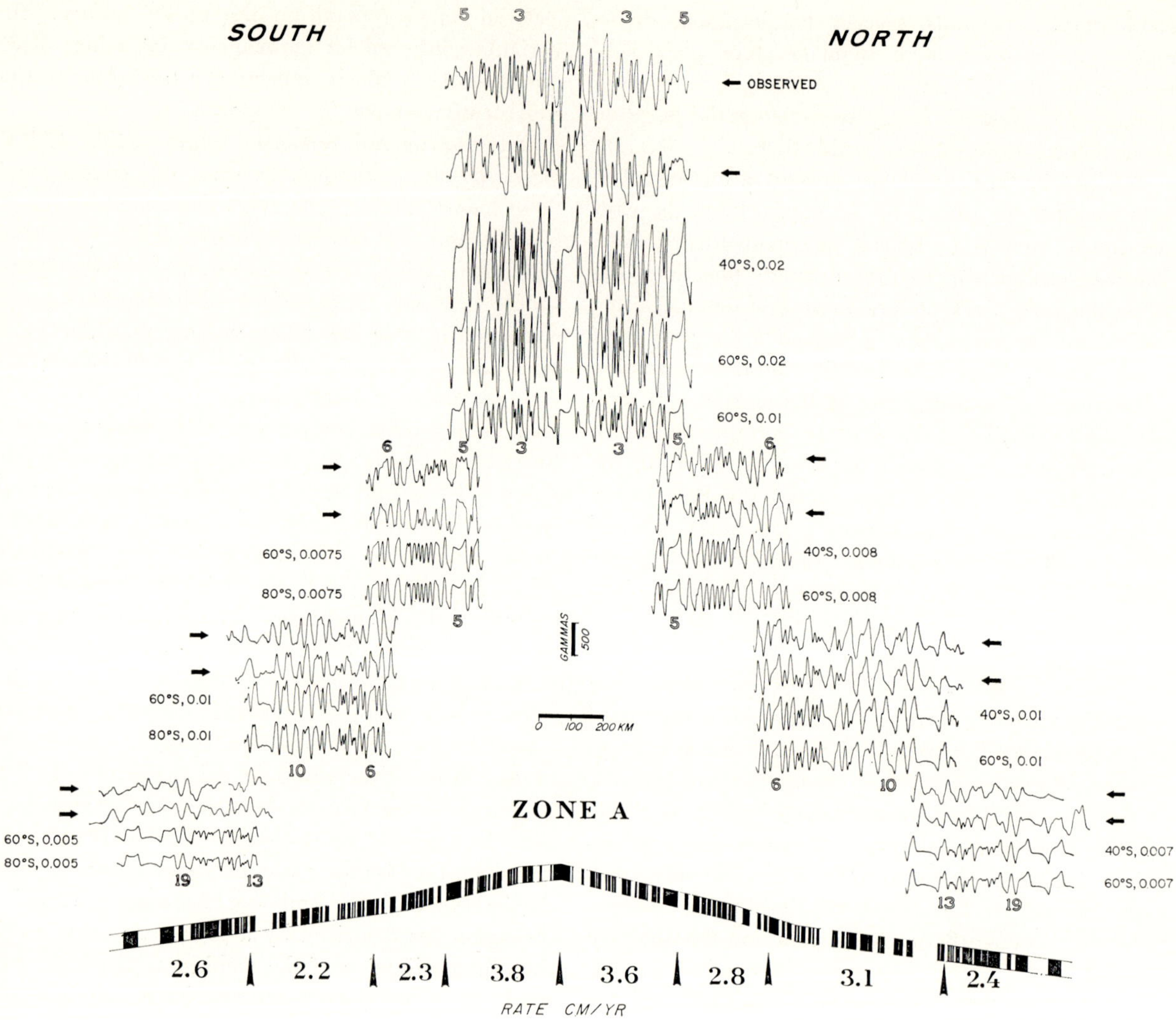

Fig. 7. Two-dimensional synthetic magnetic anomalies for zone A computed for selected sequences of anomalies. Two representative observed profiles (indicated by the arrows) were selected from all observations and are shown for comparison with the synthetic anomalies and their range of NRM parameters judged to give the best fit. Spreading rates used in the models are indicated and are not necessarily equal for the same anomaly sequences over opposite limbs. Spreading rates used were taken from *Weissel and Hayes* [1971, Table 11] as summarized in Figure 6.

occurrence of small variations in spreading rates including discrete jumps in the position of the ridge crest, (3) the magnetic effects of small topographic features not modeled, (4) errors in the assumed geomagnetic time scale, and (5) a combination of 1–4. The close over-all correspondence of the observed and the synthetic anomalies demonstrates that the effects mentioned above are generally not important for this area.

The computational procedure used here [*Talwani and Heirtzler*, 1964] differs from the method advocated by *McKenzie and Sclater* [1972] in that only the component of the anomalous field in the direction of the ambient field is actually calculated. Generally, this quantity is a close approximation to the total residual field if the magnitude of the anomalous field is small with respect to that of the ambient field (here the ratio of the two is about 1:60 or less). The more rigorous approach of McKenzie and Sclater should be used if the anomalous field is large when compared with the ambient field. The ambient field parameters used in the calculations were determined from the International Geomagnetic Reference Field coefficients [*Cain et al.*, 1968].

An examination of Figures 7, 8, and 9 reveals variations in both the magnitude and the inclination of

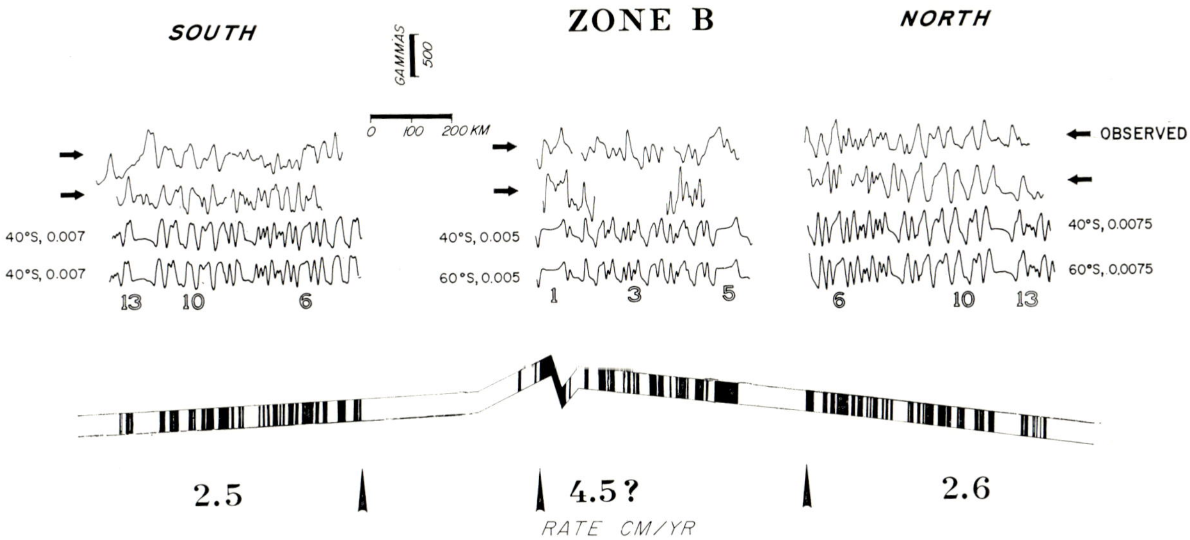

Fig. 8. Two-dimensional synthetic magnetic anomalies for zone B computed for selected sequences of anomalies. Two representative observed profiles were selected from all observations and are shown for comparison with the synthetic anomalies and their range of NRM parameters judged to give the best fit (also see Figure 7).

the inferred NRM vector for the same sequences of anomalies from the three zones (A, B, and C) and also for corresponding anomalies on opposite sides of the ridge axis but within a single zone.

The model studies of Figures 7, 8, and 9 serve to complement the data presented in Figure 5 regarding the magnitude variations of the anomalies. The near-axial anomalies of zone A require a large value of magnetization relative to the same anomalies from zone B and zone C. The central anomaly of zone A requires a value of 0.02 emu when compared with about 0.005 emu for zone B and about 0.01 emu for zone C. However, it should be remembered that the unusual morphologic fabric of zone B may be responsible for some of this difference. Within zone A, the required magnetization value falls from the high value at the ridge axis to about 0.01 emu or lower for anomaly 5. In general, the magnitude of magnetization required to reproduce anomaly amplitudes on both flanks of zone A is equal to or greater than the values for corresponding anomalies from the other two zones.

Note from Figure 8 that within zone B a greater value of magnetization characterizes anomalies older than anomaly 6 than characterizes anomalies younger than anomaly 5. The older anomalies are associated with a smoother, less disrupted ridge morphology than that associated with the younger anomalies.

For zone C (Figure 9), anomalies older than about anomaly 5B on the north flank (Indian plate) require a higher value of magnetization (0.006–0.012 emu) than their corresponding anomalies on the south flank, the antarctic plate (0.005–0.007 emu). This relationship is also clearly illustrated in Figure 5.

The estimates of the optimum inclination of the NRM vector for limited sequences of magnetic anomalies (i.e., inferred latitude of formation of the magnetic blocks) show considerable scatter as seen from a careful examination of Figures 7, 8, and 9. This scatter prevents us from unequivocally determining the movement of the Indian and antarctic plates relative to the magnetic dipole. We therefore choose to discuss these results only in general terms so as not to bias the reader in gaging the significance of the inclination differences shown.

If all the estimates of the value of inclination obtained from anomalies between 6 and 13 are averaged for zone A where the observed data are most numerous, then an average inclination of 76° is found for the south flank, whereas an average inclination of 67° is obtained for the north flank. These values of magnetic inclination correspond to average latitudes of formation for this sequence of anomalies of about 64°S for the antarctic plate and near 50°S for the Indian plate. Anomalies 5B and 5C on the Indian plate require inclinations around 60°–70° to reproduce the anomaly forms, whereas their counterparts on the antarctic plate generally require inclinations of 70°–80° or greater. This result provides an important discrepancy with the generally accepted

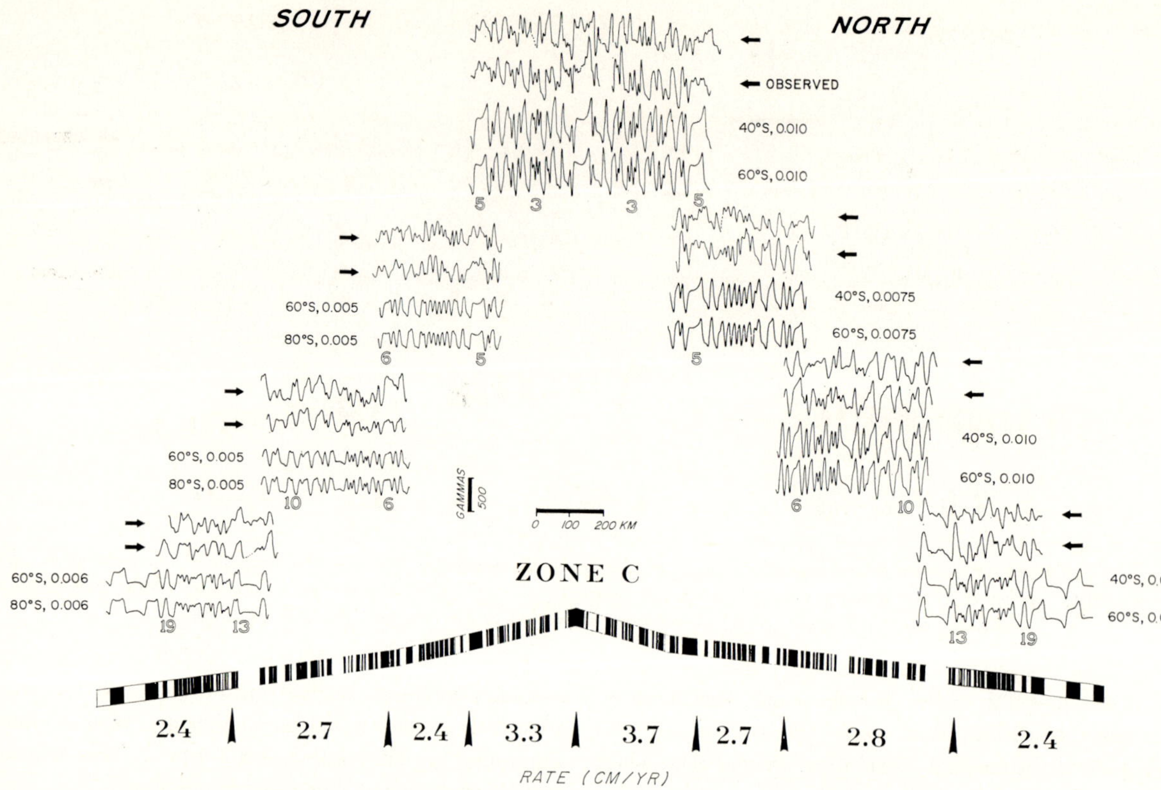

Fig. 9. Two-dimensional synthetic magnetic anomalies for zone C computed for selected sequences of anomalies. Two representative observed profiles were selected from all observations and are shown for comparison with the synthetic anomalies and their range of NRM parameters judged to give the best fit (see also Figure 7).

ideas regarding the genesis of sea floor spreading magnetic lineations. Further, we suggest that our data from zones B and C also show these kinds of discrepancies.

McKenzie and Sclater [1972] have calculated magnetic models for profiles across the Indian and antarctic plates. They present a synthetic magnetic profile that is compared with the *Conrad* 8 track that runs southwest from Fremantle. For the anomaly sequence older than about anomaly 7 on the Indian plate [*McKenzie and Sclater,* 1972, Figure 31], their model suggests that these anomalies were formed at 45°S on a ridge striking N 70°W. However, the same sequence of anomalies from the antarctic plate corresponds poorly with the model profile, this poor correspondence further suggesting that real differences in the parameters of the NRM vector are needed to model the older anomalies from opposite sides of the ridge axis. J. Sclater (personal communication, 1971) recomputed synthetic anomalies for the *Conrad* 8 track. He found that, although the anomalies on the north flank agreed with a latitude of formation of 45°S, the same sequence on the south flank could have been generated at 75°S.

For all three zones, the near-axial anomalies (younger than anomaly 5) can be modeled by using inclinations between 60° and 70° that correspond to latitudes of formation of between 41°S and 54°S. These latitudes are close to the observed positions of these anomalies, and thus the derived inclinations are not at variance with plate tectonic theory.

Some additional anomalous characteristics of the observed magnetic lineations can be seen in Figure 7. On the south flank of zone A, most observations of anomaly 5 and some of anomaly 6 show a level crest or a slight peak over the southern side that is contrary to that expected for these model blocks magnetized in the Southern Hemisphere. In fact, only one observation on the south flank of zone A shows the peak of anomaly 5 over the northern edge of the positive polarity block as predicted by the synthetic anomalies. This feature must represent some fundamental inadequacy in the assumptions implicit in our magnetic model.

It follows from the theory of plate tectonics that anomalies of the same age, now observed both on the Indian and antarctic plates, were originally formed together at a paleo-accreting plate boundary. Thus essentially the same NRM vectors should characterize the corresponding magnetic lineations on both plates. The variations in both magnitude and inclination of the NRM vectors estimated from the magnetic data south of Australia clearly do not support this important corollary. A particular consequence of this fact is that we were unable to use the derived NRM inclination values to estimate the past movement of the Indian and antarctic plates with respect to the geocentric axial dipole. The anomalies here could provide erroneous paleomagnetic pole positions if considered in a manner similar to that of *Francheteau et al.* [1970] for their seamount studies. Paleomagnetic determinations from Antarctica [*Irving*, 1964] give a pole at 86°S, 178°E for Upper Cretaceous time. By assuming that Antarctica has moved uniformly to its present position, the inferred NRM inclinations giving rise to the observed magnetic lineations on the antarctic plate should correspond to latitudes of formation close to the present positions of the lineations.

This section of the world ocean is one where the direction of the magnetic lineation fabric is almost east-west. Thus this section is unusual when compared with the composite world oceanic ridge system. *Vine* [1968] has derived anomalously low inclinations from a study of both the east-west and north-south lineations segments of the Great Magnetic Bight of the northeast Pacific. He concluded that the main Pacific plate has moved northward with respect to the axis of rotation of the earth. Our assumption that the magnetic declination is near zero within the southeast Indian Ocean is probably valid throughout the Cenozoic, and the east-west trending sea floor spreading lineations are potentially very useful parameters for studying the inclination of the remanent magnetic vector and its stability.

Another important although less striking result is that some anomalies computed by using the *Heirtzler et al.* [1968] time scale are compressed or absent from the observed profiles. The sequence of reversals representing anomalies immediately older than anomaly 5 out to about anomaly 5A is compressed on the observed profiles, and a slowdown in the spreading rate or a change in spreading geometry may be proposed. Only data older than anomaly 5A were considered in the linear least squares analysis made by *Weissel and Hayes* [1971] (see Figure 6) to determine the best fit constant spreading rates for the period 10–20 m.y. B.P. The anomalies between anomalies 6 and 7 are also compressed to a slight degree and may also reflect a real decrease in separation rate. Anomaly 14 is almost everywhere absent, and anomalies 15 and 16 are often so compressed that initial recognition of these anomalies is very difficult. Thus the period from 39–42 m.y. B.P. may also mark a period of adjustment in the relative motion between the Indian and antarctic plates.

These sequences of compressed anomalies are characterized by polarity reversals of short duration. Figures 7, 8, and 9 show that for all three zones the synthetic profiles require less magnetization for these anomalies than for the younger and older long wave length anomalies. The assumption of constant magnetization implicit in our models appears to be invalid for these anomaly sequences. The lower net magnetization inferred for these compressed blocks might be related to the contamination of material of one magnetic polarity by material of the opposite polarity during the process of accretion at the ridge crest, although this explanation is not the only feasible one. Note the indication of an 'extra' reversed event occurring in the middle of anomaly 5C (see Figure 3) not recognized in the *Heirtzler et al.* [1968] time scale. The short normal event between the Kaena and Mammoth reversed events (see *Foster and Opdyke* [1970]) also seems displaced in the models in comparison with our observations.

Perhaps these detailed discrepancies are caused by small errors in the time scale of *Heirtzler et al.* [1968]. Recently, *Talwani et al.* [1971] have presented evidence for a slight revision of the reversal time scale out to anomaly 5 time (10 m.y. B.P.). Other refinements of the time scale will probably prove necessary with the accumulation of more data.

TECTONIC HISTORY OF THE INDIAN OCEAN

As discussed by *McKenzie and Parker* [1967], *Morgan* [1968], and *Le Pichon* [1968], the relative motion of any two rigid 'plates' on a sphere can be mathematically described by a rotation around a pole. In this paper, the extensive magnetic data in conjunction with topographic and seismologic information are used to calculate poles of relative motion of the Indian and antarctic plates for the present and for five times during the Cenozoic. These times are 10 m.y. B.P. (anomaly 5 time), 21.2 m.y. B.P.

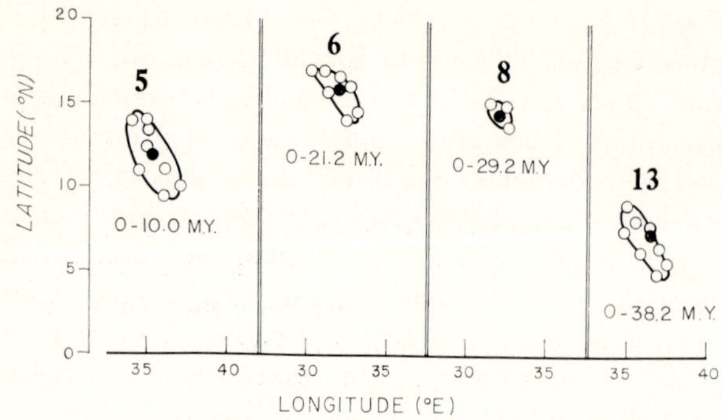

Fig. 10. The co-ordinates of acceptable trial poles of opening that allow selected anomaly lineations observed on the north flank of the Southeast Indian ridge to be rotated (about the pole) back onto anomaly lineations of the same age on the south flank. The optimum pole for each period is shown by the darkened circles. The uncertainty for each pole is indicated by the ellipse that encloses the individual determinations.

(anomaly 6 time), 29.2 m.y. B.P. (anomaly 8 time), 38.2 m.y. B.P. (anomaly 13 time), and approximately 55 m.y. B.P. when the continents were last in the position calculated by *Sproll and Dietz* [1969]. With the exception of 29.2 m.y. B.P., these times mark the end points of 'constant' spreading rate epochs previously recognized by *Weissel and Hayes* [1971]. The anomaly 8 lineation was also considered because there is a good distribution of this anomaly throughout the region and we felt that examining the history of separation of Australia and Antarctica at intervals of about 10 m.y. was desirable.

Magnetic lineations may be thought of as isochrons, and an average pole of opening that describes the relative movement of two isochrons may be calculated. The method of *Pitman and Talwani* [1972] in their study of the opening of the North Atlantic is used. Briefly, magnetic anomalies of a certain age are superimposed on anomalies of the same age on the opposite side of the ridge axis by a simple rotation about a trial pole. The position of the pole and the amount of rotation are adjusted to constrain the two sets of anomalies to lie along a smooth curve and so that anomalies generated on one side of a fracture zone would remain on the same side and at the same distance from the fracture zone after rotation to the opposite side of the ridge axis. The fracture zones that have been interpreted from offsets in magnetic lineations, seismicity, and morphologic evidence thereby act as constraints in the determination of pole positions. Anomalies 5, 6, 8, and 13 were manipulated by using this procedure.

Figure 10 shows the location of poles of opening to the various magnetic lineations and a measure of the precision of the determinations. Our estimate of the optimum pole position is shown. The scatter shown arises from uncertainties in the location of fracture zones and in the precise distribution of magnetic anomalies. We can see that the pole position of anomaly 8 and to some extent the pole position of anomaly 6 are more accurately determined than the pole positions for anomalies 5 and 13. The scatter about the anomaly 5 pole position is largely due to the proximity of the two lineations sets and thus the small arc through which the anomalies were rotated. The scatter about anomaly 13 pole position reflects the relatively small number of observations of this anomaly. As apparent from the elliptical distribution of acceptable poles shown in Figure 10, a greater uncertainty exists in latitude than in longitude with respect to the determined poles of opening.

A pole of opening that describes the movement of Australia from the *Sproll and Dietz* [1969] fit of the two continents to its present position was calculated. The opening is located at 3.4°S, 38.8°E, and the total rotation is 31°. This result is similar to that ob-

TABLE 1. Poles of Opening to Key Magnetic Lineations and Poles Assumed Operating for the Time Stages between Key Anomalies for the Indian and Antarctic Plates during the Cenozoic

Duration, m.y. B.P.	Pole of Opening: Latitude, Longitude	Pole of Opening: Rotation, °	Stage	Duration, m.y.	Stage Pole: Latitude, Longitude	Stage Pole: Rotation, °
10 (anomaly 5)	12.0°N, 35.25°E	6.75	I	0–10	12.0°N, 35.25°E	6.75
21.2 (anomaly 6)	16.0°N, 32.0°E	12.20	II	10–21.2	20.8°N, 27.8°E	5.52
29.2 (anomaly 8)	14.5°N, 32.0°E	15.90	III	21.2–29.2	9.3°N, 32.0°E	3.55
38.2 (anomaly 13)	7.3°N, 36.5°E	20.00	IV	29.2–38.2	17.2°S, 51.5°E	4.85
55* (reconstruction)	3.4°S, 38.8°E	31.20	V	38.2–55	20.6°S, 42.7°E	12.38

* *Sproll and Dietz* [1969].

tained by *Smith and Hallam* [1970] from their reconstructions of the 500-fm isobaths of the two continents. Their calculations give a pole at 3.6°S, 40°E and a rotation of 31°. Essentially equivalent results were obtained by *McKenzie and Sclater* [1972], who calculated a pole at 6°S, 40.5°E with a rotation of 31.6°. They considered the 1-km isobath of the Kerguelen plateau and Broken ridge as well as the Australian and antarctic continents.

Two new sets of centers of rotation (north and south of the ridge axis) can be obtained by addition of finite rotations using poles and amounts of rotation shown in Table 1 and Figure 10. The successive rotations define theoretical fracture zones formed during the time intervals indicated in Table 1. The hypothetical fracture zones drawn by using these two sets of centers of rotation are shown in Figure 11. Also shown in Figure 11 are the key locations of selected magnetic anomalies used in calculating the finite pole rotations. The symbols denote anomalies from the north flank that have been rotated onto their counterparts on the south flank according to the procedure outlined above. In the construction of hypothetical fracture zones, we tacitly assume that symmetrical generation of crust takes place. Segments of the hypothetical fracture zones are drawn both toward the north and toward the south and start from the present position of the accreting plate margin.

For zone C, the hypothetical fracture zones cross the Australian continental margin and predict that anomaly 21 should occur inside the continent. This result may reflect special conditions in the early rifting history that were unique to zone C. *Hayes and Conolly* [this volume] briefly discuss the possibility of intensive rifting along this portion of the margin. Anomaly 19 is the oldest observable anomaly on the north flank of zone C, and it is bounded on the north by the Diamantina fracture zone.

The most striking aspect of the hypothetical fracture zone map is that the fracture zones seem markedly inappropriate for the area east of zone A. The present ridge axis east of 140°E is displaced southward from the median position between the two continents. The result is that the hypothetical fracture zones cross deep into the antarctic continent but fail to reach the Australian continental margin south of Adelaide and west of Tasmania. The fracture zone analysis of Figure 11 suggests that the gap between the ends of the fracture zones and the continental margin in this area should be oceanic in nature. This gap is associated with the northern magnetic quiet zone as shown in Figure 2. Curiously, the southern magnetic quiet zone is associated with an area of slight overlap.

Stage poles that are defined as the poles of opening describing the relative motion of the two plates for a particular time period during the history of separation are calculated for the intervals 0–10 m.y. B.P. (stage I), 10–21.2 m.y. B.P. (stage II), 21.2–29.2 m.y. B.P. (stage III), 29.2–38.2 m.y. B.P. (stage IV), and 38.2–55 m.y. B.P. (stage V). (Note that the time intervals used by *Chase* [1971] are slightly different for stages III, IV, and V.) The individual stage poles are found by first removing the effect of subsequent rotations of the plates. For example, to find the stage II pole that is applicable from anomaly 5 time to anomaly 6 time (10–21 m.y. B.P.), the two centers of rotation describing the north and south flank portions of the fracture zones created during this interval are rotated in opposite directions about the pole of opening for the anomaly 5 lineation. This rotation brings the two centers of rotation into coincidence. The stage poles obtained are listed on the right-hand side of Table 1 and are plotted in Figure 12. Also shown in Figure 12 are the stage poles of *Chase* [1971]; the poles of opening for the reconstructions of *Sproll and Dietz* [1969], *Smith and Hallam* [1970], and *McKenzie and Sclater* [1972]; and the poles of opening and instantaneous poles of *Christoffel* [1971] and *Le Pichon* [1968].

McKenzie and Sclater [1972] used available magnetic and morphologic data from the area north and west of the southeast Indian Ocean to calculate poles and rotations that describe the relative movement of the Indian and antarctic plates during the Cenozoic. They calculated an instantaneous rotational pole at 11°N, 32°E with a rate of 6.4°/10 m.y. from fracture zone azimuth and near-axial magnetic anomalies. This pole may be applicable back to 10 m.y. B.P. If so, it should approximate the pole at 12°N, 35.25°E presented in this paper, the amount of rotation being 6.75°. The pole presented by the above authors describing the opening of anomaly 11 magnetic lineation is situated at 4.9°N, 36.4°E, the amount of rotation being 18.3°. A similar pole presented in this paper describes this opening from the time of formation of anomaly 13 (38 m.y. B.P.). This pole is located at 7.3°N, 36.5°E with 20° of rotation (see Table 1).

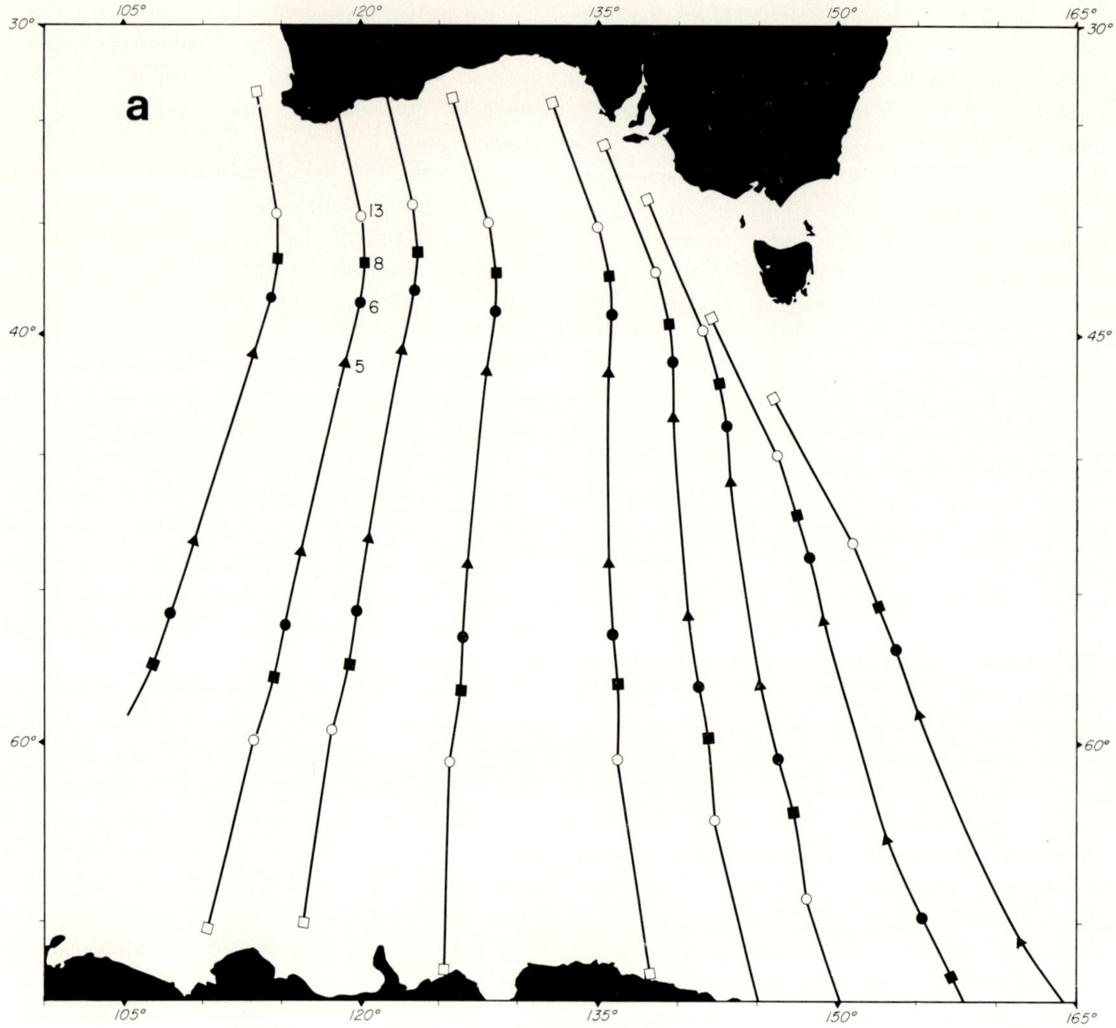

Fig. 11. Hypothetical fracture zones deduced from the poles of opening shown in Figure 10. Figure 11*a* is a Mercator projection, and these fracture zones should be compared with observed offsets of the lineation pattern shown in Figure 2 (see foldouts at back of volume). Figure 11*b* is a stereographic projection for comparison with the relative continental position shown in Figure 13. The finite rotation segments for each stage pole are indicated. The magnetic anomalies defining the end of each consecutive stage are shown on both limbs by symbols, and these coincide closely with the corresponding observed magnetic lineations.

As derived from Figure 10 and illustrated in Figure 12, the poles of opening of magnetic lineations 5, 6, 8, and 13 presented in this paper and the pole of opening calculated from the *Sproll and Dietz* [1969] reconstruction all group in northeast Africa. In Figure 12, Africa is shown in its present position for reference only and is not in any sense intended as a representation of paleogeography. The group of poles is elongated in a northwest-southeast direction with the 'oldest' pole, the Sproll and Dietz pole, at the southeastern end. There is a progressive migration of these pole positions to the northwest with decreasing age until the pole for anomaly 6 time. This trend is then reversed with the pole of opening for anomaly 5 migrating back to the southeast. Note that the stage I pole and the 0–5 anomaly pole are by definition the same. The stage poles also show the northwest-southeast grouping but to a more pronounced degree and a similar migration path, except that between the stage IV and V poles the trend is to the northeast. The stage II poles are at the northwest extremity of the group. The poles determined by *Christoffel*'s [1971] two models, considered applicable for the past 10 m.y., both lie to the north-

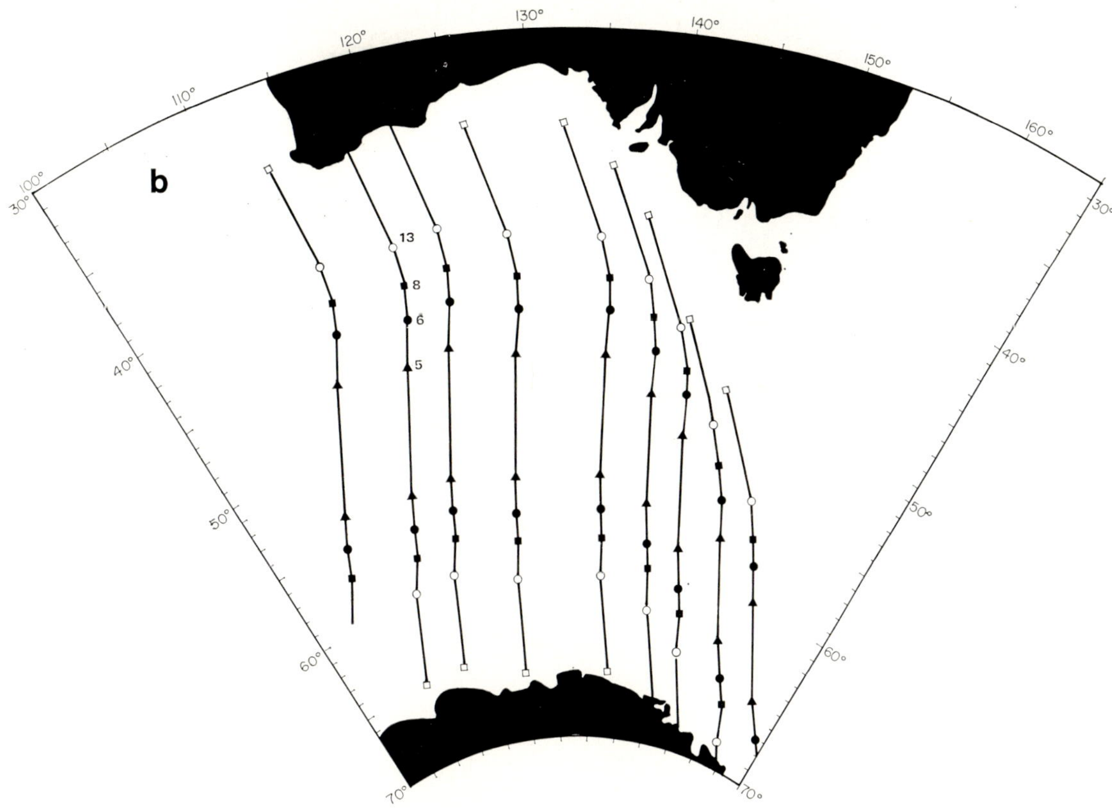

Fig. 11. (continued)

west of those calculated in this study. *Chase*'s [1971] poles for stages I–V, except for his stage II pole, lie to the southeast of the analogous group presented in this paper. His poles also show a northwest-southeast trend in grouping with the oldest stage poles at the southeast end.

The *Le Pichon* [1968] 'instantaneous' pole, considered applicable for recent rotations (~10 m.y.), lies to the south or southwest of poles of similar applicability determined above. The *Le Pichon* [1968] pole of opening of the anomaly 18 lineations lies considerably to the south of the analogous poles of *Sproll and Dietz* [1969], *Smith and Hallam* [1970], and *McKenzie and Sclater* [1972]. The pole determinations of *Christoffel* [1971] and *Le Pichon* [1968] were based on far less data than that available to us and are thus considered as subject to significant error.

We wish to stress that these poles do not necessarily have any physical significance. They are, however, a convenient manner in which one can describe the characteristics of relative plate motions and are useful in evaluating the extent of systematic motions.

The northwest-southeast grouping of the stage poles with the oldest stages at the southeast end of the group suggests that the pole defining the relative motion between the Indian and antarctic plates has responded to the absolute movement of these plates. The locus of the pole follows a path that is somewhat analogous to that of the Indian plate with respect to the antarctic plate (see Figure 13) and may be considered effectively fixed with respect to the Indian plate. The reversal of the northwest migration trend for the stage I pole and the divergence during stage V are regarded as perturbations in the general motion. These apparent perturbations may coincide in time with the observed decreases in spreading rates as seen from the compression or complete absence of anomalies by using the *Heirtzler et al.* [1968] geomagnetic time scale as previously discussed.

The close similarities between the various poles

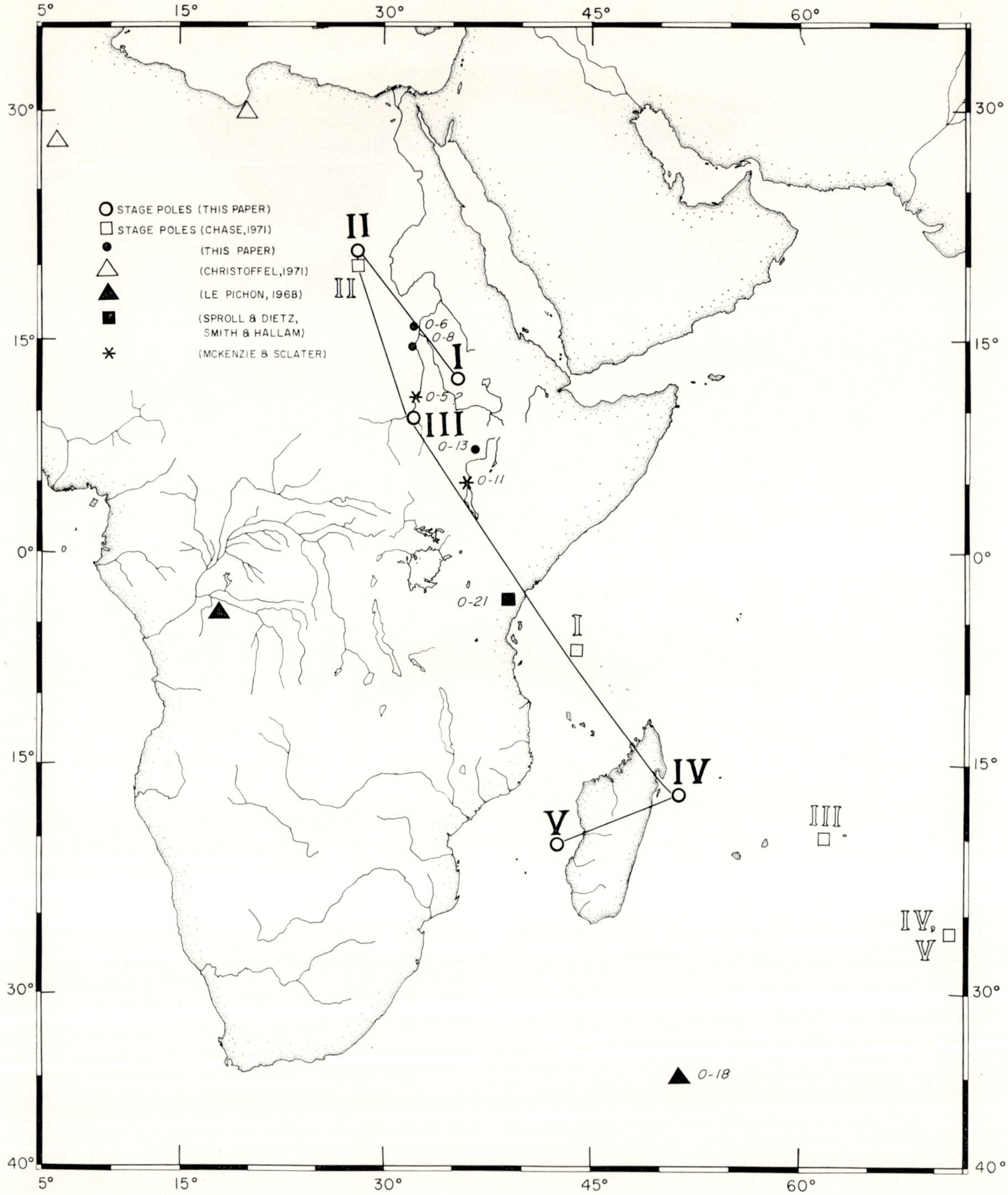

Fig. 12. Map of poles of opening and stage poles for the motion between the Indian and antarctic plates for selected time intervals. The distribution of the poles suggests a systematic migration of the pole relative to the antarctic plate. Africa is shown in its present position and is for reference only.

and rotations calculated by *McKenzie and Sclater* [1972] and the ones presented in this paper constitute a significant result of our study. Both groups of workers used sets of data from those parts of the Indian and antarctic plates lying between the Indian Ocean triple junction and the Macquarie triple junction, a distance exceeding 5000 km. However, McKenzie and Sclater used data to the west of about 105°E, whereas our data were confined to the area east of 105°E. The agreement between these inde-

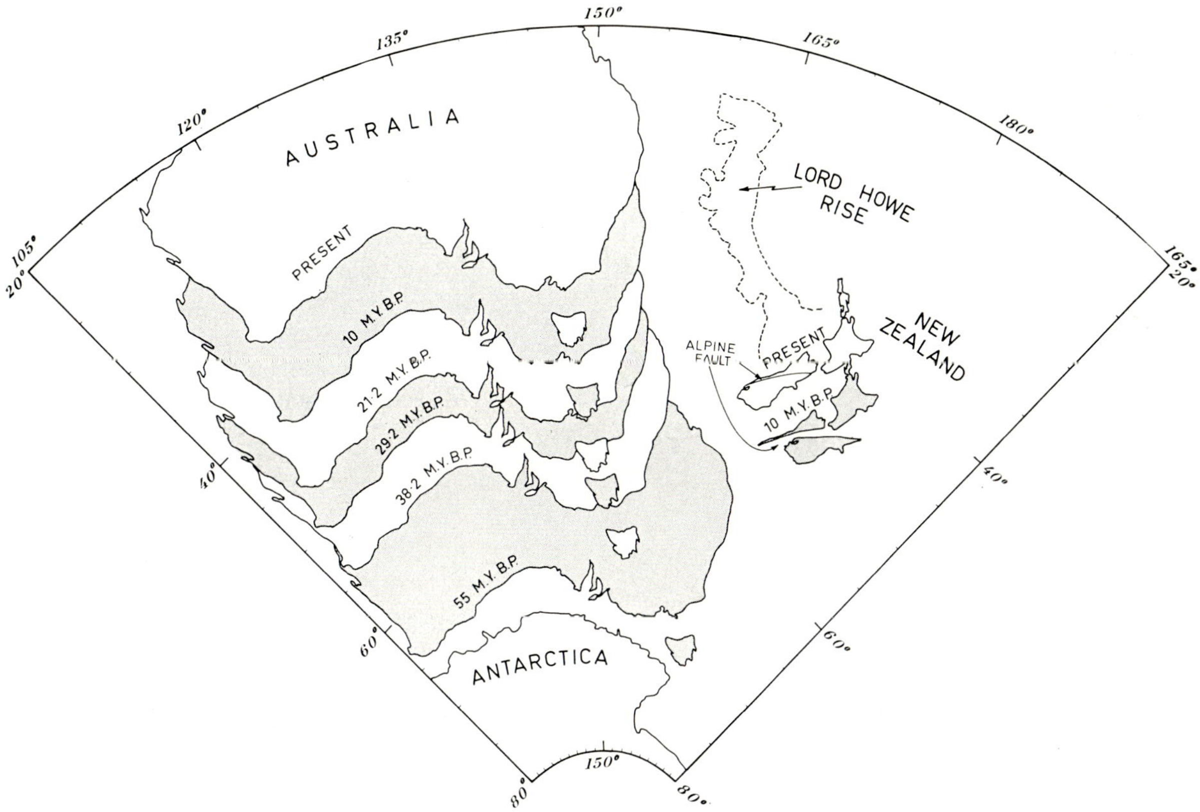

Fig. 13. The relative position of Australia and Antarctica during the past 55 m.y. as derived from the stage poles shown in Figure 12. Antarctica is assumed fixed. The configuration of the Indian-Pacific boundary (in part, the Alpine fault of New Zealand) prior to 10 m.y. B.P. is unknown.

pendent studies attests to the reliability of the techniques used and to the applicability of the assumption of the continuity of the Indian plate.

The position of Australia relative to Antarctica was calculated for the following five times in the geologic past: 10, 21.2, 29.2, 38.2, and 55 m.y. B.P. The successive stages in the evolution of the Indian Ocean are shown in Figure 13; the detail at 38.2 m.y. is shown in Figure 14. In these reconstructions, the position of Antarctica is held fixed with respect to the present geographic co-ordinates. In addition, we have assumed that the Tasman Sea basin opened prior to 55 m.y. B.P. Evidence in support of this assumption is given in *Hayes and Ringis* [1972], *Vogt and Conolly* [1971], and *Fleming* [1969].

Whereas the tectonic history of separation of Australia from Antarctica is fairly straightforward from 55 m.y. B.P. to the present, except, perhaps, for the very early rifting history, the evolution of the present geology and geography of the New Zealand area is much more difficult to deduce.

Christoffel [1971] postulated that the observed movement along the Alpine fault of New Zealand can be accounted for by 10 m.y. of relative motion of the Indian and Pacific plates. To test his hypothesis, the new stage I pole for the motion of the Indian plate with respect to the antarctic plate and *Christoffel*'s [1971] Pacific-antarctic pole at 82°S, 120°E with 10.4° of rotation in 10 m.y. was used to compute the amount of relative motion between the Indian and Pacific plates along the Alpine fault during the past 10 m.y. (see also *Hayes and Talwani* [this volume]). The amount of relative movement deduced, 580 km, is the same as obtained by Christoffel using a different Indian-antarctic pole. The pole of *Le Pichon* [1968] at 72°S, 118°E with a rotation of 10.8° in 10 m.y. was also tested with the stage I pole from Table 1. This test gives a relative displacement of 410 km in 10 m.y. across the Alpine fault (see Figure 13). Geologic evidence summarized by *Fleming* [1969] shows that about 450–480 km of movement has occurred across the Alpine fault during

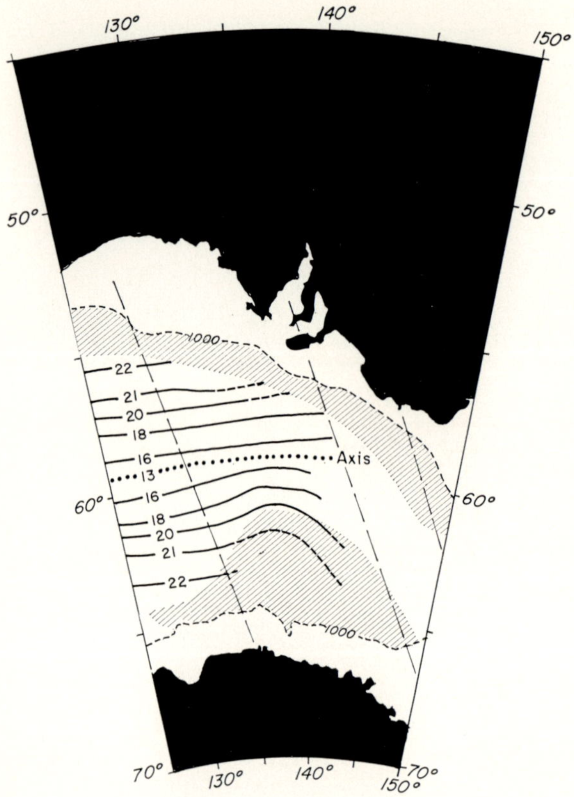

Fig. 14. Reconstruction at the time of formation of anomaly 13 (38 m.y. B.P.) showing geometry of magnetic lineations, initial direction of relative motion (broken line), and magnetic quiet zones (shaded).

or after the Rangitata Orogeny (Upper Jurassic–Lower Cretaceous). *Wellman and Cooper* [1971] recently presented radiometric age dates that indicate that the movement of the Alpine fault has occurred in two main phases: one during the Early Cretaceous and one during the last 10 m.y., the later phase involving a displacement of only about 120 km. These results are clearly in variance with those obtained by using the theories of plate tectonics. If, as the motions of the Indian and Pacific plates suggest, the total displacement along the Alpine fault can be accounted for in the last 10 m.y., what is the nature of the plate boundary between the Pacific and Indian plates prior to 10 m.y. B.P.?

It seems necessary to propose that the tectonic history of New Zealand included about 70° of anticlockwise rotation of the Campbell plateau region with respect to Northland sometime between 10 and 90 m.y. B.P. This proposal is necessary if the geologic trends on the Northland and the aeromagnetic trends over the rocks of the New Zealand geosyncline on the Campbell plateau are to be brought into alignment with each other and if they are also to be aligned with the trends of the Mesozoic geosyncline in West Antarctica. There is evidence from the occurrence of anomaly 32 and older anomalies in the Pacific basin south of the Campbell plateau that the onset of the relative movement between the Pacific and antarctic plates was in the Upper Cretaceous when the Campbell plateau started moving away from Antarctica [*Pitman et al.*, 1968; *Le Pichon*, 1968; *Christoffel and Falconer*, this volume]. However, the Cenozoic history of the New Zealand area is still very speculative, and its solution awaits more geologic and geophysical data.

Predicted Motion between the Indian and Pacific Plates for the Last 10 m.y.

Although the behavior of the boundary between the Indian and antarctic plates as a function of time is now known, the nature and history of the Pacific-Indian plate boundary has been the subject of many recent inquiries. *Christoffel* [1971] has studied the last 10 m.y. of relative motion along the Alpine fault segment of this boundary by using the theories of plate tectonics and sea floor spreading data. *Christoffel* [1971], *Hayes et al.* [1972], and *Hayes and Talwani* [this volume] all concluded that the relative motion along the Indian-Pacific plate boundary south of the Alpine fault is very complex. Subduction of the Indian plate at the Puysegur trench is predicted to have occurred during the past 10 m.y. Some extension may have taken place in the vicinity of Macquarie Island. For the southern end, the Hjort trench, some compression may have taken place, although pure strike slip motion may also characterize this section. However, the predominant motion along the plate boundary was found to be transcurrent with small components of extension and compression predicted for different points along the boundary. *Hayes and Talwani* [this volume] present additional bathymetric, gravity, seismic, and magnetic data for this area and discuss in detail the difficulties in reliably predicting the detailed relative motions at this boundary.

A test was then made to compare the predicted motions of the Indian and Pacific plates for the last 10 m.y. as calculated by *Christoffel* [1971] and *Hayes et al.* [1972] with those motions produced when the Indian-antarctic stage I (see Table 1) pole is substituted in the calculations. Two cases were considered by using different poles and rotations for the

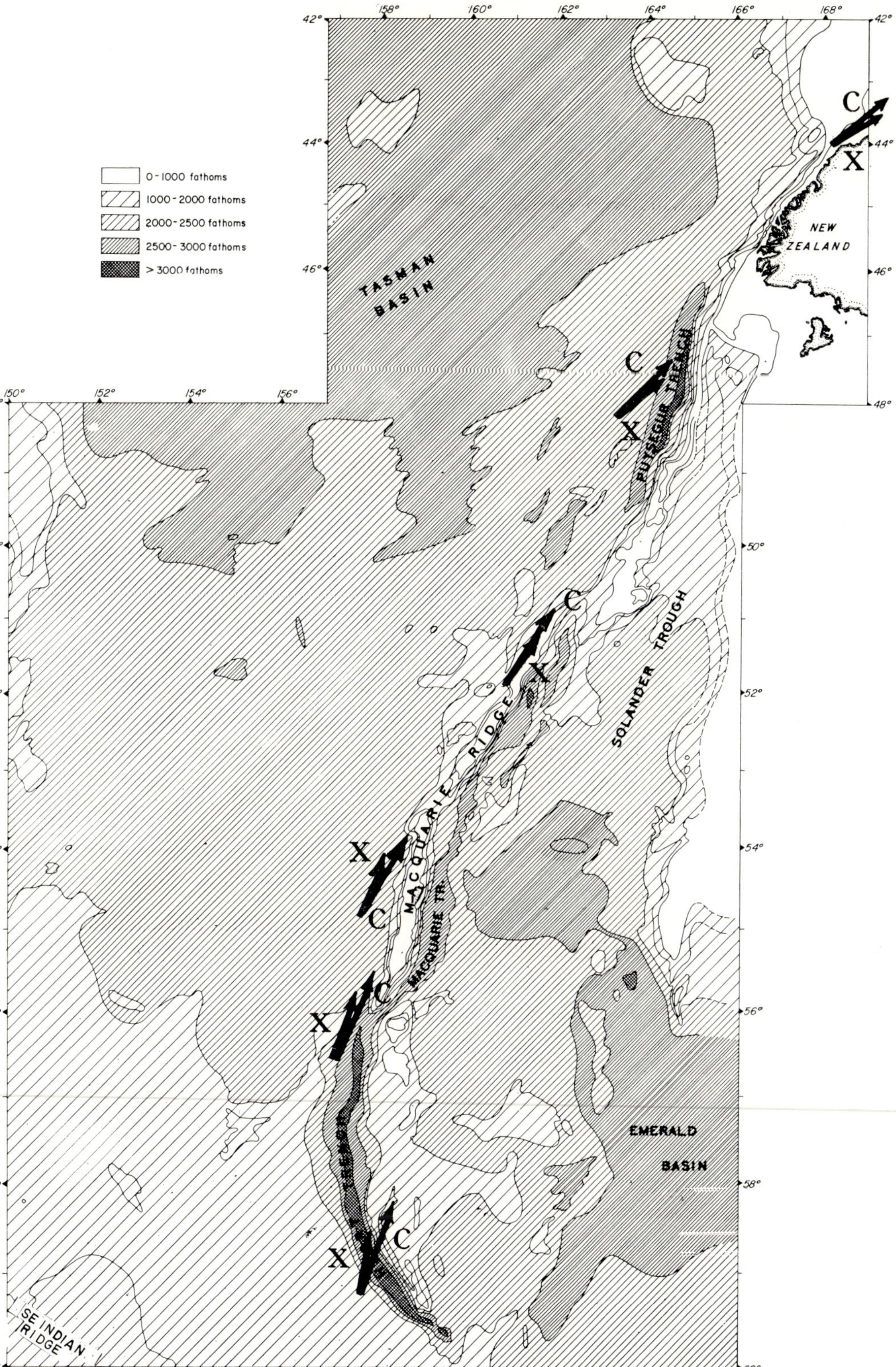

Fig. 15. The Indian-Pacific plate boundary showing the direction and amount of relative motion of the Indian plate with respect to the Pacific plate inferred for the last 2 m.y. The Indian-antarctic stage I pole of this paper was combined with the Pacific-antarctic instantaneous poles of opening of *Le Pichon* [1968] and *Christoffel* [1971] shown by × and C, respectively. The topographic base map is taken from *Hayes et al.* [1972].

relative motion between the Pacific and antarctic plates: (1) *Le Pichon*'s [1968] pole at 72°S, 118°E and a rate of 10.8°/10 m.y.; these values give a Pacific-Indian pole at 59°S, 173°W and a rate of 13.6°/10 m.y.; and (2) *Christoffel*'s [1971] pole at 82°S, 120°E and a rate of 10.4°/10 m.y.; these values give a Pacific-Indian pole at 60°S, 157°W and a rate of 13.6°/10 m.y.

By using these two calculated Indian-Pacific poles, velocity vectors were calculated at the same points along the Indian-Pacific boundary that were used by *Hayes et al.* [1972, Figure 1]. For case 1 above, the results do not show the extensional zones predicted by Christoffel and by Hayes et al. but rather show mainly strike slip motion together with compressional zones at the Puysegur and Hjort ends of the complex. For case 2 above, the same general remarks apply, but the magnitudes of the velocities are greater, and there is more compressional motion predicted for the southern end of the complex. The calculated velocity vectors are shown in Figure 15, which is modified from Figure 1 of *Hayes et al.* [1972]. Units are in kilometers per 2 m.y. to be comparable with the amounts of motion calculated by *Hayes et al.* [1972] along the boundary.

As a test of accuracy of the position and relative velocity of the stage I Indian-antarctic pole, the velocity vector between these plates was computed at the location of the Macquarie triple junction at the Southeast Indian ridge axis. The inferred separation was then compared to the observed separation of anomaly 5 on opposite sides of the ridge axis. The computation gives a velocity of 6.77 cm/yr at an azimuth of 154°, a movement 677 km in 10 m.y. This movement compares to an observed separation of 685 km between the oldest parts of anomaly 5 on both sides of the ridge axis measured from *Eltanin* cruise 27 data. A less accurate estimation of about 700 km of separation of anomaly 5 lineations can be made from a combination of *Eltanin* cruise 36 and cruise 37 traverses. Hence the stage I pole that was computed from all observations of anomaly 5 from 105°E to 160°E fits the observations at the extreme end of the area quite satisfactorily.

Predicted Motion of the Macquarie Triple Junction

The boundaries between Indian, Pacific, and antarctic plates meet at a point termed the Macquarie triple junction. The geographical position of this triple junction was taken by *Sykes* [1970] to be near 62°S, 160°E.

From the work of *Sykes* [1970], *Christoffel* [1971], and *Christoffel and Falconer* [this volume], the Pacific-antarctic boundary at the triple junction was taken to be a transform fault. The Indian-antarctic boundary was taken to be an accreting plate margin on the basis of magnetic pattern and morphology. The nature of the Indian-Pacific boundary at the triple junction is more difficult to assess. *Christoffel* [1971] and *Hayes et al.* [1972] attributed mainly strike slip motion to this boundary at the southern end, but a small amount of compression may be allowed according to *Christoffel* [1971], *Hayes and Talwani* [this volume], and the present results. The exact orientation of this boundary at the triple junction is not well determined and adds to the uncertainty of determining the motion of the triple junction with respect to the three plates. The orientation of the Pacific-Indian plate boundary at the triple junction is arbitrarily assigned to be parallel to the direction of movement between the antarctic and Indian plates (Figure 15).

The pole of relative motion used for the Indian and antarctic plates is the stage I pole shown in Table 1. We have selected three possible poles and rotations describing the recent motion between the Pacific and antarctic plates: (*a*) 70°S, 118°E, rate of 10.8°/10 m.y. [*Chase*, 1971]; (*b*) 72°S, 118°E, rate of 10.8°/10 m.y. [*Le Pichon*, 1968]; and (*c*) 82°S, 120°E, rate of 10.4°/10 m.y. [*Christoffel*, 1971].

Following the procedure of *McKenzie and Morgan* [1969], a velocity vector triangle was constructed from the information presented above, and the motion of the triple junction with respect to the three plates was obtained by the method outlined below. The results are shown in Figure 16 where the Indian, Pacific, and antarctic plates are designated by A, B, and C, respectively. The subscripts *a*, *b*, and *c* refer to the three Pacific-antarctic poles given above.

The vector triangle is in velocity space; we consider an infinitesimal area around the triple junction. The sides of the triangle are composed of the relative velocity vectors between the plates; e.g., line AC is the velocity vector between the Indian and antarctic plates. The triple junction is defined by the concurrence of three plates, and movement of the junction is thus restricted to one of three plate boundaries.

In Figure 16, the boundary between plates A and C is a ridge axis. Assuming that the spreading is symmetrical, the midpoint of line AC gives the velocity of the ridge relative to plate A when measured from point A and relative to plate C when measured from point C. In both cases, the magnitude of the velocity is ½

AC. Any line drawn perpendicular from the midpoint of AC represents the velocity of a point moving relative to the ridge axis. Therefore, the perpendicular bisector of AC, shown as a broken line in Figure 16, is the locus of all velocities of points moving along the ridge axis. The triple junction represents one such point.

Along the boundary between plates A and B, some subduction occurs with plate A being subducted by plate B. Therefore the A–B plate boundary has zero velocity relative to plate B. Any line drawn from point B parallel to the direction of the trench boundary at the triple junction yields the velocity of a point with respect to plate B. The locus of all velocities along the trench is the broken line through point B in Figure 16. It is clear that the triple junction is again one such point.

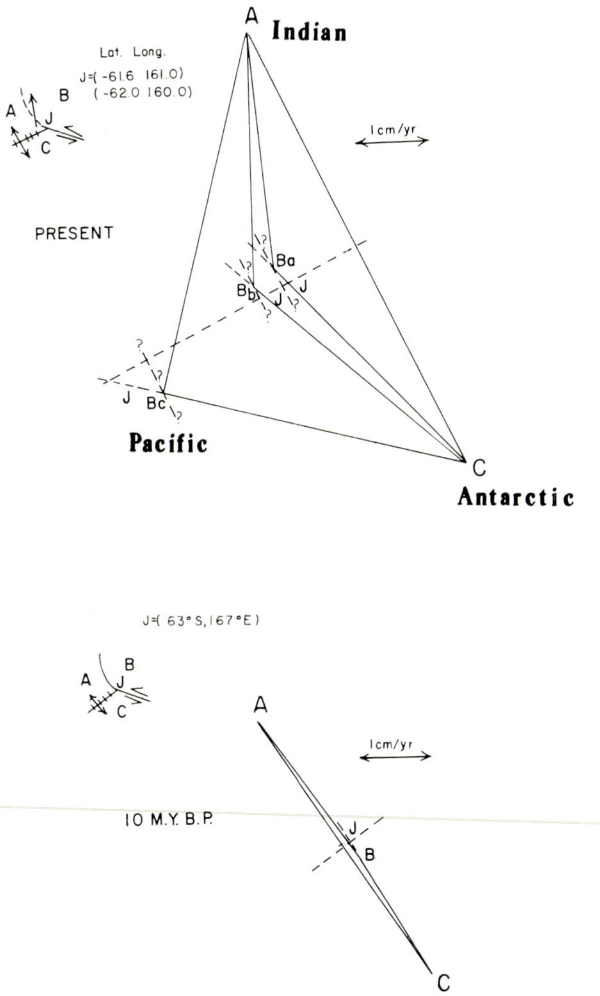

Fig. 16. Velocity vector triangles of the Macquarie triple junction (J) following the procedure of *McKenzie and Morgan* [1969]. The top diagram denotes the present time analyzed by using the Pacific-antarctic motion from (*a*) *Chase* [1971], (*b*) *Le Pichon* [1968], and (*c*) *Christoffel* [1971]. Intersections of the broken lines (with question marks) and the perpendicular bisector of line AC represent southwest velocities of 1.1, 1.5, and 3.7 cm/yr for cases *a*, *b*, and *c*, respectively. The bottom diagram denotes a possible dynamic state of the triple junction at 10 m.y. B.P. and at the stage II pole of Table 1.

For the transform fault boundary between plates B and C, the vector BC represents the slip rate between the plates. The line defined by any point on line BC (or BC extended) and by points B and C represents the velocity of a point moving along this plate boundary relative to plate B and plate C, respectively. Therefore, the locus of all velocities along the boundary is simply the extended part of line BC, which is shown broken in Figure 16.

Thus there are three 'velocity loci' that the relative motion of the triple junction must simultaneously satisfy. This situation can occur only if the broken lines in Figure 16 intersect at a point. If this condition exists, then the junction is stable and the plates defining the triple junction can maintain a fixed geometric configuration as they evolve. If the broken lines do not meet at a point, the triple junction is unstable. Either the plate boundary geometry or the relative motion between the plates is adjusted immediately so that stability is obtained.

For each of the three Pacific-antarctic poles given above, the triple junction is nearly at rest with respect to the Pacific plate. However, the top vector diagram in Figure 16 shows that the triple junction may be unstable. This instability was suspected by *Sykes* [1970] on the basis that the present location of the junction is hard to discern 'to better than a few hundred kilometers.' The effects of possible errors in the determination of the poles concerned on the apparent instability of the triple junction have not been evaluated. An interesting feature is that the geometry predicts that the most eastern segment of ridge axis between Australia and Antarctica is becoming shorter with time. In a reference frame fixed with respect to the ridge axis, the triple junction is moving southwest along the Southeast Indian ridge axis at a rate of about 1.1, 1.5, or 3.7 cm/yr for cases *a*, *b*, and *c*, respectively. This behavior was previously inferred as anomalies older than about anomaly 6 on the Indian plate increase in the along-strike length of magnetic

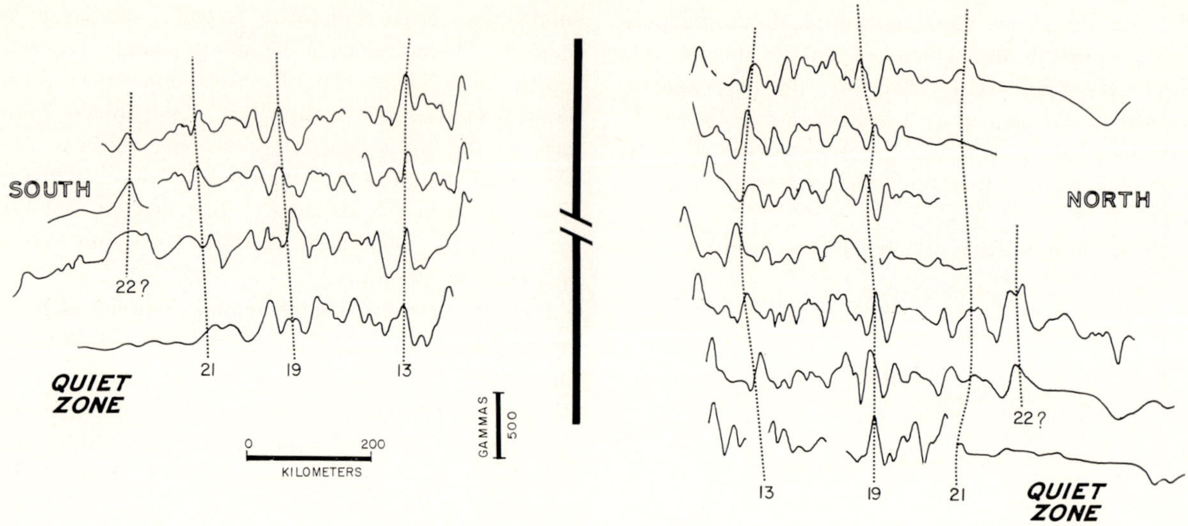

Fig. 17. Projected magnetic profile anomalies 13–21 and the magnetic quiet zone. The bottom left profile is an aeromagnetic profile. Correlations beyond anomaly 21 are tenuous.

lineations (see Figure 2). The applicability of the triple junction migration to the present time is uncertain because of the complications due to possible periods of subduction along the southern part of the plate margin at various times during the past 10 m.y.

Triple Junction at 10 m.y. B.P.

The dynamics of the Macquarie triple junction at 10 m.y. B.P. are considered by using the pole applicable for the period 10–21.2 m.y. B.P. (stage II).

At the time of formation of anomaly 5, the ancient Macquarie triple junction was in the vicinity of 63°S, 167°E as long as that motion along the Pacific-antarctic plate boundary was purely transcurrent during the last 10 m.y. If the stage II poles were acting at the time of formation of anomaly 5, then the triple junction was at rest with respect to the Pacific plate and also at rest in a reference frame fixed with respect to the Southeast Indian ridge axis (Figure 16).

Various errors and unknowns may alter the analysis of the dynamics of the Macquarie triple junction at the present time and at the time of formation of anomaly 5 (10 m.y. B.P.). These include (1) errors in the positions and rates of rotation of the poles used (particularly so in the case of the stage II pole and older poles describing the relative movement of the Pacific and antarctic plates), (2) uncertainties in the orientation of the Indian-Pacific boundary at the triple junction, and (3) unknown amounts (if any) of subduction at the above boundary that may have taken place during the last 10 m.y.

INITIAL RIFTING HISTORY

The earliest rifting history of Australia and Antarctica has been recently studied by *Smith and Kamerling* [1969], *Conolly et al.* [1970], *von der Borch et al.* [1970], *Veevers et al.* [1971], and *Griffiths* [1971]. Most of these workers conclude that initial breakup of the continents began in the Middle to Upper Jurassic. The occurrence of Jurassic dolerites in Tasmania and the Victoria Land portion of Antarctica has been cited as evidence for determining the relative position of the two areas and for the time of the first fragmentation of Australia and Antarctica.

The oldest magnetic lineation identified in the southeast Indian Ocean is anomaly 21 (or perhaps anomaly 22), the age of which is 54 m.y. B.P. (56.5 m.y. B.P.). Figure 17 illustrates the sequence of anomalies older than anomaly 18 (45 m.y. B.P.). There are only a few observations of anomaly 22, and prior to the development of anomaly 21 spreading may have been relatively slow. Consequently, an interval of about 100 m.y. exists between the first indications of continental rifting and the formation of the oceanic crust that recorded the magnetic lineations. During this long time interval, the evolution of the Australian and antarctic continental margins is poorly known. Several depositional basins located along the South Australian continental margin have accumulated large quantities of Mesozoic and Cenozoic sediment. These basins include the onshore Eucla, Otway, and Gibbsland basins; the offshore extensions of the Duntroon basin and Bass basin; and the deep Eyre, Ce-

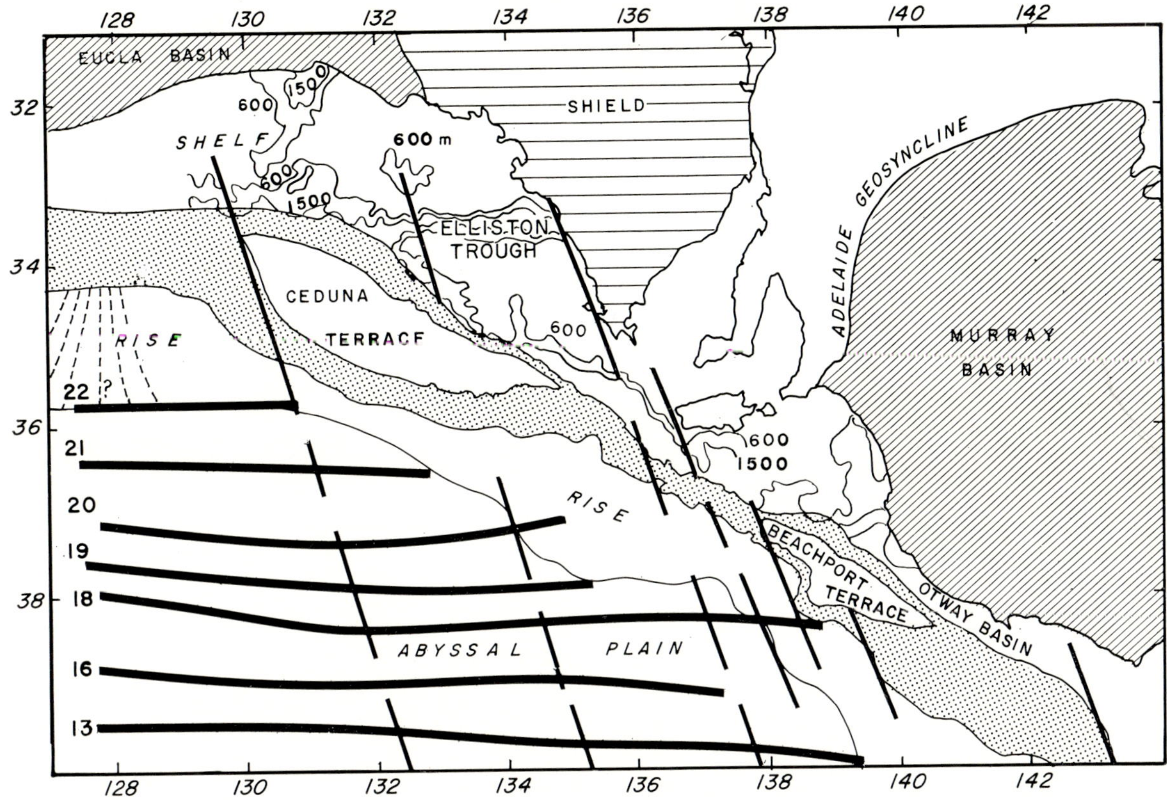

Fig. 18. The continental margin of the Great Australian Bight area after *von der Borch* [1967] showing the relationship of sedimentary basins, troughs, plateaus, and other physiographic lineaments to the inferred direction of motion for the earliest separation of the Australian and antarctic continents. The depths in meters to the magnetic basement are shown as selected portions of the continental shelf area.

duna, and Beachport marginal terraces (see Figure 18). *Conolly et al.* [1970] and *von der Borch et al.* [1970] have shown that in the vicinity of the Otway basin a marked unconformity occurs between the Mesozoic and Cenozoic sediments. The Mesozoic sediments have been disrupted by large-scale block faulting down to the basement [e.g., *von der Borch et al.*, 1970] that resulted in the landward dip of these strata. This faulting is thought to be the result of postrifting relaxation of the newly formed continental margin of Australia. *Griffiths* [1971] summarized schematically several stages in the evolution of the rifted continental margins. Above the Mesozoic-Cenozoic unconformity, which may represent an uppermost Cretaceous-Upper Eocene hiatus [*Jones*, 1971], the younger sediments have the nature of fluviatile deposits or deltaic wedges prograding seaward over the newly formed continental margin of Australia. The initial transgression of the sea occurred in the Upper Cretaceous [*Griffiths*, 1971] and progressed from west to east along the youthful plate boundary. Subsidence along the observed block faults could have continued until about 55 m.y. B.P. Thus in the Otway region, Mesozoic sediments that are block faulted and dipping landward are unconformably overlain by Eocene and younger fluvial and deltaic depositional wedges. *Conolly et al.* [1970] have presented evidence for Upper Eocene and Middle Miocene marine transgressions.

Recently, *Le Pichon and Hayes* [1971] proposed a model of early rifting for the equatorial Atlantic in which the initial direction of relative motion between two plates controls the distribution of subsiding basins and sedimentation along evolving continental margins. Relict 'transform directions' may, in many cases, be directly related to the trends of sedimentary basins and offsets of the present continental margin. *Le Pichon and Hayes* [1971] used these presumed relict transform directions along the margins of Africa and South America to compute a more accurate pole for the initial opening of the African and American plates. Using the initial pole of opening determined by *Le Pichon and Hayes* [1971], *Francheteau and Le Pichon*

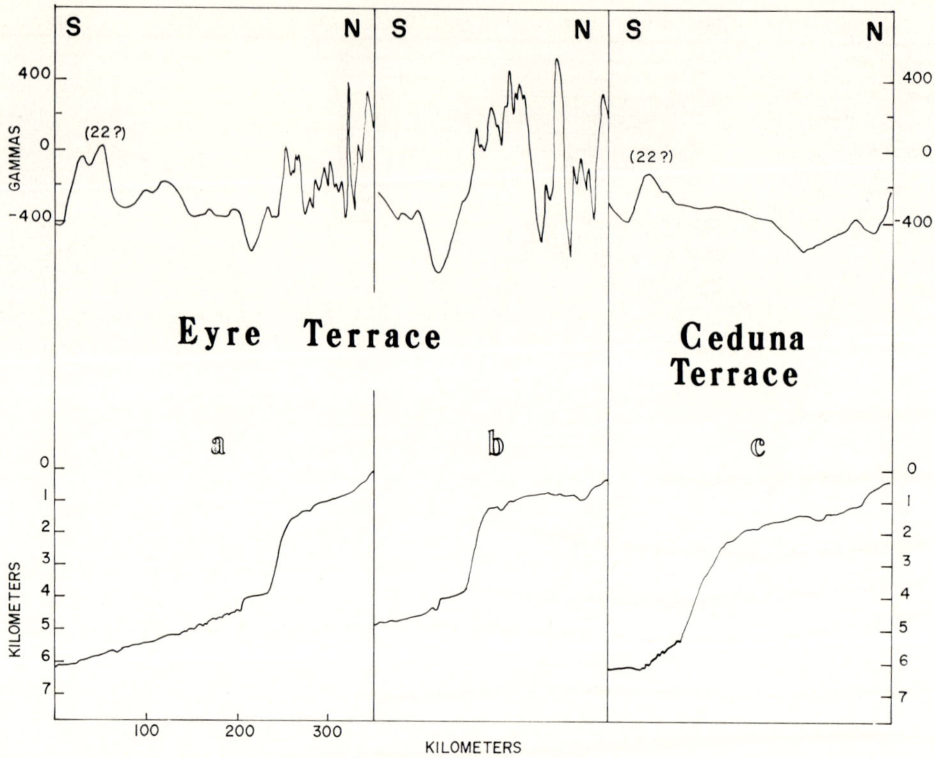

Fig. 19. Topographic and magnetic profiles across the Eyre and Ceduna terraces.

[1972] found that Mesozoic and younger sedimentary basins, offsets in bathymetry, and the trends of fracture zones abutting the continental margins of Africa and South America were then aligned parallel to hypothetical fracture zones, i.e., small circles about the pole of opening.

Since the South Australian continental margin has not been subjected to obvious subduction during its evolution, the initial transform directions may likewise have influenced the build-up of sedimentary basins along this margin. Figure 18 compares the trends in physiography and geology of the Australian margin near Adelaide with the azimuths of hypothetical fracture zones calculated from the stage V pole of Table 1. Also shown are the trends of magnetic anomaly lineations that are truncated at the continental margin. We can see that the hypothetical fracture zones are subparallel to the western margins of both the Ceduna and Beachport terraces. The northwest trend of the coast line west of Kangaroo Island (~35°S, 136°E) also seems to be consistent with the hypothetical fracture zone trend. The strike of Elliston trough is parallel to the strike of the magnetic lineations farther south (Figure 18), thus the idea that the Ellison trough was an incipient rift feature that became dormant by the end of the Mesozoic is supported [*Griffiths*, 1971].

Conolly et al. [1970] concluded from a study of geophysical data that the depth to magnetic basement beneath the Eyre and Ceduna terraces located off the continental margin of the Great Australian Bight is different. The Eyre terrace was thought to have a relative shallow magnetic basement, and the Ceduna terrace was thought to have a deeper one. Magnetic and depth profiles over the Eyre terrace and the Ceduna terrace are shown in Figure 19. Much shorter wave length anomalies occur over the Eyre terrace than over the Ceduna and support the contention that the magnetic basement is relatively shallow under the Eyre terrace. The outer slope of the Eyre terrace appears steeper than that of the Ceduna terrace, but a more extensive continental rise has been built off the Eyre terrace than off the Ceduna terrace. The distribution of the terraces and marginal basins and the contrast in their properties probably reflect the tectonic constraints imposed by the geometry of the initial break and the direction of motion of the earliest continental separation. No comparable information is available regarding sedimentary basins of the Wilkes Land continental borderland of Antarctica.

A distance of 100–250 km exists between the position of anomaly 22 and the respective continental slopes of Australia and Antarctica. Much of this 'unknown' area is coincident with the magnetic quiet zones (see Figure 17). The combined band that is 200–500 km wide probably represents a time span of 100 m.y. in the initial rifting history. The magnetic quiet zone may be the result of relatively old foundered continental blocks. The margins could thus be an ancient analogue to the Red Sea fault depression. Except over the axial trough, the Red Sea is characterized by low amplitude and long wave length magnetic anomalies [*Drake and Girdler*, 1964], i.e., a magnetic quiet zone. The seaward quiet zone boundary along the Australian and antarctic margins may correspond to the first onset of oceanic crustal formation. If so, the quiet zone boundary should define an isochron, but additional isochrons could not be extrapolated toward the continents because in this region the underlying crust has not been generated by crustal accretion. The magnetic quiet zone appears to cut across magnetic lineations and is thus inconsistent with the quiet zone boundary strictly representing an isochron. This discrepancy is closely related to the problem of explaining the pattern of the older magnetic lineations (anomalies 18–22) as they trend obliquely toward the continental margin in the Otway basin area.

DISCUSSION AND CONCLUSIONS

Weissel and Hayes [1971] suggested two explanations for the longitudinal differences in morphologic and geophysical properties along the accreting boundary between the Indian and antarctic plates. These were (1) contrasts in the mechanical processes active along the length of the accreting boundary, and (2) intraplate deformation occurring in areas away from the ridge axis. Our detailed study of all the available magnetic data leads to the following conclusions.

The existence of longitudinal zones of contrasting magnetic and morphologic properties recognized earlier is verified and additional contrasting properties are identified. Significant and probably systematic differences occur in the inferred NRM properties of the oceanic crust of the Indian and antarctic plates.

The amplitudes of certain anomaly lineations are greater over the Indian plate than amplitudes for corresponding anomalies on the antarctic plate. Figure 5 shows that this difference occurs for anomalies as young as the Gauss anomalies (2.8 m.y. B.P.) in zone A and is most marked in zone C for anomalies older than anomaly 5. Recently, *Watkins and Richardson* [1971] showed that magnetic anomaly amplitudes may be increased by a veneer of flood basalt or a localized extrusive component as long as the width of the extrusives is less than or equal to the depth to the sea floor. This effect could account for asymmetries in anomaly amplitudes from both plates over restricted areas. However, the area involved here is quite large, and the abrupt longitudinal change from large axial zone anomalies for zone A into much smaller anomalies in zone B leads us to believe the changes are related to the abrupt change in the morphology rather than to isolated extrusions of flood basalt. The contrast in amplitudes of the axial anomalies between zone A and zone C could be attributed to the effects described by Watkins and Richardson. Whatever the actual process responsible for these observed differences in axial anomaly amplitudes is, it must take place at or near the accreting plate boundary.

Calculated values of inclination of the NRM vector are also different for corresponding anomalies from both plates. The exceptions are the near-axial anomalies that show no conclusive difference. Natural remanent magnetization inclinations determined for the older parts (>10 m.y.) of the Indian plate are on the average lower by 15°–30° than inclinations determined for corresponding parts of the antarctic plate. The scatter of the individually determined inclination values does not allow us to unequivocally state that the inclination differences are systematic. However, we believe that different NRM inclinations are characteristic of the older parts of the two plates. Other plates that have an accreting plate boundary in common with the antarctic plate also show possible differences in their deduced NRM vectors. For anomalies in the sequence 25–32, W. C. Pitman (personal communication, 1971) needed different models on the South Pacific and antarctic plates. Also, anomalies older than anomaly 23 south of India (see *McKenzie and Sclater* [1972, Figures 23 and 26]) appear to record a different paleolatitude from those in the Crozet basin. Thus the differences in the NRM vectors for corresponding anomalies seemingly are not restricted to the southeast Indian Ocean but occur in other large geographical areas.

The low inclinations inferred seem to be a property of plates that have moved away from the relatively fixed antarctic plate. In such an instance, the accreting boundary must also migrate away from the fixed plate. Movement of the locus of crustal accretion with respect to the underlying asthenosphere may give rise to a mechanism of accretion substantially different from that of an accreting boundary and asthenosphere at rest. An extensive study of this problem is badly needed.

We have previously shown that differential migration of the ridge crest related to asymmetric sea floor generation in zone A has resulted in complexities of the morphology and magnetic pattern within zone B. The cause of the 'instability' of crustal accretion with zone B is unknown. Its effects are confined to areas of the ridge younger than about 15 m.y. A progressive offset of the ridge axis has been generated during the interval of asymmetric spreading. Zones of asymmetric spreading (e.g., zone A) that lie adjacent to zones of symmetric spreading (e.g., zone B) give rise to an offset magnetic lineation pattern and fracture zones. The resulting amount of offset is different for lineations of different ages.

Magnetic quiet zones are mapped along portions of the South Australian and antarctic margin. These boundaries are almost certainly not 'contemporaneous' with similar magnetic quiet zones along the margins of the North Atlantic. The quiet zones may define the areas of earliest rifting and of foundered and/or altered continental crust. Basement rock underlying the quiet zone south of Australia is less than 1 km deep and could be easily examined with the present deep-sea drilling capabilities of *Glomar Challenger*.

The history of relative motion between the Indian and antarctic plates has been documented in Table 1 and illustrated in Figure 13. Normal oceanic crust began to form between the continents of Australia and Antarctica about 55 m.y. B.P. by accretion at a mid-ocean ridge in a fashion similar to that in the South Atlantic. The initial direction of relative motion between the two plates appears to have influenced the location of sedimentary basins on the southern Australian continental margin. Counterparts to these Mesozoic and Cenozoic basins may eventually be found on the antarctic margin by following the hypothetical fracture zones from the Australian margin to the Antarctic.

Apparently, the pole of relative motion between the Indian and antarctic plates has systematically migrated with time. *Le Pichon and Hayes* [1971] suggested that, although initial offsets in continental lithosphere provide constraints on the pole of relative motion, these constraints are relaxed when evolution progresses to a stage where the offsets displace only oceanic lithosphere. The opening pole then tends to migrate to a position that minimizes the variation in spreading rates along the length of accreting boundary. The migration path of the pole for the Indian and antarctic plates shows a large shift between 38 and 29 m.y. B.P. Figure 13 shows that between these times the Tasman plateau moved past the edge of Victoria Land and thus relaxed the constraints on the pole of opening in the manner proposed by Le Pichon and Hayes. The present pole of opening has moved so that the equator of spreading passes about midway across the accreting boundary that runs from Ile Amsterdam to the Macquarie triple junction.

The suggestion by *Christoffel* [1971] that all the observed 450–480 km of strike slip displacement across the Alpine fault of New Zealand took place in the last 10 m.y. is generally supported by the new Indian–antarctic plate motions determined in this paper.

The area of extension in the vicinity of Macquarie Island on the Macquarie complex predicted by *Christoffel* [1971] and *Hayes et al.* [1972] is not predicted from new calculations. However, the contention by Christoffel and Hayes et al. that the Macquarie complex is a plate boundary showing mainly strike slip motion is supported.

An analysis of the relative velocity vectors acting at the Macquarie triple junction indicates that the length of the accreting boundary between the Indian and antarctic plates has been shortened during the last 10 m.y.

Acknowledgments. This research has been supported by the Office of Polar Programs (grants GV-23334 and GV-27472) and the Oceanography Branch (grant GA-27281) of the National Science Foundation and by the Office of Naval Research (contract N00014-67-A-0108-0004).

We wish to thank W. C. Pitman III for making much of the magnetic data used in this study available to us and for useful discussions and general support. J. K. Weissel received support for study at Lamont-Doherty from the Geophysics Section, University of New South Wales. The Geophysics Department of the University of New South Wales under the supervision of L. Hawkins co-operated with Lamont-Doherty Geological Observatory in conducting the marine geophysics program aboard the *Eltanin*. We acknowledge the important contribution of the Lamont-Doherty geophysical crews of the *Eltanin* in collecting most of the data incorporated in this study. We thank W. C. Pitman III, R. Larson, and J. G. Sclater for reviewing the manuscript.

Contribution 1778, Lamont-Doherty Geological Observatory.

REFERENCES

Banghar, A. R., and L. R. Sykes, Focal mechanisms of earthquakes in the Indian Ocean and adjacent regions, *J. Geophys. Res., 74*(2), 632–649, 1969.

Barazangi, M. W., and J. Dorman, World seismicity maps

compiled from ESSA coast and geodetic survey epicenter data, 1961–1967, *Bull. Seismol. Soc. Amer., 59*(1), 369–380, 1969.

Cain, J. C., S. Hendricks, W. E. Daniels, and J. C. Jensien, Computation of the main geomagnetic field from spherical harmonic expansions, *Tech. Rep. MSSDC 68-11*, Goddard Space Flight Center, Greenbelt, Md., 1968.

Chase, C., Tectonic history of the Fiji plateau, *Geol. Soc. Amer. Bull., 82*(11), 3087–3110, 1971.

Chase, C. G., H. W. Menard, R. L. Larson, F. G. Sharman III, and S. M. Smith, History of seafloor spreading west of Baja, California, *Geol. Soc. Amer. Bull., 81*(2), 491–498, 1970.

Christoffel, D. A., Motion of the New Zealand Alpine fault deduced from the pattern of sea floor spreading, in Recent Crustal Movements, edited by B. W. Collins and R. Fraser, *Roy. Soc. N. Z. Bull., 9*, 25–30, 1971.

Christoffel, D. A., and R. K. H. Falconer, Marine magnetic measurements in the southwest Pacific Ocean and the identification of new tectonic features, in *Antarctic Oceanology II: The Australian–New Zealand Sector, Antarctic Res. Ser.*, vol. 19, edited by D. E. Hayes, pp. 197–209, AGU, Washington, D. C., 1972.

Conolly, J. R., A. Flavell, and R. S. Dietz, Continental margin of the Great Australian Bight, *Mar. Geol., 8*, 31–58, 1970.

Drake, C. L., and R. W. Girdler, A geophysical study of the Red Sea, *Geophys. J. Roy. Astron. Soc., 8*(5), 473–495, 1964.

Drake, C. L., and J. E. Nafe, Geophysics of the North Atlantic region (abstract), *Eos Trans. AGU, 53*(2), 175–176, 1972.

Fleming, C. A., The Mesozoic of New Zealand, History of the circum-Pacific mobile belt, *J. Geol. Soc. London, 125*, 125–170, 1969.

Foster, J. H., and N. D. Opdyke, Upper Miocene to Recent magnetic stratigraphy in deep-sea sediments, *J. Geophys. Res. 75*(23), 4465–4473, 1970.

Francheteau, J., G. A. Harrison, J. G. Sclater, and M. L. Richards, Magnetization of Pacific seamounts: A preliminary polar curve for the northeastern Pacific, *J. Geophys. Res., 75*(11), 2035–2061, 1970.

Francheteau, J., and X. Le Pichon, Marginal fracture zones as a structural framework of the continental margins in the South Atlantic Ocean, *Bull. Amer. Ass. Petrol. Geol.*, in press, 1972.

Griffiths, J. R., Continental margin tectonics and the evolution of southeast Australia, *Aust. Petrol. Explor. Ass. J., 11*, 75–79, 1971.

Hayes, D. E., and J. R. Conolly, Morphology of the southeast Indian Ocean, in *Antarctic Oceanology II: The Australian–New Zealand Sector, Antarctic Res. Ser.*, vol. 19, edited by D. E. Hayes, pp. 125–145, AGU, Washington, D. C., 1972.

Hayes, D. E., and J. Ringis, Early opening of the central Tasman Sea, paper presented at the International Symposium on the Oceanography of the South Pacific, UNESCO, Wellington, New Zealand, February 1972.

Hayes, D. E., and M. Talwani, Geophysical investigation of the Macquarie ridge complex, in *Antarctic Oceanology II: The Australian–New Zealand Sector, Antarctic Res. Ser.*, vol. 19, edited by D. E. Hayes, pp. 211–234, AGU, Washington, D. C., 1972.

Hayes, D. E., M. Talwani, and D. A. Christoffel, The Macquarie ridge, in *Antarctic Geology and Geophysics*, edited by R. J. Adie, Universitets forlaget, Oslo, in press, 1972.

Heirtzler, J. R., and D. E. Hayes, Magnetic boundaries in the North Atlantic Ocean, *Science, 157*(3785), 185–187, 1967.

Heirtzler, J. R., G. E. Dickson, E. M. Herron, W. C. Pitman III, and X. Le Pichon, Marine magnetic anomalies, geomagnetic field reversals, and motions of the ocean floor and continents, *J. Geophys. Res., 73*(6), 2119–2136, 1968.

Herron, E. M., Sea-floor spreading and the Cenozoic history of the East Central Pacific, *Geol. Soc. Amer. Bull., 83*, in press, 1972.

Houtz, R. E., and R. G. Markl, Seismic profiler data between Antarctica and Australia, in *Antarctic Oceanology II: The Australian–New Zealand Sector, Antarctic Res. Ser.*, vol. 19, edited by D. E. Hayes, pp. 147–164, AGU, Washington, D. C., 1972.

Irving, E., *Paleomagnetism and Its Application to Geological and Geophysical Problems*, John Wiley, New York, 1964.

Isacks, B., J. Oliver, and L. R. Sykes, Seismology and new global tectonics, *J. Geophys. Res., 73*(18), 5855–5899, 1968.

Jones, J. G., Australia's Cenozoic drift, *Nature, 230*, 237–239, 1971.

Le Pichon, X., Sea-floor spreading and continental drift, *J. Geophys. Res., 73*(12), 3661–3697, 1968.

Le Pichon, X., and D. E. Hayes, Marginal offsets, fracture zones, and the early opening of the South Atlantic, *J. Geophys. Res., 76*(26), 6283–6293, 1971.

Le Pichon, X., and J. R. Heirtzler, Magnetic anomalies in the Indian Ocean and sea-floor spreading, *J. Geophys. Res., 73*(6), 2101–2117, 1968.

McKenzie, D. P., and W. J. Morgan, Evolution of triple junctions, *Nature, 224*, 125–133, 1969.

McKenzie, D. P., and R. L. Parker, The North Pacific: An example of tectonics on a sphere, *Nature, 216*(5122), 1276–1280, 1967.

McKenzie, D. P., and J. G. Sclater, The evolution of the Indian Ocean since the Late Cretaceous, *Geophys. J. Roy. Astron. Soc.*, in press, 1972.

Morgan, W. J., Rises, trenches, great faults, and crustal blocks, *J. Geophys. Res., 73*(6), 1959–1982, 1968.

Opdyke, N. D., Paleomagnetism, in *The Sea*, vol. 4, edited by A. E. Maxwell, part 1, pp. 157–182, Interscience, New York, 1970.

Pitman, W. C., III, and J. R. Heirtzler, Magnetic anomalies over the Pacific-antarctic ridge, *Science, 154*(3753), 1164–1171, 1966.

Pitman, W. C., III, and M. Talwani, Sea floor spreading in the North Atlantic, *Geophys. Soc. Amer. Bull., 83*(3), 1972.

Pitman, W. C., III, E. M. Herron, and J. R. Heirtzler, Magnetic anomalies in the Pacific and sea floor spreading, *J. Geophys. Res., 73*(6), 2069–2085, 1968.

Ringis, J., Magnetic lineations in the Tasman, *Geoexploration, 8*(3-4), 250, 1970.

Schlich, R., and Ph. Patriat, Profiles magnetiques sur la dorsale medioceanique 'l'Indo-Pacifique,' *Ann. Geophys., 23*, 629–633, 1967.

Schouten, J. A., A fundamental analysis of magnetic anomalies over oceanic ridges, *Mar. Geophys. Res., 1*, 111–144, 1971.

Smith, A. G., and A. Hallam, The fit of the southern continents, *Nature, 225*, 139–144, 1970.

Smith, R., and P. Kamerling, Geological framework of the Great Australian Bight, *Aust. Petrol. Explor. Ass. J., 9*, 60–66, 1969.

Sproll, W. P., and R. S. Dietz, Morphological continental drift fit of Australia and Antarctica, *Nature, 222*, 345–348, 1969.

Sykes, L. R., Seismicity of the Indian Ocean and a possible nascent island arc between Ceylon and Australia, *J. Geophys. Res., 75*(26), 5041–5055, 1970.

Talwani, M., and J. R. Heirtzler, Computation of magnetic anomalies caused by two dimensional structures of arbitrary shape, *Computers in the Mineral Industries*, pp. 464–480, School of Earth Sciences, Stanford University, Calif., 1964.

Talwani, M., C. Windisch, and M. G. Langseth, Jr., Reykjanes ridge crest: A detailed geophysical study, *J. Geophys. Res., 76*(2), 473–517, 1971.

Taylor, P. T., I. Zeitz, and C. S. Dennis, Geologic implications of aeromagnetic data for the eastern continental margin of the United States, *Geophysics, 33*, 755–780, 1968.

Tomoda, Y., K. Ozawa, and J. Segawa, Measurement of gravity and magnetic field on board a cruising vessel, *Bull. 3*, 169 pp., Ocean Res. Inst., Univ. of Tokyo, Tokyo, 1968.

Veevers, J. J., G. Jones, and J. A. Talent, Indo-Australian stratigraphy and the configuration and dispersal of Gondwanaland, *Nature, 229*(5284), 383–388, 1971.

Vine, F. J., Paleomagnetic evidence for the northward movement of the north Pacific basin during the past 100 m.y., *Eos. Trans. AGU, 49*(1), 156, 1968.

Vogt, P. R., and J. R. Conolly, Tasmantid guyots, the age of the Tasman basin, and motion between the Australian plate and the mantle, *Geol. Soc. Amer. Bull., 82*(9), 2577–2584, 1971.

Vogt, P. R., C. N. Anderson, D. R. Bracey, and E. D. Schneider, North Atlantic magnetic smooth zones, *J. Geophys. Res., 75*(20), 3955–3967, 1970.

von der Borch, C. C., Marginal plateaus: Their relationship to continental shelf sedimentary basins in southern Australia, *J. Geol. Soc. Aust., 14*(2), 309–316, 1967.

von der Borch, C. C., J. R. Conolly, and R. S. Dietz, Sedimentation and structure of the continental margin in the vicinity of the Otway basin, southern Australia, *Mar. Geol., 8*, 59–83, 1970.

Watkins, N. D., and A. Richardson, Intrusives, extrusives, and linear magnetic anomalies, *Geophys. J. Roy. Astron. Soc., 23*, 1–13, 1971.

Weissel, J. K., and D. E. Hayes, Asymmetric spreading south of Australia, *Nature, 231*(5304), 517–521, 1971.

Wellman, P., and A. Cooper, Potassium-argon age of some New Zealand lamprophyre dikes near the Alpine fault, *N. Z. J. Geol. Geophys., 14*(2), 341–350, 1971.

MARINE MAGNETIC MEASUREMENTS IN THE SOUTHWEST PACIFIC OCEAN AND THE IDENTIFICATION OF NEW TECTONIC FEATURES

D. A. Christoffel*

Lamont-Doherty Geological Observatory of Columbia University, Palisades, New York 10964

R. K. H. Falconer

Physics Department, Victoria University of Wellington, Wellington, New Zealand

Abstract. Analysis of 41 magnetic profiles in the southwest Pacific has enabled the pattern of magnetic anomaly correlations in the area to be clarified and extended. The time scale for sea floor spreading recognizable by characteristic magnetic anomalies has been increased from 76 to 81 m.y. We deduce that West Antarctica and New Zealand rifted just prior to 81 m.y. ago. Distinct periods of sea floor spreading can be identified with different poles of rotation for the motion between the Pacific and antarctic plates. South of New Zealand, the Albatross cordillera becomes a fracture zone at least 1200 km long, and we propose to name this section the Pacific-Antarctic fracture zone. A revised pole of 82°S, 120°E is suggested for the present Pacific-antarctic pole of rotation.

The pattern of magnetic lineations in the southwest Pacific Ocean was first demonstrated from two magnetic profiles by *Christoffel* [1961]. These profiles and those of *Christoffel and Ross* [1965] for just south of the Campbell plateau assisted in establishing the correlation of magnetic anomalies across the abyssal plain regions of the main oceans [*Vine*, 1966]. The global pattern of sea floor spreading and the relative poles of rotation between the Pacific and adjoining plates assigned by *Le Pichon* [1968] were shown by *Christoffel and Ross* [1970] to be not entirely compatible with the pattern of magnetic anomalies between the Campbell plateau and the shelf of the antarctic continent. The Pacific-antarctic pole of rotation of 70°S, 118°E assigned by Le Pichon does not correspond with the trend of the magnetic anomalies immediately south of the Campbell plateau. South of the Campbell plateau, poor correlation has been found between magnetic anomalies younger than number 20 on the classification of *Heirtzler et al.* [1968].

The most comprehensive analysis of magnetic and bathymetric data in the southwest Pacific Ocean has been presented by *Pitman et al.* [1968]. *Herron* [1971] has interpreted magnetic and bathymetric data in the southeastern Pacific as far west as 150°W. Much attention has recently been given to the Macquarie ridge by *Hayes et al.* [1972], *Houtz et al.* [1971], and *Christoffel* [1971].

We have obtained numerous magnetic and bathymetric measurements between New Zealand and Antarctica with our operations from the New Zealand antarctic supply ship, HMNZS *Endeavour*. We present here bathymetric and magnetic data, both from our own measurements and from previously published data, for the region between 150°W and 165°E that extends from the Campbell plateau in the north to south of the Pacific-Antarctic ridge. These data can be used to solve some of the questions raised in the interpretation of the plate tectonics of this and other regions.

DATA REDUCTION

Data Sources

For the region south of the Campbell plateau, most of the magnetic data have been obtained with our shipborne proton magnetometer towed from the *Endeavour* [*Christoffel and Ross*, 1965]. Those 18 tracks are identified in this paper by single letters. Portions of the tracks from the USNS *Eltanin* cruises 19, 23, 25, and 27 have been used [*Heirtzler et al.*, 1969; *Hayes et al.*, 1969] and are identified in the

* Now at Victoria University of Wellington, Wellington, New Zealand.

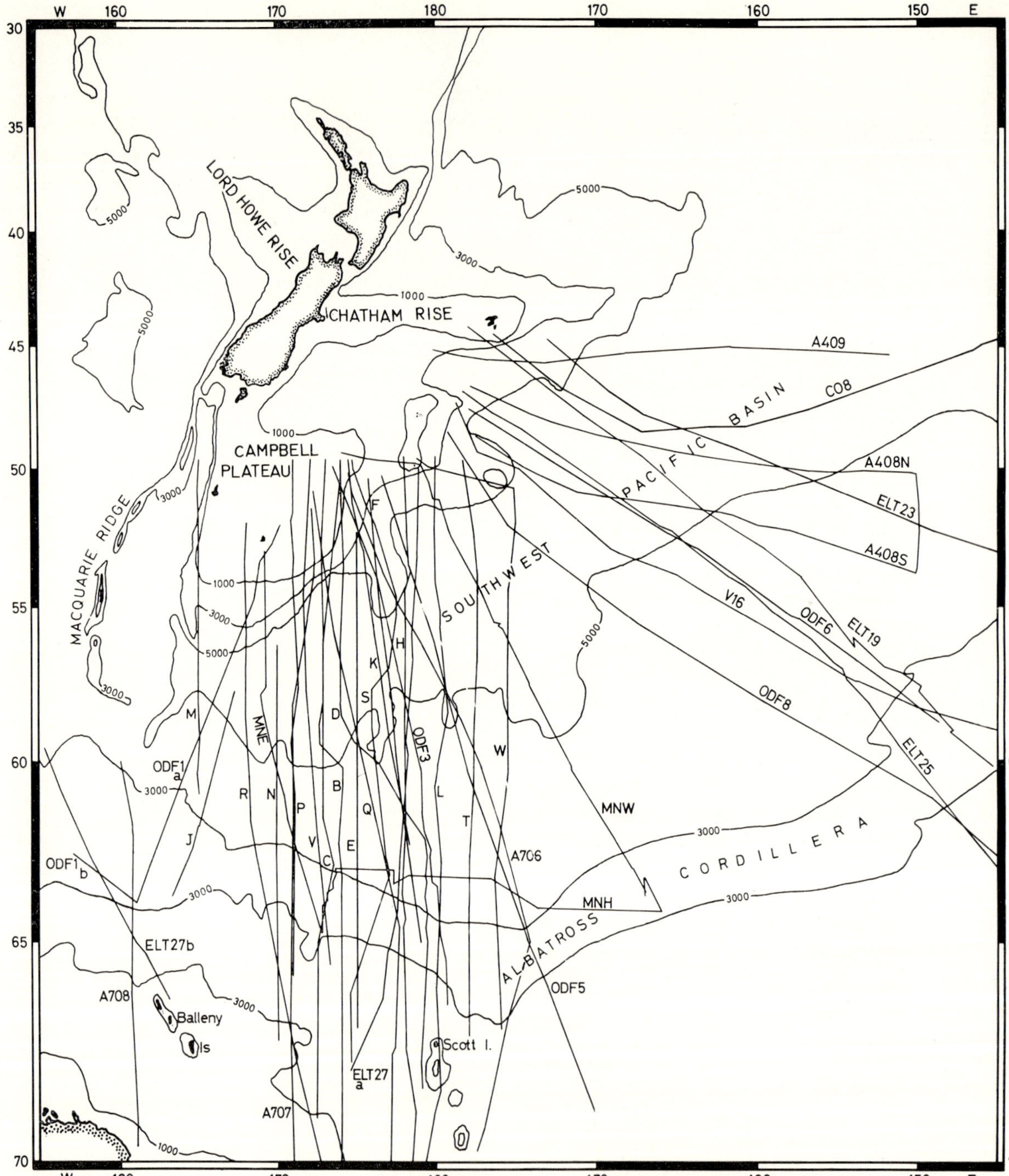

Fig. 1. Tracks used in the present study. The following abbreviations have been used: E, *Endeavour;* ELT, *Eltanin;* MN, Monsoon Expedition (RV *Argo*); ODF, Operation Deep Freeze (USNS *Staten Island* and *Burton Island*); A, aeromagnetic profiles from Project Magnet; V, *Vema;* CO, *Conrad.* Bathymetry contours are in meters.

following text and figures by ELT and the cruise number. Navigation, bathymetric, and magnetic measurements from the Operation Deep Freeze tracks 3, 5, 6, and 8 of the *Staten Island* and one track of the *Burton Island* have been obtained from the *U.S. Navy Hydrographic Office* [1962] and the *U.S. Naval Oceanographic Office* [1965]. These measurements are labeled ODF. Aeromagnetic profiles from Project

Magnet are identified by the letter A and the flight number. Portions of *Vema* cruise 16 are identified by V16, and portions of *Robert D. Conrad* cruise 8 are identified by CO8. Tracks from the Monsoon Cruise of the RV *Argo* from Scripps Institute of Oceanography are labeled MN.

Navigation

Most navigation was by conventional celestial and dead reckoning methods. The more accurate satellite navigation was available for *Eltanin* cruises 23, 25, and 27. Position errors for conventional navigation could be up to 75 km. Data from intersecting tracks show that average position errors are less than 25 km. Trends or offsets in anomaly patterns are considered established only to the latter limit unless several tracks delineate the feature.

Magnetic Data Reduction

The magnetic anomaly profiles were produced by subtracting from the recorded total field measurements a regional magnetic field calculated from the international geomagnetic reference field (IGRF) coefficients [*Fabiano and Peddie*, 1969]. The IGRF coefficients may require some revision since the anomaly profiles in this area had mean values differing considerably from zero. Most of the anomaly trends with wave lengths greater than 200 km were removed by this method. The only additional adjustment to any profile was the shifting of the zero reference line; the shift was equivalent to the removal of very long wave length components. No corrections were made for temporal variations. For the *Endeavour* cruises, the magnetic K indices and magnetic variograph recordings from Scott Base, Antarctica (77°51'S, 166°45'E), and Amberley, New Zealand (43°10'S, 172°45'E), have been checked over the cruise periods. No substantial magnetic storms occurred during the periods of measurement.

Anomalies are numbered according to the system of *Pitman et al.* [1968]. We have added the identification 32A for a characteristic positive anomaly just preceding anomaly 32.

DATA ANALYSIS

Figure 1 shows the tracks for which magnetic data are presented. Also shown are bathymetry contours that were compiled from *Lawrence* [1967], *U.S. Naval Oceanographic Office* [1969], *Udintsev et al.* [1964], and the New Zealand Oceanographic Institute (unpublished data, 1971). Figure 2 shows five bathymetric and magnetic profiles that we consider to be representative of the area. These profiles illustrate several features the identification of which simplifies the analysis.

The ELT25 magnetic profile exhibits symmetry about the ridge center (the Albatross cordillera) and contains an almost complete sequence of magnetic reversals. The corresponding bathymetric profile also shows general symmetry and is typical of bathymetry found over the East Pacific rise.

Profile ODF5 is similar to profile ELT25 in that symmetry about a spreading center is observed out to anomaly 5. The sequence of anomalies from 25 to 32 is also clearly seen, but, although this section is of similar length to that in ELT25, this section is closer to the ridge axis and the anomaly sequence 5–25 is not wholly clear.

The magnetic profile H shows the anomaly group 25–32, again with similar length to the previous profiles. No symmetrical center is obvious on either the magnetic or the bathymetric profile. High amplitude, short wave length anomalies are observed where the topography is most broken.

Profile B still shows the anomalies 25–32 but shows them closer to the 'ridge.' The section of disturbed bathymetry is rougher and broader than that to the east.

The ELT27b magnetic profile clearly shows symmetry, and the anomaly sequence agrees closely with that of the currently spreading ridges [*Vine*, 1966]. The bathymetry is also symmetrical. This profile indicates an area of active sea floor spreading similar to that observed much farther to the east along the ELT25 track.

These representative profiles illustrate the uniformity and persistence of the older magnetic anomalies throughout most of the area, the presence of active spreading in both the east and the west, and a region of disturbed bathymetry and magnetic signatures in between. In the light of these general observations, we shall divide the area into three regions and discuss them separately.

Region 1

This region includes the area containing magnetic anomalies numbered 25 and upward. The sequence of anomalies numbered from 25 to 32 is dated by *Heirtzler et al.* [1968] as covering the time span of 63–76 m.y. The sequence is identifiable on almost every track in this area.

Figure 3 shows the positions of these anomalies together with assumed trends and offsets. All the profiles

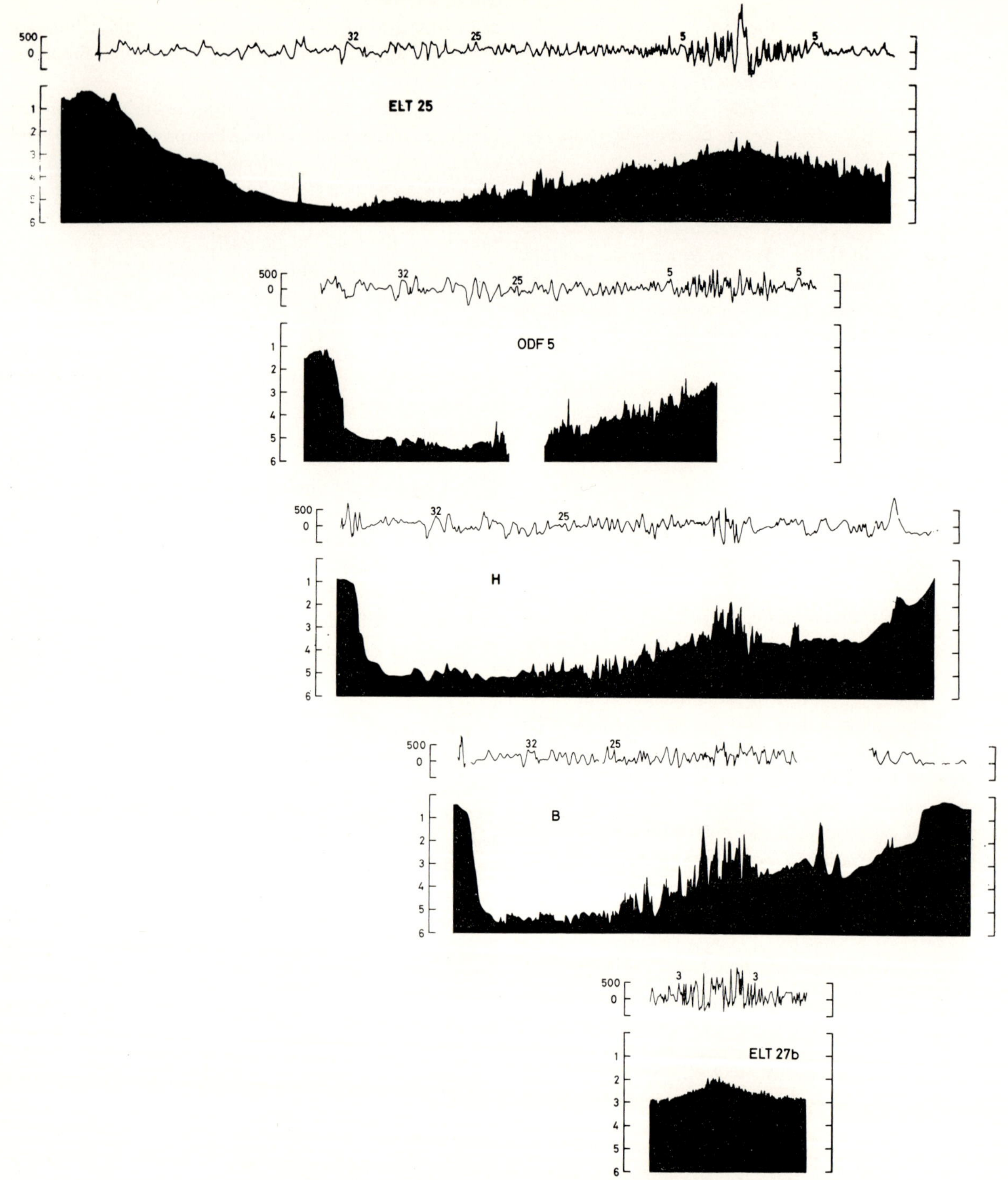

Fig. 2. Magnetic and bathymetric profiles for five tracks acrcss the Southwest Pacific basin showing characteristic features discussed in the text. The upper trace in each pair of profiles shows the magnetic anomaly and scale in gammas; the lower trace in each pair of profiles shows the bathymetry scale in kilometers.

on which this figure is based are shown in Figures 4 and 5. The azimuths of the tracks vary considerably, and the profiles have been projected to produce roughly the same length of the sequence. For this reason, the positional relationships of the anomalies should be taken only from Figure 3.

A profile from the North Pacific (labeled NPAC) is included in Figures 4 and 5. This profile from *Raff*

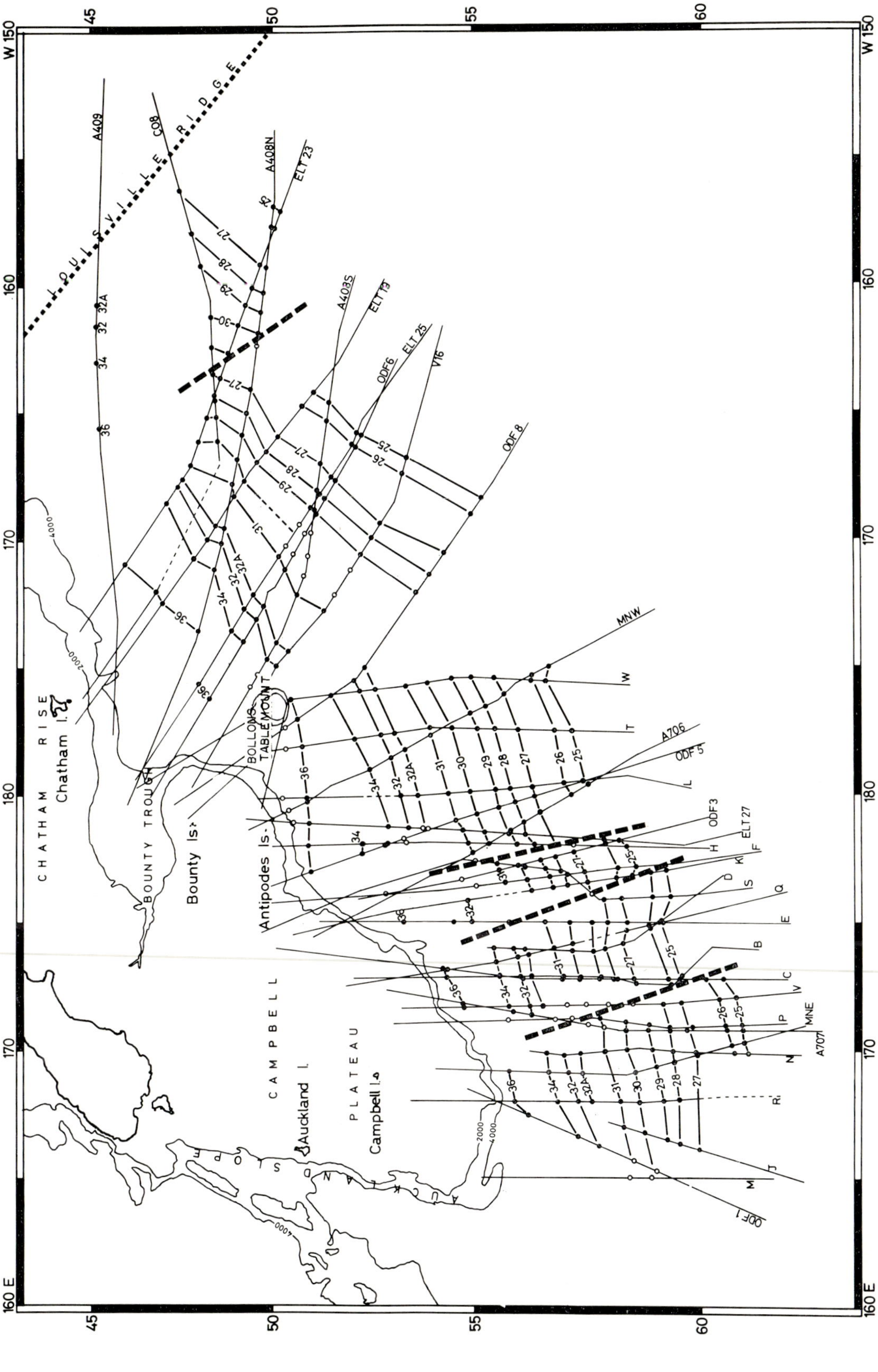

Fig. 3. Inferred correlation of magnetic anomalies in region 1 (anomalies 25–36). Fracture zones are shown as broken lines.

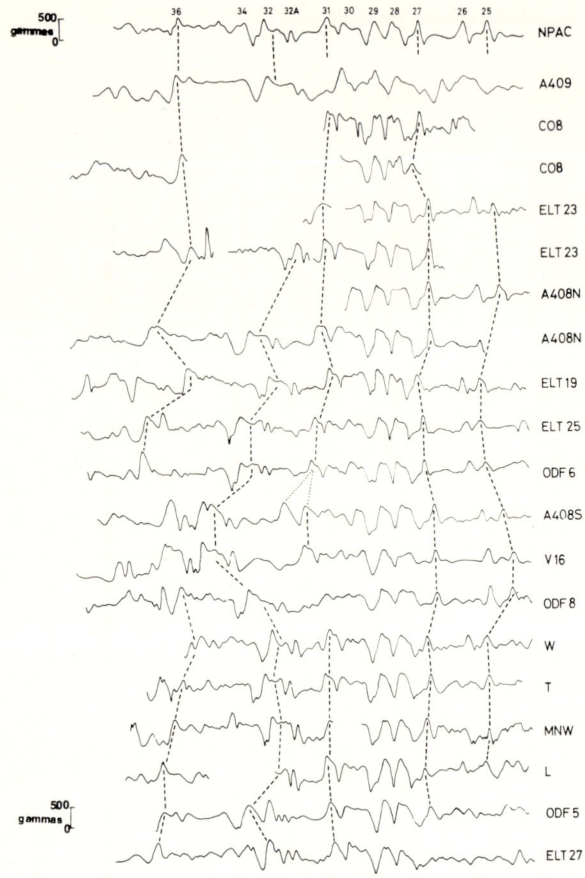

Fig. 4. Magnetic anomaly profiles from the northeastern portion of region 1. Profiles are aligned on anomaly series 27–31. Correlations of selected anomalies between tracks are shown as broken lines.

[1966] has previously been used out to anomaly 32 to establish the geomagnetic time scale [*Heirtzler et al.*, 1968]. An extension of this time scale is indicated by the fairly regular occurrence of a broad positive anomaly beyond anomaly 32. This broad anomaly, labeled by the numbers 34 and 36, is seen clearly on profiles A409, A408N, ODF8, and MNW. The anomaly can also be identified on some North Pacific profiles [*Raff*, 1966, Figure 2, tracks A, B, Q, R, and S]. By assuming the sea floor spreading rate for the section 32–36 to be the same as that immediately prior to anomaly 32, the age of anomaly 36 is approximately 81 m.y. Anomalies older than this age cannot be firmly correlated in this area.

Details of the anomaly pattern are discussed below. In general, the discussion proceeds from east to west in Figure 3. The profiles in Figures 4 and 5 are arranged with the most easterly profiles at the top.

In profile A409 along 45°S, the identification of anomalies 32A–36 is tentative but reasonable. The anomalies to the east of 32A are not very similar to those observed to the south. *Hayes and Ewing* [1971] have identified an apparently continuous aseismic ridge (the Louisville ridge) indicated in the upper right-hand corner in Figure 3. If their suggestion that this ridge is a westerly extension of the Eltanin fracture zone is correct, the sea floor to the east of the ridge could be quite different from that to the west. This suggestion would explain the irregular anomalies observed on A409 and at the eastern end of CO8.

In Figure 4, profiles CO8, ELT23, and A408N are each split into two, although the tracks are straight and continuous. We see that anomalies 27–30 are duplicated. Figure 3 shows that this duplication indicates an offset in the anomaly pattern of about 340 km

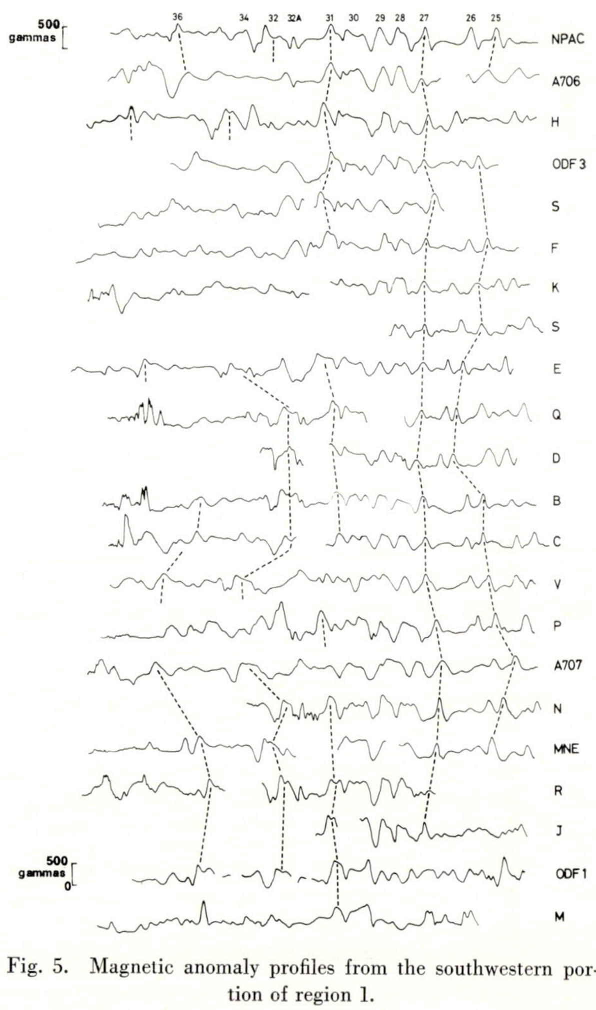

Fig. 5. Magnetic anomaly profiles from the southwestern portion of region 1.

in a dextral sense. The strike of the presumed fracture zone is probably between due north and 30° west of north. The strikes of the anomalies in this area lie between 30° and 40° east of north. Thus the fracture zone is not at right angles to the anomaly trend. *Pitman et al.* [1968, Figure 2] show that where ELT23 crosses the offset there is a blocklike bathymetric feature. The top of the block is about 220 km wide and is 1200 meters higher than the surrounding flat sea floor. Note that the western set of anomalies 27–29 are directly over this block but are not in any way irregular.

West of the fracture zone, the pattern is clear except for a disturbance between anomalies 29 and 31 at 170°W. Profiles ELT25, ODF6, A4085, and V16 indicate a broadening of anomaly 30. On ELT25, this broadening corresponds to a very slight bathymetric hump.

The anomaly pattern between 175°W and 180° is quite regular with a trend of N 70°E. This area has not been linked with that to the east for the following reasons:

1. Profiles ODF6 and V16 both show an apparent duplication of anomalies 32A and 32 and do not have anomalies 34–36.
2. The group of anomalies 32A and 32 is displaced 200 km northward with respect to the similar anomalies west of 175°W.
3. Profile ODF8 shows no anomaly 30 or 31.
4. The trend of anomalies 25–29 west of 175°W is different from that to the east. However, the eastern trend, N 30°E, is critically determined by track ODF8.

These points suggest that between 170°W and 175°W there is either a marked change of trend or one or more offsets. The data are insufficient to resolve these alternatives.

Between 175°E and 180°, *Christoffel and Ross* [1970] have mapped a fracture zone with a variable sinistral offset tending to zero at 60°S. The addition of more data in this study indicates that this area is more complex. On most profiles, anomalies 25–31 are relatively clear (e.g., A706, ODF3, E). However, the area north of anomaly 31 is best considered unclear (e.g., profiles S, F, K). At 178°E, profiles ODF5 and H indicate a dextral offset of anomalies 32A–34, but this offset cannot be identified either to the north in anomaly 36 or to the south. It is thus a localized feature. Farther west, the Endeavour fracture zone of *Christoffel and Ross* [1970] consists of two separate fractures, each with a sinistral offset and each striking normal to the anomaly trend. The displacement across each fracture zone is difficult to determine because the offsets are clear only over short lengths. The displacement across the fracture at approximately 179°E is about 45 km. The fracture at 177°E indicates a displacement of 55 km. When the fractures are considered together, the total displacement is comparable to that given by *Christoffel and Ross* [1970].

West of 175°E, the pattern is relatively clear. The complete sequence 25–36 is identifiable from several profiles, e.g., B, C, A707, MNE. A sinistral offset occurs between 173°E and 172°E. The offset of anomaly 25 is about 80 km, but the offset of anomaly 32 is only 45 km. This difference is not conclusive because profiles P and A707 are disturbed north of anomaly 30 and profile V is disturbed north of anomaly 28. Thus, possibly, the offset along this fracture decreases to the north.

West of 166°E, the pattern is confused. Anomalies 30 and 31 are tentatively identified on ODF1 and M, but the other anomalies of the sequence are not obvious. Unpublished *Eltanin* data recorded farther to the west also suggest that the regular anomaly pattern does not extend beyond about 165°E.

The anomaly sequence 25–32 shows that between 150°E and 165°W the sea floor of the Southwest Pacific basin is basically uniform in structure, although numerous offsets do occur. From Figure 3 we notice that west of 180°, anomaly 36 is close to the continental slope. The offsets on the fracture zones are such that the anomalies are successively displaced southward to the west and thus maintain a uniform distance from the steep continental slope.

East of 175°W, the continental slope is not so steep, but the anomalies still appear to maintain a constant distance from the New Zealand continent. The dextral offset on the major fault at 163°W is consistent with the eastward trend of the continent north of 45°S.

Around 50°S, 180°, the bathymetry is not well known, but a prominent bathymetric feature, the Bollons tablemount [*Lawrence*, 1967], is usually charted. The V16 track shows no sign of this feature, but it is seen on track T. The feature has thus been plotted smaller and farther to the south than is usually shown. The fact that anomaly 36, which occurs just south of this feature, is elsewhere close to the continent suggests that the Bollons tablemount may be a continental fragment. The presence of a continental fragment may explain why the magnetic

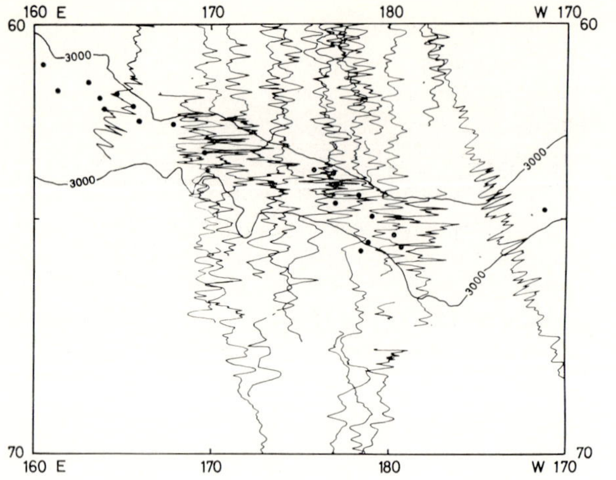

Fig. 6. Magnetic anomaly profiles over the Pacific-Antarctic ridge (region 2). Earthquake epicenters from *Barazangi and Dorman* [1969] and the 3000-meter bathymetric contour are shown.

trends are disturbed in the area. Anomaly 32 just to the east of the tablemount would be related to the continental slope well to the northwest.

The recognizable pattern of anomalies terminates to the west along 165°E; this longitude is in line with the Auckland slope that forms the western boundary of the Campbell plateau. This location further suggests that the anomalies are closely related to the continent bounding them to the north.

Region 2

The profiles in Figure 2 show that active sea floor spreading occurs east of 175°W and west of 160°E. Irregular high amplitude magnetic anomalies are observed on sections of profiles between those longitudes. Figure 6 shows that the irregular anomalies occur on many profiles and that they define a linear feature striking N 75°W. This feature coincides with an irregular bathymetric feature [*Ross*, 1967] and is associated with numerous earthquake epicenters [*Barazangi and Dorman*, 1969]. All the profiles west of 150°E that cross either the spreading ridge or the linear feature are shown in Figure 7. They have been plotted so as to align the prominent central anomaly of each profile and have been arranged in order of longitude with the most eastern profiles at the top.

If we consider the evidence for recent sea floor spreading to be the typical prominent squarish central anomaly and the characteristic anomaly sequence on both sides or even on only one side [*Vine*, 1966], then sea floor spreading is clear on only a few profiles. These profiles are ODF8, MNW, MNH, and

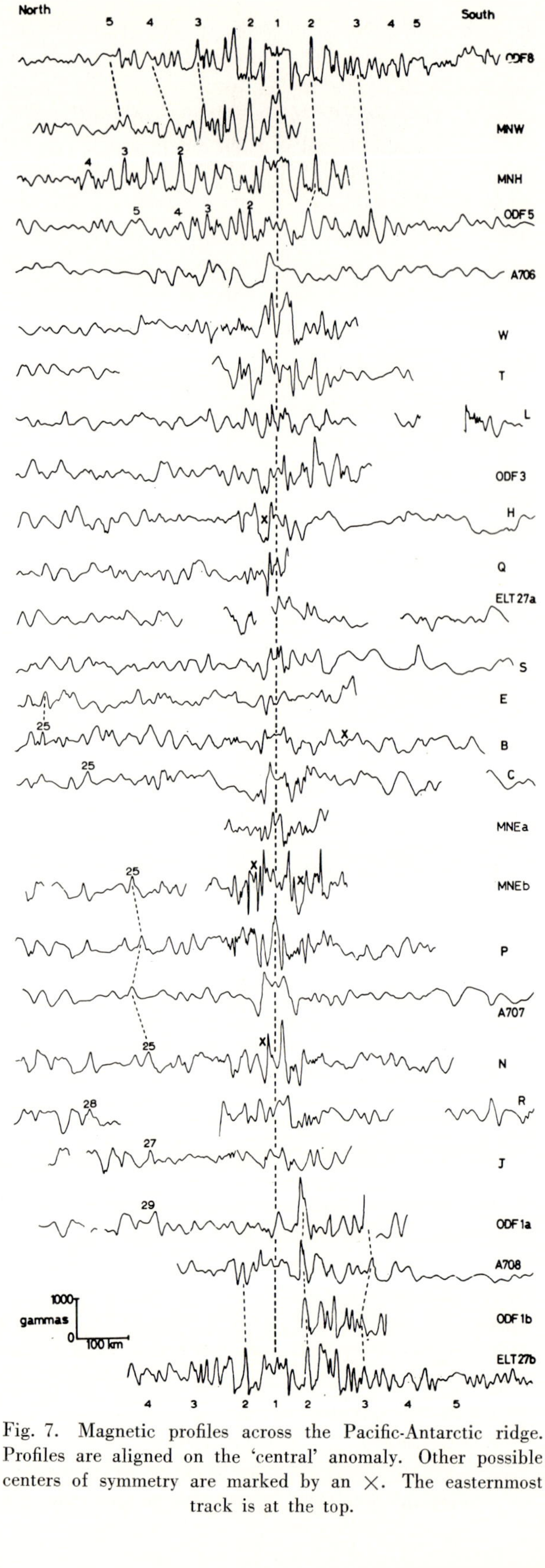

Fig. 7. Magnetic profiles across the Pacific-Antarctic ridge. Profiles are aligned on the 'central' anomaly. Other possible centers of symmetry are marked by an X. The easternmost track is at the top.

ODF5, (all east of 180°) and ELT27b, ODF1b, and A708 (all west of 161°E). Profile ODF1a indicates part of the normal sequence on the south side. The central anomaly with the peak on either side associated with the Jaramillo event can be suggested for several other profiles but can be considered clear only on T and MNEb. Limited symmetry is visible in W, ODF3, H, B, C, P, and N. In profiles H, B, MNEb, and N, possible alternative centers of symmetry are recognized. These centers are marked with an × in Figure 7.

The alignment of profiles prejudices the visual identification of the prominent anomalies, symmetry centers, and identified sequences. Tests with 50 first year university students suggest that perception of such features depends largely on the scale in which profiles are drawn, similar amplitudes for symmetric anomalies, lack of data gaps, prealignment, and the availability of tracing paper. Numerical methods being developed by R. K. H. Falconer show signs of being able to overcome the visual problems.

Ross [1967], using five bathymetric profiles and seven magnetic profiles between 163°E and 180°, has shown that the high amplitude magnetic anomalies are associated with the crest of a rough ridge. More recent data confirm that this ridge is much rougher than the sections to the east and west on which active spreading is observed (Figure 8). The ridge is also noticeably asymmetric. The Balleny basin immediately south of the crest is flat-floored with a mean depth of 3 km. There is virtually no transition zone between the ridge and this basin. North of the ridge crest, rough flanks extend to the Southwest Pacific basin at a depth of 5 km. This depth is 2 km deeper than the Balleny basin.

Magnetic data are incomplete on most profiles in the Balleny basin because pack ice is often encountered there. However, the data are sufficient to show that anomaly wave lengths are longer than those observed across the ridge flanks to the north. No clear anomaly correlations have been established. Calculations of the depths to anomaly sources by *Ross* [1967] gave a mean anomaly depth that was less than 1 km deeper than the sea floor. Seismic profile records from *Houtz et al.* [1971] and more unpublished *Eltanin* data also show that the sediment thickness in the Balleny basin is less than 1 km. The bathymetric asymmetry is therefore not due to sediment in the Balleny basin.

For the period 1961–1967, *Barazangi and Dorman* [1969] list 19 earthquake epicenters for this area between 160°E and 180°. All are on or close to the linear feature defined by the bathymetry and magnetics. For the 20° of longitude from 180° to 160°W, only three epicenters are given. Data from many sections of the mid-ocean ridge suggest that most of the activity is on fracture zones that offset the ridge crest [*Sykes*, 1970].

The magnetic, bathymetric, and epicentral data all suggest that the prominent feature striking N 75°W between 180° and 160°E is a major fracture zone. We propose to call this feature the Pacific-Antarctic fracture zone.

In Figure 6, we see that the disturbed magnetic zone, presumably the result of fracturing, is about 70 km wide. This width also corresponds to the width of the ridge shown by the bathymetry. The epicenters of *Barazangi and Dorman* [1969] and *Sykes* [1970] for this region are scattered from a line to distances greater than the stated accuracy of determination. All these results show the Pacific-Antarctic fracture zone to be a broad feature. The short sections of symmetry on some magnetic profiles suggest that the fracture zone may consist of a series of en echelon fractures.

The western end of the fracture zone is not well defined but the actively spreading ridge clearly identified from profiles ELT27b, ODF1b, A708 (Figure 7) and the unpublished *Eltanin* data strikes N 55°E and would intersect the fracture zone at about 61°S, 162°E. The form of the Macquarie ridge in this area is not accurately known, but in general the form appears to be a fracture zone rather than a ridge [*Christoffel*, 1971; *Hayes et al.*, 1972]. The Pacific, Indian, and antarctic plates therefore meet in a triple junction that is of the ridge–fault–fault type [*McKenzie and Morgan*, 1969]. This junction is at 61°S, 162°E.

A change in the bathymetry trend of approximately 90° occurs at 65°S between 180° and 175°W, the epicenter lineation terminates at 65°S, 177°W, and sea floor spreading is clearly identifiable on profile ODF5 at 175°W. These three observations suggest that the Pacific-Antarctic fracture zone terminates in the east at 65°S between 180° and 175°W. The zone is thus approximately 1100 km long.

The strike of the ridge (Albatross cordillera) east of the fracture zone is difficult to determine because spreading is clearly seen on only MNW, MNH, and ODF5. Profile MNH is important because it is the only one running predominantly east-west. Anomalies 4–2 are clear but are displaced considerably from a spreading center to the east that is identified by the

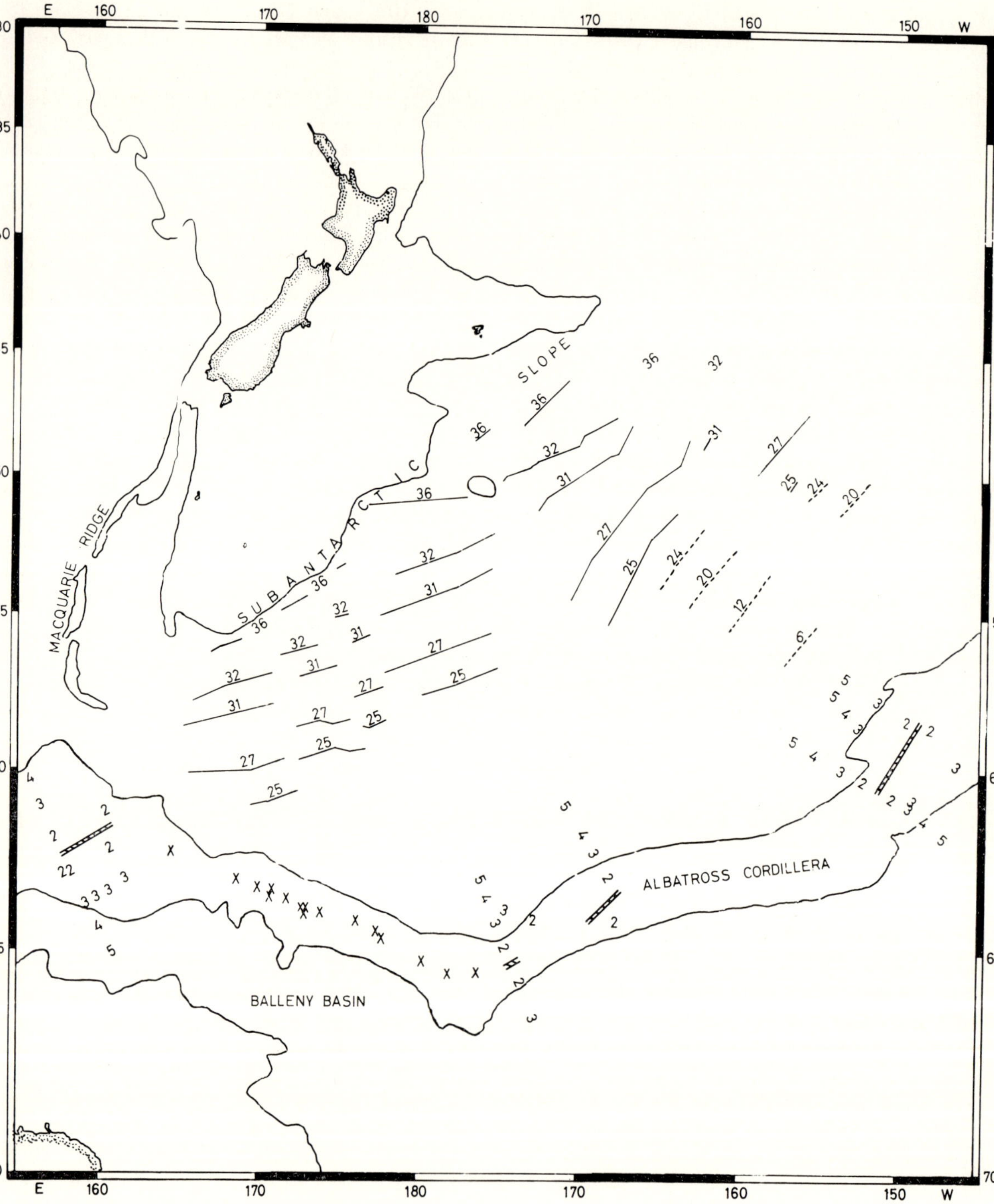

Fig. 8. Correlations of magnetic anomaly profiles in the three regions. Anomalies are numbered in the *Heirtzler et al.* [1968] convention. Along Pacific-Antarctic fracture zone, X's mark the positions of the 'central' anomaly.

central anomaly and characteristic anomalies east of it on the same profile. The general trend of the ridge is N 60°E, but probably in this area the ridge consists of short offset sections that may strike N 45°E. These ridge sections may not be normal to the direction of spreading, which should be parallel to the Pacific-Antarctic fracture zone. These sections are shown diagrammatically in Figure 8. Similar short ridge sections are observed farther east over much of the Albatross cordillera [*Herron*, 1971].

Region 3

The area between the well-identified anomalies 25–36 and the ridge and fracture zone to the south is difficult to interpret (Figure 8). East of 170°W, bathymetry profiles show relatively smooth ridge flanks that merge into the Southwest Pacific basin. *Heirtzler et al.* [1968] have shown that magnetic anomalies along some sections of ODF6, ELT19, and ELT23 can be correlated with profiles from elsewhere in the world. However, throughout the area east of 170°W, the anomalies over the flanks between the numbers 5 and 25 have amplitudes less than half of those older than 25. The wave lengths are also generally shorter. The general appearance of the profiles is one of noise rather than the more distinctive patterns of both the older anomalies and the crest region. Detailed correlations are possible only when tracks are close, and even then correlations are not unambiguous. Identified anomalies are shown in Figure 8, but the trends indicated throughout this flank region east of 170°W should be considered tentative only.

To the west of 170°W, the data coverage is better, but the interpretation is no clearer. Bathymetry profiles show that the flanks of the Pacific-Antarctic fracture zone are rougher than the more conventional ridge flanks to the east of 170°W. Also, a sharper bathymetric boundary occurs between the flanks and the basin [*Christoffel and Ross*, 1965]. Neither the magnetic anomaly amplitudes nor the wave lengths differ greatly from those in the regions to the north or south. In this respect, they contrast with the anomalies to the east of 170°W.

Correlations between the tracks could be suggested, but these correlations would be so ambiguous that they would be of little real value. The rougher bathymetry presumably is a disturbing factor, but the area appears complex. Tracks with orientations other than north-south and closer track spacings are needed if correlations are to be identified.

CONCLUSIONS

We can now extend our interpretation of the southwest Pacific Ocean in terms of plate tectonics beyond that of *Christoffel and Ross* [1970]. Figure 9 shows the main features that we intend to discuss.

On the basis of a consistent magnetic anomaly pattern south and east of the Campbell plateau, we feel justified in extending the *Heirtzler et al.* [1968] anomaly sequence from 32 to 36. Anomaly 36, calculated at 81 m.y., closely follows the 3500-meter contour at the base of the subantarctic slope. Our data extend just beyond the Albatross cordillera and allow us to recognize only one-half the magnetic anomaly pattern. However, we feel confident in predicting the mirror pattern to exist in the Southeast Pacific basin. We therefore conclude that the New Zealand continent rifted from the region of West Antarctica just prior to 81 m.y. ago. The Louisville ridge possibly formed an eastern boundary with a South American plate in a manner similar to the present-day Chile rise [*Herron and Hayes*, 1969].

At the time of anomaly 25 (i.e., 62 m.y. ago), a change in the spreading occurred. Recognition of younger anomalies becomes progressively less clear so that, west of 170°W, anomalies between 25 and 5 are unrecognized. The most probable reason for this can be seen from Figures 8 and 9. The mean strike of each of the fracture zones between anomalies 25 and 36 south of the subantarctic slope is reasonably well established from our measurements. The present strike of the Pacific-Antarctic fracture zone is also well determined, and we also assume that the strikes of the other fracture zones across this southern section of the Albatross cordillera will be determined by the same pole of relative motion. Between anomalies 25 and 5, a change in strike of over 45° has to be accommodated. From Figures 8 and 9, we see that the triangular region bounded by longitudes 170°E and 170°W is particularly affected. Considerable fracturing would be expected and is borne out by the disturbed topography. Probably only short segments of coherent spreading ridges will be present, these short segments requiring comprehensive magnetic and bathymetric surveys to establish any reasonable correlation of anomalies.

Although the measurements of *Christoffel and Ross* [1970] can be interpreted as showing a continual transition of the spreading direction, the rate of change must have considerably increased more recently than anomaly 20. The traces of the old transform faults south of the subantarctic slope have sinistral displacements. The present ones along the Albatross cordillera between 150°W and 175°W obviously have dextral displacement. We are unable to trace the old fracture zones to the south beyond anomaly 25, but there may be a pattern of continual transition of strike similar to that reported in the North Pacific Ocean by *Menard and Atwater* [1968]. If so,

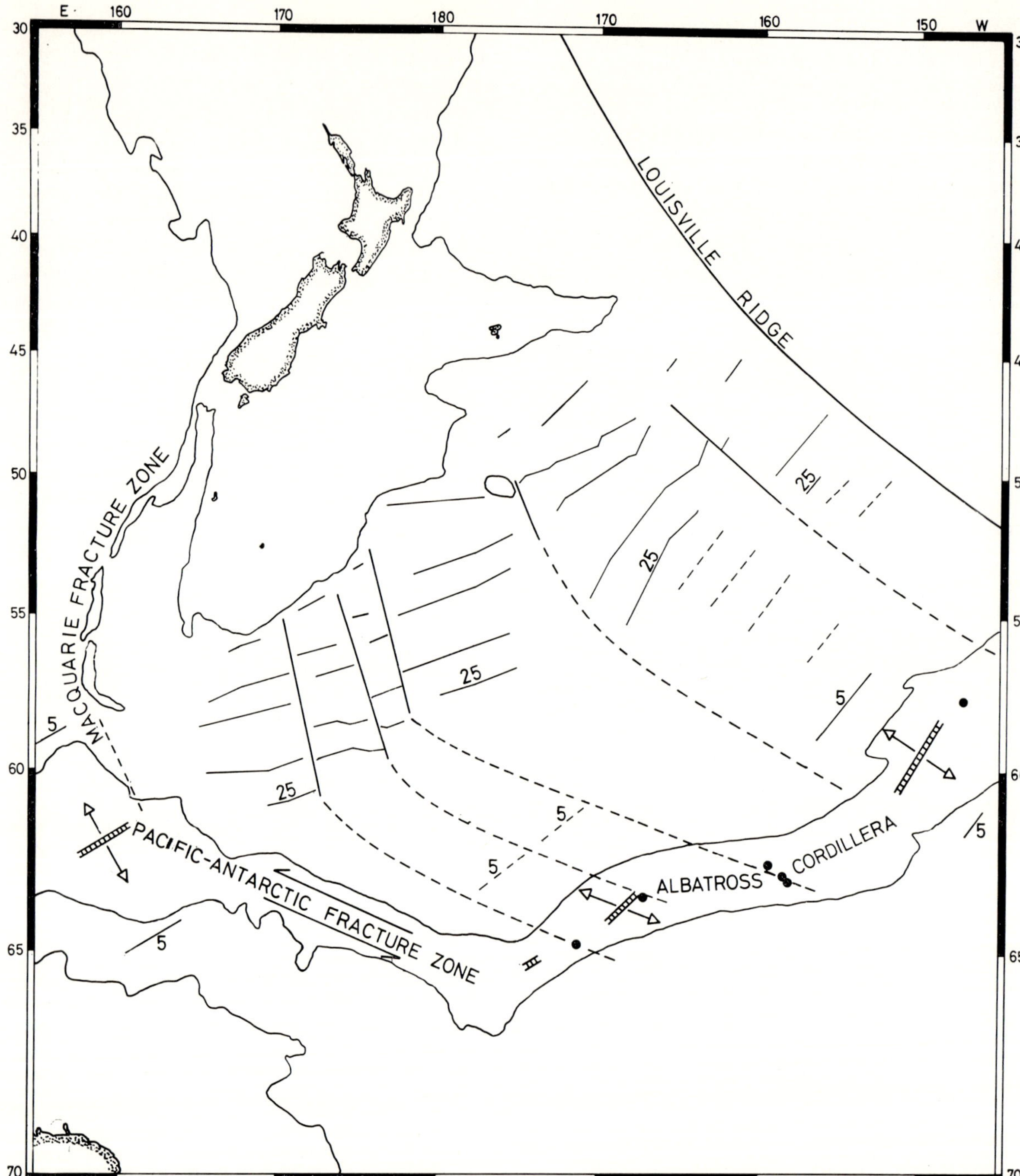

Fig. 9. Model of sea floor spreading in the southwestern Pacific. Solid lines mark established fracture zones; broken lines show inferred trends. Arrows indicate directions of motion. Closed circles on the Albatross cordillera are epicenters from *Barazangi and Dorman* [1969].

the fracture zones could continue in a manner shown by the broken line extensions in Figure 9.

Since anomaly 5 (i.e., 10 m.y. ago), the pattern of spreading seems to have persisted about the same relative pole of rotation until the present. We can identify a major fracture zone, which we propose to call the Pacific-Antarctic fracture zone, that extends at least 1200 km from approximately 65°S, 175°W to 61°S, 162°E. This fracture zone corresponds to a pole of relative motion for the Pacific and antarctic plates at 82°S, 120°E rather than to the *Le Pichon* [1968], pole at 70°S, 118°E.

Acknowledgments. We gratefully acknowledge the cooperation from the Royal New Zealand Navy and the officers and men of HMNZS *Endeavour* for making our antarctic operations possible. We thank H. P. Stockard, U.S. Naval Oceanographic Office, for supplying us with microfilm copies of the Project Magnet records. Mrs. Feriel Falconer gave valuable assistance in data reduction and interpretation and in the drawing of some of the diagrams. D. A. Christoffel is indebted to Walter B. Pitman III, to the National Science Foundation grants GA-1415 and GV-23334, and to the Office of Naval Research grant N00014-67-A-0108-0004 for making possible a period of leave at Lamont-Doherty Geological Observatory, where the aeromagnetic and much of the shipborne data reduction was carried out.

Contribution 1779, Lamont-Doherty Geological Observatory.

REFERENCES

Barazangi, M., and J. Dorman, World seismicity maps compiled from ESSA coast and geodetic survey epicenter data, 1961–1967, *Bull. Seismol. Soc. Amer., 59*(1), 369–380, 1969.

Christoffel, D. A., Total magnetic field measurements between New Zealand and Antarctica, *Nature, 190*, 776–778, 1961.

Christoffel, D. A., Motion of the New Zealand Alpine fault deduced from the pattern of sea floor spreading, in *Recent Crustal Movements*, edited by B. W. Collins and R. Fraser, *Roy. Soc. N. Z. Bull., 9*, 25–30, 1971.

Christoffel, D. A., and D. I. Ross, Magnetic anomalies south of the New Zealand plateau, *J. Geophys. Res., 70*(12), 2857–2861, 1965.

Christoffel, D. A., and D. I. Ross, A fracture zone in the south west Pacific basin south of New Zealand and its implications for sea-floor spreading, *Earth Planet. Sci. Lett., 8*, 125–130, 1970.

Fabiano, E. B., and N. W. Peddie, Grid values of total magnetic intensity, I.G.R.F. 1965, *Tech. Rep. C & GS 38*, Environ. Sci. Serv. Admin., Rockville, Md., 1969.

Hayes, D. E., and M. Ewing, The Louisville ridge–A possible extension of the Eltanin fracture zone, in *Antarctic Oceanology I, Antarctic Res. Ser.*, vol. 15, edited J. L. Reid, pp. 223–228, AGU, Washington, D. C., 1971.

Hayes, D. E., J. R. Heirtzler, E. M. Herron, and W. C. Pitman III, Preliminary report of volume 21, USNS *Eltanin* cruises 22–27, January 1966–February 1967, part A, Navigation, part B, Bathymetric and geomagnetic measurements, *Tech. Rep. 2-CU-2-69*, Lamont-Doherty Geol. Observ. of Columbia Univ., Palisades, N. Y., 1969.

Hayes, D. E., M. Talwani, and D. A. Christoffel, Macquarie ridge complex, in *Antarctic Geology and Geophysics*, edited by R. J. Adie, Universitets forlaget, Oslo, in press, 1972.

Heirtzler, J. R., G. O. Dickson, E. M. Herron, W. C. Pitman III, and X. LePichon, Marine magnetic anomalies, geomagnetic field reversals, and motions of the ocean floor and continents, *J. Geophys. Res., 73*(6), 2119–2136, 1968.

Heirtzler, J. R., D. E. Hayes, E. M. Herron, and Walter C. Pitman III, Preliminary report of volume 20 USNS *Eltanin* cruises 16–21 January 1965–January 1966, part A, Navigation, part B, Bathymetric and geomagnetic measurements, *Tech. Rep. 3-CU-3-69*, Lamont-Doherty Geol. Observ. of Columbia Univ., Palisades, N. Y., 1969.

Herron, E. M., Crustal plates and sea-floor spreading in the southeastern Pacific, in *Antarctic Oceanology I, Antarctic Res. Ser.*, vol. 15, edited J. L. Reid, pp. 229–237, AGU, Washington, D. C., 1971.

Herron, E. M., and D. E. Hayes, A geophysical study of the Chile ridge, *Earth Planet. Sci. Lett., 6*, 77–83, 1969.

Houtz, R., J. Ewing, and R. Embley, Profiler data from the Macquarie ridge area, in *Antarctic Oceanology I, Antarctic Res. Ser.*, vol. 15, edited J. L. Reid, pp. 239–245, AGU, Washington, D. C., 1971.

Lawrence, P., New Zealand region, Bathymetry, scale 1: 6,000,000, Miscellaneous Series, 15, N. Z. Oceanogr. Inst., Wellington, 1967.

Le Pichon, X., Sea-floor spreading and continental drift, *J. Geophys. Res., 73*(12), 3661–3697, 1968.

McKenzie, D. P., and W. J. Morgan, Evolution of triple junctions, *Nature, 224*, 125–133, 1969.

Menard, H. W., and T. Atwater, Changes in direction of sea-floor spreading, *Nature, 219*, 463–467, 1968.

Pitman, W. C., III, E. M. Herron, and J. R. Heirtzler, Magnetic anomalies in the Pacific and sea floor spreading, *J. Geophys. Res., 73*(6), 2069–2085, 1968.

Raff, A. D., Boundaries of an area of very long magnetic anomalies in the northwest Pacific, *J. Geophys. Res., 71*(10), 2631–2636, 1966.

Ross, D. I., Magnetic and bathymetric measurements across the Pacific-Antarctic ridge south of New Zealand, *N. Z. J. Geol. Geophys., 10*(6), 1452–1465, 1967.

Sykes, L. R., Seismicity of the Indian Ocean and a possible nascent island arc between Ceylon and Australia, *J. Geophys. Res., 75*(26), 5041–5055, 1970.

Udintsev, G. B., et al., New bathymetric map of the Pacific Ocean, *Oceanolog. Res., 9*, pp. 66–101, Inst. of Oceanolog., Acad. of Sci. of the USSR, Moscow, 1964.

U.S. Navy Hydrographic Office, Operation Deep Freeze 61, 1960–1961, Marine geophysical investigations, *Tech. Rep. 105*, Washington, D. C., 1962.

U.S. Naval Oceanographic Office, Operation Deep Freeze 62, 1961–1961, Marine geophysical investigations, *Tech. Rep. 118*, Washington, D. C., 1965.

U.S. Naval Oceanographic Office, New Zealand to Cape Adare, *H.O. 6710*, Washington, D. C., 1969.

Vine, F. J., Spreading of the ocean floor; New evidence, *Science, 154*, 1405–1415, 1966.

GEOPHYSICAL INVESTIGATION OF THE MACQUARIE RIDGE COMPLEX

Dennis E. Hayes and Manik Talwani

Lamont-Doherty Geological Observatory of Columbia University, Palisades, New York 10964

Abstract. A new topographic map of the Macquarie ridge and adjacent areas is presented and shows that the ridge is a highly complicated feature composed of three discrete elements. The ridge is bounded in places by pronounced topographic deeps that lie both to the east and to the west of the ridge. The free air gravity anomalies, total intensity magnetic anomalies, sediment distribution (from seismic profiler records), and seismicity associated with the area are examined in an effort to understand its evolution. The relative motion along this boundary, as predicted by sea floor spreading in the Pacific and Indian oceans, is examined in several models. We conclude that the Macquarie ridge complex has evolved as the consequence of the interaction of the Pacific and Indian plates and that the predominant motion has been right-lateral strike slip at about 4–5 cm/yr. Secondary components of underthrusting are indicated for the northern and perhaps southern ends of the complex. A small amount of crustal extension may have occurred within the central section (50°–57°S), although this interpretation is highly speculative. An apparent contradiction involving the predicted strike slip motion along the Alpine fault and the zone of subduction implied by intermediate earthquakes in the Fiordland region is considered in several alternative explanations. Discrete elements of the complex represent all three major tectonic regimes (extension, compression, and transcurrent motion), and the difficulty of categorizing the feature is the consequence of small components of extension and compression acting in conjunction with a major component of transcurrent motion. The complicated evolution of the area is expected in view of the proximity of the complex to the Pacific-Indian pole of relative motion.

The Macquarie ridge is a narrow topographic high that extends along the trend of New Zealand southward toward a connection with the Southeast Indian ridge–Pacific-Antarctic ridge system near 60°S, 160°E (Figure 1; see foldouts at back of volume). The existence of this ridge has been known for some time, although only recently has adequate sounding data been collected and compiled to define the details of the ridge and the many complex topographic features adjacent to it. The Macquarie ridge is seismically active. This conspicuous seismicity drew attention to the ridge and generated considerable speculations in the middle 1950's [e.g., *Ewing and Heezen*, 1956; *Carey*, 1958] on the origin, evolution, and significance of this feature as a unique morphologic element. Macquarie ridge provides the main barrier to the deep circumpolar flow of water around the Antarctic, and its growth and evolution must have had a profound influence in modifying the patterns of polar circulation and ultimately the circulation and climate of the entire world ocean [e.g., *Gordon*, this volume; *Boyer and Guala*, this volume].

The origin and evolution of the Macquarie ridge and its relationship to processes of global tectonics are controversial topics. This study will show that the morphologic expression of the Macquarie ridge is far more complicated than was previously known. Numerous trenches and elongate deeps bound various segments of the Macquarie ridge and also exhibit a complex morphology. We will examine both the ridge and the associated deeps and will attempt to provide insight into the possible origin of these features when viewed as an integral unit that we will refer to as the Macquarie ridge complex.

Several investigators have examined the Macquarie ridge complex, but in most instances their studies have either been confined to limited portions of the ridge or have treated only one or two properties of the complex. Because of this divergence in techniques and the differing areas of concentration, the conclusions reached by several key investigators appear incompatible with one another.

Summerhayes [1969] did an exhaustive study of a substantial portion of the area, and he presents many good arguments in arriving at his interpretation that the Macquarie ridge is a Tertiary-Quaternary island arc. *Summerhayes* [1967] and *McKenzie and Morgan* [1969] suggest that the Macquarie and Kermadec ridges are oppositely facing island arcs separated by the Alpine fault, a trench–trench transform fault of the type described by *Wilson* [1965]. Most of Summerhayes' investigations were confined to the central and northern part of the Macquarie ridge. His interpretation is supported by the seismic first motion studies of *Banghar and Sykes* [1969] and also by the geologic and geophysical studies of *Hatherton* [1967].

In marked contrast, *Houtz et al.* [1971] suggest that the entire Macquarie ridge was originally formed as a shear zone. However, they do not rule out the possibility that the ridge may have developed into an island arc very recently. Their study was concerned primarily with seismic profiler data concentrated over the central and southern part of the Macquarie ridge complex. They based their conclusions on the morphology, the sediment distribution, and the depth to acoustic basement in the vicinity of the narrow arcuate deep known as the Hjort trench west of the Macquarie ridge between approximately 56°S and 60°S.

Varne et al. [1969] present a third alternative for the origin of the Macquarie ridge that is based on convincing petrologic data. Pliocene oceanic crust is now exposed on the northern portion of Macquarie Island. They suggest that the Macquarie ridge was formed in a manner similar to that of other ridges that mark the sites of crustal plate separation. Moderately deepwater foraminifera were found within the basalt samples that Varne et al. studied, their presence indicating that the island has experienced considerable uplift since the Pliocene. *Ewing and Heezen* [1956] had also suggested many years earlier that the Macquarie ridge was a branch of the mid-oceanic ridge system primarily because of the narrow belt of associated seismic activity.

Jones [1971] speculates that the northwestern boundary of the Campbell plateau (that coincides with the northern part of the Macquarie ridge complex) marks the southern extension of the Alpine fault and that this boundary was formerly connected to the southeast Australia continental slope.

In general, earthquake first motion studies and the precise epicenter relocations of such workers as *Sykes* [1967, 1970], *Banghar and Sykes* [1969], and *Hamilton and Evison* [1967] are extremely valuable as independent tests and often allow the selection of a best hypothesis from among the broad spectrum of available hypotheses. For this area, however, the number of first motion studies published to date is so limited that these studies do not yield definitive answers regarding the tectonic mechanisms active along the Macquarie ridge complex. Seismic first motion studies are compatible with the models that we present. The diversity of mechanisms inferred for the limited number of solutions provides qualitative support for our contention that the evolution of the Macquarie ridge complex has been and continues to be extremely complicated.

This study is intended as a continuation of the studies presented by *Hayes et al.* [1972] and *Christoffel* [1971]. Christoffel concentrated on examining and explaining the movement along the Alpine fault in terms of the interaction of the Indian and Pacific plates as deduced from patterns of sea floor spreading in the southwest Pacific and the southeast Indian oceans. By following this approach, certain conclusions regarding the origin of the Macquarie ridge follow automatically. *Hayes et al.* [1972], on the other hand, made a preliminary examination of the marine geophysical data (seismic reflection, gravity, magnetics, and bathymetric data) and attempted to see if the complex morphology and geophysical properties of the Macquarie ridge complex could be understood as the natural consequence of the interaction of two major lithospheric plates. They attempted to reconcile the apparent incompatibility of existing hypotheses. The conclusion of Hayes et al. was that all three main types of oceanic tectonic provinces (e.g., mid-oceanic ridge, deep-sea trench and arc, and transform fault–fracture zone) may be represented as discrete elements of the Macquarie ridge complex. This complicated evolution is in part attributed to the fact that the pole of relative motion between the Indian and antarctic plates, as deduced by magnetic lineation patterns and transform faults in the southeast Indian Ocean and southwest Pacific Ocean, lies very close to the Macquarie ridge complex. If this complex is viewed as a boundary between the two plates, then because of the proximity of the pole the direction of relative motion between the two plates changes rapidly over short geographical distances. Also the Macquarie ridge complex does not have one strike direction but trends in its northern part in a generally north-northeasterly direction and has an arcuate shape in its southern portion. The complex is also disrupted by many pronounced dislocations that are transverse to the general strike of features. All these factors may combine to produce a highly variable tectonic pattern.

MORPHOLOGY

Approximately 20,000 track miles of precision echo-sounding data have been collected in this area during the last 5 years. The positions of the largest proportion of these tracks were controlled by precise satellite navigation. Many of the data in this area have been collected aboard the *Eltanin*. These data were combined with echo-sounding data from the bathymetric libraries of the Australian Hydrographic Office, the New Zealand Hydrographic Office, and the U.S. Naval Oceanographic Office. A new bathymetric map of the area was produced at a contour interval of 100 fm. The data were compiled into the standard Gebco (general bathymetric charts of the ocean) area designations. All the data from the Lamont-Doherty Geological Observatory were plotted with the aid of digital computer and plotting systems. The units of fathoms used were corrected for the variation of sound velocity in sea water according to the tables of *Matthews* [1939]. A diversity of opinion exists regarding the appropriate units in the preparation of bathymetric maps. In this instance, the units of corrected fathoms were chosen strictly from practical considerations (see *Hayes and Conolly* [this volume]). Ship tracks along which bathymetric data were collected and used in this study are shown in Figure 2. Figure 1 (see foldouts at back of volume) shows the detailed bathymetry of the Macquarie ridge complex and the surrounding areas as described by 100-fm contours.

The over-all description of the Macquarie ridge is not radically different from that which appeared on the earlier maps (e.g., *Brodie and Dawson* [1965], *Summerhayes* [1969]) and the Oceanic Series charts of the New Zealand Oceanographic Institute in Wellington. Our map shows many previously unknown details of the ridge. The most striking difference between the map shown in Figure 1 and earlier maps is the existence of numerous trenches and troughs that bound both sides of the Macquarie ridge. There are three very prominent such features.

1. The Puysegur trench lies west of the Macquarie ridge between about 46°S and 49°S. It is continuous at the 2500-fm isobath over a distance of 200 miles.

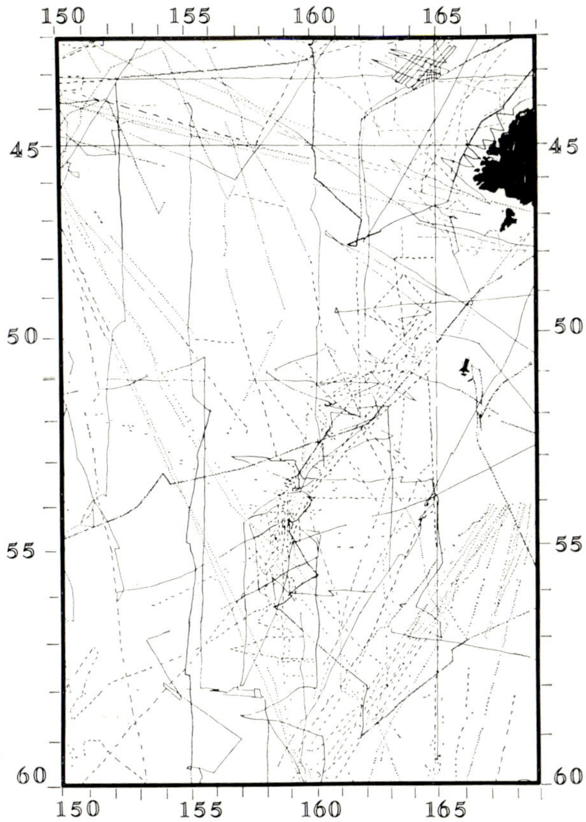

Fig. 2. Ship tracks along which sounding data were used in contouring the map of Figure 1. *Eltanin*, *Vema*, and *Robert D. Conrad* tracks are represented by solid lines. Data from Australian Hydrographic Office and New Zealand Oceanographic Office sources are indicated by broken lines. Dotted lines represent data compiled by the U.S. Naval Oceanographic Office. Spot soundings are not shown.

2. The Macquarie trench lies east of the Macquarie ridge between approximately 51°S and 56°S. The Macquarie trench is a continuous deep over a distance of approximately 300 miles. It reaches a maximum depth of more than 3000 fm and is typically deeper than 2600 fm.

3. The Hjort trench between approximately 56°S and 60°S is a narrow arcuate deep, is convex to the west, and is adjacent to the western flank of the Macquarie ridge. It is continuous over a distance of more than 240 miles, reaches a maximum depth of more than 3300 fm, and is typically deeper than 3000 fm over its entire extent.

The Hjort trench and the Macquarie trench are connected by a narrow, deep passage that bisects the Macquarie ridge near 56°S. These two features are connected at the 2500-fm isobath level. This deep passage and other deep passages are extremely important in terms of the circumpolar flow of water across the Macquarie ridge and are discussed in some detail by *Gordon* [this volume] and *Boyer and Guala* [this volume]. The second pronounced break in the ridge occurs at about 53.5°S, although the passage is not deeper than 1900 fm according to our bathymetric data. There is no evidence that a narrow deep exists on the west side of the Macquarie ridge near this passage. At 50.5°S, another passage is indicated, although the ridge appears to be continuous to the 1500-fm isobath level. A pronounced right-lateral offset (~50 miles) of the ridge is indicated. This passage also marks the northernmost limit of the Macquarie trench. North of this area are pronounced elongate troughs trending northerly in en echelon fashion. Each of the major passages through the Macquarie ridge appears to coincide with an offset or dislocation of the ridge, although the offset is fairly small at the 56°S passage.

Except for the major breaks just discussed and several minor disruptions, the Macquarie ridge is shoaler than 1000 fm over most of its length between approximately 47°S and 57°S. South of 57°S, it becomes much less regular and branches into two prominent segments. The entire complex intersects the Pacific-Antarctic ridge system near 61°S and 162°E.

The Solander trough is a narrow, flat depression in the sea floor that extends southward and parallel to the Macquarie ridge approximately from the latitude of Stewart Island (47°S) to the vicinity of 53°S where it merges with the Emerald basin. The general morphology of the Emerald basin is much more irregular than previously defined. The basin is characterized by numerous large seamounts and pronounced depressions especially in the vicinity of 55°S–59°S. Several deeps occur approximately where the edge of the Campbell plateau and the Hjort trench bend sharply to the east.

The northerly trend of some isobaths lying west of the Puysegur trench may in part be the effect of the orientation of the track lines. However, the general tendency for these lineaments to be aligned in a northerly or northwesterly direction is consistent with the observed and predicted directions of fracture zone morphology resulting from the separation of the Indian and antarctic plates. The Puysegur trench is terminated on the north at about 46°S where it is interrupted by a northeasterly trending ridge system, Resolution ridge (see also *van der Linden and Hayes* [1972]).

Near its southernmost limit, the Hjort trench is bounded on both flanks by relatively shallow depths of less than 1500 fm. The southwestern elevation is still fairly far removed from the main topographic high associated with the crest of the Pacific-Antarctic ridge [*Hayes and Conolly,* this volume]. The Hjort trench does not continue southward as a pronounced deep beyond about 60°S and cannot be traced as a discrete feature into the axis of the Pacific-Indian-Antarctic ridge system. The basin lying to the west of Macquarie ridge is characterized by relatively low relief of generally less than 200 fm. Abrupt changes occur in the thickness of sediment present on the sea floor just west of the southern and central sections of the Macquarie ridge complex [*Houtz et al.,* 1970]. Consequently, the true relief of the oceanic basement is obscured.

GRAVITY ANOMALIES

Gravity measurements over the Macquarie ridge complex have been obtained principally aboard the *Eltanin* during cruises 34, 36, 37, 42, and 44 and aboard the *Robert D. Conrad* during cruise 8. The measurements were obtained by the Graf Askania Gss 2-25 sea gravimeter mounted on the Anschutz gyrotable. Real time cross-sampling correction was provided by an analogue computer designed at Lamont-Doherty Geological Observatory.

Free air gravity anomalies are shown on Figure 3 (see foldouts at back of volume). These anomalies generally reflect the major topographic features. However, there are several important areas where this relationship does not hold. The anomaly wave lengths are of the order of 30–60 miles (50–100 km). Extremely steep gradients persist throughout the area and are the consequence of large positive anomalies (more positive than 150 mgals) associated with the Macquarie ridge and adjacent large negative anomalies (more negative than −125 mgals) associated with the trenches that bound alternate sides of the Macquarie ridge. The gravity control lines are also shown on Figure 3. Six cruises are represented by these data and numerous crossings of the tracks were available to evaluate the reliability of the free air gravity measurements. For 45 crossings examined, the root mean square (rms) discrepancy in free air anomalies was about 8 mgals. In many instances (~15), track crossings occur in areas of steep gravity gradients (>5 mgals/n. mi.). Discrepancies there may be due largely to errors in navigation. If the remainder of the crossings are treated separately, the rms discrepancy reduces to 6 mgals but is still somewhat larger than the rms discrepancy typical of other Lamont-Doherty surveys.

There is much less gravity data in this area than topographic data. Free air anomaly contours are shown as solid lines where there is a good control and as broken lines in areas where the contours have been extrapolated or interpolated primarily on the basis of the underlying topography.

The Macquarie ridge coincides with a long, continuous free air gravity maximum that extends from approximately 47°S to approximately 56°S. However, the most positive values within this belt are broken into three major segments and coincide almost exactly with the three topographic elements of the ridge crest defined by major offsets. Most of the area lying west of the complex is characterized by free air anomalies that are slightly positive and that average about +10–+15 mgals. One exception is near 46°S where larger positive anomalies (greater than +25 mgals) are associated with the northeasterly trending Resolution ridge (see *van der Linden and Hayes* [1972]). In contrast, the area east of the complex shows generally negative anomalies (more negative than about −25 mgals) that are associated with the Solander trough and the northern extension of the Emerald basin. Large positive anomalies are also associated with the Campbell slope near 54°S. These positive anomalies are probably in part due to edge effects along the steep scarp. South of 56°S, the Macquarie ridge is less well defined topographically and branches into at least two prominent spurs. Although the gravity data control between 56°S and 58°S is not very good, we believe that this southern portion of the Macquarie ridge is characterized by positive anomalies much smaller than those of its counterparts to the north. We have no indication that anomalies over this portion of the ridge ever exceed +75 mgals.

Perhaps the most remarkable free air anomalies lie between 44°S and 46°S, adjacent to the Fiordland coast of New Zealand (see also Figure 8). The Puysegur trench cannot be traced north of 46°S as a topographic feature. The continental slope west of the Fiordland coast grades directly into the continental rise. There is no topographic depression at the base of the slope. Free air anomalies at the base of the slope in this region are more negative than −200 mgals. This gravity minimum is bounded on the north by the seaward extension of the Alpine fault trace and on the south by the southwest coast of New Zealand and by the Resolution ridge. The Puysegur trench gravity minimum is not more negative than −125 mgals, even though it is situated over a pronounced topographic trench. Although there is a general continuity of gravity trends between the Fiordland slope minima and the Puysegur trench minima, these two gravity minima are possibly related to somewhat different structures.

Note that the contours of zero milligals that enclose the negative anomalies associated with the Hjort trench, Macquarie trench and Puysegur trench never cross over Macquarie ridge itself. In other words, they define discrete closures associated with the individual features. The major gravity anomalies reflect the major topographic features including their discontinuities and serve to define the same basic elements as defined by the topography. The influence of topography alone can be approximated very well by applying a simple two-dimensional Bouguer correction that assumes a density contrast of +1.64 g/cm^3 between crustal rocks and water.

Large Bouguer gradients are encountered in going from the deep ocean to land (or very shallow topography). Therefore considering local isostatic anomalies is more useful in deducing deep crustal structure. Local isostatic anomalies were computed along simulated profiles (Figure 6*b*) simplified and taken from the topographic map of Figure 1 and the free air gravity anomaly map of Figure 3 (see foldouts at back of volume). Results of this analysis are discussed in a later section.

MAGNETIC ANOMALIES

The magnetic anomalies associated with the Macquarie ridge complex are not yet well understood. As *Hatherton* [1967] first pointed out, the ridge is characterized by a large positive magnetic peak that is often but not always associated with several short wave length peaks on the main magnetic positive.

Magnetic anomalies associated with the northerly spreading of the Southeast Indian ridge have been mapped west of the complex (see Figure 6*d*). They can be traced as far north as about 53°S where anomaly 13 (38 m.y. B.P.) lies very close to the complex but trends roughly east-west, a trend consistent with the predicted southeast Indian Ocean spreading (see *Weissel and Hayes* [this volume]). Older anomalies may lie adjacent to the complex somewhat farther

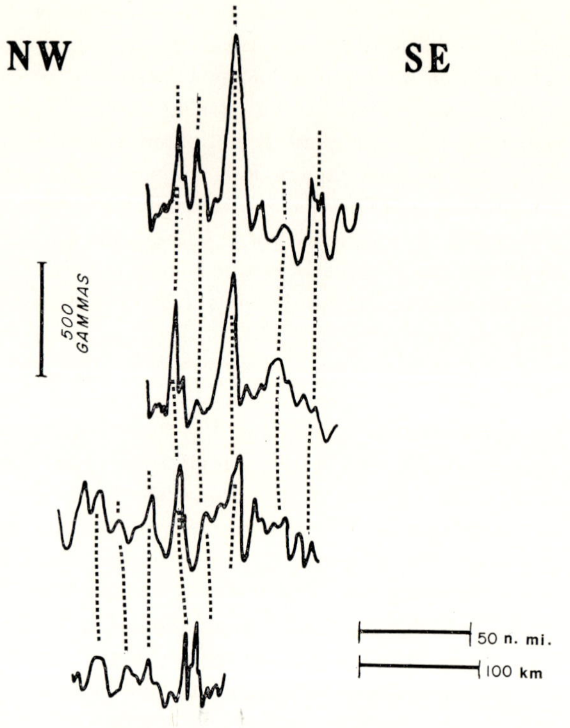

Fig. 4. Tentative magnetic anomaly correlations east of the Macquarie ridge. Figure 6*d* shows the inferred lineation trends just east of the central portion of the Macquarie ridge.

to the north, but these have not been identified with confidence. The magnetic character in the central and southeast Tasman basin changes at about 48°S. The apparent wave lengths along the traverses oriented in a northerly direction change to much longer values north of this latitude. *Ringis* [1970] and *Hayes and Ringis* [1972] show that magnetic lineations in the northern and central Tasman Sea are oriented about N 20°W, but these lineations cannot be traced as far to the southeast as where they would intersect the Macquarie ridge complex.

Southeast of the complex, magnetic lineations from about anomaly 25 to anomaly 32 and beyond have been described and are adjacent to the southern slope of the Campbell plateau [*Christoffel and Falconer*, this volume]. However, the nature of the magnetic anomalies in the area of the Macquarie triple junction is still unknown. Magnetic anomalies throughout most of the Emerald basin and Solander trough appear to be fairly long in wave length (of the order of 50–100 km), but trends have not been established in this area. Similar long wave length anomalies are typical of the Campbell plateau except for isolated cases, e.g., northeast of Campbell Island and near the Auckland Islands. Solander Island, composed largely of hornblende andesites, is the only known source of Quaternary volcanism in the area [*Summerhayes*, 1967].

Although the topographic trenches associated with the complex are not typified by the magnetic quiet zones that are characteristic of many deep ocean trenches [*Hayes and Ewing*, 1971], the borderland along the New Zealand fiordland region does exhibit this characteristic (see also *Christoffel and van der Linden* [this volume]).

Evidence suggests that Macquarie Island had once been formed by a process of crustal accretion associated with sea floor spreading (see *Varne et al.* [1969]). Our preliminary analysis of relative plate motions [*Hayes et al.*, 1972] suggested the possibility that some extension had occurred along this boundary within the last 10 m.y. It is therefore of prime importance to determine whether or not anomaly lineations could be identified and correlated with spreading along the central part (or any part) of the Macquarie ridge complex. A short sequence of magnetic anomalies recorded in the vicinity of 51.5°S, 162.5°E have been tentatively correlated (Figure 4). If the correlations are correct, the lineations trend in a direction roughly parallel to the Macquarie ridge in this area (Figure 6*d*). The lineations lie to the southeast of the ridge. Within this region, the bottom topography is extremely rough. We have not yet been able to discount the possibility that these magnetic lineations are associated with the structure of the basement or magnetic contrasts within it. The need to invoke a model of alternating polarity blocks to account for the lineated anomaly trends has not been demonstrated. The average depth to the upper surface of the magnetic layers is a minimum of about 1800 fm (3.3 km) in this region, and it is unlikely that the topographic relief exhibited there could give rise to the anomalies correlated in Figure 4. However, we wish to stress that the relationship of these anomaly lineations to sea floor spreading models and, in fact, the correlations themselves are questionable.

There is no pronounced magnetic signature associated with the Resolution ridge. The magnetic high associated with Macquarie ridge changes markedly at various points along the feature. The magnetic high is of the highest amplitude in the northern section and is very subdued in the southern section, particularly in that area south of 56°S where the ridge becomes broken up.

Although magnetic lineations have not been identified over most of the area of the Emerald basin, the orientation of tracks and the apparent wave lengths displayed suggest that if lineations are present they are more nearly oriented in a north-northeasterly direction than in a northwest or westerly direction. In other words, if lineations are eventually found, their trend will most likely be quite different from the trend associated with the Southeast Indian ridge spreading or with the Pacific-Antarctic ridge spreading.

SEDIMENT DISTRIBUTION

There is great variability in the over-all sediment distribution in the region surrounding and including the Macquarie ridge complex [*Houtz et al.*, 1971; *Houtz and Markl*, this volume]. The Macquarie ridge provides the major barrier to the circumpolar flow of water, one of the major systems of transport in the world ocean. The gaps within the ridge greatly affect the path that the flow takes. Upwelling and increased productivity occur in some areas, and bottom scour and nondeposition occur in others. Extremely large quantities of sediment are present within the Tasman Sea north of about 46°S. These sediments are essentially acoustically transparent and frequently exceed 1 km in thickness. They are presumably hemipelagic.

If there has been recent subduction along the Puysegur trench region, we would expect these thick Tasman Sea sediments to be transported into the trench. Figure 5*a* (profiles 2–6) shows that there are essentially no sediments in the bottom of the narrow, V-shaped Puysegur trench. Either no sediments have been brought into the trench, or they have been swept away by bottom currents, or they have been thrust onto the Macquarie ridge and disturbed beyond recognition. As there are virtually no measurable quantities of sediment in the trench now, vigorous bottom currents are likely to be presently active there. The possibility that older, distorted sediments are present on the eastern flank of the Puysegur trench cannot be discounted. North of Resolution ridge, thick quantities of sediment characteristic of those of the entire northern Tasman basin are present along the base of the Fiordland continental slope. As nearly as we can tell, the basement horizon does not bend downward as the continental slope is approached. If a buried trench structure is present near the Fiordland margin, it must lie as a seismically masked feature beneath the continental slope there.

Figures 5*a*, *b*, and *c* illustrate the bottom morphology and sediment disposition for the three main segments of the complex. Weak reflectors have been highlighted only in instances where they are clearly present in the original records. The Macquarie trench is the only major deep of the complex that contains a measurable thickness of sediment. These sediments occur as undisturbed flat-lying layers and are not observed on all crossings.

Thick accumulations of well-layered sediments are characteristic of a small, nearly enclosed basin near 59°S, 160°E (Figure 5*c*; profiles 13, 14; see also *Houtz et al.* [1971]). The only other area of pronounced sediment accumulation is the Solander trough (Figures 5*a* and *b*, profiles 4, 5, 7). Here the sediments are disturbed by high angle postdepositional faults, and the trough appears to be a grabenlike feature.

The Macquarie ridge is essentially devoid of measurable thicknesses of sediments. *Summerhayes* [1969] has reported a variety of mafic crystalline rock types dredged from the flanks of the ridge.

INFERRED PLATE MOTIONS

As mentioned in the introduction, the relationship of the Macquarie ridge complex to the standard tectonic regimes (extension–accreting zones, compression–subduction zones, transcurrent motion–transform faults) is unknown. If the ridge can be related to one or several of these features, a variety of inferences can be drawn from studies of other areas of the ocean. In keeping with the concepts of global tectonics, we assume that the boundary between the Pacific and the Indian plates (both major crustal plates) is defined by the zone of seismic activity along the Tonga-Kermadec trench system, the seismicity belt of New Zealand, and the seismically active Macquarie ridge complex. We can examine the relative motion at this boundary as defined by the pole of relative motion between the Indian and Pacific plates to see if it is consistent with the style of tectonic deformation observed at the present boundary.

In examining the Macquarie ridge complex, the success of this technique is limited primarily because of the uncertainty in determining the precise position of the pole that describes the 'instantaneous' relative motion between the Indian and Pacific plates. The uncertainty is critical in this area principally because the pole of relative motion lies very close to the Macquarie ridge complex, and therefore small changes in the pole position can give rise to large

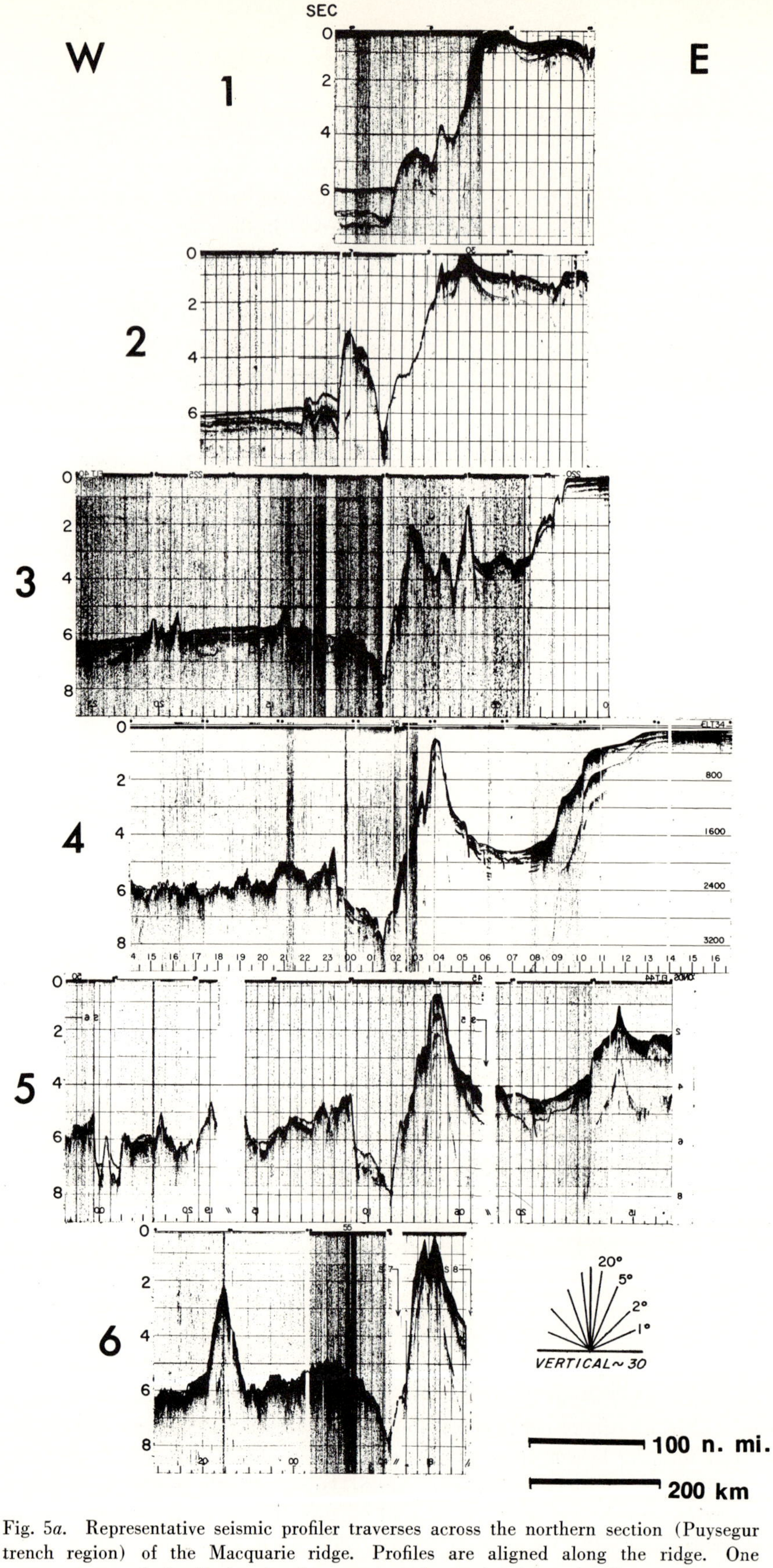

Fig. 5*a*. Representative seismic profiler traverses across the northern section (Puysegur trench region) of the Macquarie ridge. Profiles are aligned along the ridge. One second of reflection time is approximately equal to 1 km of sediment.

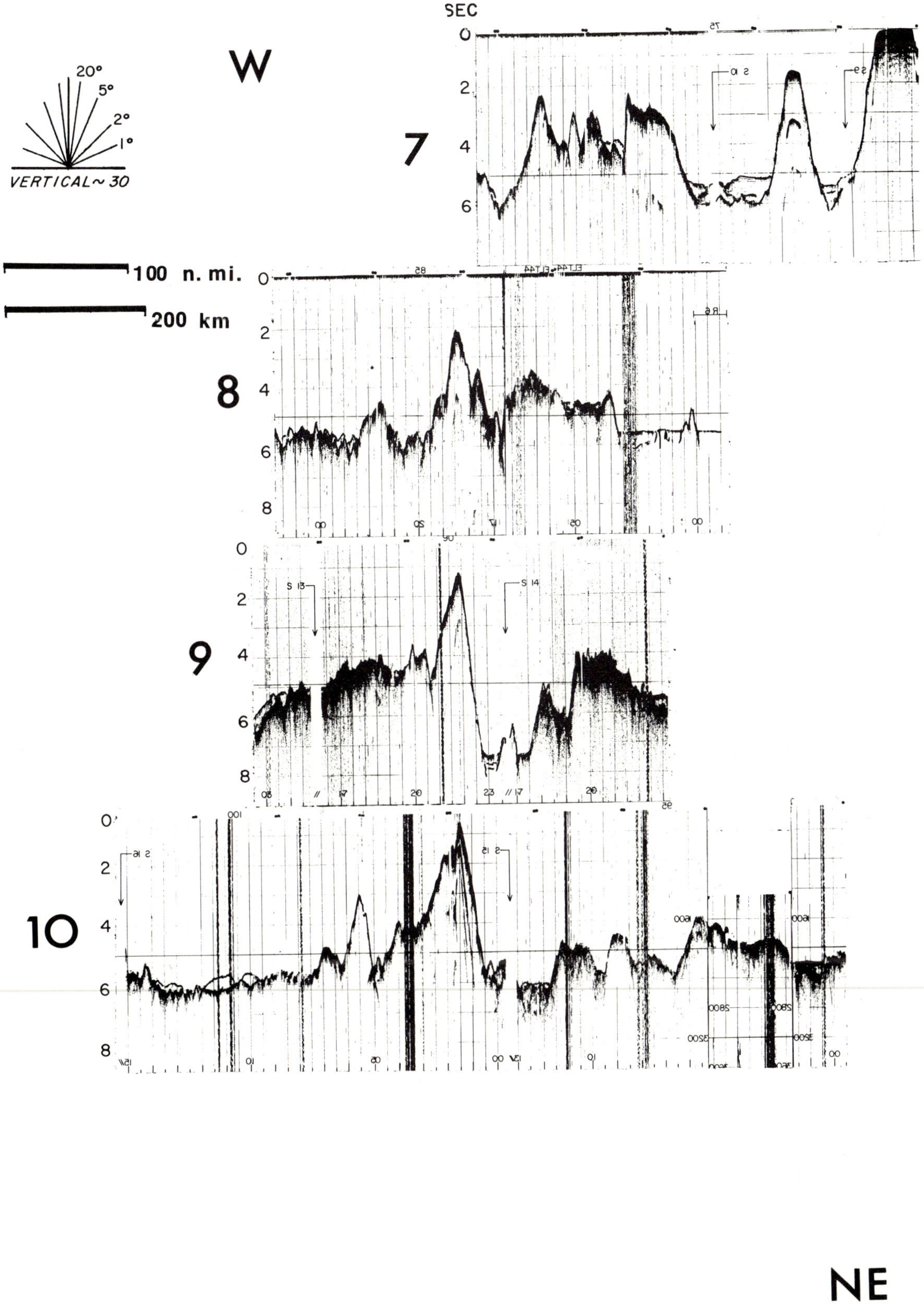

Fig. 5*b*. Representative seismic profiler traverses across the central section of the Macquarie ridge. Profiles are aligned along the ridge. One second of reflection time is approximately equal to 1 km of sediment.

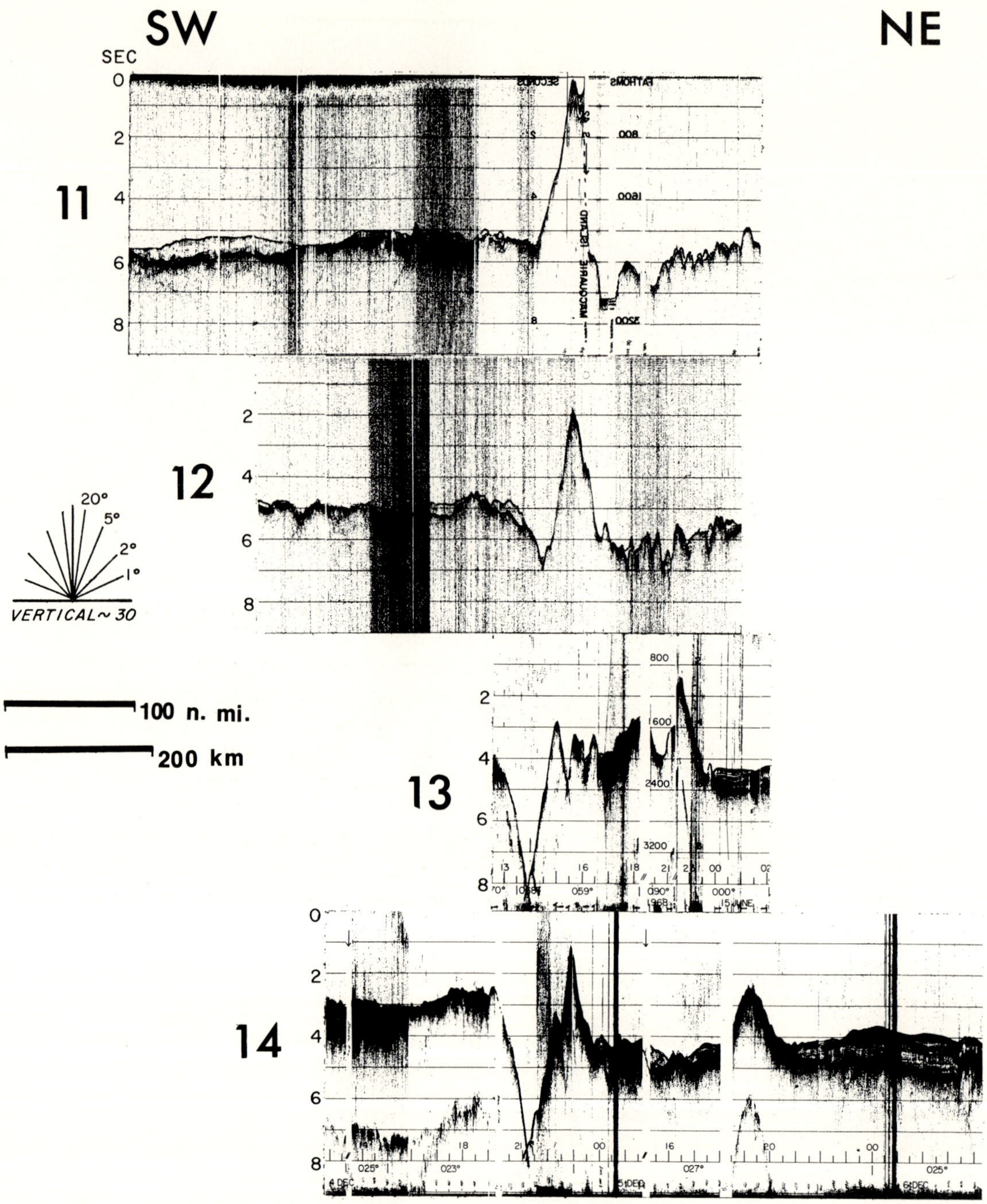

Fig. 5c. Representative seismic profiler traverses across the southern section (Hjort trench region) of the Macquarie ridge. Profiles are aligned along the ridge. One second of reflection time is approximately equal to 1 km of sediment.

differences in the direction of relative motion at various points along the complex.

The relative motion derived from a variety of models is shown for this area (Figure 6*e*) and the area to the north (Figure 7). Table 1 identifies the pole positions used in deriving each model. The range of azimuths of the vectors is an indication of the effects of the uncertainty in determining the relative motion at various points along this boundary.

In the area of the Tonga-Kermadec trench, the range of azimuths is significantly large, but the inference regarding the mode of tectonism active there

TABLE 1. Instantaneous Poles of Relative Motion between the Indian and Pacific Plates

Pacific-antarctic			Indian-antarctic			Indian-Pacific		
Source	Latitude, Longitude	Rate, 10^{-7} deg/yr	Source	Latitude, Longitude	Rate, 10^{-7} deg/yr	Model	Latitude, Longitude	Rate, 10^{-7} deg/yr
Le Pichon, 1968	70.0°S, 118.0°E	10.8	*Le Pichon*, 1968	4.7°S, 7.3°E	5.7	1	51.2°S, 161.1°E	12.4
Christoffel, 1971	82.0°S, 120.0°E	10.4	*Christoffel*, 1971	30.0°S, 160.0°W	10.5	2	58.7°S, 168.7°W	18.2
Le Pichon, 1968	...	...	*Christoffel*, 1971	21.0°S, 167.0°W	10.4	3	51.0°S, 174.5°E	17.9
Le Pichon, 1968	...	...	*Christoffel*, 1971	...	...	4	56.3°S, 179.2°E	18.5
Le Pichon, 1968	...	...	*Weissel and Hayes*, this volume	12.0°S, 145.0°W	6.8	5	59.0°S, 173.0°W	13.6
Christoffel, 1971	...	...	*Weissel and Hayes*, this volume	...	...	6	60.0°S, 157.0°W	13.6
Le Pichon, 1968	...	...	*McKenzie and Sclater*, 1972	11.0°S, 148.0°W	6.4	7	58.2°S, 179.5°W	13.4
Christoffel, 1971	...	...	*McKenzie and Sclater*, 1972	...	...	8	61.0°S, 161.1°W	13.2

The calculations for the Indian-Pacific zone were obtained by vector addition ($\omega_{\text{Pacific-antarctic}} + \omega_{\text{Indian-antarctic}} = \omega_{\text{Indian-Pacific}}$).

is not dependent on the selection of a particular model. All models are compatible with the relative motion deduced along the boundary from analyses of earthquake hypocenters and sea floor spreading in the South Pacific.

In the vicinity of the Alpine fault in South Island, New Zealand, all models except model 1 predict a dominant or exclusive strike slip motion with respect to the fault trace. (The *Le Pichon* [1968] Indian-antarctic pole of motion was deduced from the summation of several rotation vectors from other oceans. It is subject to large errors as pointed out by *Le Pichon* [1968] and is considered in model 1 only for reference purposes; see Table 1.) The right-lateral sense of motion across the fault is consistent with the observed offset of geologic features on opposite sides of the fault. However, the amount of displacement as predicted by the various models is different by almost a factor of two. The two extreme examples are model 2 and model 3 (Figure 7), which indicate motions of 660 km and 280 km, respectively, over a period of 10 m.y. Using geologic field investigations, *H. Wellman* [1964] suggests that 450 km of displacement has occurred on the Alpine fault since the late Miocene. However, *P. Wellman and Cooper* [1971] believe the movement to be as little as 120 km of displacement since the late Cretaceous. None of the models presented here are compatible with the *P. Wellman and Cooper* [1971] hypothesis.

The relative motion vectors deduced farther south along the Macquarie ridge complex diverge considerably, and the circle of uncertainty for the pole determinations is much more critical here than to the north. The strike of the Macquarie ridge complex is quite variable over its 1800-km length. This factor, coupled with the proximity of the relative motion pole to the complex, gives rise to major differences in the deduced relative motion along this boundary.

In the Puysegur trench region, most of the models predict a primary component of strike slip motion and a secondary component of compressive dip slip motion. The relative motion vectors are in good agreement with the deduced slip vectors and earthquake mechanism solutions for the two events studied [*Banghar and Sykes*, 1969; *Sykes*, 1967] for this area (Figure 6*a*).

In the Hjort trench region (the extreme southern portion), the relative motions are again primarily transcurrent but with a small component of compression indicated for most models. In both extremities of the complex (the Puysegur and Hjort trenches), all models except model 1 exclude any significant component of extension.

The central portion of the complex that includes Macquarie Island at 54.6°S, 158.8°E provides the greatest enigma, for here the various models predict not only pure strike slip motion (models 5 and 7) but also significant components of both extension (models 1 and 3) and compression (models 2, 6, and 8).

Certain geophysical parameters are associated with the different kinds of relative motion of the crust. Areas of extension are normally associated with shallow seismicity, high heat flow, small gravity anomalies, and recognizable magnetic anomaly linea-

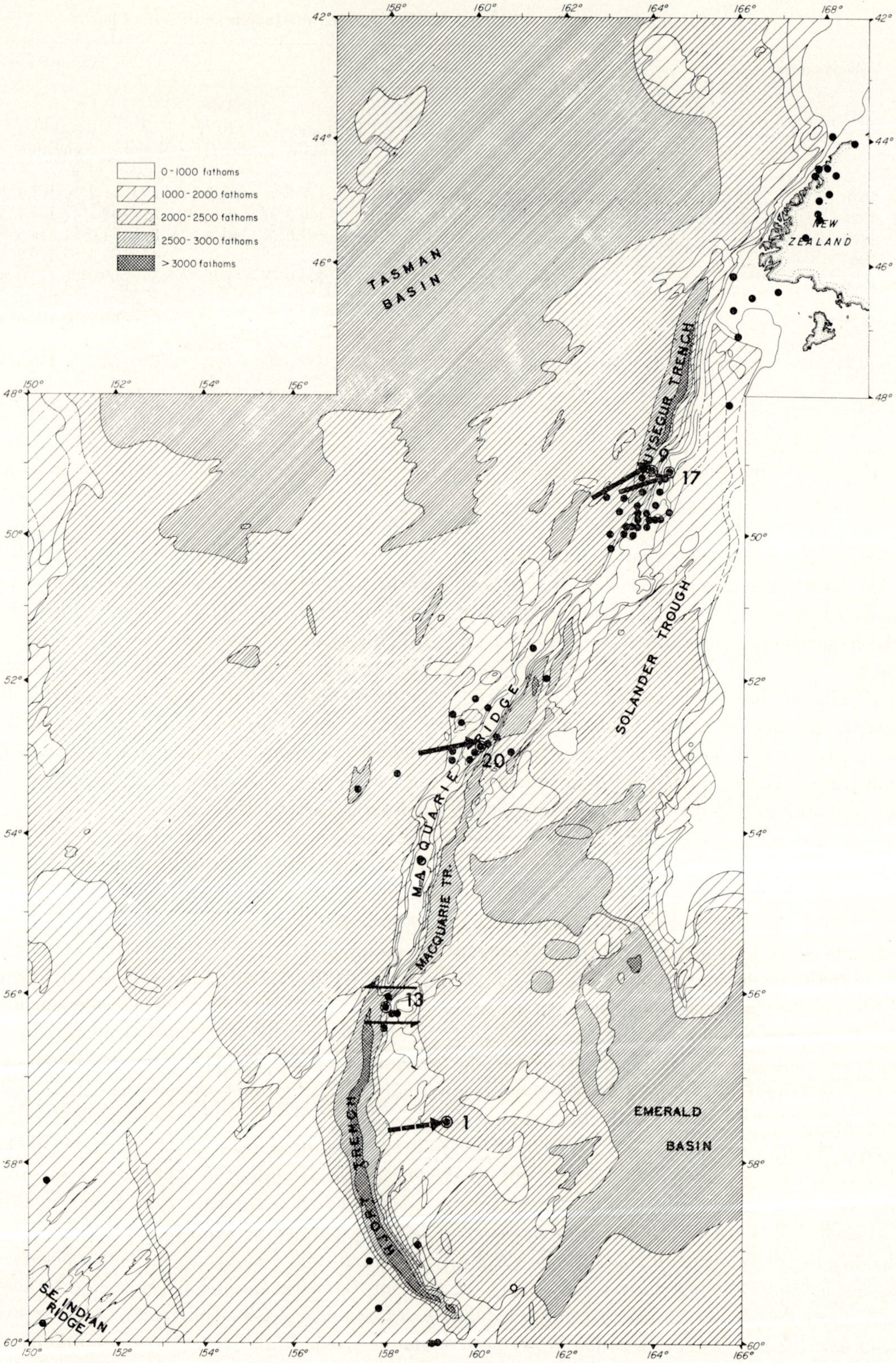

Fig. 6*a*. Generalized topography as taken from Figure 1. Earthquake epicenters are after *Barazangi and Dorman* [1969]. Slip vectors and their identification numbers for events studied by *Banghar and Sykes* [1969] and *Sykes* [1967] are indicated.

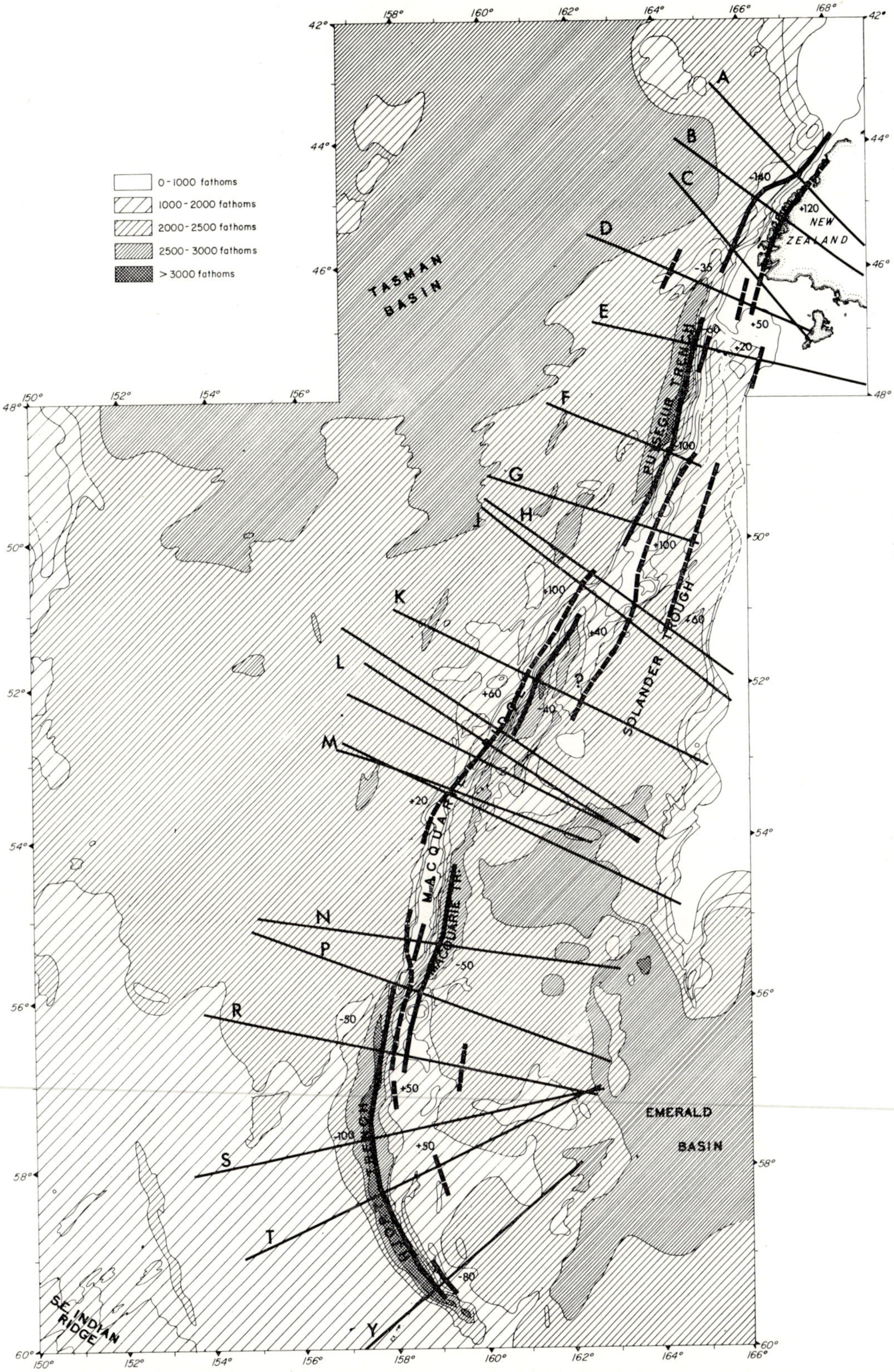

Fig. 6*b*. Generalized topography as taken from Figure 1. Axes of maxima (broken lines) and minima (solid lines) of local Airy isostatic anomalies. Representative values at the axis are shown in milligals. Lines A–Y show the locations of profiles examined and key profiles of Figure 8.

tions. Areas of transcurrent motion are also characterized by shallow seismicity and are easily identified on land by measured displacements of once contiguous geologic elements. Areas of transcurrent motion are not always obvious for water-covered areas especially if a component of compression is also involved. Large transcurrent faults (e.g., the Alpine fault and San Andreas fault) often do not have a significant gravity signature, although in such areas as the Romanche fracture zone large gravity anomalies are present that are attributed to great water depths. Regions of compression are generally related to zones of deep and intermediate seismicity. Large gravity anomalies are characteristic of these regions as well as major topographic deeps. The zones of compression are likewise difficult to identify if a significant transcurrent motion is present. The Cayman trench, the Puerto Rico trench, and western extremities of the Aleutian trench are good examples of features that provide difficulty in evaluating the relative importance of transcurrent and compressive motions. The morphologic expression of these features is not in itself a good criterion for determining their mode of evolution.

We must now briefly review the known geophysical characteristics of the three segments of the complex in light of the relative motions predicted by our models. In the Puysegur trench region, the morphology and the large gravity anomalies suggest that this is a zone of at least some compression. The earthquake mechanism solutions also support this interpretation. However, the seismicity here is confined to shallow depths. The apparent grabenlike structure of the nearby Solander trough is puzzling in view of the inference of compressive stresses acting at the Puysegur trench.

In the Macquarie Island region, a topographic trench lies east of the ridge and contains undeformed sediments. The scattered seismic activity here is presumed to be shallow, although the possibility of the occurrence of unidentified intermediate events cannot be completely discounted. No heat flow measurements are available for this area. The gravity anomalies over the central section are smaller than at the extremities but are still considered to be large.

Varne et al. [1969] proposed that Macquarie Island is composed in part of typical oceanic basalt formed in moderately deep water. They suggest that the petrology and magnetic properties of these rocks can be related to the formation of basalt and associated magnetic lineations during the process of sea floor spreading. The fact that the island is now emergent testifies to its subsequent uplift of about 3 km. *Williamson and Rubenach* [this volume] question the conclusions of *Varne et al.* [1969]. An unequivocal identification of the sea floor spreading type of magnetic lineations is the only evidence likely to resolve this issue. As we have shown in an earlier section, magnetic lineations are likely to be present within a small zone to the east of the central section of the complex. Their orientation is consistent with that expected for sea floor spreading anomalies associated with crustal extension in model 3. We have not yet identified any of the correlated anomalies as part of the Cenozoic spreading pattern; in view of the small component of extension predicted (<1 cm/yr), such identifications would be very unlikely.

The strike of the dyke swarms and ultramafic provinces in the northern part of Macquarie Island (see *Varne et al.* [1969]; *Williamson and Rubenach* [this volume]) is consistent with the direction of extension and hence the hypothetical fracture zone azimuths of model 3 (Figure 6*e*). However, local uplift of the Macquarie Island basalts by more than 3 km since Pliocene time is difficult to relate to extensional tectonics in this region. One earthquake first motion study has been done for this area and indicates a large component of thrusting (Figure 6*a*, event 20).

The Hjort trench is a pronounced arcuate deep. The trace of the deep is nearly coincident with a small circle about the pole of motion of model 4. Gravity anomalies are large over the deep but relatively small over the adjacent section of the Macquarie ridge. The seismic activity is low and is probably confined to shallow depths. These observations are compatible with the models that indicate a predominantly transcurrent motion that acts in combination with a secondary, compressive component.

The offshore Fiordland region presents a special problem. *Smith* [1971] has examined new seismic evidence, obtained from local New Zealand seismic stations, and has demonstrated the presence of intermediate depth earthquakes lying within a narrow area near the Fiordland region of South Island, New Zealand. Smith proposes that this activity defines a narrow Benioff zone steeply dipping to the southeast. This dipping zone is confined on the north by the trace of the Alpine fault and terminates to the south about 100 km from the New Zealand coast. This narrow region coincides closely with the area of the large negative gravity anomalies offshore (the free air nega-

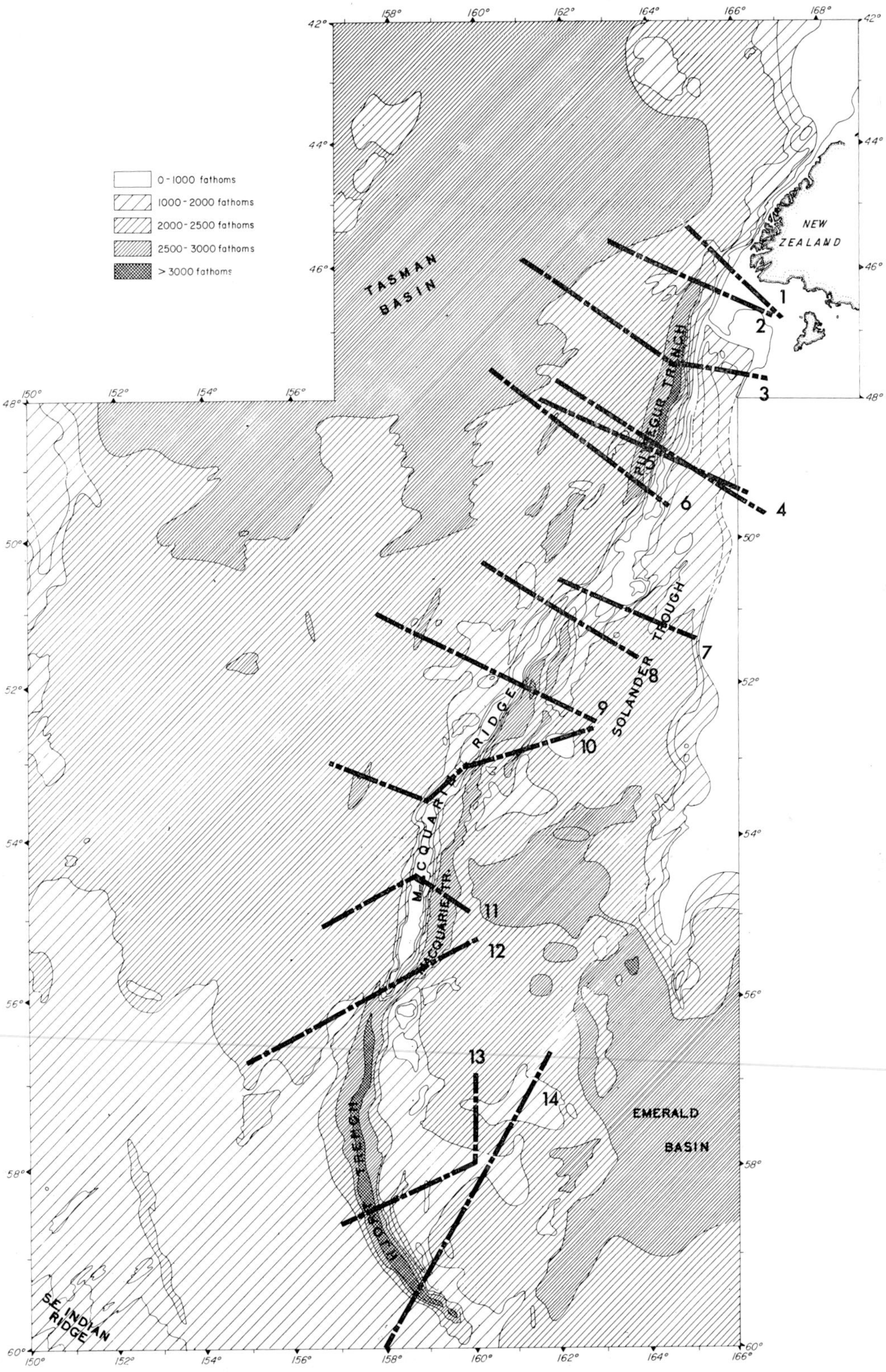

Fig. 6*c*. Generalized topography as taken from Figure 1. Locations of seismic profiler sections that were shown in Figure 5.

tive anomaly we have earlier described lying north of Resolution ridge; see Figure 3) and the large positive anomalies on land described by *Reilly* [1965] (see Figure 9).

On first inspection, it appears that the sense of motion deduced by sea floor spreading and plate tectonics (i.e., motion essentially parallel to the Alpine fault trace) and the southeasterly dipping Benioff zone presumed to be indicative of crustal subduction are incompatible. This problem is also discussed by *Christoffel and van der Linden* [this volume]. The most difficult aspect of explaining the intermediate depth of the earthquake activity is that its boundaries are sharp and define a rather small area. The very large isostatic gravity anomalies (negative offshore, positive onshore) are also restricted to a small zone adjacent to the area of the intermediate earthquake activity. Anomalies here are a factor of two or more larger than the anomalies of the northern Puysegur trench. There is no obvious reason to predict a drastic change in the tectonic mode coincident with these abrupt seismic and gravity boundaries.

We can ask the question, Has the relative motion along the Alpine fault been exclusively strike slip motion during the period of time recorded by these earthquakes? Sea floor spreading information in the adjacent oceans strongly suggests that motion has been essentially strike slip along the presently well-defined Alpine fault trace of South Island for a period of at least 10 m.y. (Figure 6). There is evidence of some vertical motion near the fault trace during the Quaternary [*Suggate*, 1963]. Suggate suggests that this vertical motion is about the same as the component of strike slip motion each amounting to about 20 km in 10 m.y. The relative importance of vertical versus horizontal movements on the Alpine fault in the time frame of *Smith*'s [1971] earthquakes is unresolved.

We can also question the reliability with which the attitude and extent of the South Island Benioff zone are defined. The existence of intermediate earthquake activity cannot be questioned. However, the demonstration of a unique distribution of that activity and of the shallow activity is not straightforward. If we accept the premise that the intermediate depth earthquakes are indicative of subduction, then the lack of similar activity behind the Puysegur trench is even more puzzling.

We offer several alternative explanations that may account for the apparent conflict of observations.

1. The primary difference between the Fiordland coast and the Puysegur trench is that motion is between a continent–ocean boundary for the former and between an ocean–ocean boundary for the latter. The intermediate depth earthquakes may not be related to subduction but rather to pure transcurrent motion. The motion gives rise to intermediate seismic activity by virtue of a contrast in the properties of crustal and upper mantle rocks that extend to considerable depth on opposite sides of the fault. The large gravity anomalies may also reflect significant density differences associated with the contrast in properties near the boundary of the two lithospheric plates. Similar intermediate earthquake activity is absent along the Alpine fault farther north and along other major transcurrent faults (e.g., the San Andreas fault) because motion mainly occurs across a continent–continent boundary. The near-vertical dip of the Fiordland seismic zone lends support to this interpretation.

The area where oceanic crust is moving past continental crust is confined to the south Fiordland coast, and this confinement can thus account for the limited areal extent and the sharp limits of the intermediate seismic zone and the large isostatic anomalies.

We prefer this explanation and suggest that this area may serve to demonstrate that different types of boundaries (i.e., ocean–ocean, continent–continent, and ocean–continent) react differently. Although we cannot cite other examples where intermediate earthquakes are clearly associated with major strike slip faults at ocean–continent boundaries, the southern extension of the Philippine fault may be a good place to look for such a relationship.

2. The large gravity minimum north of Resolution ridge and adjacent to the Fiordland coast may not be oriented parallel to the coast line and the continental slope. Only a short number of traverses were run north of 43°S, and in contouring the free air anomalies the topography was used to interpolate and extrapolate trends. We assumed that the major gravity trends were parallel to the continental margin. Thus a possible bias is built into the free air anomaly contour map and cannot be eliminated without further survey. If the isostatic anomalies trend more nearly north-south than northeast-southwest, a trend suggesting a corresponding trend to the deep crustal structure, then motion along the Alpine fault would also give a component of compression relative to the deep structure consistent with the southeasterly dipping Benioff zone. We think that a major error in the

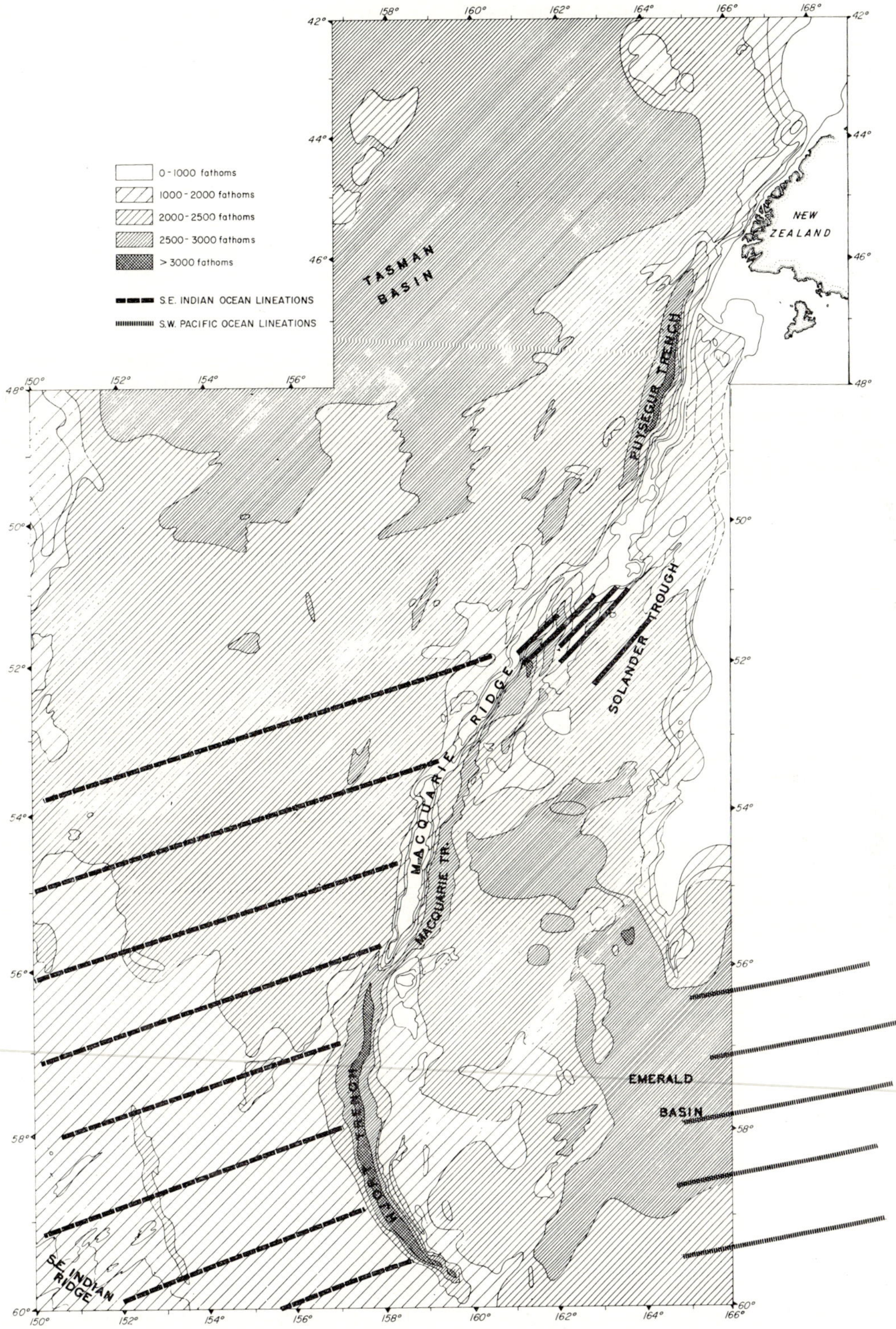

Fig. 6*d*. Generalized topography as taken from Figure 1. Magnetic lineation trends deduced for the area east of the central Macquarie ridge and based on the correlations shown in Figure 4. The orientations of magnetic lineations identified as part of the Cenozoic spreading pattern are schematically shown for the area west of the complex [*Weissel and Hayes*, this volume] and the area southeast of the complex [*Christoffel and Falconer*, this volume].

definition of the gravity low is unlikely, and hence this explanation is probably not valid.

3. The strike of the Benioff zone of *Smith* [1971] is not well determined. If only events deeper than 33 km are considered (see Figure 9), an alternative inferred strike of northwest-southeast is not unreasonable. If the Benioff zone dips steeply to the northeast, then strike slip motion parallel to the Alpine fault could also accommodate a component of underthrusting beneath the Fiordland coast of South Island. Although this explanation could reconcile the problem of conflicting types of relative motion in this area, it does not help explain the distribution of the gravity anomalies or the narrow limits of the seismic activity as does our first explanation.

4. *Weissel and Hayes* [this volume] have suggested that an ancient boundary exists that runs across the central Tasman Sea from the South Tasman rise to the tip of South Island. This hypothetical boundary coincides closely with the Resolution ridge near New Zealand. The boundary may represent the dividing line in the Tasman Sea that separates the southern portion relating to spreading in the southeast Indian Ocean from the northern portion relating to an earlier and entirely different pattern of spreading. Therefore the large gravity anomalies of the Fiordland area may in fact reflect a tectonic history and deep crustal structure not directly related to processes active in the last few million years. The region as a whole is in close isostatic balance and may be able to sustain large local gravity anomalies over a long period of time. This explanation implies that the close correspondence between the distribution of the gravity anomalies and the intermediate seismicity is simply coincidental, and for this reason we disfavor this alternative.

5. The large gravity anomalies and intermediate seismic activity might relate to a detached portion of the lithosphere that is sinking gravitationally. The detached lithospheric segment may have been underthrust along a part of the plate boundary the configuration of which is different from that presently defined. The configuration of the plate boundary 3–5 m.y. ago is purely conjectural, and an evaluation of the plausibility of the above explanation must await further investigations here and elsewhere.

6. In arriving at the models for relative motion across the Indian-Pacific boundary (Figure 7), we have assumed a continuity for the major plates. If smaller plates are moving as discrete units, our analysis is invalid.

GENERAL DISCUSSION AND CONCLUSIONS

To evaluate the possible differences in deep crustal structure along the complex, two-dimensional local Airy isostatic anomalies were computed. Locations of the profiles examined are shown in Figure 6*b*, and representative profiles are given in Figure 8. These profiles were computed from simulated topographic and gravity profiles taken from the maps of Figures 1 and 3 (see foldouts at back of volume). The axes of the isostatic gravity maxima and minima are also shown in Figure 6*b*. These gravity axes are discontinuous, although they collectively define trends parallel to the physiographic features. The distribution of isostatic anomalies suggests that the three segments (Puysegur, Central, Hjort) of the complex defined by other parameters probably can also be identified with significant differences in the deep crustal structure. Large negative isostatic anomalies are present over the Puysegur and Hjort trenches, and their minima are displaced about 25 km toward Macquarie ridge from their free air anomaly minima (Figure 8, profiles F and T). The central section of the complex (Figure 8, profile N) has negative isostatic anomalies that are much smaller than those to the north or to the south (see Figure 6*b*). The magnitude of the positive isostatic maxima associated with Macquarie ridge is relatively small (about +50 mgals) south of 56°S (profile T) and also in areas where the ridge is offset (profile D). The maxima are characteristically broad in these regions.

Although Macquarie ridge is offset near 51°S and the bounding trenches change relative position with the ridge, the trend of the axis of the negative isostatic anomaly minimum is uninterrupted. The change in the position of the trench (Puysegur trench) from west of the ridge to east (Macquarie trench) of the ridge may be only superficial to the deeper structure.

Studies of fault mechanisms using the first motions have been made for five events in the Macquarie ridge complex area [*Banghar and Sykes*, 1969; *Sykes*, 1970, 1967]. In all but one case, a large component of thrusting is inferred, although there is uncertainty in determining the strikes of the fault planes and auxiliary planes.

The events that have been studied are indicated in Figure 6*a*. The slip vectors deduced from three events (9, 20, 17) are consistent with the relative motion vectors in models 1 and 5–8. The southernmost event (1) lies on one spur of the southern part

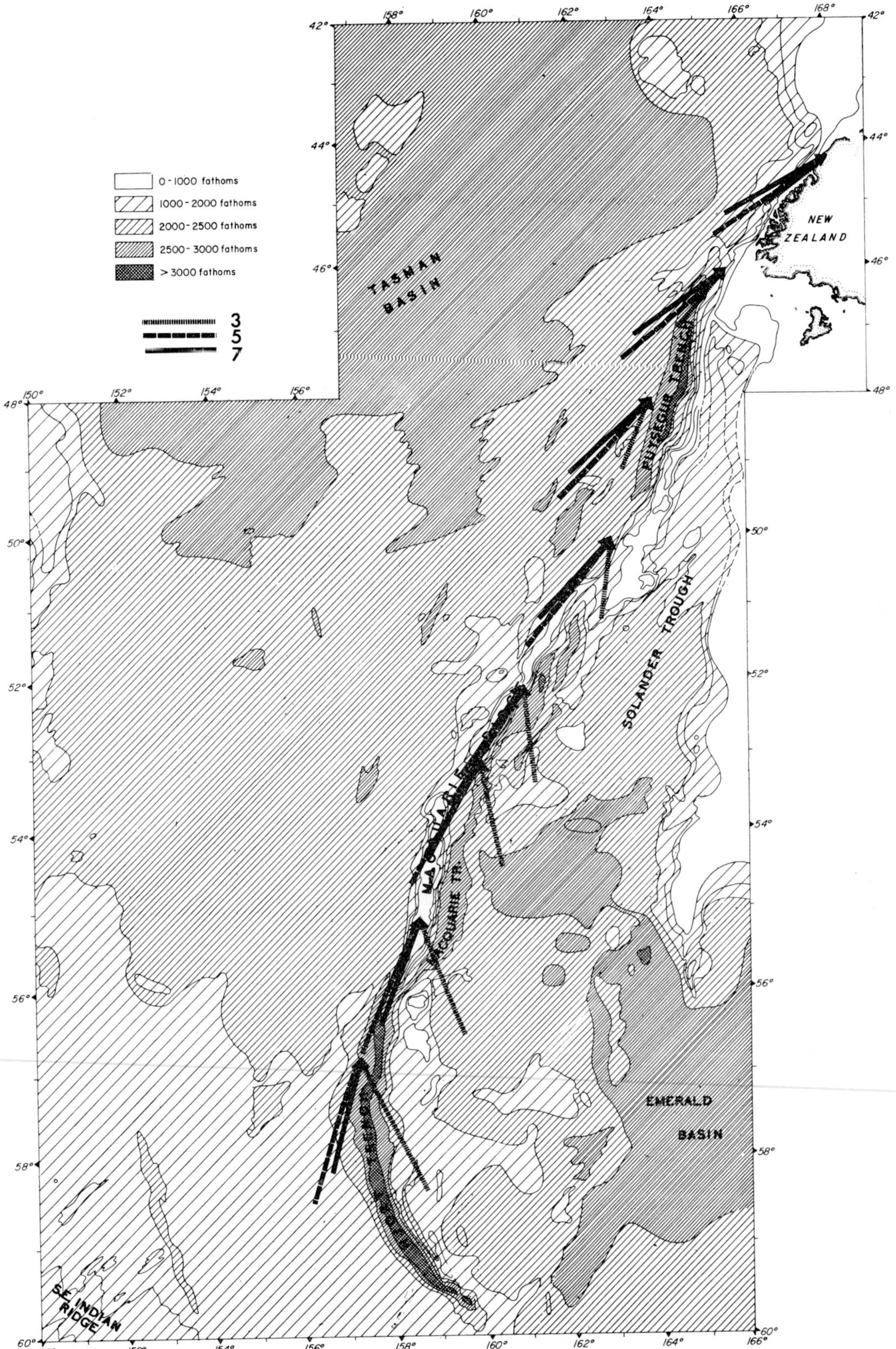

Fig. 6*e*. Generalized topography as taken from Figure 1. Relative motion of the Indian plate with respect to the Pacific plate as deduced from models 3, 5, and 7 of Table 1. We view model 5 as representing the most reliable Indian-Pacific pole position that was determined. The length of the vectors represents the amount of motion deduced for the last 5 m.y.

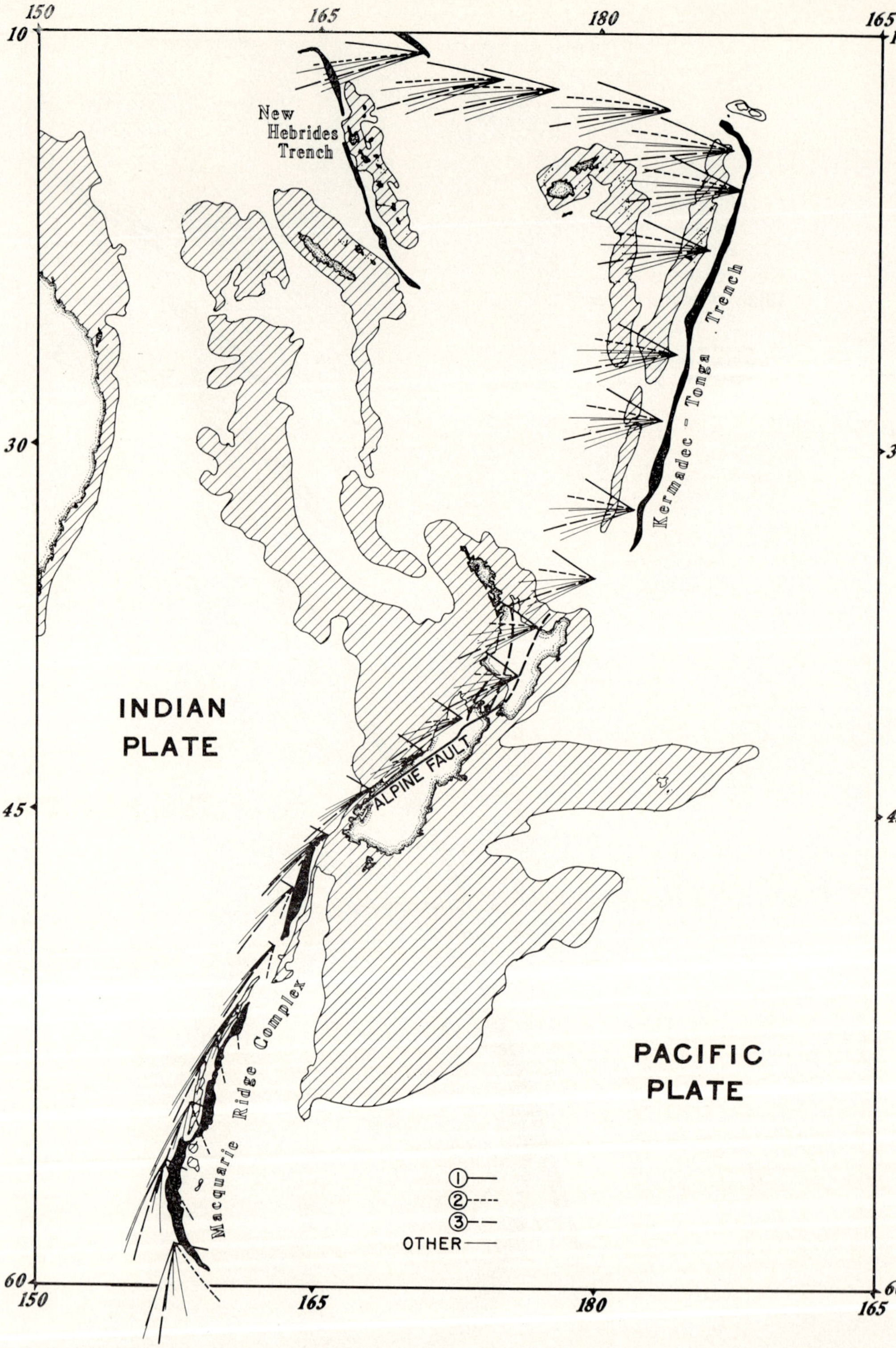

Fig. 7. Relative motion of the Indian plate with respect to the Pacific plate including the Alpine fault and the Tonga-Kermadec trench. Only models 1, 3, and 5 of Table 1 are identified. The other models are somewhat similar to one another except near the southern extremity of the map. The focal point of vectors represents the 'point' of the vector, and the lengths represent the amount of motion deduced for the past 5 m.y. The scale for these vectors is the same as the sliding distance scale for any Mercator projection. The major plateaus and ridges are cross-hatched.

of the Macquarie ridge, and the mechanism solution is very poor. One event (13) located right in the 56°S gap indicates a predominantly strike slip mechanism. Relative motion along the fault plane selected by *Banghar and Sykes* [1969] is shown in Figure 6*a*. If the auxiliary plane were chosen as the fault plane, the relative motion is consistent with models 2 and 4–8.

Although the epicenters for the limited period 1961–1967 (see *Barazangi and Dorman* [1969]) clearly define a seismic belt associated with the Macquarie ridge complex, the total activity of this zone is not great when compared with most areas of the circum-Pacific. This relative lack of seismic activity may in part be the result of an inadequate distribution of seismic stations, but probably the true activity of this area is relatively low. Only about 100 events are recorded for the area for the period from 1961 to 1967.

In addition to the fact that there are relatively few epicenters defining the Macquarie seismic zone, these earthquakes are not uniformly distributed along the length of the complex. There are 2 or 3 areas where the greatest proportion of the earthquake activity is concentrated (Figure 6*a*). One area is near 49°S, 164°E, another occurs near 53°S, and a third minor one is at 55°–57°S near 158°E. There are only a few events in the vicinity of the complex south of about 56°S. Similarly, to the north there is an apparent hiatus in the seismic activity between approximately 49°S and 46°S (the tip of South Island). These events have not occurred uniformly in time. For example, in the 49°S region, 10 events occurred within a 1-year period (not aftershocks) followed by a 2-year period when only three events occurred. The seismic activity of the complex is far from uniform in either time or space. For this reason, we are hesitant to consider that the earthquake studies provide very strong support for any particular hypothesis in a geologic time frame.

Although we have essentially no evidence of a component of extension for the northern part of the Macquarie ridge complex, the ridge here is bounded on the east by the Solander trough. This broad depression contains at least one and probably several kilometers of well-layered sediment (Figure 5) [*Houtz et al.*, 1971]. These sediments have been disturbed by postdepositional high angle faulting. Both flanks of the Solander trough are rectilinear and essentially parallel with the Macquarie ridge. We note the proximity of the Puysegur trench, a feature presumably formed by strike slip motion and compression, and

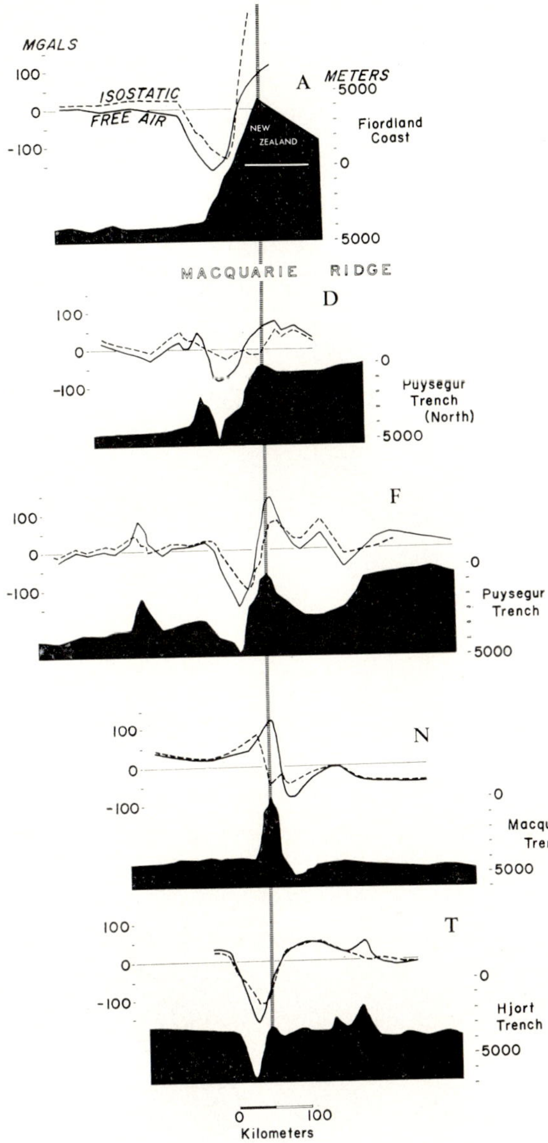

Fig. 8. Selected profiles of generalized topography, free air anomaly, and local Airy isostatic anomaly. The isostatic anomaly was computed by assuming a density contrast of 0.53 g/cc between crustal and mantle rocks. The location of these profiles is given in Figure 6*b*.

the Solander trough, another major feature of large geographical extent but that has apparently been formed by crustal extension. This part of the complex may be analogous to areas of interarc extension as proposed by *Karig* [1970].

The Macquarie ridge complex has probably evolved as the consequence of the interaction of the Indian and Pacific plate motions along a pre-existing zone of weakness. The dominant motion along the boundary

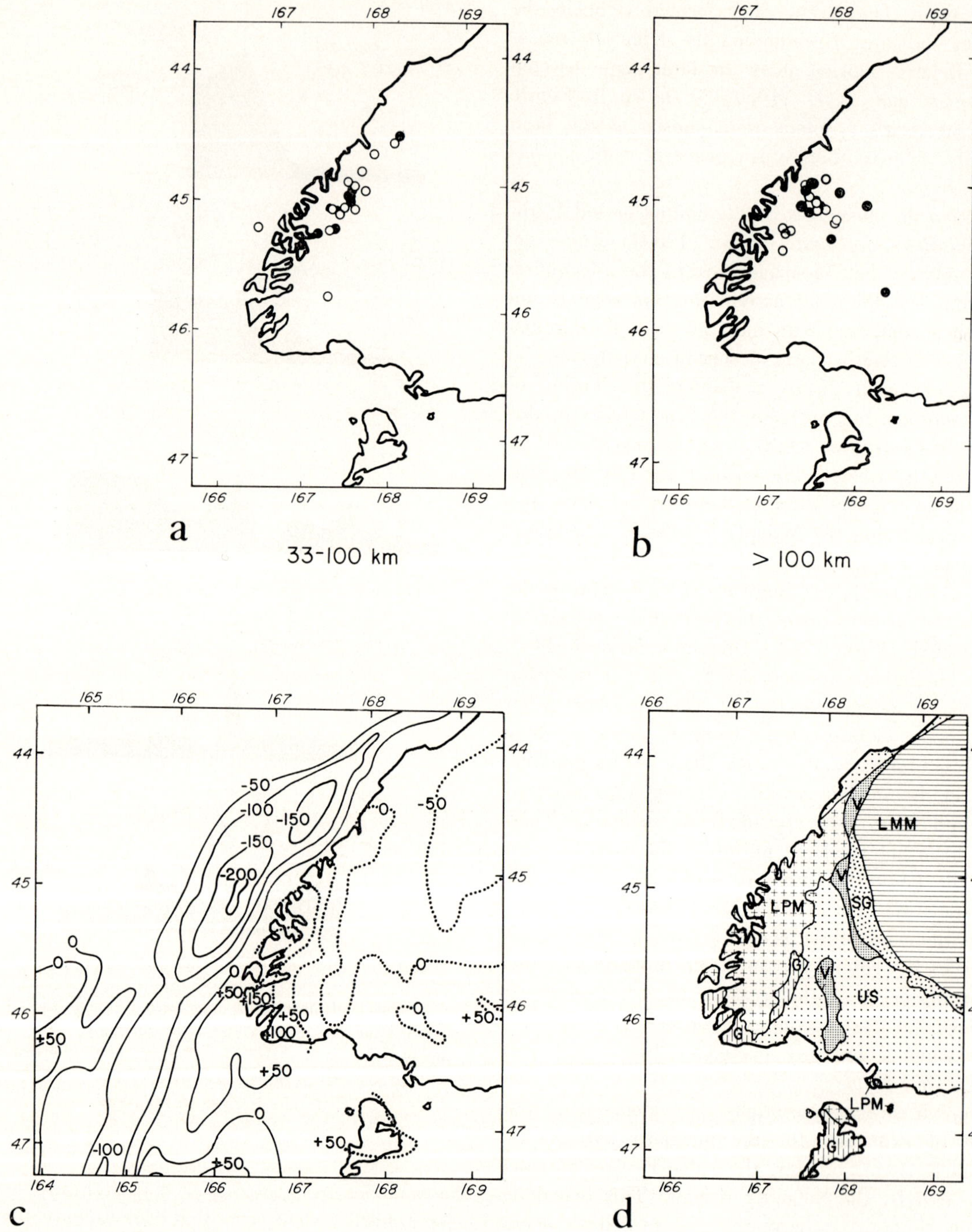

Fig. 9. (*a, b*) Epicenter locations of intermediate earthquake activity modified from *Smith* [1971]. Open circles represent less reliable locations. (*c*) Free air anomalies offshore from the Fiordland coast given in this paper and Bouguer anomalies on land taken from *Reilly* [1965]. (*d*) Generalized geology of Fiordland and nearby areas modified from the geological map of New Zealand [*New Zealand Department of Scientific and Industrial Research*, 1958]. The following abbreviations have been used: US, Paleozoic to Quaternary sediments; LPM, Lower Paleozoic metamorphics; LMM, Lower Mesozoic metamorphics; SG, Southland geosyncline and Upper Paleozoic sediments; G, granite batholiths; and V, volcanics.

has been right-lateral strike slip motion. A smaller component of compressive dip slip motion has been active in the northern Puysegur trench region and possibly in the southern Hjort trench region. The central portion of the complex is the only area where a small amount of extension is indicated. Extension has not yet been firmly established. Our inability to categorize the composite features stems from the likelihood that a combination of tectonic modes are active along the boundary and that the relative importance of these modes changes along the complex. This manner of tectonic evolution is to be expected along boundaries that are close to the pole that describes the relative motion across the boundary.

It would be incorrect to classify the Macquarie ridge complex as an island arc trench system, a zone of crustal accretion, or a transform fault. Elements of each of these types of features can be observed along discrete segments of the complex.

Acknowledgments. This research has been supported by the Office of Polar Programs (grants GV-23334 and GV-27472) and the Oceanography Branch of the National Science Foundation (grant GA-27281) and by the Office of Naval Research (contract N00014-67-A-0108-0004). We thank W. C. Pitman III and R. Houtz for access to unpublished magnetics and seismic data. D. A. Christoffel was involved in many useful discussions in the early stages of this project. We thank C. Scholz, P. Rabinowitz, and G. Thompson for the critical reviews of the manuscript.

Contribution 1780, Lamont-Doherty Geological Observatory.

REFERENCES

Banghar, A. R., and L. R. Sykes, Focal mechanisms of earthquakes in the Indian Ocean and adjacent regions, *J. Geophys. Res., 74*(2), 632–649, 1969.

Boyer, D. L., and J. R. Guala, Model of the antarctic circumpolar current in the vicinity of the Macquarie ridge, in *Antarctic Oceanology II: The Australian–New Zealand Sector, Antarctic Res. Ser.*, vol. 19, edited by D. E. Hayes, pp. 79–93, AGU, Washington, D. C., 1972.

Brodie, J. W., and E. W. Dawson, Morphology of north Macquarie ridge, *Nature, 207*, 844–845, 1965.

Carey, S. W., The tectonic approach to continental drift, paper presented at Continental Drift Symposium, Aust. Acad. of Sci., Hobart, Tasmania, 1958.

Christoffel, D. A., Motion of the New Zealand Alpine fault deduced from the pattern of sea floor spreading, in Recent Crustal Movements, edited by B. W. Collins and R. Fraser, *Roy. Soc. N. Z. Bull., 9*, 25–30, 1971.

Christoffel, D. A., and R. K. H. Falconer, Marine magnetic measurements in the southwest Pacific Ocean and the identification of new tectonic features, in *Antarctic Oceanology II: The Australian–New Zealand Sector, Antarctic Res. Ser.*, vol. 19, edited by D. E. Hayes, pp. 197–209, AGU, Washington, D. C., 1972.

Christoffel, D. A., and W. J. M. van der Linden, The Macquarie Ridge–New Zealand alpine fault transition, in *Antarctic Oceanology II: The Australian–New Zealand Sector, Antarctic Res. Ser.*, vol. 19, edited by D. E. Hayes, pp. 235–242, AGU, Washington, D. C., 1972.

Ewing, M., and B. C. Heezen, Some problems of antarctic submarine geology, *Antarctica in the International Geophysical Year, Geophys. Monogr.*, vol. 1, edited by A. P. Crary et al., pp. 75–81, AGU, Washington, D. C., 1965.

Gordon, A. L., On the interaction of the antarctic circumpolar current and the Macquarie ridge, in *Antarctic Oceanology II: The Australian–New Zealand Sector, Antarctic Res. Ser.*, vol. 19, edited by D. E. Hayes, pp. 71–78, AGU, Washington, D. C., 1972.

Hamilton, R. M., and F. F. Evison, Earthquakes at intermediate depths in southwest New Zealand, *N. Z. J. Geol. Geophys., 10*(5), 1319–1329, 1967.

Hatherton, T., Total magnetic force measurements over the north Macquarie ridge and Solander trough, *N. Z. J. Geol. Geophys., 10*(5), 1204–1211, 1967.

Hayes, D. E., and J. R. Conolly, Morphology of the southeast Indian Ocean, in *Antarctic Oceanology II: The Australian–New Zealand Sector, Antarctic Res. Ser.*, vol. 19, edited by D. E. Hayes, pp. 125–145, AGU, Washington, D. C., 1972.

Hayes, D. E., and M. Ewing, The Louisville ridge—A possible extension of the Eltanin fracture zone, in *Antarctic Oceanology I, Antarctic Res. Ser.*, vol. 15, edited by J. L. Reid, pp. 223–228, AGU, Washington, D. C., 1971.

Hayes, D. E., and J. Ringis, Early opening of the central Tasman Sea, paper presented at the International Symposium on the Oceanography of the South Pacific, UNESCO, Wellington, New Zealand, February 1972.

Hayes, D. E., M. Talwani, and D. Christoffel, The Macquarie ridge, in *Antarctic Geology and Geophysics*, edited by R. J. Adie, Universitets forlaget, Oslo, in press, 1972.

Houtz, R., J. Ewing, and R. Embly, Profiler data from the Macquarie ridge area, in *Antarctic Oceanology I, Antarctic Res. Ser.*, vol. 15, edited by J. L. Reid, pp. 239–245, AGU, Washington, D. C., 1971.

Houtz, R. E., and R. G. Markl, Seismic profiler data between Antarctica and Australia, in *Antarctic Oceanology II: The Australian–New Zealand Sector, Antarctic Res. Ser.*, vol. 19, edited by D. E. Hayes, pp. 147–164, AGU, Washington, D. C., 1972.

Jones, J. G., Australia's Cenozoic drift, *Nature, 230*, 237–239, 1971.

Karig, D. E., Ridges and basins of the Tonga-Kermadec island arc system, *J. Geophys. Res., 75*(2), 239–254, 1970.

Le Pichon, X., Sea-floor spreading and continental drift, *J. Geophys. Res., 73*(12), 3661–3697, 1968.

Matthews, D. J., *Tables of the Velocity of Sound in Pure Water and Sea Water for Use in Echo-Sounding and Sound-Ranging*, 2nd ed., 52 pp., H. M. Stationery Office, London, 1939.

McKenzie, D. P., and W. J. Morgan, Evolution of triple junctions, *Nature, 224*, 125–133, 1969.

New Zealand Department of Scientific and Industrial Research, Geological map of New Zealand, Wellington, 1958.

Reilly, W. I., Gravity map of New Zealand, scale 1:4,000,000, *Isostatic Anomalies*, 1st ed., New Zealand Department of Scientific and Industrial Research, Wellington, 1965.

Ringis, J., Magnetic lineations in the Tasman, paper presented at International Conference of the Earth and the Ocean, Sydney, Australia, January 1970.

Smith, W. D., Earthquakes at shallow and intermediate depths in Fiordland, New Zealand, *J. Geophys. Res., 76*(20), 4901–4907, 1971.

Suggate, R. P., The Alpine fault, *Trans. Roy. Soc. N. Z. Geol., 2*(7), 105–129, 1963.

Summerhayes, C. P., Marine geology of the New Zealand subantarctic sea floor, *N. Z. Dep. Sci. Ind. Res. Bull., 190,* 92 pp., 1969.

Summerhayes, C. P., Note on Macquarie ridge and the Tonga-Kermadec complex–Are they parts of the mid-ocean ridge system?, *N. Z. J. Sci., 10,* 808–812, 1967.

Sykes, L. R., Seismicity of the Indian Ocean and a possible nascent island arc between Ceylon and Australia, *J. Geophys. Res., 75*(26), 5041–5055, 1970.

Sykes, L., Mechanism of earthquakes and nature of faulting on the mid-oceanic ridges, *J. Geophys. Res., 72*(8), 2131–2153, 1967.

van der Linden, W. J. M., and D. E. Hayes, Resolution bathymetry scale 1:1,000,000, Oceanic Series, N. Z. Oceanogr. Inst., Wellington, in press, 1972.

Varne, R., R. D. Gee, and P. C. Quilty, Macquarie Island and the cause of oceanic linear magnetic anomalies, *Science, 166*(3902), 230–232, 1969.

Weissel, J. K., and D. E. Hayes, Magnetic anomalies in the southeast Indian Ocean, in *Antarctic Oceanology II: The Australian–New Zealand Sector, Antarctic Res. Ser.,* vol. 19, edited by D. E. Hayes, pp. 165–196, AGU, Washington, D. C., 1972.

Wellman, H. W., Age of the Alpine fault, New Zealand, in the Proceedings of Section 4 of the Report of the 22nd International Geological Congress, India, *Rock Deformation and Tectonics,* edited by R. K. Sundaram, 22nd International Geological Congress, New Delhi, 1964.

Wellman, P., and A. Cooper, Potassium-argon age of some New Zealand lamprophyre dikes near the Alpine fault, *N. Z. J. Geol. Geophys., 14*(2), 341–350, 1971.

Williamson, P., and M. Rubenach, Preliminary report on geophysical studies on Macquarie Island, in *Antarctic Oceanology II: The Australian–New Zealand Sector, Antarctic Res. Ser.,* vol. 19, edited by D. E. Hayes, pp. 234–249, AGU, Washington, D. C., 1972.

Wilson, J. T., A new class of faults and their bearing on continental drift, *Nature, 207*(4995), *343–347,* 1965.

MACQUARIE RIDGE–NEW ZEALAND ALPINE FAULT TRANSITION

D. A. Christoffel

Physics Department, Victoria University of Wellington, Wellington, New Zealand

W. J. M. van der Linden

New Zealand Oceanographic Institute, Department of Scientific and Industrial Research Wellington, New Zealand

Abstract. In the Fiordland-Puysegur trench region, there are discrepancies between observation and the present concepts of plate movements. The southeast dipping Benioff zone strikes parallel to the transform Alpine fault and therefore appears to be in the wrong direction. To overcome this apparent difficulty, we propose an inverted plough model for the subduction zone. In our model, the Benioff zone is broken into relatively narrow sections by a number of transform faults. Each section rotates as it sinks. The leading edges incline toward the vertical and run in the transform direction. The depth of the earthquakes is related to the width of the plough furrow.

Recently, an analysis of the Macquarie ridge based on calculations of the relative motions between the Pacific-antarctic and Indian-antarctic plates was presented [*Christoffel*, 1971]. The analysis showed that the entire Macquarie ridge and the New Zealand Alpine fault very nearly, but not quite, formed a single continuous transform fault around the computed pole of relative motion between the Indian and Pacific plates (model 1) at 58°S, 169°W. An alternative pole was then calculated by assuming the New Zealand Alpine fault to be a pure transform fault (Figure 1, model 2). This pole at 53°S, 180°E seemed to be in best agreement with the observed displacements. Bathymetric, seismic, and magnetic data from the Macquarie ridge presented by *Houtz et al.* [1971] and *Hayes et al.* [1972] are compatible with the model.

We intend to present newly available details for the Alpine fault and the New Zealand Fiordland and Puysegur trench regions at the northern end of the Macquarie ridge to see whether these details also agree with the model. For the region, the model predicts subduction along the Puysegur trench during the last 10 m.y. As seen in Figure 1, a section of crust has to be consumed at the site of the Puysegur trench to arrive at the present position of the plate boundary.

THE ALPINE FAULT

The New Zealand Alpine fault as we know it today [*Suggate*, 1971] was first recognized 30 years ago as a continuous South Island feature measuring 500 km long from Lake Rotoroa to Milford Sound [*Wellman and Willett*, 1942]. Scarcity of data and conflicting interpretations make it difficult to determine unambiguously the statistics of the fault.

Notwithstanding this observation, we can safely say that the Alpine fault is predominantly transcurrent, has a dextral displacement of 500 km, and is most probably post-Eocene in age [*Mason*, 1961; *Wellman*, 1964; *Christoffel*, 1971].

The absence of a large gravity anomaly across the main South Island fault trace [*Hatherton and Hunt*, 1968] may at first seem conspicuous when taking into account an uplift of 10–20 km and a strike slip motion of several hundred kilometers. The lack of a gravity signature (also true for the San Andreas fault) only means that the displaced rocks show no appreciable density contrast.

The central South Island part of the fault traverses an area that at present has low seismic activity [*Eiby*, 1971]. The extension of the Alpine fault northward through the Marlborough province and North Island is ambiguous. Perhaps we can best follow *Kupfer* [1964] and H. W. Wellman (personal communication, 1970) and combine all the New Zealand active long faults (which according to H. W. Wellman happen to be all dextral faults) into an Alpine fault. This action essentially combines the Alpine fault,

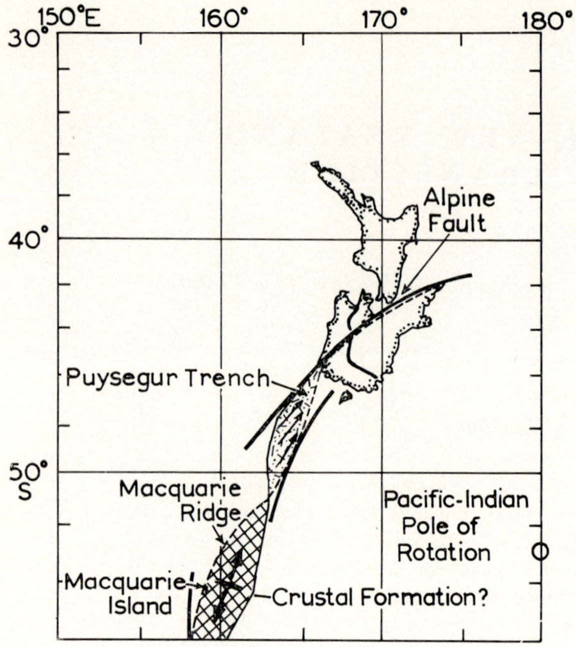

Fig. 1. Computed motions along the north Macquarie ridge and the New Zealand Alpine fault. The broken line shows the present Pacific-Indian plate boundary. The thin solid line and the outline of New Zealand show the inferred position 10 m.y. ago (after *Christoffel* [1971]).

sensu stricto, with the Marlborough faults of South Island and the East Coast faults of North Island into one zone that is 1100 km long (Figure 2). Note, however, that the age of the North Island faults is most likely Upper Jurassic [*Wellman*, 1955]. Since the strike of the North Island faults is influenced by the North Island Benioff zone, we shall consider the South Island section only.

FIORDLAND AND PUYSEGUR TRENCH REGION

New bathymetric and geophysical data in the eastern Tasman Sea west and south of Fiordland have been obtained during USNS *Eltanin* cruise 44 in 1970. We shall discuss these and other results in turn.

Bathymetry

The bathymetry is given in Figure 3 that also shows the *Eltanin* cruise 44 track where continuous gravity, seismic, reflection, and magnetic data were obtained.

From an exceptionally wide continental shelf platform in the Cook Strait region where the Lord Howe rise joins New Zealand, the shelf width decreases rapidly southward. The western Fiordland coast is the top of the continental slope. In the south, the Puysegur Bank is a fingerlike north-south elongated appendix of New Zealand. The bank is not in line with the Macquarie ridge and is separated from the continental shelf to the east by the Solander trough. South of 47°S, the Macquarie ridge is clearly defined as a singular feature with steep slopes on both sides.

The continental slope off Fiordland is cut by a number of deep, incised valleys and canyons. Presumably, these are all erosional features associated with the Fiordland drainage pattern, although primary tectonic control cannot be excluded. At the foot of the slope is a wedge-shaped zone of elongated relatively sharply crested ridges and flat-floored basins. The alignment of the highs and deeps in this zone is between 44° and 52°E of north, and the zone widens toward the southwest to several hundred kilometers. To the northwest, this borderland is bounded by the Tasman abyssal plain; to the southeast, it is bounded by the 400-km-long Puysegur trench at the western foot of the north Macquarie ridge. In the northeast at about 44°S, the zone merges with the New Zealand continental slope. South of 49°S, the lack of bathymetric information in this area prevents us from defining a southern boundary.

Seismic Reflection and Magnetic Profiles

The Fiordland continental borderland, crossed by the profiles of Figures 4 and 5, shows sedimentary cover varying in thickness and increasing outward. In the abyssal plain, sediment thickness is at least 2 km corresponding to a travel time of about 2 sec. Especially in the borderland, the profiles show rough bottom topography and rough acoustic basement.

There is no magnetic expression over the Resolution ridge, which therefore must be of sedimentary, metamorphic, or perhaps granitic composition. The pronounced ridge in the middle of profile 16 also is probably nonmagnetic since the large positive anomaly is displaced westward by 20 km. Over both the borderland and abyssal plain, the wave length of the magnetic anomalies is of the order of 20 km, and with only one exception the amplitude does not exceed 200 γ. These calculations indicate a deep-seated magnetic basement. The most striking feature is the quiet magnetic zone approximately 90 km wide in the south. This feature suggests an even further deepening magnetic basement approaching the Puyse-

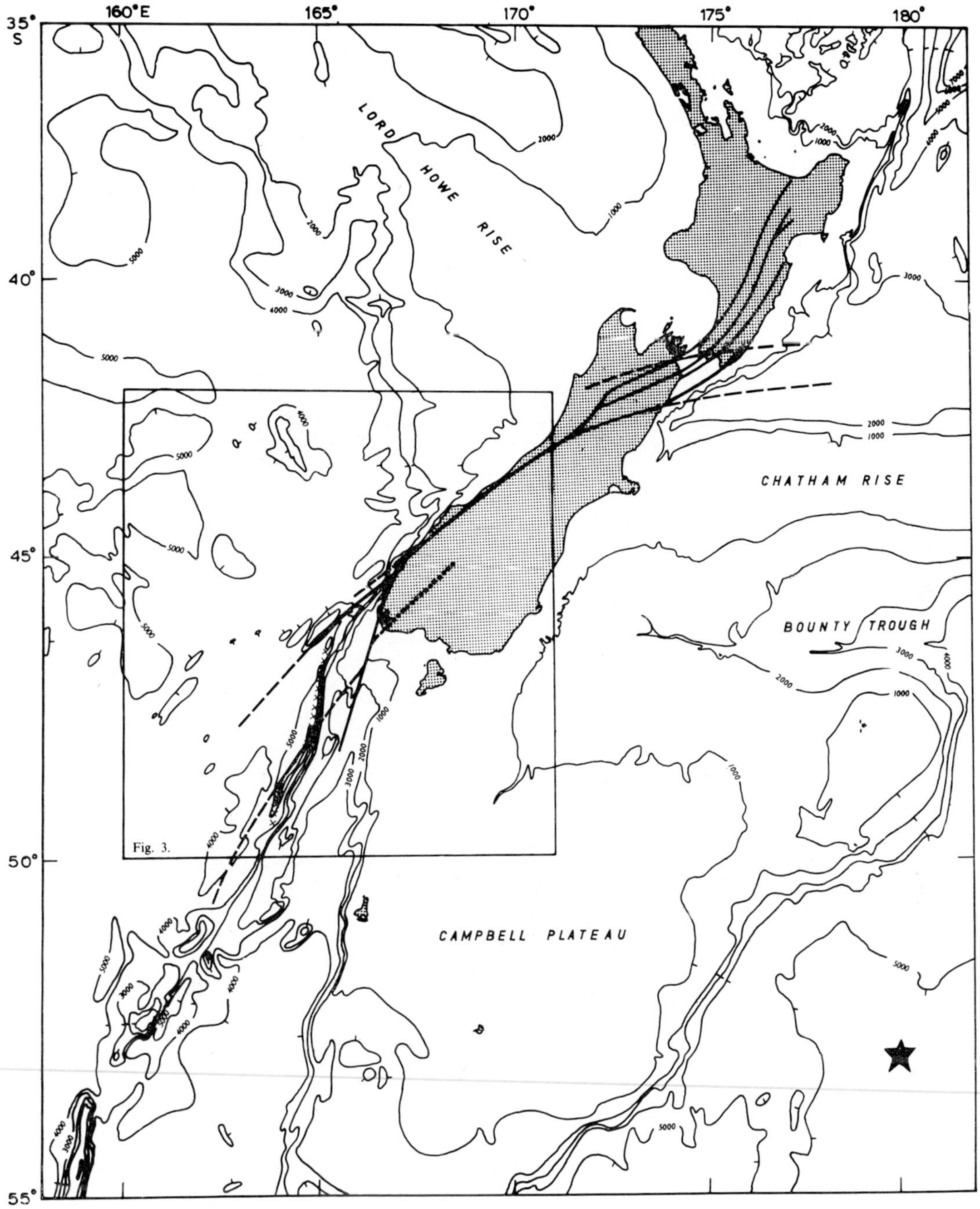

Fig. 2. The Campbell plateau and the southeast Tasman basin showing the Alpine fault zone. Depth contours are at intervals of 1000 meters. The broken lines are circles of latitude around the starred Pacific-Indian pole of Figure 1.

gur trench. The above observations suggest that the ocean floor in the area is at least mid-Tertiary in age.

Seismicity

Recent analyses of earthquakes in the Fiordland region established that, in addition to shallow shocks, intermediate depth earthquakes also occur [*Hamilton and Evison,* 1967; *Smith,* 1971] and extend down to 140 km. According to Smith, the envelope of all shallow and intermediate quakes in the Fiordland region may be a Benioff zone dipping southeastward at an angle of about 80° to the horizontal. Smith

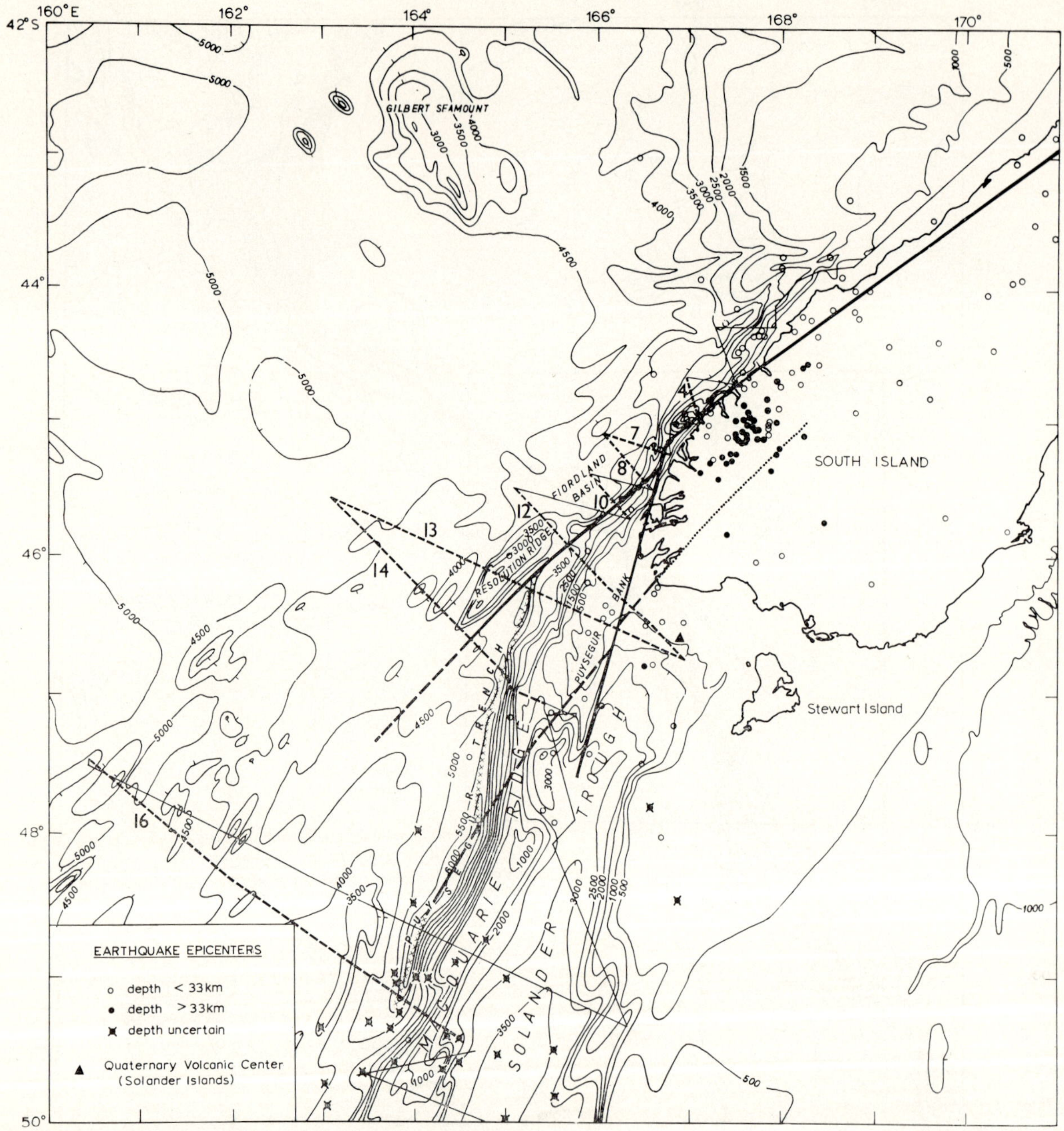

Fig. 3. Detail of the southern end of the Alpine fault and Puysegur trench showing earthquake epicenters, bathymetry (after *van der Linden and Hayes* [1972]), and the *Eltanin* 44 track.

remarks that it may be significant that the Alpine fault lies close to the surface projection of the Benioff zone. He further suggests the possibility that the seismic zone may be a surface rather than a volume.

Gravity

The *Eltanin* cruise 44 survey revealed an elongated negative isostatic gravity anomaly, a 'Meinesz' zone, with a minimum of −150 mgals parallel to the coast (D. Woodward, personal communication, 1971). This anomaly and the anomaly of +150 mgals measured over Fiordland [*Reilly*, 1965] appear to be a typical gravity anomaly pair as observed over plate margins, island arc, and trench or continental margin and trench. There are, however, some differences. The gravity anomalies are very localized, and although

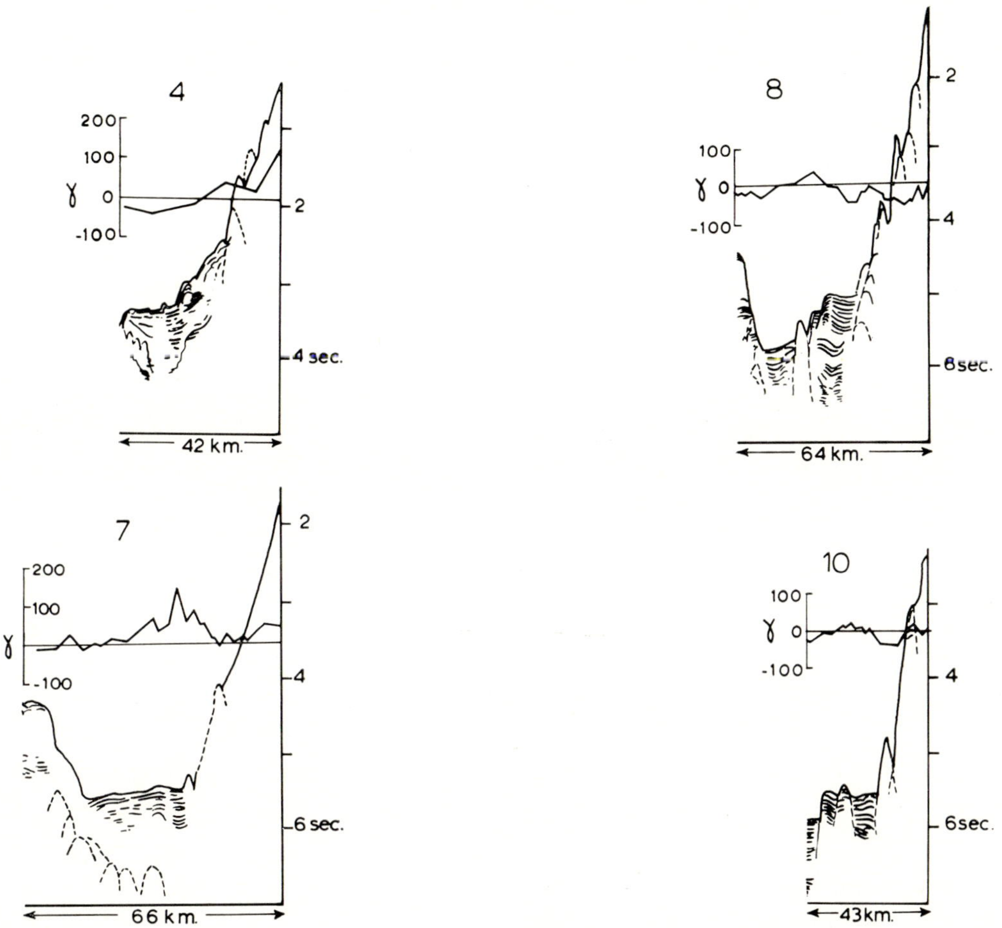

Fig. 4. Seismic and magnetic profiles for selected *Eltanin* cruise 44 tracks along the Fiordland coast.

elongated parallel to the coast, they are restricted between Milford Sound and Foveaux Strait. The gravity gradient is unusually steep (12.5 mgals/km), and this steep gradient suggests a near-surface cause for the anomaly.

INTERPRETATION

The data presented on the Fiordland-Puysegur trench region show it to be an active plate margin where the Indian plate subducts the Pacific plate. We have a very narrow Benioff zone at least 140 km deep; we have a trench; there is a characteristic gravity anomaly; and Quaternary volcanicity occurs behind the trench on the landward side (the Solander Islands).

The northeast strike of the earthquake plane (i.e., in the transform direction of our pole of relative motion, a direction that should therefore preclude subduction) appears to contradict the established plate tectonics model. We propose a model that overcomes this apparent difficulty without shifting our pole. In our model, the subduction zone follows a surface that is shaped like an inverted plough share. The process envisioned is similar to ploughing a furrow except that the furrow is ploughed down instead of up. The model (Figure 6) fits the observations in remarkable detail. To clearly show the mechanism, we have drawn the model by looking from the north.

The Puysegur trench north of 48°S is the surface expression of the subduction zone, whereas the movement along the Alpine fault and central Puysegur trench is transcurrent. As can be seen, the deeper earthquakes lie to the northeast (Figure 3), and although they dip at 80° [*Smith,* 1971], these deeper earthquakes are as much as 500 km away from the trench. The leading edge of the Indian plate is thus most inclined toward the vertical, and the deepest earthquakes occur in this region. The maximum

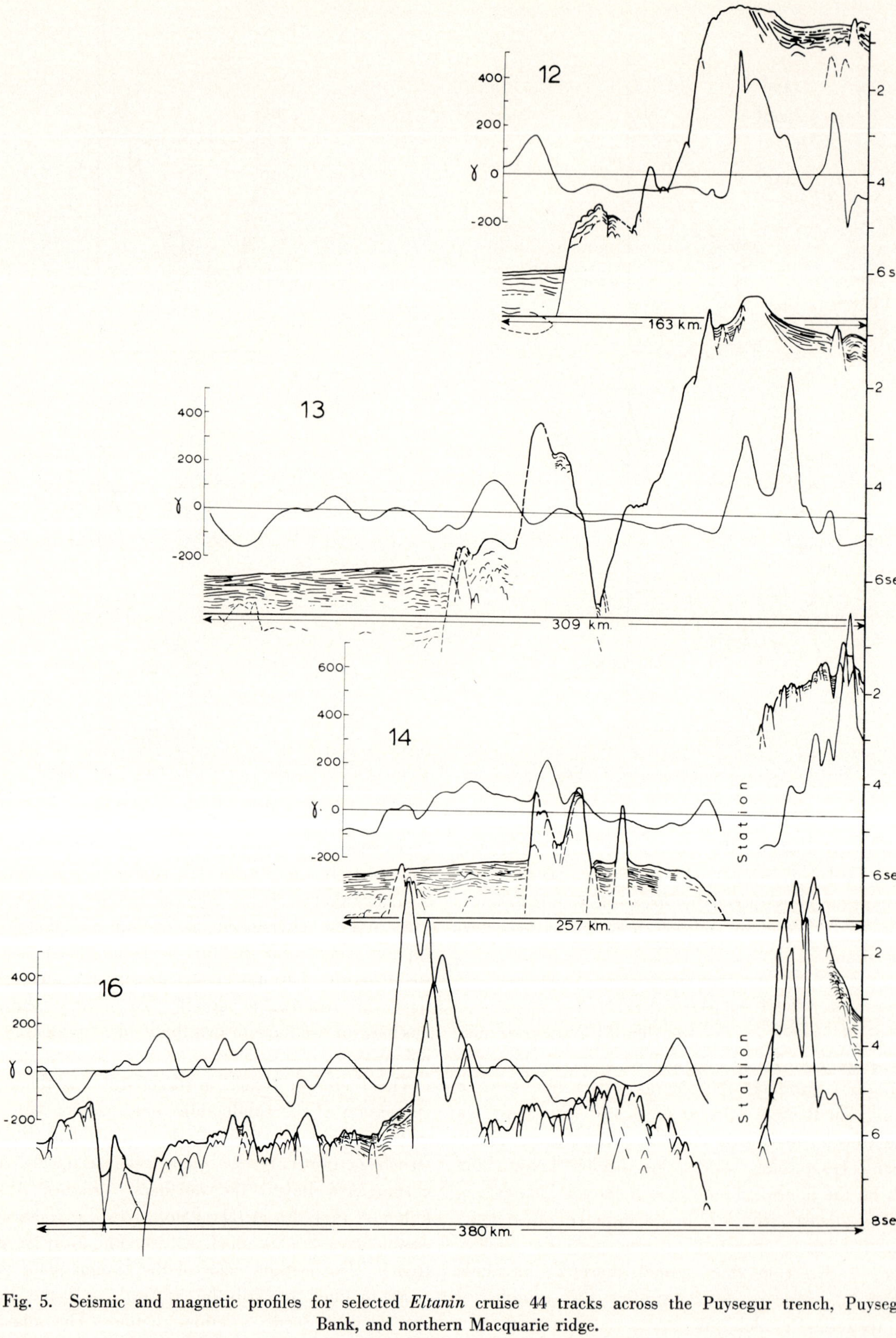

Fig. 5. Seismic and magnetic profiles for selected *Eltanin* cruise 44 tracks across the Puysegur trench, Puysegur Bank, and northern Macquarie ridge.

depth of the earthquakes is determined not only by the rate and duration of the relative plate motions but also by the distance between the bounding fracture zones. For the south Fiordland zone, the distance between the Alpine fault trace and the central Puysegur trench (see Figure 3) is 120 km, whereas the maximum earthquake depth so far reliably determined is 140 km.

Figure 6 represents only one segment of the Indian plate as it moves latitudinally around the pole of relative motion with respect to the Pacific plate. The south Puysegur trench is another surface expression of a similar sliver diving and rotating in a northeasterly direction under the encroaching Pacific. The north and south Puysegur trench subduction segments are thus separated en echelon by a transform fault through the central part of the trench. Similar large and small slivers may be expected in other places along the Indian-Pacific margin, e.g., to the north under New Zealand and along the Tonga-Kermadec arc. This enlarged model thus adequately explains the separation of zones of low and high seismicity [*Eiby*, 1971].

So far, we have concentrated our attention on the Indian plate. Figure 7, a west-east cross section over

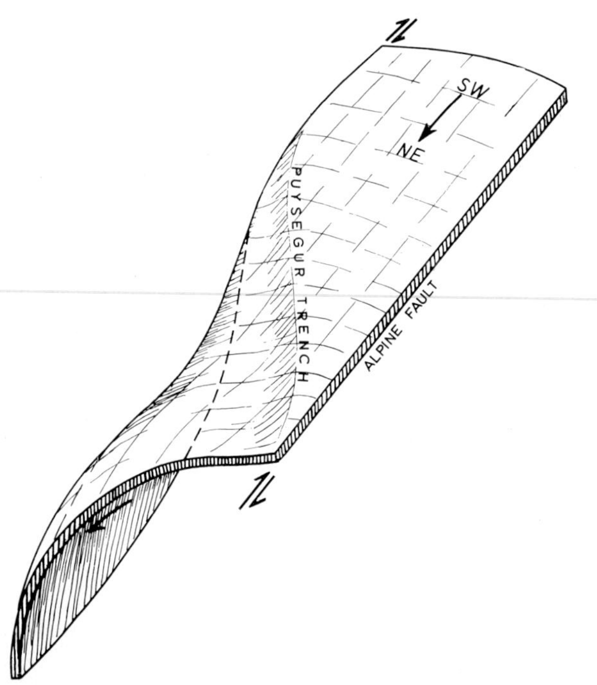

Fig. 6. Subduction model shaped like a plough share for the Puysegur-Fiordland region. Note that the model is viewed from the north.

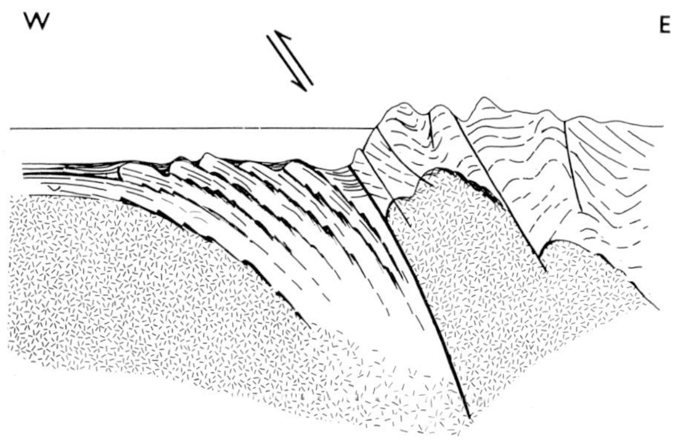

Fig. 7. West-east section across the Indian-Pacific margin in the Fiordland region.

the Fiordland area, shows that subduction of the Indian plate and overthrusting of the Pacific plate are two aspects of one and the same process. On the basis of our observations and as confirmed by the gravity anomaly pattern, the magnetic basement is shown to deepen as one approaches the plate boundary from the west. On the other side, the opposite effect can be observed and thus produces the gravity high in western Fiordland. Clearly from our model (Figure 6), this gravity high must be restricted in a northeast-southwest direction.

Both the ridge and valley topography offshore and the breakup of the Fiordland ranges are shown in Figure 7 as largely caused by reversed faulting. As must be realized, this figure is only a schematic representation, and crustal shortening can be produced by faulting and/or folding. Deciding from the profiles (Figure 5, profiles 12–16) which process prevails is impossible.

CONCLUSION

A subduction zone has been identified in the Fiordland region by the occurrence of a zone of intermediate earthquakes dipping 80° to the southeast. According to *Christoffel*'s [1971] earlier model, this Benioff zone strikes in the transcurrent direction that seems to be in direct conflict with the model.

Rather than moving the relative pole of motion over a long distance, we here propose that the subduction zone rotates as it descends. We thus arrive at a new concept for the Fiordland region in which the subduction zone is shaped like an inverted plough share. This concept establishes the Puysegur trench as the surface expression of two subduction ploughs separated by a transform fault.

Although our model agrees with earlier plate motion concepts (e.g., *Oliver* [1970]), it explains how the edges of the encroaching plates can be broken up. Our model satisfactorily explains the separation of the plates into zones of high and low seismic activity along the margin. It also explains differences in the depth of the Benioff zones along a common margin and shows this depth difference to be largely influenced by the width of the plough furrows. Observation of earthquake epicenter distributions indicates that our model may have general application to other active margins.

Acknowledgments. D. E. Hayes and R. Houtz kindly made available *Eltanin* cruise 44 magnetic and seismic profiler data. Participation by D. A. Christoffel in the *Eltanin* cruise 44 is gratefully acknowledged.

REFERENCES

Christoffel, D. A., Motion of the New Zealand Alpine fault deduced from the pattern of sea floor spreading, in *Recent Crustal Movements*, edited by B. W. Collins and R. Fraser, *Roy. Soc. N. Z. Bull.*, *9*, 25–30, 1971.

Eiby, G. A., Seismic regions of New Zealand, in *Recent Crustal Movements*, edited by B. W. Collins and R. Fraser, *Roy. Soc. N. Z. Bull.*, *9*, 161–165, 1971.

Hamilton, R. M., and F. F. Evison, Earthquakes at intermediate depths in south-west New Zealand, *N. Z. J. Geol. Geophys.*, *10*(6), 1319–1329, 1967.

Hatherton, T., and T. M. Hunt, Gravity profiles across the Alpine fault, New Zealand, *J. Geophys. Res.*, *73*(16), 5343–5352, 1968.

Hayes, D. E., M. Talwani, and D. A. Christoffel, Macquarie ridge complex, in *Antarctic Geology and Geophysics*, edited by R. J. Adie, Universitets forlaget, Oslo, in press, 1972.

Houtz, R., J. Ewing, and R. Embley, Profiler data from the Macquarie Ridge area, in *Antarctic Oceanology I*, *Antarctic Res. Ser.*, vol. 15, edited by J. L. Reid, pp. 239–245, AGU, Washington, D. C., 1971.

Kupfer, D. H., Width of the Alpine fault zone, New Zealand, *N. Z. J. Geol. Geophys.*, *7*(4), 685–701, 1964.

Mason, B. H., Potassium-argon ages of metamorphic rocks and granites from Westland, New Zealand, *N. Z. J. Geol. Geophys.*, *4*(4), 352–356, 1961.

Oliver, J., Structure and evolution of the mobile seismic belts, *Phys. Earth Planet. Interiors*, *2*, 350–362, 1970.

Reilly, W. I., Gravity map of New Zealand, scale 1:4,000,000, *Isostatic and Bouguer Anomalies*, 1st ed., N. Z. Dep. Sci. and Ind. Res., Wellington, 1965.

Smith, W. D., Earthquakes at shallow and intermediate depths in Fiordland, New Zealand, *J. Geophys. Res.*, *76*(20), 4901–4907, 1971.

Suggate, R. P., The Alpine fault, *Trans. Roy. Soc. N. Z. Geol.*, *2*(7), 105–129, 1971.

van der Linden, W. J. M., and D. E. Hayes, Resolution bathymetry, scale 1:1,000,000, Oceanic Series, N. Z. Oceanogr. Inst., Wellington, in press, 1972.

Wellman, H. W., New Zealand quaternary tectonics, *Geol. Rundsch.*, *43*(1), 248–257, 1955.

Wellman, H. W., Age of the Alpine fault, New Zealand, in the Proceedings of Section 4 of the Report of the 22nd International Geological Congress, India, *Rock Deformation and Tectonics*, pp. 148–162, 1964.

Wellman, H. W., and R. W. Willett, The geology of the west coast from Abut Head to Milford Sound, 1, *Trans. Roy. Soc. N. Z.*, *71*(4), 282–306, 1942.

PRELIMINARY REPORT ON GEOPHYSICAL STUDIES ON MACQUARIE ISLAND

P. Williamson

Department of Geology, University of New South Wales, New South Wales, Australia

M. J. Rubenach

Department of Geology, University of Tasmania, Tasmania, Australia

Abstract. A simple Bouguer gravity map of Macquarie Island shows a gradient of approximately +5 mgals/km in a westerly direction. No locally significant Bouguer anomalies were observed within the confines of the island. The amplitudes and wave lengths of magnetic anomalies on Macquarie Island are similar to those observed near the ocean floor from deep-tow magnetometer studies. The amplitude of anomalies and the noise level of the magnetic profiles are related to the underlying rock types and to their degree of metamorphism. Relatively high noise levels are observed on magnetic profiles over volcanic and ultramafic rocks. Our interpretation relates the magnetic noise to observed variations in low grade metamorphism and to serpentinization. The dyke swarms, which are mainly restricted to the northern part of the island and are metamorphosed to greenschist and amphibolite facies, are associated with relatively low noise level magnetic signatures. Correlations between magnetic profiles in the central and southern parts of the island indicate a magnetic fabric striking northwest to north-northwest, and this trend coincides with the dominant regional strike of geologic features. Results from this study do not support the hypothesis that Macquarie Island was formed by crustal extension at the Macquarie ridge.

Macquarie Island is located on the Macquarie ridge, which extends from New Zealand to the Southeast Indian ridge. The island morphology consists primarily of a narrow coastal plain and a steep-sided plateau with an average elevation of about 350 meters. The bathymetry indicates a possible structural extension of the island feature along the crest of the ridge. Macquarie Island is believed to be an uplifted block of sea floor material [*Varne et al.*, 1969]. Serpentinized harzburgites and gabbros on the island have been metamorphosed to amphibolite and greenschist facies grade and thus may have formed in deeper levels within the ocean crust; the relatively unmetamorphosed pillow lavas and volcanic breccia also present may represent initially higher levels in the oceanic crust.

An age of early to middle Miocene has been determined for coccoliths in globigerina ooze (P. G. Quilty and J. A. Wilcoxon, personal communication, 1972). The samples were obtained from oozes occurring in interstices in pillow lavas in the extreme north of the island. The island thus offers a unique opportunity for direct geophysical studies on outcropping sea floor rocks and for comparison of these results with marine geophysical studies.

This paper presents the preliminary results of geophysical studies carried out by P. Williamson in the summer of 1970–1971. The field work consisted of a regional gravity survey and reconnaissance magnetic traverses. Oriented samples of paleomagnetic work were also collected, but these results are not yet available. In this paper, the observed geophysical parameters are related to the geology and petrology of the rocks of Macquarie Island. These rocks are described in more detail by *Varne and Rubenach* [this volume].

GRAVITY

The gravity survey consists of measurements made at 59 localities on the island by using a Worden gravity meter. The absolute gravity value was obtained by a tie-in to the gravity stations on the Australian continent by using a La Coste-Romberg gravity meter. The locations of the gravity stations were obtained by reference to mapped geographical features. The errors in locations are considered to be less than 100 meters. The station altitudes were obtained by comparison

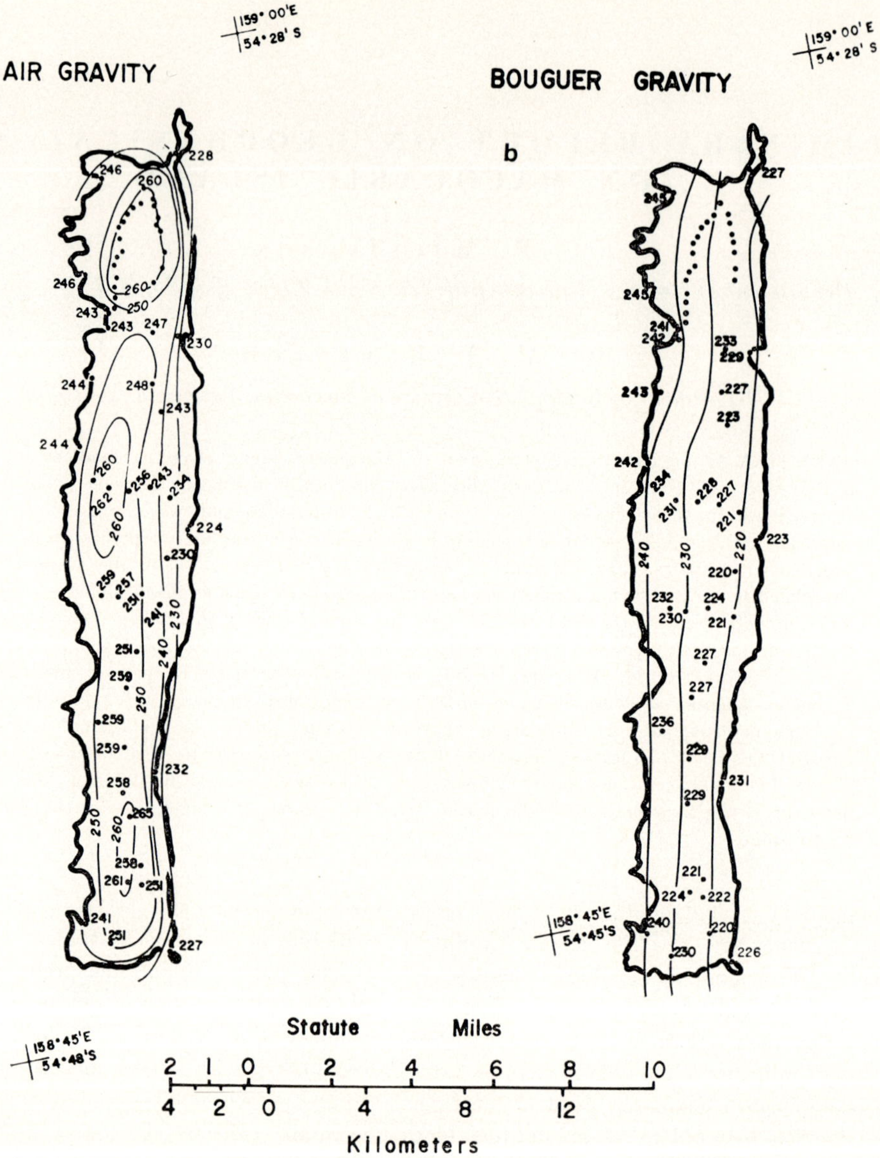

Fig. 1. (*a*) Free air gravity map of Macquarie Island. (*b*) Simple Bouguer gravity map of Macquarie Island.

between the pressure readings on a portable microbarometer and those on a base microbarometer and are considered correct to 10 meters. The errors correspond to an accuracy of 3 mgals in the free air anomaly.

A simple Bouguer correction was applied to the free air values; the free air gravity and Bouguer gravity results are shown in Figure 1. The gravitational effects of the topography are only approximate in the simple Bouguer correction. A more rigorous topographic correction using the method described by *St. John and Green* [1967] is in progress.

The free air gravity results are consistent with other free air anomalies obtained over the Macquarie ridge [*Hayes and Talwani*, this volume] and reflect the island topography. In the northern part of the island, the plateau is cut by a deep valley that runs north-northwest and that coincides with a major fault and appears as a minimum in the free air gravity.

The density used in the Bouguer correction was

2.83 g/cm^3, which is the average of the measured densities of the pillow lavas from the island. Bouguer anomalies show a gradient of +5 mgals/km in a westerly direction. Preliminary calculations indicate that this gradient could be largely the result of the mass deficiency off the east coast caused by the deep bathymetric trench (2600 fm) in that area [*Hayes and Talwani*, this volume]. The smooth westerly gradient is partly obscured in the Bouguer gravity map by an apparent discontinuity between the coastal and plateau station values. This discontinuity is in part the result of applying a simple Bouguer correction. No significant local Bouguer anomalies are observed within the confines of the island, and, although some of the rock types in the north of the island probably represent petrologies characteristic of deeper levels within the oceanic crust than those rocks elsewhere on the island, only a small increase in the Bouguer gravity value is observed.

MAGNETICS

The magnetic data presented are the results of a number of vertical field magnetic traverses that used a Sharpe flux-gate magnetometer. The location of these traverses is shown in Figure 2. Readings were obtained at intervals of about 30 meters. The data have been corrected for diurnal variations and instrumental drift, and the zero level on the traverses corresponds to the theoretical vertical field strength [*IAGA Commission 2, Working Group 4*, 1969].

A preliminary examination of these results yields two useful kinds of information. First, a comparison between the rock types traversed and the character of the magnetic profiles can be made. Second, the correlation of anomalies between traverses yields possible insights into the structure and possible origin of the island.

Comparisons of Magnetic Profiles over Different Rock Types

The magnetic properties of the rocks forming oceanic crust depend on a number of factors that have been discussed in detail in *Watkins and Paster* [1970]. The magnetic properties are a function of rock type and in particular the percentage of magnetite or titanomagnetite in the rock. The properties also depend on the cooling history, and differences are expected in the magnetic profiles over pillow lavas, volcanic breccias, and massive lavas even if they are of the same chemical composition. Alteration and metamorphism generally have a pronounced effect on the magnetic properties of a rock [*Watkins and Paster*, 1970]. Metamorphism of basic rocks is commonly magnetically destructive, whereas the magnetic effects of serpentinization are variable. The character of the magnetic profiles on Macquarie Island could then be expected to depend on such factors as the composition of the rocks, the physical characteristics of the rock types, the amount of metamorphism, and the structural relations between rock types (e.g., faulting, relative tilting of blocks, igneous intrusions; see *Varne and Rubenach* [this volume]), as well as variations in the ambient magnetic field of the earth.

The volcanic rocks of Macquarie Island consist of pillow lavas, volcanic breccias, and some massive lavas. They are hypersthene normative and are composed of augite, plagioclase, and titanomagnetite. This suite of rocks is described in *Varne and Rubenach* [this volume]; the rocks are believed to be ocean floor basalts. The average noise level over the volcanics is about 500 γ. In this text, noise level refers to the average difference between the filtered and unfiltered values. This noise is superimposed on longer wave length anomalies with an average amplitude of about 1500 γ.

No differences are obvious in the character of the magnetic profiles over pillow lavas and volcanic breccias or in the character of profiles over aphyric and feldsparphyric lavas. This situation is true even where distinctive units of lavas can be traced over several kilometers of strike length.

The volcanics show varying degrees of metamorphism [*Varne and Rubenach*, this volume], and these variations would certainly influence the magnetic properties of the rocks. Some samples show only slight alteration, but in the majority the metamorphism is indicated by the replacement of interstitial glass by clay minerals. Some of the volcanics have been metamorphosed to greenschist facies, and the original titanomagnetite laths have been recrystallized to granular iron oxides and sphene. Although the detailed relationship between the magnetic data and petrology remains to be ascertained, it seems likely that the magnetic characteristics observed over the lavas reflect the degree of metamorphism. Local variations in the degree of alteration could account for much of the noise in the magnetic profiles.

The profiles over the dyke swarms in the north of the island and the pillow lavas on part of the traverses LM and LN (Figure 2) show a noise level of about 50 γ superimposed on longer wave length anomalies with amplitudes of less than 200 γ. The

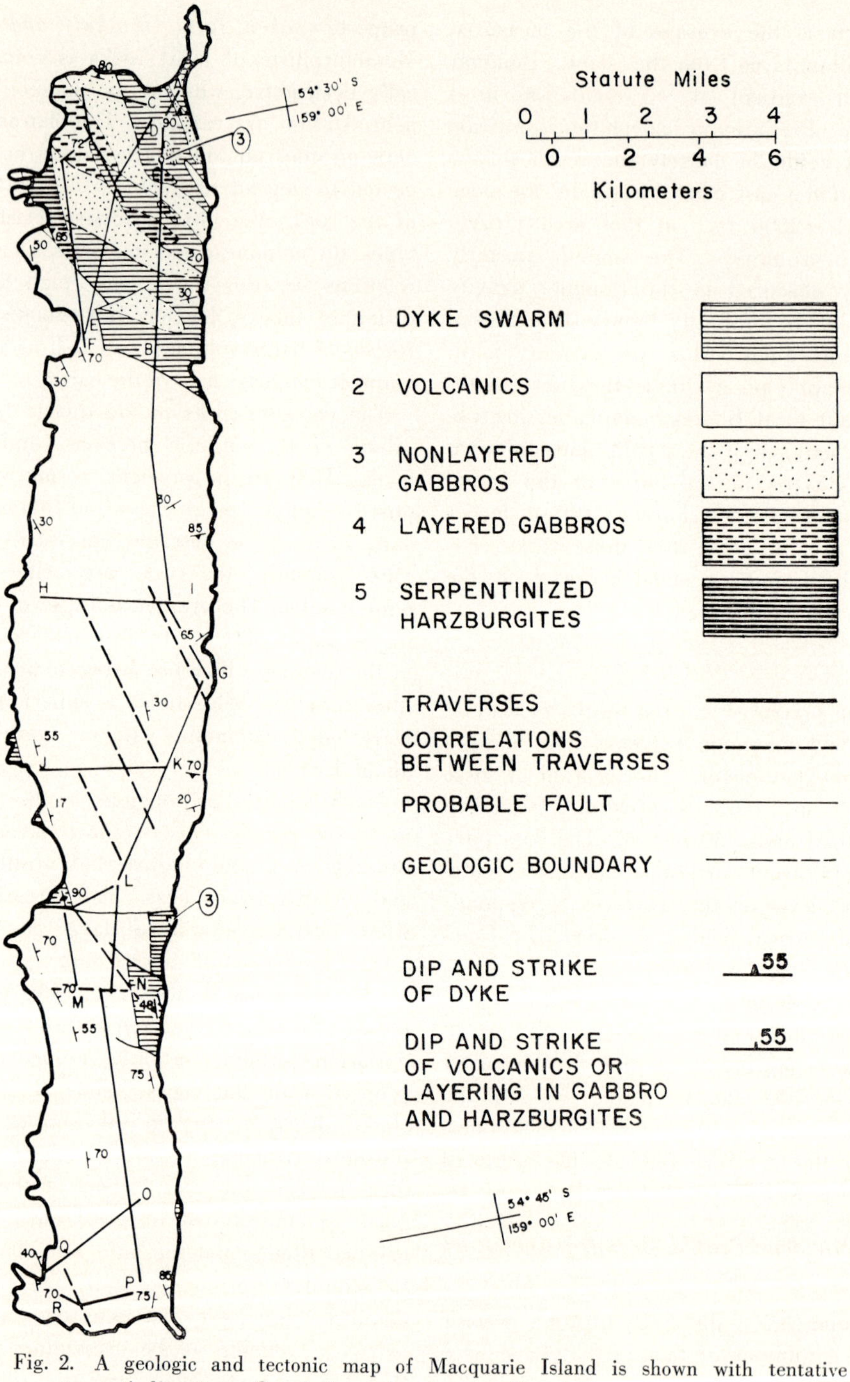

Fig. 2. A geologic and tectonic map of Macquarie Island is shown with tentative magnetic lineations. Correlations taken from profiles of Figure 3 are also shown.

dolerites from dyke swarms are metamorphosed, mainly to greenschist and amphibolite facies [*Varne and Rubenach,* this volume]. A greater degree of metamorphism is consistent with the dyke swarms being formed at a deeper level in the oceanic crust than the pillow lavas and with the proposed models for the formation of oceanic crust at a mid-ocean ridge [*Cann,* 1968]. The original titanomagnetite laths have been recrystallized during metamorphism of the dykes, and this alteration would presumably have to have been uniformly magnetically destructive to account for the very low noise level of the anomalies over the dyke swarms. This hypothesis will be checked by laboratory studies on the magnetic properties of the dolerites.

The noise level over the serpentinized harzburgites is about 300 γ superimposed on longer wave length anomalies having an average amplitude of 2000 γ. The character of the profiles over the serpentinized plagioclase dunites, which form the basal zone of a layered gabbro complex in the northwestern corner of the island, is similar to that of the serpentinized harzburgites. Both these rock types contain only accessory chromite as the primary spinel phase. However, abundant fine-grained magnetite has formed during the serpentinization, and the character of the magnetic profiles over the rocks is probably largely a result of this secondary magnetite.

In the layered gabbros, except where plagioclase dunite layers are abundant, the noise in the magnetic profiles is relatively low. This low noise level is probably due to the extremely low spinel content in these rocks. The character of the profiles is in general similar to that over the metamorphosed dyke swarms. The noise level in the gabbros is higher in some localities, but this higher level is possibly due to the presence of relatively unmetamorphosed dolerite dykes and larger xenoliths and screens of serpentinized harzburgites, all of which have been observed.

Correlation of Magnetic Profiles

The large magnetic anomaly obtained over the Macquarie ridge [*Hayes and Talwani,* this volume] is probably due in part to the presence of harzburgite in the ridge [*Watkins and Gunn,* 1970] as well as to the plagioclase dunite. However, large magnetic anomalies are observed at some localities in the center and south of the island without any apparent relationship with the above rock types.

To remove local noise from the magnetic profiles, a three-point convolution filter was used, and it removed wave lengths of less than 60 meters. The filtered data and examples of the unfiltered data are presented in Figure 3. The observed correlations between profiles are shown in Figures 2 and 3. The only correlations indicated are those that seem to be most reliable since no objective assessment of the correlations has as yet been carried out. The main trend in the correlations is north-northwest.

Since measurements are made by standing on the outcrop, the amplitudes of the anomalies are much higher and the wave lengths much shorter than those measured at some distance above the outcrop. Thus the anomaly wave lengths from such a survey cannot be compared directly with those over oceanic rocks taken from survey ships. The most common anomaly wave length on the island is about 400 meters and has an amplitude of about 1500 γ. Anomaly wave length in this text refers to the distance between adjacent magnetic peaks.

Anomalies of similar wave lengths and amplitudes have been observed over the East Pacific rise from deep tow magnetic studies [*Larson and Spiess,* 1969]. Anomaly widths of 500 meters and average amplitudes of 1000 γ were observed there. An anomaly with a width of about 2 km and an amplitude of 2000 γ is correlated between profiles IH and GL. Anomaly widths and amplitudes of this order (anomaly widths of 1–5 km and amplitudes of up to 1500 γ) have also been observed in deep-tow magnetometer studies [*Luyendyk,* 1969].

The trend of the magnetic correlations in the center of the island corresponds to the regional strike (northwest to north-northwest) of the pillow lavas and dykes in this locality. The most easterly magnetic correlation line in the center of the island falls, however, within a zone in which the pillow lavas and intruding tholeiitic dykes strike east-northeast. Prominent north-northwest trending faults have been located on the island, and some magnetic correlations could correspond to faults.

Correlations lines between tracks LM and LN transgress the strike of the lavas in the locality. The northern contact with the magnetically smooth lavas strikes northwest and coincides with a fault indicated by surface geology. The southern contact with this rock type strikes west and may also be associated with faulting.

In the southern part of the island, correlations trending north-northwest have been made, and these lines also correspond to the local strike of the pillow lavas. In the northern part of the island, the trend of the geologic units and the faults is dominantly northwest to north-northwest, and the southern contact of the magnetically flat area in the northeast section of the island also strikes north-northwest. Good correlations were not obtained in the north part of the island in spite of the presence of a pronounced lithologic striping. The geology of this area is quite complex [*Varne et al.,* 1969; *Varne and Rubenach,* this volume], rock types sometimes varying across adjacent traverses.

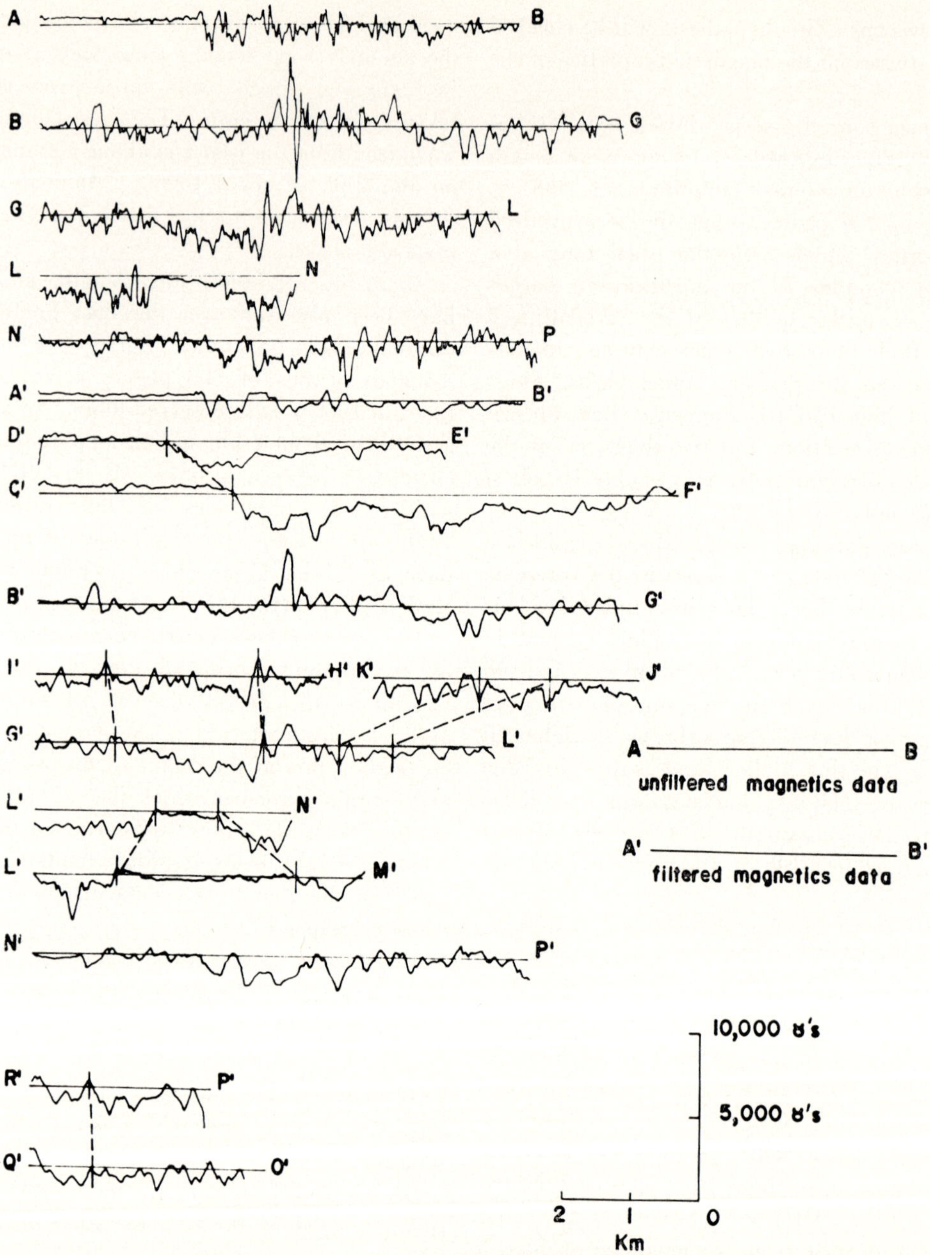

Fig. 3. Magnetic anomaly profiles across Macquarie Island. Profile locations are given in Figure 2.

CONCLUSIONS

The early to middle Miocene age suggested for rocks from Macquarie Island and the observed dominant northwest to north-northwest strike of the geology and magnetics anomaly correlations do not support the hypothesis that Macquarie Island was formed by crustal extension at the Macquarie ridge [*Christoffel*, 1971]. Christoffel implies an age of less than 10 m.y. for the island and a dominant east-northeast strike direction for hypothetical magnetic lineations. Although a strip of rocks on the east coast does have this trend, there is no geologic reason to suspect that these rocks are younger than those in the remainder of the island. Further consideration of the relative motion between the Indian and Pacific plates also indicates that the Macquarie Island region is one where the major motion is a strike slip motion accompanied by minor compression and not extension [*Weissel and Hayes*, this volume].

This report is preliminary, and the conclusions from the geophysical studies are tentative. Further magnetic data are necessary to establish the validity of suggested correlations. Also, additional gravity surveys and paleomagnetic studies may help to solve the problem of the evolution of Macquarie Island and the significance of this area in evaluating the present theories on the nature of sea floor material.

Acknowledgments. Thanks are due to ANARE which hosted the authors when they carried out the above work; to R. Varne, J. Major, and B. Long, who assisted in some phases of the data collecting; to B. Johnson, J. Weissel, and D. Falvey for aid and discussion in putting together this report. The facilities of the University of Tasmania, Macquarie University, and the University of New South Wales were used in processing the data and compiling this report.

REFERENCES

Cann, J. R., Geological processes at mid-ocean ridge crests, *Geophys. J., 15*, 331–341, 1968.

Christoffel, D. A., Motion of the New Zealand Alpine fault deduced from the pattern of sea floor spreading, in Recent Crustal Movements, edited by B. W. Collins and R. Fraser, *Roy. Soc. N. Z. Bull., 9*, 25–30, 1971.

Hayes, D. E., and M. Talwani, Geophysical investigation of the Macquarie ridge complex, in *Antarctic Oceanology II: The Australian–New Zealand Sector, Antarctic Res. Ser.*, vol. 19, edited by D. E. Hayes, pp. 211–234, AGU, Washington, D. C., 1972.

IAGA Commission 2, Working Group 4, International geomagnetic reference field 1965.0, *J. Geophys. Res., 74*(17), 4407–4448, 1969.

Larson, R. L., and F. N. Spiess, East Pacific rise crest: A near-bottom geophysical profile, *Science, 163*, 68–70, 1969.

Luyendyk, B. P., Origin of short-wavelength magnetic lineations observed near the ocean bottom, *J. Geophys. Res., 74*(20), 4869–4881, 1969.

St. John, V. P., and R. Green, Topographic and isostatic corrections to gravity surveys in mountainous areas, *Geophys. Prospect., 15*, 151–162, 1967.

Varne, R., and M. J. Rubenach, Macquarie Island and its relationship to oceanic crust, in *Antarctic Oceanology II: The Australian–New Zealand Sector, Antarctic Res. Ser.*, vol. 19, edited by D. E. Hayes, pp. 251–266, AGU, Washington, D. C., 1972.

Varne, R., D. Gee, and P. G. J. Quilty, Macquarie Island and the cause of oceanic linear magnetic anomalies, *Science, 166*, 230–233, 1969.

Watkins, N. D., and B. M. Gunn, Petrology, geochemistry and magnetic properties of some rocks dredged from the Macquarie Ridge, *N. Z. J. Geol. Geophys., 14*, 153–168, 1970.

Watkins, N. D., and T. P. Paster, The magnetic properties of igneous rocks from the ocean floor, *Phil. Trans. Roy. Soc. London, Ser. A, 268*, 507–550, 1970.

Weissel, J. K., and D. E. Hayes, Magnetic anomalies in the southeast Indian Ocean, in *Antarctic Oceanology II: The Australian–New Zealand Sector, Antarctic Res. Ser.*, vol. 19, edited by D. E. Hayes, pp. 165–196, AGU, Washington, D. C., 1972.

GEOLOGY OF MACQUARIE ISLAND AND ITS RELATIONSHIP TO OCEANIC CRUST

R. VARNE AND M. J. RUBENACH

Department of Geology, University of Tasmania, Hobart, Tasmania 7001

Abstract. Macquarie Island is formed of fault-bounded blocks from different crustal levels. The northern part of the island is composed of gabbro and serpentinized peridotite masses and dolerite dyke swarms metamorphosed under greenschist and amphibolite facies conditions. This northern part appears to have been derived from deeper crustal levels than the southern part of the island where the rocks are mainly fresh or slightly metamorphosed lavas and volcanic breccias that are cut by many dykes. The basalts and dolerites more closely resemble ocean floor rocks than four representative groups of basalts and andesites from island arc and oceanic volcano environments. The rocks of the island could have been formed at a spreading zone in an oceanic environment. A hypothetical vertical section would probably consist of an upper layer of slightly metamorphosed extrusive rocks overlying a lower level of amphibolite facies dolerite dyke swarms, gabbros, and serpentinized peridotites.

It has recently been proposed that Macquarie Island is formed of uplifted oceanic crust originally generated at a spreading plate boundary in a mid-oceanic ridge environment [*Varne et al.*, 1969]. In this progress report, we present new maps and geologic data gained from recent studies of the island and its rocks, and we briefly consider their implications. This report is of an interim nature, and much more work, particularly of a petrologic nature, remains to be undertaken. Elsewhere in this volume, the preliminary results of gravity and magnetic surveys of the island are set out and discussed by P. Williamson and M. J. Rubenach.

Macquarie Island lies on the Macquarie ridge (Figure 1) about 700 miles south-southwest of the southern tip of New Zealand, the nearest major continental land mass. The island is elongated north-northwest along the ridge axis and is about 21 miles long and as much as 3 miles wide. This alignment of the long axis of the island parallel to the ridge axis and the linearity of the east coast, and less markedly, the west coast led *Mawson* [1943] to suggest that the island was fault bounded. Rocks similar to those that occur on the island have been recovered both north and south of the island in dredge hauls from the crest of the Macquarie ridge [*Summerhayes*, 1969; *Watkins and Gunn*, 1971]. *Cullen*'s [1969] bathymetric data suggest that morphologically the island is part of the ridge crest that breaks the surface. A preliminary Bouguer gravity map of the island [*Williamson and Rubenach*, this volume] shows no pronounced local gravity anomalies.

GEOLOGY OF MACQUARIE ISLAND

The island stands about 1000 feet above sea level. Topographically, it is defined by an elevated and undulating plateau cut off seaward by steep cliffs and studded with many lakes of apparently glacial or periglacial origin. In places, particularly in the northwest, the island is fringed by a coastal platform with relict sea stacks and beach deposits standing slightly above sea level. Good exposures of country rock occur along the coasts, but the inland is concealed by vegetation and superficial deposits. Both raised beach and fluvioglacial deposits have been found. Consequently, how much of the topography of the island was shaped by marine or glacial processes is not known. *Mawson* [1943] proposed that an ice sheet, which supposedly originated from a now foundered land mass to the west, in the past overrode and partially planated the island, but this concept should be abandoned [*Varne et al.*, 1969].

Faulting is widespread, and essentially the island is a number of fault-bounded blocks, probably on all scales. Most of the island is formed of basaltic lava flows, volcanic breccias, and sediments that are cut by many dolerite dykes (Figure 2). Gabbros and serpentinized peridotites are restricted to the northern part of the island. In the geologic map of the island (Figure 2), these rocks are represented as one group, but in a more detailed geologic map of the northern part of the island (Figure 3), we have distinguished serpentinized peridotite from both layered and

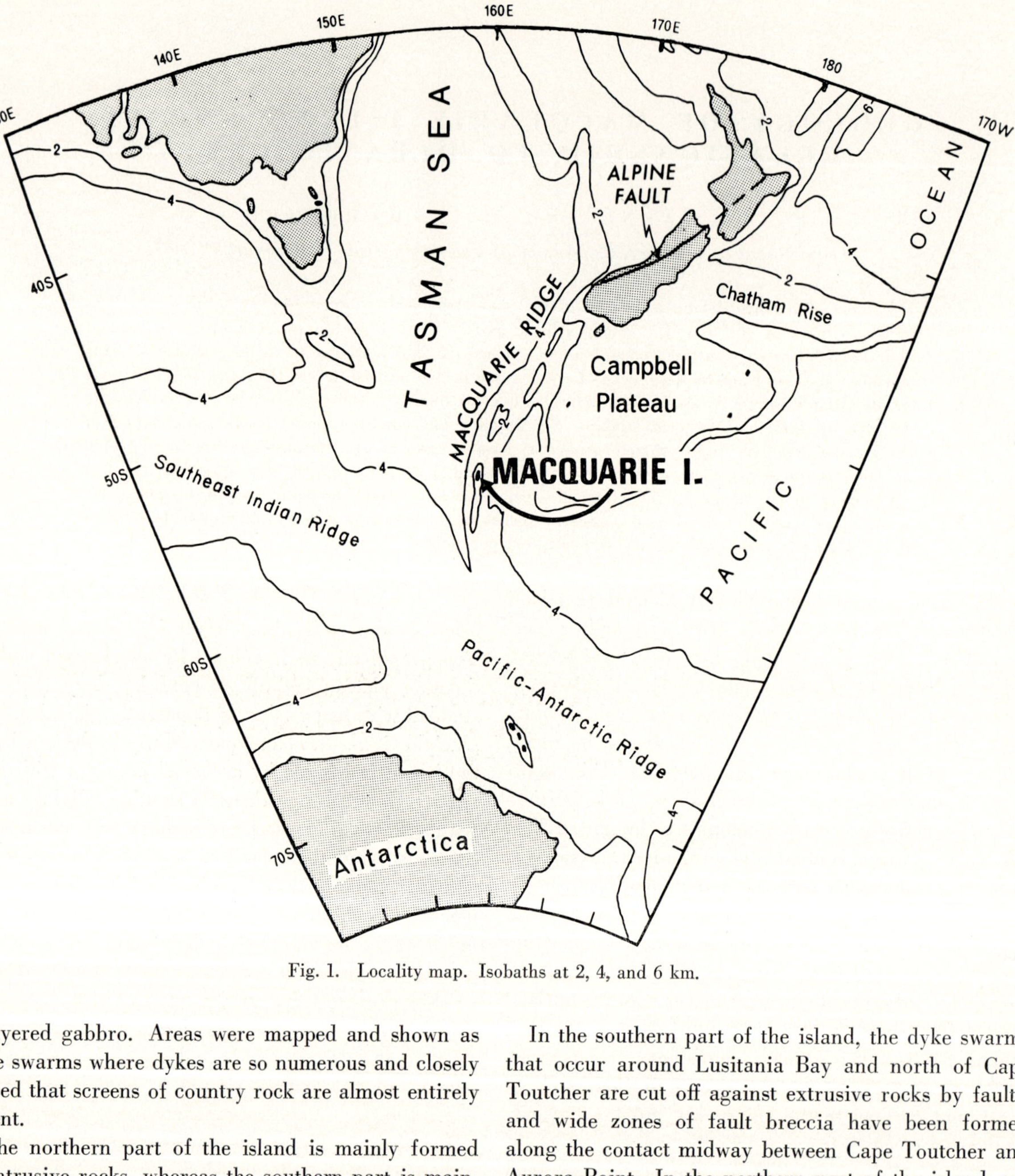

Fig. 1. Locality map. Isobaths at 2, 4, and 6 km.

unlayered gabbro. Areas were mapped and shown as dyke swarms where dykes are so numerous and closely spaced that screens of country rock are almost entirely absent.

The northern part of the island is mainly formed of intrusive rocks, whereas the southern part is mainly formed of extrusive rocks. The contact between these two parts of the island runs between Eagle Point and Sandy Bay (Figure 3); where this contact is exposed, it is faulted. East of Douglas Point, dyke swarms and extrusive rocks have been thrown against one another along a complex fault zone, and on Eagle Point along a continuation of this fault zone, extrusive rocks are in contact with sheared serpentinized peridotite.

In the southern part of the island, the dyke swarms that occur around Lusitania Bay and north of Cape Toutcher are cut off against extrusive rocks by faults, and wide zones of fault breccia have been formed along the contact midway between Cape Toutcher and Aurora Point. In the northern part of the island, extrusive rocks occur only at North Head and as fault-bounded blocks enclosed by dyke swarms north of Nuggets Point(Figure 3).

Extrusive Rocks and Associated Sediments and Dykes

The extrusive rocks occur both as basaltic lava flows and as basaltic volcanic breccias. The bulk of them are flows of pillow lava, although massive lavas are also found. The volcanic breccias are formed of pillow fragments, basaltic blocks, and ash and in places are clearly transitional downward into pillow lavas. Ropy tops are preserved on some of the massive lavas.

Series of parallel dolerite dykes that range in thickness up to about 8 feet cut the lavas, volcanic breccias, and associated sediments in all parts of the island. The dykes form up to 40% of the coastal exposures for 6 miles to the south of Brothers Point (Figure 2) but are less abundant elsewhere. There is not a continuous gradation from these areas of extrusive rocks cut by many dykes into the areas of dyke swarms.

Globigerina ooze occurs as interstitial material in pillow lavas. Although widely distributed, this ooze is abundant only in a few localities. Usually the ooze is baked and silicified, but poorly preserved coccoliths have been found in two samples from North Head. A few reworked Cretaceous forms are present in one sample, but most forms are Miocene and are probably middle or early rather than late Miocene (J. A. Wilcoxon and P. G. Quilty, personal communication, 1971).

The structure and stratigraphy of the extrusive rocks of the island are complex and as yet poorly understood. In the central part of the island, sediments are interbedded with the lavas and breccias. Beds of red mudstone, commonly less than a foot thick, occur at Brothers Point and at various localities on the west coast between Mawson Point and Cape Toutcher. Thicker beds of mudstone, graywacke, and paraconglomerate are interbedded with massive lava flows at Mawson Point. At Brothers Point, a near-horizontal unconformity is exposed over a small area. Below the unconformity, brecciated aphyric pillow lavas with irregular dips are cut by many minor faults. Above the unconformity, a thin bed of grits is overlain by feldsparphyric pillow lavas with shallow dips. North of Mawson Point, the pillow lavas dip steeply and strike north-northwest, but to the south of the point, the lavas and breccias dip shallowly south. Equally abrupt changes in the dips of the extrusive rocks occur elsewhere on the island, but no extensive unconformities have been found.

In the Mawson Point area and in coastal exposures for 6 miles to the south of Brothers Point, the dominant strike trends are westerly. Elsewhere, the dominant strike trends are in the northwest quadrant. Dips on lavas range from shallow to steep and are variable even in areas of constant strike. The pillow forms of some pillow lavas imply that the lavas have been tilted to near-vertical attitudes, but the steepest

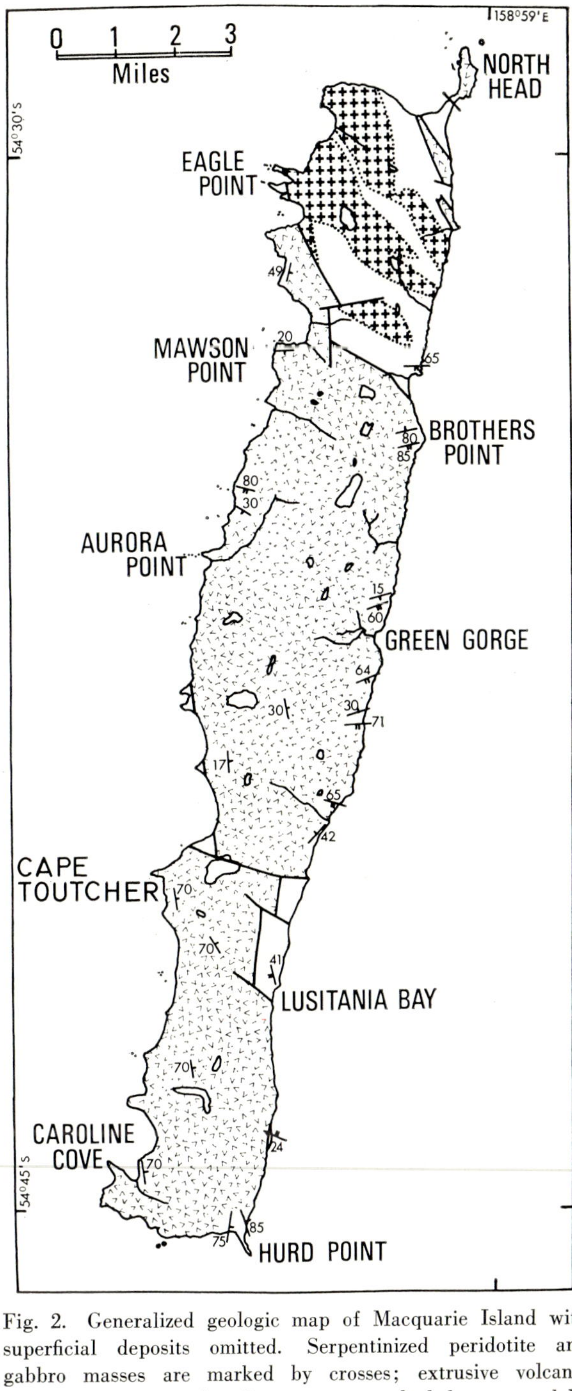

Fig. 2. Generalized geologic map of Macquarie Island with superficial deposits omitted. Serpentinized peridotite and gabbro masses are marked by crosses; extrusive volcanic rocks and associated sediments are marked by vees; dyke swarms are blank. Strikes on lavas are shown with a single tick, and strikes on dykes are shown with a double tick. Faulted contacts are drawn as heavy lines, and gradational or uncertain contacts are drawn as dotted lines.

reliable dips, measured on lava flows with interbedded mudstone, are about 45°. These flows were presum-

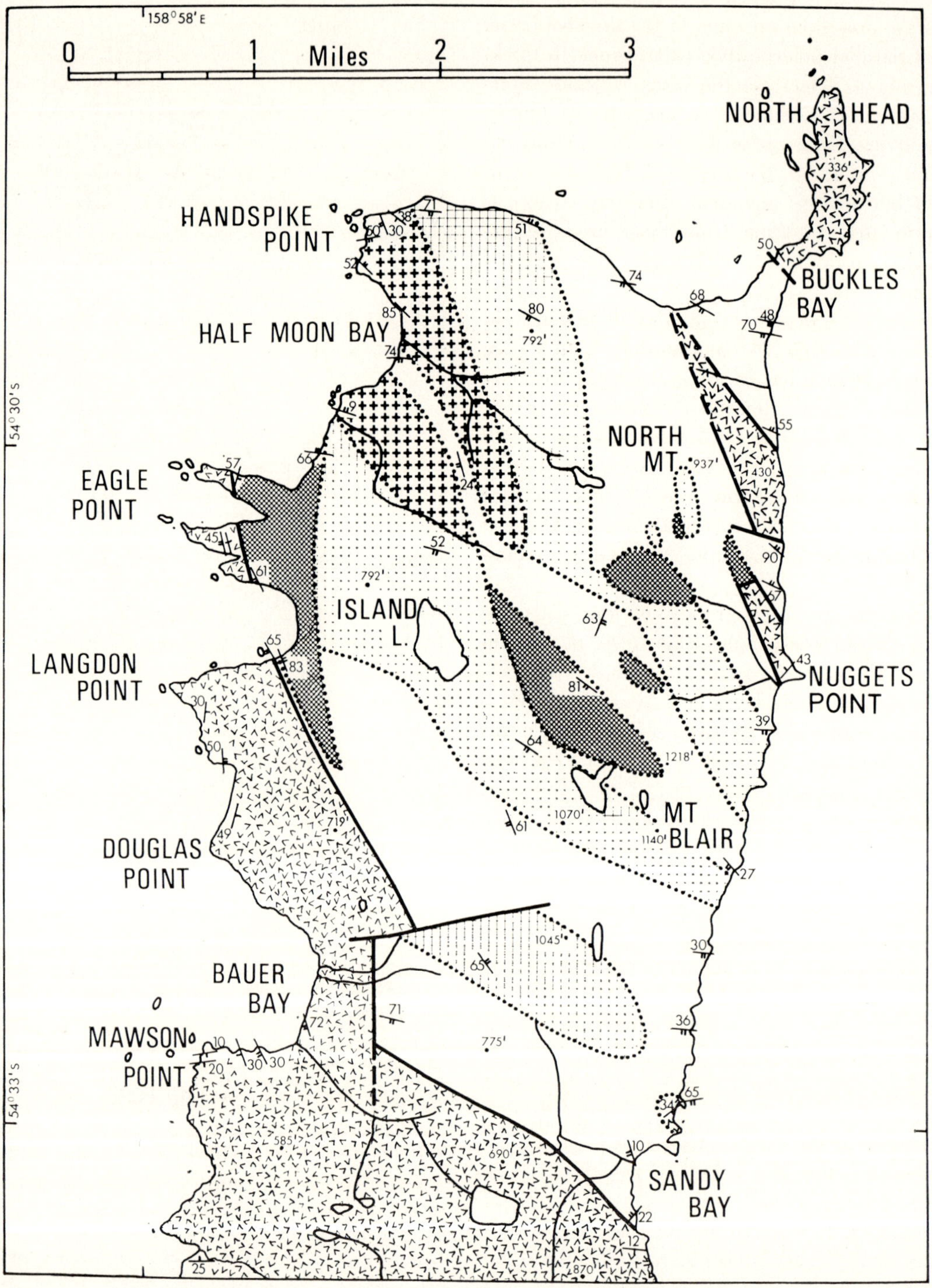

Fig. 3. Geologic map of the northern part of Macquarie Island with superficial deposits omitted. Serpentinized peridotite bodies are marked by heavy diagonal cross-hatching; layered gabbro complex is marked by crosses; other gabbro masses are marked by light cross-hatching; extrusive volcanic rocks and associated sediments are marked by vees; dyke swarms are blank. Other symbols correspond with those in Figure 1. Spot heights in feet above sea level.

ably horizontal or near-horizontal at the time of eruption. Many high and low angle faults cut the extrusive rocks, and it is not yet possible to estimate the thickness of the volcanic succession on the island or to determine if a true superimpositional stratigraphic succession exists.

Dykes cut the interbedded lava flows, breccias, and sediments as series of parallel sheets and do not intersect one another. Acute angles between the planes of the dykes and the bedding planes of the extrusive rocks are almost always large no matter how steep the inclinations of the bedding planes. This relationship implies that the dykes were originally steeply inclined intrusions into relatively flat-lying extrusive rocks and that the tilting occurred subsequently.

The strikes of these tilted lava flows, breccias, and sediments are also parallel or near-parallel to the strikes of the dykes that cut them. Seemingly, therefore, the axes about which the tilting took place must have been approximately horizontal and parallel to the dyke-bedding plane intersections. No fold closures have been found. Any folding that may have occurred was probably about the same axes as the tilting, because folding about different axes would distort the geometric relationships between the dykes and the lava flows, breccias, and sediments that these dykes have intruded. These geometric relationships also remain the same regardless of the strike of the rocks and thus the variations of regional strike may be the result of later rotation about vertical axes of the tilted rocks.

Intrusive Rocks

Where intrusive relationships are exposed, the serpentinized peridotites are older than and are intruded by gabbros, microgabbros, and dolerites. Gabbros are also intruded by dolerites. Blocks of peridotite are enclosed by gabbro east of Island Lake.

Screens of gabbro and serpentinized peridotite are common in the dyke swarms, particularly near gradational transitions from gabbro and serpentinized peridotite masses intruded by few dykes into areas of numerous and closely spaced dolerite dykes. The poorly exposed contact between dyke swarms and the eastern margin of the strip of gabbro west of Nuggets Point and North Mountain may be faulted as the contact is not gradational and the strikes of the dykes are oblique to it. We do not know, therefore, whether intrusive or tectonic processes have been the dominant factor in the formation of the lithologic striping and outcrop pattern of the northern part of the island where the coarse-grained intrusive rocks are exposed.

The serpentinized peridotites are usually massive rocks. The only obvious compositional layering occurs at Eagle Point where adjacent layers differ in their proportions of serpentinized olivine and orthopyroxene. The style and regularity of this layering are similar to that found in cumulate rocks, and the layers are continuous over the length of the outcrop (Figure 4). Elsewhere, irregular layering is found occasionally on weathered surfaces. Individual layers have indistinct boundaries and are not continuous. Anastamosing veins of cross-fiber asbestos occur adjacent to fault zones.

A layered gabbro complex disturbed by faulting and dyke swarm emplacement occurs in the northwest section of the island (Figure 3). Large outcrops showing spectacular layering are exposed at Handspike Point and Half Moon Bay (Figure 5). The ultramafic zone of the complex is represented only by plagioclase dunites. Orthopyroxenites and harzburgites, commonly found in the ultramafic zone of other layered intrusions, are here absent.

The southern part of the layered gabbro complex is composed of troctolite with occasional bodies of plagioclase dunite and anorthosite, but at Handspike Point and to its southeast, the layered gabbros are norites and hypersthene gabbros grading into anorthosites with scattered layers of plagioclase pyroxenites, plagioclase dunites, and troctolites. Rhythmic layering is best developed in the troctolites but is also present in the other gabbroic rocks; other layers, though not rhythmic, differ from adjacent ones in the proportions of minerals present.

The intrusive relationship of the layered gabbros to the other gabbros on Macquarie Island is not known. Most of these gabbros show no layering, are coarse-grained, and are composed of olivine, orthopyroxene, clinopyroxene, and plagioclase. Spinels are very rare. Gabbro pegmatites occur at various localities on the plateau, but their relationship to the main gabbro masses is obscured by lack of exposures. Other smaller gabbro masses are formed of clinopyroxene and plagioclase alone.

The dyke swarms are composed of series of dykes lying parallel or subparallel to one another (Figure 6). The majority of the dykes are between 1 and 4 feet thick, and only a few are as thick as 10 feet. However, exposures within the swarms showing several chilled edges that all face the same way exemplify

Fig. 4. Serpentinized layered harzburgite from Eagle Point (see Figure 2). The style and regularity of this layering is reminiscent of that found in cumulates. Nearby outcrops of serpentinized harzburgites show the effects of deformation.

the difficulties of measuring the thickness of a particular dyke. In general, dykes intrude within earlier dykes rather than at their margins.

Screens of gabbro, serpentinized peridotite, and brecciated and veined dolerite have been identified within the dyke swarms, but over large areas no screens are apparent. Large, fault-bounded masses of pillow lava occur within the dyke swarms, but screens of pillow lava have been identified at only two localities. Dykes from the dyke swarms in the northern part of the island are distributed about a horizontal axis trending 300° [*Varne et al.*, 1969].

Petrography of Some Macquarie Island Rocks

The lavas and dykes are basaltic, but many have been altered and metamorphosed. They range from coarsely porphyritic to aphyric types and from almost wholly crystalline to almost wholly glassy. Petrographic descriptions of some Macquarie Island rocks have earlier been given by *Mawson* [1943]. Practically all the types that he mentions are represented among our rock collections as well as others from the southern parts of the island that had not been previously sampled.

When fresh or only slightly altered, the porphyritic lavas and dykes carry phenocrysts of plagioclase that are usually accompanied by phenocrysts of altered magnesian olivine. Small octahedra of reddish-brown spinel also occur, sometimes enclosed by plagioclase but frequently as discrete crystals, and large but rounded and faceless crystals of clinopyroxene are

Fig. 5. Layered troctolite from the southern point of Half Moon Bay (see Figure 2).

found more rarely. The plagioclase phenocrysts are calcic in the bytownite-labradorite compositional range. Glomerocrysts of plagioclase and clinopyroxene occur as well as rare xenoliths of gabbro and dolerite. The aphyric lavas and dykes resemble the ground masses of the porphyritic rocks and are characteristically composed of plagioclase laths, clinopyroxene grains, and opaque minerals in intergranular or subophitic textural relationships with intersertal patches of altered or devitrified glass. With increasing proportions of glassy or originally glassy material, there is a gradation to rocks that are largely turbid altered glass with subvariolitic textures, although in three localities we have collected fresh

Fig. 6. Dyke swarm from one-half mile north of Nuggets Point (see Figure 3).

honey-brown glass containing crystals of unaltered olivine and plagioclase. Olivine has not been definitely observed as a ground mass mineral, and so far neither orthopyroxene nor pigeonite has been identified in the basaltic lavas and dykes, although both are present in the gabbros. Amphibole occurs in the ground mass of some of the lavas, particularly those from Green Gorge. This amphibole is pleochroic from reddish-brown to fawn tints and occurs either as small discrete crystals or as rims on clinopyroxene; amphibole does not appear to be a secondary alteration product. These lavas have turbid interstitial material in the ground mass and may have slight alkaline affinities. A small number of the lavas and dykes so far examined are petrographically more clearly alkaline; one dyke is composed of olivine pseudomorphs and large phenocrysts of titanaugite that are mantled by thin rims of aegirine-augite set in a ground mass of oligoclase, biotite, and brown hornblende together with a little K-feldspar and possibly analcite with sheaves of zeolites after glass.

Dykes from dyke swarms differ petrographically from the dykes emplaced in the extrusive rocks. The dykes from dyke swarms are commonly porphyritic and are more coarsely grained with doleritic textures, whereas the other dykes are infrequently porphyritic and are more finely grained with basaltic textures. Also, all the dykes so far examined from the dyke swarms appear metamorphosed. They have a matrix composed essentially of hornblende, plagioclase, chlorite, and opaque minerals. The hornblende has replaced clinopyroxene that is on occasion preserved unaltered as relicts within it. The hornblende is fibrous or forms ragged plates and commonly is pleochroic, usually in tints of bluish-green and greenish-brown, although some is almost colorless. Hornblende that is pleochroic from light to dark brown rarely occurs together with the more common greenish hornblende and may represent a product of magmatic crystallization. The plagioclase is calcic and appears unaltered. The phenocrysts and textures of these apparently metamorphosed dolerite dykes are largely unaffected. Narrow veins of quartz, quartz and calcite, chlorite and epidote, and also zeolitic material have been noticed to transect the rocks.

The lavas and the dykes that intrude them show varying degrees of alteration. One pillow lava possesses a similar mineral assemblage to the dykes of the dyke swarms, but in many the ground masses are composed of a fibrous actinolite, commonly associated or intergrown with chlorite, together with some sphene and plagioclase patchily replaced by albite. The actinolite and chlorite appear to have replaced ferromagnesian silicates, mainly pyroxenes, and intersertal glass; and the sphene occurs as tiny granules associated with the opaque minerals of the rocks. These rocks are also cut by epidote, calcite, quartz, and chlorite-bearing veins.

Slight alteration of lavas that are otherwise fresh in appearance is represented either by the appearance of clay minerals pleochroic in greens and yellow-browns that replace glassy or cryptocrystalline parts of the rock, or by the ground mass of the rock taking on a murky and turbid appearance.

Serpentinized peridotites examined briefly by us have a primary mineralogy of magnesian olivine + magnesian orthopyroxene + chromian spinel and are therefore harzburgites, although *Mawson* [1943] mentions boulders of dunite from glacial deposits. Orthopyroxene commonly poikilitically encloses olivine, a texture that may be cumulate in origin. Some orthopyroxenes are kink-banded, but the rarer fresh olivines show only weak undulose extinction and have no deformation lamellae. Therefore, although the peridotites suffered some deformation before serpentinization took place, there is no petrographic evidence of plastic flow. Studies made by X-ray diffraction show that olivine is largely or entirely replaced by lizardite and antigorite with many small magnetite grains, whereas orthopyroxene, although often unaltered, is partly pseudomorphosed by bastite (lizardite). Chromian spinels commonly have fresh cores but opaque rims. Postserpentinization deformation that has affected some of the rocks has led to the development in mesh-textured serpentinite of a distinct foliation that wraps around orthopyroxene augen.

The gabbroic rocks include a considerable range of types. The gabbro mass around Island Lake is composed of olivine + orthopyroxene + clinopyroxene + plagioclase. Protoclastic textures are common, and spectacular flaser textures are developed at some localities. The clinopyroxene is characterized by very small optic axial angles and may be pigeonite. Coronites and symplectites involving olivine, pyroxenes, and plagioclase are also present. Other gabbros described by *Mawson* [1943] include norites, olivine gabbros, troctolites, eucrites, olivine eucrites, and allivalites. An equivalent of the alkaline dolerites may be represented by gabbro boulders from Buckles Bay that contain subophitic clinopyroxene and plagioclase together with titanomagnetite, olivine, and biotite.

As a group, the gabbros are little altered. In the

plagioclase dunites, olivine is replaced in part by lizardite and antigorite, and interstitial plagioclase is replaced by hydrogrossular. Incipient 'uralitization' of the gabbros is fairly common, pyroxene being rimmed by hornblende or nests of tremolite and chlorite replacing pyroxene. Epidote and prehnite replace plagioclase in some gabbros. Some of the dolerites intruded into serpentinized peridotite and gabbro are now wholly composed of hornblende and plagioclase, but others have a fresh appearance. Abundant veins and irregular dykes of pegmatitic gabbro emplaced within the peridotites have been altered to prehnite + actinolite + hydrogarnet assemblages, probably during serpentinization. In the flaser gabbros, hornblende has replaced clinopyroxene postdeformation as it has formed along cracks that cut clinopyroxene augen.

Metamorphism

Similar metamorphic assemblages are developed throughout groups of rock samples collected from particular areas. All the dykes sampled from the dyke swarm exposed on the east coast north of Nuggets Point are characterized by strongly pleochroic hornblende + plagioclase + opaques assemblages, whereas in the dyke swarms east and southeast of Mount Blair, the primary minerals are replaced by weakly pleochroic hornblende + chlorite + plagioclase + opaques assemblages. This difference suggests that the dyke swarms from these different areas have been metamorphosed at differing grades within the amphibolite facies.

In the Bauer Bay area, the pillow lavas have been in part replaced by an actinolite + albite + sphene assemblage that is taken to indicate metamorphism under greenschist facies conditions. In the Green Gorge area, however, the lavas are little altered, although some show replacement of mesostasis by clay minerals. This slight alteration is interpreted to suggest that if these rocks have been metamorphosed, the metamorphism has taken place at very low grades. Dykes within extrusive rocks exhibit mineralogic changes and assemblages similar to those of the extrusive rocks.

A start has been made on the construction of a metamorphic map of the island (Figure 7*a*). We have made a preliminary two-fold division of the island into (1) areas where greenschist and/or amphibolite facies metamorphism appears to have taken place and (2) areas where the basaltic rocks are little affected by reconstitution of their original mineralogy.

Fig. 7. (*a*) An outline map of Macquarie Island to show the distribution of metamorphism. Cross-hatching indicates areas where the rocks are metamorphosed at greenschist and/or amphibolite facies grade; diagonal lines indicate areas where metamorphism is either at lower grade or has not occurred. Areas have only been included where many samples have been collected and examined petrographically. (*b*) The localities of Macquarie Island basalts and dolerites that have been analyzed for Ti, Rb, Sr, Y, Zr, and Nb. Those that on other grounds (see text) resemble 'typical ocean floor basalt' are plotted as stars; those that are 'unlike typical ocean floor basalt' are plotted as darkened circles.

In addition, only those parts of the island where many rocks have been collected and examined have been included so that the map is limited in extent. Provisionally, the peridotites and gabbros have been included within the areas of greenschist and/or amphibolite facies metamorphism, as the peridotites and gabbros are closely associated in the field with metamorphosed dolerites.

The preponderance of coarse-grained intrusive rocks and dolerite dyke swarms in the northern part of the island implies that this part represents a deeper crustal level than the southern part where extrusive rocks are abundant. The character and distribution of the metamorphosed basalts and dolerites support this conclusion. The dolerite dyke swarms were metamorphosed at higher grades than the majority of the extrusive rocks and associated basaltic dykes.

Notably, the obvious tectonic contacts between dyke swarms and extrusive rocks in the Sandy Bay and Bauer Bay areas do not also mark a change in the metamorphic grade. The amphibolite and/or greenschist facies metamorphism that has affected the dyke swarms extends across the contacts to the extrusive rocks. Similarly, the transition from dyke swarms to extrusive rocks in the Buckles Bay–North Head area in the extreme north of the island does not correspond to a lessening in metamorphic grade. Passages from rocks that have suffered amphibolite and/or greenschist facies metamorphism to rocks that are only slightly metamorphosed and are associated with fresh tachylites occur within extrusive rocks. Presumably, the metamorphosed extrusive volcanic rocks have also been at deeper crustal levels than those that are relatively fresh.

The lack of schistosity or penetrative cleavage in the metamorphosed basalts and dolerites and the survival of igneous textures suggest that the metamorphism was not accompanied by deformation, and also could be taken to imply that the metamorphism was similar in nature to burial metamorphism. However, the partial metamorphic reactions that took place in the rocks involve hydration, and *Turner* [1968] has suggested that reactions of this type are exothermic and that once begun the metamorphism of basalts and dolerites to greenschist facies assemblages could even be self-perpetuating. The extent of the reaction in the metamorphosed Macquarie Island basalts and dolerites is likely to have been limited mainly by the amounts of water and heat available. Water would have been supplied from the sea and in smaller amounts from the crystallizing magmas themselves [*Cann*, 1970*b*]. In a wholly igneous terrain such as Macquarie Island, the magmatic heat introduced during the emplacement of lavas and dykes could have been sufficient to hold rocks at shallow depths at temperatures high enough to sustain the metamorphic reactions observed. It does not seem necessary at this stage to assume that the rocks of Macquarie Island have undergone deep burial to account for the grades of metamorphism that they have suffered.

Some Analytical Data for Macquarie Island Rocks

Eighteen major element analyses of gabbros, dolerites, basalts, and tachylites from Macquarie Island are presented by *Mawson* [1943]. He does not always make clear whether the dolerite analyses are of dykes from dyke swarms or of dykes intruding extrusive rocks. However, of the 14 analyzed dolerites, basalts, and tachylites, all but one contain normative Ol and Di, and the majority also contain normative Hy. One of the basalts contains a small amount (0.56%) of normative Q, and three dolerites and one basalt contain normative Ne. The marked resemblance of the mean chemical composition of the 9 Ol, Di, and Hy normative dolerites, basalts, and tachylites to the mean chemical composition of ocean floor basalts has already been noted [*Varne et al.*, 1969], and two aphyric rocks of this group also plot in *Kuno*'s [1960] compositional field for high-alumina basalts. On the basis of petrographic comparisons with the analyzed rocks, the majority of the Macquarie Island dolerites and basalts may prove to be similar to rocks dredged from the deep ocean floors, although subsidiary amounts of alkaline types may also be present. However, according to *Mawson*'s [1943] petrographic description, all the analyzed rocks are altered or metaphorphosed. As these processes commonly lead to changes of bulk composition in basaltic rocks [e.g., *Smith*, 1968; *Cann*, 1969; *Matthews*, 1971], our analytical studies of Macquarie Island rocks have so far been directed toward establishing in a preliminary way by using abundances of Ti, Y, Zr, and Nb any chemical affinities that may exist between these rocks and groups of volcanic rocks from various known environments. These elements were chosen because *Cann* [1970*a*] has shown that their abundances in ocean floor basalts appear to be unaffected even by severe secondary alteration processes such as weathering and metamorphism and because *Pearce and Cann* [1971] have successfully used Ti, Y, and Zr abundances to discriminate between groups of rocks from various environments.

Analyses of Ti, Rb, Sr, Y, Zr, and Nb abundances in 31 Macquarie Island dolerites and basalts have been carried out by using the methods of *Cann* [1970] and *Pearce and Cann* [1971]. Copies of the analytical results are available on request.

The group of Macquarie Island rocks has been compared with a group of (1) ocean floor basalts from the Atlantic and Indian oceans and the Gulf of Aden, (2) tholeiitic basalts from Japan, (3) tholeiitic basalts from Hawaii, (4) alkali basalts from Flores, Azores, and (5) andesites and basaltic andesites from Japan and New Zealand. Data for group 1 are from *Cann* [1970] and *Pearce and Cann* [1971]; data for groups 2, 3, and 4 are from *Pearce and Cann* [1971]; and data for group 5 are from *Taylor and White*

[1966]. These five groups of rocks are felt to provide samples of those volcanic rocks that the Macquarie Island rocks may be most reasonably expected to resemble. They include representatives of *Kuno*'s [1968] island arc pigeonitic and hypersthenic series lavas, representatives of tholeiitic and alkaline lavas from oceanic volcanoes, and representatives of ocean floor basalts.

A plot of Ti versus Zr (Figure 8) after *Pearce and Cann* [1971] shows that the five groups can be separated effectively. The alkali basalts from Flores, Azores, were not included, but all these basalts have higher Ti and Zr abundances than the Hawaiian tholeiites. In Figure 9, the Macquarie Island rocks are plotted with the fields of the four groups plotted on Figure 8 transferred for comparison. On this basis, the rocks from Macquarie Island plainly resemble the ocean floor basalts most closely.

Pearce and Cann [1971] have also shown that the five groups of rocks listed above can be well discriminated statistically by using Ti, Zr, and Y abundances. Discriminant analysis was used to calculate three discriminant functions (F_1, F_2, and F_3) of the form $F = a_1$ Ti + a_2 Zr + a_3 Y. In projection, the five groups each form well-defined fields with no overlap.

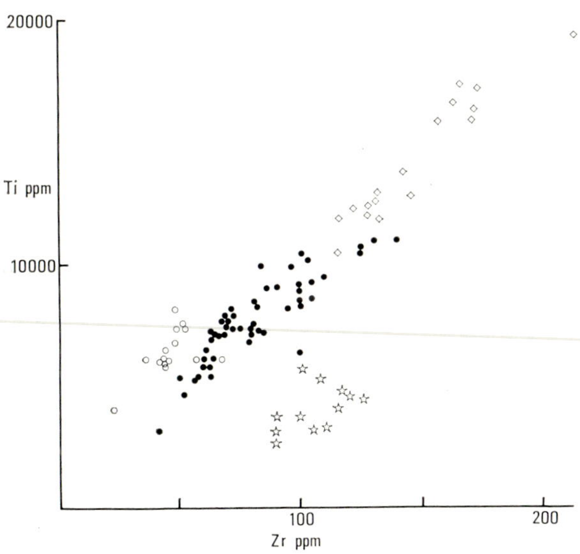

Fig. 8. A plot, after *Pearce and Cann* [1971], of Ti versus Zr in parts per million for four groups of rocks referred to in the text. Open circles represent Japanese tholeiites; darkened circles represent ocean floor basalts; stars represent andesites and basaltic andesites from Japan and New Zealand; and open rhombs represent Hawaiian tholeiites. A fifth group of alkali basalts from Flores, Azores, is characterized by higher abundances of both Ti and Zr than Hawaiian tholeiites but has not been included. Data sources are given in the text.

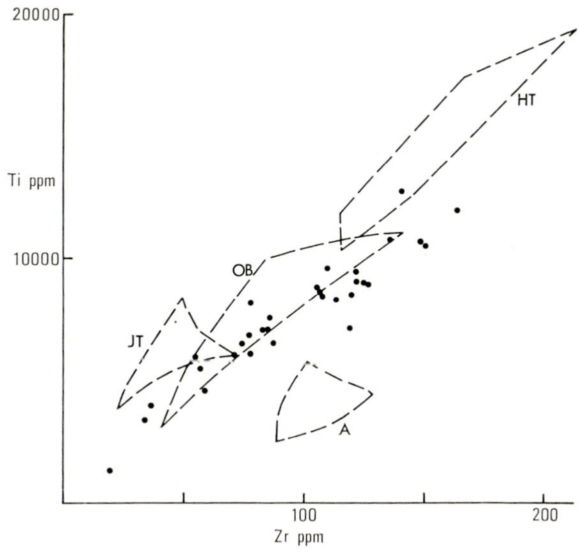

Fig. 9. A plot of Ti versus Zr in parts per million for Macquarie Island basalts and dolerites (darkened circles) with the fields of the four groups of Figure 7 transferred for comparison. The following abbreviations are used: JT, Japanese tholeiites; OB, ocean floor basalts; A, andesites and basaltic andesites from Japan and New Zealand; and HT, Hawaiian tholeiites.

The statistical probability of the five groups not being significantly different is very low. Figure 10 is a plot of $F_1 = +0.000042$ Ti + 0.0024 Zr − 0.0192 Y versus $F_2 = +0.000057$ Ti − 0.0055 Zr + 0.0032 Y. The fields of each group are outlined after *Pearce and Cann* [1971]. Evidently, in this projection, ocean floor basalts and island arc andesites are well separated, but some overlap exists between ocean floor basalts and island arc tholeiites. Volcanic rocks from oceanic islands are clearly well separated from the other groups. The Macquarie Island rocks as a group show some overlap into the fields of ocean floor basalts and island arc tholeiites, but in part the Macquarie Island rocks show no affinities with any of the five groups used for comparison.

Figure 11 is a plot of F_2 versus $F_3 = 0.000012$ Ti + 0.00050 Zr + 0.0224 Y. The fields of ocean floor basalts and of island arc tholeiites are well separated in this projection. Here, the Macquarie Island rocks as a group show the most overlap with the field of the ocean floor basalts, but again some of the data points fall outside the fields of the groups plotted for comparison. No statistical assignment of the Macquarie Island rocks to the groups used for comparison has been attempted as yet. However, it is concluded that the study of *Pearce and Cann* [1971] shows that

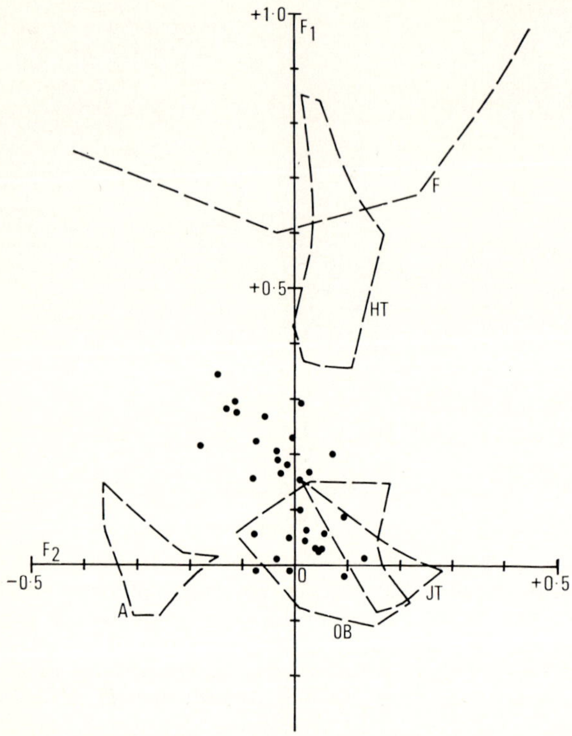

Fig. 10. Plot of F_1 — F_2 after *Pearce and Cann* [1971]. Macquarie Island basalts and dolerites (darkened circles) compared in projection with four groups of volcanic rocks labeled like those in Figure 8 with an additional group of alkali basalts from Flores, Azores, labeled F.

groups of volcanic rocks can be well discriminated by using Ti, Zr, and Y abundances and that the Macquarie Island rocks on this basis show the closest affinities to the ocean floor basalts from the rocks used for comparison.

A further step has been tentatively taken in an attempt to assess the chemical affinities of Macquarie Island rocks with other volcanic rocks from known ocean floor environments. Figure 12 is a plot of Y versus Nb for some ocean floor basalts [*Cann*, 1971], some alkaline and tholeiitic basalts dredged from seamounts southwest of the Hawaiian Islands and from the Cocos ridge in the eastern Pacific Ocean [*Engel and Chase*, 1965; *Engel and Engel*, 1966], and some Macquarie Island rocks. The ocean floor basalts and the seamount basalts occupy different fields that can be separated either on the basis of abundances of Y and Nb or on Nb/Y. The Macquarie Island rocks here form a group falling in part into the ocean floor basalt field and in part into an area of the plot occupied by neither of the two groups used for comparison. On this basis, we have carried out a simple and arbitrary classification of the Macquarie Island rocks into 'typical ocean floor basalt' types and 'unlike typical ocean floor basalt' types and have plotted these two types according to their localities on Figure 7*b*. There is a tendency for these two types to occur in different parts of the island.

Analytical studies of the Macquarie Island rocks have only just begun and there is no evidence that chemically the rocks do not form a continuum. Also, although ocean floor basalts so far seem to form a convergent group on the basis of their major element compositions, large parts of the ocean floors remain unsampled. From our point of view, there is a serious lack of analyzed ocean floor basalts for comparative purposes from the southwest Pacific and the areas south of Australia and from the Macquarie ridge. In addition, *Nicholls and Islam* [1971] have shown that variations in trace element contents of ocean floor basalts are more marked than variations in major element compositions and that geochemical provinces exist in oceanic areas. Nevertheless, the observation here that the Macquarie Island rocks show closer affinities with ocean floor basalts from the Atlantic and Indian oceans and the Gulf of Aden than with tholeiites and andesites from island arc environments in Japan and

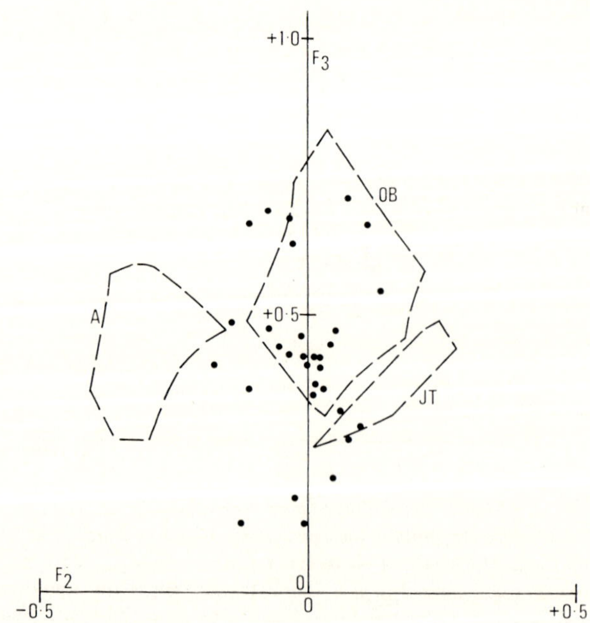

Fig. 11. Plot of F_2 — F_3 after *Pearce and Cann* [1971]. Macquarie Island basalts and dolerites (darkened circles) compared in projection with three groups of volcanic rocks labeled as in Figure 8.

New Zealand, which are spatially more closely related to the Macquarie ridge area, strongly supports the original suggestion that Macquarie Island represents uplifted oceanic crust.

There are also clear petrographic similarities between the Macquarie Island rocks and ocean floor basalts. Petrographic descriptions of ocean floor basalts [e.g., *Nicholls et al.*, 1964; *Muir et al.*, 1964] frequently refer to phenocrysts or xenocrysts of calcic plagioclase commonly accompanied by magnesian olivine, more rarely by chromian spinel, and infrequently by clinopyroxene. Similar xenocryst associations are present in Macquarie Island basalts and dolerites. Similarly, orthopyroxene or pigeonite appears to be rare or absent, particularly as phenocrysts, in ocean floor basalts. The Macquarie Island rocks share this feature, and it seems to set them off from *Kuno*'s [1968] pigeonitic and hypersthenic series volcanics of island arc environments.

The similarity in mineralogy between fresh rocks dredged from the ocean floors and those from Macquarie Island extends to their metamorphosed equivalents. Metamorphosed basalts and dolerites have been dredged from various parts of the Atlantic and Pacific oceans [e.g., *Miyashiro et al.*, 1971; *Cann*, 1971], commonly from fracture zones or regions of irregular topography. The rocks are mineralogically similar to Macquarie Island metamorphosed lavas and dykes. They have been metamorphosed at grades up to amphibolite facies but also lack schistosity or penetrative cleavage and retain their original igneous textures. By analogy with Macquarie Island, we suspect that the amphibolites with doleritic textures from the ocean floors may also be from dyke swarms exposed at shallow levels by fault movements that have brought up originally deeper parts of the oceanic crust.

Studies of the xenocryst assemblages of some of the lavas and dykes could prove valuable in interpretations of the genesis of high-alumina olivine tholeiite magma compositions of the ocean floor and Macquarie Island. *Clarke et al.* [1962] showed that some effects of increasing pressure on liquidus relations in the diopside-anorthite-silica system are that the clinopyroxenes become more aluminous, the stability field of clinopyroxene + liquid is enlarged, and the field of anorthite + liquid is diminished. A liquid in equilibrium with clinopyroxene and anorthite at high pressures will be relatively much richer in anorthite component than a liquid in equilibrium with clinopyroxene and anorthite at low pressures. As *Clarke et al.* [1962] point out, if a cotectic liquid at high pressures were decompressed suddenly, perhaps by rapid upward movement, the liquid would become superheated and its composition would at low pressures be within the anorthite field. The aluminous clinopyroxene would tend to be resorbed, and the liquid would precipitate large quantities of anorthite. An analogous although more complex mechanism may account for the observed xenocryst mineralogies of ocean floor and Macquarie Island basalts.

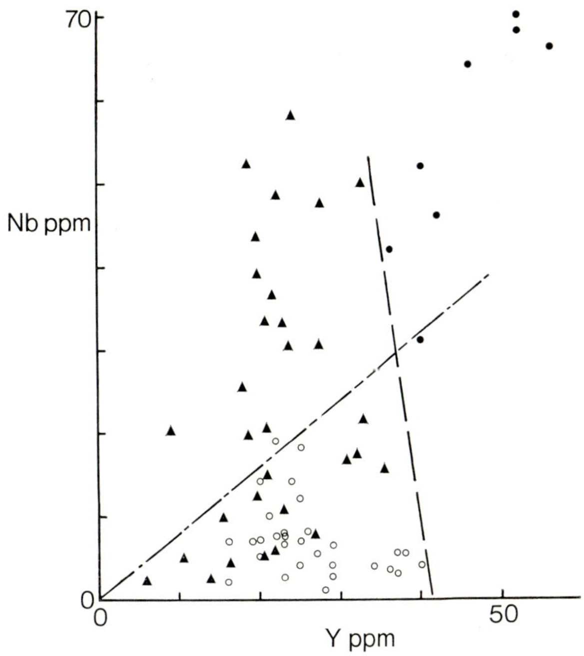

Fig. 12. A plot of Nb versus Y for basalts and dolerites from Macquarie Island (triangles), ocean floor basalts (open circles) from *Cann* [1970], and some basalts from Pacific seamounts (darkened circles) from *Engel and Chase* [1965] and *Engel and Engel* [1966]. The dash-dot line divides the group of ocean floor basalts from seamount basalts on the assumption that Nb/Y is higher in seamount rocks. Macquarie Island rocks plot across this line, but their Y abundances are clearly lower than those of seamount rocks. The broken line is inserted to emphasize this difference and to show that it may also be generally true for ocean floor basalts.

The experimental study of *Green et al.* [1967] of the liquidus and near-liquidus phases from high-alumina olivine tholeiite compositions shows that in the composition most similar to ocean floor basalts, clinopyroxene is on the liquidus at 9 kb and is joined at lower temperatures by plagioclase, whereas plagioclase is on the liquidus in the pressure range 1 atm–6.8 kb and is joined at lower temperatures by olivine and clinopyroxene. These phase relations are consistent with the suggestion outlined above. Rapid eruption of magma of this composition at about its liquidus from depths equivalent to 9 kb pressure could lead to a situation where aluminous clinopyroxene stable at higher pressures would react with the magma, and plagioclase and olivine would be precipitated as the magma fractionally crystallized to move to a cotectic relationship at lower pressures. The rare clinopyroxene xenocrysts of ocean floor basalts should prove to be highly aluminous if this suggestion is valid.

Orthopyroxene appears unlikely to crystallize as a near-liquidus phase from olivine tholeiite magma at pressures below 5 kb on the basis of experimental studies so far carried out. This relationship may account for the absence of orthopyroxene from the plagioclase dunite ultramafic zone of the Macquarie Island layered gabbro complex. In the Stillwater complex, orthopyroxene appears as a cumulus phase very low in the ultramafic zone, whereas in the ultramafic zone of the Bushveld complex, orthopyroxene appears as a postcumulus phase at lower levels and as a cumulus phase at higher levels [*Wager and Brown*, 1967]. Whereas the relative order of the first appearances of cumulus and postcumulus phases is certainly in part a function of bulk composition and other compositional variables such as fO_2 [*Hamilton et al.*, 1964], the order is also a function of pressure and temperature. Possibly, the Macquarie Island layered gabbro complex crystallized at much lower pressures than the Stillwater and Bushveld complexes. The Island Lake gabbro with its mineralogy of olivine + orthopyroxene + possibly low calcium clinopyroxene + plagioclase and the coronite and symplectite structures may have crystallized at higher pressures. Studies of the gabbros may provide clues to assessing the depths at which they crystallized.

ORIGIN OF MACQUARIE ISLAND

Varne et al. [1969] earlier pointed out that the extrusive rocks and associated sediments of Macquarie Island are characteristic of deposition on the ocean floor at depths between 2000 and 4000 meters. The additional studies reported here have confirmed that basalts and dolerites from the island closely resemble rocks dredged from the ocean floors both in their petrography and their chemistry. In summary, the island is formed of fault-bounded blocks. The northern part of the island is composed mainly of serpentinized peridotites, gabbros, and dolerite dyke swarms metamorphosed under greenschist and amphibolite facies conditions. These intrusive rocks were derived from greater crustal depths than the fresh or slightly metamorphosed extrusive rocks that make up most of the southern part of the island, although some of the extrusive rocks may have been once buried and metamorphosed under greenschist facies conditions.

We suggest that the island represents a faulted section through different levels of oceanic crust. A hypothetical vertical section through the island, undisturbed by faulting, has at the top a layer of unmetamorphosed extrusive rocks overlying greenschist facies extrusive rocks. This layer of extrusive rocks is intruded by many dykes. Beneath this layer is a layer of greenschist and amphibolite facies dolerite dyke swarms together with gabbro and serpentinized peridotite masses.

Miyashiro et al. [1970] have based their two alternative models for Mid-Atlantic ridge oceanic crust mainly on petrographic studies of dredge hauls. In one model, the oceanic crust is composed principally of serpentinites accompanied by smaller amounts of metamorphosed basalts and gabbros and overlain by a thinner layer of relatively unmetamorphosed basalts. In the other model, serpentinites are less voluminous and the oceanic crust is formed mainly of basalts and gabbros. Macquarie Island oceanic crust fits either of these models equally well. However, the presence of the dolerite dyke swarms seems to require 100% crustal extension to account for their emplacement, a mode of formation amounting to sea floor spreading [*Moores and Vine*, 1971].

Cann [1970*b*] has used sea floor spreading and plate tectonics theory to infer the nature of geologic processes that could be taking place at spreading mid-oceanic ridges. He suggests that in the active spreading zone the oceanic crust consists of a layer of pillow lavas fed by dykes that overlies a zone of dykes without lavas that become more coarsely grained downward toward a basaltic magma chamber at the base of the crust. As a consequence of this process, the final crustal structure envisaged by *Cann* [1970*b*] is layered. Unmetamorphosed pillow lavas overlie pillow metamorphosed at low grades (greenschist facies and below). The lavas are cut by feeder dykes. This layer of extrusive rocks overlies a layer composed of dyke swarms, largely metamorphosed under amphibolite facies conditions and grading downward through more coarsely grained dykes into massive layered gabbro forming the lower oceanic crust. Serpentinite diapirs from the uppermost part of the upper mantle intrude these rocks. Cann's model layered oceanic crust is very similar to Macquarie Island oceanic crust.

Although Macquarie Island forms part of the crest of the Macquarie ridge, the rocks of the island were not necessarily formed by a process related to the tectonic evolution of the ridge. In plate tectonics theory, the ridge is assumed to mark the boundary between the Indian-Australian and the Pacific plates. *Hayes and Talwani* [this volume] conclude that the ridge has evolved as a consequence of the interaction of these two plates. They suggest that the proximity of the ridge to the pole of relative motion between the plates has caused a combination of tectonic regimes (extension, compression, and transcurrent motion) to be active along the plate boundary. They also tentatively suggest that a small amount of extension may have occurred within the central part of the ridge. If the rocks of Macquarie Island were formed as the result of such an episode, the angular separation between the concordant northwest direction of narrow linear magnetic anomalies and structural trends in areas of extrusive rocks on the island [*Williamson and Rubenach*, this volume] and the north-northeast strike of this part of the ridge suggests that sea floor spreading did not occur perpendicular to the ridge axis. However, although tectonic activity along the ridge has caused the island to appear, the tectonic regimes in the Macquarie ridge region may not have been the same when the rocks of the island were formed.

Magnetic profiles spanning the period from 50 to 10 m.y. ago give half-spreading rates of 2.65 cm/yr between Australia and Antarctica [*Weissel and Hayes*, 1971] and 1.4–1.65 cm/yr between the Campbell Plateau and Antarctica [*Pitman et al.*, 1968]. From 10 m.y. ago to the present, the half-spreading rate on the Southeast Indian ridge is 3.7 cm/yr [*Weissel and Hayes*, 1971] and on the Pacific-Antarctic ridge about 3.0–3.5 cm/yr [*Pitman et al.*, 1968].

The implications of these changes in spreading rates about 10 m.y. ago have been discussed by *Griffiths and Varne* [1972], who conclude that the present plate geometry in the region may not have been established for more than 10 m.y.

Some of the freshest volcanic rocks on Macquarie Island are associated with sediments that appear to be middle or early Miocene in age. These rocks could therefore have formed before the existence of the present tectonic regime, perhaps during the evolution of the south Tasman Sea.

Wherever the rocks of Macquarie Island were formed, it seems likely that they originated at a spreading zone in an oceanic environment. We suggest that the various geologic relationships we have described between the different rock types on the island could also exist between their counterparts beneath the oceans.

Acknowledgments. The Antarctic Division of the Commonwealth of Australia Department of Supply generously arranged transport to and supported field work on the island. We are grateful to Dr. G. Johnstone, P. Williamson, and R. Gould for carrying many rock specimens in the field, and to Dr. J. R. Cann and C. Winter for providing the analytical data. Dr. Cann also offered much valuable advice and made available for comparison the collection of ocean floor rocks housed at the University of East Anglia. Our research was supported financially by the University of Tasmania, the Australian Research Grants Committee, and the grant of a Royal Society Nuffield Foundation Commonwealth Bursary to R. Varne.

REFERENCES

Cann, J. R., Spilites from the Carlsberg ridge, Indian Ocean, *J. Petrology, 10*, 1–19, 1969.

Cann, J. R., Rb, Sr, Y, Zr and Nb in some ocean floor basaltic rocks, *Earth Planet. Sci. Lett., 10*, 7–11, 1970*a*.

Cann, J. R., New model for the structure of the oceanic crust, *Nature London, 226*, 928–930, 1970*b*.

Cann, J. R., Petrology of basement rocks from Palmer ridge, N.E. Atlantic, *Phil. Trans. Roy. Soc. London, Ser. A, 268*, 605–617, 1971.

Clark, S. P., Jr., J. F. Schairer, and J. de Neufville, Phase relations in the system $CaMgSi_2O_6$-$CaAl_2SiO_6$-SiO_2 at low and high pressure, *Carnegie Inst. Washington Yearb., 61*, 59–68, 1962.

Cullen, D. J., Macquarie Island, provisional bathymetry, scale 1:200,000, Island Series, N. Z. Oceanogr. Inst., Wellington, 1969.

Engel, C. G., and T. E. Chase, Composition of basalts dredged from seamounts off the west coast of central America, *U.S. Geol. Surv. Prof. Pap. 525C*, 161–163, 1965.

Engel, C. G., and A. E. J. Engel, Volcanic rocks dredged southwest of the Hawaiian Islands, *U.S. Geol. Surv. Prof. Pap. 550-D*, 104–108, 1966.

Green, T. H., D. H. Green, and A. E. Ringwood, The origin of high-alumina basalts and their relationships to quartz tholeiites and alkali basalts, *Earth Planet Sci. Lett., 2*, 41–51, 1967.

Griffiths, J. R., and R. Varne, Evolution of the Tasman Sea, Macquarie ridge, and Alpine fault, *Nature London, 235*, 83–86, 1972.

Hamilton, D. L., C. W. Burnham, and E. F. Osborn, The solubility of water and effects of oxygen fugacity and water content on crystallization in mafic magmas, *J. Petrology, 5*, 21–39, 1964.

Hayes, D. E., and M. Talwani, Geophysical investigation of the Macquarie ridge complex, in *Antarctic Oceanology II: The Australian–New Zealand Sector, Antarctic Res. Ser.,* vol. 19, edited by D. E. Hayes, pp. 211–234, AGU, Washington, D. C., 1972.

Kuno, H., High-alumina basalt, *J. Petrology, 1,* 121–145, 1960.

Kuno, H., Differentiation of basalt magmas, in *Basalts, Poldervaart Treatise on Rocks of Basaltic Composition,* vol. 2, edited by H. H. Hess and A. Poldervaart, pp. 623–688, Interscience, New York, 1968.

Matthews, D. H., Altered basalts from Swallow Bank, an abyssal hill in the NE Atlantic, and a nearby seamount, *Phil. Trans. Roy. Soc. London, Ser. A, 268,* 551–571, 1971.

Mawson, D., Macquarie Island, its geography and geology, *Australasian Antarctic Expedition 1911–1914, Sci. Rep. Sec. A:5,* Govt. Print. Office, Sydney, 1943.

Miyashiro, A., F. Shido, and M. Ewing, Petrologic models for the Mid-Atlantic ridge, *Deep Sea Res., 17,* 109–123, 1970.

Miyashiro, A., F. Shido, and M. Ewing, Metamorphism in the Mid-Atlantic ridge near 24° and 30°N, *Phil. Trans. Roy. Soc. London, Ser. A, 268,* 589–603, 1971.

Moores, E. M., and F. J. Vine, The Troodos Massif, Cyprus and other ophiolites as oceanic crust: Evaluation and implications, *Phil. Trans. Roy. Soc. London, Ser. A, 268,* 443–466, 1971.

Muir, I. D., C. E. Tilley, and J. Scoon, Basalts from the northern part of the rift zone of the Mid-Atlantic ridge, *J. Petrology, 5,* 409–434, 1964.

Nicholls, G. D., and M. R. Islam, Geochemical investigations of basalts and associated rocks from the ocean floor and their implications, *Phil. Trans. Roy. Soc. London, Ser. A, 268,* 469–486, 1971.

Nicholls, G. D., A. J. Nalwark, and E. E. Hays, The nature and composition of rock samples dredged from the Mid-Atlantic ridge between 22°N and 52°N, *Mar. Geol., 1,* 333–343, 1964.

Pearce, J. A., and J. R. Cann, Ophiolite origin investigated by discriminant analysis using Ti, Zr, and Y, *Earth Planet. Sci. Lett., 12,* 339–349, 1971.

Pitman, W. C., III, E. M. Herron, and J. R. Heirtzler, Magnetic anomalies in the Pacific and sea floor spreading, *J. Geophys. Res., 73*(6), 2069–2085, 1968.

Smith, R. E., Redistribution of major elements in the alteration of some basic lavas during burial metamorphism, *J. Petrology, 9,* 191–219, 1968.

Summerhayes, C. P., Marine geology of the New Zealand subantarctic sea floor, *N. Z. Dep. Sci. Ind. Res. Bull. 190,* 94 pp., 1969.

Taylor, S. R., and A. J. R. White, Trace element abundances in andesites, *Bull. Volcanol., 29,* 177–194, 1966.

Turner, F. J., *Metamorphic Petrology,* 403 pp., McGraw-Hill, New York, 1968.

Varne, R., R. D. Gee, and P. G. J. Quilty, Macquarie Island and the cause of oceanic linear magnetic anomalies, *Science, 166,* 230–233, 1969.

Wager, L. R., and G. M. Brown, *Layered Igneous Rocks,* 558 pp., Oliver and Boyd, Edinburgh, 1968.

Watkins, N. D., and B. M. Gunn, Petrology, geochemistry, and magnetic properties of some rocks dredged from the Macquarie ridge, *N. Z. J. Geol. Geophys., 14,* 153–168, 1971.

Weissel, J. K., and D. E. Hayes, Asymmetric seafloor spreading south of Australia, *Nature London, 231,* 518–522, 1971.

Williamson, P., and M. Rubenach, Preliminary report on geophysical studies on Macquarie Island, in *Antarctic Oceanology II: The Australian–New Zealand Sector, Antarctic Res. Ser.,* vol. 19, edited by D. E. Hayes, pp. 243–249, AGU, Washington, D. C., 1972.

3

Marine Sediments

INTRODUCTION: MARINE SEDIMENTS OF THE SOUTHEAST INDIAN OCEAN

JOHN R. CONOLLY

Department of Geology, University of South Carolina, Columbia, South Carolina 29208

Abstract. Up to 5 km of sediment has been deposited off the antarctic continent since the separation of the Australian-antarctic continents; this sediment has built up the surface of the deep South Indian plain some 1 km higher than the Australian counterpart, the South Australian plain. Pelagic sediment forms a blanket less than 100 meters thick across the young mid-oceanic ridge, and Pleistocene sedimentation rates range from 0.5 to 2 cm/1000 yr. Interbedded calcareous and siliceous ooze layers, 10–50 cm thick, record striking changes in the Pleistocene planktonic record and the Tasman manganese pavement formed between Tasmania and the mid-oceanic ridge in the past 2.5 m.y.

The five papers in this section discuss the sediments of the southeast Indian Ocean. The basic data for all the papers came from about 300 piston cores. Some of the piston cores were taken during Lamont-Doherty Geological Observatory expeditions of the *Vema* and *Robert D. Conrad,* but most of the cores were taken during cruises of the *USNS Eltanin* [*Conolly and Payne,* this volume, Figure 1]. All the papers except that of *Glass* [this volume] have drawn extensively from the *Eltanin* cores. Glass has reviewed the occurrence of Australasian microtektites, and his review is based on *Vema* and *Conrad* cores. *Eltanin* cores taken from the southeast Indian Ocean have been routinely sampled and examined for paleomagnetic stratigraphy and faunal stratigraphy [*Watkins* and *Kennett,* this volume] and have been described and sampled for sediment variation [*Conolly and Payne,* this volume]. In addition, specific cores have been sampled for sediment chemistry studies [*Scott et al.,* this volume].

SEDIMENT DISTRIBUTION AND SEDIMENTATION RATES

Houtz and Markl [this volume] describe the distribution of sediment thickness over the oceanic basement as seen from seismic profiler records. Their results show that large areas of the mid-oceanic ridge are generally covered with less than 100 meters of sediment (Figure 1). Direct sampling and analysis of the uppermost few meters of sediment provide no surprises. Sedimentation rates of 0.5–2 cm/1000 yr calculated from paleomagnetic magnetic stratigraphy [*Watkins and Kennett,* this volume] compare well with sedimentation rates calculated by using α and γ spectrometry of excess ^{230}Th [*Scott et al.,* this volume].

Ooze measuring 3–5 meters thick was deposited across the mid-oceanic ridge during the Brunhes period, and approximately the same amount of sediment was deposited during the Matuyama period. Thus a total of 6–10 meters of sediment has been deposited during the past 2 m.y. [*Conolly and Payne,* this volume]. If ooze sedimentation rates have remained the same during the past 30 m.y. of ridge development [*Weissel and Hayes,* this volume], then up to 150 meters of ooze may be expected to cover the ridge where it is 30 m.y. old. Since most of the ridge is less than 30 m.y. old and since this part of the ridge is covered with less than 100 meters of sediment, then sedimentation rates may well have been the same or lower than those observed for the past 2 m.y.

As the continents are approached from the mid-oceanic ridge, the sediment pile thickens markedly to greater than 2 km near the base of the continental rise off both Antarctica and Australia [*Houtz and Markl,* this volume, Figure 3]. We must presume that even higher average sedimentation rates occur on the continental rise and slope off both continents, but thus far there is insufficient core data to show that these higher rates exist. *Scott et al.* [this volume] have made some useful calculations of the rate of clay accumulation from Al_2O_3 analyses. They have found that terrigenous clay sedimentation rates are much higher close to the antarctic continent, whereas accu-

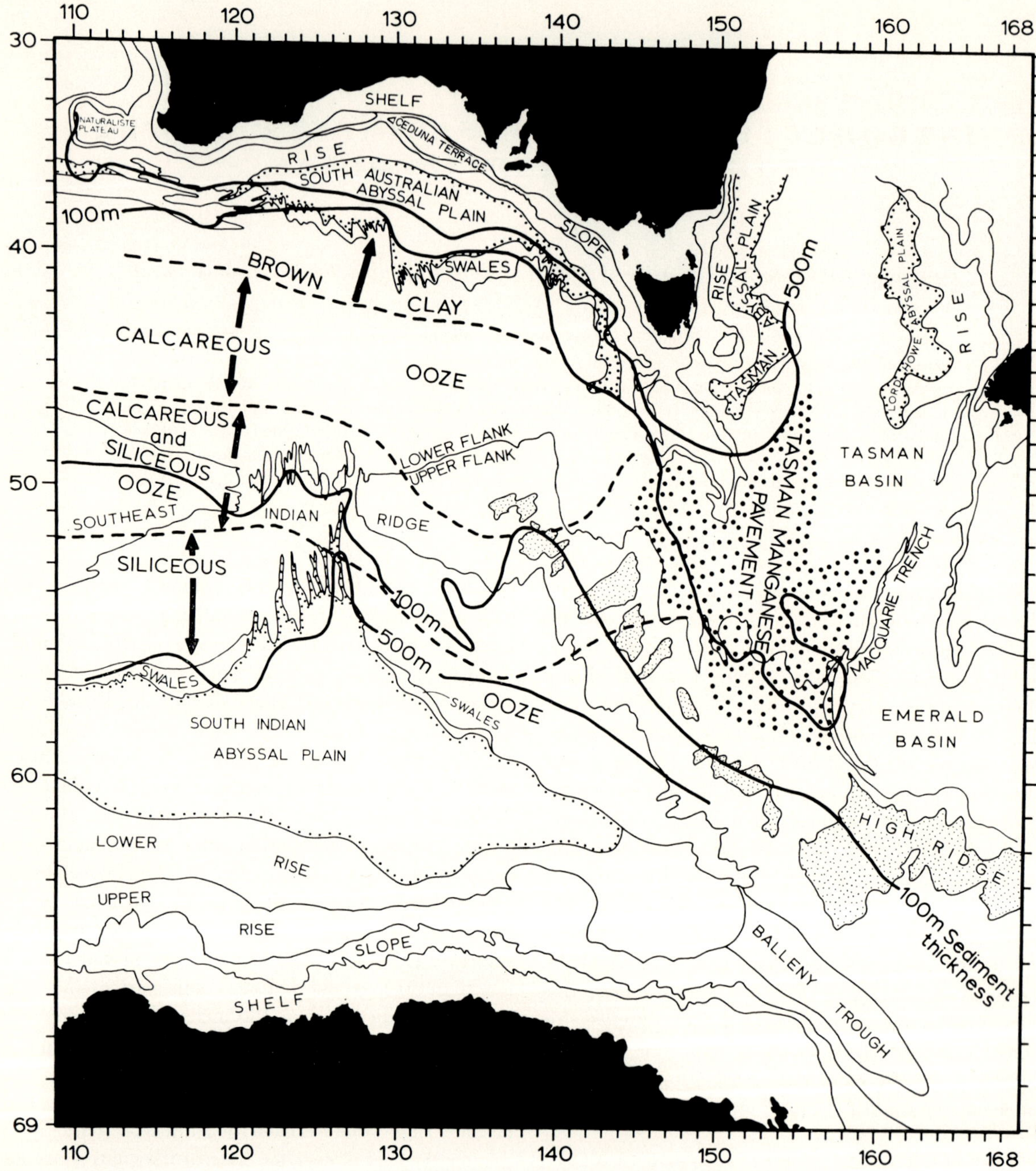

Fig. 1. Physiographic diagram of the southeast Indian Ocean after *Hayes and Conolly* [this volume]. The diagram shows sediment thickness above oceanic basement after *Houtz and Markl* [this volume], the Tasman manganese pavement, and the location of major Brunhes ooze types across the mid-oceanic ridge south of Australia after *Conolly and Payne* [this volume].

mulation rates based on Fe, Mn, and U in the ridge sediments are not significantly different from those found for sediments close to the antarctic continent.

Scott et al. [this volume] also show a strong positive correlation between the excess ^{230}Th content in surface sediments and the Fe/ (Al + Fe + Mn) oxide ratios; the correlation suggests that ^{230}Th is removed from sea water in association with Fe and Mn colloids. Scott et al. also suggest that the relatively low ^{230}Th content found close to the continent may be caused by the transportation of ^{230}Th to the north by currents leaving the continent.

PLEISTOCENE CLIMATE HISTORY

Little is known about the details of the changes in antarctic ocean climate as registered in the deep-sea ooze record. *Kennett*'s [1970] climate analysis of cores from the subantarctic sector of the South Pacific Ocean suggested that at least 12 major climate changes occurred during the Brunhes magnetic epoch. Preliminary faunal data based on a combination of diatom, radiolarian and foraminiferal data by *Conolly and Payne* [this volume] support Kennett's conclusions. However, the most striking feature of the Brunhes and Matuyama ooze preserved in the southeast Indian Ocean is an interbedding and mixing of siliceous and calcareous ooze (Figure 1).

Most ooze recovered from the region of the mid-oceanic ridge south of Australia consists of layers of calcareous ooze that was presumably deposited when the antarctic convergence lay in a more southerly position and of siliceous ooze that was deposited when the antarctic convergence lay in a more northerly position. At least 10–12 layers of different ooze composition occur in most Brunhes ooze sections, and a smaller number of layers occur in the Matuyama ooze [*Conolly and Payne,* this volume].

Detailed faunal studies of long cores that penetrate this ooze are currently needed to fully document these changes before any refinement of our paleoclimatic interpretations can be made.

CONTINENTAL MARGIN SEDIMENTATION PATTERNS

A review of continental margin sedimentation is given in two papers [*Conolly and Payne,* this volume; *Payne and Conolly,* this volume]. Total deposition from the Australian continent to the deep ocean is strikingly less than that from the antarctic continent. This difference is evidenced by the high level of the abyssal plain off Antarctica (2500 fm) as compared with the level of the abyssal plain south of Australia (3000 fm). This change in level implies that terrigenous sedimentation off the antarctic continent has been much more extensive during the past 50 m.y. than that off Australia.

The relative lack of large feeding rivers from the flat interior of Australia is in great contrast with the ice-covered antarctic continent during the past 5–10 m.y., and this contrast may be the major reason for the difference in the sedimentation observed. However, much is still to be learned from deep-sea drilling in the antarctic continental margin, particularly with reference to the climatic and tectonic history of the antarctic continent.

Patterns of continental margin sedimentation off both continents are consistent with those observed off other continents. Graded beds of sand, silt, and clay characterize the continental rise and proximal abyssal plain, and sedimentation rates are relatively high (20 meters during the Brunhes magnetic epoch). The deep distal margins of the abyssal plain are commonly characterized by swale topography composed of beds of silt and clay presumably representing the tail end of major sediment-carrying flows that originate on the continental margin.

AUSTRALASIAN MICROTEKTITES

A review of our present knowledge of microtektite distribution by *Glass* [this volume] shows that approximately 10^{14} grams of microscopic glassy particles were deposited on the ocean floor south of Australia and the adjacent Indian Ocean at about the time of the magnetic field reversal of the earth 0.7 m.y. ago. Glass presents physical and chemical data that show that these tektites are part of the Australasian tektite-strewn field and that they represent a remarkable chronostratigraphic event.

TASMAN MANGANESE PAVEMENT AND REGIONAL UNCONFORMITIES IN SOUTH TASMAN SEA

The paleomagnetic and faunal studies of *Watkins and Kennett* [this volume] have led to the discovery of a large area of nondeposition or low deposition in the south Tasman Sea region (Figure 1). Watkins and Kennett show that 'normal' ooze of Brunhes to middle Gauss age (0–3.0 m.y.) is largely absent over this region. They theorize that this spectacular area of nondeposition was created by a major change in water circulation that started about 3.5 m.y. ago because of an increase in the production of antarctic bottom water due to a critical development of antarctic sea and shelf ice.

Conolly and Payne [this volume] also describe this region and establish that it consists of a pavement of manganese nodules and/or reworked or 'relict' foraminiferal and manganese sand that overlies firm olive-gray siliceous ooze. The distribution of this pavement, herein called the Tasman manganese pavement, follows the deep troughs of the Tasman fracture zone that separate Tasmania from the mid-oceanic ridge. From these deep troughs, the pavement extends eastward into the adjacent deep Tasman sea floor. These authors are of the opinion that the manganese pavement was created by bottom water movement through the Tasman fracture zone into the adjacent Tasman basin. Whether a change in topography due to sea floor spreading or a change in the antarctic climate caused the onset of the new hydrologic regime that occurred about 3 m.y. ago is still difficult to assess.

REFERENCES

Conolly, J. R., and R. R. Payne, Sedimentary patterns within a continent–mid-oceanic ridge–continent profile: Indian Ocean south of Australia, in *Antarctic Oceanology II: The Australian–New Zealand Sector, Antarctic Res. Ser.*, vol. 19, edited by D. E. Hayes, pp. 295–315, AGU, Washington, D. C., 1972.

Glass, B. P., Australasian microtektites in deep-sea sediments, in *Antarctic Oceanology II: The Australian–New Zealand Sector, Antarctic Res. Ser.*, vol. 19, edited by D. E. Hayes, pp. 335–348, AGU, Washington, D. C., 1972.

Hayes, D. E., and J. R. Conolly, Morphology of the southeast Indian Ocean, in *Antarctic Oceanology II: The Australian–New Zealand Sector, Antarctic Res. Ser.*, vol. 19, edited by D. E. Hayes, pp. 125–145, AGU, Washington, D. C., 1972.

Houtz, R. E., and R. G. Markl, Seismic profiler data between Antarctica and Australia, in *Antarctic Oceanology II: The Australian–New Zealand Sector, Antarctic Res. Ser.*, vol. 19, edited by D. E. Hayes, pp. 147–164, AGU, Washington, D. C,. 1972.

Kennett, J. P., Pleistocene paleoclimates and foraminiferal biostratigraphy in subantarctic deep-sea cores, *Deep Sea Res.*, *17*, 125–140, 1970.

Payne, R. R., and J. R. Conolly, Turbidite sedimentation off the antarctic continent, in *Antarctic Oceanology II: The Australian–New Zealand Sector, Antarctic Res. Ser.*, vol. 19, edited by D. E. Hayes, pp. 349–364, AGU, Washington, D. C., 1972.

Scott, M. R., J. K. Osmond, and J. K. Cochran, Sedimentation rates and sediment chemistry in the South Indian basin, in *Antarctic Oceanology II: The Australian–New Zealand Sector, Antarctic Res. Ser.*, vol. 19, edited by D. E. Hayes, pp. 317–334, AGU, Washington, D. C., 1972.

Watkins, N. D., and J. P. Kennett, Regional sedimentary disconformities and Upper Cenozoic changes in bottom water velocities between Australasia and Antarctica, in *Antarctic Oceanology II: The Australian–New Zealand Sector, Antarctic Res. Ser.*, vol. 19, edited by D. E. Hayes, pp. 273–293, AGU, Washington, D. C., 1972.

Weissel, J. K., and D. E. Hayes, Magnetic anomalies in the southeast Indian Ocean, in *Antarctic Oceanology II: The Australian–New Zealand Sector, Antarctic Res. Ser.*, vol. 19, edited by D. E. Hayes, pp. 165–196, AGU, Washington, D. C., 1972.

REGIONAL SEDIMENTARY DISCONFORMITIES AND UPPER CENOZOIC CHANGES IN BOTTOM WATER VELOCITIES BETWEEN AUSTRALASIA AND ANTARCTICA

N. D. WATKINS AND J. P. KENNETT

Graduate School of Oceanography, University of Rhode Island, Kingston, Rhode Island 02881

Abstract. Paleomagnetic and micropaleontologic dating of 126 sedimentary cores collected during *Eltanin* cruises 16, 26, 27, 34, 35, 36, and 39 south of Australia and New Zealand reveals the existence of an area of the order of 3×10^6 km^2 centered at the junction of the Southeast Indian ridge, the south Tasman basin, and the Macquarie ridge, where sediments of Brunhes to middle Gauss age (t = 0–3.0 m.y.) are largely absent. Flanking this zone toward the Australian shelf, the South Tasman rise, the northern Tasman Sea, and the antarctic continental shelf, the cores are mostly of Brunhes age (t = 0–0.69 m.y.) throughout. Despite the widespread nature of the sedimentary hiatus detected, only five cores have sediments older than 6 m.y. The preferred interpretation for these data is an increase at some time since t = 3.5 m.y. in the velocity of the antarctic bottom water from dominantly less than 10 cm sec^{-1} to dominantly more than 10 cm sec^{-1}. This increase enhances the removal of finer sedimentary fractions and the corresponding formation of residual sands, increases in micromanganese nodule concentrations, and development of manganese nodules. Available bottom water velocity data, bottom photograph features, and limited observations of the sediment type all support this interpretation. Local factors, such as topographic variations, are subordinate to bottom water velocity in creating the coherent regional sedimentary regime. Cores of probable accretionary sands are identified in the scour zone, presumably in pockets possessing local velocity minima. The proposed increase in average bottom current velocity probably resulted from a marked and persistent increase in the production of antarctic bottom water.

During the last 4 years, the paleomagnetic method has become established as an important tool for dating deep-sea sedimentary cores of Pliocene and Pleistocene age. The method is based on recognition of the major features of the known geomagnetic polarity history (Figure 1) as summarized by *Cox* [1969]. These polarity changes, which are global synchronous phenomena, are readily recorded in most deep-sea sediments, although fine details may become blurred by faunal redeposition [Watkins, 1968] and other means.

A constant sedimentation rate would ideally lead to the reproduction of the recognizable polarity history (Figure 1) in the sedimentary cores; in the absence of such ideal circumstances, the identification of polarity boundaries in sedimentary cores is clearly susceptible to ambiguities. Nevertheless, for most of the ocean, if normal polarity sediment is found at the surface of or throughout any core, the probability is very high that Brunhes epoch (t = 0–0.69 m.y.) material is represented.

Disconformities and rapid changes in sedimentation rates may be relatively common in deep-sea cores and may thus make the identification of specific polarity changes difficult, if not impossible, without independent micropaleontologic dating methods. Almost all deep-sea cores from water depths shallower than 5000 meters contain one or more microfossil groups that can be used in age determinations. Plio-cene-Pleistocene micropaleontologic zonations have now been established for most of the major water masses and are being used in conjunction with paleomagnetic stratigraphy. This combination of disciplines is leading to considerable advances in the understanding of faunal, climatic, and other changes as recorded in deep-sea sediments [*Berggren et al.*, 1967; *Hays et al.*, 1969; *Kennett et al.*, 1971].

The methods are, by definition, appropriate to the definition of at least gross changes in sedimentation rates for the past 5 m.y. Changes may be expected, for instance, with distance from continental source areas and as a function of variations in biogenic productivity with time. The most drastic change in the sedimentation rate is, of course, a halt in sedimen-

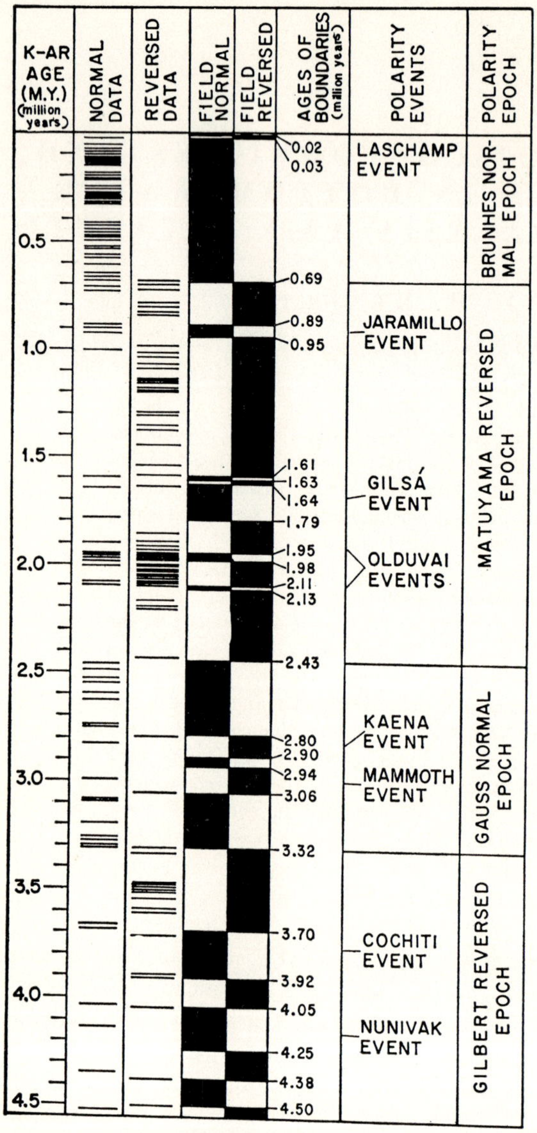

Fig. 1. The known geomagnetic polarity scale for the period $t = 0$–4.5 m.y. The diagram includes constituent data points, inferred periods of normal (darkened areas) and reversed (open areas) polarity, and established nomenclature of the polarity events and epochs. Doubt now exists about the reality of the Laschamp event as a global polarity change. Figure taken from *Cox* [1969].

tation or even a negative sedimentation rate (or scour). When present in the deeper parts of oceanic regions, such extreme variations in sedimentation rates, if persistent over large regions with different topographic regimes, are most readily understood to be functions of changes in bottom water velocity.

In this communication, we present the paleomagnetic data for the *Eltanin* collection of deep-sea sedimentary cores from the region between Australia, New Zealand, and the antarctic continent (Figure 2, Table 1). Micropaleontologic control has been added for all cores. We shall examine the distribution of ages in the core collection for the purpose of defining and interpreting the causes of regional sedimentation rate variations.

PREVIOUS WORK

Paleomagnetic determinations of sedimentation rates have been made for cores from all the oceans. In a broad analysis of sedimentation rates in the Indian Ocean, *Opdyke and Glass* [1969] included two cores from south of Western Australia. The Brunhes epoch sediments are thin in these two cores possibly because of some recent erosion.

The only regional studies of paleomagnetically defined sedimentation rates so far made are by *Goodell and Watkins* [1968] and *Opdyke and Foster* [1970]. The former study used 159 cores from the *Eltanin* collection; the cores were taken from between 20°W and 160°E in the South Atlantic and South Pacific oceans. This survey did not incorporate micropaleontologic control, but the construction of isopachs on the Brunhes epoch sediment thickness led to the detection of an east to west zone of low deposition rate across the Southeast Pacific basin into the Scotia Sea. The zone is most reasonably explained as due to the inhibiting effect of the circumantarctic current on normal depositional patterns. *Hollister and Heezen* [1967] have defined, by using *Eltanin* sea bottom photographs, a zone in the same region where bottom currents are active.

Conolly and von der Borch [1967], *von der Borch et al.* [1970], and *Conolly et al.* [1970] have compiled seismic, aeromagnetic, gravity, and bottom photograph data and have analyzed a number of carbonate cores from the continental shelf to the abyssal plain off South Australia. No age data were given. *Conolly* [1969] has also described the general sediment types in eight piston cores from the south and central Tasman basin. Two of these are of early Miocene age.

Since the circumantarctic current is probably the world's major current in terms of volume of water transported per second [*Callahan,* 1971] and since the dynamics of ocean currents and bottom water transport are likely to be associated to some extent, the consideration of bottom water velocity as a factor relevant to the sedimentary character in the study area (Figure 2) is clearly appropriate.

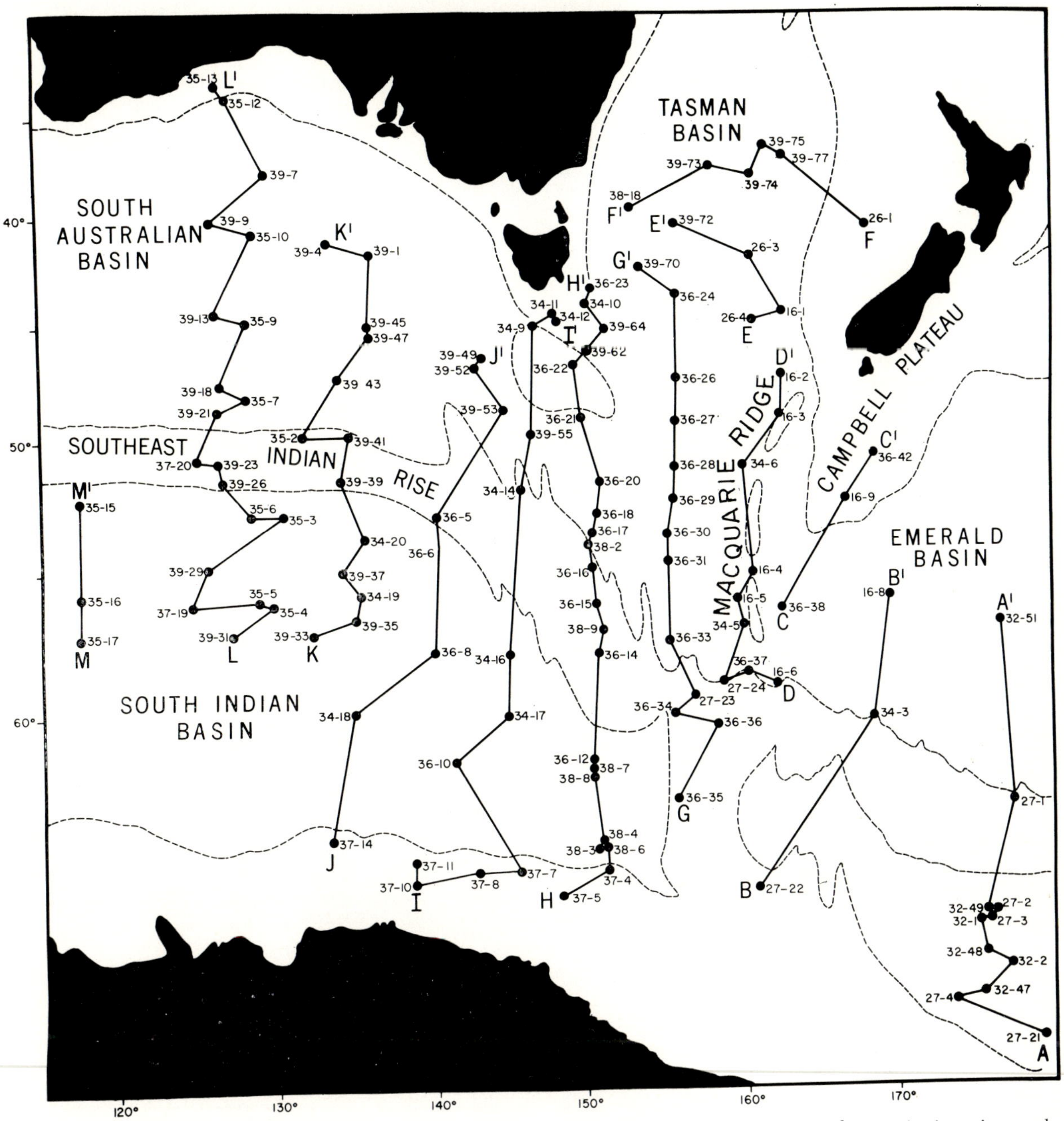

Fig. 2. Map showing locations of cores and selected traverses A–A′ to M–M′. Number next to each core site is cruise number followed by core number. Bathymetric contours are 2000 meters in depth. For exact locations and water depth at each core site, see Table 1.

METHODS

Shipboard methods have been described previously [*Watkins and Goodell*, 1967]. The cores, which were collected in plastic liners, were 6 cm in diameter. The liners were split along a diameter, and the sediment was separated by using a nonmagnetic fiber. Specimens were taken for the paleomagnetic study by inserting oriented 10.7 cm^3 plastic boxes into the sediment face. Those cores showing signs of sediment disturbance and very limited recovery were not examined. All specimens were demagnetized in alternating magnetic fields of 100 or 150 oersteds to minimize unstable components. The direction and intensity of remanent magnetism were measured by using a 5-cycle/sec spinner magnetometer [*Foster*, 1966]. In several cases, the natural remanent magnetism (which existed

TABLE 1. Location, Length, and Water Depth for Each Core

Cruise	Core	Length, cm	Latitude, S	Longitude, E	Water Depth, meters
			Traverse A–A′		
27	21	1584	69°02′	179°50′	3456
32	47	1089	68°06′	176°10′	3429
27	4	1968	68°02.9′	174°35.5′	3449
32	2	882	67°22′	178°44′	3674
32	48	1562	67°00′	176°16′	3526
27	3	1689	66°04′	176°30′	3537
32	1	240	66°10′	175°30′	3486
32	49	1264	66°00′	176°25′	3502
27	2	1694	66°01.5′	176°25.8′	3490
27	1	484	63°00′	177°42.5′	3356
32	51	731	56°50′	176°52′	4868
			Traverse B–B′		
27	22	808	64°58.2′	160°37.4′	2944
34	3	565	60°01.81′	167°21.88′	4517
16	8	292	56°08′	169°42′	5117
			Traverse C–C′		
36	38	1062	56°27′	161°45.7′	4129
16	9	520	52°25′	166°44′	1051
36	42	294	51°00′	167°15′	460
			Traverse C–C′		
16	6	613	58°59′	161°55′	4636
36	37	517	58°42′	159°29′	3877
27	24	569	59°05.2′	157°02.9′	3100
34	5	595	57°23.25′	159°59.64′	3795
16	5	290	56°05′	159°05′	3850
16	4	636	55°36′	160°12′	4151
34	6	530	51°21.84′	159°56.43′	3877
16	3	420	49°09′	162°00′	3919
16	2	694	47°08′	162°07′	4466
			Traverse E–E′		
26	4	1630	45°03.4′	160°04′	4810
16	1	586	44°12′	162°01′	4755
26	3	400	41°57′	160°02.5′	4872
39	72	1462	40°37.27′	154°36′	4645
			Traverse F–F′		
26	1	547	39°2.2′	172°09′	695
39	77	941	37°03.38′	161°56.9′	3946
39	75	923	36°29.31′	161°12.9′	4007
34	7	1205	38°17.1′	160°14.2′	4535
39	73	1592	37°39.35′	157°53′	4759
38	18	575	40°03′	152°14.5′	4535
			Traverse G–G′		
36	35	584	62°44′	154°58′	3493
36	36	572	60°23′	157°32′	2816
36	34	452	59°59′	155°02′	2794
27	23	955	59°37.1′	155°14.3′	3182
36	33	503	57°45′	154°53′	3433
36	31	461	55°00′	155°00′	4270
36	30	551	54°04′	155°00′	4087
36	29	500	53°00′	155°10′	3904
36	28	584	51°35′	155°09′	4426
36	27	1153	49°40′	155°00′	4535
36	26	1224	47°51′	155°08′	4691
36	24	569	44°00′	155°00′	4548
39	70	1673	42°23.8′	153°10′	4663

TABLE 1. (continued)

Cruise	Core	Length, cm	Latitude, S	Longitude, E	Water Depth, meters
			Traverse H–H′		
37	5	684	65°31′	147°26.3′	2990
37	4	584	64°49′	150°30′	3273
38	6	559	64°17.5′	150°11′	3456
38	3	943	64°14.5′	150°01.3′	3493
38	4	731	64°13.9′	150°3.8′	3658
38	8	544	61°48.6′	149°54.2′	3292
38	7	599	61°49.3′	149°53′	3658
36	12	603	61°45′	149°33′	4056
36	14	601	58°06′	150°10′	3054
38	9	610	57°27.7′	150°06.3′	3173
36	15	207	56°34′	150°15′	3517
36	16	590	55°07′	150°00′	3822
38	2	423	54°14.1′	149°57.7′	4060
36	17	612	54°02′	150°05′	3950
36	18	582	53°00′	150°00′	3877
36	20	541	51°47′	150°27′	3862
36	21	493	49°27′	149°08′	3846
36	22	422	47°32′	148°01′	1102
39	62	284	46°56.8′	149°32.6′	3219
39	64	453	45°33.6′	150°21′	4652
34	10	1186	44°31.6′	149°58.4′	2853
36	23	542	43°53′	150°02′	2533
			Traverse I–I′		
37	11	433	64°30′	137°59′	3118
37	10	414	65°14′	137°51′	2249
37	8	241	64°57′	142°31′	3036
37	7	844	65°00′	145°00′	3155
36	10	559	61°37′	140°13′	4206
34	17	582	60°11.8′	144°45′	3853
34	16	441	58°07′	144°56.5′	3548
34	14	519	52°11.9′	145°03.9′	3264
39	55	862	49°56.9′	145°55.8′	4810
34	9	588	45°20.09′	146°06.2′	2743
34	11	490	45°12.6′	147°48.2′	3932
34	12	2262	45°11.6′	147°47.7′	3932
			Traverse J–J′		
37	14	497	63°54′	132°27′	3288
34	18	922	60°00.03′	134°52.26′	4437
36	8	299	58°04′	139°56′	4371
36	6	299	54°32′	140°03′	3356
36	5	282	53°00′	140°00′	2962
39	53	411	48°48.5′	144°32′	3987
39	52	432	47°34.8′	142°59.7′	4581
39	49	392	47°06′	142°33.6′	4678
			Traverse K–K′		
39	33	968	57°29.9′	129°56.2′	4691
39	35	595	57°30.6′	133°58.5′	4627
34	19	262	56°39.90′	135°00′	4016
39	37	522	55°03′	133°58.10′	4298
34	20	606	54°10.5′	135°12.8′	3987
39	39	283	52°05.1′	133°56′	3447
39	41	393	49°53.12′	134°00.6′	3255
35	2	581	50°00′	131°36′	3365
39	43	453	47°26.35′	134°00.6′	4023
39	47	421	45°47.7′	136°48.1′	4298
39	45	471	45°09′	133°57.9′	4521
39	1	221	41°56′	136°50′	4810
39	4	582	41°26′	133°01.2′	4170

TABLE 1. (continued)

Cruise	Core	Length, cm	Latitude, S	Longitude, E	Water Depth, meters
			Traverse L–L′		
39	31	672	57°37′	126°15′	4663
35	4	1110	56°50′	129°35′	4594
35	5	520	56°02′	128°10′	4490
37	19	475	56°04′	124°54′	4572
39	29	1620	55°05.8′	126°04.4′	4700
35	3	410	53°08′	130°29′	3767
35	6	600	53°12′	128°06′	3996
39	26	933	52°58.5′	126°08.2′	4782
39	23	464	51°07.28′	126°10.60′	4791
37	20	234	51°10′	124°42′	4032
39	21	245	48°51.7′	126°01′	4261
35	7	530	48°58′	128°04′	3877
39	18	585	48°01.975′	126°07.8′	4700
35	9	523	45°02′	128°00′	5176
39	13	572	45°00.62′	125°58.9′	4663
35	10	985	41°58′	128°00′	4865
39	9	580	40°00.07′	126°02.7′	5121
39	7	991	38°13.11′	130°15.07′	5596
35	12	1113	34°30′	127°06′	4096
35	13	481	33°16′	126°24′	186
			Traverse M–M′		
35	17	675	58°40′	117°00′	4471
35	16	1212	56°00′	117°07′	4298
35	15	420	52°56′	117°00′	3764

prior to demagnetization) was measured initially. All data were processed and plotted by computer. Since the purpose of this study was to identify and use the positions of the major polarity changes in the core collection, the fine detail of the magnetic data reflecting noise or possible undiscovered short polarity events was suppressed by subjecting most results to a five point running average. Polarities were then nonsubjectively assigned to each data point by using the magnetic inclination results: inclinations above the horizontal were normal and those below the horizontal were reversed polarity in these high latitude Southern Hemisphere cores.

Because of the reconnaissance nature of this study, no detailed micropaleontologic investigations were carried out. In the majority of cores, examination of only three samples was necessary to check the age assignments based on magnetic stratigraphy; this examination was accomplished by taking specimens from near the top, middle, and bottom of these cores. Disconformities in many cores, however, made micropaleontologic examination necessary at closer intervals. These cores required study of up to 10 samples per core for age assignment of the magnetic stratigraphy. Intervals examined were rarely less than 100 cm and in no case were closer than 50 cm; thus the position of disconformities not marked by a polarity change may be as much as 50 cm from their exact position.

The correlation lines for the major polarity boundaries have been placed by applying the radiolarian zonation of *Opdyke et al.* [1966] and *Hays and Opdyke* [1967] and the foraminiferal zonation of *Kennett* [1970]. The radiolarian species that were particularly valuable in our correlations were *Triceraspyris* sp. and *Lychnocanium grande* in the early and middle Gilbert; *Desmospyris spongiosa*, the extinction of which coincided approximately with the Gauss-Matuyama boundary; *Eucyrtidium calvertense* and *Clathrocyclas bicornis*, the extinction of which occurred in the middle Matuyama; and *Saturnulus planetes*, which in subantarctic waters disappeared near the Brunhes-Matuyama boundary (t = 0.69 m.y.). The foraminiferal species that were particularly valuable in our correlations were *Globorotalia puncticulata* the disappearance of which is close to the Brunhes-Matuyama boundary and *Globorotalia inflata* and *Globorotalia crassaformis* the appearance of which in subantarctic waters is closely associated with the Brunhes-Matuyama boundary. Paleontologic control for the shorter duration magnetic polarity events (Figure 1) is often not available, especially for those events within the Gilbert epoch where detailed paleontologic control has yet to be established. Correlations of events are therefore made subjectively by using the magnetic data alone and comparing these data with the established geomagnetic polarity scale [*Cox*, 1969]. Several cores consisting primarily of glacial marine material and red clay contain little biogenic material and hence have also been dated by using magnetic data alone. Of the cores dated entirely as Brunhes epoch, 14 were dated by micropaleontologic means only, and the magnetic stratigraphy is inferred.

RESULTS

The location, length, and water depth for each core are given in Table 1. The paleomagnetic properties of a total of over 10,000 specimens from 126 cores have been measured. The results are given in Figure 3. Polarities for all data points are included. Intensities of magnetization range from 10^{-5} to 10^{-7} emu g^{-1}. Magnetic stabilities, which were inferred from the series of cores measured before and after demagnetization, are generally very high.

INTERPRETATION

Paleomagnetism

Inspection of the polarity data shows that cores with normal polarity throughout (and thus probably restricted to the Brunhes epoch) commonly occur in those parts of the traverses close to the continental shelves of Antarctica and Australia and in the northern Tasman Sea. Since the 0.7-m.y. isochron is not reached, only minimum sedimentation rates (i.e., up to 2 cm/1000 yr) can be derived.

The polarity data for the cores in the central part of the area, especially in the vicinity of the south Tasman Sea and the region midway between Tasmania and Antarctica (Figure 3, traverses G–G′, H–H′, I–I′, J–J′, K–K′, L–L′) possess an entirely different magnetic character with substantial segments of reversed polarity. Although this finding is clear evidence that sediments older than 0.7 m.y. are encountered, the ages cannot be fixed with any certainty by using the magnetic data alone.

Precision of Combined Paleomagnetic and Micropaleontologic Methods

Despite the dual methods employed for determination of time ranges several imperfections still exist, as discussed below. In almost all cases, cores restricted to the Brunhes epoch (t = 0–0.69 m.y.) have not been assigned to a more specific time interval. Such cores are therefore arbitrarily assigned to reach an age of $t = 0.6$ m.y.

Sedimentation rate extrapolation has been used to estimate the age ranges of many cores. Figure 4 shows six examples of the principle employed. Linear or nearly linear sedimentation rates exist in all these examples and range from 1 cm/1000 yr in core E32-49 to 0.25 cm/1000 yr in core E34–17. In the six examples shown, unambiguous identification of major polarity boundaries was possible in all cores except E36-16. When such unambiguous identification is available for at least two boundaries (including, if necessary, the t = 0 m.y. boundary that is assumed to exist at the top of a Brunhes epoch sedimentary sequence), then a sedimentation rate can be calculated. This calculation is not possible if only one boundary exists in the core or if the available polarity boundaries cannot be confidently identified as a known polarity change. A confidence of only ±0.4 m.y. in the age assignment results in many instances.

A major difficulty in the accurate delineation of the age range is the existence of disconformities and variable sedimentation rates in the cores. A subjective judgment made by using the magnetic data alone is then more commonly necessary to estimate the probable age ranges. Micropaleontologic data, however, make possible the recognition of disconformities and general dating of core sections to reduce subjectivity in the determination of polarity boundaries. Future detailed work on individual cores will in all probability result in more accurate identification of the polarity events, especially those in the Gilbert epoch.

We must stress the fact that if we assume that sedimentation rates have varied throughout a core, then any magnetic data can be forced into equivalence with virtually any part of the polarity scale. This degree of freedom clearly cannot be legitimately adopted. Linear extrapolation of sedimentation rates between assigned polarity boundaries must be used consistently, even if the resulting assigned ages result in a conflict between measured polarity and the details in the known polarity time scale. The possibility of the magnetic data being spurious in part and/or the polarity time scale being imperfectly known cannot be overlooked. In our opinion, a precision of ±0.4 m.y. can still be maintained for the Pliocene and Pleistocene.

Age Ranges of Each Core

Figure 5 shows the assigned age ranges of all cores in each traverse. The limitations in our confidence of these age ranges have been discussed above. Combining these age ranges with the core lengths given in Figure 3 and Table 1 yields the sedimentation rates. The following additional notes on individual cores shown in Figure 5 are given in the order of the traverses.

Traverse A–A′. Four of the cores (E27-21, E27-2, E27-3, E27-4) appear to have sedimentation rates that are nearly linear for the past 2.5 m.y., but four cores (E32-48, E32-49, E32-51, E31-1) feature definite disconformities. Three of these (E32-48, E32-49, and E32-1) that were taken in a restricted area have disconformities in which all or most of the Matuyama epoch material is missing. This characteristic suggests that either erosion took place at this location during much of the Matuyama or a slump took place in late Matuyama prior to Brunhes deposition. All the cores in the traverse except E27-1 contain intervals of high volcanic glass shard concentration, the distribution of which has been described by *Kennett and Watkins* [1970]. Radiolarians are abundant throughout the two northern cores (E32-51 and E27-1) and are common enough at specific horizons in the more southern

Fig. 3. Paleomagnetic polarities for each core in traverses A–A' to M–M' as a function of depth below the top of each core. For locations of traverses and cores see Figure 2 and Table 1. In each log, the darkened area shows normal polarity and the open area shows reversed polarity. See text for specimen intervals in each core and methods of interpretation of data.

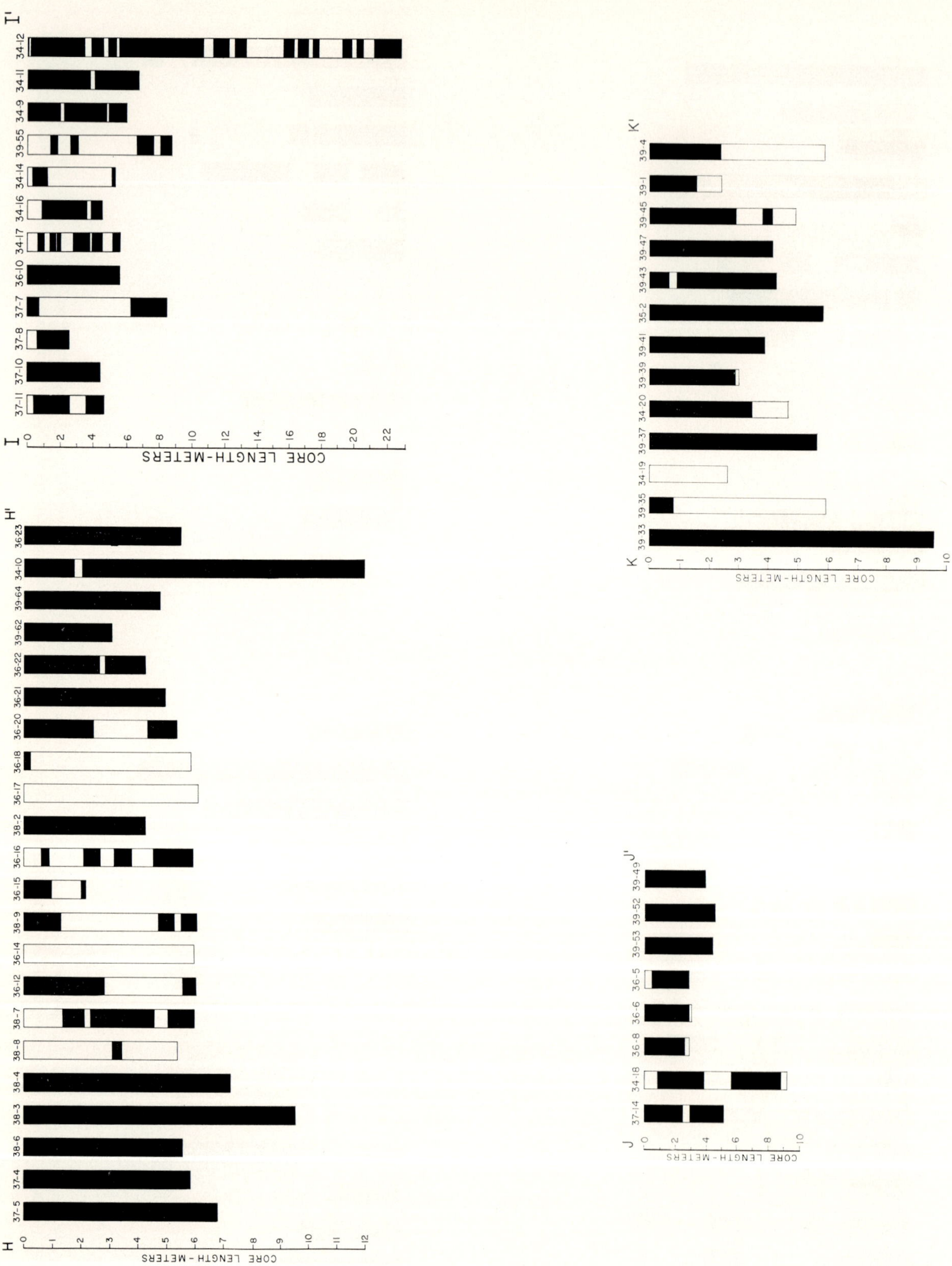

Fig. 3. (continued)

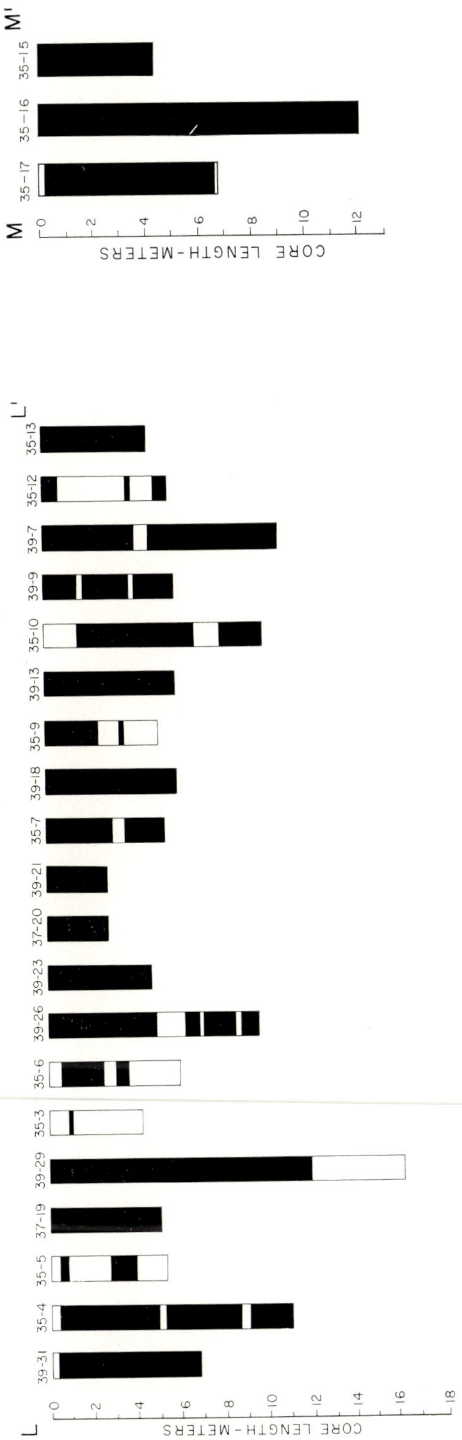

Fig. 3. (concluded)

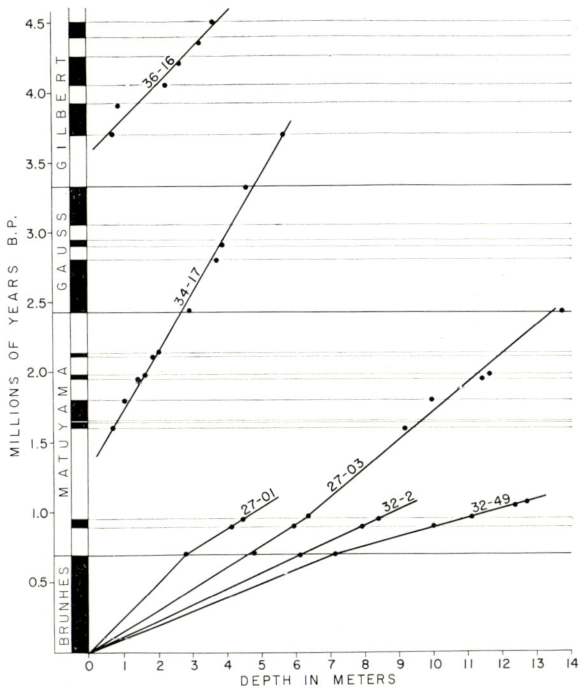

Fig. 4. Sedimentation rate diagram for six selected cores from traverse A–A′, traverse H–H′, and traverse I–I′ (Figure 3). The polarity time scale at the left is after *Cox* [1969] as in Figure 1. The depths and assigned ages of the observed polarity boundaries (traverses A–A′ and I–I′) are plotted and yield sedimentation rates from 1 cm/1000 yr (core 32-49) to 0.25 cm/1000 yr (core 36-16). See text for further discussion.

cores to offer adequate age control. Even at these southern latitudes, Gauss sediments commonly contain abundant *Desmospyris spongiosa*, *Eucyrtidium calvertense*, and *Clathrocyclas bicornis*.

Traverse B–B′. All three cores in this traverse contain disconformities. Micropaleontologic control of E27-22 is poor. Radiolarian assemblages more typical of the Gauss age occur at several intervals, although several horizons contain rare Gilbert radiolarian species that may, however, have been reworked. Core E16-8, located in the western part of the Emerald basin, contains a high proportion of coarse rock and biogenic debris throughout that is strongly abraded and thus almost certainly reworked. The core appears to be largely an accretionary deep-sea sand.

Traverse C–C′. All cores are rich in biogenic material of Brunhes age. Core E36-38 has abundant radiolarian assemblages throughout, and E16-9 and E36-42 have abundant planktonic foraminiferal assemblages. The magnetic data in E36-42 suggest an older age, but both *Globorotalia inflata* and *Globoro-*

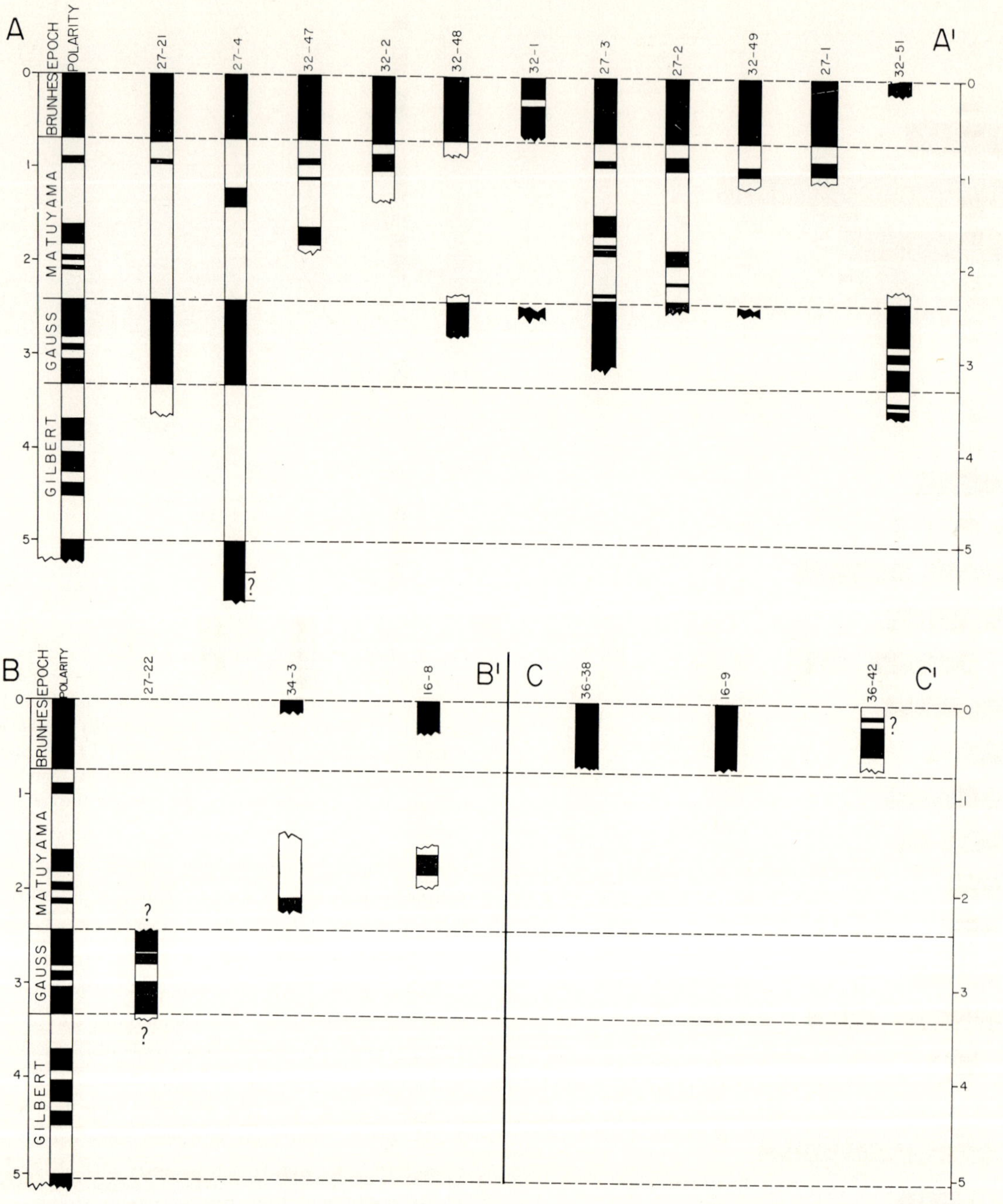

Fig. 5. Diagrams showing assigned time ranges and measured polarities of all cores in traverses A–A′ to M–M′ of Figure 2. The results are compared with the known polarity history (after *Cox* [1969] as shown in Figure 1), which is at the left of each traverse. Darkened area shows normal polarity; open area shows reversed polarity. See text for discussion of methods and limitations of age range assignments and for discussion of data from each traverse.

talia truncatulinoides are present, and at these latitudes their presence suggests a Brunhes age [*Kennett*, 1970].

Traverse D–D′. Disconformities exist in six of the cores. In E34-5, a rich radiolarian fauna of pre-Gilbert age occurs below 150 cm. A sparse foraminiferal assemblage contains *Globigerina* cf. *bulloides* and *Globorotalia* cf. *scitula* but no *Globigerina pachy-*

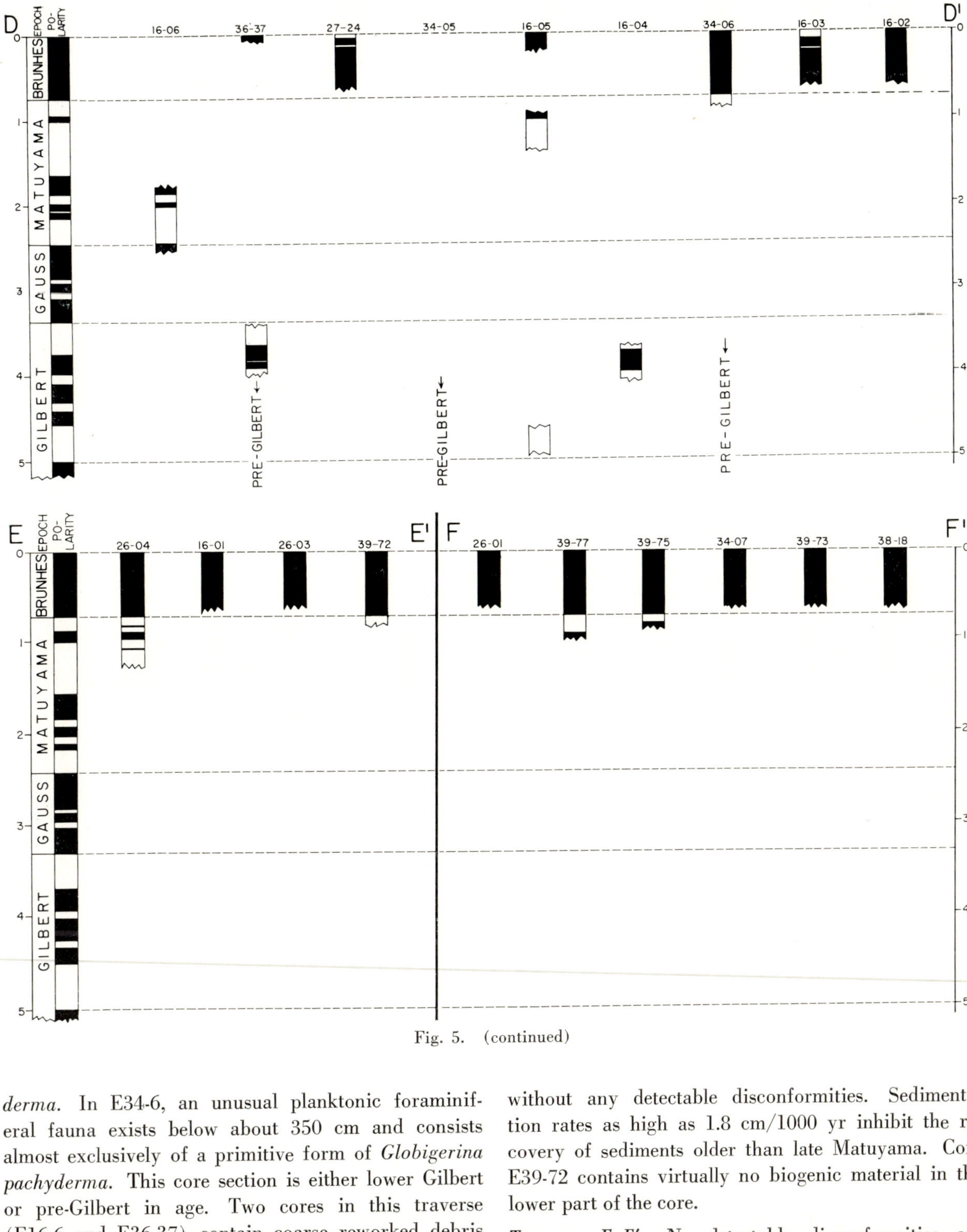

Fig. 5. (continued)

derma. In E34-6, an unusual planktonic foraminiferal fauna exists below about 350 cm and consists almost exclusively of a primitive form of *Globigerina pachyderma*. This core section is either lower Gilbert or pre-Gilbert in age. Two cores in this traverse (E16-6 and E36-37) contain coarse reworked debris at various horizons.

Traverse E–E′. This traverse at the southern edge of the Tasman basin appears to be composed of cores without any detectable disconformities. Sedimentation rates as high as 1.8 cm/1000 yr inhibit the recovery of sediments older than late Matuyama. Core E39-72 contains virtually no biogenic material in the lower part of the core.

Traverse F–F′. No detectable disconformities are present. Sedimentation rates reach as much as 2 cm/1000 yr. Because of a lack of biogenic material in E39-73, no paleontologic age could be determined.

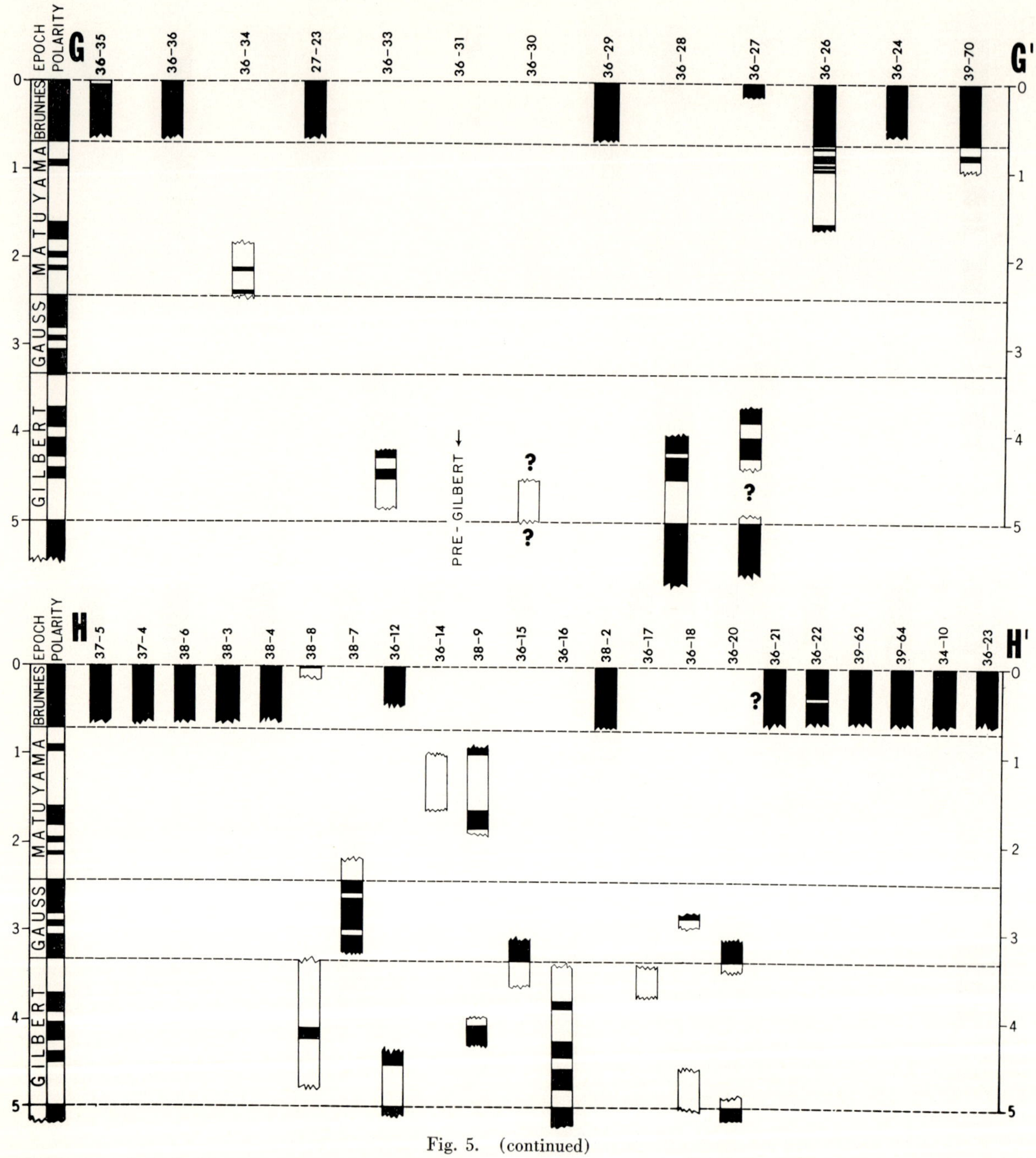

Fig. 5. (continued)

The magnetics, however, indicate a Brunhes age. The 4 easternmost cores are rich in foraminifera throughout, and E38-18 contains intervals rich in foraminifera in addition to levels of terrigenous sand. All cores are Brunhes or Brunhes to late Matuyama age.

Traverse G–G′. Five of the cores in this traverse are composed of dominantly Gilbert epoch material. All the cores but E36-34 occur in the central part of the traverse. Substantial difficulty exists in confidently assigning age ranges within this time interval. Micropaleontologic control is sufficient to restrict core segments to upper or lower Gilbert in most cases. This restriction is sufficient, we believe, to yield a ±0.4-m.y. precision, although we cannot be certain that the normal polarity events identified are in all cases exact equivalents of the known polarity scale features (see

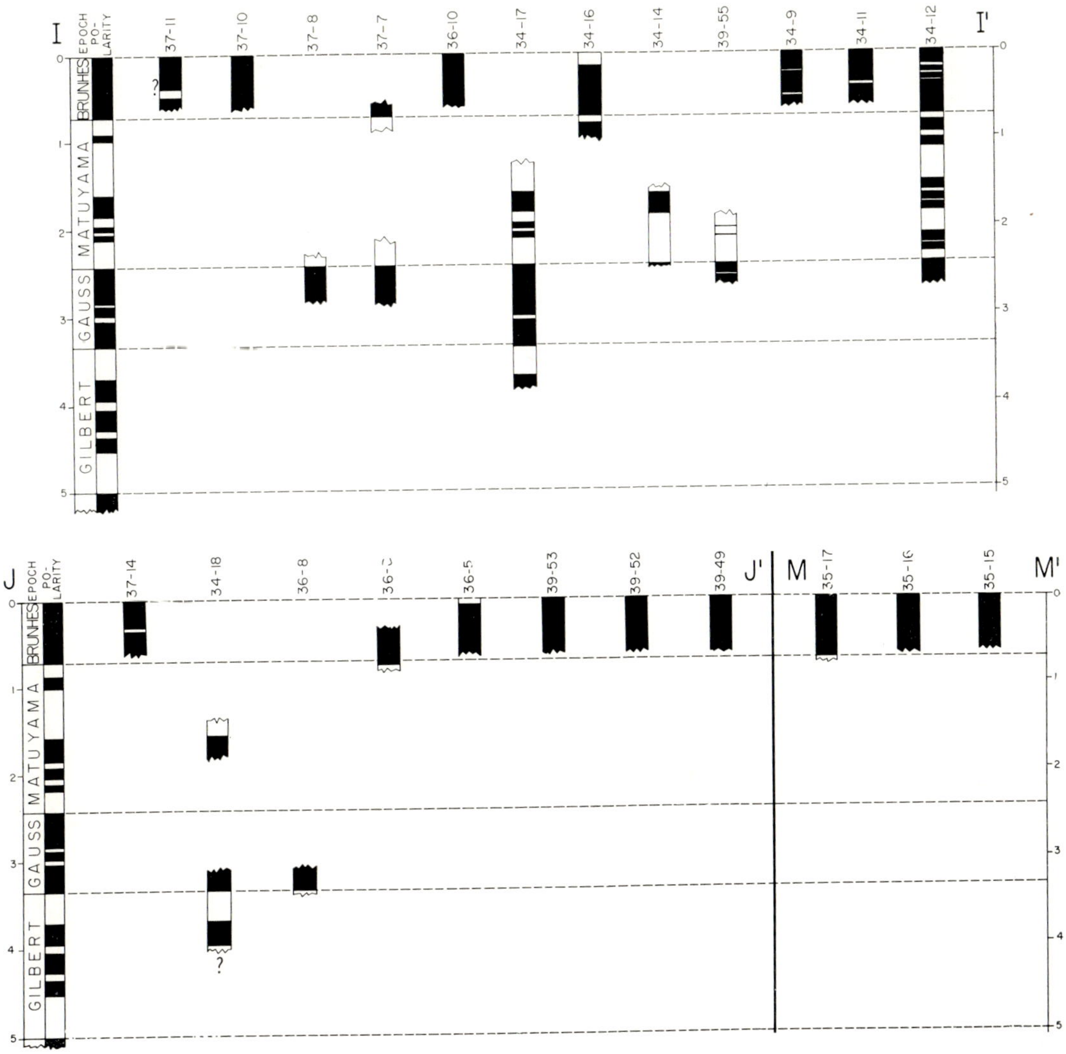

Fig. 5. (continued)

polarity scale on left-hand side of traverse G–G′). Core E36-28 has a high proportion of coarse debris and manganese micronodules throughout and a radiolarian assemblage that appears to have been highly reworked. Likewise, core E36-27 contains concentrations of coarse debris suggesting substantial reworking. The anomalous presence of *G. truncatulinoides* in the lower part of the core associated with possible pre-Gilbert assemblages has almost certainly resulted from reworking. Core E36-29 is a relatively fine-grained siliceous ooze of Brunhes age. This core is noteworthy in that it represents normal Brunhes biogenic sedimentation and yet is surrounded by cores of much greater age. In core E36-26, *G. truncatulinoides* is associated with abundant *E. calvertense* at 710 cm in sediments whose magnetic data suggest a late Matuyama age.

Traverse H–H′. This traverse closely resembles G–G′ in that 10 cores feature definite disconformities. All these cores are restricted to the central part of the traverse. Core E38-8 has a small reversed polarity sequence at the top of the core; radiolarians suggest that this core is Brunhes in age. A disconformity separating Gilbert from Brunhes sediments is marked by a slight winnowed horizon that is between 100 and 150 cm and that contains a mixture of Brunhes and Gilbert radiolarians. Core E38-2 of Brunhes age is very coarse throughout and contains micromanganese

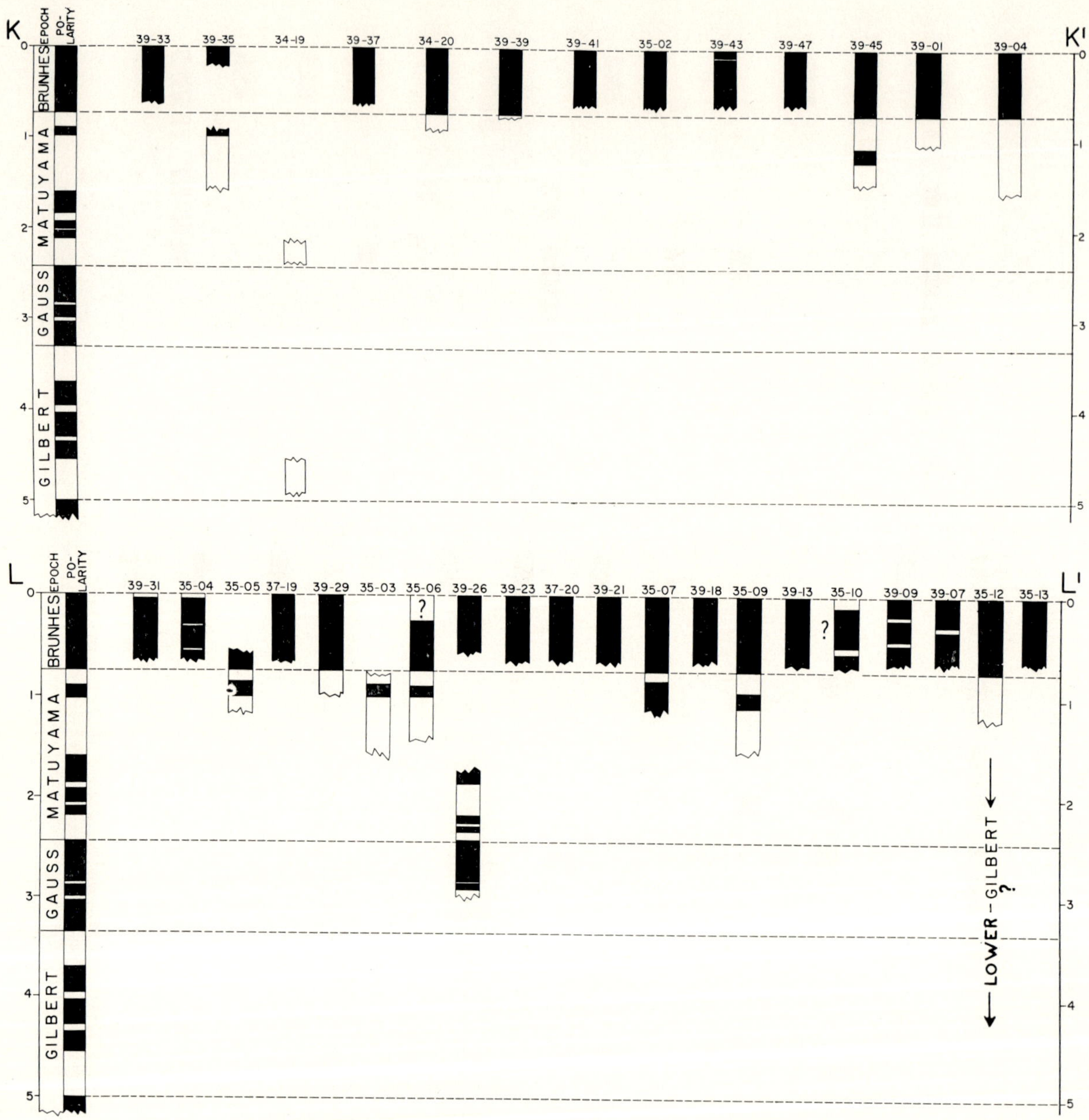

Fig. 5. (concluded)

nodules, obsidian fragments, and robust foraminifera and radiolarians, most of which show signs of abrasion. The core appears to be an accretionary sandy silt. Micropaleontologic age dating of E36-21 was not possible below about 200 cm because no biogenic material was observed. Normal polarity throughout suggests, however, a Brunhes age.

Traverse I–I'. Five cores possess disconformities. A glacial marine core, E37-11, contains at its base a sparse, unusual planktonic foraminiferal fauna consisting of *Globigerina* cf. *bulloides* and *Globigerina* cf. *woodi* but no *Globigerina pachyderma.* The foraminiferal specimens are heavily abraded and may have been reworked. A sparse Brunhes radiolarian fauna occurs, and hence the core is probably of this age despite the unusual foraminiferal fauna. The two northernmost cores in this traverse (E34-11 and E34-12) contain little biogenic material, and their ages are determined from the magnetic data. Core E34-16 contains a latest Matuyama through Brunhes foraminiferal sequence.

Traverse J–J′. Three cores in this traverse contain disconformities. The absence of *Globorotalia truncatulinoides* at the top of E36-6 suggests that the late Brunhes is missing. The southernmost core (E37-14) consists of glacial marine sediments of Brunhes age, whereas the four northernmost cores are foraminiferal oozes of Brunhes age.

Traverse K–K′. Core E34-19, which has reversed polarity throughout, is part Matuyama and part Gilbert epoch. Only one additional core (E39-35) contains a disconformity. Core E39-43, which has normal polarity throughout, is possibly late Matuyama in age at its base because of the occurrence at this level of abundant *Globorotalia puncticulata.* Core E39-39 contains an anomalous association of *G. truncatulinoides* and *G. puncticulata* at 134 cm. The magnetic data suggest a Brunhes age for this core, except for the lowest 10 cm of this core where the data suggest a late Matuyama age. Both E39-1 and E39-4 did not contain enough biogenic sediment to obtain a micropaleontologic age.

Traverse L–L′. Most of E35-12 is composed of sediment rich in foraminifera and bryozoa of lower Pliocene age. Four other cores contain definite disconformities. Cores E35-10, E39-9, and E39-7 do not contain enough biogenic material for paleontologic age determinations, but the magnetic stratigraphy suggests that these cores are Brunhes in age. No biogenic material occurs in the five specimens examined below 550 cm in E39-26.

Traverse M–M′. Sedimentation rates up to 1 cm/1000 yr without hiatuses are found in three cores of this traverse, the cores consisting primarily of siliceous biogenic sediment of Brunhes age.

Trend Surface Analysis of Surface Sediment Ages

Examination of the traverses in Figure 5 reveals a well-defined zone in the mid-oceanic area where sediment younger than Gauss or Gilbert epoch age is largely absent. This disconformity is far from random in its occurrence and is clearly a feature of substantial size with some east-west extent, but there are obvious conflicting local data. For example, E36-29 (traverse G–G′), E38-2 (traverse H–H′), and E36-10 (traverse I–I′) are Brunhes throughout, whereas much older sediments are found in other cores in the same region. The suspected reasons for this anomalous situation will be discussed later. We present here a mathematical expression of the regional hiatus through the technique of trend surface analysis.

A surface of limited curvature, or order, is fitted to the data that in this case is the duration in millions of years of the sediment that is missing (or the age of the sediment at the top of each core). The configuration of the surface is controlled by minimizing the root mean square residual. *Watkins and Self* [1971] have discussed the rationale and methods employed in selecting limits of surface order choice. A low order surface has been selected to attempt delineation of only the very broad controlling factors involved in the creation of the hiatus. We stress our awareness of the fact that several factors, especially topography, can create variable sedimentation rate characteristics, and this fact reinforces our motive for employing a limited curvature surface. The results are shown for the whole area of study (Figure 2) in Figure 6*a*. Since the possibility exists that the trend surface may be more meaningfully examined in restricted segments of the study area where sedimentary regimes are more likely to be controlled by a single major factor rather than diverse factors, we present additional trend surface maps of selected regions in Figures 6*b* and 6*c*.

From the trend surface, a major zone clearly exists and is centered in the south Tasman Sea and southeast end of the Southeast Indian ridge where sediment of t = 0–3.5 m.y. age is largely absent. The actual age of the surface toward the center of the region (Figures 6*a*, 6*c*) is likely to be younger than would be the case if the cores were generally longer because we can compute only minimum sedimentation rates in cores of normal polarity (Brunhes epoch age). Therefore the resulting trend surface is elevated toward a younger age on the north and south flanks (where such cores are more common), this elevation leading to an inevitable artificial depression of the surface in the central part of the area. If much longer cores were available and the 0.69-m.y. isochron was reached in all cases, we would expect somewhat steeper trend surface gradients. Nevertheless, the general pattern is unlikely to change. Therefore, maximum ages of the sea floor are centered in a region that includes the south Tasman basin, the southeastern end of the Southeast Indian ridge and the southern extent of the Macquarie ridge. When the total area (Figure 6*a*) and the most central area of maximum ages (Figure 6*c*) are considered, ages decrease away from this central region in a concentric fashion. When an area of more westerly extent is considered (Figure 6*b*), the ages tend to increase from west to east. This tendency shows that

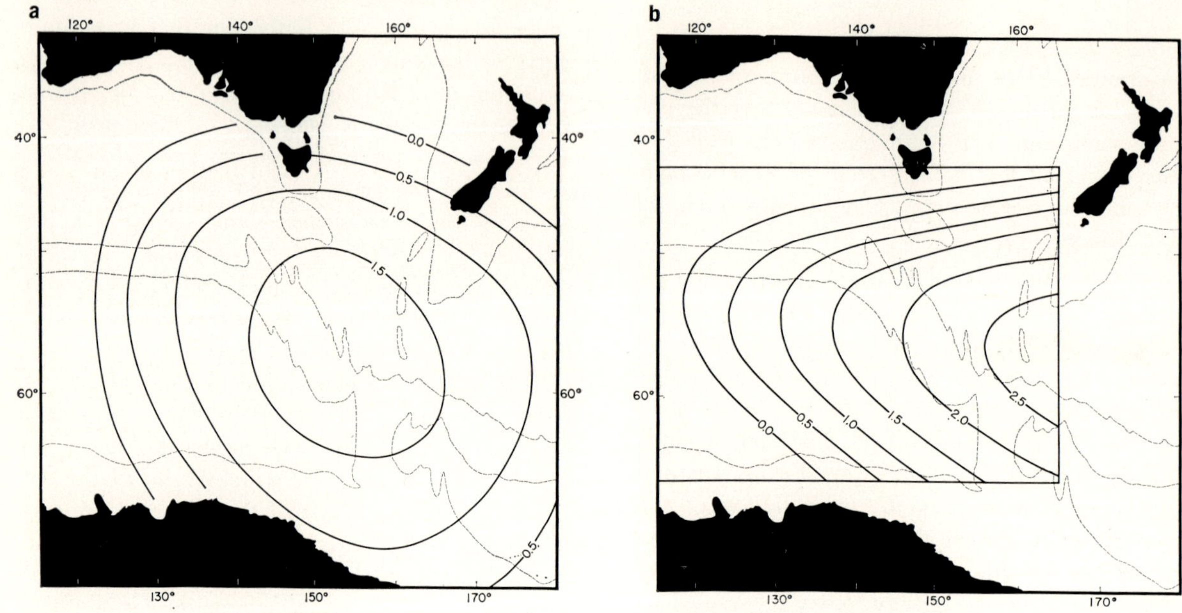

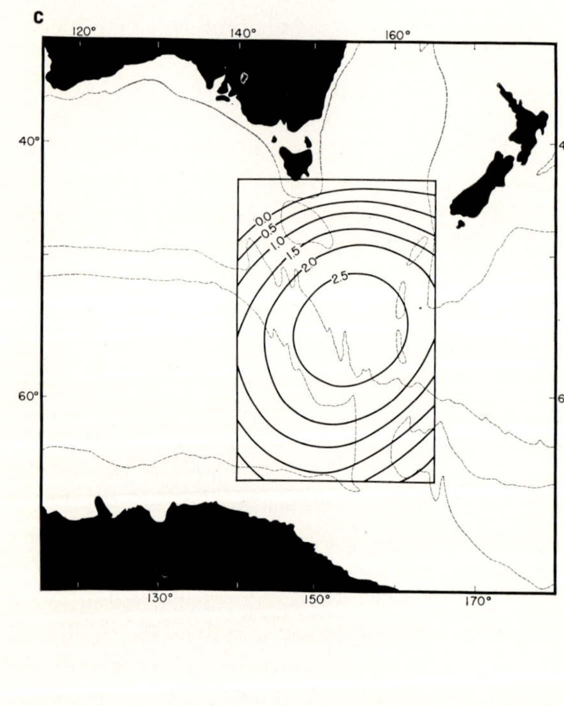

Fig. 6. Maps of second order trend surfaces on the age (in millions of years) of the sediment at the base of the regional erosional disconformity, as detected in some cores (Figures 3 and 5). Note that if two disconformities are present, only the oldest horizon is included in the analysis; and if only a small upper Brunhes epoch sedimentary fraction is present above an older core segment, then the age of the top of the older segment is employed in the analysis.

the factor (or factors) causing the hiatus increases from the west toward the east.

Cause of Hiatus

The absence of sediment can result from one of two possible causes: a lack of deposition through complete reduction in biogenic productivity or a removal of sediment by processes of regional extent.

For several obvious reasons, the possibility of a complete reduction in biogenic productivity over such

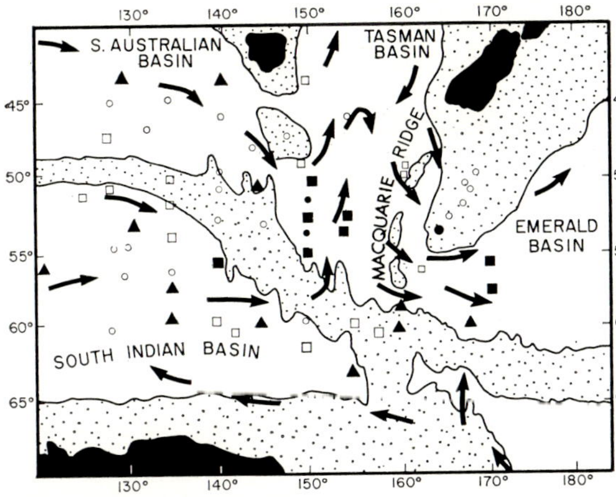

Fig. 7. Map of the ocean between Australia, New Zealand, and Antarctica showing circulation patterns of bottom waters and bottom photograph [*Jacobs et al.*, 1970] characteristics. Arrows represent bottom current directions from *Gordon* [1971*a*]. Submarine ridge system is stipled. Bathymetric contour is 2000 meters in depth. Bottom photograph symbols are shown at camera station sites and indicate the following features: open squares, abundant bioturbation; open circles, bioturbation common or absent; closed triangles, manganese nodules common; closed squares, manganese nodules abundant; closed circles, ripples and scour marks present.

a large area as the cause of the hiatus is considered untenable. We propose that much evidence exists to support the possibility of removal of sediment by current activity. Deposition at apparently normal rates has nevertheless taken place in regions closer to the continents, in the South Australian basin, and the north and central Tasman basin. Evidence of the removal of the sediments by sea bottom current activity comes from three main sources discussed below.

Bottom current velocities. The several available laboratory experimental data relating current velocities, sedimentary size, and erosion, transportation, and deposition have been summarized by *Heezen and Hollister* [1964, Figure 18]. Although some doubts must exist about the direct and exact applicability of the data to in situ sea bottom sediments, the order of the velocities suggested for removal of sediment of a given particle size is unlikely to be greatly different from those suggested. Currents greater than 15 cm sec^{-1} are required for erosion of noncohesive particles of fine sand and finer sizes [*Heezen and Hollister*, 1964] but may be effective on coarser sand-sized sedimentary fractions [*Southard et al.*, 1971]. Erosion of cohesive particles requires velocities between two and ten times greater depending on the grain size. Once in motion, substantially lower velocities are required to maintain transportation. Sedimentation can be expected when currents drop to approximately 1 cm sec^{-1} for fine sand and 0.1 cm sec^{-1} for silt fractions. Lutite and fine silt may be expected to remain in transportation until virtually stationary water velocities (less than 0.01 cm sec^{-1}) or natural physiographic barriers are encountered.

Figure 7 shows the proposed pattern of circulation of bottom waters inferred from salinity, oxygen, and temperature data in the study area [*Gordon*, 1971*a*]. Irregular currents measured in situ in the area of the hiatus (Figures 5 and 6) reach sufficiently high velocities [*Eittreim et al.*, 1971] to initiate erosion and readily maintain transportation. General current directions (Figure 7) are parallel to the length of the hiatus, with a substantial northern distortion in the vicinity of the southern end of the Southeast Indian ridge. In the area of the hiatus, intersections of currents from the South Australian basin, the South Indian basin, and the Tasman basin (Figure 7) occur. Bottom velocity measurements according to *Callahan* [1971, Figure 1] show the asymmetrical distribution of velocities with respect to the ridge; the velocities are highest just north of the ridge. All available current data therefore show that erosion and transportation by bottom currents are to be expected in the study area.

Sea bottom photographs. The sea bottom photographs taken during *Eltanin* cruises 34 and 36 have been published [*Jacobs et al.*, 1970]. The relationship between features that can be recognized in deep-sea photographs and current velocities has been summarized by *Heezen and Hollister* [1964], *Hollister and Heezen* [1967], and others. These features range from surfaces with distinct faunal tracks at the lowest velocities, surfaces with elongate sedimentary deposits behind positive features, ripple marks, scour around high density objects such as manganese nodules, and rocks swept bare of sediment at the highest velocities. This qualitative approach to an understanding of current velocities has been successfully applied by several authors including *Heezen and Hollister* [1964], *Heezen et al.* [1966], and others. In Figure 7, we include a compilation of inferences that can be drawn from the available *Eltanin* sea bottom photographs.

An inspection of Figure 7 shows that those features most readily associated with the higher current velocities and erosion are common in the area of sediment hiatus. In the vicinity of the south Tasman basin in particular, ripple marks, manganese nodules, and scour are characteristic. Areas where Brunhes sedimentation has taken place show, on the other hand, bottom features more consistent with low current velocities.

Sedimentology. Our interpretation leads to the inference that the surface sediments in the proposed scour region may be dominated by coarser and high density fractions. These will be mainly residual types, but transported concentrations can be expected. The development of thin residual sands [*Heezen and Hollister*, 1964] is now occurring in parts of the scoured area. These sands are in most cases residual because only older biogenic components are frequently present and the original (frequently reversed) polarities appear to be maintained. (Normal polarity can be expected in modern transported sediments deposited during the Brunhes epoch.) A high concentration of manganese and micromanganese nodules is also predictable. Our observations of common micromanganese nodules within the sand-sized fraction is consistent with these predictions. Cores of Brunhes age throughout, when found in the scour zone, are very likely to be composed of dominantly coarse transported sediments. Such a core is, for example, E38-2 in traverse H–H′. Deviations from these simple situations do exist (for example, E36-29 from the scour zone consists of siliceous ooze of Brunhes age), but we believe that future sedimentologic studies will substantiate our predictions. A core (E55-3) from the center of the scour region and recently collected by Japanese oceanographers in fact features coarse clastics at the top of the core [*Kobayashi and Kitazawa*, 1970], and two early Miocene cores from the same area described by *Conolly* [1969] contain manganese nodules in their upper parts.

We suspect that echogram characteristics may possibly substantiate and further delineate the scour area containing residual sands, since *Heezen et al.* [1959, 1966] have shown bottom reflectivity to be poor in areas of high sedimentation rate and stronger echoes to be associated with coarse sands. Our results show that a scour area with a surface of residual sands can be clearly diachronous, and yet probably the surface would appear as a single reflector in any seismic or echo sounder profiles. Because of this difficulty in distinguishing diachronous from nondiachronous horizons, such methods cannot be expected to assist in fine resolution of sedimentary history, although major sediment thickness variation can readily be defined by such methods [e.g., *Jones et al.*, 1970].

Regional Sedimentation Rate Variation

Combining core lengths (Figure 3, Table 1) with estimated time ranges yields average sedimentation rates for each core. In the cores that do not reach the Brunhes-Matuyama boundary, minimum sedimentation rates can be calculated. Where such cores are less than 5 meters, however, such minimum sedimentation rate estimates may be a factor of five too small, and so such estimates are not included in any regional analysis.

In Figure 8*a*, we present second order trend surface maps of the calculated average sedimentation rates for the whole area for the period t = 0–2.5 m.y. Figure 8*b* shows the same surface for a restricted part of the study area for reasons similar to the earlier analyses of the age of the disconformity surface.

We must realize that sedimentation rates for the period t = 0–2.5 m.y. cannot be estimated from many of the cores in the region of the disconformity defined above. The inherent nature of trend surface analyses, however, facilitates estimates of what the sedimentation rates would be for the central region if no excessive scour were occurring by simple interpolation and extrapolation between cores north and south of the scour region. In similar fashion to the trend surface on the age of the disconformity surface, the sedimentation rate surface is to some extent artificially diminished toward the flanks (because of the use of minimum sedimentation rate estimates only) and therefore probably artificially elevated slightly in the center of the region. According to our analysis, sedimentation rates for the region vary on the average from less than 0.4 cm/1000 yr to about 1.6 cm/1000 yr. When a trend surface for the entire area is considered (Figure 8*a*), the expected influence on sedimentation rates of the Southeast Indian ridge is apparent. The sedimentation rate for the region is therefore coherent and is proposed to be evidence that the local variations in sedimentation rate are subordinate to the regional controls.

IMPLICATIONS OF THE LIMITED AGE OF THE ERODED SEDIMENT

A major west to east scour zone exists in the study area. As shown in Figure 5, the age of the cores exceeds Gilbert epoch or uppermost Miocene in only six of 126 cores despite the very large aerial coverage of the core collection. In other words, according to our trend surface analyses of sedimentation rates, no more than 15–20 meters of sediments has been removed or

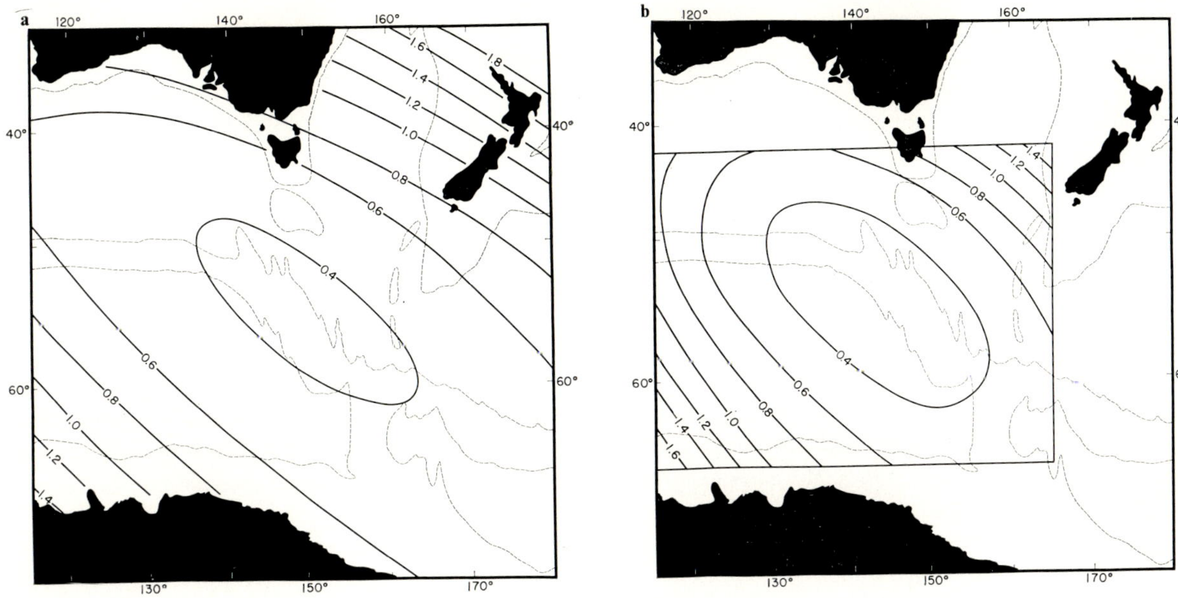

Fig. 8. Maps of second order trend surface on the sedimentation rate (in cm/1000 yr) for the period $t = 0$–2.5 m.y. as calculated for all cores (Figures 2, 3, and 5) except for those cores of less than 5 meters in length that are Brunhes epoch age throughout. See text for discussion of limitations of the method, the effect of the use of minimum sedimentation rates in cores restricted to the Brunhes epoch, and the role of extrapolation in extending the surface over a region where sediment of 0–2.5 m.y. of age is largely absent.

inhibited from normal deposition at any of the core sites. The Brunhes epoch material has been very readily removed over a very large area. Why is it that much Gauss and particularly Gilbert epoch material still exists in the area of scour? Much of this sediment is fine-grained biogenic sediment with little evidence of winnowing action by currents. Therefore the current system responsible for the scour could not have been in existence throughout much of the Gilbert epoch, although spasmodic erosion did occur (Figure 5, traverses G–G′ and H–H′).

Examination of Figure 5 shows that Matuyama epoch sediments are less commonly preserved than Gilbert epoch sediments. We therefore favor initiation of a period dominated by erosion in the area during the Gauss. The scouring of Brunhes age material is more easily accomplished by a given bottom current velocity relative to the older materials because of the probable increase of cohesiveness with depth below the surface sediment. As *Heezen and Hollister* [1964] summarized, a substantial increase in current velocity is required for erosion of a given cohesive (as opposed to incohesive) sediment. Therefore scouring efficiency can be expected to decrease with the age of sediment. According to our data, a burial depth of only 10 meters or so may cause a critical increase in cohesiveness, although this figure will undoubtedly vary with sediment type. The creation of a protective residual sand and/or manganese nodule pavement at the sediment surface would also decrease the erosion rate for a given current velocity.

CAUSE OF THE UPPER CENOZOIC BOTTOM CURRENT VELOCITY CHANGE

It is appropriate to speculate on the cause of the marked change in bottom current velocity that we believe occurred during the Gauss epoch in the South Atlantic, Pacific, and Indian oceans. The change required is (from examination of Figure 18 of *Heezen and Hollister* [1964]) from a system with currents very dominantly less than 10 cm sec^{-1} to dominantly greater than 10 cm sec^{-1}. We doubt that a 'threshold' concept would apply to an inherently variable system of bottom currents, so a small absolute increase in average current velocity is not envisaged. A 5-cm sec^{-1} average velocity increase is proposed. This increase is unlikely to have occurred as a single event. Changes may have occurred spasmodically over a long period of time before reaching a more sustained higher velocity. The presence of disconformities with associated winnowed sediment horizons within the Gilbert epoch indicates that increased current activity did not commence suddenly but was active at times during the Gilbert. Despite the obvious relation between topography and current systems, insufficient time has elapsed since deposition for changes in currents to result from substantial changes in submarine topography. Instead we propose that this increase in average current velocity is associated with a large change in the production of antarctic bottom water from antarctic ice. There is strong evidence for extensive glaciation of Antarctica in the late Cenozoic since at least the beginning of the Gilbert epoch (t = 0–5.0 m.y.) and for world-wide pre-Pleistocene cool climatic cycles assumed to be associated with increased glaciation of Antarctica [*Kennett,* 1967; *Margolis and Kennett,* 1971; *Bandy,* 1967; *Ingle,* 1967]. Periods of nonglaciation or partial glaciation of Antarctica result in decreased circulation of antarctic waters and a decrease in the magnitude of bottom circulation [*Gordon,* 1971*b*]. Possibly, therefore, ice development on the antarctic continent reached a critical level during post-Gilbert or Gauss time. Conceivably, extensive floating ice shelves developed at this time, and such shelves are likely to be related to the production of antarctic bottom water [*Gordon,* 1971*b*]. Also deep circulation may have experienced periods of even more highly intensified flow in response to more extreme climatic conditions associated with glacial periods, and that interglacial deep circulation is now relatively quiescent [*Johnson,* 1971]. We suspect that the number of factors involved in our postulated change in current velocity is likely to be large. Changes in latitudinal temperature gradients [*Hendy and Wilson,* 1971] and corresponding wind system changes [*Munk,* 1950] would be substantial contributors to differences in current velocity with time.

SPECULATION ON MID-TERTIARY SEDIMENT CHARACTER IN THE STUDY AREA

Our study of the *Eltanin* piston cores recovered from the Australasian-antarctic region, although restricted to age determinations and very limited sedimentologic examinations, has given strong emphasis to the intimate relation between current velocities and sedimentary character. We therefore speculate on the nature of the mid-Tertiary and older sediments in the area between Australia and Antarctica.

Dating of the oceanic crust by conventional recognition of the classic marine magnetic anomaly pattern [*Le Pichon and Heirtzler,* 1968; *Dietz and Holden,* 1970] suggests that oceanic crust between Australia and Antarctica is dominantly younger than early Eocene (t = 50 m.y.), this suggestion meaning that this part of the oceanic crust is spreading as rapidly on average as any part of the world ridge system.

During the middle Tertiary (t = 35–20 m.y.), the oceanic crust of the study area can only have been half as wide as it is now. Probably, therefore, the total current flow of the circumantarctic current between Australia and Antarctica now of the order of 10^8 m^3 sec^{-1} was much higher during the middle Tertiary. In all likelihood, bottom currents would also have been higher. We therefore propose that coarse fractions will be found to dominate in the mid-Tertiary parts of the cores recovered during the *Joides* drilling in the region, particularly to the south of the ridge system where sources of such coarser fractions are more likely to be available. Since manganese nodule growth is enhanced by rapid currents and minimal deposition of finer sediments, we also suggest that manganese nodule concentrations will be correspondingly high. We also reasonably suspect that, as is the case for the present, hiatuses of regional scale may be encountered. If the Macquarie ridge (Figure 1) was at any time outcropping above sea level, these arguments would become applicable to the early Tertiary sedimentary regime of the Tasman Sea.

Acknowledgments. The performance of the various *Eltanin* crews is greatly appreciated. The careful supervision and curating of the sediment cores by Dennis Cassidy of Florida State University are a critical part of any study of the core collection. We thank Rosmarie Raymond for drafting assistance; and Dennis Cassidy for photographic assistance.

These studies have been supported by the antarctic branch of the National Science Foundation and by grants GA-1678, GA-13132, and GV-25400 to N. D. Watkins and grant GV-28305 to J. P. Kennett. The initial paleomagnetic phases of this work were carried out at Florida State University and we gratefully acknowledge that institution for the provision of much time on their CDC 6400 computer.

REFERENCES

Bandy, O. L., Problems of Tertiary foraminiferal and radiolarian zonation, circum-Pacific area, in *Tertiary Correlations and Climate Changes in the Pacific, Symposium 25,* Sasaki, Sendai, Japan, 1967.

Berggren, W. A., J. D. Phillips, A. Bertels, and D. Wall, Late Pliocene-Pleistocene stratigraphy in deep-sea cores from the south-central north Atlantic, *Nature, 216,* 253–255, 1967.

Callahan, J. E., Velocity structure and flux of the antarctic circumpolar current south of Australia, *J. Geophys. Res., 76*(24), 5859–5864, 1971.

Conolly, J. R., Western Tasman Sea floor, *N. Z. J. Geol. Geophys., 12,* 310–343, 1969.

Conolly, J. R., and C. C. von der Borch, Sedimentation and physiography of the sea floor south of Australia, *Sediment. Geol., 1,* 181–220, 1967.

Conolly, J. R., A. Flavelle, and R. S. Dietz, Continental margin of the Great Australian Bight, *Mar. Geol., 8,* 31–58, 1970.

Cox, A., Geomagnetic reversals, *Science* (AAAS), *163,* 237–245, 1969.

Dietz, R. S., and J. C. Holden, Reconstruction of Pangea: Breakup and dispersion of continents, Permian to present, *J. Geophys. Res., 75*(26), 4939–4956, 1970.

Eittreim, S., A. L. Gordon, M. Ewing, E. Thorndike, and P. Bruchhausen, The nepheloid layer and observed bottom currents in the Indian-Pacific antarctic sea, in *Studies in Physical Oceanography—A Tribute to Georg Wüst on His 80th Birthday,* edited by A. L. Gordon, Gordon and Breach, New York, in press, 1971.

Foster, J. H., A paleomagnetic spinner magnetometer using a fluxgate gradiometer, *Earth Planet. Sci. Lett., 1,* 463–466, 1966.

Goodell, H. G., and N. D. Watkins, The paleomagnetic stratigraphy of the Southern Ocean: 20° West to 160° East longitude, *Deep Sea Res., 15,* 89–112, 1968.

Gordon, A. L., Spreading of antarctic bottom waters, 2, in *Studies in Physical Oceanography—A Tribute to Georg Wüst on His 80th Birthday,* edited by A. L. Gordon, Gordon and Breach, New York, in press, 1971*a*.

Gordon, A. L., Oceanography of antarctic waters, in *Antarctic Oceanology I, Antarctic Res. Ser.,* vol. 15, edited by J. L. Reid, pp. 169–203, AGU, Washington, D. C., 1971*b*.

Hays, J. D., and N. D. Opdyke, Antarctic Radiolaria, magnetic reversals and climatic change, *Science, 158,* 1001–1011, 1967.

Hays, J. D., T. Saito, N. D. Opdyke, and L. H. Burckle, Plio-Pleistocene sediments of the equatorial Pacific: Their paleomagnetic, biostratigraphic, and climatic record, *Geol. Soc. Amer. Bull., 80,* 1481–1514, 1969.

Heezen, B. C., and C. D. Hollister, Deep-sea current evidence from abyssal sediments, *Mar. Geol., 1,* 141–174, 1964.

Heezen, B. C., M. Tharp, and M. Ewing, The floors of the ocean, 1, The North Atlantic, *Geol. Soc. Amer. Spec. Pap., 65,* 122 pp., 1959.

Heezen, B. C., C. D. Hollister, and W. Ruddiman, Shaping of the continental rise by deep geostrophic contour currents, *Science, 152,* 502–508, 1966.

Hendy, C. H., and A. T. Wilson, Changes in latitudinal thermal gradient during the last 50,000 years (abstract), *Eos Trans. AGU, 52*(4), 229, 1971.

Hollister, C. D., and B. C. Heezen, The floor of the Bellingshausen Sea, in *Deep-Sea Photography, Johns Hopkins Oceanogr. Stud. 3,* edited by J. B. Hersey, pp. 177–189, Johns Hopkins Press, Baltimore, Md., 1967.

Ingle, J. C., Foraminiferal biofacies variation and the Miocene-Pliocene boundary in Southern California, *Bull. Amer. Paleontol., 52*(236), 217–394, 1967.

Jacobs, S. S., P. M. Bruchhausen, and E. B. Bauer, Hydrographic stations, bottom photographs, and current measurements, USNS *Eltanin* cruises 32–36, 460 pp., Lamont-Doherty Geol. Observ. of Columbia Univ., Palisades, N. Y., 1970.

Johnson, D. A., A model of intensified sea floor erosion during glacial stages (abstract), *Eos Trans. AGU, 52*(4), 224, 1971.

Jones, E. J. W., J. I. Ewing, and S. L. Eittreim, Influences of Norwegian Sea overflow water on sedimentation in the northern North Atlantic and Labrador Sea, *J. Geophys. Res., 75*(9), 1655–1680, 1970.

Kennett, J. P., Recognition and correlation of the Kapitean stage (Upper Miocene, New Zealand), *N. Z. J. Geol. Geophys., 10,* 1051–1063, 1967.

Kennett, J. P., Pleistocene paleoclimates and foraminiferal biostratigraphy in subantarctic deep-sea cores, *Deep Sea Res., 17,* 125–140, 1970.

Kennett, J. P., and N. D. Watkins, Geomagnetic polarity change, volcanic maxima and faunal extinction in the South Pacific, *Nature, 227,* 930–934, 1970.

Kennett, J. P., N. D. Watkins, and P. Vella, Paleomagnetic chronology of Pliocene-Early Pleistocene climates and the Plio-Pleistocene boundary in New Zealand, *Science, 171,* 276–279, 1971.

Kobayashi, K., and K. Kitazawa, Piston coring, in *Preliminary Report of the Hakuho Maru Cruises KH-68-4,* pp. 36–62, Ocean Res. Inst. Univ. of Tokyo, Tokyo, 1970.

Le Pichon, X., and J. R. Heirtzler, Magnetic anomalies in the Indian Ocean and sea-floor spreading, *J. Geophys. Res., 73*(6), 2101–2117, 1968.

Margolis, S. V., and J. P. Kennett, Cenozoic paleoglacial history of Antarctica recorded in subantarctic deep-sea cores, *Amer. J. Sci., 271,* 36, 1971.

Munk, W. H., On the wind-driven ocean circulation, *J. Meteorol., 7,* 79–93, 1950.

Opdyke, N. D., and J. Foster, The paleomagnetism of cores from the North Pacific, *Geol. Soc. Amer. Mem., 126,* 83–120, 1970.

Opdyke, N. D., and B. P. Glass, The paleomagnetism of sediment cores from the Indian Ocean, *Deep Sea Res., 16,* 249–261, 1969.

Opdyke, N. D., B. Glass, J. D. Hays, and J. Foster, Paleomagnetic study of antarctic deep-sea cores, *Science, 154,* 349–357, 1966.

Southard, J. B., R. A. Young, and C. D. Hollister, Experimental erosion of calcareous ooze, *J. Geophys. Res., 76*(24), 5903–5909, 1971.

von der Borch, C. C., J. R. Conolly, and R. S. Dietz, Sedimentation and geology of the continental margin off the Otway Basin, *Mar. Geol., 8,* 31–58, 1970.

Watkins, N. D., Short period geomagnetic polarity events in deep-sea sedimentary cores, *Earth Planet. Sci. Lett., 4,* 341–349, 1968.

Watkins, N. D., and H. G. Goodell, Confirmation of the reality of the Gilsa geomagnetic polarity event, *Earth Planet. Sci. Lett., 2,* 123–129, 1967.

Watkins, N. D., and R. Self, An examination of the *Eltanin* dredged rocks from the Scotia Sea, in *Antarctic Oceanology I, Antarctic Res. Ser.,* vol. 15, edited by J. L. Reid, pp. 331–343, AGU, Washington, D. C., 1971.

SEDIMENTARY PATTERNS WITHIN A CONTINENT–MID-OCEANIC RIDGE–CONTINENT PROFILE: INDIAN OCEAN SOUTH OF AUSTRALIA

John R. Conolly and Robert R. Payne

Department of Geology, University of South Carolina, Columbia, South Carolina 29208

Abstract. The southeast Indian Ocean between the Australian and antarctic continents is represented by a mid-ocean ridge that, according to magnetic data and sea floor spreading hypotheses, has grown between the drifting continents during the past 50 m.y. Analyses of nearly 300 piston cores from this region provide us with a model for sedimentation during the Brunhes within a typical continent–ocean–continent profile. Up to 20 meters of graded beds were deposited during the past 700,000 years within the South Australian and South Indian abyssal plains but as little as 1 meter of ooze was deposited in some regions on the crest of the ridge during the same period. Generally, the mid-oceanic ridge is covered by 3–6 meters of Brunhes calcareous ooze on the northern ridge flank, siliceous ooze on the southern ridge flank, and a mixture in a 600 km wide zone across the center of the ridge. Sedimentation patterns in these areas were similar during the Matuyama magnetic epoch. The interbedding of calcareous ooze with siliceous ooze throughout the past 2 m.y. is thought to be caused by fluctuations in antarctic climate. Six warm and six cold cycles are recognized for the Brunhes magnetic epoch. On the deep ridge flanks at depths greater than about 4500 meters, manganese nodules commonly cover the sea floor. In the Tasman fracture zone and southern Tasman basin, the sea floor is covered with a manganese nodule and sandy pavement that overlies compact gray siliceous ooze of early Gauss or Gilbert age. The spatial arrangement of this lithologic disconformity suggests that it may be caused by a major change in sedimentary regime that began about 3 m.y. ago. Since that time, this region has been kept clean of sediment presumably by bottom currents that flow through the Tasman fracture zone into the south Tasman basin.

During the years 1968–1971, the USNS *Eltanin* took over 300 piston cores from the southeast Indian ocean between Australia and Antarctica (Figure 1). We are most concerned with problems of ocean floor sediment dispersal and the processes that control the sediment accumulation in the deep ocean. To understand more fully these problems, we intend to use all available sea floor sediment information to help to explain the patterns of sediment cover found across the Australian-antarctic profile. The density of piston coring in the ocean south of Australia (Figure 1) is as good as any collected from the deep ocean floor. As well as using bathymetric data [*Hayes and Conolly*, this volume], we have drawn on the results of a complimentary paper on the same suite of deep-sea cores by *Watkins and Kennett* [this volume]; these results provide paleomagnetic and faunal control.

Previous investigations of the deep-sea sediments of the oceans bounding Antarctica were mainly based on studies of piston cores collected from the USNS *Eltanin* from the sea floor between South America and New Zealand and have been summarized by *Goodell and Watkins* [1968]. They show that during the Brunhes epoch up to 16 meters of sediment was deposited along the continental margin of Antarctica but that less than 2 meters of sediment was deposited in the deep ocean basin north of the antarctic continent. They also show that within this area of low deposition is an area characterized by pavements of manganese nodules and located presumably where strong currents keep deposition rates to a minimum [*Hollister and Heezen*, 1967]. Another suite of cores taken from research vessels of Lamont-Doherty Geological Observatory in the Indian Ocean and south of Australia was discussed by *Opdyke and Glass* [1969] and *Glass* [1967].

Apart from the studies on the Lamont-Doherty cores in the Indian Ocean and the *Eltanin* cores in the South Pacific sector, little work has been done on studies of piston cores from the circumpolar ocean. Studies of faunal changes on isolated cores have been made by numerous authors [*Hays*, 1965, 1967; *Hays and Op-*

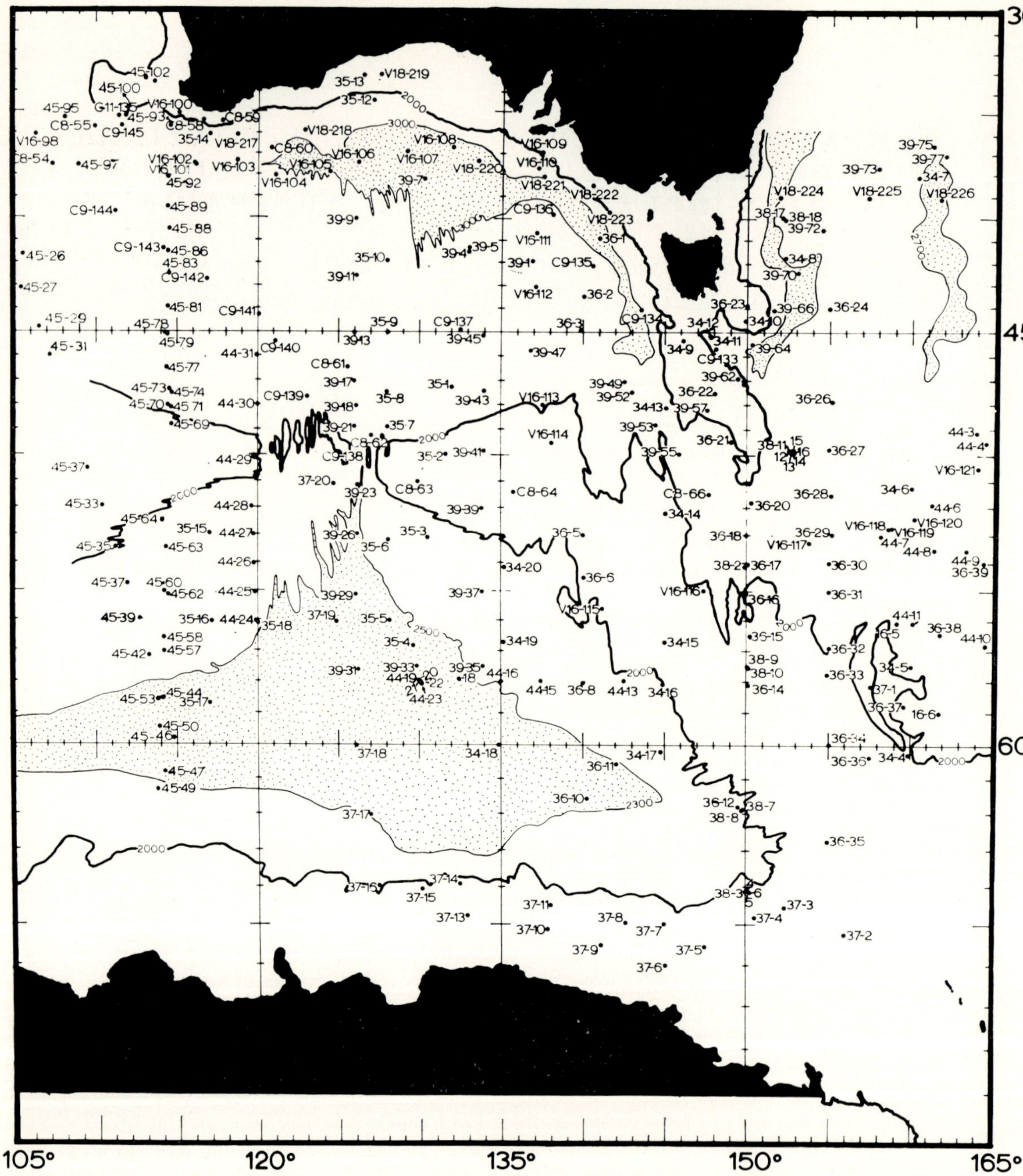

Fig. 1. The southeast Indian Ocean showing locations of piston-coring stations and an outline physiography. *Eltanin* cores are shown without a prefix. Lamont-Doherty cores are shown with the prefix V for the *Vema* and the prefix C for the *Robert D. Conrad.* Deep basin areas are stippled, and the 2000-fm contour outlines the mid-oceanic ridge and the edge of the continents.

dyke, 1967; *Kennett and Geitzenauer*, 1969; *Geitzenauer*, 1969; *Margolis and Kennett*, 1971; *Bandy et al.*, 1971]. Sedimentation along the continental margin of Australia was discussed by *Conolly* [1967, 1968], *Conolly and von der Borch* [1967], *Conolly et al.* [1970], *von der Borch et al.* [1970], and *Wass et al.* [1970]. *Conolly* [1969] reported on sediments from the southern part of the Tasman Sea.

In this report, we bring up to date the continental margin sedimentation studies of *Conolly and von der Borch* [1967] by using the *Eltanin* cores taken in the past 4 years. A companion paper [*Payne and Conolly*, this volume] describes the sedimentation pattern of the antarctic continental margin and more fully discusses turbidite sedimentation in the deep ocean.

PHYSIOGRAPHIC SETTING AND GENERAL OUTLINE

The Australian and antarctic continents are separated by an east-west trending mid-oceanic ridge, most of which has been emplaced due to sea floor spreading processes in the last 50 m.y. [*Weissel and Hayes*, 1971]. The mid-oceanic ridge is displaced by a series of major right-lateral fracture zones as it trends southeastward from south of Tasmania to the Balleny Islands where this ridge is joined by the Macquarie trench trend in a triple plate junction.

Our data is presented in a series of profiles (Figures 2, 5, 6, 8) illustrating the major lithologies and ages of the sediment cored from a strip of ocean that runs from the Australian-antarctic discordance east to the south Tasman basin. We intend to discuss the sedimentary patterns of the upper 10 meters of sediment by starting with the Australian continental margin, then continuing with the much fractured eastern arm of the ridge and the south Tasman basin, and finally concluding with the central ridge or discordance.

SOUTHERN AUSTRALIAN CONTINENTAL MARGIN

The continental shelf in the Great Australian Bight is 30–100 miles wide and 600 miles long and is covered by calcareous Pleistocene sands [*Conolly and von der Borch*, 1967]. The shelf is commonly 20–60 fm deep, and evidence of strong bottom currents is present in most bottom photographs of rippled coarse sand dominated by a living fauna of sponges and bryozoans.

Many Tertiary and even Cretaceous fossils are reworked from the Cretaceous to Tertiary sediments of the Eucla basin that form 200-foot-high cliffs along the Great Australian Bight coast line. These cliffs are being actively eroded today by the washing of older fossil debris onto the inner shelf plain. As no rivers drain into the Great Australian Bight today, the only detritus supplied to the shelf comes from the reworking of the older Eucla basin cliff sediments with a smaller contribution from living epifauna [*Wass et al.*, 1970].

East and west of the Great Australian Bight, the shelf narrows. Small coastal rivers feed onto the shelf south of the Pre-Cambrian shield of Western Australia. Larger rivers drain from inland Australia and from the geologically younger and geographically higher eastern highlands onto the shelf to the east of the Great Australian Bight and add contributions of mineral and rock detritus to the calcareous shelf sediment.

Section A–A′ (Figure 2) illustrates the style of sedimentation found off the southern Australian continent. Cores C8-57, C8-58, V18-219, and E35-13 come from the upper continental slope and consist of reworked bryozoan and molluscan shelf sands mixed with small amounts (10–30%) of planktonic foraminifera. This coarse detritus is reworked downhill along submarine canyons and mixed with slope calcareous ooze (coccolith and foraminiferal ooze). Cores from the wide continental rise (cores E35-12, V18-217, V18-218) consist mainly of graded beds of calcareous sand (Figure 3). The upper parts of the graded sand beds are sometimes covered with silt and clay, but sand makes up more than 50% of the total sediment, and apparently silt and clay particles have mainly bypassed this area of coarse deposition.

The continental rise merges with the wide and flat South Indian abyssal plain. Piston cores from this region are characterized by sand similar in composition to that from the continental rise but with a larger admixture of 'lighter' particles such as planktonic foraminifera. The sands commonly grade upward into thick (20 cm to 1 meter) clay zones, the upper parts of which may be burrow mottled. Core V16-110 described by *Conolly and von der Borch* [1967, Figure 12] contains a number of graded beds of slightly different color and composition and is typical of beds cored from the abyssal plain and continental rise (Figure 4). Core E39-7 from the central part of the abyssal plain contains only thin (10 cm) fine sand layers at the base of thick (up to 4 meters) homogeneous clay beds. Core 39-5 from the deep zone of swale topography that flanks the abyssal plain consists of discreet thin beds of homogeneous clay. Since these beds of clay muds are similar in appearance (color and mineralogy) to the clay beds in core E39-7, we assume that they probably represent deposition during episodic flooding of the abyssal plain by large turbidity currents. We assume that the density currents do not deposit just on the flat abyssal surface but that a considerable portion of the suspended load is deposited on the higher abyssal hills that flank the outer margin of the abyssal plain. Continued deposition of this kind would eventually smooth this flanking region and would thus produce a zone of swale topography.

The thin region of swale topography merges ridgeward with deep (2500–2900 fm) ridge flank topog-

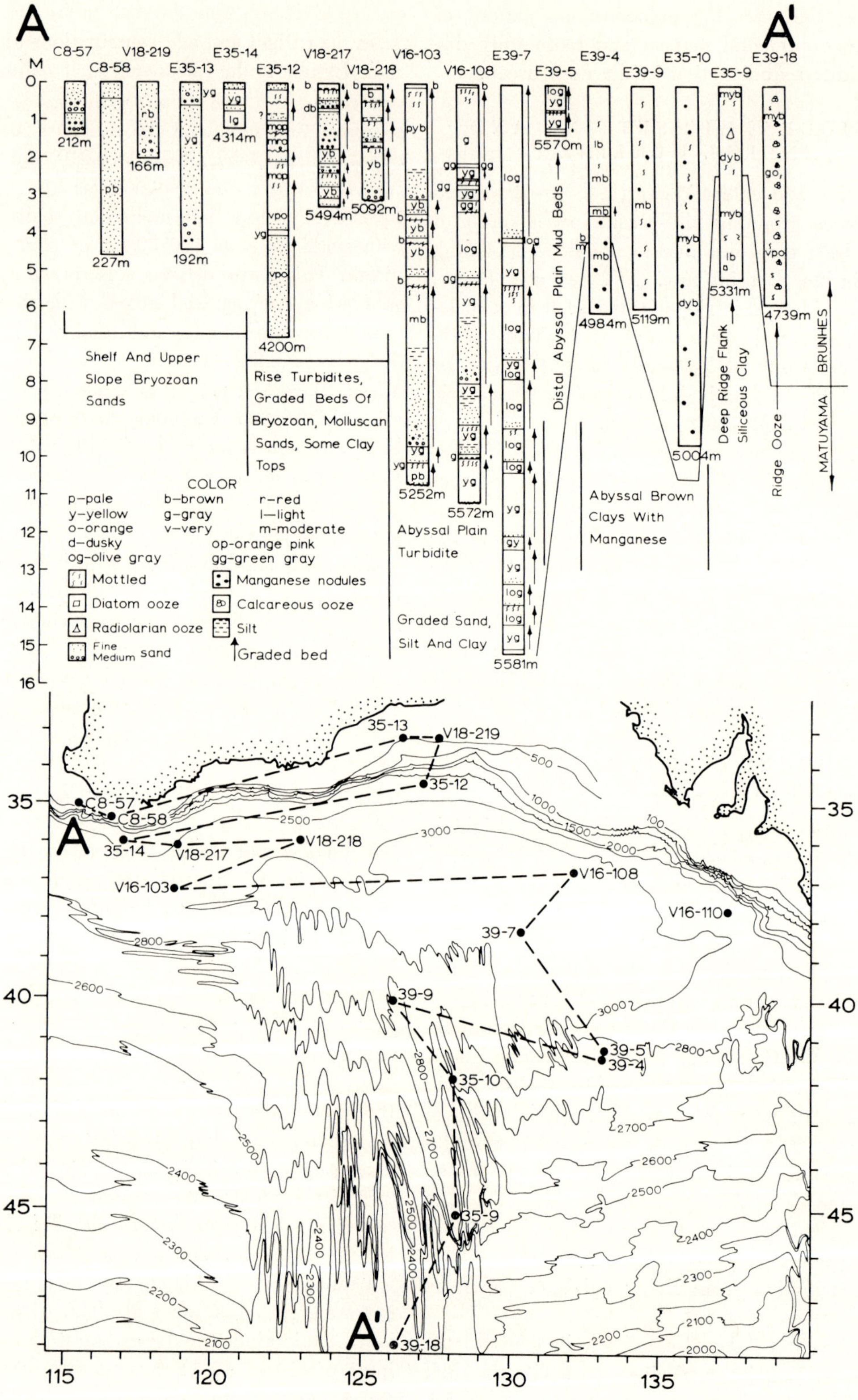

Fig. 2. Continental margin of southern Australia. Traverse A–A′ represents a typical cross section from continental shelf sands to abyssal plain turbidites and then to deep ridge muds and flank ooze. Bathymetry in corrected fathoms is after *Hayes and Conolly* [this volume].

Fig. 3. Photograph of sand from the sieved fraction greater than 500 μ from a turbidite sand in core V18-221 at a depth of 105 cm in the South Australian continental rise. The sand consists mostly of bryozoan fragments with scattered benthic foraminifera.

raphy. The deep ridge hills are covered with homogeneous abyssal brown clays composed mainly of mineral silt and clay and characterized by manganese micronodules that tend to become more common with depth in the core. These cores come from a zone that lies below the depth where both calcareous and siliceous skeletal material remains preserved in the sediment. Paleomagnetic stratigraphy (Figure 2) suggests, however, that the sedimentation rate is quite high (greater than 1 cm/1000 yr in core E35-10), a high rate indicating that much terrigenous detritus deriving from the Australian continent has been deposited over this region. As the central part of the mid-ocean ridge is approached, the brown abyssal muds tend to retain a relatively higher proportion of siliceous plankton (diatoms and radiolarians), and these siliceous clays merge with the calcareous oozes of the northern flank of the ridge.

TASMAN FRACTURE ZONE OF THE MID-OCEANIC RIDGE AND THE SOUTH TASMAN BASIN

The mid-oceanic ridge extends southeastward from about 138°E to the Balleny Islands in a series of blocks broken by several major fractures. The Tasman fracture zone forms the major fracture zone. The Tasman basin lies to the east of the mid-ocean ridge in this region. Five core traverses can be used to illustrate sedimentary patterns of the upper 10 meters of sediment from the mid-ocean ridge and south Tasman basin area.

The positions of paleomagnetic boundaries were fixed by combining the data of *Watkins and Kennett* [this volume] with our lithologic data. A traverse on the western margin of this region near 140°E (Figure 5, traverse B–B′) illustrates that the ridge is covered by 2–6 meters of Brunhes sediment overlying, perhaps, a similar thickness of Matuyama sediment. There are no obvious disconformities in this sequence, and the Brunhes sediment is essentially the same composition as the underlying Matuyama sediment.

Cores from the deep flanks of the mid-oceanic ridge commonly recover brown abyssal clays and/or manganese nodules. Calcareous ooze covers the ridge north of 52°S. Calcareous oozes mix with siliceous oozes between 52°S and about 55°S, and siliceous ooze is the dominant sediment south of 55°S.

A similar pattern of latitudinal sediment variation occurs for Brunhes and Matuyama sediment along traverse C–C′. Below the Matuyama boundary, however, the sedimentation pattern may be different. For instance, in core 39-55, Matuyama siliceous-calcareous ooze overlies Gauss siliceous ooze containing very few calcareous fossils. The upper 1.5 meters of this core consists almost entirely of calcareous ooze. Hence there is a trend toward a more siliceous ooze from the top to the bottom of core 39-55.

Profile C–C′ is located along the major fracture zone trends that offset the mid-oceanic ridge south of Australia [*Hayes and Conolly*, this volume]. Cores such as 39-55 that come from the deep fracture zone valleys of the offset blocks of the mid-oceanic ridge tend to contain more manganese and shortened intervals of Brunhes and Matuyama sediment relative to cores taken at shallower depths on the ridge.

The association of very little sedimentation with abundant manganese nodules becomes even more pronounced in the three traverses east of traverse C–C′ (Figure 6). These three traverses illustrate the startling disconformity that exists between a manganese pavement and underlying siliceous oozes of Gauss or Gilbert age within the deep troughs of the Tasman fracture zone and their extensions into the south Tasman basin. Associated with the manganese nodule pavement in most instances are reworked manganese nodule and foraminiferal sands. The foraminiferal sands and reworked oozes are commonly 1–2 meters thick as shown in core 36-21 on traverse D–D′. Bottom photography throughout the region (Figures 7 and 12) documents the prevailing existence of manganese nodules associated with current-rippled sands [*Jacobs et al.*, 1970].

The manganese nodule–sand pavement rests on a firm and compact pale olive to light olive-gray siliceous ooze or on a compact brown calcareous-siliceous ooze (Figure 6). This lithology is very distinctive and easy to identify as contrasted with the less firm Brunhes and Matuyama oozes deposited to the north or south of the manganese pavement region.

Profile D–D′ (Figure 6) illustrates the general sedimentary situation in the most striking fashion. Core 34-13 in the deep basin immediately south of Tasmania cored only manganese nodules, whereas core 36-22 recovered 5 meters of calcareous sands and core 39-57 recovered 5 meters of calcareous ooze deposited during the Brunhes. South of about 49°S, cores penetrate a sand-manganese pavement that overlies firm gray Gauss and Gilbert siliceous oozes. Hence siliceous ooze of Gilbert age can now be recovered from latitudes as far north as 50°S. This latitude is considerably (2°–5°) farther north than the deposition of modern siliceous oozes, and this change in latitudes

Fig. 4. Photo of core V16-110 from the South Australian plain at a depth of 5287 meters. This core consists of many graded beds with calcareous sands (such as that at 216–190 cm) at the base that grade upward into calcareous silt and mud and that then grade into darker hemipelagic brown or greenish clays. Many of the turbidite events have deposited white calcareous silts (such as at 663–650 cm or 566–560 cm) that grade upward into darker burrow-mottled pelagic muds.

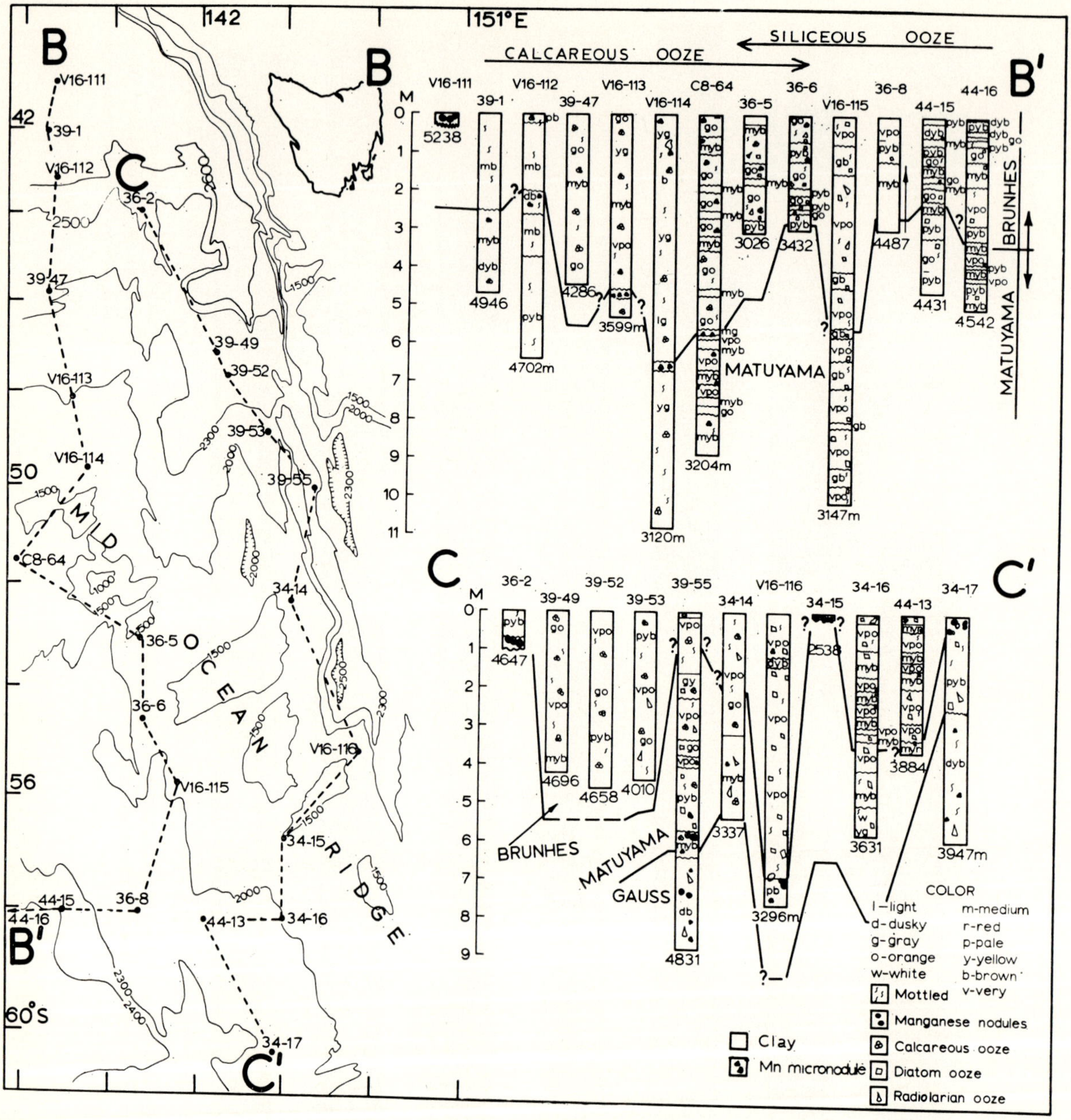

Fig. 5. Core profiles B–B′ and C–C′ from the mid-ocean ridge south of Australia. Outline bathymetry is after *Conolly and Hayes* [this volume].

indicates a change in the ooze sedimentation pattern since Gilbert time.

Core 38-9 recovered 3 meters of Brunhes foraminiferal sand. About another 1 meter of similar sand was deposited during the upper part of the Matuyama. These sands also lie on compact gray siliceous oozes of Gauss or older age.

The data from the cores taken at the edges of the manganese pavement region (cores 36-21, 36-20, and 38-9) suggest that currents became active sometime near the end of the Gauss (2.5 m.y. ago), for ooze of Gauss age is preserved in all these cores. Conceivably, sediment of Gauss age has been eroded from the central manganese pavement region represented by cores 36-17, 38-2, and 36-16 on traverse D–D′. This theory substantiates the conclusions of *Watkins and Kennett* [this volume], who advocate a period of intense current activity and erosion in the past 3 m.y. Rates of

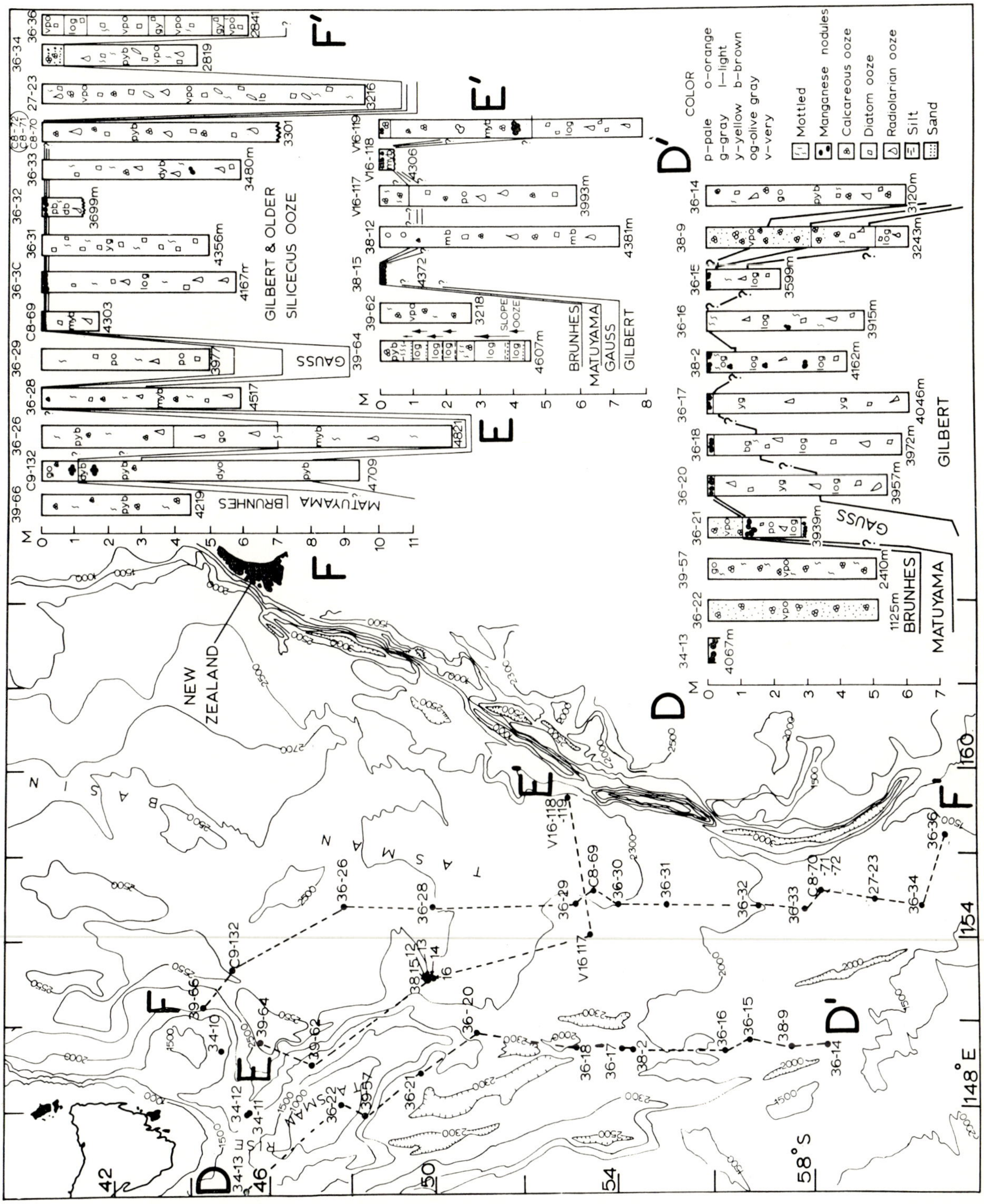

Fig. 6. Core profiles from the eastern flank of the mid-oceanic ridge south of Australia and the Tasman basin. Traverses D–D', E–E', and F–F' show features characteristic of the upper 10 meters of sediment cover from this region. Bathymetry is after *Hayes and Conolly* [this volume].

pelagic deposition are higher (1 cm/1000 yr) for some siliceous ooze zones south of the current-affected region (Figures 5 and 6).

MID-OCEAN RIDGE AND AUSTRALIAN-ANTARCTIC DISCORDANCE

Two traverses (Figure 8, G–G′ and H–H′) across the mid-oceanic ridge and the Australian-antarctic discordance illustrate the sedimentary patterns of the upper 10 meters of sediment that covers the east-west trending ridge south of Australia. The less fractured ridge section (traverse G–G′) is covered by 3–6 meters of Brunhes ooze that is calcareous on the northern ridge flank, siliceous on the southern ridge flank, and a mixture of both in the central ridge zone.

The interbedding of mixtures of calcareous and siliceous ooze provides us with an excellent opportunity to analyze major changes in paleoclimate in the circumpolar ocean, since we can derive climate curves for all major fossil plankton groups. Our investigations of these changes by using these techniques is still incomplete; however, our preliminary information on some cores suggests that the following changes can be observed (Figures 9, 10, 11).

Core C8-61 (Figures 9 and 10) is a typical siliceous-calcareous ooze from the mid-oceanic ridge south of Australia. It is mottled throughout and consists of zones of differently colored sediment. Changes in color correspond to changes in composition (Figure 10). The percentage of foraminifera within the total sediment, as approximated by weighing the greater than 62 μ fraction that is mostly composed of foraminifera, varies from 2 to 30%. Zones of low foraminiferal content are commonly marked by zones of different color; for instance, the grayish-orange sediment from 100 to 140 cm contains less than 10% foraminifera, and the darker brown zones above and beneath this zone contain about 20% foraminifera. Further analysis of this core indicates that such grayish-orange zones contain low percentages of foraminifera but high percentages of diatoms, whereas the reverse is true for the dark-colored zones. However, the core is not a simple mixture of light-colored siliceous ooze with dark-colored calcareous ooze, for there are many zones intermediate in color and composition. If it were true that the more siliceous ooze zones represent a more northerly incursion of upwelling cold antarctic bottom waters and hence a colder antarctic climate, then the climate interpretation shown on Figure 10 could be approximate to the truth. However, we need to examine carefully the sedimentation changes over this region to be sure of our interpretations.

Our preliminary analysis of the 12 meters of Brunhes sediment in core C8-63 (Figure 11) yields a climate curve similar to that derived by *Kennett* [1970, Figure 8], and the similarity suggests that there are at least six major warm periods during the Brunhes (including the Holocene). Our analysis is based on changes in warm and cold water assemblages of foraminifera, diatoms, and radiolarians [*Abbott*, 1971; *Payne*, 1971].

Representative foraminifera and radiolarian samples from the sand fractions (>62 μ) were counted at 50-cm intervals. These data were then compared with known distributions of cold and warm fauna in the Pacific and Indian Ocean sectors of the Southern Hemisphere oceans [*Kennett*, 1970; *Bé*, 1969; *Kustanowich*, 1963; *Boltovsky*, 1969; *Riedel*, 1958; *Hays*, 1965; *Petrushevskaya*, 1968].

An index for each group is shown in Figure 11. In each case, the curves are produced by a simple ratio of warm species as compared with cold. With the foraminifera index, a value of 1 means an equal number of cold and warm species. Values of less than 1 mean cold. Values of more than 1 mean warm. The radiolarian and diatom indices are normalized by factors of 50 and 100, respectively, owing to the dominance of colder species.

Fig. 7. Bottom photography of manganese pavement region south of the South Tasman rise and in the south Tasman basin. In this area, cores recover reworked foraminiferal sands and manganese nodules that lie on a firm and compact gray siliceous ooze of Gauss or Gilbert age. (*a*) Manganese pavement taken on *Eltanin* cruise 36, camera station 19, and frame 7, at 56°33′S, 150°06′E, and at a depth of 3757 meters. Manganese nodules are generally about 4–8 cm in diameter and similar in size to those shown in the other plates. (*b*) Manganese pavement and manganese-covered rocks taken on *Eltanin* cruise 36, camera station 20, and frame 17, at 55°08′S, 150°01′E, and at a depth of 3930 meters. (*c*) Manganese pavement patches within manganese and foraminiferal sand with some rocks taken on *Eltanin* cruise 36, camera station 21, and frame 3, at 54°04′S, 150°04′E, and at a depth of 4008 meters. (*d*) Manganese pavement within sandy ooze sediment taken on *Eltanin* cruise 36, camera station 22, and frame 5 and at a depth of 3930 meters. (*e*) Manganese and rock pavement in current-rippled sand taken on *Eltanin* cruise 36, camera station 23, and frame 15 and at a depth of 4430 meters. (*f*) Manganese pavement with extremely tightly packed nodules taken on *Eltanin* cruise 36, camera station 24, and frame 7 and at a depth of 4071 meters.

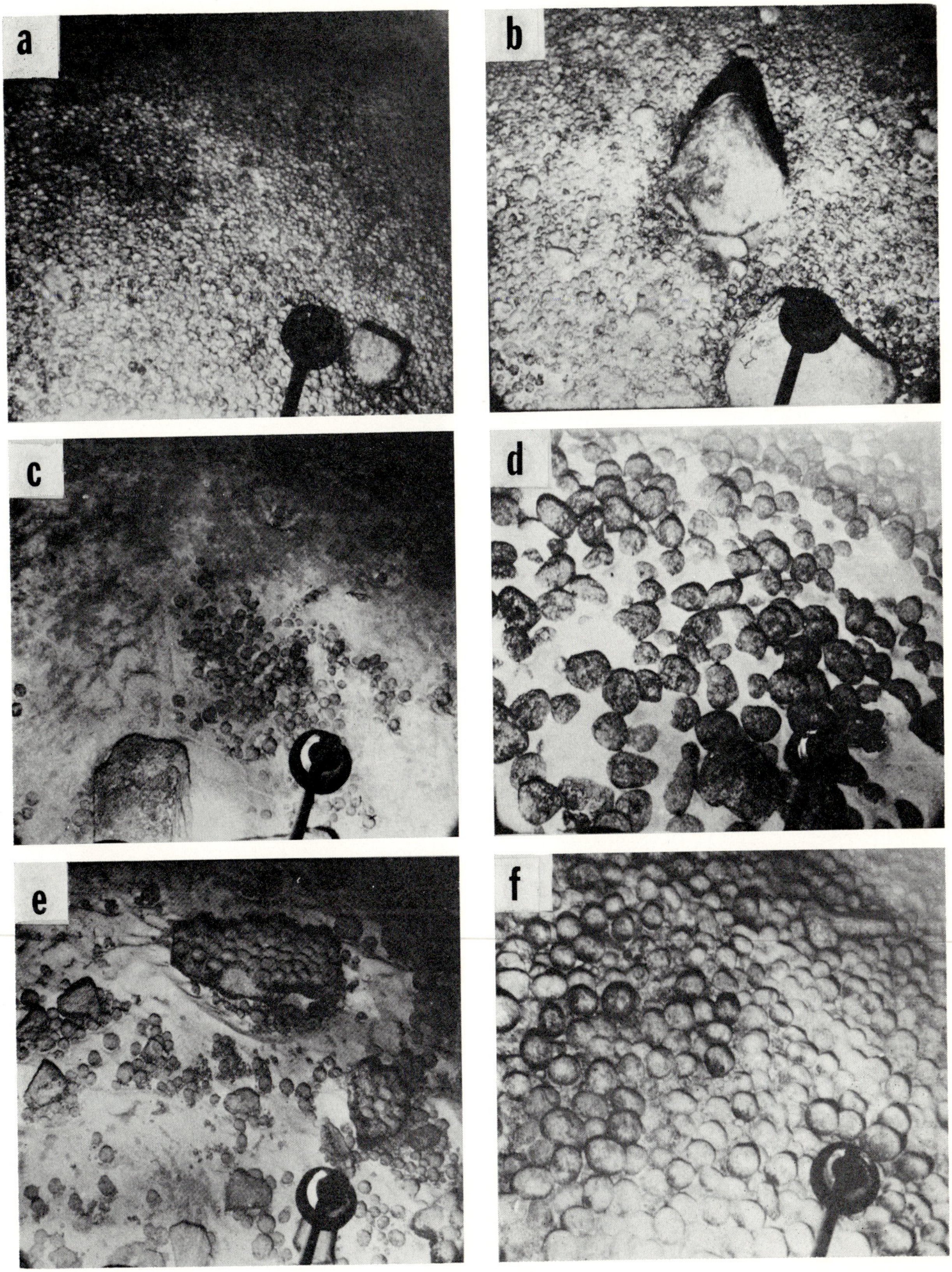
a
b
c
d
e
f

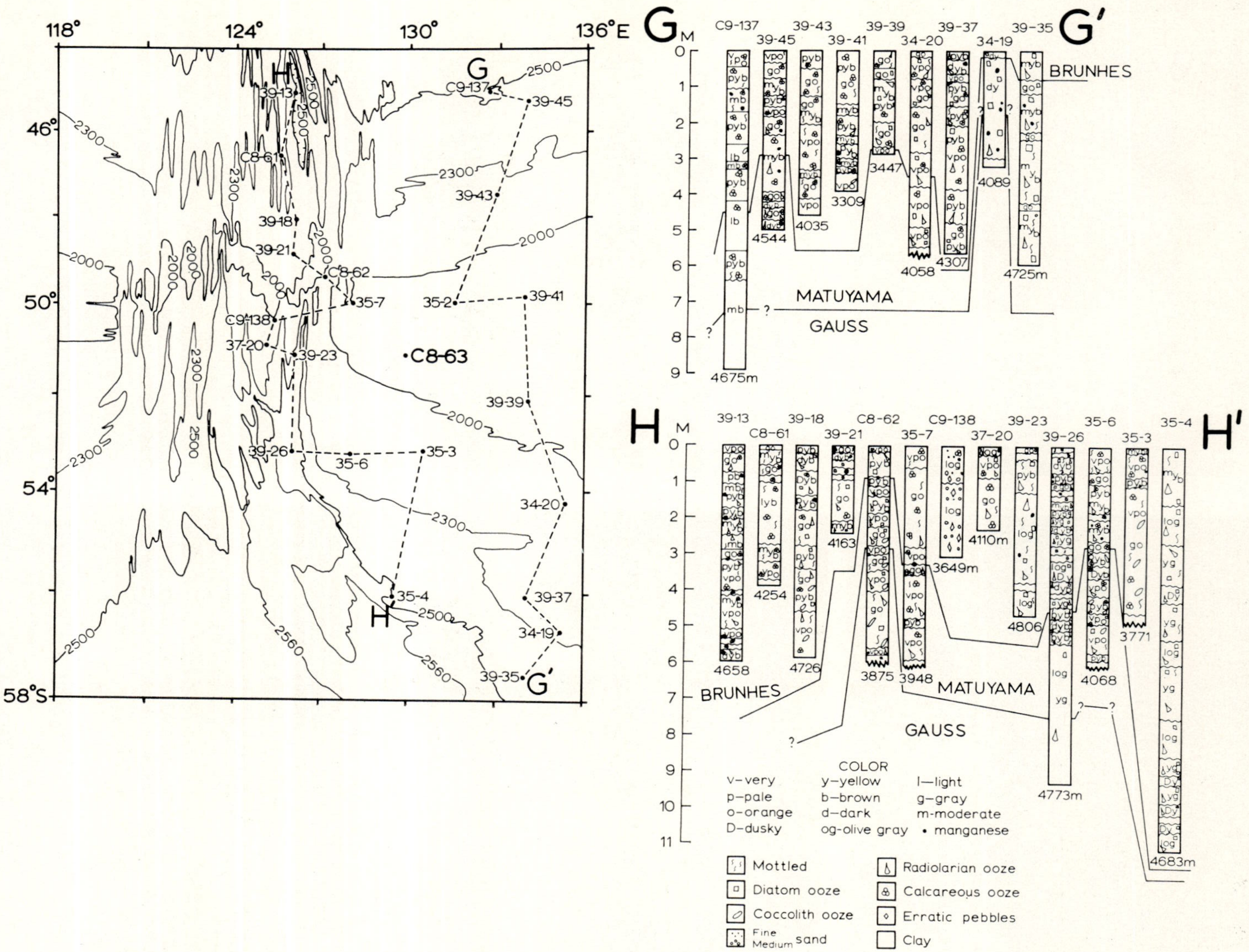

Fig. 8. Core profiles G–G′ and H–H′ from the Australian discordance region. Bathymetry after *Hayes and Conolly* [this volume].

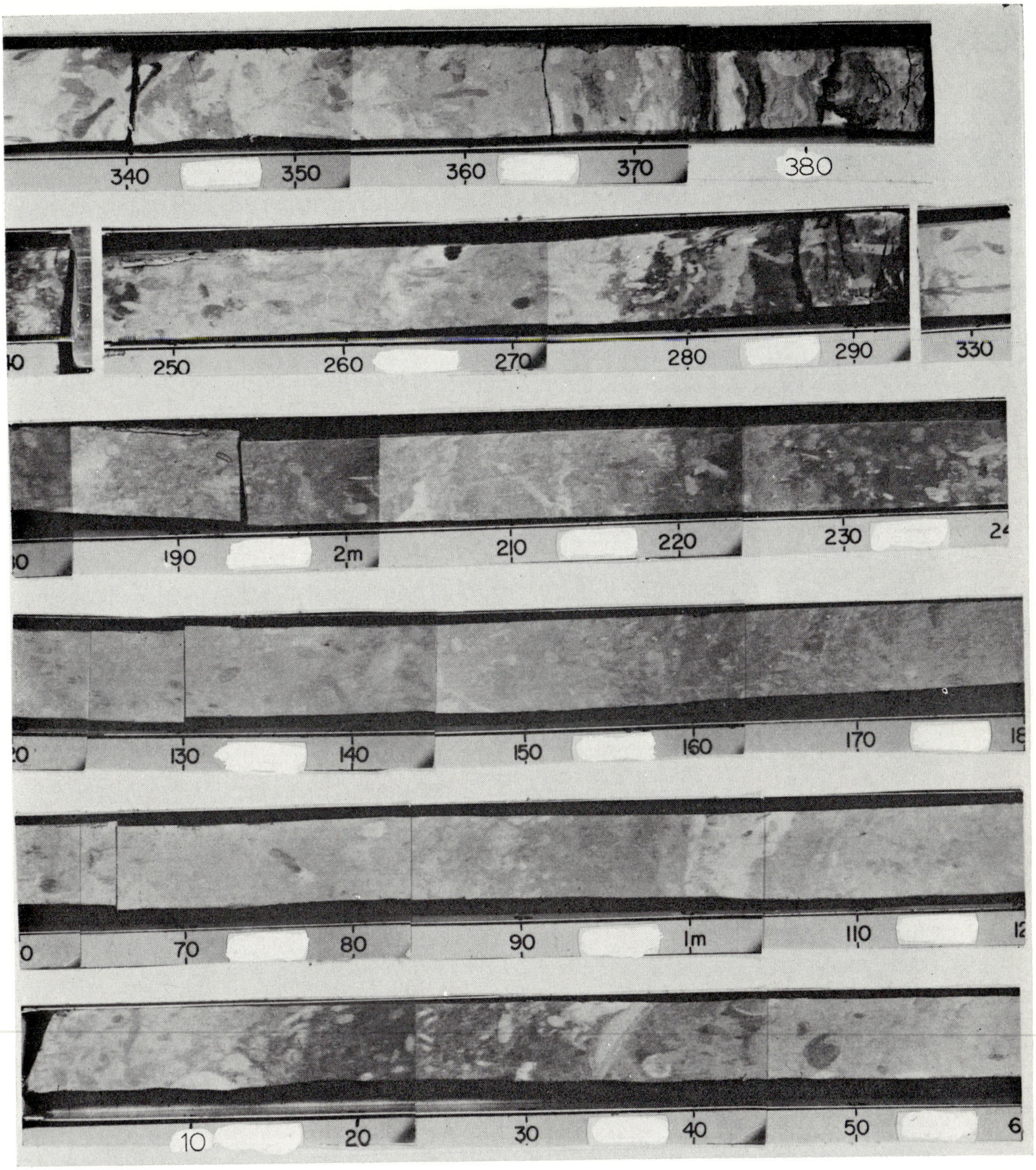

Fig. 9. Core C8-61 from the zone of interbedded siliceous and calcareous ooze on the mid-oceanic ridge south of Australia. This core is typical of oozes from the mid-oceanic ridge. It is burrow-mottled throughout and consists of differently colored zones of different composition as shown on Figure 10.

During examination of the radiolarian and diatom assemblages, known index species (e.g., the radiolarian *Clathrocyclas bicornis* and the diatom *Actinocyclus ingens* [*Hays*, 1965; *Donahue*, 1970]) from pre-Brunhes biostratigraphic zones were noted in the interval between 240 and 460 cm. The occurrence of these older species may indicate a slump of older sediment from an adjacent high area near the coring

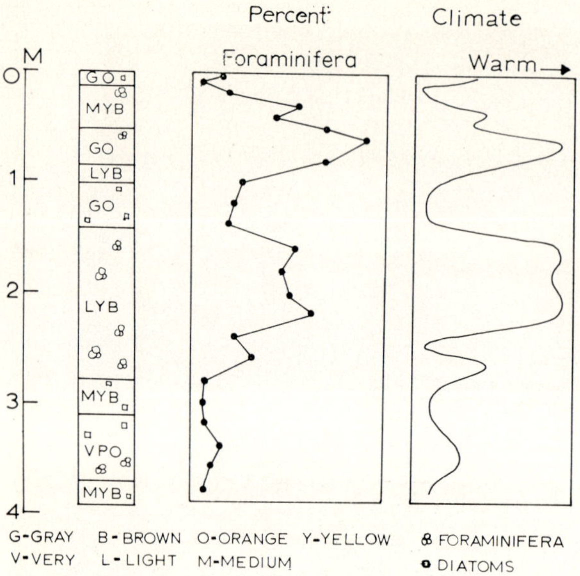

Fig. 10. Analysis of core C8-61 showing the variation in the total percentage of foraminifera with depth in the core. Warm zones are characterized by 20–30% foraminifera. Cold zones are mostly diatom ooze and contain low percentages of foraminifera.

location. As a result, this interval is omitted from Figure 11. Six major warm intervals showing fairly good correlation with light-colored zones are compared with a climate curve based on foraminiferal assemblages in subantarctic cores in the South Pacific [*Kennett*, 1970] on Figure 11.

We notice that all the Brunhes oozes shown in Figure 8 contain at least 6–12 major color zones. Each of these zones denotes a change in composition and hence probably a change in climate similar to the changes shown to exist in core C8-63. We are confident that further detailed faunal work on these cores should provide sufficient data to show that circumpolar waters have responded to major changes in climate at least 12 times during the past 700,000 years. A similar pattern of sedimentation may well be true for a major part of the Matuyama epoch, since it too is characterized by interbedded siliceous and calcareous ooze.

DISTRIBUTION OF SURFACE MANGANESE, BOTTOM CURRENTS, AND BOTTOM PHOTOGRAPHY

Many bottom camera stations have been set up in the region discussed above. The bottom photography and direct current measurements made during *Eltanin* cruises 32–36 were summarized by *Jacobs et al.* [1970]. These data and additional data from more recent cruises are summarized in Figure 12.

At most camera stations away from the manganese pavement area south of Tasmania (Figures 12 and 13), the bottom is muddy and shows traces of animal life in the form of tracks, trails, mounds, burrow pits, fecal pellets. In some instances, the camera has caught the organisms in the process of walking, crawling, or burrowing on the sea floor. At most of these stations with abundant sea floor life (Figure 7), the current meter has measured bottom currents of 2–10 cm/sec or greater, but there is little evidence of any of the mud becoming entrained in this current. Conversely, in areas of sandy bottoms, the sand is commonly rippled by bottom currents of the same intensity, this occurrence indicating that a muddy bottom is com-

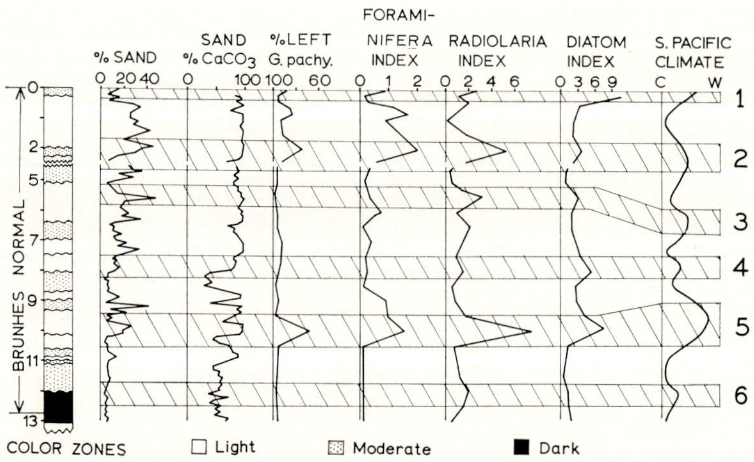

Fig. 11. Analysis of core C8-63 from the central part of the mid-oceanic ridge south of Australia. This core consists of variable mixtures of siliceous and calcareous ooze deposited over the past 0.8 m.y.

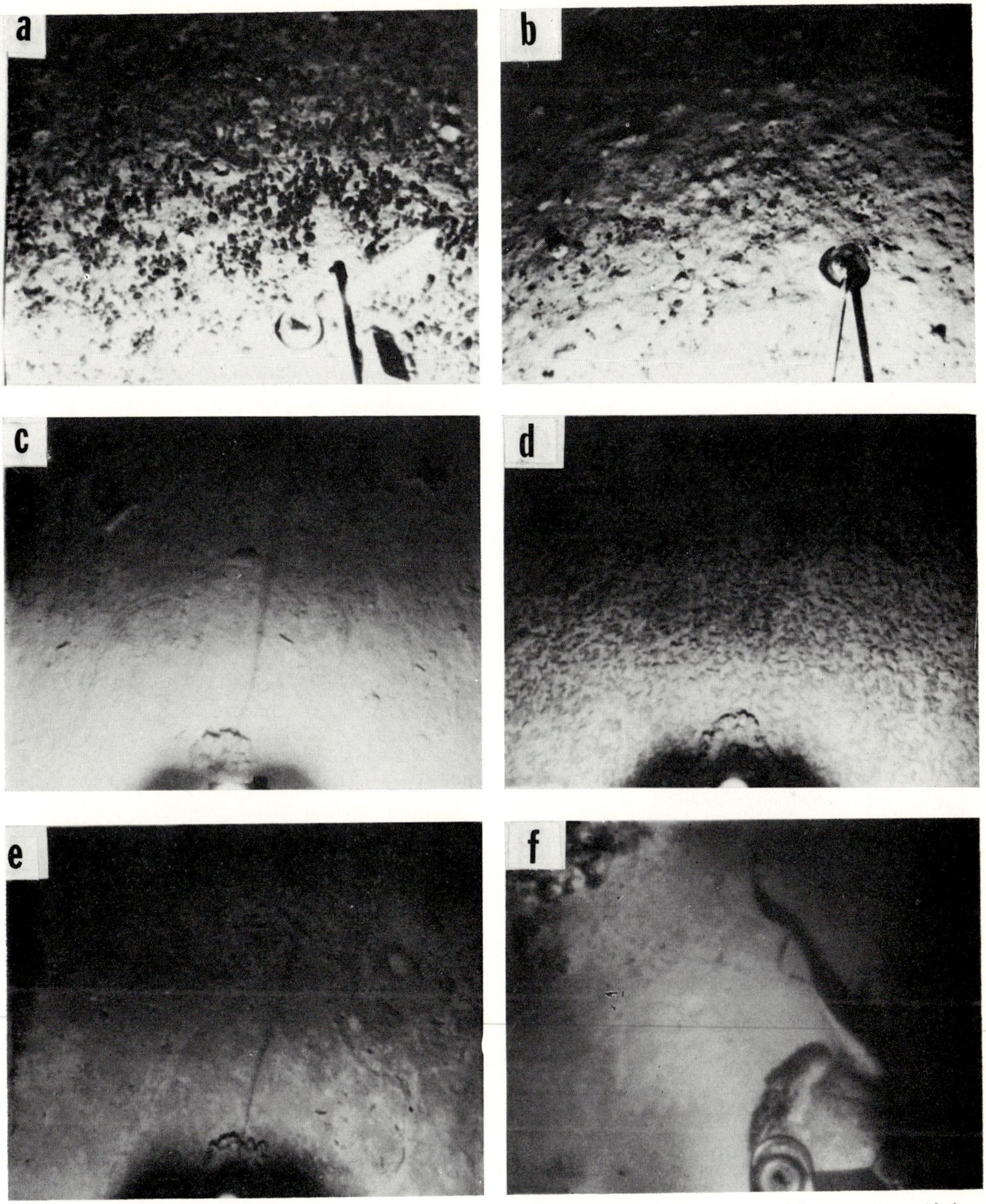

Fig. 12. Bottom photography of the mid-oceanic ridge south of Australia. (*a*) A manganese nodule pavement amid foraminiferal sand taken on Eltanin cruise 37, station 29, frame 21, at 50°55′S, 125°10′E, and at a depth of 2200 fm. (*b*) Same station as above showing sands with a sparse manganese nodule distribution. (*c*) Muddy bottom with holothurian tracks and fecal pellets, pits, and mounds taken on *Eltanin* cruise 44, station 26, frame 4, at 58°S, 132°30′E, and at a depth of 2490 fm. (*d*) Thin manganese nodular crust in ooze taken on *Eltanin* cruise 44, station 24, frame 4, at 57°58′S, 146°58′E, and at a depth of 1745 fm. (*e*) Muddy bottom with abundant tracks, trails, pits, and mounds taken on *Eltanin* cruise 44, station 25, frame 2, at 58°03′S, 139°58′E, and at a depth of 1862 fm. (*f*) Muddy bottom, rocks in lower right corner and a deep-sea fish taken on *Eltanin* cruise 39, station 4, frame 24, at 52°S, 134°E, and at a depth of 1470 to 1625 fm.

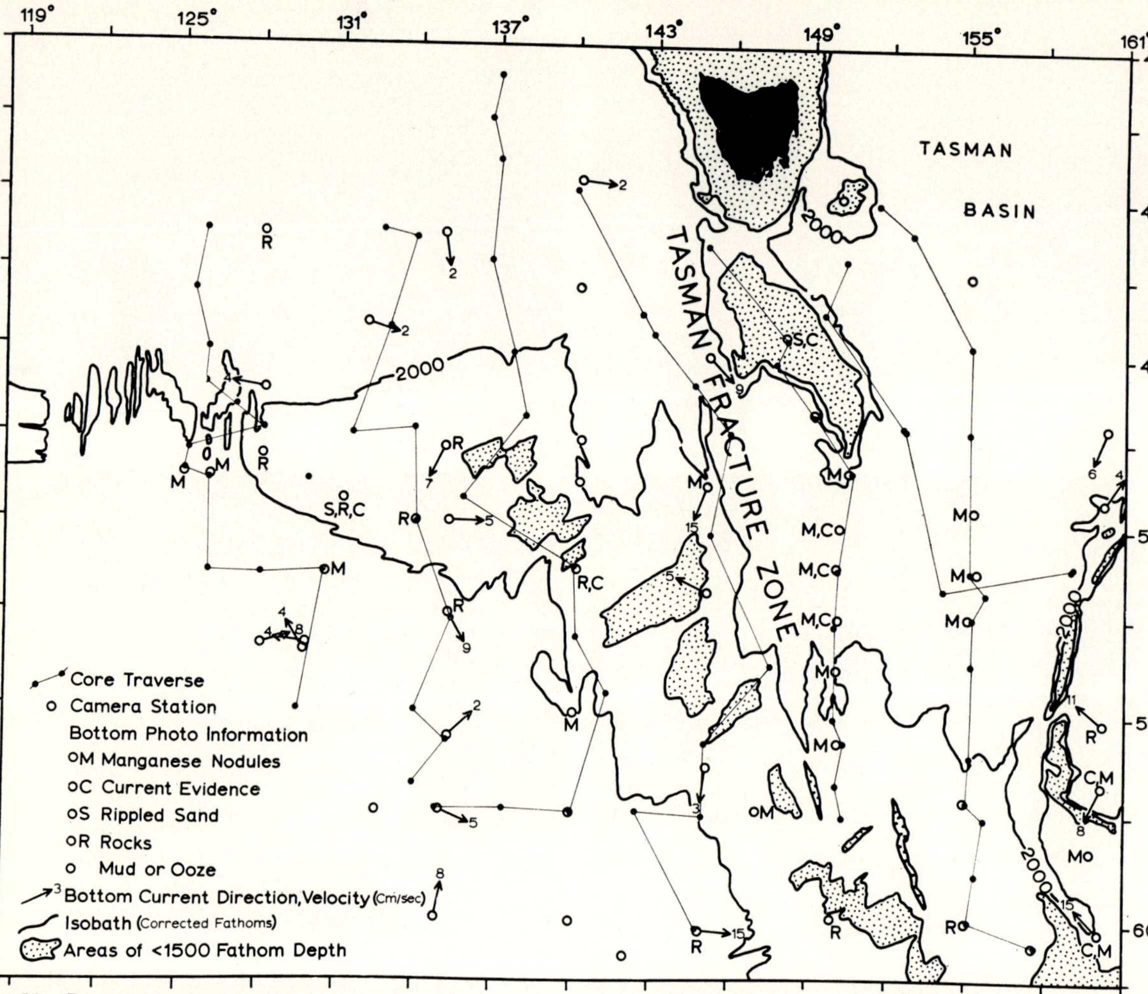

Fig. 13. Bottom photography, upper sediment cover, and bottom current data from the mid-oceanic ridge south of Australia.

monly more resistant to bottom currents than a sandy floor.

The region of deep topography south of the South Tasman rise and bordered to the south by the mid-oceanic ridge is characterized by a manganese nodule pavement (Figure 12). Not only do most camera stations show manganese nodules (Figure 13), but most cores from this region recover either manganese nodules or reworked foraminiferal sands (Figure 14). Studies by *Gordon* [1971*a*, 1971*b*, this volume] show that at least two different types of antarctic bottom water occur in the region south of Australia. Both are cold and high in oxygen, but they have different salinities, the low salinity variety (type B) being more widespread in the South Indian basin. The high salinity variety (type C) is derived from the Ross ice shelf water and is diluted in the southeastern sector of the South Indian basin.

The basic current direction predicted by Gordon is west to east with a westerly flow along the antarctic continental margin. A northward movement of water is predicted to occur across the mid-ocean ridge near 145°–150°E to introduce cold bottom water into the Tasman basin. This movement must occur along the Tasman fracture zone, which forms a series of deep troughs oriented north-northwest–south-southeast at 145°–150°E (Figure 12). Gordon predicts that this cold water then moves eastward out of the Tasman fracture zone across the south Tasman basin and eventually crosses the Macquarie ridge into the Southwest Pacific basin.

The bottom camera and piston-coring data certainly substantiate Gordon's predictions (Figures 12 and 14) and show that most of the Tasman fracture zone is floored by manganese nodules and reworked sands. This manganese pavement area extends into the south Tasman basin to the Macquarie ridge. Bottom photographs of the deep-sea floor east of the Macquarie

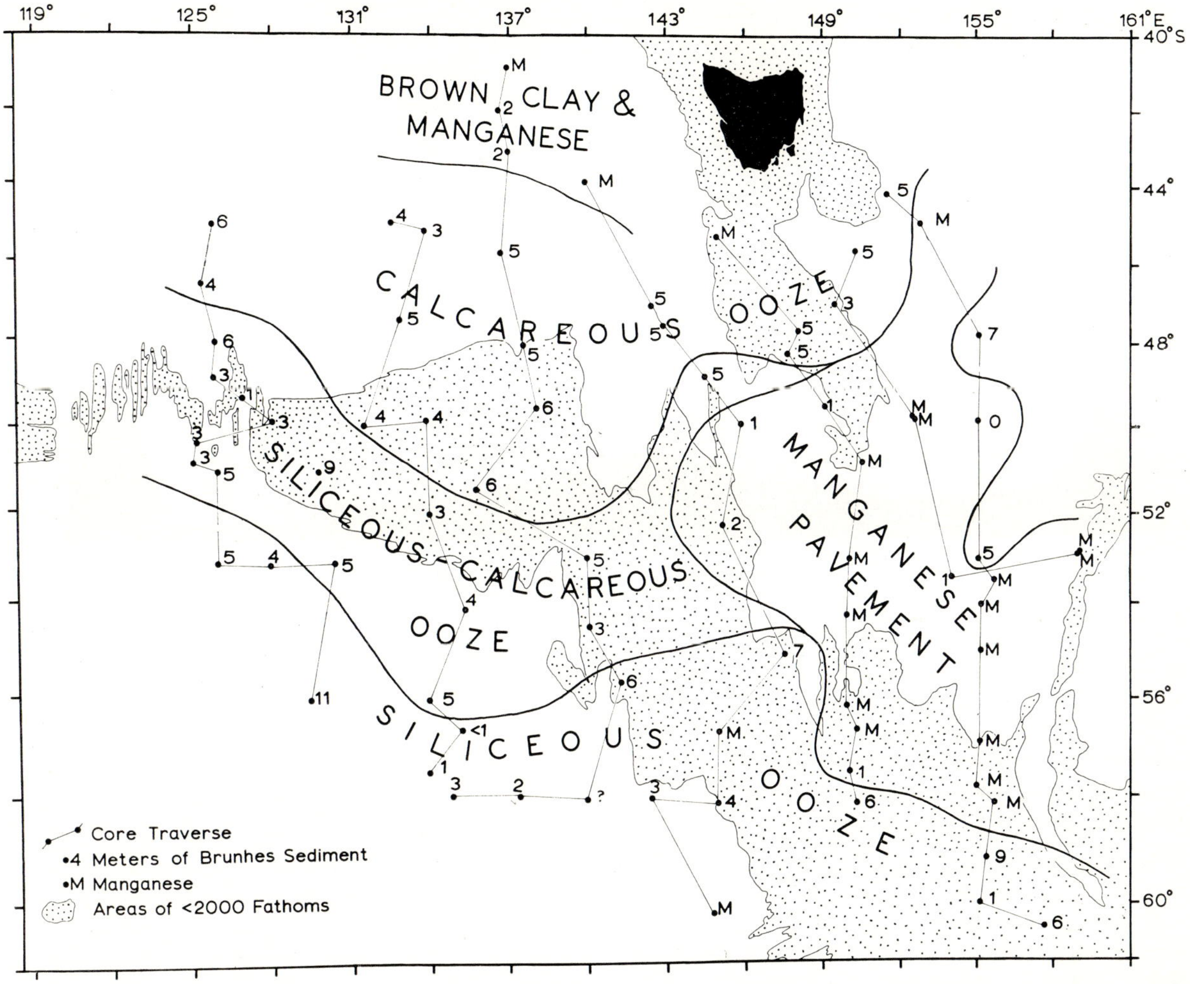

Fig. 14. Sedimentation patterns during the Brunhes magnetic epoch across the mid-oceanic ridge south of Australia. Thickness of Brunhes sediment is shown in meters.

ridge also show manganese pavements [*Jacobs et al.*, 1970] that indicate the flow of the cold bottom water through the ridge into the Southwest Pacific basin. *Eittreim et al.* [1971] describe a similar distribution for a thick nepheloid layer in the Indian-Pacific antarctic sea. They also show that their short duration current measurements (those readings shown on Figure 12) have an irregular motion with a mean clockwise flow in the South Indian basin.

SUMMARY

The composition and thickness of sediment deposited during the Brunhes epoch across the mid-oceanic ridge south of Australia is shown on Figure 14. By using the data shown on profile A–A′ (Figure 2) and published in *Conolly and von der Borch* [1967] for the continental margin of Australia and by using the data of *Payne and Conolly* [this volume] for the antarctic continental margin, a profile for sedimentation across the entire southeast Indian Ocean is shown on Figure 15.

Continental Margin Sedimentation

Processes of continental margin sedimentation are now well documented from other oceans, particularly from the western coast of the Atlantic Ocean [*Emery*, 1965; *Emery et al.*, 1970; *Heezen and Hollister*, 1964; *Uchupi*, 1963], the southern coast of California [*Normark*, 1970; *Shepard et al.*, 1969], and the northwestern coast of the United States [*Griggs and Kulm*, 1970; *Griggs et al.*, 1968], as well as from other miscellaneous regions around the world. It is generally accepted that sediment is moved down the continental

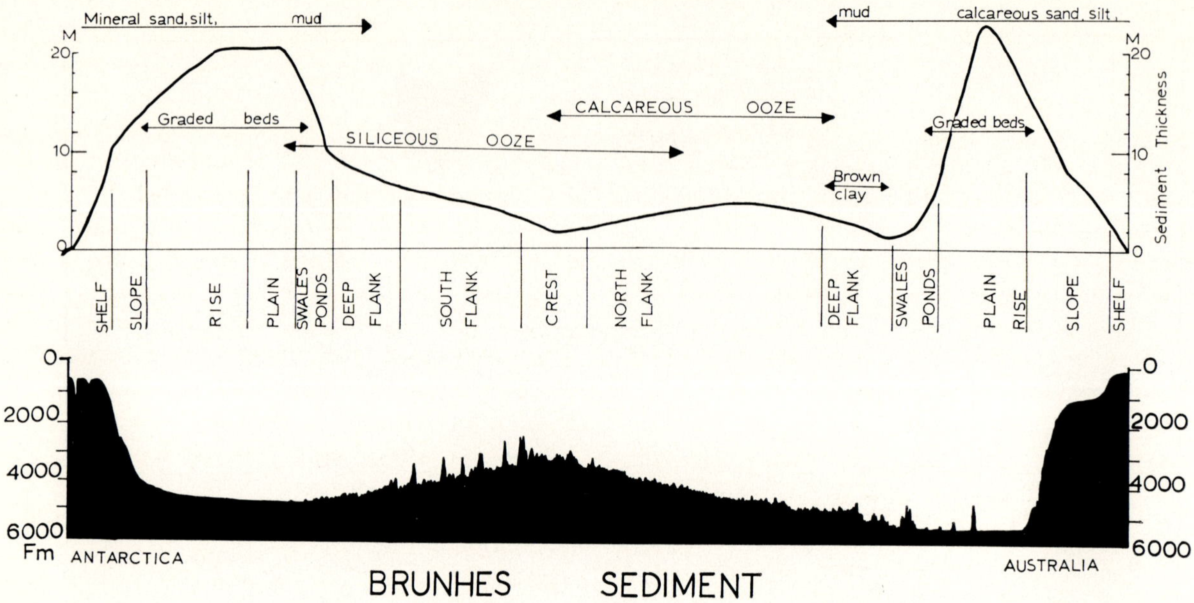

Fig. 15. A model for sedimentation patterns across a typical continent–mid-oceanic ridge–continent profile. This profile is drawn at about 133°E from Australia to Antarctica across the mid-oceanic ridge and shows the distribution of the major sediment types for each physiographic province. The average thickness of sediment deposited in the Brunhes in each province is also shown.

slope by way of submarine canyons that build submarine 'deltas' or deep-sea fans and deltas [*Normark*, 1970; *Conolly and Cleary*, 1971].

The flat abyssal plains of the deep ocean are characterized by graded beds of sand, silt, and clay that merge laterally into beds of clay. These beds of clay were presumably deposited as the fine tail of turbidity currents and form clay ponds at the distal margin of the plains [*Needham and Conolly*, 1970].

Our analyses of the sedimentation patterns across the continental rise and abyssal plain off Antarctica [*Payne and Conolly*, this volume] show the same sedimentation pattern. A similar pattern exists off the Australian continental margin (Figure 2). The fan valleys of the continental rise and the proximal region of the abyssal plain are characterized by graded beds of sand and silt. The percentage of sand decreases away from the feeding canyons of the continental margin. Any profile across the abyssal plain will show graded beds of sand and silt with clay tops that merge distally into silt and clay beds.

Up to 20 meters of sediment was deposited during the Brunhes in the central part of the South Indian abyssal plain [*Payne et al.*, 1971, 1972], and just as much sediment was deposited in the South Australian abyssal plain. The swale and ponded areas of mud that occur as a distal apron to the abyssal plain generally contain up to 15 meters of sediment in the Brunhes. Sediment thickness in the distal abyssal plain region, judged from seismic reflection profiles, is from 300 meters to 1 km in the distal portions of the South Indian abyssal plain, this thickness giving average sedimentation rates of 1–3 cm/1000 yr in that part of the ocean basin, if it is about 30 m.y. old. These rates are in keeping with the rates of sedimentation calculated for the Brunhes (Figure 15).

Mid-Oceanic Ridge Sedimentation

Both siliceous and calcareous ooze cover the mid-oceanic ridge south of Australia. The siliceous oozes occur on the south ridge flank and the calcareous oozes on the north ridge flank (Figure 15). A zone of mixed calcareous-siliceous oozes 600–1000 km wide covers the central mid-oceanic ridge and must be caused by the mixing of two contrasting types of surface waters, cold upwelling antarctic bottom water from the south and warmer Indian Ocean water coming in from the northwest.

There is a southward bulge of the mixed ooze zone between about 130° and 140°E that may be caused by a more southward movement of the warmer Indian Ocean water in this region. There does not seem to be any other way to explain this bulge. It must be caused by changes in water circulation close to the surface, for apparently most pelagic skeletal remains fall to the sea floor close to the geographical location in which they thrived in the surface waters.

Ooze sedimentation rates across the mid-oceanic ridge vary from 0.1 to 1 cm/1000 yr for the Brunhes and were probably similar or slightly less for the Matuyama. For the most part, ooze rates average about 0.5 cm/1000 yr, and hence if the central ridge zone is only 10–20 m.y. old [*Weissel and Hayes*, this volume], only 50–100 meters of sediment can be expected to be deposited over it. This amount is in keeping with the extremely thin (< 100 meters) blanket of sediment recognized on seismic reflection profiles [*Houtz and Markl*, this volume].

Pleistocene Climate Changes

Preliminary studies of interbedded siliceous-calcareous oozes in the mid-oceanic ridge region south of Australia provide interesting paleoclimate data. Detailed studies of both siliceous and calcareous fossils in 12 meters of Brunhes ooze in one core (C8-63) suggest that there have been six major warm periods in the last 700,000 years; this suggestion is in keeping with *Kennett*'s [1970] climate assessment for the Brunhes for cores from the subantarctic South Pacific Ocean. Core C8-63, like most other ooze cores from the mixed ooze region of the mid-oceanic ridge south of Australia, consists of siliceous ooze zones presumably deposited during colder times and calcareous ooze zones deposited during warmer periods. Many cores contain up to 12 different ooze zones in the Brunhes. The zones are easily distinguished by their differences in color and lithology and substantiate the hypothesis that there have been major changes in the surface waters during the past 700,000 years. The same pattern of sedimentation extends back into the Matuyama.

The only Gauss and Gilbert sediment extensively cored occurs beneath the manganese pavement in the Tasman fracture zone and south Tasman basin. In this region, the Gauss and Gilbert sediment consists of compact and gray siliceous ooze. The northernmost limit of this ooze lies near 49°S, and this limit is 100 km or more farther north than the location of Brunhes or Matuyama siliceous ooze. This location suggests that there has been a major change in the sedimentation pattern since Gauss time.

The timing of this change is difficult to calculate, but one traverse of cores along the Tasman fracture zone suggests that it may have occurred about halfway into the Gauss, or about 3 m.y. ago, since all sediment older than this age can be found missing from cores taken in this region. However, given sea floor spreading rates of up to 5 cm/yr [*Weissel and Hayes*, 1971] over periods of 2–3 m.y., the geographical position of one core could be displaced northward 100–150 km because of the northward movement of the Australian continent away from a 'fixed' Antarctica [*Vogt and Conolly*, 1971]. Hence the location of siliceous ooze that is 3 m.y. old 100 km northward of the present-day siliceous ooze limit is to be expected.

Recent studies by *Denton et al.* [1970] have shown that Antarctica has suffered at least four major glaciations during the last 5 m.y. and that a large ice sheet has occupied most of West Antarctica since the late Miocene [*Rutford et al.*, 1968]. The existence of an ice sheet across Antarctica prior to 5 m.y. ago still remains an unanswered question. Currently, most world-wide evidence comes from sea level data [*Tanner*, 1968].

Note here that a major glacier episode registered in Antarctica may not necessarily be caused by a world-wide cooling, since *Hollin* [1962] has shown that when some glaciers are advancing in Antarctica, others are retreating. It is possible to derive a climatic curve from glaciations in Antarctica that does not record world-wide climate changes. Notwithstanding this possibility, the indication that probably at least six major warm and six major cold climatic cycles occurred during the Brunhes is in keeping with the 12 major cycles recognized for the same period from cores from the subantarctic Pacific Ocean [*Kennett*, 1970], though these 12 cycles are not similar to the 16 cycles recognized by *Hays et al.* [1969] from the eastern equatorial Pacific.

The Tasman Manganese Pavement

The extensive manganese pavement of the Tasman fracture zone consists of reworked foraminiferal sands that are interbedded with manganese nodules that range in size from a few millimeters to several centimeters. The zone must owe its formation to a combination of many factors, such as (1) the massive movement of water along the deep troughs of the Tasman fracture zone into the south Tasman basin, (2) extremely low rates of pelagic deposition on adjacent parts of the ridge, and (3) the fact that this region of the sea floor does not receive sediment from continental sources being located in a position away from the influence of sedimentation from continental margins.

Acknowledgments. We thank Mr. D. Cassidy and Dr. L. Frakes for their assistance with the core facility at Florida State University, Mr. William Abbott for help with the diatom assemblages on core C8-63, Mrs. M. Galloway for typing and editorial assistance, and Mr. S. Salley for drafting the figures. This research was funded by National Science Foundation grant GV-28803 for work on antarctic cores. Dr. Norman Watkins provided paleomagnetic stratigraphy for the cores, and we appreciate his cooperation.

REFERENCES

Abbott, W. H., Diatom investigations of southern ocean deep-sea cores, *Antarctic J. U.S.*, *6*(5), 171–172, 1971.

Bandy, O. L., R. E. Casey, and R. C. Wright, Late Neogene planktonic zonation, magnetic reversals, and radiometric dates, antarctic to the tropics, in *Antarctic Oceanology I*, *Antarctic Res. Ser.*, vol. 15, edited by J. L. Reid, pp. 1–26, AGU, Washington, D. C., 1971.

Bé, A. W. H., Planktonic foraminifera, in *Antarctic Map Folio Ser.*, folio 11, pp. 9–12, Amer. Geogr. Soc., New York, 1969.

Boltovsky, E., Living planktonic foraminifera at the 90°E meridian from the equator to the antarctic, *Micropaleontology*, *15*(2), 237–255, 1969.

Conolly, J. R., Postglacial-glacial change in climate in the Indian Ocean, *Nature*, *214*, 873–875, 1967.

Conolly, J. R., Geology and submarine canyons, East Bass Strait, Australia, *Mar. Geol.*, *6*, 449–461, 1968.

Conolly, J. R., Western Tasman sea floor, *N. Z. J. Geol. Geophys.*, *12*, 310–343, 1969.

Conolly, J. R., and W. J. Cleary, Braided deep sea deltas and the origin of turbidite sands (abstract), *Eos Trans. AGU*, *52*(4), 244, 1971.

Conolly, J. R., and C. C. von der Borch, Sedimentation and physiography of the sea-floor south of Australia, *Sediment. Geol.*, *1*, 181–220, 1967.

Conolly, J. R., A. Flavelle, and R. S. Dietz, Continental margin of the Great Australian Bight, *Mar. Geol.*, *8*, 31–58, 1970.

Denton, G. H., R. L. Armstrong, and Minze Stuiver, Late Cenozoic glaciation in Antarctica: The record in the McMurdo Sound region, *Antarctic J.*, *5*(1), 15–21, 1970.

Donahue, J. G., Diatoms as quaternary biostratigraphic and paleoclimatic indicators in high latitudes of the Pacific Ocean, Ph.D. thesis, Columbia University, New York, 1970.

Eittreim, S., A. L. Gordon, M. Ewing, E. M. Thorndike, and P. Bruchhausen, The nepheloid layer and observed bottom currents in the Indian-Pacific antarctic sea, in *Studies in Physical Oceanography—A Tribute to Georg Wüst on His 80th Birthday*, edited by A. L. Gordon, Gordon and Breach, New York, in press, 1971.

Emery, K. O., Characteristics of continental shelves and slopes, *Bull. Amer. Ass. Petrol. Geol.*, *49*(9), 1379–1384, 1965.

Emery, K. O., et al., The continental rise off eastern North America, *Amer. Ass. Petrol. Geol. Bull.*, *54*(1), 1970.

Geitzenauer, K. R., The Pleistocene calcareous nannoplankton of the subantarctic Pacific Ocean, *Contrib. 32*, Sediment Res. Lab., Fla. State Univ., Tallahassee, 1969.

Glass, B., Microtektites in deep-sea sediments, *Nature*, *214*, 372–374, 1967.

Goodell, H. G., and N. D. Watkins, The paleomagnetic stratigraphy of the Southern Ocean: 20° West to 160° East longitude, *Deep Sea Res.*, *15*, 89–112, 1968.

Gordon, A. L., Oceanography of Antarctic Waters, in *Antarctic Oceanology I*, *Antarctic Res. Ser.*, vol. 15, edited by J. L. Reid, pp. 169–203, AGU, Washington, D. C., 1971*a*.

Gordon, A. L., Spreading of antarctic bottom waters, 2, in *Studies in Physical Oceanography — A Tribute to Georg Wüst on His 80th Birthday*, edited by A. L. Gordon, Gordon and Breach, New York, in press, 1971*b*.

Gordon, A. L., On the interaction of the antarctic circumpolar current and the Macquarie ridge, in *Antarctic Oceanology II: The Australian–New Zealand Sector*, *Antarctic Res. Ser.*, vol. 19, edited by D. E. Hayes, pp. 71–78, AGU, Washington, D. C., 1972.

Griggs, C. B., and L. D. Kulm, Sedimentation in Cascadia deep-sea channel, *Geol. Soc. Amer. Bull.*, *81*, 1361–1384, 1970.

Griggs, G. B., A. G. Carey, and L. D. Kulm, Deep-sea sedimentation and sediment-fauna interaction in Cascadia channel and on Cascadia abyssal plain, *Deep Sea Res.*, *16*, 157–170, 1968.

Hayes, D. E., and J. R. Conolly, Morphology of the southeast Indian Ocean, in *Antarctic Oceanology II: The Australian–New Zealand Sector*, *Antarctic Res. Ser.*, vol. 19, edited by D. E. Hayes, pp. 125–145, AGU, Washington, D. C., 1972.

Hays, J. D., Radiolaria and Late Tertiary and Quaternary history of antarctic seas, in *Biology of the Antarctic Seas II*, *Antarctic Res. Ser.*, vol. 5, edited by G. Llano, pp. 125–184, AGU, Washington, D. C., 1965.

Hays, J. D., Quaternary sediments of the antarctic ocean, *Progress in Oceanography*, vol. 4, edited by M. Sears, pp. 117–131, Pergamon, New York, 1967.

Hays, J. D., and N. D. Opdyke, Antarctic radiolaria, magnetic reversals and climatic change, *Science*, *158*, 1001–1011, 1967.

Hays, J. D., T. Saito, N. D. Opdyke, and L. H. Burckle, Plio-Pleistocene sediments of the equatorial Pacific: Their paleomagnetic, biostratigraphic, and climatic record, *Geol. Soc. Amer. Bull.*, *80*, 1481–1514, 1969.

Heezen, B. C., and C. Hollister, Deep-sea current evidence from abyssal sediments, *Mar. Geol.*, *1*, 141–174, 1964.

Hollin, J. T., On the glacial history of Antarctica, *J. Glaciol.*, *4*(32), 173–195, 1962.

Hollister, C. D., and B. C. Heezen, The floor of the Bellingshausen Sea, in *Deep-Sea Photography*, *Johns Hopkins Oceanogr. Stud. 3*, edited by J. B. Hersey, pp. 177–189, Johns Hopkins Press, Baltimore, Md., 1967.

Houtz, R. E., and R. G. Markl, Seismic profiler data between Antarctica and Australia, in *Antarctic Oceanology II: The Australian–New Zealand Sector*, *Antarctic Res. Ser.*, vol. 19, edited by D. E. Hayes, pp. 147–164, AGU, Washington, D. C., 1972.

Jacobs, S. S., P. M. Bruchhausen, and E. B. Bauer, Hydrographic stations, bottom photographs, and current measurements, USNS *Eltanin* cruises 32–36, 460 pp., Lamont-Doherty Geol. Observ. of Columbia Univ., Palisades, N. Y., 1970.

Kennett, J. P., Pleistocene paleoclimates and foraminiferal biostratigraphy in subantarctic deep-sea cores, *Deep Sea Res.*, *17*, 125–140, 1970.

Kennett, J. P., and K. R. Geitzenauer, Pliocene-Pleistocene boundary in a South Pacific deep-sea core, *Nature*, *224* (5222), 899–901, 1969.

Kustanowich, S., Distribution of planktonic foraminifera in surface sediments of the south-west Pacific Ocean, *N. Z. J. Geol. Geophys.*, *6*(4), 534–565, 1963.

Margolis, S. V., and J. P. Kennett, Cenozoic paleoglacial history of Antarctica recorded in subantarctic deep-sea cores, *Amer. J. Sci., 271*, 1–36, 1971.

Needham, H. D., and J. R. Conolly, Sedimentation in modern abyssal plains, paper presented at annual meeting, Geol. Soc. of Amer., Milwaukee, Wis., 1970.

Normark, W. R., Growth patterns of deep-sea fans, *Amer. Ass. Petrol. Geol. Bull., 54*(11), 2170–2195, 1970.

Opdyke, N. D., and B. P. Glass, The paleomagnetism of sediment cores from the Indian Ocean, *Deep Sea Res., 16*, 249–261, 1969.

Payne, R. R., Paleoclimatic changes in antarctic Pelagic sediments, *Antarctic J. U.S., 6*(5), 181–182, 1971.

Payne, R. R., and J. R. Conolly, Turbidite sedimentation off the antarctic continent, in *Antarctic Oceanology II: The Australian–New Zealand Sector, Antarctic Res. Ser.*, vol. 19, edited by D. E. Hayes, pp. 349–364, AGU, Washington, D. C., 1972.

Payne, R. R., J. R. Conolly, and W. H. Abbott, Variation between *Eltanin* piston core sites: An intensive coring station in the Wilkes abyssal plain, *Antarctic J. U.S., 6*(5), 169–170, 1971.

Payne, R. R., J. R. Conolly, and W. H. Abbott, Turbidite muds within diatom ooze off Antarctica: Pleistocene sediment variation defined by closely-spaced piston cores, *Geol. Soc. Amer. Bull., 83*, 481–486, 1972.

Petrushevskaya, M. G., Radiolaria of orders Spumellaria and Nassellaria of the antarctic region (from material of the Soviet antarctic expedition), in *Biological Reports of the Soviet Antarctic Expedition (1955–1958)*, vol. 3, edited by A. P. Andriyashev and P. V. Ushakov, pp. 2–186, Israel Program for Scientific Translations, Jerusalem, 1968.

Riedel, W. R., Radiolaria in antarctic sediments, *B.A.N.Z.A.R.E. Rep., Ser. B., 6*(10), 217–255, 1958.

Rutford, R. H., C. Craddock, and T. W. Bastien, Late Tertiary glaciation and sea-level changes in Antarctica, *Paleogeogr. Paleoclimatol. Paleoecol., 5*(1), 15–39, 1968.

Shepard, F. P., R. F. Dill, and U. Von Rad, Physiography and sedimentary processes of La Jolla submarine fan and fan-valley, California, *Amer. Ass. Petrol. Geol. Bull., 53*(2), 390–420, 1969.

Tanner, N. F., Tertiary sea level symposium introduction, *Paleogeogr. Paleoclimatol. Paleoecol., 5*(1), 7–14, 1968.

Uchupi, E., Sediments on the continental margin off eastern United States, *U.S. Geol. Surv. Prof. Pap. 475-C*, C132–C137, 1963.

Vogt, P. R., and J. R. Conolly, Tasmantid guyots, the age of the Tasman basin, and motion between the Australia plate and the mantle, *Geol. Soc. Amer. Bull., 82*, 2577–2584, 1971.

von der Borch, C. C., J. R. Conolly, and R. S. Dietz, Sedimentation and geology of the continental margin off the Otway basin, *Mar. Geol., 8*, 31–58, 1970.

Wass, R. E., J. R. Conolly, and R. J. Macintyre, Bryozoan carbonate sand continuous along southern Australia, *Mar. Geol., 9*, 63–73, 1970.

Watkins, N. D., and J. P. Kennett, Regional sedimentary disconformities and Upper Cenozoic changes in bottom water velocities between Australasia and Antarctica, in *Antarctic Oceanology II: The Australian–New Zealand Sector, Antarctic Res. Ser.*, vol. 19, edited by D. E. Hayes, pp. 273–293, AGU, Washington, D. C., 1972.

Weissel, J. K., and D. E. Hayes, Asymmetric sea-floor spreading south of Australia, *Nature, 231*(5304), 518–522, 1971.

Weissel, J. K., and D. E. Hayes, Magnetic anomalies in the southeast Indian Ocean, in *Antarctic Oceanology II: The Australian–New Zealand Sector, Antarctic Res. Ser.*, vol. 19, edited by D. E. Hayes, pp. 165–196, AGU, Washington, D. C., 1972.

SEDIMENTATION RATES AND SEDIMENT CHEMISTRY IN THE SOUTH INDIAN BASIN

MARTHA R. SCOTT*, J. KENNETH OSMOND, AND J. KIRK COCHRAN

Department of Geology, Florida State University, Tallahassee, Florida 32306

Abstract. Sedimentation rates and sediment chemistry have been studied on a series of piston cores and trigger cores from the South Indian basin and the Southeast Indian ridge. Sedimentation rates were measured by α and γ spectrometry of ^{230}Th excess in the sediment; general agreement was observed between the two methods, and these data are generally consistent with the rates determined from paleomagnetic reversal data. Continental slope and rise sediments are accumulating at 0.7–1.6 cm/10^3 yr. The sedimentation rates generally decrease northward across the South Indian basin but are locally high in biogenic oozes on and near the ridge; biogenic sediment accumulation rates are commonly 1–2 cm/10^3 yr. The clay accumulation rates, estimated from Al_2O_3 analyses, decrease from 1.4 cm/10^3 yr on the continental slope to 0.1 cm/10^3 yr or less on the ridge. Ridge sediments have high Fe/(Al + Fe + Mn) and Mn/(Al + Fe + Mn) oxide ratios relative to basin sediments. Most of the uranium in the sediments is detrital in origin; however, $^{234}U/^{238}U$ activity ratios approaching the 1.15 value of sea water occur in ridge sediments, where the U/Al ratios are also highest. The increased U/Al ratios and uranium isotope ratios occur in biogenic sediments under the antarctic convergence, their occurrence implying possible removal of uranium from sea water by sediments rich in organic matter; one sample, however, has a $^{234}U/^{238}U$ ratio of 1.29 ± 0.07, which is suggestive of volcanic origin. There is a positive correlation between the excess ^{230}Th of surface sediments and the Fe/(Al + Fe + Mn) oxide ratios. The correlation suggests that ^{230}Th is being removed from sea water by or in association with Fe (and Mn) colloids. Accumulation rates of Fe, Mn, and U in ridge sediments analyzed in this study are not significantly larger than those in nonridge sediments.

Sediments of the South Indian basin grade northward from glacial marine deposits on the antarctic continental shelf, to sands, silts, and clays on the continental slope and rise, to siliceous oozes in the vicinity of the antarctic convergence, and to predominantly calcareous oozes on the Southeast Indian ridge. The large number of sediment cores taken by the USNS *Eltanin* during the last 4 years have made extensive research on this area possible for the first time. This paper reports the results of a study of sediment accumulation rates and related sediment chemistry of piston cores and surface samples from the South Indian basin and the Southeast Indian ridge (Figure 1, Table 1). The cores were collected on cruises 34–44 of the USNS *Eltanin*.

SEDIMENTATION RATES

Analysis of marine sedimentation rates by disequilibrium in the uranium decay series has been carried out mostly by leaching or chemical solution of the samples and measurement of the α activity of ^{230}Th and sometimes ^{231}Pa [*Goldberg*, 1968; *Goldberg and Koide*, 1962, 1963; *Ku*, 1966; *Ku and Broecker*, 1966, 1967; *Ku et al.*, 1968; *Rosholt et al.*, 1961, 1962; *Sackett*, 1964, 1965]. These isotopes are present in the sediments in excess of equilibrium values with their parent isotopes ^{234}U and ^{235}U. The excess ^{230}Th and ^{231}Pa are generated in sea water by dissolved uranium and are removed almost immediately to the sediments, where these isotopes are unsupported by their parent uranium isotopes [*Sackett et al.*, 1958; *Sackett*, 1960; *Moore and Sackett*, 1964]. The decrease with depth of ^{230}Th excess is measured by α spectrometry and yields accumulation rates for the sediment. Comparisons of sedimentation rates measured by paleomagnetic reversal studies and by α determination of excess ^{230}Th in the sediments have shown good agreement between the methods [*Ku et al.*, 1968; *Goldberg*, 1968]; the usefulness of the method has been well demonstrated.

* Now at the Department of Oceanography, Texas A & M University, College Station, Texas 77843.

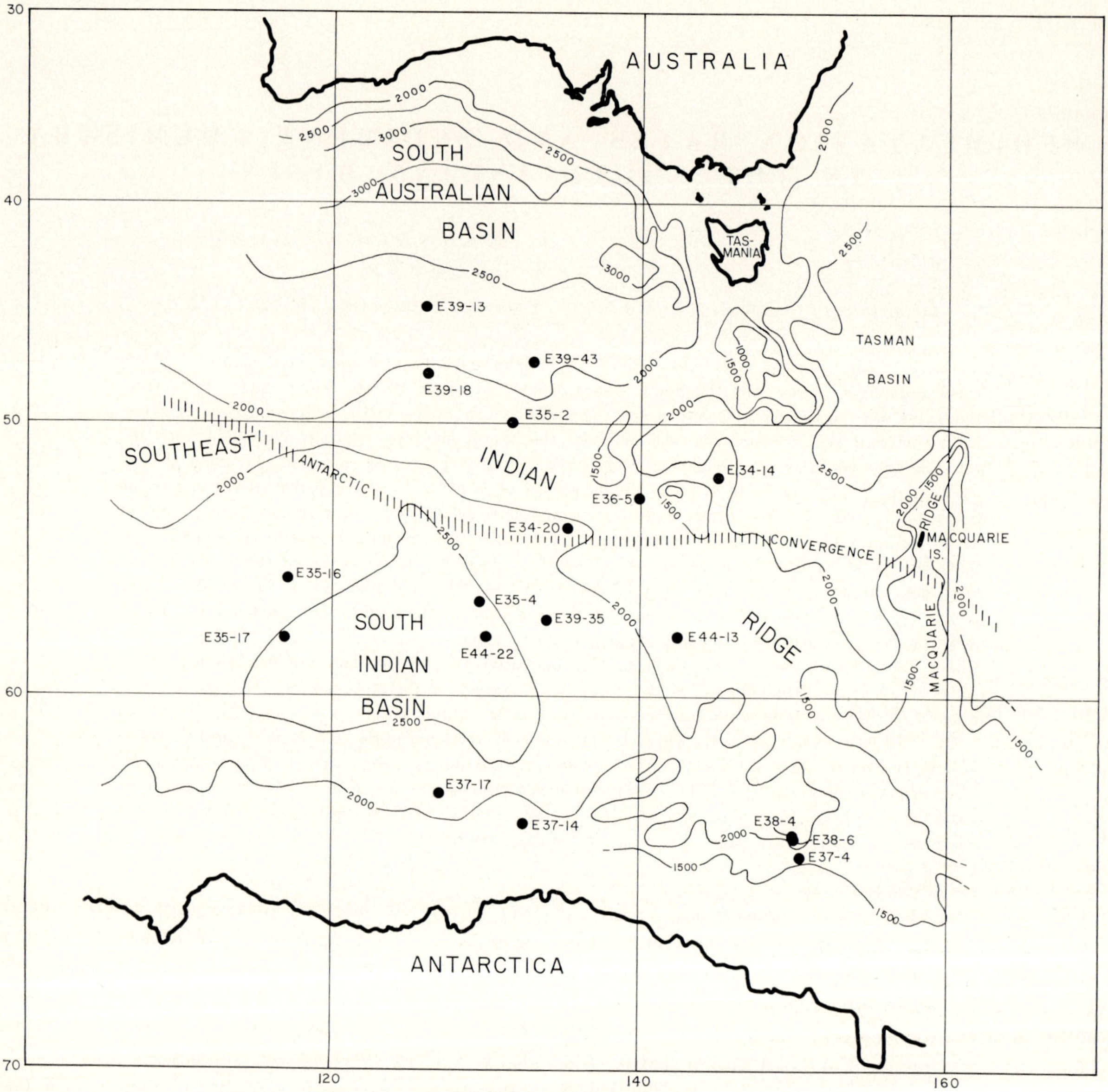

Fig. 1. Sample location map. Contours are in fathoms.

In this study, the sedimentation rates of five cores were measured by α spectrometry of ^{230}Th in the sediments (Figure 2, Tables 2*a* and 2*b*). Uranium and thorium isotopes were also measured in top samples from 11 trigger cores (Table 2*c*). Samples were prepared for α spectrometry in a manner similar to that used by *Ku* [1966]. Samples from cores E35-16, E38-6, E39-13, and E39-18 were completely dissolved by HF and aqua regia treatments, spiked with a ^{232}U-^{228}Th tracer, and were separated into purified uranium and thorium fractions with ion exchange resins. Samples from one core (E37-14) were treated by successive leaching with HCl and total dissolution of the residue from four of the leached samples. The uranium and thorium were plated from TTA (thenoyltrifluoro acetone in benzene) solutions onto stainless steel planchets and were counted by Ortec surface barrier detectors and a multichannel analyzer system. The data are presented in Table 2; errors listed in this Table represent 1 σ values computed from counting statistics. Data on $CaCO_3$ per cent, also given in Tables 2*a* and 2*c*, were determined by using an adaptation of the method described by *Turekian* [1956]. The decay curves with depth for the α method are reported as ^{230}Th excess per gram noncarbonate (Figure 2). The activity ratio (^{230}Th excess/^{232}Th) can be used as a means of normalizing ^{230}Th activity to correct for fluctuating amounts of biogenic silica and carbonate in the sediments; decay curves from this ratio are not shown in Figure 2 because the resulting sedimentation rates were usually the same as those from ^{230}Th excess.

Sedimentation rates were also measured on these cores and on several others by an adaptation of the γ-ray analysis method used by *Osmond and Pollard* [1967]. The core halves were counted intact in their plastic core liners by moving them sequentially along a track under a lead-shielded NaI crystal attached to a multichannel analyzer (Table 3). The method hinges on the detection of ^{214}Bi, a daughter of the ^{238}U-^{230}Th decay series, and on the assumption that the ^{214}Bi is in equilibrium with the ^{230}Th. The decrease in ^{214}Bi down the core represents the decay of ^{230}Th excess with time. One can also derive a ^{230}Th/^{232}Th ratio from the γ activities of the ^{230}Th daughter, ^{214}Bi, and the ^{232}Th daughter, ^{208}Tl.

Sedimentation rates based on γ-ray measurements by *Osmond and Pollard* [1967] were estimated by noting the point at which ^{214}Bi activity levels off to an apparent equilibrium value; this point was taken to be about 300,000 years old, or about four half lives of ^{230}Th. However, the equilibruim point is difficult to identify in cores with variable sediment type and density, and it is not reached at the bottom of cores taken in areas of high sedimentation rates. To avoid these problems and to achieve more direct comparison with α analysis, the γ-ray data have been computed as ^{230}Th excess activity in disintegrations per minute per gram (Table 3); sedimentation rates were determined from the slopes of the lines drawn through plots of ^{230}Th excess versus depth (Figures 2 and 3). The ^{230}Th/^{232}Th activity ratio derived from the ^{214}Bi/^{208}Tl ratio was not used because of the large error generated by the low ^{208}Tl activity in most of these cores. The best estimates of the counting rate errors are about $\pm 10\%$ of the total count rates; uncertainty values for ^{230}Th and ^{230}Th/%K were computed from this value. Ratios of ^{230}Th/%K were computed only where ^{40}K activities exceeded 2 net counts per minute. Data from four cores rich in clay (E37-4, E-37-14, E37-17, and E38-6) fit this criterion and were computed in terms of both ^{230}Th excess and ^{230}Th excess/%K (Table 3, Figure 2). The ratio is comparable to the α activity ratio ^{230}Th/^{232}Th, as both ^{232}Th and K are associated mainly with the detrital component of the sediment.

TABLE 1. Cores Analyzed in This Study

Core	Location	Water Depth, meters	Description
E34-14	52°12′S, 145°4′E	3267	foraminiferan, radiolarian ooze
E34-20	54°10′S, 135°13′E	4019	radiolarian, diatom ooze
E35-2	49°59′S, 131°36′E	3367	foraminiferan ooze
E35-4	56°52′S, 129°38′E	4957	radiolarian ooze
E35-16	56°00′S, 117°7′E	4300	diatom, radiolarian ooze
E35-17	58°5′S, 117°1′E	4474	clay, silt
E36-5	53°2′S, 139°59′E	2965	foraminiferan, radiolarian diatom, ooze
E37-4	64°50′S, 150°29′E	3276	sandy, diatom, radiolarian clay
E37-14	63°54′S, 150°30′E	3276	clay, silt, sand
E37-17	63°4′S, 127°6′E	4297	radiolarian, diatom ooze and clay
E38-4	64°14′S, 150°4′E	3486	sandy, radiolarian, diatom ooze
E38-6	64°18′S, 150°11′E	3459	silt, sand
E39-13	45°1′S, 125°59′E	4538	foraminiferan, radiolarian ooze
E39-18	48°1′S, 126°6′E	4615	foraminiferan, radiolarian ooze
E39-35	57°31′S, 133°58′E	4630	sandy, radiolarian, diatom ooze
E39-43	47°26′S, 133°1′E	3947	foraminiferan ooze
E44-13	58°1′S, 142°27′E	3795	foraminiferan, radiolarian, diatom ooze
E44-22	58°5′S, 130°8′E	4597	radiolarian, diatom clay

Because the method of determining limits of error is somewhat arbitrary, a very conservative approach has been taken to the data. Some of the data eliminated by low count rates yield apparently reasonable ^{230}Th excess/%K and ^{230}Th/^{232}Th decay curves with depth in spite of large uncertainties. A detailed discussion of the method of γ-ray analysis used in this study will appear elsewhere.

Osmond and Pollard [1967] have noted a lower ^{230}Th excess at the very top of sediment cores relative to deeper levels when the ^{230}Th was measured by γ activity of ^{214}Bi. This effect was attributed to the diffusion of the intermediate daughter products, ^{226}Ra and ^{222}Rn, out of the sediment. A few cores exhibit an apparent loss of radium near their tops (Figures 2 and 3). However, the phenomenon appears to be less common in these cores than was observed by Osmond and Pollard in their study. We do not know whether this difference is real and due to differences in sedimentary regimes in the two cases, or whether this difference represents an improvement in technique (in-liner counting) in the present study.

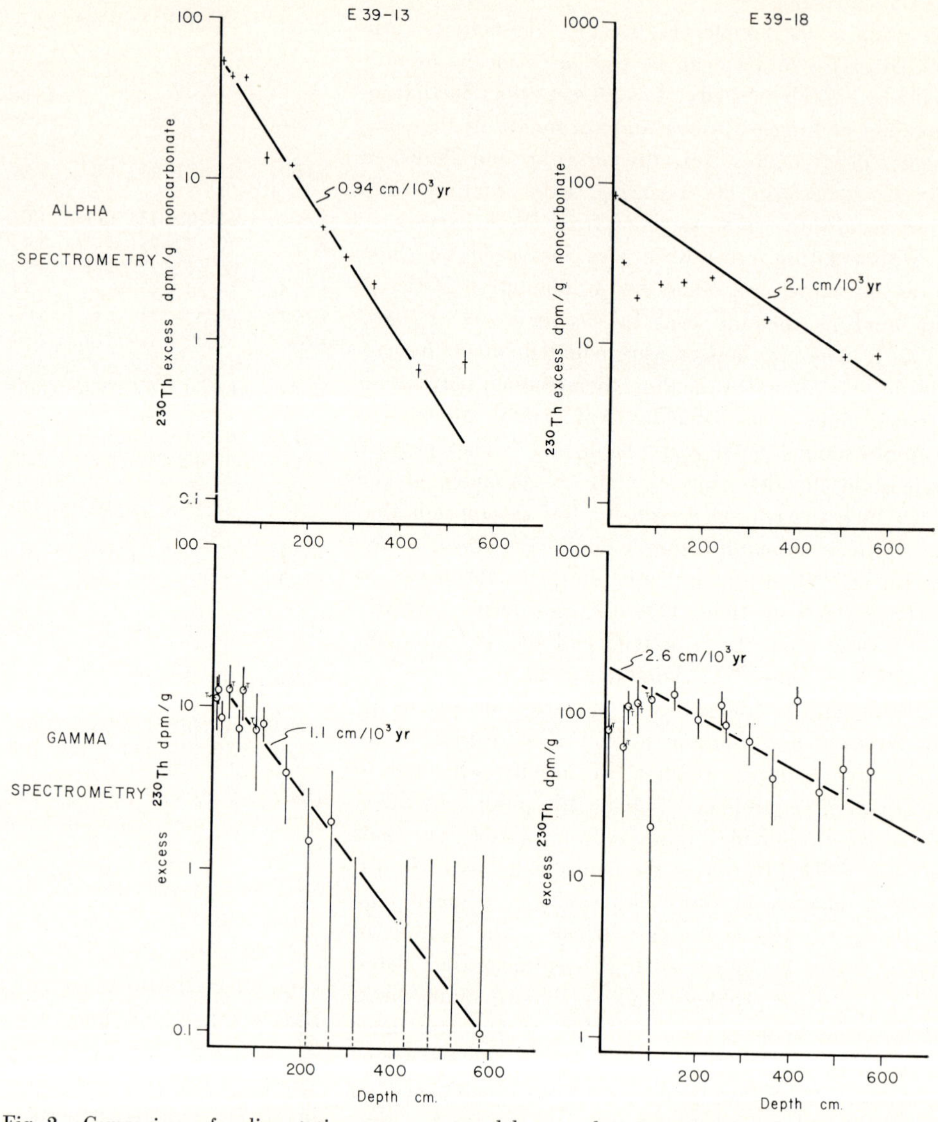

Fig. 2. Comparison of sedimentation rates measured by α and γ spectrometry. Crosses represent ^{230}Th excess measured by α spectrometry; open circles represent ^{230}Th excess measured by γ spectrometry. Vertical lines through all points represent error from counting statistics. Leaches A and B for core E37-14 are described in Table 2. The letter T designates trigger core samples; all other data are from piston cores.

There is general agreement between sedimentation rates of the same core determined by the α and γ spectrometry of ^{230}Th excess (Figure 2, Table 4). The sedimentation rates obtained from paleomagnetic data on the position of the 700,000-year-old Brunhes-Matuyama reversal boundary (J. P. Kennett and N. D. Watkins, personal communication, 1971) also show general agreement with the two ^{230}Th decay methods, except for the data for core E35-16 (Table 4). The sedimentation rate for this siliceous ooze core based on ^{230}Th excess per gram noncarbonate is high, i.e., about 4 cm/10^3 yr. This rate is not conclusively supported by the average rate determined by the location of the 700,000-year-old reversal boundary; the boundary gives a minimum accumulation rate of 1.71 cm/10^3 yr. Disparate sedimentation rates were also found for siliceous antarctic cores by ^{32}Si and ^{230}Th excess methods [*Kharkar et al.*, 1969]. The shorter time range method (^{32}Si with a half life of about 500 years) gave very high sedimentation rates relative to the rates measured by the long-range method (^{230}Th with a half life of 75,200 years). In the present

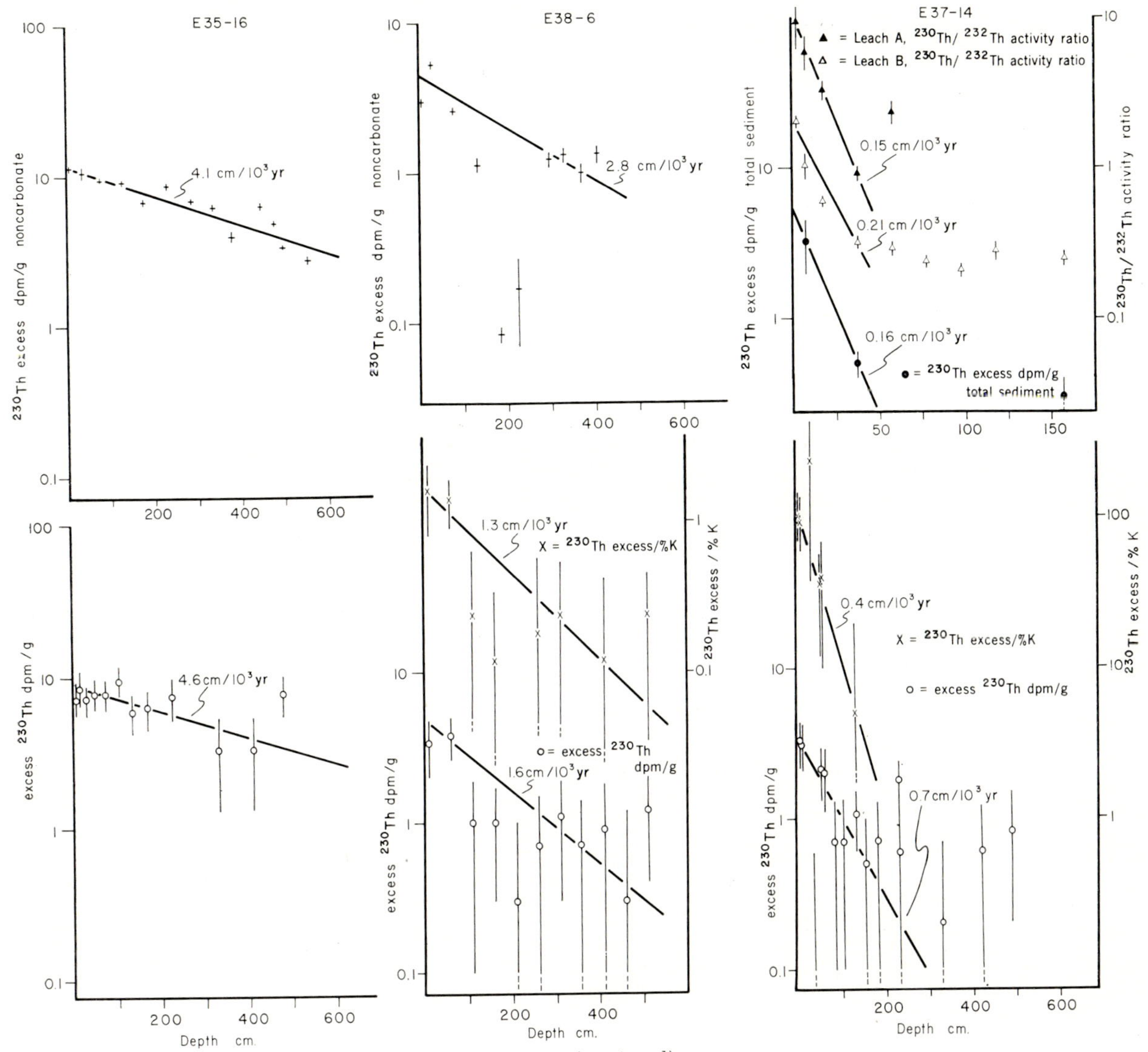

Fig. 2. (continued)

study, the ^{230}Th excess method with a total range of about 300,000 years gives a rate three to four times as great as that based on the 700,000-year-old paleomagnetic horizon for this core. The differences in rates with different amounts of penetration into the past may reflect either the increased probability of sediment loss by slumping and current transport with time or an increase in the accumulation rate of the remains of siliceous organisms in relatively recent times.

Four of the cores studied are located on the continental slope and rise (E37-4, E37-14, E37-17, and E38-6), a sediment regime of reworked material and turbidity current deposits [*Payne and Conolly*, this volume]. Despite the lack of particle by particle deposition of the sediment, the sedimentation rates measured by ^{230}Th excess methods are reasonable when compared with paleomagnetic rates (Table 4). Low ^{230}Th activity levels in the middle of core E38-6 (130–230 cm) coincide with a turbidite sand layer (Figure 2); the low activity may result from older material added to the sediment or from the lower surface area per gram for adsorption of ^{230}Th by the sand.

Watkins and Kennett [1971] have demonstrated large-scale Cenozoic erosion of surface sediments in the south Tasman basin. The erosion is attributed to increased velocity of the antarctic bottom water, the increase possibly caused by a change in the production of bottom water from antarctic ice. In some

TABLE 2a. Alpha Spectrometric Analysis of Sediments

Depth in Core, cm	U, ppm	$^{234}U/^{238}U$	Th, ppm	^{230}Th excess, dpm/g	^{230}Th excess/^{232}Th	$CaCO_3$, %
			Core E39-13			
0–5	0.346 ± 0.022	0.97 ± 0.08	3.07 ± 0.26	14.9 ± 0.63	19.7 ± 1.57	71.3
25–30	0.478 ± 0.020	0.99 ± 0.06	4.67 ± 0.37	19.6 ± 0.92	17.0 ± 0.97	52.6
55–60	0.442 ± 0.026	0.96 ± 0.08	5.82 ± 0.33	19.3 ± 0.64	13.5 ± 0.67	54.3
105–110	0.407 ± 0.019	1.04 ± 0.06	4.23 ± 0.42	5.84 ± 0.38	5.84 ± 0.56	57.6
155–160	0.851 ± 0.042	1.07 ± 0.07	6.78 ± 0.26	9.60 ± 0.23	5.75 ± 0.23	21.9
225–230	0.769 ± 0.052	1.00 ± 0.09	8.18 ± 0.29	4.53 ± 0.14	2.25 ± 0.09	10.5
275–280	0.772 ± 0.041	0.97 ± 0.07	7.03 ± 0.39	3.14 ± 0.16	1.82 ± 0.12	9.60
335–340	0.421 ± 0.032	0.80 ± 0.09	4.23 ± 0.19	1.18 ± 0.06	1.13 ± 0.07	47.2
435–440	0.442 ± 0.022	0.80 ± 0.06	3.07 ± 0.16	0.350 ± 0.04	0.16 ± 0.06	47.8
535–540	0.302 ± 0.019	1.03 ± 0.09	2.96 ± 0.20	0.347 ± 0.04	0.48 ± 0.07	69.4
			Core E39-18			
0–5	0.373 ± 0.019	1.13 ± 0.08	2.80 ± 0.19	31.0 ± 1.02	44.9 ± 2.7	62.9
25–30	0.511 ± 0.036	1.02 ± 0.10	4.63 ± 0.24	17.3 ± 0.50	15.2 ± 0.73	44.8
52–57	0.833 ± 0.040	0.96 ± 0.06	5.77 ± 0.28	17.3 ± 0.45	12.2 ± 0.58	9.32
105–110	0.624 ± 0.039	1.14 ± 0.09	5.73 ± 0.40	20.0 ± 0.69	14.2 ± 0.81	14.6
155–160	0.545 ± 0.029	0.94 ± 0.07	3.93 ± 0.28	14.5 ± 0.54	15.0 ± 1.0	40.0
215–220	0.382 ± 0.021	0.98 ± 0.07	2.80 ± 0.11	11.7 ± 0.22	17.0 ± 0.60	54.5
275–280	...	...	...	...	...	58.9
335–340	0.420 ± 0.029	1.10 ± 0.10	3.76 ± 0.24	10.8 ± 0.37	11.7 ± 0.73	23.8
411–416	...	...	...	...	...	61.5
504–509	0.261 ± 0.020	1.04 ± 0.08	2.20 ± 0.09	3.31 ± 0.08	6.13 ± 0.25	61.5
575–580	0.304 ± 0.015	0.93 ± 0.08	2.25 ± 0.14	3.38 ± 0.12	6.10 ± 0.41	61.0
			Core E35-16			
0–5	0.810 ± 0.05	1.06 ± 0.07	5.31 ± 0.26	10.6 ± 0.32	8.09 ± 0.41	8.4
34–39	1.06 ± 0.02	1.06 ± 0.10	8.50 ± 0.34	10.3 ± 0.95	4.93 ± 0.20	2.0
74–79	1.05 ± 0.05	0.953 ± 0.06	8.42 ± 0.47	9.53 ± 0.09	4.60 ± 0.26	1.3
123–128	0.592 ± 0.057	1.14 ± 0.11	6.44 ± 0.45	9.12 ± 0.42	6.90 ± 0.41	1.7
175–180	0.953 ± 0.066	0.869 ± 0.08	5.00 ± 0.41	6.79 ± 0.35	5.66 ± 0.48	1.3
224–229	1.06 ± 0.05	0.842 ± 0.06	8.13 ± 0.52	8.74 ± 0.40	3.84 ± 0.27	1.3
283–288	0.709 ± 0.051	1.02 ± 0.10	4.86 ± 0.28	6.73 ± 0.25	5.66 ± 0.32	1.1
334–339	0.652 ± 0.039	1.01 ± 0.08	3.87 ± 0.24	6.20 ± 0.29	6.50 ± 0.56	0.80
385–390	0.532 ± 0.043	1.08 ± 0.12	2.09 ± 0.14	3.98 ± 0.31	7.70 ± 0.56	0.98
444–449	0.610 ± 0.060	1.17 ± 0.14	3.75 ± 0.37	6.13 ± 0.34	6.62 ± 0.71	0.70
475–480	0.635 ± 0.036	1.11 ± 0.08	3.85 ± 0.28	4.84 ± 0.22	5.09 ± 0.41	0.80
495–500	0.462 ± 0.032	1.17 ± 0.11	2.47 ± 0.19	3.37 ± 0.15	5.57 ± 0.49	0.40
554–559	0.432 ± 0.032	1.13 ± 0.11	1.56 ± 0.20	2.72 ± 0.16	7.10 ± 1.0	0.30
			Core E38-6			
0–5	1.82 ± 0.11	1.00 ± 0.06	11.0 ± 0.58	3.00 ± 0.21	1.11 ± 0.09	0.93
26–31	2.10 ± 0.10	0.95 ± 0.05	13.7 ± 0.70	5.20 ± 0.30	1.55 ± 0.12	1.20
75–80	2.07 ± 0.11	0.88 ± 0.05	10.0 ± 0.47	2.58 ± 0.17	1.05 ± 0.08	1.08
131–136	2.11 ± 0.10	0.94 ± 0.04	13.6 ± 0.64	1.05 ± 0.13	0.32 ± 0.04	1.46
185–190	1.23 ± 0.06	0.94 ± 0.04	5.80 ± 0.31	0.085 ± 0.006	0.06 ± 0.05	0.70
223–228	2.27 ± 0.10	0.95 ± 0.03	17.8 ± 0.84	1.8 ± 0.10	0.04 ± 0.02	1.46
292–297	2.17 ± 0.11	0.92 ± 0.05	13.1 ± 0.56	1.25 ± 0.13	0.39 ± 0.04	1.75
325–330	2.02 ± 0.12	0.91 ± 0.06	12.7 ± 0.60	1.34 ± 0.14	0.43 ± 0.05	1.80
365–370	1.92 ± 0.10	0.94 ± 0.05	14.9 ± 0.80	1.02 ± 0.14	0.28 ± 0.04	1.51
403–408	1.93 ± 0.18	0.94 ± 0.10	12.9 ± 0.60	1.35 ± 0.17	0.42 ± 0.06	1.93

$^{234}U/^{238}U$ and ^{230}Th excess/^{232}Th are activity ratios; dpm/g, disintegrations per minute per gram.

areas, the sediments have been eroded down to the level of the Gilbert paleomagnetic epoch. Disconformities have been noted in many of the cores. The authors selected cores free from evidence of large-scale erosion on the basis of data from *Watkins and Kennett* [1971, also personal communication, 1971].

A summary of the sedimentation rate data from both α and γ methods is given in Figure 4. Accumulation rates of continental margin sediments range from 0.7 to 1.6 cm/10^3 yr. Cores in the South Indian basin and on the Southeast Indian ridge have a large biogenic component and locally high sedimentation

TABLE 2*b*. Alpha Spectrometric Analysis of Sediments

	$^{230}Th/^{232}Th$		Total Sediment Data				
Depth in Core, cm	Leach A	Leach B	U, ppm	$^{234}U/^{238}U$	Th, ppm	^{230}Th excess, dpm/g	^{230}Th excess/^{232}Th
Core E37-14							
0–5	9.54 ± 3.37	2.03 ± 0.18	...	...	...	...	...
5–10	5.98 ± 1.60	1.04 ± 0.20	1.23 ± 0.30	1.06 ± 0.12	15.5 ± 5.0	3.24 ± 1.25	0.89 ± 0.30
15–20	3.30 ± 0.53	0.60 ± 0.05	...	...	...	...	...
35–40	0.92 ± 0.11	0.32 ± 0.03	1.22 ± 0.20	1.02 ± 0.20	8.2 ± 2.0	0.20 ± 0.08	0.12 ± 0.5
55–60	2.38 ± 0.40	0.29 ± 0.03	...	...	...	...	...
75–80	...	0.24 ± 0.02	1.80 ± 0.25	0.85 ± 0.10	11.3 ± 2.0	...	...
95–100	...	0.21 ± 0.02	...	...	...	...	...
115–120	...	0.28 ± 0.04	...	...	...	...	...
155–160	...	0.25 ± 0.03	0.99 ± 0.12	1.04 ± 0.08	10.6 ± 2.5	0.30 ± 0.10	0.11 ± 0.04

$^{230}Th/^{232}Th$, $^{234}U/^{238}U$, and ^{230}Th excess/^{232}Th are activity ratios. Leach A is a solution of 0.1 *N* HCl for 30 minutes; leach B is a solution of hot 10 *N* HCl for 30 minutes. The total sediment data are the computationally combined data from leach A and leach B and the analysis of the residue left from the two leach treatments.

rates, which are up to 4.3 cm/10^3 yr in the case of E35-16. However, because much of the sediment is composed of siliceous or calcareous organisms, the accumulation rate for clay is much lower than that for the whole sediment. Estimates of the clay accumulation rates are shown in Table 4. The bulk rates are highest in the vicinity of the Southeast Indian ridge where biogenic sediments are deposited. The clay sedimentation rates, however, are highest near the antarctic continent, and these rates decrease northward away from the major terrigenous sediment source (Figure 4).

SEDIMENT CHEMISTRY

Analyses for Fe, Mn, and Al were made by atomic absorption spectrophotometry on piston core samples from cores E34-14, E35-16, E39-13, E39-18, E38-6, E37-14, and from the top portions of 11 trigger cores (Table 5). The atomic absorptions analyses were based on U.S. Geological Survey igneous rock standards and U.S. Bureau of Standards sediment standards; appropriate amounts of MnO were added to some of the sediment standards to bring the concentrations within the range expected for marine sediments. The analyses were performed on a Perkin-Elmer model 303 atomic absorption spectrophotometer. Precision is estimated to be ±3% for Fe, ±3% for Mn, and ±4% for Al.

Recent work has demonstrated that sediments on and near active ocean ridges are characteristically low in Al content and high in Fe, Mn, and trace metals, presumably as the result of additions of these metals by volcanic emanations [*Bonatti and Joensuu*, 1966; *Bostrom*, 1970; *Bostrom and Fisher*, 1971; *Bostrom and Peterson*, 1966, 1969; *Bostrom et. al.*, 1969; *Bostrom and Valdes*, 1969]. *Bostrom et al.* [1969] have shown that sediments from the crest of the East Pacific rise and the Mid-Indian ridge have character-

TABLE 2*c*. Alpha Spectrometric Analysis of Sediments

Trigger Cores	Depth in Core, cm	U, ppm	$^{234}U/^{238}U$	Th, ppm	^{230}Th excess, dpm/g	^{230}Th excess/^{232}Th	$CaCO_3$, %
E34-20	0–5	0.218 ± 0.013	1.29 ± 0.07	1.30 ± 0.08	14.0 ± 0.31	43.8 ± 2.7	70.0
E35-2	0–5	0.140 ± 0.009	1.19 ± 0.07	0.581 ± 0.041	7.87 ± 0.17	55.0 ± 3.9	89.2
E35-4	0–6	1.05 ± 0.04	0.99 ± 0.04	4.50 ± 0.35	15.1 ± 0.68	13.6 ± 1.0	1.42
E35-17	0–5	2.22 ± 0.09	0.88 ± 0.04	22.7 ± 1.62	1.61 ± 0.25	0.29 ± 0.04	1.71
E36-5	0–6	0.216 ± 0.023	1.20 ± 1.5	3.15 ± 0.22	22.3 ± 0.83	28.7 ± 1.8	78.1
E38-4	0–5	1.77 ± 0.07	0.94 ± 0.04	8.10 ± 0.52	9.23 ± 0.49	4.64 ± 0.28	0.97
E39-13	0–5	0.480 ± 0.03	1.02 ± 0.08	2.98 ± 0.24	29.2 ± 1.2	39.8 ± 2.9	66.1
E39-18	0–6	0.600 ± 0.041	0.98 ± 0.5	4.96 ± 0.24	17.9 ± 0.49	14.7 ± 0.64	22.2
E39-35	0–6	1.10 ± 0.06	1.10 ± 0.08	8.21 ± 0.50	14.2 ± 0.61	7.03 ± 0.39	1.39
E39-43	0–5	0.169 ± 0.009	0.91 ± 0.07	0.984 ± 0.050	12.2 ± 0.24	50.4 ± 2.4	54.6
E44-22	0–6	1.16 ± 0.06	1.02 ± 0.07	5.62 ± 0.42	13.3 ± 0.63	9.64 ± 0.68	1.15

$^{234}U/^{238}U$ and ^{230}Th excess/^{232}Th are activity ratios.

TABLE 3. Gamma Spectrometric Analysis of Sediment Cores

Depth, cm	KN	RN	TN	^{230}Th excess, dpm/g	^{230}Th excess/ %K
Core E34-14					
5	0.75	0.66	0.76	4.7 ± 2.0	
10	1.80	0.63	0.43	3.6 ± 2.0	
30	0.75	0.62	0.50	4.3 ± 2.0	
50	0.74	0.51	0.51	3.5 ± 1.8	
60	0.70	0.40	0.34	2.5 ± 1.8	
70	1.02	0.34	0.56	1.8 ± 1.6	
90	9.63	0.16	0.18	0.4 ± 1.4	
135	0.54	0.30	0.52	1.7 ± 2.0	
175	1.15	0.55	0.76	3.5 ± 2.0	
190	0.99	0.56	0.65	3.7 ± 2.0	
225	1.28	0.57	0.87	3.5 ± 2.2	
270	0.81	0.33	0.73	1.8 ± 2.0	
330	1.10	0.56	0.80	3.5 ± 1.8	
410	0.22	0.00	0.18	0.0 ± 1.6	
510	1.06	0.22	0.30	0.8 ± 1.6	
Core E35-2					
10*	0.15	0.59	0.10	10.4 ± 4.4	
10	0.17	0.64	0.11	5.2 ± 3.2	
30	0.46	0.92	0.16	7.4 ± 2.4	
35*	0.00	0.37	0.16	6.4 ± 4.2	
50	0.12	0.50	0.14	3.8 ± 2.0	
105	0.19	0.78	0.12	6.5 ± 2.2	
195	0.03	0.31	0.12	2.1 ± 1.2	
310	0.00	0.00	0.17	0.0 ± 1.2	
370	0.02	0.05	0.15	0.0 ± 1.2	
505	0.00	0.00	0.19	0.0 ± 1.2	
Core E35-16					
5	2.1	1.2	0.57	7.4 ± 1.7	
15	3.6	1.65	0.47	8.5 ± 1.9	
30	2.6	1.26	0.50	7.2 ± 1.6	
50	3.8	1.60	0.42	7.9 ± 1.8	
75	2.9	1.40	0.48	7.8 ± 1.8	
105	3.2	1.76	0.56	9.6 ± 2.1	
135	2.4	1.04	0.42	5.9 ± 1.7	
170	2.1	1.04	0.48	6.3 ± 1.8	
225	2.1	1.22	0.57	7.5 ± 2.3	
330	0.76	0.50	0.66	3.3 ± 2.0	
410	0.63	0.49	0.78	3.3 ± 2.0	
480	1.2	1.10	0.92	7.7 ± 2.2	
Core E37-4					
10	10.9	1.02	0.98	2.1 ± 0.8	0.48 ± 0.26
40	9.8	1.21	1.17	3.0 ± 1.0	0.80 ± 0.35
60	9.1	1.27	0.89	3.4 ± 1.1	1.1 ± 0.40
100	7.2	0.62	0.86	1.5 ± 1.0	0.48 ± 0.40
160	11.4	0.75	1.00	1.2 ± 0.08	0.24 ± 0.60
160	12.6	0.57	1.11	0.6 ± 0.7	0.0 ± 0.25
210	12.9	0.54	1.20	0.5 ± 0.7	0.0 ± 0.20
260	10.7	0.60	1.00	0.9 ± 0.8	0.12 ± 0.28
300	3.7	0.38	0.54	1.2 ± 1.2	0.6 ± 0.80
315	9.0	0.85	0.88	2.0 ± 0.9	0.5 ± 0.33
365	9.4	0.63	0.95	1.1 ± 0.7	0.24 ± 0.30
415	7.2	0.29	0.69	0.2 ± 0.8	0.0 ± 0.32
465	10.2	0.39	0.97	0.2 ± 0.7	0.0 ± 0.30
515	10.1	0.32	0.97	0.0 ± 0.7	0.0 ± 0.24
565	9.2	0.35	0.81	0.2 ± 0.8	0.0 ± 0.32
Core E37-14					
5	7.9	1.14	1.38	3.3 ± 1.1	1.1 ± 0.4
10	10.2	1.29	1.26	3.1 ± 1.0	1.0 ± 0.4

TABLE 3. (continued)

Depth, cm	KN	RN	TN	^{230}Th excess, dpm/g	^{230}Th excess/ %K
35	9.7	0.24	0.41	0.0 ± 0.60	2.4 ± 2.2
55	13.3	1.07	1.69	2.1 ± 0.80	0.37 ± .25
60	15.1	1.25	1.77	2.0 ± 0.90	0.40 ± 0.3
85	13.1	0.63	0.86	0.7 ± 0.60	0.0 ± 0.2
105	13.5	0.66	1.04	0.7 ± 0.6	0.0 ± 0.2
135	22.0	1.18	1.64	1.1 ± 0.4	0.50 ± 0.15
135	17.2	0.91	1.54	1.0 ± 0.4	0.04 ± 0.15
155	17.1	0.71	1.09	0.5 ± 0.5	0.0 ± 0.20
185	14.8	0.71	1.33	0.7 ± 0.6	0.0 ± 0.20
235	16.3	1.21	2.04	1.8 ± 0.6	0.29 ± 0.20
235	16.3	0.71	1.00	0.6 ± 1.2	0.0 ± 0.35
330	19.0	0.61	1.32	0.2 ± 0.5	0.0 ± 0.16
425	17.7	0.77	1.40	0.6 ± 0.6	0.0 ± 0.14
495	18.2	0.86	1.47	0.8 ± 0.6	0.0 ± 0.20
Core E37-17					
5	1.63	0.47	0.26	2.5 ± 1.5	2.9 ± 2.0
10	3.18	0.93	0.41	4.6 ± 1.8	2.9 ± 0.9
25	1.79	0.44	0.29	2.2 ± 1.6	2.4 ± 1.8
60	2.22	0.30	0.29	1.0 ± 1.4	1.1 ± 1.2
95	1.98	0.88	0.37	5.3 ± 1.8	4.8 ± 2.0
110	4.62	0.27	0.72	0.4 ± 0.9	0.12 ± 0.5
160	10.5	0.43	1.16	0.3 ± 0.7	0.0 ± 0.25
210	10.5	0.49	1.09	0.5 ± 0.7	0.0 ± 0.40
260	8.1	0.46	1.07	0.7 ± 0.7	0.12 ± 0.35
310	2.5	0.33	0.36	1.2 ± 1.2	1.0 ± 1.1
325	5.3	1.06	0.59	4.1 ± 1.7	1.8 ± 1.2
360	4.7	0.21	0.72	0.1 ± 0.8	0.0 ± 0.5
410	1.8	0.18	0.23	0.3 ± 1.2	0.6 ± 1.0
460	9.7	0.32	0.48	0.6 ± 1.2	0.24 ± 0.6
510	8.8	0.52	0.68	0.9 ± 0.8	0.12 ± 0.28
560	7.7	0.18	1.10	0.0 ± 0.9	0.0 ± 0.30
Core E38-6					
10	4.9	0.90	0.50	3.4 ± 1.4	1.6 ± 0.8
60	7.2	1.20	0.70	3.8 ± 1.2	1.4 ± 0.5
110	7.1	0.50	0.70	1.0 ± 0.9	0.24 ± 0.4
160	12.0	0.70	1.30	1.0 ± 0.7	0.12 ± 0.24
210	13.5	0.50	0.30	0.3 ± 0.7	0.0 ± 0.20
260	6.2	0.40	0.60	0.7 ± 0.9	0.18 ± 0.40
310	9.1	0.60	1.00	1.1 ± 0.8	0.24 ± 0.32
355	11.8	0.60	1.00	0.7 ± 0.7	0.0 ± 0.24
410	8.0	0.50	0.80	0.9 ± 0.9	0.12 ± 0.30
460	5.9	0.30	0.00	0.3 ± 0.9	0.0 ± 0.18
510	10.2	0.70	1.00	1.2 ± 0.8	0.24 ± 0.26
Core E39-13					
5*	0.14	0.63	0.16	11.2 ± 4.0	
10	0.80	1.60	0.25	12.7 ± 2.5	
20	1.93	1.30	0.23	8.3 ± 1.8	
35*	0.38	0.74	0.13	12.9 ± 4.5	
60	1.00	1.00	0.12	7.3 ± 2.1	
65*	0.53	0.76	0.16	12.8 ± 4.5	
95*	0.08	0.41	0.14	7.2 ± 4.3	
110	2.80	1.40	0.52	7.8 ± 1.8	
160	1.10	0.60	0.24	3.9 ± 2.0	
210	1.20	0.32	0.12	1.5 ± 1.7	
260	1.30	0.38	0.22	2.0 ± 2.0	
310	0.50	0.09	0.68	0.0 ± 1.2	
420	1.10	0.05	0.18	0.0 ± 1.2	
470	0.60	0.10	0.15	0.0 ± 1.2	
520	0.40	0.04	0.14	0.0 ± 1.2	
580	1.20	0.14	0.20	0.1 ± 1.2	

TABLE 3. (continued)

Depth, cm	KN	RN	TN	^{230}Th excess, dpm/g	^{230}Th excess/ %K
			Core E39-18		
10*	0.41	0.49	0.16	7.9 ± 4.0	
40*	0.26	0.39	6.22	6.3 ± 4.0	
50	1.90	1.70	0.14	11.1 ± 2.4	
70*	0.37	0.68	0.12	11.8 ± 4.2	
100*	0.0	0.30	0.05	2.0 ± 2.0	
100	2.10	1.90	0.18	12.2 ± 2.6	
150	1.90	2.00	0.13	13.3 ± 2.8	
200	1.20	1.30	0.22	9.2 ± 2.0	
250	1.40	1.60	0.11	11.3 ± 2.5	
260	1.80	1.30	0.19	8.6 ± 1.8	
310	1.20	1.00	0.06	6.9 ± 2.0	
360	1.00	0.61	0.20	4.1 ± 2.0	
410	1.20	1.70	0.04	12.4 ± 2.6	
460	0.90	0.53	0.20	3.4 ± 1.7	
510	0.80	0.66	0.14	4.7 ± 2.0	
570	0.97	0.67	0.26	4.6 ± 2.0	
			Core E39-35		
5*	0.49	0.36	0.18	5.6 ± 4.4	
5	3.31	1.08	0.42	5.4 ± 1.7	
25	2.29	0.45	0.40	2.1 ± 1.6	
30*	0.98	0.40	0.23	5.7 ± 3.5	
40	1.95	0.32	0.40	1.3 ± 1.6	
55	2.18	0.02	0.40	0.0 ± 1.2	
75	3.05	0.05	0.49	0.0 ± 1.2	
105	0.54	0.00	0.19	0.0 ± 1.2	
			Core E44-13		
7	0.24	0.60	0.14	4.1 ± 2.0	
88	0.00	0.81	0.23	7.1 ± 2.4	
115	1.08	0.48	0.22	2.6 ± 1.6	
198	0.48	0.07	0.22	0.0 ± 1.2	
270	0.01	0.04	0.17	0.0 ± 1.2	
328	0.0	0.02	0.19	0.0 ± 1.2	
345	0.86	0.17	0.22	0.3 ± 1.2	

RN, net counts per minute ^{214}Bi; KN, net counts per minute ^{40}K; TN, net counts per minute ^{208}Tl; dpm/g, disintegrations per minute per gram.

The ^{230}Th, ^{230}Th excess and %K were calculated from the raw data on the basis of empirical equations from α and γ analyses of U series isotopes, atomic absorption and γ analyses for K, and measurements of the sediment density. Trigger core data were multiplied by 2 to correct for the smaller volume of trigger cores relative to piston cores.

* Trigger core sample.

istically low Al/(Al + Fe + Mn) ratios. They have predicted that low Al sediments describe a continuous trend from the Mid-Indian ridge crest along the Southeast Indian ridge south of Australia that joins the trend of low Al sediments on the East Pacific rise. Sediment analyses done as part of this study support their suggestion (Table 5, Figure 5). Figure 5 shows the variation in the oxide ratios Al/(Al + Fe + Mn) and Fe/(Al + Fe + Mn) across the South Indian basin from Antarctica to the ridge. The figure was constructed with latitude as the abscissa so that an effective traverse across the basin is described; this presentation was chosen because the major boundaries of the basin (Southeast Indian ridge and antarctic continent) fall generally along lines of latitude. The data in Figure 5 clearly show that low Al sediments are concentrated on the ridge; as expected, the Fe/(Al + Fe + Mn) oxide ratio in the sediments shows the inverse relationship, the ridge sediments having relatively high Fe contents. The oxide ratio Mn/(Al + Fe + Mn) is not shown in Figure 5 because it has a distribution across the basin and ridge parallel to that of the Fe/(Al + Fe + Mn) oxide ratio. For the purposes of this discussion the nondetrital Fe added to the sediment is referred to as 'authigenic'; the oxide ratio Fe/(Al + Fe + Mn) is taken as a relative indicator of authigenic versus continentally derived Fe and Mn in the sediment.

Most of the uranium in the sediments analyzed for this study is associated with the detrital portion of the sediment. This relationship is illustrated in Figure 6, which shows positive correlation between the uranium and aluminum content of the sediment. However, the $^{234}U/^{238}U$ isotope ratios measured in the surface sediments are commonly greater than 1 (Table 2), whereas the isotope ratio found in most soils and detrital material is usually equal to or less than 1 [*Moore*, 1967; *Rosholt et al.*, 1966; *Scott*, 1968]. The conclusion is that part of the uranium in these sediments is nondetrital in origin.

There are two potential sources for authigenic uranium in marine sediments: uranium removed from sea water and uranium added by volcanic emanations. *Veeh* [1967] has demonstrated that uranium is being deposited from sea water into anaerobic sediments in areas of upwelling. *Turekian and Bertine* [1971] have suggested that uranium is being removed from sea water into anaerobic sediments along the ocean ridge system; they propose that the anaerobic sediments form partly in ephemeral basins along the ridges and partly as a result of the large influx of organic matter to the relatively shallow ridge areas. *Ku* [1969] has cited a case of removal of uranium from sea water in the Red Sea by the precipitation of large amounts of volcanogenic iron hydroxides. The test for a sea water origin of the authigenic uranium is the $^{234}U/^{238}U$ activity ratio, which should approach the sea water value of 1.15 ± 0.02 [*Koide and Goldberg*, 1965; *Thurber*, 1962; *Veeh*, 1968] as a limit.

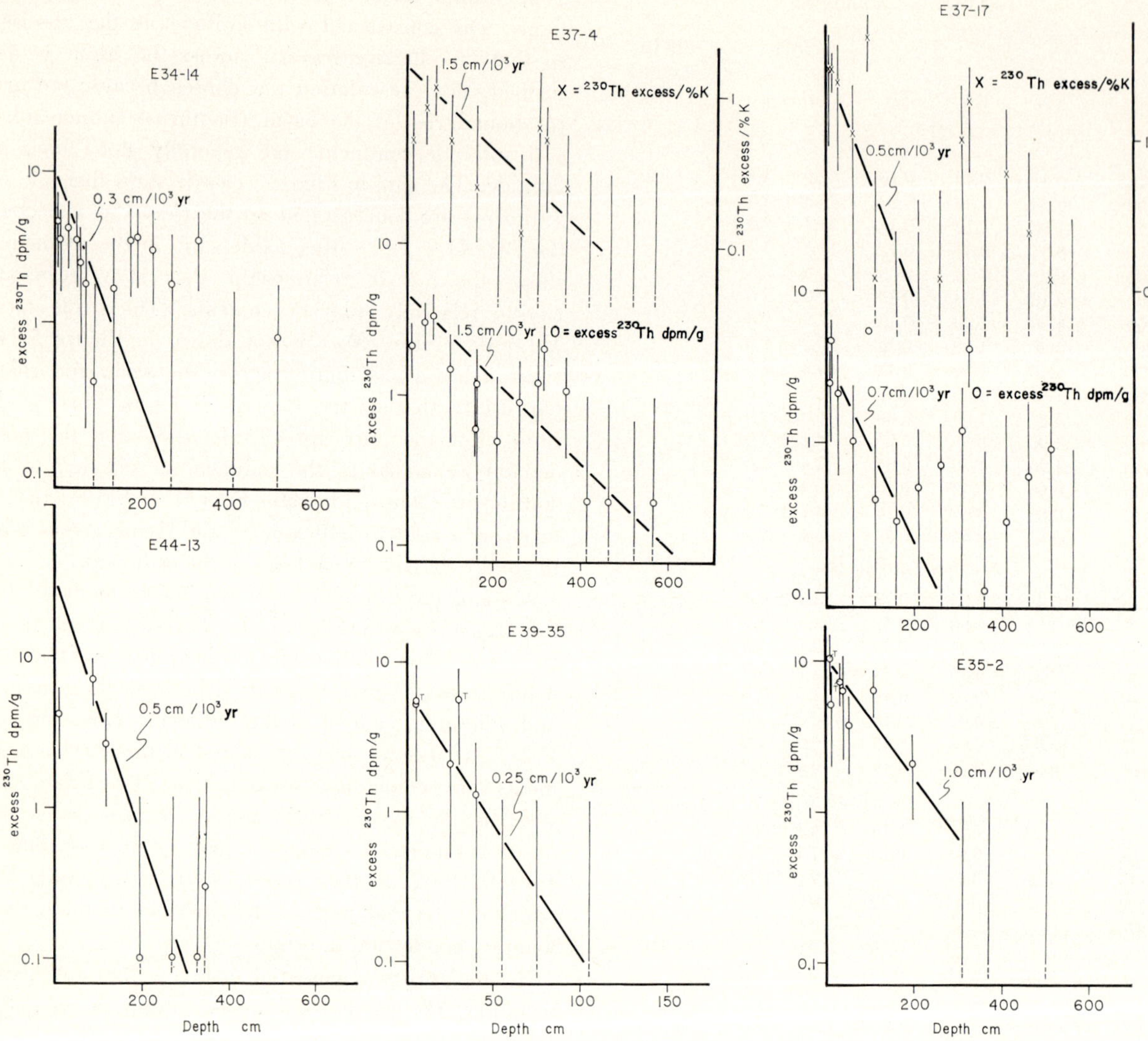

Fig. 3. Sedimentation rates measured by γ spectrometry. Open circles represent ^{230}Th excess; $\times$ represents ^{230}Th excess/%K; the letter T designates trigger core samples; all other data are from piston cores. Vertical lines through all points represent error from counting statistics.

A volcanic origin for uranium in Indian Ocean sediments has been proposed by *Bostrom and Fisher* [1971]. Their data show a correlation between high iron contents and high uranium contents in the sediment; they interpret the correlation as the result of volcanic input of both elements. *Veeh and Bostrom* [1971] cite a ^{234}U/^{238}U activity ratio of 1.2 $\pm$ 0.02 as evidence of the volcanic origin of uranium in an iron crust from the East Pacific rise.

An illustration of the distribution of U/Al and ^{234}U/^{238}U activity ratios across the basin and ridge is given in Figures 7 and 8. Both these ratios clearly increase in the general vicinity of the Southeast Indian ridge. One of the samples, E34-20, has a ^{234}U/^{238}U activity ratio of 1.29 $\pm$ 0.07, which is suggestive of a volcanic origin; however, all the other samples exhibit ratios that approach the sea water ratio of 1.15 as a maximum within 1σ from counting statistics. If uranium is being removed from sea water by the authigenic Fe hydroxides, a correlation would be expected between the uranium isotope ratios and the Fe/(Al + Fe + Mn) oxide ratio. Figure 9 shows that such a correlation appears to exist, although it is not pronounced. This correlation does not provide a unique test for a sea water origin, however, as volcanogenic uranium might also be added to the sedi-

TABLE 4. Comparison of Sedimentation Rates by α and γ Methods

Core	α Rate, cm/10^3 yr	γ Rate, cm/10^3 yr	Paleomagnetic Rate, cm/10^3 yr	Value Used in This Paper	Clay Accumulation Rates, cm/10^3 yr
E39-13	0.94	1.10	1.5 maximum	1.0	0.09
E39-18	2.10	2.60	2.9 maximum	2.3	0.15
E35-16	4.10	4.60	1.71 minimum	4.3	0.36
E38-6	2.30	1.3 (1.6)	0.78 minimum	1.6	1.43*
E37-14	0.20	0.4 (0.7)	0.73 minimum	0.7	0.37*
E34-14	...	0.3	...	0.3	0.01
E35-2	...	1.0	0.82 minimum	1.0	0.01
E37-4	...	1.5 (1.5)	0.84 minimum	1.5	...
E37-17	...	0.5 (0.7)	0.71 minimum	0.7	...
E39-35	...	0.25	0.11 minimum	0.25	0.04
E44-13	...	0.5	0.47 minimum	0.5	...

The paleomagnetic rates are taken from *Watkins and Kennett* [1971, also personal communication, 1971]. The clay accumulation rates are calculated from the Al_2O_3 (or Fe) concentration of the sediment (Al_2O_3 = 34% of the clay component).

* Calculated from Fe analyses in Table 5, the average Fe/Al ratio of 0.51 from antarctic pelagic sediments [*Angino*, 1966], and the correction factor of 34% Al_2O_3 in the nonbiogenic fraction of the sediment.

ment in proportion to the amount of volcanogenic iron.

The apparent maxima in the U/Al trend (Figure 8) and the $^{234}U/^{238}U$ trend (Figure 7) are located slightly to the south of the ridge under the position of the antarctic convergence. The rapid transport of diatoms to the sediments by the sinking surface current produces abundant biogenic oozes, a situation described by *Turekian and Bertine* [1971] as likely to promote the deposition of uranium from sea water. The high $^{234}U/^{238}U$ ratio of sample E34-20 (1.29 ± 0.07) makes it difficult to propose that the removal

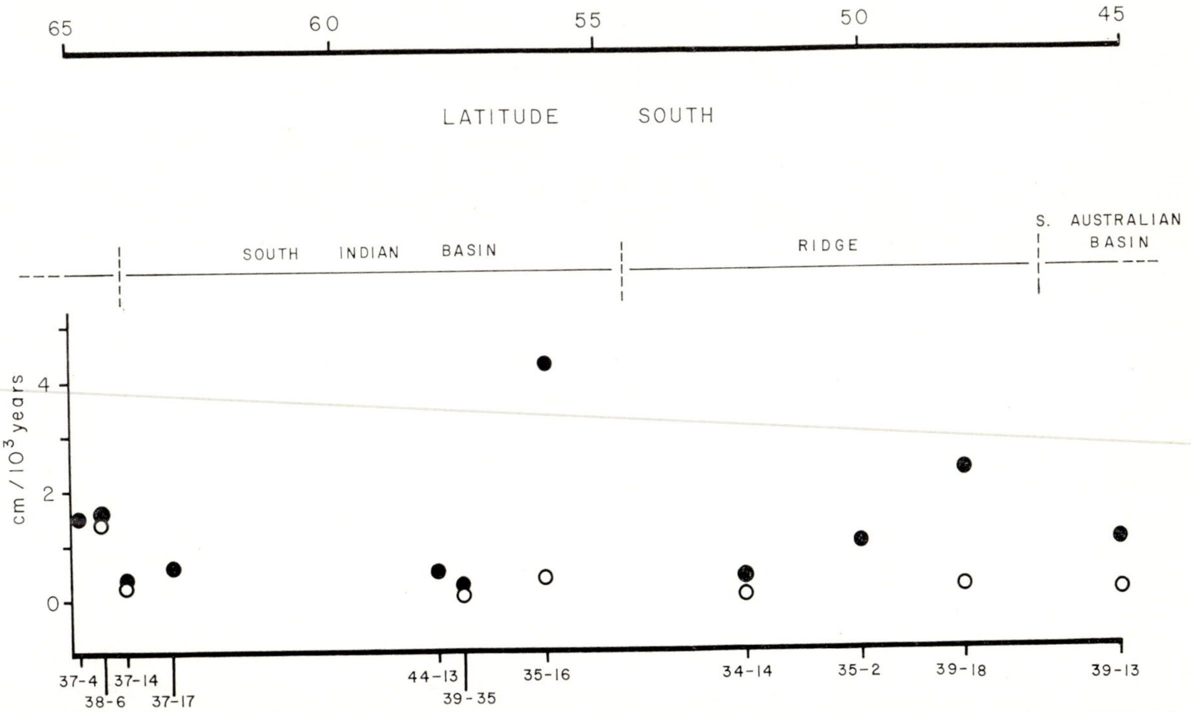

Fig. 4. Sedimentation rates across the South Indian basin. Closed circles represent bulk rates, open circles represent clay accumulation rates calculated from the average Al_2O_3 content (Table 5) of cores where this parameter is known. Where only one Al_2O_3 analysis from the top of a core was made, that value was used to represent the core average. Al_2O_3 was taken to be 34% of the nonbiogenic mineral component of the sediment [*Arrhenius*, 1963]. This estimate of clay content was used rather than per cent $CaCO_3$ because of the high content of biogenic silica in some of these cores.

TABLE 5. Chemical Analyses

Depth interval, cm	Fe_2O_3, %	MnO, %	Al_2O_3, %	Dry Bulk Density, g/cm³
		Core E39-13		
0–5	2.10	0.191	1.97	0.634
25–30	3.30	0.257	3.12	0.493
55–60	3.80	0.205	2.82	0.426
105–110	2.95	0.279	2.68	0.491
155–160	6.30	0.740	5.04	0.304
225–230	5.80	0.545	5.34	0.308
275–280	7.00	0.795	5.53	0.291
335–340	2.80	0.133	2.68	0.513
435–440	2.75	0.216	2.54	0.490
535–540	2.00	0.219	1.64	0.628
		Core E39-18		
0–5	2.85	0.241	1.77	0.458
25–30	3.05	0.275	2.61	0.425
52–57	3.90	1.23	4.30	0.325
105–110	3.60	0.720	4.00	0.314
155–160	2.70	0.324	2.49	0.391
215–220	2.10	0.284	1.74	0.418
275–280	2.20	0.118	1.90	0.480
335–340	2.75	0.072	2.56	0.322
411–416	2.00	0.202	1.70	0.469
504–509	1.95	0.113	1.65	0.456
575–580	2.10	0.106	1.71	0.466
		Core E35-16		
0–5	2.00	0.132	2.84	0.332
34–39	3.84	0.321	5.76	0.376
74–79	3.35	0.047	5.43	0.399
123–128	2.55	0.048	4.00	0.390
175–180	2.30	0.186	4.12	0.354
224–229	3.35	0.087	5.32	0.365
283–288	2.60	0.245	3.87	0.367
334–339	1.80	0.041	2.69	0.335
385–390	0.85	0.022	1.18	0.353
444–449	1.45	0.041	2.02	0.304
475–480	1.40	0.366	1.96	0.376
495–500	0.80	0.041	1.21	0.353
554–559	0.60	0.041	0.83	0.356
		Core E38-6		
0–5	9.16	0.097		0.880
26–31	12.30	0.101		0.790
75–80	10.3	0.129		0.618
131–136	12.9	0.125		0.710
185–190	5.54	0.072		1.46
223–228	17.2	0.310		0.648
292–297	12.3	0.146		0.724
325–330	12.4	0.116		0.699
365–370	12.9	0.220		0.734
403–408	12.2	0.106		0.690
		Core E37-14		
0–5	6.56	0.080		
5–10	7.40	0.074		
15–20	6.23	0.057		
35–40	7.60	0.054		
55–60	7.98	0.059		
75–80	8.74	0.063		
95–100	9.00	0.068		
115–120	8.96	0.067		
155–160	8.42	0.064		

TABLE 5. (continued)

Depth Interval, cm	Fe_2O_3, %	MnO, %	Al_2O_3, %	Dry Bulk Density, g/cm³
		*Core E34-14**		
0–5	0.95	0.103	. . .	0.730
		*Core E34-20**		
0–5	0.60	0.075	0.46	0.510
		*Core E35-2**		
0–5	0.70	0.092	0.23	0.819
		*Core E35-4**		
0–6	1.75	0.118	2.20	0.290
		*Core E35-17**		
0–5	7.05	0.159	13.2	0.548
		*Core E36-5**		
0–6	0.65	0.100	0.60	0.587
		*Core E38-4**		
0–5	4.10	0.103	7.76	0.506
		*Core E39-13**		
0–5	2.90	0.274	1.84	0.502
		*Core E39-18**		
0–6	3.65	0.446	3.28	0.361
		*Core E39-35**		
0–6	3.40	0.280	5.00	0.297
		*Core E39-43**		
0–5	0.55	0.088	0.51	0.794
		*Core 44E-22**		
0–6	2.30	0.126	3.62	0.287

* Sample from trigger core; all other samples are from piston cores.

of sea water uranium by organically rich sediments is the sole source of authigenic uranium in these sediments.

The data reported here do not document an unequivocal origin for the small amount of authigenic uranium in the sediments. The general association of higher U/Al and higher $^{234}U/^{238}U$ ratios with the ridge implies either a volcanic or sea water origin; the association of the largest amounts of authigenic uranium with biogenic sediments under the antarctic convergence implies the removal of uranium from sea water by reducing sediments or possibly the direct removal of uranium by siliceous organisms.

There is a strong positive correlation between the ^{230}Th excess/^{232}Th activity ratio and the authigenic metal content of the sediments (Figure 10). This correlation emphasizes the fact that much of the Fe and Mn and presumably all of the ^{230}Th excess are added to the sediment after it enters the marine environment; the ^{230}Th excess is added from decay of

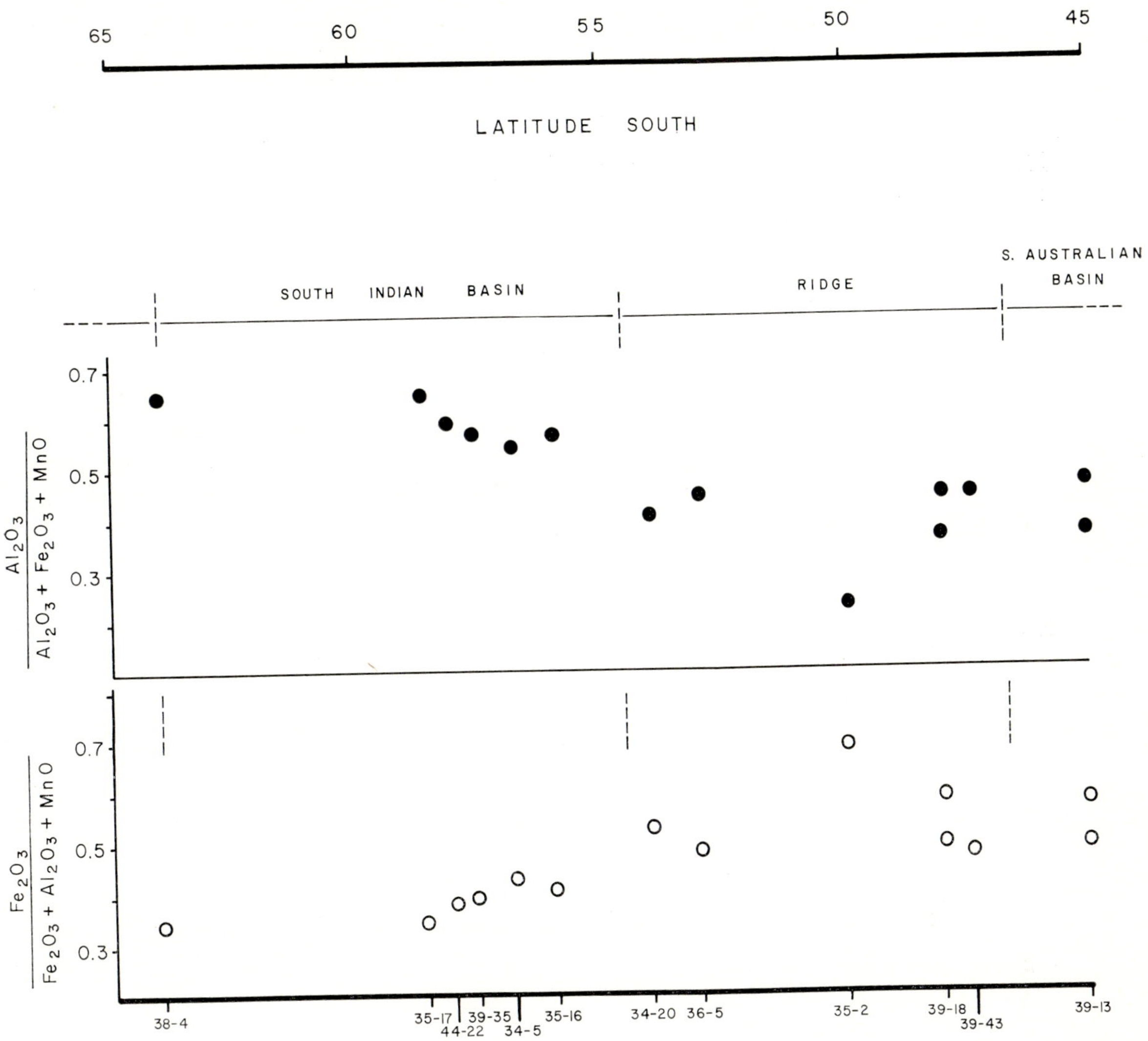

Fig. 5. Variation in the oxide ratios Al/(Al + Fe + Mn) and Fe/(Al + Fe + Mn) in surface sediment samples.

^{234}U in sea water, and the Fe and Mn apparently are added from submarine volcanic sources. There is also a clear correlation between the total ^{230}Th excess in disintegrations per minute per gram of sediment and the Fe/(Al + Fe + Mn) oxide ratio of the sediments (Figure 11). This correlation implies that the more authigenic Fe (and Mn) a sediment contains relative to its continental components, the more unsupported ^{230}Th it is also likely to contain. The conclusion that ^{230}Th is being removed from sea water by or in association with the Fe and Mn precipitates is clearly supported. The association of ^{230}Th in marine sediments with colloidal Fe and Mn has been suggested by *Goldberg* [1954], *Baranov and Kuzimina* [1958], and *Kuznetsov et al.* [1966].

There is no way to determine from these data whether the mechanism of removal involves coprecipitation or whether the ^{230}Th is adsorbed on the surface of the colloidal precipitates. The very small particle size of the volcanogenic colloidal Fe and Mn has been pointed out by *Bostrom and Peterson* [1966]; such material would be expected to have a very large surface area per unit weight and to be able to adsorb a relatively large amount of ^{230}Th from sea water. Continentally derived sediment of very small particle size would also be expected to have a large surface area per weight, and *Turekian* [1968] has suggested that the unequal distribution of such material in marine sediments may cause variation in the concentration of trace elements inherited from the weathering history of the material. *Windom* [1970] has emphasized the addition of trace elements to

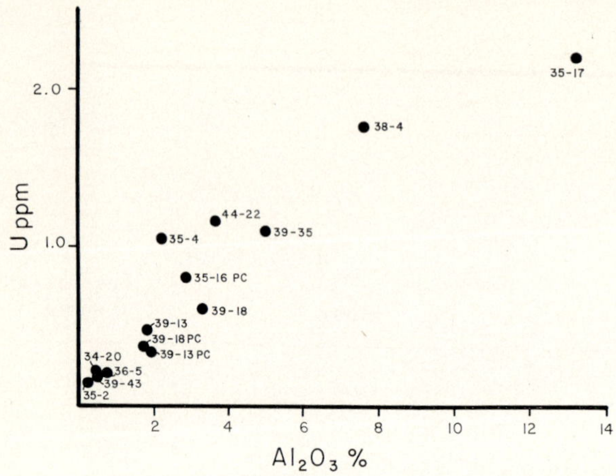

Fig. 6. Plot of uranium versus aluminum in surface sediment samples.

marine sediments via atmospherically borne dust. This material contains relatively large proportions of trace metals and may adsorb additional amounts from sea water. The distribution of trace elements adsorbed by fine particulate matter from sea water would be controlled by the physical processes of sedimentation of the adsorbing material regardless of its origin.

Ku and Broecker [1967] have suggested that the low Σ ^{230}Th excess in an arctic core may be the result of a lack of scavenging agents to remove the ^{230}Th from sea water to the ocean floor; a similar deficiency has been reported in antarctic cores [*Ku et al.*, 1968]. On the basis of 3 ppb of uranium in sea water, the $^{234}U/^{238}U$ activity ratio of 1.15, and an average water depth of 3800 meters, the expected amount of ^{230}Th excess in marine sediment cores is calculated to be 950 dpm/cm^2. This value represents the amount of ^{230}Th in equilibrium with its parent uranium in sea water; the thorium is removed to the sediment almost as soon as it is formed, as evidenced by its short residence time in sea water [*Sackett et al.*, 1958, *Moore and Sackett*, 1964]. The sediments analyzed here (see Table 6) also show a low Σ ^{230}Th excess in glacial marine sediments (core E38-6, 511 dpm/cm^2), higher Σ ^{230}Th excess in the basin center (core E35-16, 1405 dpm/cm^2) and on the ridge flank (core E39-13, 1436 dpm/cm^2), and the highest Σ ^{230}Th excess on the ridge (core E39-18, 2763 dpm/cm^2). Water depth variations for these cores parallel the Σ ^{230}Th variations but are not large enough in magnitude to cause the Σ ^{230}Th variations (Tables 1 and 6). The highest total Σ ^{230}Th excess values are found in sediments with the highest Fe/(Al + Fe + Mn) oxide ratios (Figure 5); this finding again implies an association between ^{230}Th derived from sea

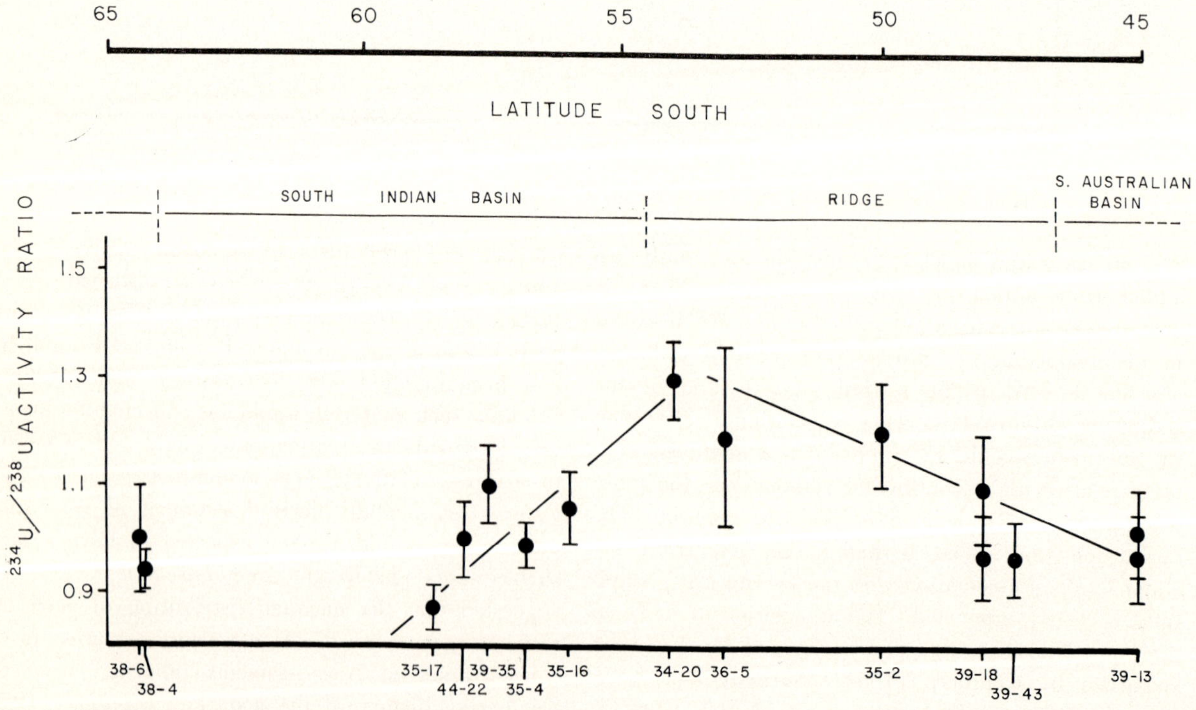

Fig. 7. Variation in $^{234}U/^{238}U$ activity ratios in surface sediment samples. Vertical lines represent 1 σ errors from counting statistics.

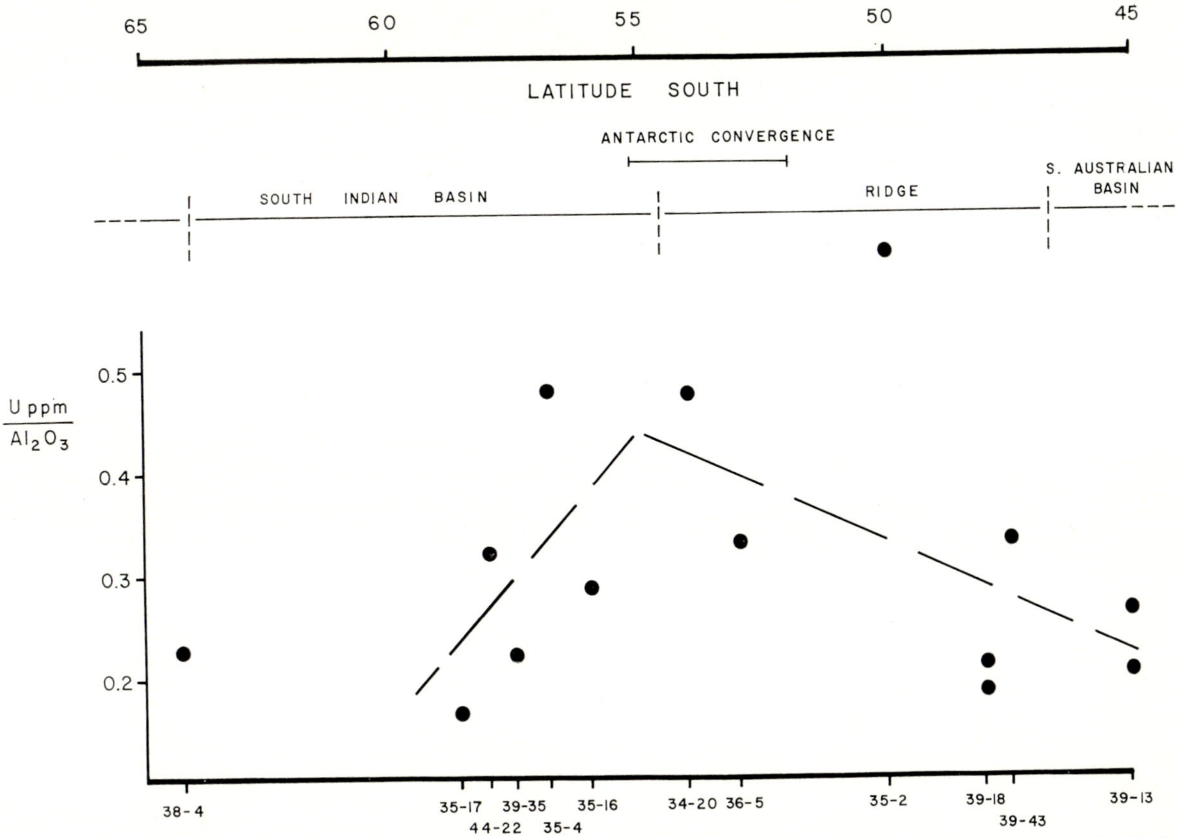

Fig. 8. Variation in the U/Al ratio in surface sediment samples. The aluminum content of core E35-2 was near the limit of detection, so the U/Al ratio for this sample has a large error.

water and the authigenic Fe and Mn content of the sediment. Where this ratio is low, as in the continental margin sediments, there appears to be an inefficient removal of excess ^{230}Th to the sediment; where the ratio is high, more than the predicted amount of ^{230}Th excess occurs in the sediments. Higher than predicted amounts of ^{230}Th excess probably represent the sum of ^{230}Th generated in sea water above the deposition site plus ^{230}Th generated in other parts of the ocean and transported some distance to the deposition site.

If ^{230}Th generated by ^{234}U in some parts of the ocean is transported to other areas before it is deposited, then some water masses should be observed to contain measurably more ^{230}Th than others. Moreover, in the case of the South Indian basin, a mechanism would have to exist for removal of ^{230}Th from the area of deposition of continental margin sediments northward toward regions of biological ooze and volcanogenic sediment deposition. *Lynn and Reid* [1968] have cited evidence that the antarctic bottom water flows northward from all sides of Antarctica; the ^{230}Th may be transported away from the continent by this water mass either in solution or more probably adsorbed on particulate matter. Analyses of the ^{230}Th content of antarctic bottom water would be expected to show high values if the suggested mechanism is operating. A relatively high ^{230}Th value found in North Atlantic bottom water may be related to a similar process [*Moore and Sackett,* 1964]. The process of removal probably involves extremely fine continental debris with high Fe and trace metal content as well as colloidal Fe and Mn of volcanic origin. Thus the movement of sediment down the continental slope via turbidity currents [*Payne and Conolly,* this volume] may transport ^{230}Th adsorbed to the clay fraction northward and affect the observed distribution of the isotope. Because the sediments with high ^{230}Th excess are complemented by ^{230}Th deficient sediments near the antarctic continent, a volcanic source for ^{230}Th is not being suggested here.

DEPOSITION RATES OF TRACE METALS

The accumulation rates of Fe, Mn, and U are given in Table 6. The rates were calculated on the basis of

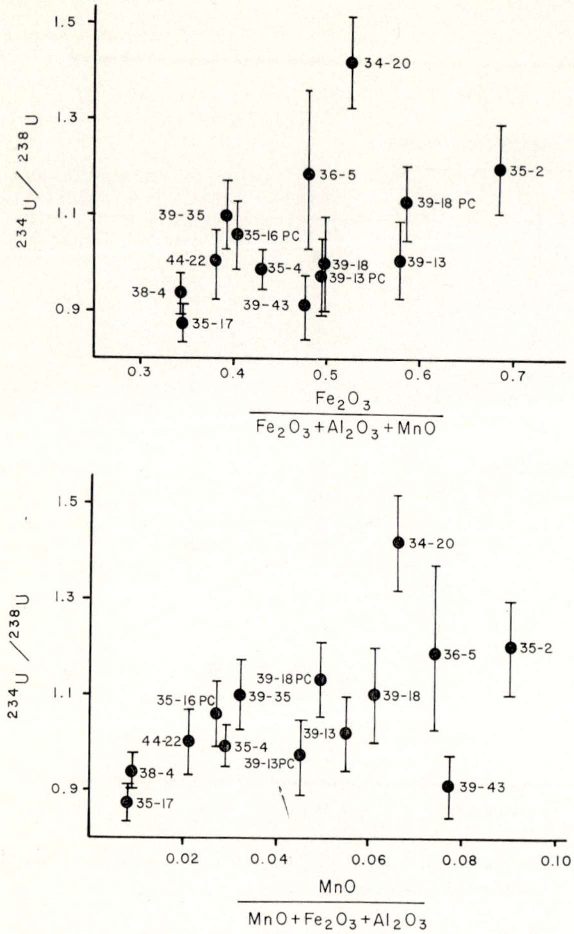

Fig. 9. Plot of $^{234}U/^{238}U$ activity ratios in surface sediment samples against the oxide ratios Fe/(Al + Fe + Mn) and Mn/(Al + Fe + Mn).

the total sediment accumulation rates determined by α and/or γ spectrometry, the Fe, Mn, and U contents given in Tables 2 and 5, and the dry bulk density of the sediment that was determined for each sample (Table 5). Note that the dry bulk densities measured on these samples commonly were as low as 0.3 g/cm^3 (Table 5), or less than half of the 0.7–1.0 densities often assumed in the literature.

The high rate of Fe and U accumulation in the continental rise core E38-6 is the result of rapidly accumulating continental debris. Rapid deposition rates of detrital versus authigenic Fe have been noted by *Somayajulu et al.* [1971]. The rapid influx of metals into core E35-16 is related to the very fast total deposition rate of 4 cm/10^3 yr measured for this core; this rate is not the average for the core based on paleomagnetic data (J. P. Kennett and N. D. Watkins, personal communication, 1971) and may not typify the area at present.

These data show no large-scale increases in Mn, Fe, and U deposition rates in ridge cores E39-18 and E34-14 relative to the other cores. Such an increase would be expected over the ridge from the work of *Bostrom* [1970] and *Bender et al.* [1970]. Possibly other ridge cores from this area would show accelerated trace element accumulation rates; the cores analyzed for this study may not represent the average conditions for the ridge crest sediments.

CONCLUSIONS

Sedimentation rates measured on sediment cores from the South Indian basin by α and γ spectrometry of ^{230}Th excess are generally consistent with the rates obtained from paleomagnetic data. Accumulation rates of continental margin sediments vary from 0.7 to 1.6 cm/10^3 yr. The sedimentation rates generally decrease northward across the South Indian basin but are locally high in biogenic oozes on and near the ridge; biogenic sediment accumulation rates are commonly 1–2 cm/10^3 yr. The rate of accumulation of

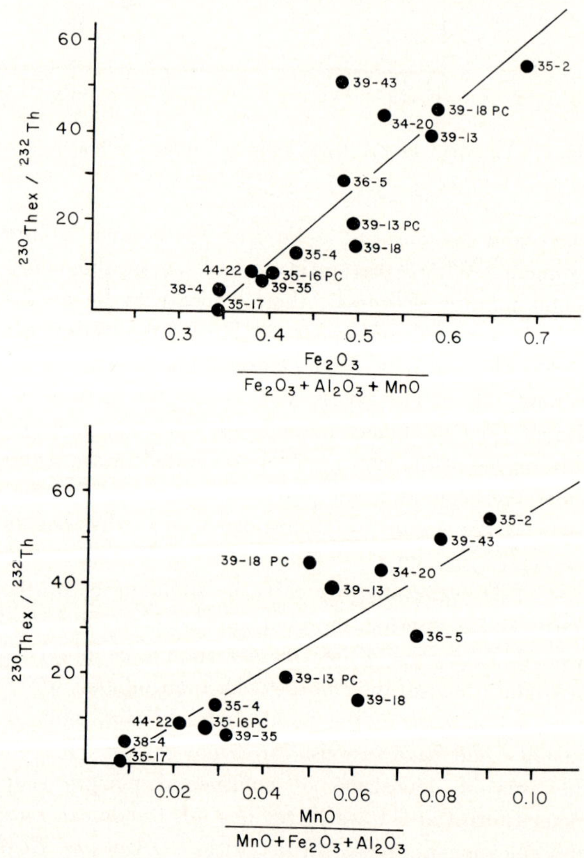

Fig. 10. Plot of ^{230}Th excess/^{232}Th activity ratios in surface sediment samples against the oxide ratios Fe/(Al + Fe + Mn) and Mn/(Al + Fe + Mn).

the clay fraction, estimated from analyses of Al_2O_3, is highest near the antarctic continent and lowest in the biogenic sediments near the ridge.

The regional study of the paleomagnetic stratigraphy of a large number of cores by *Watkins and Kennett* [this volume] shows that bulk sedimentation rates decrease northward across the South Indian basin and are lowest on the ridge. Our high bulk rate values for some of the biogenic sediments on and near the ridge are attributed to the shorter time range for the ^{230}Th method and to local geographical influences on the relatively small number of ridge cores used in this study.

Analyses of Fe, Mn, and Al show that low Al, high Fe, and Mn sediments are present on the Southeast Indian ridge. Most of the uranium in the sediments is detrital in origin, but $^{234}U/^{238}U$ ratios and U/Al ratios increase in the vicinity of the ridge; the increase implies addition of authigenic uranium to the sediments. The proximity of sediments with the highest U/Al ratios to the zone of siliceous oozes associated with the convergence implies the scavenging of uranium from sea water by sediments rich in organic matter or the direct removal of uranium by siliceous organisms. The presence of one sample with a $^{234}U/^{238}U$ activity ratio of 1.29 ± 0.07 implies a possible volcanic input of uranium. The data do not suggest an unequivocal origin for the authigenic portion of the uranium. Accumulation rates of Fe, Mn, and U in ridge sediments analyzed in this study are not significantly larger than those in nonridge sediments.

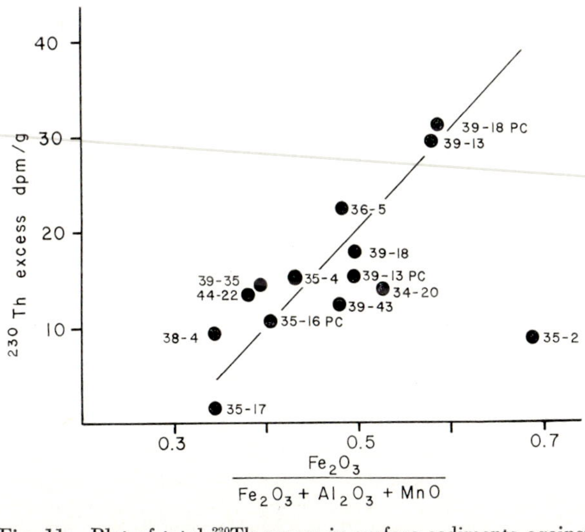

Fig. 11. Plot of total ^{230}Th excess in surface sediments against the Fe/(Al + Fe + Mn) oxide ratios. The plot of ^{230}Th excess against Mn/(Al + Fe + Mn) oxide ratios (not shown) is parallel to this line.

TABLE 6. Accumulation Rates of Trace Metals in Sediments of the South Indian Basin

Core	Fe, mg/cm²/10³ yr	Mn, mg/cm²/10³ yr	U, μg/cm²/10³ yr	Σ ²³⁰Th excess, dpm/cm²
E35-16	31	1.9	1.15	1405
E38-6	149	1.8	2.49	511
E39-13	18	1.7	0.248	1436
E39-18	26	3.2	0.457	2763
E34-14	1.9	0.21	...	...
E35-2	5.7	0.74	0.115	...
E37-14	42	0.35	0.688	...
E39-35	2.5	2.1	0.082	...

Rates calculated on the basis of the total sediment accumulation rates (Table 4), Fe, Mn, and U concentrations, (Tables 2 and 5), and dry bulk density of the samples (Table 5). For piston core samples, average values of the parameters were computed for each core. For cores represented by chemical analyses of a trigger core top only, the analysis of that sample was used to represent the entire core.

There is a strong positive correlation between ^{230}Th excess content of the surface sediments and the Fe/(Al + Mn + Fe) oxide ratios. The data show that ^{230}Th is removed from sea water by or in association with Fe (and Mn) colloids. Low Σ ^{230}Th in glacial marine sediments is complemented by high Σ ^{230}Th in sediments to the north, in the basin, and on the ridge. The ^{230}Th generated in sea water is possibly being transported northward for some distance before being deposited.

Acknowledgments. This research was supported by National Science Foundation grant GV-25786. Dr. Robert B. Scott provided standards for the atomic absorption analyses and advised the authors on the use of the technique. Dr. James P. Kennett and Dr. Norman D. Watkins allowed the use of paleontologic and paleomagnetic data on sediment cores to provide reasonable limits on conclusions made in this paper. The authors appreciate the advice of Dr. W. M. Sackett, Dr. B. J. Presley, and Dr. Richard Ku and the helpful review by Dr. J. R. Connolly.

REFERENCES

Angino, E. E., Geochemistry of antarctic pelagic sediments, *Geochim. Cosmochim. Acta, 30*, 939–962, 1966.

Arrhenius, G., Pelagic sediments, in *The Sea*, vol. 3, edited by M. N. Hill, pp. 655–727, John Wiley, New York, 1963.

Baranov, V. I., and L. A. Kuzmina, The rate of silt deposition in the Indian Ocean, *Geochemistry, 6*, 131–140, 1958.

Bender, M., W. Broecker, V. Gornitz, and U. Middel, Accumulation rates of manganese and related elements in the sediments from the East Pacific Rise (abstract), *Eos Trans. AGU, 51*(4), 327, 1970.

Bonatti, E., and O. Joensuu, Deep-sea iron deposit from the South Pacific, *Science, 154*, 643–645, 1966.

Bostrom, K., Submarine volcanism as a source for iron, *Earth Planet. Sci. Lett.*, *9*, 348–354, 1970.

Bostrom, K., and D. Fisher, Volcanogenic uranium, vanadium and iron in Indian Ocean sediments, *Earth Planet. Sci. Lett.*, *11*, 95–98, 1971.

Bostrom, K., and M. N. A. Peterson, Precipitates from hydrothermal exhalations on the East Pacific Rise, *Econ. Geol.*, *61*, 1258–1265, 1966.

Bostrom, K., and M. N. A. Peterson, Origin of aluminum-poor ferromanganoan sediments in areas of high heat flow on the East Pacific Rise, *Mar. Geol.*, *7*, 427–447, 1969.

Bostrom, K., M. N. A. Peterson, O. Joensuu, and D. E. Fisher, Aluminum-poor ferromanganoan sediments on active ocean ridges, *J. Geophys. Res.*, *74*(12), 3261–3270, 1969.

Bostrom, K., and S. Valdes, Arsenic in ocean floors, *Lithos*, *2*, 351–360, 1969.

Goldberg, E., Marine geochemistry, 1, Chemical scavengers of the sea, *J. Geol.*, *62*, 249–265, 1954.

Goldberg, E. D., Ionium/thorium geochronologies, *Earth Planet. Sci. Lett.*, *4*, 17–21, 1968.

Goldberg, E. D., and M. Koide, Geochronological studies of deep-sea sediments by the ionium-thorium method, *Geochim. Cosmochim. Acta*, *26*, 417–450, 1962.

Goldberg, E. D., and M. Koide, Rates of sediment accumulation in the Indian Ocean, *Earth Sciences and Meteoritics*, edited by J. Geiss and E. D. Goldberg, pp. 90–102, North-Holland, Amsterdam, 1963.

Kharkar, D. P., K. K. Turekian, and M. R. Scott, Comparison of sedimentation rates obtained by Si-32 and uranium decay series determinations in some siliceous antarctic cores, *Earth Planet. Sci. Lett.*, *6*, 61–68, 1969.

Koide, M., and E. D. Goldberg, Uranium-234/uranium-238 ratios in sea water, in *Progress in Oceanography*, vol. 3, edited by M. Sears, pp. 173–177, Pergamon, London, 1965.

Ku, T.-L., Uranium series disequilibrium in deep-sea sediments, Ph.D. dissertation, Columbia University, New York, 1966.

Ku, T.-L., Uranium series isotopes in sediments from the Red Sea hot-brine area, in *Hot Brines and Recent Heavy Metal Deposits in the Red Sea*, edited by E. T. Degens and D. A. Ross, pp. 512–524, Springer-Verlag, New York, 1969.

Ku, T.-L., and W. S. Broecker, Atlantic deep-sea stratigraphy: Extension of absolute chronology to 320,000 years, *Science*, *151*, 448–450, 1966.

Ku, T.-L., and W. S. Broecker, Rates of sedimentation in the Arctic Ocean, in *Progress in Oceanography*, vol. 4, edited by M. Sears, pp. 94–104, Pergamon, New York, 1967.

Ku, T.-L., W. S. Broecker, and N. Opdyke, Comparison of sedimentation rates measured by paleomagnetic and the ionium methods of age determination, *Earth Planet. Sci. Lett.*, *4*, 1–16, 1968.

Kuznetsov, Yu. V., A. N. Elizarova, and M. S. Frenklikh, Investigation of the mechanism of the precipitation of Pa-231 and Th-230 from ocean waters, *Radiokhimiya*, *8*, 459–468, 1966.

Lynn, R. J., and J. L. Reid, Characteristics and circulation of deep and abyssal waters, *Deep Sea Res.*, *15*, 577–598, 1968.

Moore, W. S., Amazon and Mississippi river concentrations of uranium, thorium, and radium isotopes, *Earth Planet. Sci. Lett.*, *2*, 231–234, 1967.

Moore, W. S., and W. M. Sackett, Uranium and thorium series in equilibrium in sea water, *J. Geophys. Res.*, *69*(24), 5401–5405, 1964.

Osmond, J. K., and L. D. Pollard, Sedimentation rate determination in deep-sea cores by gamma-ray spectrometry, *Earth Planet. Sci. Lett.*, *3*, 476–480, 1967.

Payne, R. R., and J. R. Conolly, Turbidite sedimentation off the antarctic continent, in *Antarctic Oceanology II: The Australian–New Zealand Sector*, *Antarctic Res. Ser.*, vol. 19, edited by D. E. Hayes, pp. 349–364, AGU, Washington, D. C., 1972.

Rosholt, J. N., C. Emiliani, J. Geiss, F. F. Koczy, and P. J. Wangersky, Absolute dating of deep-sea cores by the Pa-231/Th-230 method, *J. Geol.*, *69*, 162–185, 1961.

Rosholt, J. N., C. Emiliani, J. Geiss, F. F. Koczy, and P. J. Wangersky, Pa^{231}/Th^{230} dating and O^{18}/O^{16} temperature analysis of core A254-BR-C, *J. Geophys. Res.*, *67*(7), 2907–2911, 1962.

Rosholt, J. N., B. R. Doe, and N. Tatsumoto, Evolution of the isotopic composition of uranium and thorium in soil profiles, *Geol. Soc. Amer. Bull.*, *77*, 987–1004, 1966.

Sackett, W. M., Protactinium-231 content of ocean water and sediments, *Science*, *132*, 1761–1762, 1960.

Sackett, W. M., Measured deposition rates of marine sediments and implications for accumulation rates of extraterrestrial dust, *Ann. N. Y. Acad. Sci.*, *119*, 339–346, 1964.

Sackett, W. M., Deposition rates by protactinium method, in *Symposium on Marine Geochemistry*, *Occas. Publ. 3*, edited by D. R. Schink and J. T. Corless, pp. 29–40, Univ. of R. I., Kingston, 1965.

Sackett, W. M., H. A. Potratz, and E. D. Goldberg, Thorium content of ocean water, *Science*, *128*, 204–205, 1958.

Scott, M. R., Thorium and uranium concentrations and isotope ratios in river sediments, *Earth Planet. Sci. Lett.*, *4*, 245–252. 1968.

Somayajulu, B. L. K., G. R. Heath, T. C. Moore, Jr., and D. S. Cronan, Rates of accumulation of manganese nodules and associated sediment from the equatorial Pacific, *Geochim. Cosmochim. Acta*, *35*, 621–624, 1971.

Thurber, D. L., Anomalous U^{234}/U^{238} in nature, *J. Geophys. Res.*, *67*(11), 4518–4530, 1962.

Turekian, K. K., Rapid technique for determination of carbonate content of deep-sea cores, *Amer. Ass. Petrol. Geol. Bull.*, *40*, 2507–2509, 1956.

Turekian, K. K., Deep-sea deposition of barium, cobalt, and silver, *Geochim. Cosmochim. Acta*, *32*, 603–612, 1968.

Turekian, K. K., and K. Bertine, Deposition of molybdenum and uranium along the major ocean ridge systems, *Nature*, *229*, 250–251, 1971.

Veeh, H. H., Deposition of uranium from the ocean, *Earth Planet. Sci. Lett.*, *3*, 145–151, 1967.

Veeh, H. H., U-234/U-238 in the East Pacific sector of the antarctic ocean and in the Red Sea, *Geochim. Cosmochim. Acta*, *32*, 117–119, 1968.

Veeh, H. H., and K. Bostrom, Anomalous U-234/U-238 on the East Pacific Rise, *Earth Planet. Sci. Lett.*, *10*, 372–373, 1971.

Watkins, N. D., and J. P. Kennett, Antarctic bottom water: Major change in velocity during the Late Cenozoic between Australia and Antarctica, *Science*, *173*, 813–817, 1971.

Windom, H., Contribution of atmospherically transported trace metals to South Pacific sediments, *Geochim. Cosmochim. Acta*, *34*, 509–514, 1970.

AUSTRALASIAN MICROTEKTITES IN DEEP-SEA SEDIMENTS

BILLY P. GLASS

Department of Geology, University of Delaware, Newark, Delaware 19711

Abstract. Coincident with the reversal of the magnetic field of the earth 0.7 m.y. ago, approximately 10^{14} grams of microscopic glassy particles were deposited on the ocean floor south of Australia and over much of the rest of the Indian Ocean. These glassy particles, called microtektites, represent a portion of the Australasian tektite-strewn field as indicated by their appearance, geographical occurrence, age of deposition, age of formation (fission track age), petrography, physical properties, and chemical compositions. Present theories of tektite origin suggest that tektites were formed by impact and thus deposited over the entire strewn field instantaneously (geologically speaking). Thus the Australasian microtektite layer provides a chronostratigraphic horizon throughout much of the Indian Ocean including the area south of Australia. The microtektite horizon appears to be coincident with several paleontologic boundaries. The amount of sediment reworking that has taken place since the microtektites fell is indicated by the vertical dispersion of the microtektites in the cores between 35 and 90 cm.

Tektites are small, greenish to jet black natural glass bodies found in large but limited areas, or strewn fields, on the surface of the earth (Figure 1). They are generally a few centimeters in diameter, and many are characteristically in the form of a spheroid, lense, button, disc, teardrop, dumbbell, or cylinder. In addition, some show a type of surface sculpturing indicative of aerodynamic ablation that is unique to tektites. Megascopically, tektites are often easily confused with obsidian, but petrographically they can be distinguished from all other natural glasses such as obsidian by (1) the universal presence of flow structure (schlieren) accompanied by strain birefringence, (2) the general presence of siliceous glass inclusions (such as lechatelierite particles), and (3) the complete absence of microlites. The youngest and largest strewn field (called the Australasian strewn field) covers the northern part of the Philippines, Indochina, Indonesia, southern Australia, and the island of Tasmania (Figure 1).

In 1966, microscopic (<1 mm) glassy objects were found in five deep-sea sediment cores taken south of Australia and Sumatra. Based on their appearance, physical properties, age of deposition (~700,000 years ago), and geographical location (adjacent to the Australasian tektite-strewn field), *Glass* [1967] concluded that the glassy objects were microtektites and that they constituted a portion of the Australasian strewn field. Since 1967, much evidence supporting this conclusion has been published in many widely scattered papers. In addition, microtektites have been found in the mid-Atlantic Ocean that are related to the Ivory Coast strewn field [*Glass,* 1968*a,* 1969*a*].

The purpose of this paper is to bring together all the data dealing with the Australasian microtektites with special reference to the area south of Australia.

GEOGRAPHICAL DISTRIBUTION

Microtektites were first discovered in cores RC8-52 and RC8-53 taken on the north flank of the mid-oceanic ridge southwest of Australia [*Glass,* 1967]. The microtektites were generally found to be restricted to a layer approximately 50 cm thick immediately above the Brunhes-Matuyama geomagnetic reversal boundary. To date, approximately 60 cores from many parts of the oceans of the world (Figure 1) have been investigated for the presence or absence of microtektites associated with the Brunhes-Matuyama reversal boundary. Microtektites have been found in 12 cores (Table 1, Figure 1). The cores containing Australasian microtektites define an area extending from just south of the Japanese Islands through Sumatra to south of Australia and westward to just southeast of Madagascar. Nearly half the cores known to contain Australasian microtektites are from the South Australian basin south and southwest of Australia. The distribution of the Australasian microtektites is similar to that of the Australasian tektites (Fig-

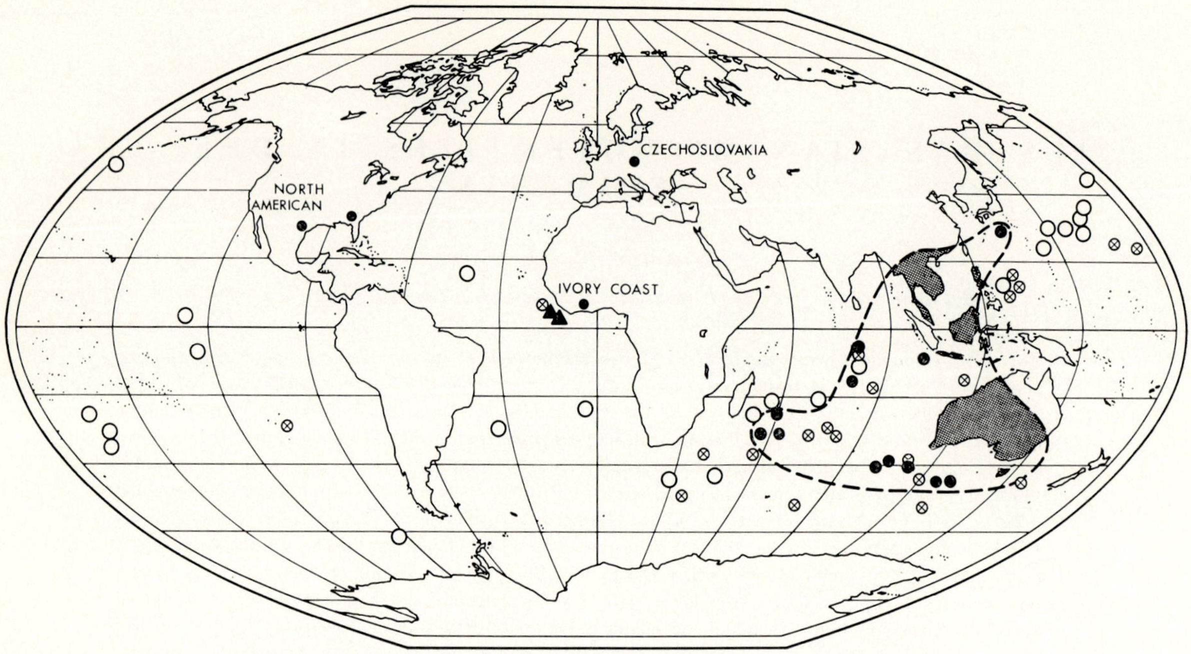

Fig. 1. Tektite-strewn fields and core locations. Areas where Australasian tektites have been found are stippled. Locations of cores containing Australasian microtektites are indicated by darkened circles. Provisional boundary of the Australasian strewn field is indicated by the broken line. Cores not containing Australasian microtektites associated with the Brunhes-Matuyama reversal boundary are indicated by an open circle. Cores not containing Australasian microtektites but with confusing magnetic stratigraphies are indicated by an open circle with an $\times$ in it. Locations of the two cores known to contain Ivory Coast microtektites are indicated by darkened triangles. North American tektites have been found in Texas and Georgia, and one has been found on Martha's Vineyard. The North American, Czechoslovakian, Ivory Coast, and Australasian tektites have been dated (by K-Ar and fission track dating) at ~35, ~15, ~1, and ~0.7 m.y., respectively.

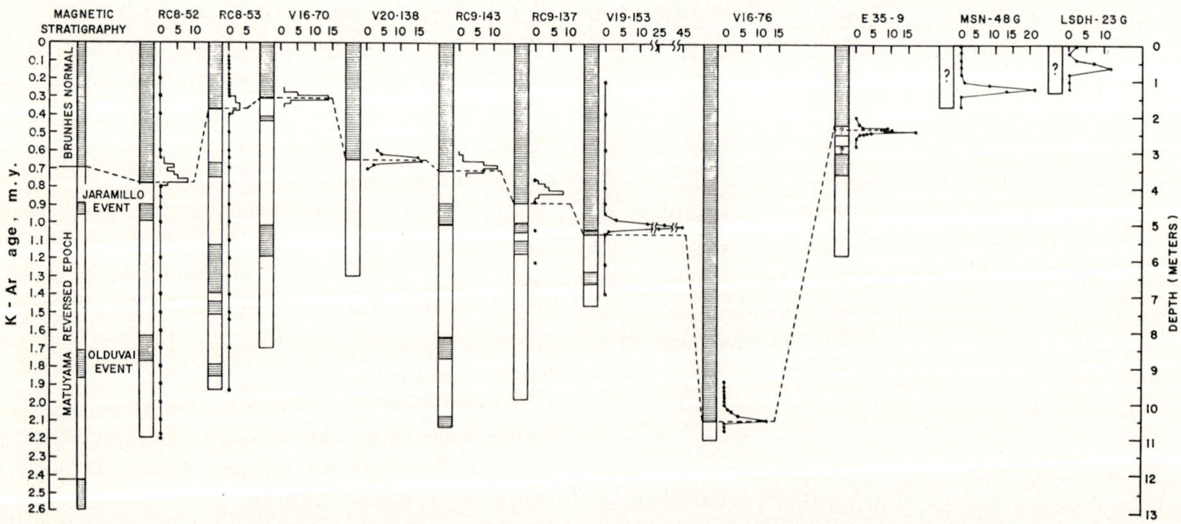

Fig. 2. Correlation of microtektite horizon with paleomagnetic stratigraphy. The paleomagnetic stratigraphies of most of the cores are discussed by *Opdyke and Glass* [1969]. Shaded areas represent normal polarity, and unshaded areas represent reversed polarity. The broken line shows the location of the Brunhes-Matuyama reversal boundary. Vertical distribution of microtektites within the cores is indicated by a scale to the right of each core showing the number of microtektites, greater than 125 μm in diameter (>62 μm in diameter for cores E35-9 and LSDH-23G), recovered from approximately 2-cm^3 samples.

ure 1), except that the microtektites are known to occur across the Indian Ocean almost to Madagascar. The location of the southern and eastern boundary of the strewn field south of Australia is not well established at the present time.

The maximum number of microtektites (>125 μm in diameter) per 2-cm^3 sample recovered from a core varies from 3 in core RC8-53 to 45 in core V19-153. The total number of microtektites recovered from each core does not seem to show any geographical pattern, except that the largest number of microtektites was recovered from core V19-153 (Table 1), which is near the center of the area where the microtektites are known to occur.

The average number of microtektites (>125 μm in diameter) per square centimeter for the cores investigated is approximately 50. By using this average for the entire strewn field, the mass of microtektites for the Australasian strewn field is calculated to be 10^{14} grams. A previous estimate of the total mass of tektites in the Australasian strewn field was 10^9 grams [*Urey*, 1962].

AGE OF DEPOSITION AND FISSION TRACK AGE

Paleomagnetic stratigraphies have been determined for 10 of the 12 cores found to contain Australasian microtektites. In each core, the microtektites are concentrated in a relatively thin layer (20–40 cm thick) that occurs on or just above the Brunhes-Matuyama geomagnetic reversal boundary (Figure 2). In all but one core (RC9-137), the peak abundance occurs within 10 to 20 cm of the Brunhes-Matuyama boundary. If the microtektites were deposited on the ocean floor prior to or after the Brunhes-Matuyama reversal, then the distance between the microtektite layer and the Brunhes-Matuyama boundary should be proportional to the sedimentation rate of the core; i.e., the greater the sedimentation rate, the greater the distance between the microtektite layer and the Brunhes-Matuyama boundary. This effect is not observed (see Figure 2); thus the microtektite fall must have been closely associated with the Brunhes-Matuyama geomagnetic reversal. Since the reversal occurred 0.69 m.y. ago [*Cox*, 1969], the microtektites must have fallen on the ocean floor approximately 0.69 m.y. ago.

The Australasian microtektites have a fission track age of 0.71 ± 0.1 m.y. [*Gentner et al.*, 1970]. Likewise, Australasian tektites have K-Ar ages [*Zähringer*, 1963] and fission track ages [*Fleischer and Price*, 1964; *Wagner*, 1966; *Gentner et al.*, 1969] of ap-

TABLE 1. Cores Containing Australasian Microtektites

Core No.	Latitude	Longitude	Core Depth, m	Sedimentation Rate, cm/1000 yr	Depth of B-M Boundary, cm	Lower Limit of Microtektite Zone, cm	Length of Microtektite Zone, cm	Duration of Microtektite Zone, yr	Total No. of Microtektites >125 mm
E35-9	45°02′S	128°00′E	5176	0.33	233	250	30	90,000	10
LSDH-23G	31°14′S	62°58′E	4910	~0.10‡	~ 70‡	60–80	60–80	~700,000‡	7†
MSN-48G	15°15′S	81°08′E	4996	~0.20‡	~140‡	140–175	60–95	~390,000‡	44†
RC8-52	41°06′S	101°25′E	4393	0.56	390	400	80	145,000	58
RC8-53*	39°23′S	104°22′E	4429	0.26*	185	190	40*	155,000*	17
RC9-137	45°01′S	123°45′E	4675	0.64	445	420	40	63,000	40
RC9-143	41°21′S	114°08′E	4396	0.51	355	360	50	98,000	71
V16-70	29°27′S	56°06′E	4649	0.22	155	170	70	320,000	54
V16-76	25°09′S	59°54′E	5316	1.51	1045	1050	~50	~ 33,000	21†
V19-153	8°51′S	102°07′E	5433	0.75	530	520	35	47,000	151†
V19-171	7°04′S	80°46′E	5053	0.50	350	...	...	...	11†
V20-138	28°52′N	135°33′E	4244	0.46	325	~345	≧40	≧ 87,000	37†

Sedimentation rates based on depth to Brunhes-Matuyama (B-M) reversal boundary.

*Core RC8-53 seems to have a large hiatus near the Brunhes-Matuyama boundary. Thus calculations of sedimentation rates and length and duration of microtektite zone may not be meaningful.

†Number of microtektites recovered from cores not sampled continuously through the microtektite layer.

‡Paleomagnetic stratigraphy not determined for these cores. Depth of Brunhes-Matuyama boundary based on peak abundance of microtektite occurrence.

proximately 0.7 m.y. Thus both the age of deposition and fission track age of the Australasian microtektites are the same (as closely as can be determined) as the K-Ar and fission track ages of the Australasian tektites.

Because of the close association of the microtektites with the Brunhes-Matuyama reversal boundary, *Glass and Heezen* [1967] speculated that the two phenomena might have had a common cause. They stated that an investigation of the paleomagnetic stratigraphy of the other known tektite deposits and a search throughout the world for tektites in sediments deposited during other reversals are necessary steps to rule out the possibility of an accidental coincidence. A search in the Atlantic Ocean for microtektites associated with the Ivory Coast strewn field resulted in the discovery of the Ivory Coast microtektites [*Glass*, 1968*a*, 1969*a*]. Ivory Coast microtektites have been found in two cores. Paleontologic and paleomagnetic data [*Glass*, 1968*a*] and fission track dating [*Gentner et al.*, 1970; *Durrani and Khan*, 1971] suggest that the Ivory Coast microtektite fall coincided with the Jaramillo magnetic event (0.89–0.95 m.y. ago) [*Cox*, 1969]. Thus two out of the four known tektite-strewn fields appear to be associated with reversals of the magnetic field of the earth.

PHYSICAL CHARACTERISTICS

Based on their physical properties and chemical compositions, the Australasian microtektites can be divided into two groups: 'normal' microtektites and bottle-green microtektites [*Glass*, 1967, 1968*b*; *Cassidy et al.*, 1969]. The normal Australasian microtektites have physical properties and chemical compositions similar to those of the Australasian tektites. However, the bottle-green microtektites have physical properties and chemical compositions different from those of most Australasian tektites but similar to those of a rare group of Australasian tektites characterized by their high magnesium (Mg) content [*Chapman and Scheiber*, 1969].

Normal Australasian Microtektites

More than 500 normal Australasian microtektites greater than 125 μm in diameter have been recovered. The largest spherical ones have a diameter of ~800 μm, and some teardrop and dumbbell forms are nearly 1 mm long. The number of microtektites increases with a decrease in diameter (down to 38 μm), whereas the weight percentage decreases with a decrease in diameter [*Glass*, 1968*b*].

The normal microtektites are mostly spherical in shape, but disc, rod, pear, dumbbell, teardrop, and irregular shapes have also been found (Figure 3). In most of the cores, only spherical forms were recovered. Approximately 90% of the normal microtektites recovered from cores taken south and southeast of Australia (in the South Australian basin) are spherical. The remaining 10% are mostly disc, teardrop, pear, or dumbbell shaped. In contrast to this percentage, only about 35% of the microtektites from core V19-153 taken south of Sumatra (in the Wharton basin) are spherical, and nearly 40% are irregular shapes or fragments. The remainder are pear, teardrop, or dumbbell shapes, many of which are flattened. Except for a few of the microtektites from core V19-153, all the normal microtektites are remarkably symmetrical in shape and, unlike many volcanic spherules and artificial glass beads, do not have protrusions.

The normal microtektites from the Wharton basin core (V19-153) vary from opaque (<1%) to transparent and are generally dark brown to yellowish brown in color. Most of the normal microtektites recovered from the other cores are transparent colorless to yellowish green, but yellow to yellowish-brown ones were also found. Most of the microtektites have a vitreous (glassy) luster, but some have a dull luster.

The surface texture varies from perfectly smooth to slightly pitted to frosted (Figure 3). In contrast to tektites (especially australites), none of the microtektites shows any signs of having undergone atmospheric ablation. We do not know if the microtektites are ablation droplets from the tektites or merely small particles that passed through the atmosphere unablated.

The refractive index of 47 normal microtektites was determined by using the oil immersion method. The normal microtektites had refractive indices ranging from 1.497 to 1.540 with an average of 1.518. The specific gravity of seven of the larger normal microtektites was determined by the sink or float method by using heavy liquids. The specific gravity was found to range from 2.37 to 2.55. The range in refractive index and specific gravity is similar to that reported by *Chapman and Scheiber* [1969] for Australasian tektites (Table 2). In a plot of refractive index versus specific gravity, the normal microtektites fall on the same trend as Australasian tektites [*Glass*, 1968*b*].

Senftle et al. [1969] measured the magnetic susceptibility, magnetization, and Curie constants for 11 normal Australasian microtektites. They concluded that the magnetic properties of the normal microtektites closely parallel those of tektites.

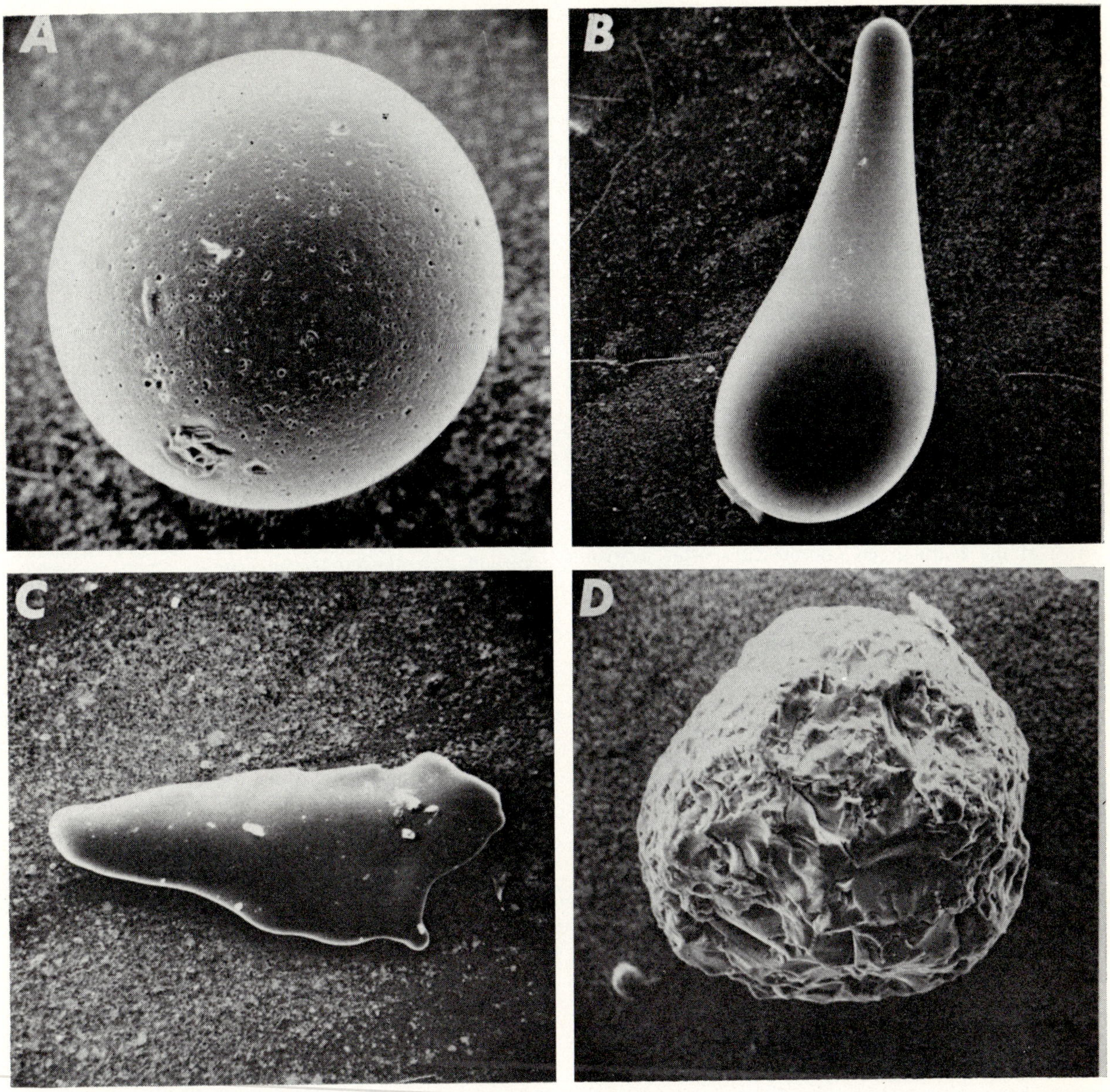

Fig. 3. Scanning electron microscope photographs of Australasian microtektites. (*A*) Finely pitted yellow-green spherule that is 470 μm in diameter. (*B*) Teardrop that is 880 μm long from core V19-153. (*C*) Smooth irregular microtektite that is 535 μm long from core V19-153. (*D*) Bottle-green microtektite that is 250 μm in diameter from core V19-153; note the irregular surface that is in contrast to the normal microtektites in *A*, *B*, and *C*.

Bottle-Green Microtektites

Transparent bottle-green glassy objects have been found in three cores (V19-153, RC9-143, and V19-171) associated with the normal microtektites. The bottle-green microtektites vary in shape from spherical to irregular (disc, teardrop, and dumbbell forms are exceedingly rare) and generally have a deeply corroded surface and granular appearance (Figure 3). The refractive index of seven of the bottle-green microtektites ranges from 1.580 to 1.620. A group of bottle-green microtektites was picked from core V19-153 and separated into specific gravity fractions by using heavy liquids. Two hundred fifteen bottle-green microtektites were recovered that had specific gravities ranging from less than 2.4887 up to 2.8622 with a modal abundance at about 2.79. Thus the bottle-green microtektites have distinctly higher refractive indices and specific gravities than most Australasian tektites, but

TABLE 2. Physical Properties and Fission Track Age of Australasian Microtektites and Australasian Tektites

	Microtektites				Tektites			
	Normal	No. of Analyses	Bottle-Green	No. of Analyses	All Except High Mg	No. of Analyses	High Mg	No. of Analyses
Refractive index	1.497–1.540	47	1.580–1.620	7	1.487–1.542*	50	1.501–1.544*	9
Specific gravity	2.37–2.55	8	2.489–2.87	215	2.361–2.590*	38	2.381–2.591*	9
Magnetic susceptibility, cmu/g $\times 10^{-6}$	4.7–11.7	9	5.1–14.3	10	5.6–11.2	...	...	...
Fission track age, m.y.	0.71±0.10†	...	...	...	0.7‡	17	...	...

Magnetic susceptibility data from *Senftle et al.* [1969].
* Data from *Chapman and Scheiber* [1969].
† Data from *Gentner et al.* [1970].
‡ Data from *Gentner et al.* [1969].

there is some overlap with the high Mg group of Australasian tektites described by *Chapman and Scheiber* [1969] (Table 2).

Senftle et al. [1969] found that the magnetic properties of the bottle-green microtektites after a strong acid leach are considerably different from those of the tektites. They suggested that the bottle-green microtektites may have partially reacted with the sea water, the reaction thus altering their original magnetic properties. They found, however, that, like the normal microtektites, many of the bottle-green microtektites have a positive temperature-independent paramagnetism indicating the presence of Fe_3O_4 or of possibly metallic iron microspherules, which were found for tektites.

PETROGRAPHY

The normal microtektites are optically isotropic except for a few that have a weakly birefringent outside layer that may be the result of strain. Many contain spherical bubble cavities. No crystalline inclusions have been observed in any of the normal microtektites, but lechatelierite particles have been identified by electron microprobe analysis [*Cassidy et al.*, 1969]. The lechatelierite particles often occur as hemispherically shaped bodies at the surface of the microtektite. The rounded portion of the hemisphere is toward the center of the microtektite, and the flat side is at the surface of the spherule. In some cases, differential solution has caused the flat surface of the hemispherical particle to be slightly raised above the surface of the spherule to form a slightly elevated circular area. Thin sections of a few of the larger microtektites show flow lines (due to slight variations in refractive index) similar to those observed in tektites. In contrast to the normal Australasian microtektites, the bottle-green microtektites do not contain bubble cavities, lechatelierite particles, or crystalline inclusions.

CHEMICAL COMPOSITION

Normal Australasian Microtektites

Cassidy et al. [1969] determined the major element compositions for 60 Australasian microtektites (42 normal microtektites, 7 bottle-green microtektites, and 11 with intermediate compositions) by electron microprobe analysis. Many of the normal microtektites had compositions indistinguishable from Australasian tektites, and variation diagrams of the major oxides versus silica (SiO_2) show that the microtektites fall on the same composition trends as the Australasian tektites. However, the microtektites were found to have a much greater range in composition than that reported for tektites. For example, the lowest silica content reported for Australasian tektites was about 68%, but the silica content for the normal Australasian microtektites was as low as 59%. Cassidy et al. concluded from their investigation that the normal Australasian microtektites did indeed represent a portion of the Australasian tektite-strewn field and that Australasian tektites have a greater range in composition than that previously reported.

Chapman and Scheiber [1969] chemically analyzed 530 Australasian tektites. They reported specimens with silica contents as low as 62%; thus only a few normal Australasian microtektites have SiO_2 contents lower than that now reported for Australasian tektites (Table 3). An even more convincing demonstration of the chemical similarity between the normal Australasian microtektites and Australasian tektites was achieved by crushing Australasian tektite specimens and separating the grains into various specific gravity fractions by using heavy liquids. Representative grains

TABLE 3. Chemical Composition in Per Cent of Australasian Microtektites, Australasian Tektites, and Other Microforms of Silicate Glasses

	Australasian Microtektites									Australasian Tektites	
	Normal					Bottle-Green				Normal	High Mg
	Sample 53	Sample A2	Sample 1F	Sample A18	Range of 121 Analyses	Sample 13F	Sample 84-1	Sample 93-3	Range of 55 Analyses	Range of 48 Analyses	Range of 10 Analyses
SiO_2	55.0	72	73	81	55.0–81	46.2	51.6	70.3	46.2–74.2	62.2–79.7	64.76–77.0
TiO_2	1.05	0.8	0.73	0.3	0.3–2.09	0.7	0.73	0.71	0.57–1.09	0.43–1.00	0.66–0.77
Al_2O_3	23.4	13.6	12.1	7.5	7.4–23.4	13.4	14.7	12.4	8.0–21.6	8.2–17.7	10.7–13.3
FeO	4.92	4.9	5.3	3.1	2.78–9.68	6.41	4.44	6.48	3.42–10.8	3.09–6.88	3.85–8.63
MnO	0.09	0.08	0.08	0.03	0.03–0.15	0.09	0.10	0.11	0.06–0.22	0.058–0.22	0.085–0.139
MgO	12.7	2.4	2.5	1.2	1.2–12.7	25.6	21.4	6.4	3.47–27.3	1.13–4.68	1.83–7.95
CaO	2.76	3.6	2.4	0.5	0.5–6.18	5.8	4.2	2.1	1.35–6.2	0.73–9.77	1.79–3.73
Na_2O	0.32	0.6	0.7	1.1	0.32–2.8	0.19	0.06	0.62	0.02–0.73	0.91–3.91	0.62–1.38
K_2O	0.34	1.9	1.9	2.2	0.19–3.7	0.08	0.06	1.2	0.01–2.11	0.90–2.81	1.34–2.56
P_2O_5	...	...	0.01	0.00	0.01–0.17	0.04	...	...	<0.01–0.04	...	...
BaO	...	...	0.04	0.01	<0.01–0.06	0.04	...	...	0.04–0.06	0.033–0.053	0.039–0.046
Cr_2O_3	...	...	...	...	...	...	0.22	0.09	0.07–0.24	0.009–0.056	0.016–0.064
Total	100.58	99.9	98.8	96.9	...	98.6	97.5	100.4	...	...	...

	Other Microforms of Silicate Glasses									
						Lunar Glasses				
	Pele's Tears, Range of 3 Analyses	Hay Silica Glass	Antarctic Glass Spherule	Tunguska Silicate Spherules		Apollo 11 Glasses, Range of 36 Analyses	Apollo 12 Glasses, Range of 125 Analyses	Anorthositic, Range of 34 Analyses	Soda-Lime Glass Spherules	
				Sample 753	Range of 6 Analyses				Sample V94	Sample 376
SiO_2	46.91–50.82	61.7	73.0	61.3	41.0–54.8	37.7–50.0	38.1–57.1	37.1–51.9	76.5	73.6
TiO_2	2.74–2.75	0.19	...	0.31	0.23–0.91	0.08–11.2	0.39–8.6	0.19–1.15	0.07	0.09
Al_2O_3	9.14–13.30	1.16	12.5	18.1	11.9–18.6	5.1–30.9	4.12–18.6	20.3–36.0	1.21	0.77
FeO	10.87–13.63	1.03	4.5	3.2	1.5–8.6	1.85–23.6	7.36–29.8	0.67–8.41	0.16	1.94
MnO	0.08–0.38	0.03	...	0.23	0.45–0.85	0.04–0.31	0.06–0.23	0.01–0.24	0.02	0.01
MgO	7.22–9.19	4.88	2.0	2.9	6.2–12.2	2.0–16.2	5.42–14.9	3.40–11.1	3.71	3.01
CaO	11.27–11.63	6.77	3.0	6.7	14.0–30.9	7.5–16.7	7.28–15.0	13.1–22.0	9.32	9.64
Na_2O	1.02–2.64	8.73	1.5	0.8	0.54–1.3	0.04–1.09	0.05–1.84	0.04–1.02	10.3	13.1
K_2O	0.59–3.06	8.53	2.5	5.5	0.9–3.2	0.04–1.15	0.04–1.27	0.02–0.35	0.34	0.29
P_2O_5	...	5.66	...	...	...	0.09–0.87	0.02–0.9	0.01–0.04	...	...
BaO	...	...	...	...	...	...	...	...	...	...
Cr_2O_3	...	...	...	...	...	0.05–0.63	0.3–0.9	0.1–0.3	...	...
Total	...	98.95	99.0	99.0	...	...	...	...	101.6	102.45

Sixty of the analyses used for determining the range in composition for Australasian microtektites are taken from *Cassidy et al.* [1969]; the remainder are unpublished analyses. The data for the Australasian tektites are compiled from *Chapman and Sheiber* [1969]. The range of the three analyses for Pele's Tears is taken from the literature. The data for hay silica glass are taken from *Baker* [1968]. The data for the antarctic glass spherule are taken from *Shima* [1966]. The data for Apollo 11 glasses are compiled from *Chao et al.* [1970]. The data for anorthositic glasses are taken from *Glass* [1971]. For the soda-lime glass spherules, sample V94 is an artificial glass spherule used in reflecting paint, and sample 376 was supposedly recovered from South Pacific sediments.

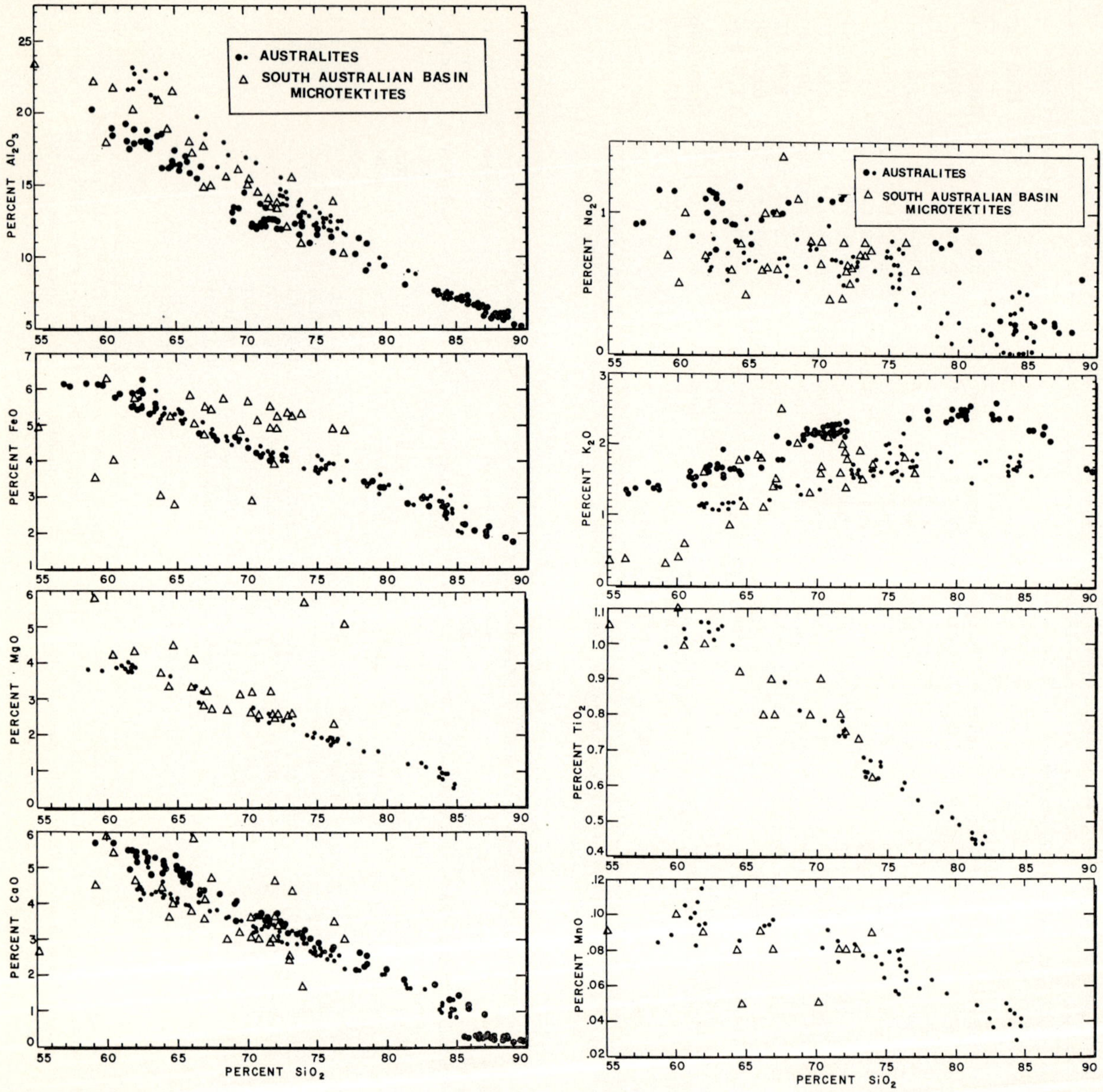

Fig. 4. Variation diagrams of major oxides versus SiO_2 for two australites and Australasian microtektites from the South Australian basin. Australite data (represented by darkened circles) show chemical variation within two australite specimens obtained by electron microprobe analysis. South Australian basin microtektite data (represented by triangles) were also obtained by electron microprobe analysis. (*Left*) Al_2O_3, FeO, MgO, and CaO versus SiO_2; (*right*) Na_2O, K_2O, TiO_2, and MnO versus SiO_2.

recovered from each specific gravity fraction were then analyzed with the electron microprobe. This method has shown that the chemical variation and trends within single tektite specimens are similar to those observed among the normal Australasian microtektites (Figure 4).

Frey et al. [1970] analyzed a composite group of normal Australasian microtektites (five specimens totaling 1.5 mg) for 20 trace elements and 3 major elements by instrumental neutron activation analysis (Table 4). They concluded that the close correspondence of elemental abundances among elements with different geochemical characteristics indicates that the normal Australasian microtektites are directly related to the Australasian tektites.

Basing their work on a study of the magnetic properties of the microtektites, *Senftle et al.* [1969] found that the bulk iron in the normal Australasian microtektites is primarily ferrous iron as it is in tektites.

Bottle-Green Microtektites

Cassidy et al. [1969] pointed out that the bottle-green microtektites overlap the normal microtektites in silica content, but that they have distinctly lower Al_2O_3 and K_2O contents and higher MgO contents than the normal microtektites of similar silica contents. The composition trends for the seven bottle-green microtektites were different from those for the normal Australasian microtektites. In their article reporting the analyses of 530 Australasian tektites, *Chapman and Scheiber* [1969] defined a high magnesium group of tektites. They pointed out that the seven bottle-green microtektites analyzed by *Cassidy et al.* [1969] overlap the high magnesium tektites in content and continue smoothly all major element variations found for the high magnesium tektites. An additional 44 bottle-green microtektites from core V19-153 have been analyzed by electron microprobe analysis, and these data show the relationship between the bottle-green microtektites and the high magnesium Australasian tektites even better (Figure 5). There can be little doubt that the bottle-green microtektites are related to the high magnesium Australasian tektites. The bottle-green microtektites have a rather high Cr_2O_3 content (up to ~0.24% according to the electron microprobe analysis), which may account for their bottle-green color.

TABLE 4. Trace Element Composition of Australasian Microtektites (Determined by Instrumental Neutron Activation Analysis) and Australasian Tektites (Compiled from the Literature)

Element	Composite of 5 Australasian Microtektites	'Normal' Australasian Tektites
Rb	66	62–144
Cs	2.3	2–4
Sc	11.9	8.2–14
Cr	81	72–105
Mn	820	670–830
Co	9.9	11–14
Ba	530	330–425
La	40	37–60
Ce	93	60–88
Nd	36	22–35
Sm	7.1	4.1–5.6
Eu	1.37	1.1–1.5
Gd	5.4	3.9–7.7
Tb	0.9	0.6–1.0
Dy	5.1	3.8–4.5
Yb	2.2	1.6–2.6
Lu	0.45	0.36
Hf	6.4	3.7–5.0
Ta	2.4	. . .
Th	14.5	9.0–14.5
Na	0.45*	0.97–1.16*
K	1.42*	1.93–2.17*
Fe	3.26*	3.54–3.98*

Table taken from *Frey et al.* [1970].

* These values are given in per cent. All others are in parts per million.

COMPARISON WITH OTHER MICROSCOPIC GLASSY PARTICLES

Igneous Glass

Several authors [*Martin*, 1967; *Baker*, 1968; *von Koenigswald*, 1968; *Dymond*, 1969] have suggested that some microtektites (especially the bottle-green ones) might be volcanic glass. *Barnes* [1940] has shown that tektites and igneous glasses fall on two different trends when the data are plotted on variation diagrams (major oxides versus silica). In particular, for a given SiO_2 content, tektites have higher MgO and lower Na_2O contents than igneous glasses. This same relationship is true for microtektites and igneous glass [*Glass*, 1969*a*]. The similarity between Australasian tektites and Australasian microtektites and their contrast with igneous glasses is well illustrated in a ternary diagram in which the chemical analyses are plotted in terms of weight percentage of MgO, Na_2O, and K_2O recalculated to 100% (Figure 6). The bottle-green microtektites with high MgO contents and low alkali contents are at the top of the diagram (~100% MgO) and continue smoothly down into the normal Australasian microtektites. Most of the tektites and microtektites plot more or less along a line joining 100% MgO with 65% K_2O and 35% Na_2O. The tektites and microtektites plot on the left of the diagram because they have Na_2O/K_2O ratios of less than 1. The igneous glasses plot to the lower right (low MgO contents) of the tektites and microtektites. Thus the microtektites (including the bottle-green ones) are more similar to the Australasian tektites than to igneous glasses.

It has also been pointed out that tektites differ from igneous glasses in having a higher FeO/Fe_2O_3 ratio. *Senftle et al.* [1969] have shown that microtektites also have a high FeO/Fe_2O_3 ratio, which again demonstrates the difference between microtektites and igneous glasses. Besides the chemical difference, the normal microtektites are distinguished from volcanic glass by their lack of microlites and the presence of lechatelierite particles.

Martin [1967], *Baker* [1968], and *von Koenigswald* [1968] point out that the microtektites are simi-

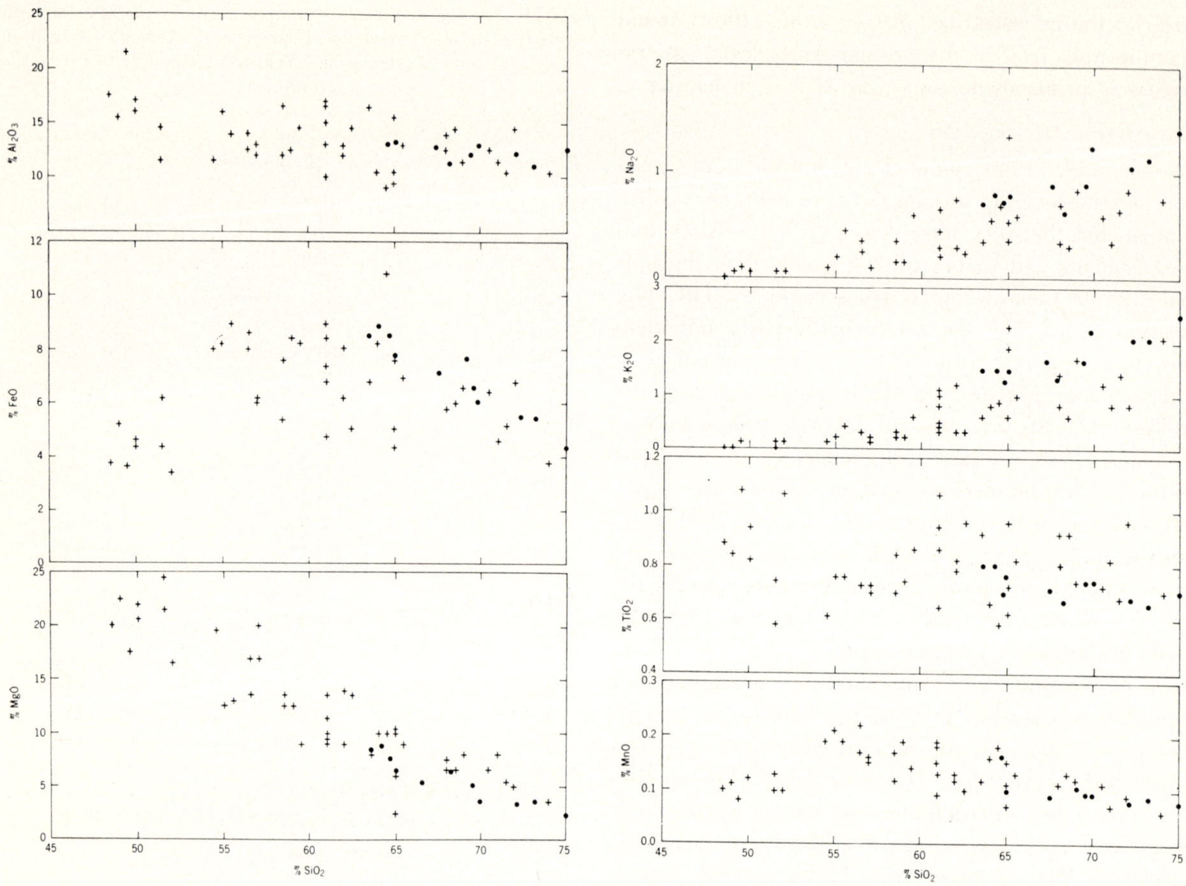

Fig. 5. Variation diagrams (Harker diagrams) of major oxides versus SiO_2 for bottle-green microtektites (crosses) and high magnesium Australasian tektites (darkened circles). Tektite data from *Chapman and Scheiber* [1969]. (*Left*) Al_2O_3, FeO (total iron), and MgO versus SiO_2; (*right*) Na_2O, K_2O, TiO_2, and MnO versus SiO_2.

lar in size and morphology to the so-called 'Pele's Tears' (small glassy bodies originating as spray from basaltic lava fountains). Pele's Tears are basaltic and generally have silica contents of about 50% or less. Only the bottle-green microtektites have silica contents this low, and the bottle-green microtektites can be distinguished from basaltic igneous glasses by their higher MgO and lower Na_2O, CaO, and TiO_2 contents (Table 3). Pele's Tears are generally dark brown and contain numerous vesicles and microlites; the bottle-green microtektites are transparent pale green and do not contain bubble cavities or crystalline inclusions. Thus even the bottle-green microtektites show no resemblance to terrestrial igneous glasses.

Hay Silica Glass

Baker [1968] suggested that some of the microtektites (those with low refractive indices) might be microbodies of glass generated by fusing opal that was contained in silica accumulator plants during prehistoric forest, bush, and grass fires in the Australian and New Guinea region and that was dispersed toward the southwest and south over the Australian basin. According to Baker, the microbodies of glass formed in the haystack fire include spheres, spheroids, discs, ovoids, ellipsoids, dumbbells, teardrops, pear shapes, flask shapes, canoe shapes, aberrant forms, distorted and nondescript shapes, and irregular fragments. Colors range from colorless through gray, pale smoky brown gray, brownish, and yellowish green and less commonly from gray through black. Spherical forms range from 10 to 320 μm; several contain numerous microbubbles, whereas others are completely free of bubbles. The microforms of hay silica glass are somewhat similar in appearance to microtektites. However, in detail, the microforms of hay silica glass are distinct from microtektites. The microforms of hay silica glass are not as symmetrical in shape as the microtektites and often show tapering ends (probably the proximal ends of long, drawn-out, filamentous glass that connected one lenticular form with another) not found on microtektites. The composition of hay silica glass according to *Baker* [1968] is given in Table 3. The low Al_2O_3 content, which is characteristic of glass produced by incineration of opal-bearing vegetation, is in contrast to the Al_2O_3 content of the normal Australasian microtektites, which contain approximately 10 times as much Al_2O_3. In addition, glass produced in the above manner contains more MnO, CaO, Na_2O, K_2O (approximately five times as much), and P_2O_5 and less total iron and TiO_2 than the normal microtektites (Table 3).

Antarctic Glass Particles

Microscopic glassy bodies have been found in antarctic ice [*Schmidt*, 1963; *Shima*, 1966]. These glassy particles are mostly spherical, but teardrop bodies are also found. They are generally transparent and colorless. Electron microprobe analysis of one teardrop-shaped glass particle [*Shima*, 1966] indicates that at least some of these particles have compositions similar to the Australasian microtektites (Table 3). The greatest difference between the microtektites and the antarctic glassy particles is one of age. The Australasian microtektites were recovered from sediments deposited 0.7 m.y. ago, whereas the antarctic glassy particles were recovered from modern snow. Thus the relationship between these two groups of glassy microparticles is not clear.

Tunguska Silicate Spherules

Microscopic silicate glass spherules have been recovered from soil samples taken in the area devastated by the explosion of the Tunguska meteorite in 1908 [*Kirova*, 1964]. The silicate spherules were found in association with black opaque magnetic spherules that have a Ni:Fe ratio indicative of a cosmogenic origin [*Florenskiy et al.*, 1969]. *Glass and Heezen* [1967] suggested that microtektites (and tektites) might have been produced in a similar manner. However, these spherules are less symmetrical than the microtektites, and many spherules contain numerous inclusions that give them a turbid appearance. Electron microprobe analysis shows that these spherules are more heterogeneous than the microtektites, and their compositions are not generally similar to microtektites. However, one (sample 753) does have a composition similar to an Australasian microtektite, except for its somewhat higher K_2O content [*Glass*, 1969*b*] (Table 3).

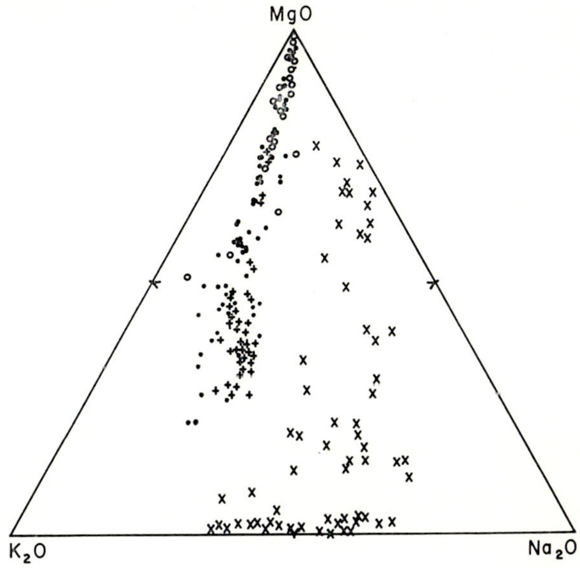

Fig. 6. Ternary diagram of MgO–Na_2O–K_2O. Weight percentage of MgO, Na_2O, and K_2O (recalculated to 100%) plotted for Australasian tektites (represented by crosses), Australasian normal microtektites (represented by darkened circles), bottle-green microtektites (represented by open circles), and igneous glasses (represented by ×'s). Tektite data from *Chapman and Scheiber* [1969]. Igneous glass data are mostly compiled from literature, but also include electron microprobe analyses of volcanic glasses from deep-sea cores (two analyses of Pele's Tears also included).

Lunar Glasses

One of the big surprises for many lunar investigators was the presence of numerous microscopic glassy particles in the lunar 'soil.' Many of the lunar spherules are similar to the microtektites in appearance. However, the lunar glasses are much more varied in color, and many contain crystalline inclusions including nickel-iron and sulfide beads. The greatest difference between the lunar glass beads and the microtektites is their composition (Table 3). Several hundred of the lunar glass spherules have been analyzed. They are generally 'basaltic' or 'anorthositic' in composition. Some of the lunar glass spherules overlap some of the bottle-green microtektites in SiO_2 content, but even these are chemically distinct (Table 3). However, the lunar glass spherules were produced by meteorite impact [*Chao et al.*, 1970], an origin that most tektite investigators favor for the tektites and microtektites.

Transparent colorless glass fragments have been found in Apollo 14 soil samples. Their compositions are similar to terrestrial igneous glasses except for their low Na_2O content. The average composition of six of these glass grains is as follows: 71.5% SiO_2, 0.4% TiO_2, 14.2% Al_2O_3, 1.8% FeO, 0.7% MgO, 2% CaO, 0.9% Na_2O, and 6.5% K_2O [*Apollo Soil Survey*, 1971]. The silica content ranges as high as 79%. Although the Na_2O content is low for terrestrial igneous glasses, it is similar to the Na_2O content of tektites. However, the K_2O content is too high for any known tektites. No lunar spherules of this composition have been reported.

Artificial Glass

A few investigators have recovered glass microspheres from sediments that they thought were similar to microtektites. These microforms of silica glass are about the same size as microtektites. They are transparent and colorless and have smooth vitreous surfaces. However, electron microprobe analysis (Table 3) shows that these glass bodies are an artificial soda-lime silica glass. The source of these particles is unknown, but they have been appearing in several investigators' samples.

ORIGIN

The data presented in the preceding sections show that the microtektites found in deep-sea sediments south of Australia and in other parts of the Indian Ocean and in the Philippine Sea represent a portion of the Australasian tektite-strewn field. In brief, the main evidence supporting this conclusion is (1) general appearance, (2) occurrence of the microtektites in deep-sea sediments adjacent to where tektites have been found on land, (3) age of deposition (and fission track age) similar to K-Ar and fission track ages of Australasian tektites (i.e., 0.7 m.y.), (4) absence of microlites and presence of lechatelierite particles, and (5) similarity in physical properties and chemical compositions between the Australasian tektites and microtektites.

Thus the origin of the microtektites is related to that of the tektites. Unfortunately, the origin of tektites has been debated since they were first described in scientific literature over 180 years ago, and there is still not unanimous agreement. However, most tektite investigators agree that tektites were formed by impact (by comet or meteorite). The evidence for an impact origin includes the presence of nickel-iron spherules in philippinites and indochinites [*Chao et al.*, 1962] and the presence of coesite [*Walter*, 1965] and shocked crystalline inclusions [*Glass*, 1970] in Muong Nong type (layered) tektites. The two principal hypotheses of tektite origin involve collisions of a cosmic body with the earth or moon. Recent investigations of returned lunar samples have not given much support to a lunar origin of tektites because of the scarcity of lunar material with compositions similar to that of tektites. However, studies of glass particles recovered from the lunar soil have shown that homogeneous glass bodies similar in size and shape to the microtektites can be produced by meteorite impact. Except for their composition and correspondingly higher refractive index and specific gravity, some of the lunar glass particles appear to be indistinguishable from microtektites.

Although controversy over the origin of the tektites has not been resolved with the discovery of the microtektites, new information about the Australasian strewn field must be explained in any hypothesis of tektite origin. For example, we now know that the Australasian strewn field extends nearly all the way across the Indian Ocean and is thus nearly twice as large as previously indicated. The total mass of microtektites is calculated to be 10^{14} grams. This calculation is about 100,000 times greater than previous estimates of the mass of tektite material in the Australasian strewn field. Also, chemical analyses of the microtektites have shown that tektite material in the Australasian strewn field has a wider range in composition than that indicated by published analyses of Australasian tektites (Table 3).

Leake [1970] used Niggli numbers in an attempt to compare Australasian microtektite data presented by *Cassidy et al.* [1969] with australite analyses and the trends given for igneous and sedimentary rocks and with glasses that have been vapor fractionated. He stated that there was a general consistency of trend from the tektites through the microtektites (making reasonable allowance for analytical scatter and sampling) but that one or more of the trends were inconsistent with a sedimentary origin, igneous fractionation, and vapor fractionation trends. Thus he concluded that unless a number of curious coincidences have occurred, the tektites and microtektites have a common history; what this history is remains an enigma.

CORRELATION WITH OTHER DEEP-SEA STRATIGRAPHIC BOUNDARIES AND REWORKING OF DEEP-SEA SEDIMENTS

Since the age of the Australasian microtektite horizon is well established (based on paleomagnetic stratigraphy and fission track dating), the horizon provides a chronostratigraphic boundary for much of the Indian Ocean including the area south of Australia. In addition, the microtektite horizon roughly correlates with several paleontologic boundaries.

Several radiolarian boundaries have been established in antarctic deep-sea sediments [*Hays*, 1965]. *Opdyke et al.* [1966] and *Watkins and Goodell* [1967] have shown that one of these boundaries (X-Ψ) is closely associated with the Brunhes-Matuyama reversal boundary. Also, there is a marked upward decrease in abundance of the foraminifera *Sphaeroidinella dehiscens* at or just above the Brunhes-Matuyama boundary [*Glass et al.*, 1967; *Hays et al.*, 1969; *Ruddiman*, 1971]. Although these boundaries have not been directly correlated with the Australasian microtektite horizon, their close association with the Brunhes-Matuyama boundary shows that they are apparently synchronous with the microtektite horizon.

Hays et al. [1969] found that below the Brunhes-Matuyama boundary there is a general drop in the level of carbonate content. They state that the generally higher concentration of calcium carbonate above the boundary suggests a cooler average climate, and they believe that the decrease in abundance of the tropical species *S. dehiscens* at the Brunhes-Matuyama boundary is further evidence of this cooling. Thus the Australasian microtektite horizon not only is closely associated with the Brunhes-Matuyama magnetic reversal boundary, but also is closely associated with at least two paleontologic boundaries.

The amount of reworking that has taken place since the microtektites fell (0.7 m.y. ago) is indicated by the vertical distribution of microtektites within the cores [*Glass*, 1969*c*] (Figure 2, Table 1). The microtektites are distributed through a vertical zone that ranges from 35 to 80 cm in length. This zone represents from 33,000 to possibly as much as 700,000 years of sedimentation based on the rates of deposition (the sedimentation rates are based on the depth of the Brunhes-Matuyama reversal boundary; see Table 1). In all but two of the cores (RC8-52 and RC8-53), there is a level of peak abundance. Scattered microtektites are found as high as 50 cm above this level, whereas microtektites are found only to a depth of 5–20 cm (10-cm average) below this level. Thus, on the average, the microtektites extend 2.5 times higher above this level than below it. This characteristic indicates that the first appearance of a fossil species provides a better chronostratigraphic horizon than the extinction of a species. In many of the cores, the upper half of the microtektite zone contains a larger percentage of large microtektites than the lower half of the zone. This occurrence suggests that the apparent time of extinction of two fossil species of different size might be significantly different because of reworking, even though these species became extinct at the same time.

SUMMARY AND CONCLUSION

The appearance, geographical distribution, age of deposition, fission track age, petrography, physical properties, and chemical compositions of microscopic glassy particles found in deep-sea sediments south of Australia and throughout much of the Indian Ocean and south of Japan in the Philippine Sea show that they are related to the Australasian tektites. We estimate that approximately 10^{14} grams of these glassy objects, called microtektites, fell on the ocean floor ~0.7 m.y. ago. The microtektites are distinct from other microforms of natural silica glass such as Pele's Tears and hay silica glass. The microtektite horizon appears to have the same age (0.7 m.y.) as Hays' (X-Ψ) boundary. A marked decrease in the abundance of the foraminifera *Sphaeroidinella dehiscens* also occurred about 0.7 m.y. ago. In addition, there seems to have been a general cooling of the climate about this time. There is a vertical dispersion of the microtektite horizon in every core in which the microtektites have been found. The average time span over which the microtektites are dispersed is greater than 100,000 years. This span indicates that the reworking of bottom sediments in the area south of Australia (and over much of the rest of the Indian Ocean) is significant and widespread.

Acknowledgments. The work reported in this article covers research carried out over the last 5 years. During those years I have received help and encouragement from numerous colleagues, mostly those at Lamont-Doherty Geological Observatory and at Goddard Space Flight Center (first when I was a captain in the U.S. Army Corps of Engineers and later as a National Research Council post-doctoral resident research associate). People who have been helpful in this research are acknowledged in my articles listed in the references. I would especially like to acknowledge the help of Drs. Bruce C. Heezen and N. D. Opdyke of Lamont-Doherty and Drs. J. A. O'Keefe and L. S. Walter of Goddard without whose assistance and encouragement this work would not have been done. My thanks to Dr. W. F. Ruddiman and Dr. J. R. Conolly who critically reviewed the manuscript.

REFERENCES

Apollo Soil Survey, Apollo 14: Nature and origin of rock types in soil from the Fra Mauro formation, *Earth Planet. Sci. Lett.*, *12*, 49–54, 1971.

Baker, C., Micro-forms of hay-silica glass and volcanic glass, *Mineral. Mag., 36,* 1012–1023, 1968.

Barnes, V. E., North American tektites, Contributions to geology, 2, *Publ. 3945,* pp. 477–582, Univ. of Tex., Austin, 1940.

Cassidy, W. A., B. Glass, and B. C. Heezen, Physical and chemical properties of Australasian microtektites, *J. Geophys. Res., 74*(4), 1008–1025, 1969.

Chao, E. C. T., I. Adler, E. J. Dwornik, and J. Littler, Metallic spherules in tektites from Isabela, Philippine Islands, *Science, 135,* 97–98, 1962.

Chao, E. C. T., J. A. Boreman, J. A. Minkin, and O. B. James, Lunar glasses of impact origin: Physical and chemical characteristics and geologic implications, *J. Geophys. Res., 75* (35), 7445–7479, 1970.

Chapman, D. R., and L. C. Scheiber, Chemical investigation of Australasian tektites, *J. Geophys. Res., 74*(27), 6737–6776, 1969.

Cox, A., Geomagnetic reversals, *Science, 163,* 237–245, 1969.

Durrani, S. A., and H. A. Khan, Ivory Coast microtektites: Fission track age and geomagnetic reversal, *Nature, 232,* 320–323, 1971.

Dymond, J., Age determinations of deep-sea sediments: A comparison of three methods, *Earth Planet. Sci. Lett., 6,* 9–14, 1969.

Fleischer, R. L., and P. B. Price, Fission track evidence for the simultaneous origin of tektites and other natural glasses, *Geochim. Cosmochim. Acta, 28,* 755–760, 1964.

Florenskiy, K. P., A. V. Ivanov, N. P. Ilyin, M. N. Petrikova, and L. Ye. Loscva, Chemical composition of meteoritic dust from the Tunguska fall and differentiation of material of cosmic bodies, *Geokhimiya,* no. 10, 1163–1173, 1969.

Frey, F. A., C. M. Spooner, and P. A. Baedecker, Microtektites and tektites: A chemical comparison, *Science, 170,* 845–847, 1970.

Gentner, W., D. Storzer, and G. A. Wagner, New fission track ages for tektites and related glasses, *Geochim. Cosmochim. Acta, 33,* 1075–1081, 1969.

Gentner, W., B. P. Glass, D. Storzer, and G. A. Wagner, Fission track ages and ages of deposition of deep-sea microtektites, *Science, 168,* 359–361, 1970.

Glass, B., Microtektites in deep-sea sediments, *Nature, 214,* 372–374, 1967.

Glass, B. P., Glassy objects (microtektites?) from deep-sea sediments near the Ivory Coast, *Science, 161,* 891–893, 1968*a.*

Glass, B. P., Microtektites and the origin of the Australasian tektite strewn field, *Publ. 6,* 22 pp., Center for Meteorite Stud., Ariz. State Univ., Tempe, 1968*b.*

Glass, B. P., Chemical composition of Ivory Coast microtektites, *Geochim. Cosmochim. Acta, 33,* 1135–1147, 1969*a.*

Glass, B. P., Silicate spherules from Tunguska impact area: Electron microprobe analysis, *Science, 164,* 547–549, 1969*b.*

Glass, B. P., Reworking of deep-sea sediments as indicated by the vertical dispersion of the Australasian and Ivory Coast microtektite horizons, *Earth Planet. Sci. Lett., 6,* 409–415, 1969*c.*

Glass, B. P., Zircon and chromite crystals in a Muong Nong-type tektite, *Science, 169,* 766–769, 1970.

Glass, B. P., Investigation of glass recovered from Apollo 12 sample 12057, *J. Geophys. Res., 76*(23), 5649–5657, 1971.

Glass, B. P., and B. C. Heezen, Tektites and geomagnetic reversals, *Nature, 214,* 372–374, 1967.

Glass, B. P., D. B. Ericson, B. C. Heezen, N. D. Opdyke, and J. A. Glass, Geomagnetic reversals and Pleistocene chronology, *Nature, 216,* 437–442, 1967.

Hays, J. D., Radiolaria and late Tertiary and Quaternary history of antarctic seas, in *Biology of the Antarctic Seas II: Antarctic Res. Ser.,* vol. 5, edited by G. Llano, pp. 125–184, AGU, Washington, D. C., 1965.

Hays, J. D., T. Saito, N. D. Opdyke, and L. H. Burckle, Pliocene-Pleistocene sediments of the equatorial Pacific: Their paleomagnetic, biostratigraphic, and climatic record, *Geol. Soc. Amer. Bull., 80,* 1481–1514, 1969.

Kirova, O. A., Scattered matter from the area of fall of the Tunguska cometary meteorite, *Ann. N. Y. Acad. Sci., 119,* 235–249, 1964.

Leake, B. E., Some paradoxes in Australasian microtektite compositional trends, *J. Geophys. Res., 75*(2), 349–356, 1970.

Martin, R., Tektites, impactites and volcanic glass, paper presented at 30th Annual Meeting of the Meteoritical Society, Moffet Field, Calif., October 25–27, 1967.

Opdyke, N. D., and B. P. Glass, The paleomagnetism of sediment cores from the Indian Ocean, *Deep Sea Res., 16,* 249–261, 1969.

Opdyke, N. D., B. P. Glass, J. D. Hays, and J. Foster, Paleomagnetic study of antarctic deep-sea cores, *Science, 154,* 349–357, 1966.

Ruddiman, W. F., Pleistocene sedimentation in the Equatorial Atlantic: Stratigraphy and faunal paleoclimatology, *Geol. Soc. Amer. Bull., 82,* 283–302, 1971.

Schmidt, R. A., Microscopic extraterrestrial particles from the Antarctic Peninsula traverse, *Res. Rep. Ser., G3-3,* 65 pp., Geophys. and Polar Res. Center, Univ. of Wis., Madison, 1963.

Senftle, F. E., A. N. Thorpe, and S. Sullivan, Magnetic properties of microtektites, *J. Geophys. Res., 74*(27), 6825–6833, 1969.

Shima, M., Glassy spherules (microtektites?) found in ice at Scott Base, Antarctica, *J. Geophys. Res., 71*(14), 3595–3596, 1966.

Urey, H. C., Origin of tektites, *Science, 137,* 746–747, 1962.

von Koenigswald, G. H. R., The relationship of shape, size and texture in Asiatic tektites, *Proc. Kon. Ned. Akad. Wetensch., Ser. B, 61,* no. 1, 1968.

Wagner, G. A., Altersbestimmungen an tektiten und anderen naturlichen glasern mittels spuren der spontanea kernspaltung des Urans-238, *Z. Naturforsch. A, 21,* 733, 1966.

Walter, L. S., Coesite discovered in tektites, *Science, 147,* 1029–1032, 1965.

Watkins, N. D., and H. G. Goodell, Geomagnetic polarity change and faunal extinction in the southern ocean, *Science, 156,* 1083–1087, 1967.

Zähringer, J., K-Ar measurements of tektites, in *Radio-active Dating,* p. 289, International Atomic Energy Agency, Vienna, 1963.

TURBIDITE SEDIMENTATION OFF THE ANTARCTIC CONTINENT

ROBERT R. PAYNE AND JOHN R. CONOLLY
Department of Geology, University of South Carolina, Columbia, South Carolina 29208

Abstract. The ocean floor off the antarctic continent south of Australia is characterized by a deep (300–500 meters) continental shelf covered with glacial marine sediment, a canyon-incised continental slope covered by gray muds and sands, and an extensive fan valley abyssal plain complex covering an area of 1000 by 1900 km (the South Indian plain). This abyssal plain is covered by terrigenous sediment up to 5 km thick along the continental margin; this sediment thins to sediment ponds (200–500 meters thick) filled with interbedded terrigenous clays and pelagic ooze along the southern flanks of the mid-oceanic ridge. Analyses of 34 piston cores taken from the USNS *Eltanin* in the continental margin–abyssal plain–distal pond region indicate that Pleistocene sedimentation is dominated by three major processes. The first process is the continual reworking and movement of terrigenous material down continental slope canyons; this material has been dropped by glacial streams or ice onto the continental shelf. The second process is the redeposition of this terrigenous material by turbidity currents in the rise fan valley complex and for hundreds of kilometers across the abyssal plain. The redisposition deposits consist of graded beds of terrigenous sand, silt, and clay mixed with reworked radiolarians and diatoms, and these deposits are interbedded with siliceous ooze. The third major process is the continual deposition of siliceous skeletons of planktonic organisms, the volume and eventual disposition of which are controlled by productivity in surface waters and deep ocean circulation. These pelagic oozes accumulate at rates of 2–5 cm/1000 yr over the southern flanks of the mid-ocean ridge. The analysis of closely spaced piston cores in the deep central part of the South Indian plain allows the correlation of turbidite and pelagic ooze sequences 1–3 meters thick over distances of more than 50 km. Sedimentation rates in this distal area of the South Indian plain are 1 cm/1000 yr for pelagic ooze and 5 cm/1000 yr overall. Such sedimentation rates account for the 500 meters of sediment that has been deposited during the past 20 m.y.

Few studies have been made of the sedimentation on the extensive continental margin surrounding the antarctic continent. Early studies by *Schott* [1939] and *Hough* [1950] indicated that the antarctic continent is surrounded by a zone of glacial marine sediment, presumably deposited by ice rafting and glacial outwash and consisting of poorly sorted muds or till. More recent studies by *Hough* [1956] and *Lisitzin* [1962] have shown that the glacial marine sediments merge seaward into diatom oozes. However, these studies have included only samples from the antarctic continental margin in the region south of Australia and none from the deep ocean basin or mid-ocean ridge flanks. *Von der Borch and Oliver* [1968] found that heavy mineral assemblages in shelf samples correlated roughly with heavy mineral assemblages in the shield rocks of the adjacent coast line. A few studies of sedimentation in the deep sea basins around the antarctic have been made [*Heezen et al.*, 1966; *Ewing et al.*, 1966], but to date the authors do not know of any studies on the South Indian plain.

CORE DESCRIPTIONS AND SAMPLING TECHNIQUES

Each core has been described (Table 1) with particular attention paid to significant color and lithologic and textural changes. Major lithologies have been examined with the microscope by using smear slides, and numerous selected samples have been separated into fractions of several sizes by using sieves and settling methods and have been examined in detail for both detrital and biogenic components.

Thin sections of epoxy-impregnated sand fractions were examined for most sandy terrigenous beds. Several long piston cores taken in the deep central (distal) portion of the South Indian abyssal plain have received special attention. Six cores, footnoted in Table 1, were examined at 10-cm intervals by using

TABLE 1. Location and Depths of USNS *Eltanin* Piston Cores Used in This Study

Core No.	Latitude S	Longitude E	Depth	Length, cm
34-18	60°00.0′	134°52.3′	4591	910
35-5	56°02′	128°10′	4573	860
35-17	58°40′	117°01′	4560	920
36-10	61°36.9′	140°14.6′	4276	559
37-2	65°15.5′	155°58.0′	3223	1056
37-4	64°49.7′	150°29.2′	3306	590
37-5	65°30.6′	147°23.0′	3018	660
37-6	66°04.9′	145°01.1′	199	114
37-7	65°01.1′	144°57.2′	3186	856
37-8	64°51.5′	142°25.0′	3065	260
37-9	65°33.1′	141°05.7′	1306	165
37-10	65°13.4′	137°52.8′	2261	580
37-11	64°31.2′	138°00.0′	3149	428
37-13	64°40.3′	132°58.6′	1332	152
37-14	63°55.1′	132°27.5′	3321	518
37-15	64°02.9′	130°15.2′	3306	282
37-16	63°58.2′	127°26.8′	3895	280
37-17	63°03.2′	127°06.0′	4314	495
37-18	60°00.7′	127°27.5′	4548	483
38-4	64°13.9′	150°03.8′	3524	746
39-29	55°05.8′	126°04.4′	4708	1626
39-31	57°37.0′	126°15.0′	4675	678
39-33*	57°29.9′	129°56.2′	4685	960
44-19*	57°59.6′	130°00.3′	4683	1020
44-20*	58°00.5′	130°03.8′	4683	1570
44-21*	58°03.5′	130°06.5′	4683	2232
44-22*	58°05.3′	130°08.2′	4683	2330
44-23*	58°06.2′	130°11.0′	4687	1920
45-44	58°28.5′	114°07.3′	4528	410
45-46	59°44.6′	114°56.8′	4493	201
45-47	60°45.3′	114°14.4′	4362	545
45-49	61°21.1′	113°44.9′	4305	430
45-50	59°25.2′	113°54.0′	4491	450
45-53	58°28.8′	113°55.2′	4528	450

Depth is given in corrected meters. Lengths given are for undisturbed part of core; flow-in is not included. In disturbed cores, lengths are estimated.

* Piston cores from deep central part of South Indian plain examined at 10-cm intervals.

smear slides. Sand, silt, and clay analyses were made on most of the turbidite beds in the six cores, and epoxy impregnations of sand fractions (62 μ) from 40 of the more sandy beds were made and examined. In addition, over 100 samples of 3–5 grams from pelagic zones were size fractionated, and counts of 200 individuals were made to determine the characteristic diatom assemblages. One core in this group (core 44-20) penetrated several meters of sediment before sediment entered the core; this core contains a different sequence of deposits from those found in neighboring cores. This difference will be discussed below. Preliminary results of this work have already been reported [*Payne et al.*, 1971*a*, 1971*b*, 1972].

GEOLOGY OF THE EAST ANTARCTIC COAST

Several isolated studies have been made of the relatively rare outcrops of Pre-Cambrian shield along the coast of East Antarctica [*Klimov and Solov'ev*, 1964; *Ravich*, 1964; *McLeod and Gregory*, 1967]. Because of the extensive pack ice and heavy snow cover, large outcrops are rare. However, from limited available data, most authors agree that the coast line in this area consists mainly of metamorphosed rocks of the granulite gneiss facies. Other rock types exposed include biotite schists, hypersthene amphibolites, charnockites, and granites [*Klimov et al.*, 1964*a*, 1964*b*]. *Evteev* [1964] has determined that valley glaciers with a zone of rock-laden ice 10–20 meters thick at their bases disgorge as much as 0.2 km^3 of glacial debris per kilometer of coast line per year. Ages of east antarctic shield rocks range from 700 to 1200 m.y. [*Craddock*, 1970].

SEDIMENTATION ON THE SHELF AND SLOPE

The lithology of piston cores (traverses A–A′ and B–B′, Figure 1) taken on the margins of the shelf and upper slope are shown schematically in Figure 2. Because of the difficulty of penetrating coarse sediments with the conventional piston corer, most of the cores are short. However, they show several distinctive features that characterize a glacial marine depositional environment.

Core 37-6, taken in 199 meters of water on the outer edge of the deep shelf, was mixed with water in the core barrel during recovery and shows a high percentage of rock fragments in its lower half including perthitic granite, gneiss, and biotite schist fragments along with sand rich in quartz (Table 2). Similarly, core 37-9 contains rock fragments. This core and core 37-13 are from the upper slope and contain poorly sorted to weakly laminated muds of an olive-gray color. Zones of high sand or rock fragments may represent increased ice rafting in a particular area.

Occasionally, cores were taken near submarine canyons cutting down through the slope. Core 37-10, taken near the west side of a canyon near 138°E, contains three graded beds of light olive-gray color composed dominantly of clays with laminated silts and thick graded sands at their bases. Below these is a poorly sorted muddy sand with many erratic pebbles. Core 37-16, taken on the lower slope to upper rise, was also disturbed during recovery, but its high sand

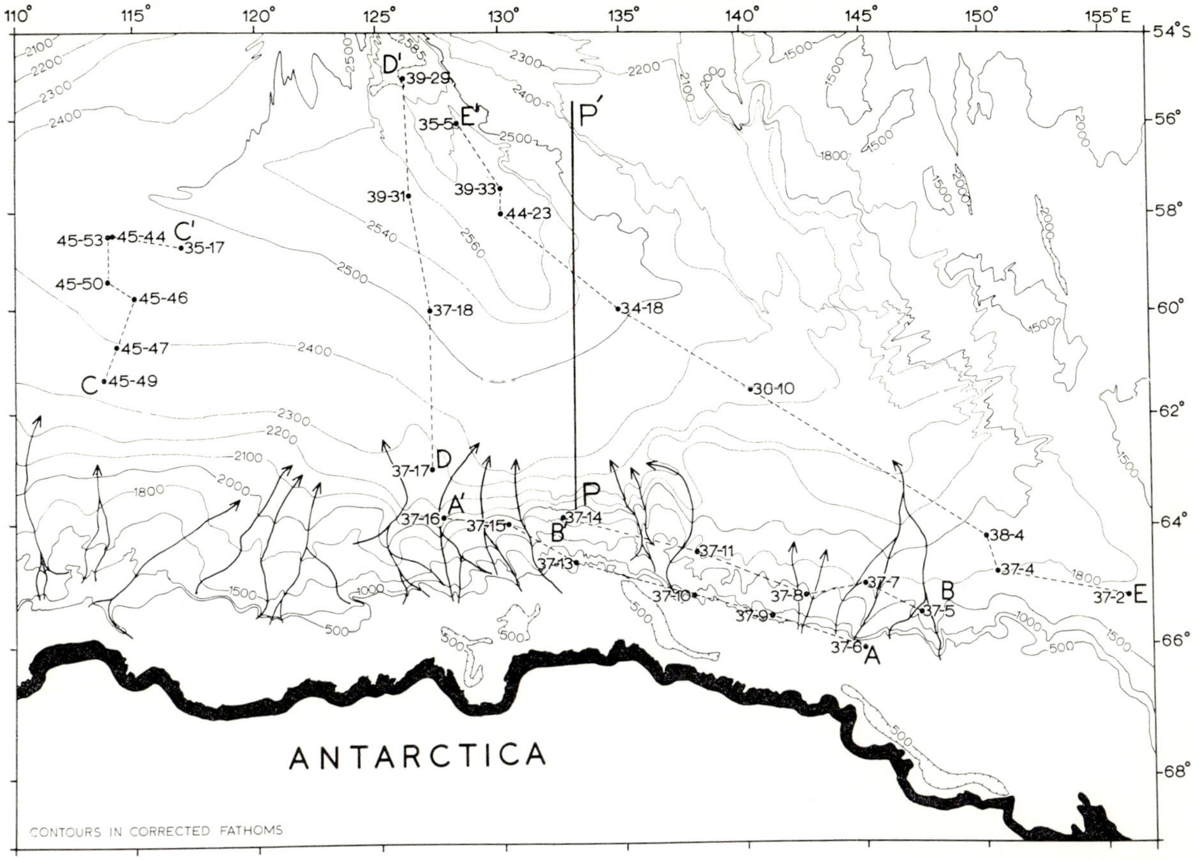

Fig. 1. East antarctic continental margin showing location of piston cores taken by the USNS *Eltanin* along the continental margin and within the South Indian abyssal plain. Arrows show major submarine canyons along the continental slope. Core locations are listed in Table 1. Traverses A–A′ and B–B′ are shown in Figure 2. Traverses C–C′, D–D′, and E–E′ are shown in Figure 5. Line P–P′ shows location of seismic profile used in Figure 9. Bathymetry after *Hayes and Conolly* [this volume].

and rock fragment content suggests that this core may have been taken from an active canyon channel.

The cores shown on traverse B–B′ (Figure 2) were all taken at similar depths from the lower slope. Whereas a few cores contain well-developed graded beds (such as core 37-5 as shown in Figures 2 and 3), most of the cores show typical slope muds with numerous dark silt laminae and small sand beds. Most contain zones rich in ice-rafted erratic pebbles. *Gordon* [1971*b*] and *Gordon and Tchernia* [this volume] have shown that Ross Sea bottom water moves westward between the Balleny Islands and Antarctica and sinks along the Adélie Coast. These bottom waters may redistribute sediments along regional contours. The occurrence of a thick nepheloid layer along the lower slope and rise substantiates this finding [*Eittreim et al.*, 1971, this volume].

One process that must be occurring on the slope is scour. A few cores (core 37-15 on traverse A–A′ and core 37-8 on B–B′) contain Tertiary firm dusky yellow clays near their tops, although neither of these cores are located in active canyons where scour might be expected.

Pelagic sediments are generally rare in these cores. Occasionally, as in core 37-7, reworked foraminifera occur in sand beds. Table 2 summarizes the mineralogy of some of the cores, and Figure 3 shows more detail of the sand, silt, and clay content of core 37–5. Figure 4 shows photomicrographs of some of the sands and rock fragments of slope and rise muds.

ABYSSAL PLAIN SEDIMENTATION

Cores from along three traverses (Figures 1 and 5) have been selected to show details of sedimentation from the lower rise and proximal abyssal plain to the distal sediment ponds 1000 km from the antarctic continent. *Houtz and Markl* [this volume] have found a large accumulation of sediment along the lower edge

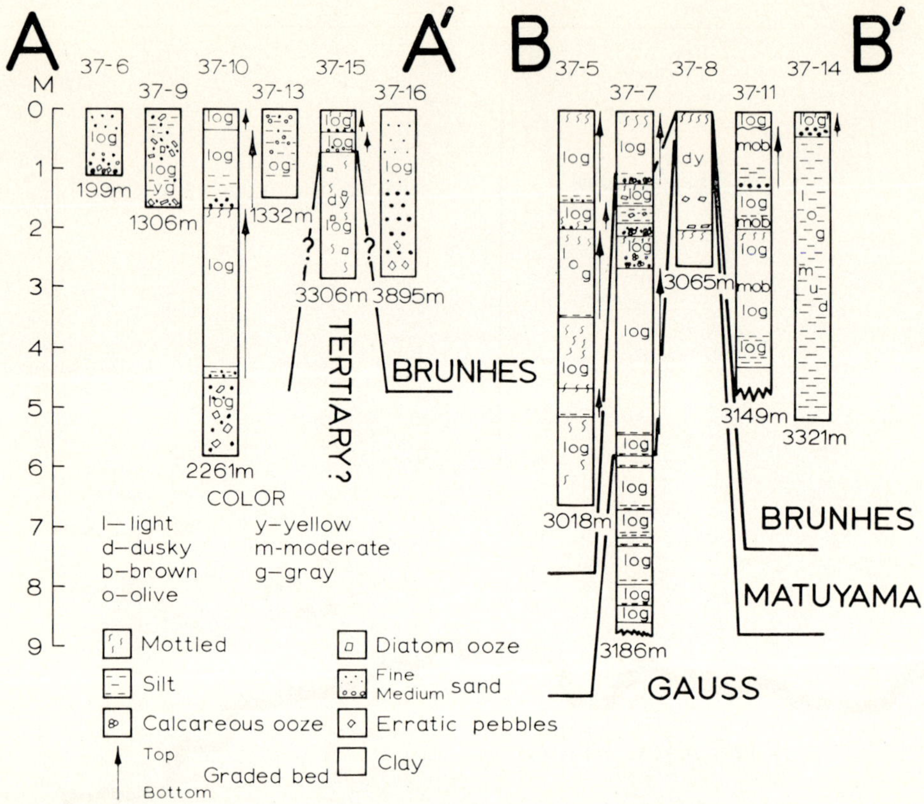

Fig. 2. Core profiles along two lines of section along the continental shelf and slope of East Antarctica. Profile locations are shown on Figure 1. Core depths are shown in corrected meters. Core lengths are in meters. Paleomagnetic stratigraphy after *Watkins and Kennett* [this volume].

of the antarctic continental margin from seismic studies. This accumulation, which is up to 5 km thick, thins northward to 200–500 meters in sediment ponds on the flanks of the mid-ocean ridge.

Extensive fan systems cover much of the lower rise off the antarctic continent, and their braided channels (Figure 1) feed sediment onto the gently sloping abyssal plain. Cores taken in the proximal areas of the abyssal plain (e.g., core 37-17 on traverse D–D′ and cores 37-2, 37-4, and 38-4 of traverse E–E′) contain graded sandy and silty clay turbidites with well-defined medium to fine sand layers and silt laminae at their bases. During periods when the influx of terrigenous material does not reach across the abyssal plain, accumulations of siliceous ooze are common. However, most of the ooze found in the cores is mixed with clay and shows evidence of transport as we find the co-occurrence of Tertiary and Recent diatom species [*Payne et al.*, 1972] and broken diatom frustules. The dominant feature of the proximal turbidites is their thick sandy and silty clay beds.

Traverse C–C′ (Figure 5) shows the sediments deposited in the western parts of the abyssal plain (Figure 1). Note that there is only a 250-meter increase in water depth from one end of this traverse to the other. Sediments found in the cores are all very similar. Clay turbidites 50–150 cm thick are the dominant deposits. Their bases often contain laminated silts or fine sands with sharp lower contacts. Only the upper 10–20 cm of the turbidites are disturbed by burrowing organisms, and their dominant color is light olive gray. Clays are illites that are rich in quartz, and the sands and silts are always well sorted with quartz and albite, the most abundant minerals present. Common accessory minerals are green hornblende and glauconite. Diatom and radiolarian oozes are found atop many of the terrigenous beds and are commonly burrow mottled throughout. Color changes in the ooze zones may reflect changes in climate [*Conolly and Payne*, this volume].

Turbidites were cored at several stations in the deep central (or distal) portions of the South Indian

TABLE 2. Sand Petrology

Core No.	Sample Depth, cm	Quartz	Feldspar	Rock Fragments	Opaques	Glauconite	Spicules	Radiolaria	Foraminifera
				Continental Slope and Rise					
37-6	50	73	21	2	2	2	...	...	...
37-6	100	72	14	11	2	1	...	...	...
37-10	32	59	31	4	4	2	...	...	...
37-10	500	54	19	26	1	...	...	...	...
37-10	570	48	28	22	2	...	...	...	...
37-16	50	54	20	16	8	2	...	...	...
37-14	31	52	11	15	15	5	2	...	...
37-15	40	32	6	3	4	2	3	...	50
37-15	65	32	7	5	4	1	2	...	49
				Abyssal Plain					
36-10	500	67	17	10	4	1	1	...	...
44-19	910	64	18	8	9	1	...	...	...
44-20	260	55	13	2	8	1	1	20	...
44-20	300	60	17	4	3	1	...	15	...
44-20	320	53	18	1	11	1	3	13	...
44-20	340	56	18	2	5	1	...	18	...
44-20	790	50	27	3	6	1	...	13	...
44-20	820	50	19	2	6	1	...	22	...
44-20	840	42	23	1	10	1	2	21	...
44-21	1060	34	15	2	13	...	3	33	...
44-21	1070	24	12	4	4	...	6	50	...
44-21	1340	50	16	2	8	2	2	20	...
44-21	1460	48	25	3	9	1	...	14	...
44-21	1670	52	35	3	7	2	...	1	...
44-22	790	50	22	1	10	1	2	14	...
44-22	970	50	19	1	13	1	...	16	...
44-22	1100	49	23	2	7	...	1	18	...
44-23	672	4	1	...	2	...	...	93	...
44-23	910	17	6	4	6	...	10	57	...
44-23	920	32	17	5	3	...	3	40	...
44-23	1243	46	23	3	9	1	2	16	...
44-23	1580	55	34	3	4	1	1	2	...
44-23	1600	52	28	2	15	1	1	1	...

abyssal plain. Cores 39-31 and 39-29 on traverse D–D′ show typical deposits. Terrigenous material is deposited in homogeneous soupy clay beds of light olive-gray to yellowish-gray color. These beds always have sharp basal contacts, and only a few contain very fine sand or silt at their bases. The tops of the beds are often burrowed for 10–20 cm. Thicker pelagic zones are found, and these are burrow mottled as well. Paleomagnetic stratigraphy [*Watkins and Kennett*, this volume] shows 12 meters of sediment deposited in core 39-29 in the past 0.7 m.y.

Details of Distal Abyssal Plain Sedimentation

Core 44-23 is one of five long cores taken at a drift station in the deepest part of the abyssal plain by the USNS *Eltanin* [*Gordon*, 1971*a*]. A sixth core (39-33) was taken on an earlier cruise at a station 55 km to the north. Figure 6 shows lithologic changes in the cores based on megascopic descriptions. These cores, representing 100 meters of cored sediment, contain clay turbidites interbedded with burrow-mottled siliceous oozes. Several features characterize the terrigenous beds.

1. Terrigenous beds thicker than 20 cm are marked by sharp basal contacts. (Beds thinner than this measurement may be reworked by burrowing organisms.)

2. Detritus grades from coarse silt or fine sand (50–300 μ) at the base upward into clay.

3. The upper 10–20 cm is generally burrow mottled.

4. In general, beds 40–50 cm thick are composed of silt and fine sand-sized detritus with low clay content (10–40%), whereas thicker beds (up to 2.5 meters) are dominantly composed of clay with only minor (less than 5%) well-sorted silt at their bases.

5. The upper portions either merge upward into burrow-mottled siliceous oozes containing clay

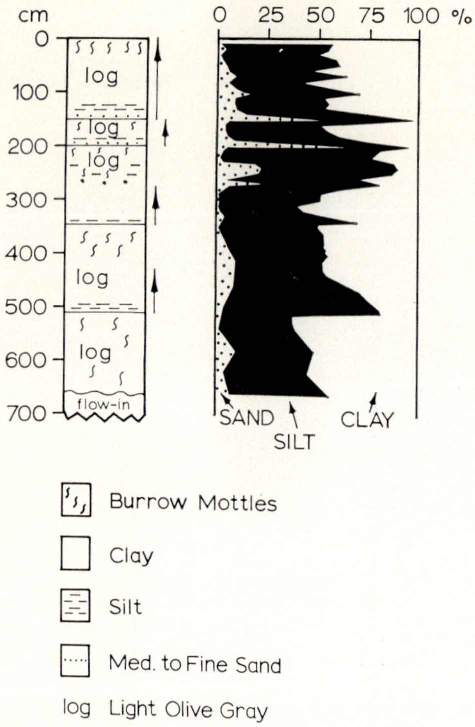

Fig. 3. Analysis of *Eltanin* core 37-5 from the continental slope off Antarctica. Core location is shown on Figure 1.

worked up into these oozes from below by benthonic burrowers, or are truncated sharply by younger turbidite beds.

6. Low amounts (5–10%) of reworked siliceous microfossils are generally incorporated in these deposits, and many show evidence of transport, such as a high percentage of breakage, relatively high abundances of species associated with the shallow waters of the antarctic shelf and ice pack, and mixtures of species from Tertiary and Recent biostratigraphic zones.

Pelagic sediment interbedded with the turbidites is quite variable in both color (a reflection of climate change [*Conolly and Payne*, this volume]) and thickness (a function of the influx of terrigenous material). Several features characterize these siliceous oozes.

1. They are composed of diatom and radiolarian skeletons (90–95%) with only very minor detrital silt or clay. Most of the South Indian abyssal plain lies below the carbonate compensation depth; planktonic foraminifera, therefore, are absent.

2. These oozes are generally burrow mottled, but occasionally contain laminations caused by local reworking by bottom currents.

3. Contacts with underlying units are always gradational, whereas contacts with overlying beds are sharp only if overlain by terrigenous beds.

4. Thin homogenous beds consisting of very well sorted and often broken diatom and radiolarians found below the base of some terrigenous beds may represent a reworked fossil pavement similar to those described by *Kuenen* [1964], e.g., core 44-23 at 672 cm (Figures 6 and 8*e*).

The original megascopic description of the six cores allowed this subdivision into terrigenous and pelagic zones as outlined above, but it was obvious that individual turbidite events or even pelagic intervals could not be recognized in adjacent cores. To make these correlations, closer examination was required. Smear slides taken at 10-cm intervals were examined systematically and relative proportions of five major terrigenous (sand, silt, and clay) and pelagic (diatoms and radiolarians) constituents were recorded. All these data were used to construct Figure 7. Diatom assemblages were examined from pelagic zones at selected intervals to help establish correlative pelagic sequences.

Correlations of Distal Turbidites

The megascopic descriptions revealed differences in the core tops that became more clear with the detailed examination of the sediment. Core 44-20, which contained a thick ooze bed at the top, had a series of sandy turbidites that appeared very different from the uppermost clay turbidite found in the balance of the cores. Core 44–20, therefore, is shown in a somewhat lower position on Figure 7 than the rest of the cores.

Features that were found in most of the cores are summarized at the right of Figure 7. At the top of all but two of the cores is a zone of Recent siliceous ooze

---

Fig. 4. Photomicrographs of continental slope and rise sediments off Antarctica. (*a*, *b*) These photomicrographs are from core 37-9 at 70 cm showing biotite schist pebble. (*c*, *d*) These are from core 37-14 at 30 cm. (*e*) This is from core 37-15 at 68 cm. Photomicrographs *d* and *e* show basal sands from proximal turbidites. Photomicrographs *a*, *c*, and *e* are in plane light; *b*, *d*, and *f* are under crossed nicols. Upper bar scale for *a* and *b*; lower bar scale for *c*, *d*, *e*, and *f*. Core locations are given in Table 1.

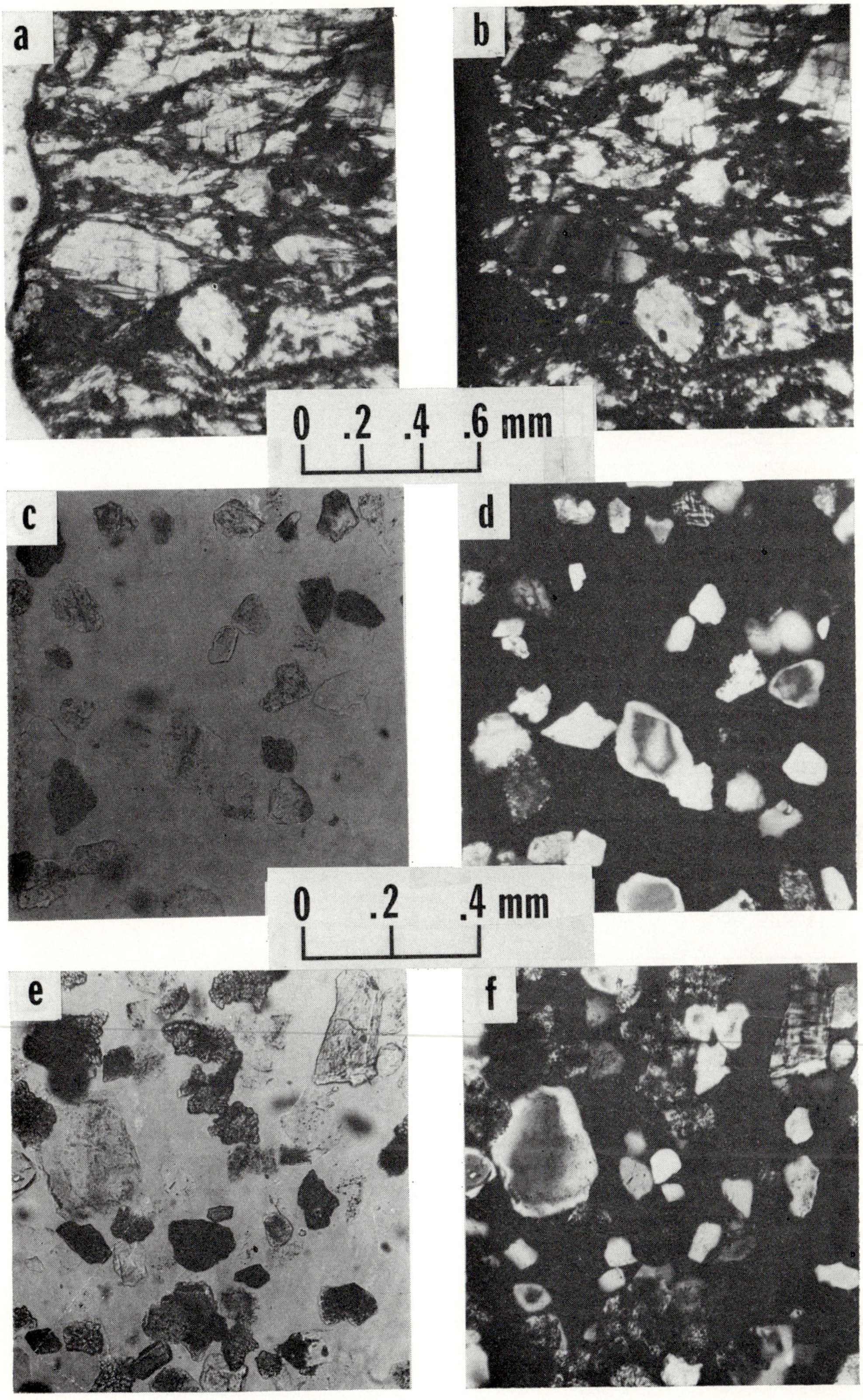
a
b
0 .2 .4 .6 mm
c
d
0 .2 .4 mm
e
f

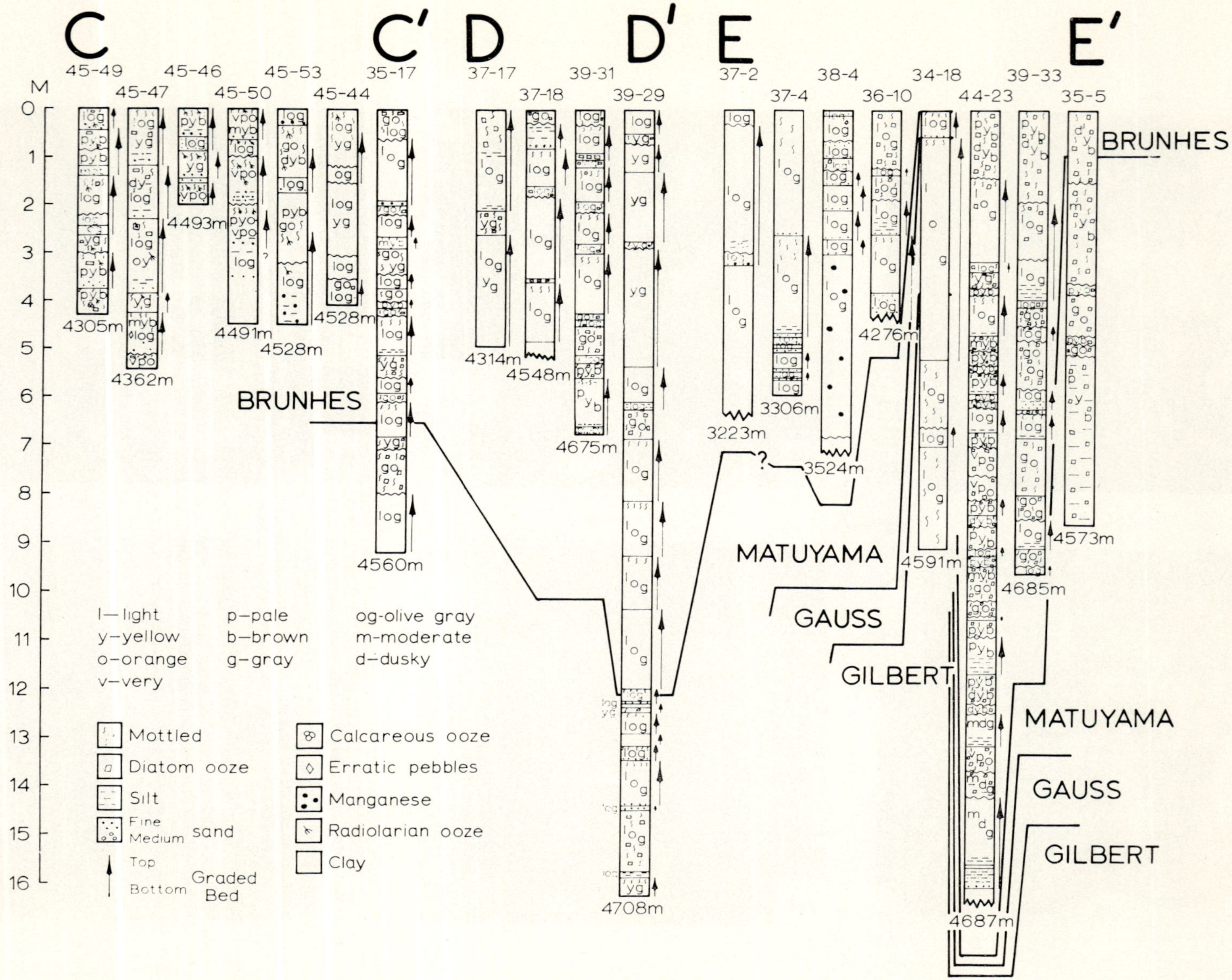

Fig. 5. Core profiles along three lines of section across the South Indian abyssal plain off Antarctica. Profile locations are shown on Figure 1. Core depths are shown in corrected meters. Core lengths in meters. Paleomagnetic stratigraphy after *Watkins and Kennett* [this volume].

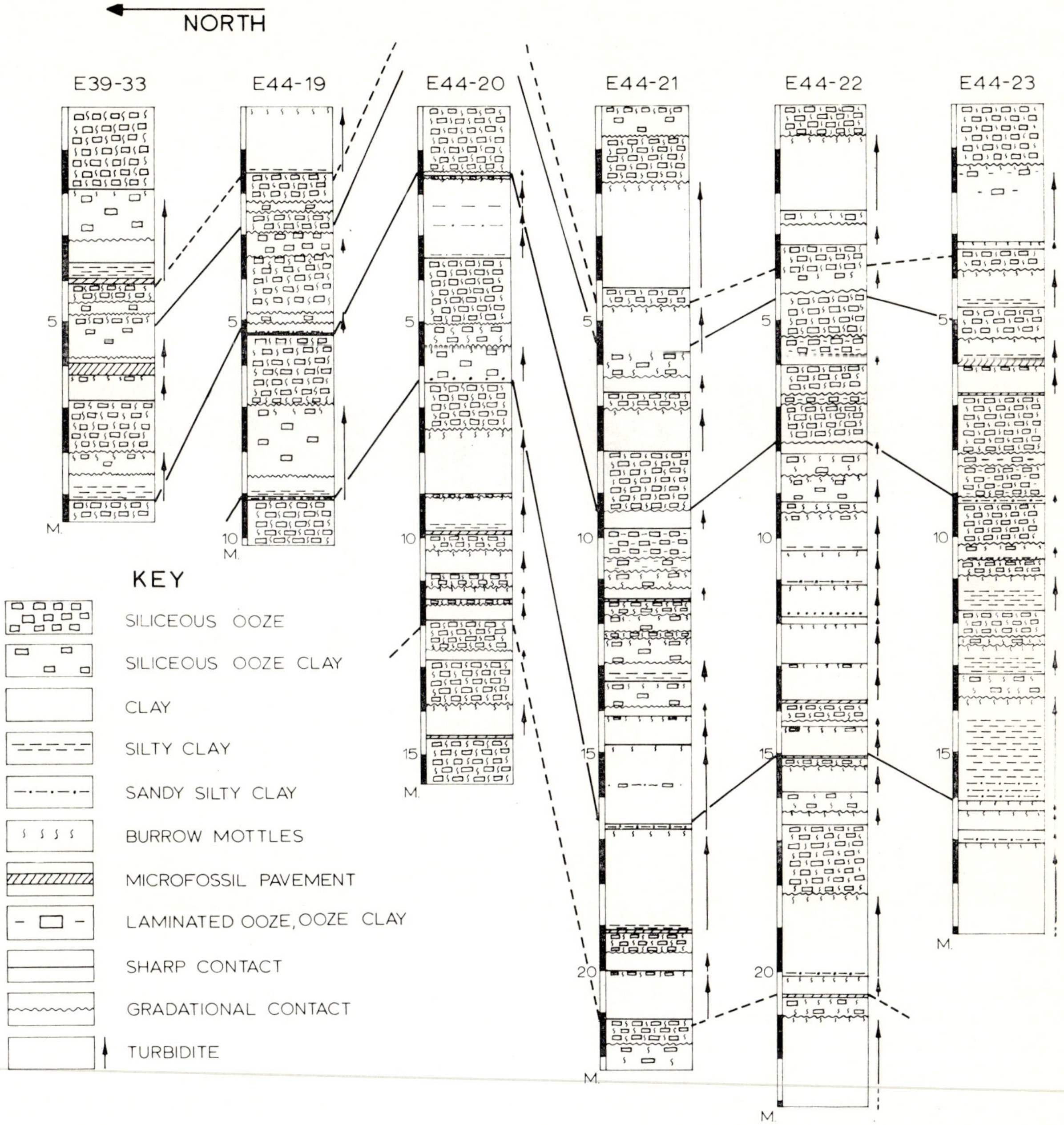

Fig. 6. Lithology of six *Eltanin* cores from the deep central part of the South Indian abyssal plain. Locations of cores 39-33 and 44-23 are shown in Figure 1. Locations of cores 44-19, 44-20, 44-21, and 44-22 are shown in Table 1. Lines between cores are boundaries shown on Figure 7 and described in text. Core lengths are in meters.

0.5–1 meter thick. Below this zone is an easily distinguished zone of clay turbidite (bed A) 2 meters thick of light olive-gray color with a small percentage of silt at its base. A diatom zone (A′) 30–50 cm thick underlies this clay bed. Another light olive-gray clay bed with a low percentage of basal silt follows (bed B) and marks a major change in lithology. Below this distinctive sandwich of clay beds and ooze lies a sequence of clayey ooze and ooze beds of mixed gray color. These beds are generally thin (10–40 cm) and lack sand. Contacts are often burrowed; this burrowing makes correlation of discreet horizons from core to core difficult. This sequence is roughly 3.5 meters thick in most of the cores, and the lower

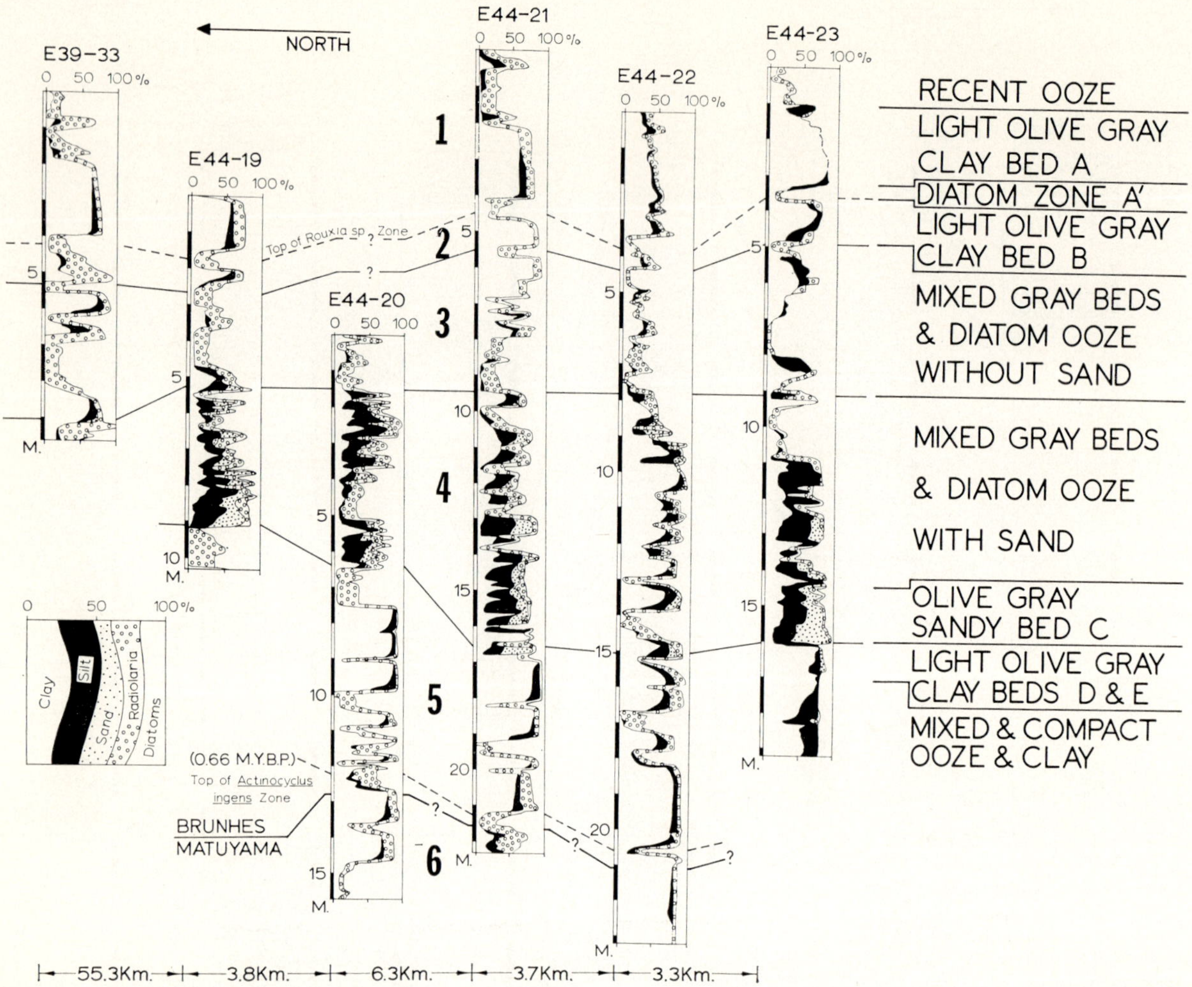

Fig. 7. Detailed microscopic and floral analyses of the six cores shown on Figure 6 showing correlations based on lithology and diatom assemblages as described in text.

1.5 meters form the uppermost sediment of the problematic core 44–20.

Below this zone of mixed gray beds and ooze lies a similar sequence with one major exception: the terrigenous beds are thinner and contain fine sands and much more silt. Many of these thin sandy silty turbidites can be recognized in individual cores (e.g., in core 44-22 from 10–13 meters; see Figure 6), but they cannot be correlated with confidence from core to core. This zone thins to the north from about 6 meters in cores 44-21, 44-22, and 44-23 to about 3.5 meters in core 44-19.

A thick olive-gray to medium dark-gray sandy turbidite zone (bed C) lies next in the sequence. This bed with its high sand content and distinctive color marks the lowermost occurrence of sand in all but one of the cores, the most southerly, core 44-23. There appears to be a general decrease in the thickness and in the silt content of this bed to the north.

Below this distinctive bed lies a pair of clay turbidites, beds D and E, which are most easily recognized in cores 44-20, 44-21, and 44-22. These beds are represented by a pair of sandy and silty beds in core 44-23 and apparently were not cored in core

Fig. 8. Photomicrographs of sediment from distal abyssal plain turbidite cores. (*a*, *b*) These photomicrographs are from core 44-19 at 910 cm showing basal sand. (*c*, *d*) These are from core 44-23 at 1580 cm (bed C of Figure 7) showing basal sand. (*e*) This is from core 44-23 at 672 cm showing radiolarian pavement (see Figure 6). Bar scale for all photographs. Photomicrographs *a*, *c*, and *e* are in plane light; *b* and *d* are under crossed nicols. Core locations are given in Table 1.

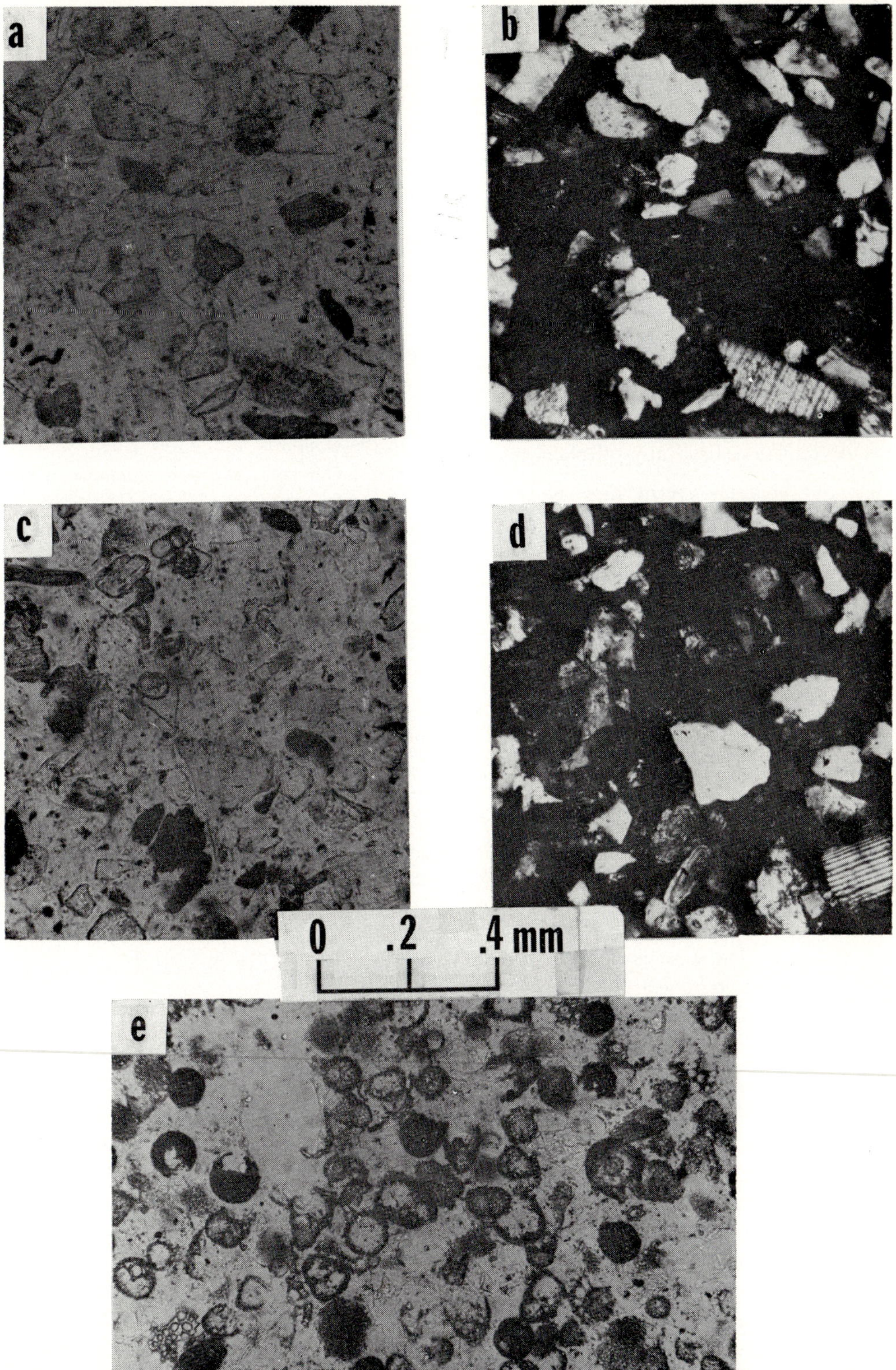
a
b
c
d
0 .2 .4 mm
e

44-19. Below this pair of light olive-gray beds, the sediment in the cores becomes more compact and consists of dark-brown diatom clays and oozes.

To augment the lithologic correlations, diatom assemblages from pelagic zones were examined. Three major diatom assemblage zones, correlative with zones established in the South Pacific by *Donahue* [1970], were identified. From youngest to oldest, these zones are the *Nitzschia kerguelensis* zone, the *Rouxia* sp. zone, and the *Actinocyclus ingens* zone [Abbott, 1971]. Tentative ages for the bases of the upper two zones are 0.35 and 0.66 m.y. B.P. These time horizons are shown as broken lines on Figure 7.

On the basis of the recognizable lithologic zones and the additional correlations of diatom assemblage zones, six major units, vertically aligned on Figure 7, were delineated for these distal abyssal plain cores.

Unit 1. This unit contains recent ooze and clay bed A.

Unit 2. The top of this unit is defined by the base of the *Nitzschia kerguelensis* zone and includes ooze A′ and clay bed B.

Unit 3. Mixed gray turbidites and ooze with no sand make up this unit. The top is defined by a prominent diatom-radiolarian ooze.

Unit 4. Thin sandy silty terrigenous beds of gray color with thin diatom oozes compose this unit. Most contacts are burrowed. Sandy bed C is included. The top of the zone is marked by the uppermost sandy bed, and the base is marked by bed C. Photomicrographs of some of the sands in this interval are shown in Figure 8.

Unit 5. Olive-gray clay beds and dark-colored ooze make up this unit. The base is defined by the top of the *Actinocyclus ingens* zone.

Unit 6. Compact dark diatom oozes and firm clays make up this unit.

The examination of smear slides showed diatoms mixed in with the clays of the turbidite beds. Many of these diatoms are species that are indigenous to shallow water and colder environments or from older (Tertiary) biostratigraphic zones. Some of these diatoms may have been scoured from areas within the abyssal plain itself (e.g., from the core 34-18 area) or from areas adjacent to the western flanks of the ridge in the center right of Figure 1 [*Conolly and Payne*, this volume].

At several places in the cores, thin (5–10 cm) homogenous diatom and radiolarian 'oozes' were found under the sandy or silty base of an overlying turbidite (e.g., 500 cm in core 44-23). Analysis of these narrow beds and the incorporation of radiolarian skeletons in the lower portions of terrigenous beds suggest that these may be microfossil pavements similar to those described by *Kuenen* [1964].

The most recent data to be added to Figure 7 are the paleomagnetic stratigraphies (N. D. Watkins, personal communication, 1971). Cores 39-33, 44-19, and 44-23 are all normal, whereas the Brunhes-Matuyama boundary in core 44-20 is at 1277 cm. The same boundary is not clearly evident in the other two cores.

Sediments of the Distal Ponds and Lower Ridge Flanks

Cores 39-29 and 35-5 of traverses D–D′ and E–E′ (Figures 1 and 5) represent typical sediment columns of the distal ponds and lower ridge flanks. Core 39-29, a 16-meter core, shows a sequence of slick yellowish-gray to light olive-gray clay turbidites with rare traces of silt or sand at the bases of these turbidites. Like the mud beds in the cores from the deepest part of the abyssal plain discussed above, these beds have sharp basal contacts and show only burrowing in the upper 10–20 cm. Diatoms are commonly incorporated in the upper two thirds of the beds in low amounts (5–15%). Sedimentation rates are high but do not appear to be as high as rates farther south.

Core 35-5 represents a typical lower ridge flank ooze. Taken from a local high at the distal abyssal plain margin, this core is dominantly a burrow-mottled siliceous ooze but contains up to 15% clay. Nephelometer measurements in the South Indian basin [*Eittreim et al.*, 1971] show a layer up to a kilometer thick in most of the basin. Hemipelagic clays from this layer may be incorporated in these lower ridge flank oozes.

SEDIMENT ACCUMULATION RATES AND TOTAL SEDIMENT THICKNESS

Paleomagnetic stratigraphy [*Watkins and Kennett*, this volume] has been included for all cores where it has been determined. Unfortunately, most cores taken on the antarctic shelf and slope are too short to penetrate the Brunhes sediment column. In a few cores (cores 37-7, 37-8, and 37-15), erosion by bottom currents (or low deposition rates) has allowed penetration of older sediment. We believe, based on seismic data by *Houtz and Markl* [this volume], that shelf-slope sedimentation rates are quite high because of the great thickness of sediment off this continental margin.

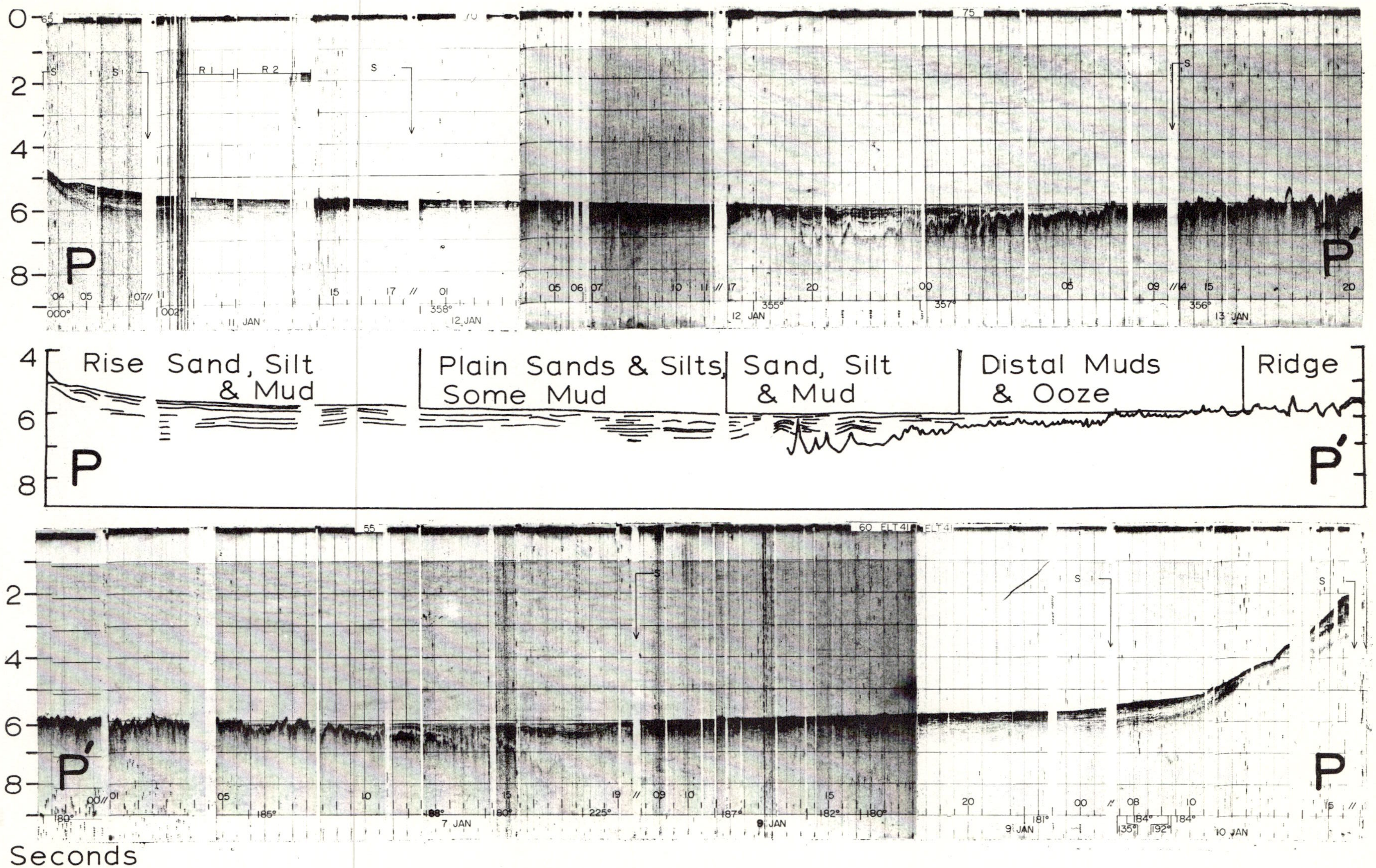

Fig. 9. Seismic profiles from *Eltanin* cruise 41 across the South Indian abyssal plain. Line of profile is shown on Figure 1.

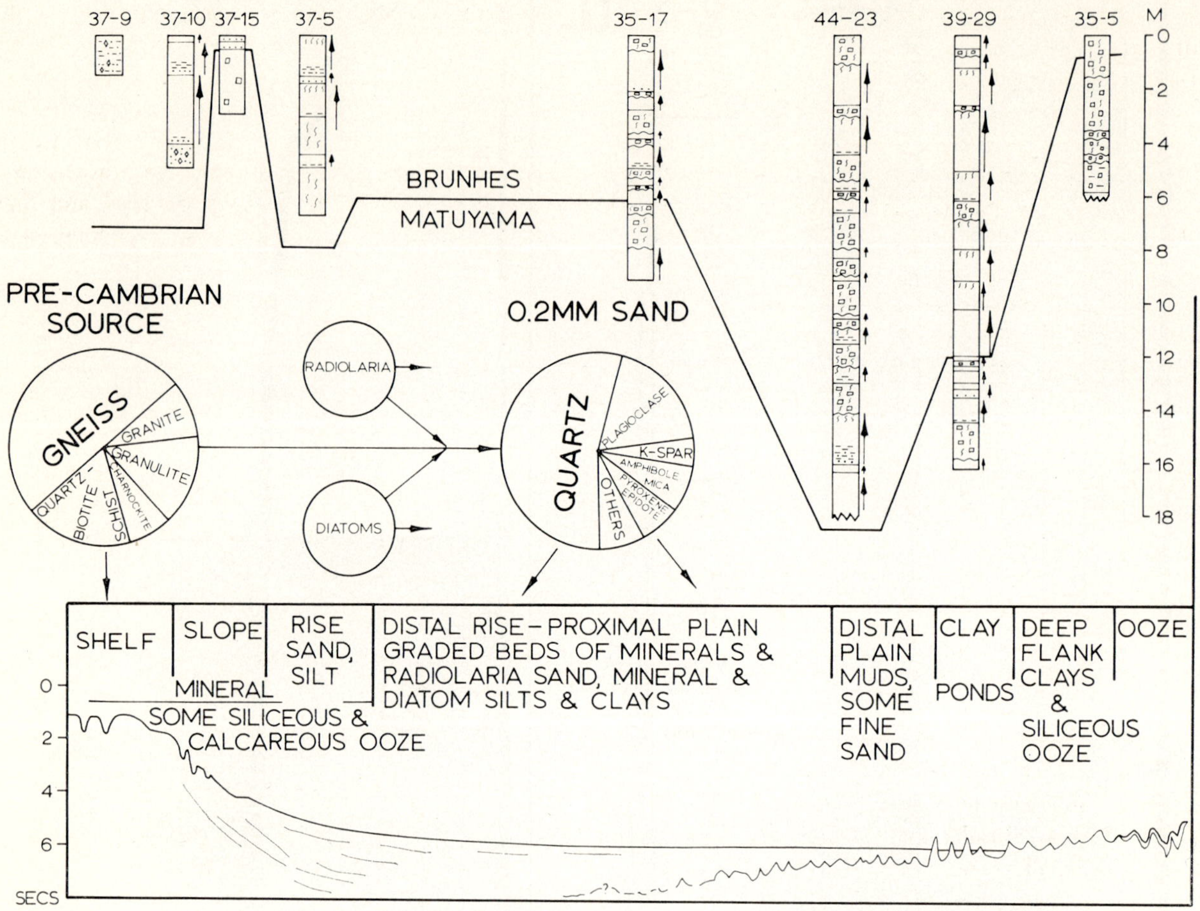

Fig. 10. Model showing sedimentation patterns across the South Indian abyssal plain.

In the proximal and distal abyssal plain, sedimentation rates are more reliably established. In core 35-17, the rate is less than 1 cm/1000 yr, whereas rates for cores 44-21, 44-22, or 44-23 are on the order of 5 cm/1000 yr. Distal pond rates are 1.8 cm/1000 yr (core 39-29). Rates in the lower ridge flanks drop appreciably to less than 1 cm/1000 yr [*Conolly and Payne*, this volume, Figure 14].

REGIONAL PROFILE

A seismic profile from *Eltanin* cruise 41 along 133°E (Figure 1) is shown in Figure 9. A line drawing interpretation of the major reflectors is also shown. In the original profile, numerous thin reflectors cross the rise and proximal abyssal plain and merge rather abruptly (near the middle of the upper profile) into much thicker reflecting horizons visible above the basement reflector. From our knowledge of sediments revealed in piston cores, we believe this boundary change represents the basinward limit of major sand-carrying turbidity currents. Farther away from Antarctica are the turbidite muds rich in clay of the distal abyssal plain.

MINERALOGY AND PROVENANCE OF ABYSSAL PLAIN SANDS

Table 2 summarizes the mineralogy of sand samples from cores in the study area. In coarse-grained sandstones, the rock fragments are mostly gneissic and schistose varieties, vein quartz, and other polycrystalline quartz grains and feldspar. Fine-grained sandstones (0.2-mm modal grain size) consist mainly of quartz and feldspar with relative high (5–40%) percentages of heavy and/or opaque minerals. These minerals consist mainly of amphibole and biotite with minor amounts of hypersthene, clinopyroxene, and epidote. The sand composition remains the same for sands sampled from all physiographic provinces, and this consistency indicates that the source area is the shelf and adjacent hinterland.

On the rise and abyssal plain, sands rich in minerals are mixed with radiolarians and sponge spicules because of the reworking of slope siliceous oozes. Fluctuations in sand composition are caused by variation in the admixture of these reworked radiolarian skeletons. 'Rock fragments' in abyssal plain sands are dominantly clay pellets that failed to disaggregate during sample preparation. Photomicrographs of typical sands from the abyssal plain are shown in Figure 8. A microfossil pavement from 672 cm in core 44–23 (Figures 6 and 8*e*) contains over 90% reworked radiolarian skeletons (Table 2).

PATTERNS OF SEDIMENTATION IN THE SOUTH INDIAN PLAIN

Figure 10 schematically depicts the patterns found in a profile across the South Indian plain from the antarctic continental margin to the ridge flank. The shelf, slope, and rise are characterized by poorly sorted gray mineral muds (core 37-9) and sands that are often graded, as well as laminated mineral silts and sands in mud that show evidence of reworking by bottom currents. In some places on the continental margin, cores penetrate calcareous and siliceous oozes that may be locally reworked in beds.

The distal rise–proximal abyssal plain contains well-developed, graded-bed sequences. The total amount of sand decreases to 10–20% of the sediment volume, whereas the distal abyssal plain sediments consist dominantly of turbidite muds with less than 1% fine sand. Interbedded with these distal turbidite muds are siliceous oozes.

The distal ponds of the lower ridge flanks are characterized by clay beds interbedded with siliceous oozes. These clay beds are missing from the deep ridge flanks that contain siliceous oozes with only minor admixtures of clay.

The general pattern of sedimentation is a simple one of decreasing grain size away from the antarctic continent, the primary sediment source. Large percentages of sand (20%) are only found on the continental slope, rise, and proximal abyssal plain. This finding means that the remainder of the deeper sea floor receives only silt and clay that can be deposited during single depositional episodes (as discrete beds with burrow-mottled upper portions), or that can be deposited from the sediment-carrying (nepheloid layers) bottom waters.

Note that individual depositional events now preserved as thick (up to 2 meters) clay beds can be correlated over distances of up to 55 km across the central abyssal plain. These events represent an interruption of ordinary pelagic sedimentation. About 20 such turbidity currents have deposited clay beds in the central South Indian abyssal plain in the last 700,000 years. These events also tend to cluster together; the clustering suggests that these events may be triggered at times of lowered sea level and that, under contemporary conditions, extensive slumping is probably a rare occurrence.

Acknowledgments. The authors express thanks to Dr. Larry Frakes and Mr. Dennis Cassidy of Florida State University for help in sampling the cores, to Mr. Daniel Grinnell and Mr. Steven Alcorn for sand analyses of some of the continental margin cores, to Mr. William Abbott for the diatom investigations of the distal abyssal plain cores, to Mr. Samuel Salley for drafting many of the figures, and to Mrs. Malcolm Galloway for typing and editorial assistance. This research was funded by National Science Foundation grant GV-28803 for work on antarctic cores.

REFERENCES

Abbott, W. H., Diatom investigations of southern ocean deep-sea cores, *Antarctic J. U.S.*, *6*(5), 171–172, 1971.

Conolly, J. R., and R. R. Payne, Sedimentary patterns within a continent–mid-oceanic ridge–continent profile: Indian Ocean south of Australia, in *Antarctic Oceanology II: The Australian–New Zealand Sector, Antarctic Res. Ser.*, vol. 19, edited by D. E. Hayes, pp. 295–315, AGU, Washington, D. C., 1972.

Craddock, C. (Ed.), Geologic maps of Antarctica, *Antarctic Map Folio Ser.*, folio 12, Amer. Geogr. Soc., New York, 1970.

Donahue, J. G., Diatoms as Quaternary biostratigraphic and paleoclimatic indicators in high latitudes of the Pacific Ocean, Ph.D. thesis, Columbia University, New York, 1970.

Eittreim, S., A. L. Gordon, M. Ewing, E. M. Thorndike, and P. Bruchhausen, The nepheloid layer and observed bottom currents in the Indian-Pacific antarctic sea, in *Studies in Physical Oceanography—A Tribute to Georg Wüst on His 80th Birthday*, edited by A. L. Gordon, Gordon and Breach, New York, in press, 1971.

Eittreim, S., P. M. Bruchhausen, and M. Ewing, Vertical distribution of turbidity in the South Indian and South Australian basins, in *Antarctic Oceanology II: The Australian–New Zealand Sector, Antarctic Res. Ser.*, vol. 19, edited by D. E. Hayes, pp. 51–58, AGU, Washington, D. C., 1972.

Evteev, S. A., Determination of the amount of morainic material carries by glaciers of the East Antarctic coast, in *Soviet Antarctic Expedition, Information Bulletin*, vol. 2, pp. 7–9, Elsevier, Amsterdam, 1964.

Ewing, M., J. I. Ewing, R. E. Houtz, and R. Leyden, Sediment distribution in the Bellinghausen basin, *Symposium on Antarctic Oceanography*, pp. 89–100, Scott Polar Research Institute, Cambridge, 1966.

Gordon, A. L., Eltanin cruise 44, *Antarctic J. U.S.*, *6*(1), 16–18, 1971*a*.

Gordon, A. L., Spreading of antarctic bottom waters, 2, in *Studies in Physical Oceanography—A Tribute to Georg Wüst on His 80th Birthday*, edited by A. L. Gordon, Gordon and Breach, New York, in press, 1971*b*.

Gordon, A. L., and P. Tchernia, Waters of the continental margin off Adélie Coast, Antarctica, in *Antarctic Oceanology II: The Australian–New Zealand Sector, Antarctic Res. Ser.*, vol. 19, edited by D. E. Hayes, pp. 59–69, AGU, Washington, D. C., 1972.

Hayes, D. E., and J. R. Conolly, Morphology of the southeast Indian Ocean, in *Antarctic Oceanology II: The Australian–New Zealand Sector, Antarctic Res. Ser.*, vol. 19, edited by D. E. Hayes, pp. 125–145, AGU, Washington, D. C., 1972.

Heezen, B. C., M. Tharp, and C. D. Hollister, Illustrations of the marine geology of the southern ocean, *Symposium on Antarctic Oceanology*, pp. 101–109, Scott Polar Research Institute, Cambridge, 1966.

Hough, J. L., Pleistocene lithology of antarctic ocean bottom sediments, *J. Geol.*, *58*(3), 254–260, 1950.

Hough, J. L., Sediment distribution in the southern oceans around Antarctica, *J. Sediment. Petrology*, *26*(4), 301–306, 1956.

Houtz, R. E., and R. G. Markl, Seismic profiler data between Antarctica and Australia, in *Antarctic Oceanology II: The Australian–New Zealand Sector, Antarctic Res. Ser.*, vol. 19, edited by D. E. Hayes, pp. 147–164, AGU, Washington, D. C., 1972.

Klimov, L. D., and D. S. Solov'ev, Preliminary report on geological observations in East Antarctica, in *Soviet Antarctic Expedition, Information Bulletin*, vol. 1, pp. 15–18, Elsevier, Amsterdam, 1964.

Klimov, L. D., M. G. Ravich, and D. S. Solov'ev, Charnockites of East Antarctica, in *Antarctic Geology*, edited by R. J. Adie, pp. 455–462, North-Holland, Amsterdam, 1964*a*.

Klimov, L. D., M. G. Ravich, and D. S. Solov'ev, Geology of the antarctic platform, in *Antarctic Geology*, edited by R. J. Adie, pp. 681–691, North-Holland, Amsterdam, 1964*b*.

Kuenen, Ph. H., The shell pavement below oceanic turbidites, *Mar. Geol.*, *2*, 236–246, 1964.

Lisitzen, A. P., Bottom sediments of the Antarctic, *Antarctic Research, Geophys. Monogr. Ser.*, vol. 7, edited by H. Wexler et al., pp. 81–88, AGU, Washington, D. C., 1962.

McLeod, I. R., and C. M. Gregory, Geological investigations along the antarctic coast between longitudes 108°E and 166°E, *Rep. 78*, pp. 1–54, Commonwealth of Aust. Dep. of Nat. Develop., Bur. of Miner. Resour., Geol., and Geophys., Canberra, 1967.

Payne, R. R., J. R. Conolly, and W. H. Abbott, Distal turbidite mud in the Wilkes abyssal plain, temporal correlations in some closely spaced piston cores, paper presented at the 8th International Sedimentological Congress, Int. Ass. of Sedimentol., Heidelberg, Aug. 30–Sept. 4, 1971*a*.

Payne, R. R., J. R. Conolly, and W. H. Abbott, Variation between *Eltanin* piston core sites: An intensive coring station in the Wilkes abyssal plain, *Antarctic J. U.S.*, *6*(5), 169–170, 1971*b*.

Payne, R. R., J. R. Conolly, and W. H. Abbott, Turbidite muds within the diatom ooze off Antarctica: Pleistocene sediment variation defined by closely-spaced piston cores, *Geol. Soc. Amer. Bull.*, *83*, 481–486, 1972.

Ravich, M. G., Some results of geological investigations by the Soviet antarctic expedition, in *Soviet Antarctic Expedition, Information Bulletin*, vol. 1, pp. 64–67, Elsevier, Amsterdam, 1964.

Schott, W., Deep-sea sediments of the Indian Ocean, Recent Marine Sediments—A Symposium, *Spec. Publ. 2*, edited by P. D. Trask, pp. 396–408, Soc. of Econ. Paleontol. and Mineral., Tulsa, Okla., 1939.

von der Borch, C. C., and R. L. Oliver, Comparison of heavy minerals in marine sediments with mainland outcrops along the coast of Antarctica between longitudes 40°E and 150°E, *Sediment. Geol.*, *2*, 77-80, 1968.

Watkins, N. D., and J. P. Kennett, Regional sedimentary disconformities and Upper Cenozoic changes in bottom water velocities between Australasia and Antarctica, in *Antarctic Oceanology II: The Australian–New Zealand Sector, Antarctic Res. Ser.*, vol. 19, edited by D. E. Hayes, pp. 273–293, AGU, Washington, D. C., 1972.